Signal Processing for Intelligent Sensor Systems with MATLAB®

Second Edition

Signal Processing for Intelligent Sensor Systems with MATLAB®

Second Edition

David C. Swanson

CRC Press
Taylor & Francis Group
Boca Raton London New York

CRC Press is an imprint of the
Taylor & Francis Group, an **informa** business

CRC Press
Taylor & Francis Group
6000 Broken Sound Parkway NW, Suite 300
Boca Raton, FL 33487-2742

First issued in paperback 2017

© 2011 by Taylor & Francis Group, LLC
CRC Press is an imprint of Taylor & Francis Group, an Informa business

No claim to original U.S. Government works

ISBN-13: 978-1-4200-4304-4 (hbk)
ISBN-13: 978-1-138-07545-0 (pbk)

Visit the Taylor & Francis Web site at
http://www.taylorandfrancis.com

and the CRC Press Web site at
http://www.crcpress.com

This book is dedicated to all who aspire to deeply understand signal processing for sensors, not just enough to pass an exam or assignment, or to complete a project, but deep enough to experience the joy of natural revelation. This takes more than just effort. You have to love the journey. This was best said by one of America's greatest inventors, George Washington Carver, in the quote "Anything will give up its secrets if you love it enough..."

Contents

Part I   Fundamentals of Digital Signal Processing

Part II   *Frequency Domain Processing*

Part III   Adaptive System Identification and Filtering

Part IV   *Wavenumber Sensor Systems*

Part V   Signal Processing Applications

Preface

The second edition of *Signal Processing for Intelligent Sensor Systems* enhances many of the unique features of the first edition with more answered problems, web access to a large collection of MATLAB® scripts used throughout the book, and the addition of more audio engineering, transducers, and sensor networking technology. All of the key algorithms and development methodologies have been kept from the first edition, and hopefully all of the typographical errors have been fixed. The addition of a chapter on Digital Audio processing reflects a growing interest in digital surround sound (5.1 audio) techniques for entertainment, home theaters, and virtual reality systems. Also, new sections are added in the areas of sensor networking, use of meta-data architectures using XML, and agent-based automated data mining and control. This later information really ties large-scale networks of intelligent sensors together as a network of thin file servers. Intelligent algorithms, either resident in the sensor/file-server nodes, or run remotely across the network as intelligent agents, can then provide an automated situational awareness. The many algorithms presented in *Signal Processing for Intelligent Sensor Systems* can then be applied locally or network-based to realize elegant solutions to very complex detection problems.

It was nearly 20 years ago that I was asked to consider writing a textbook on signal processing for sensors. At the time I typically had over a dozen textbooks on my desk, each with just a few small sections bookmarked for frequent reference. The genesis of this book was to bring together all these key subjects into one text, summarize the salient information needed for design and application, and organize the broad array of sensor signal processing subjects in a way to make it accessible to engineers in school as well as those practicing in the field. The discussion herein is somewhat informal and applied and in a tone of engineer-to-engineer, rather than professor-to-student. There are many subtle nuggets of critical information revealed that should help most readers quickly master the algorithms and adapt them to meet their requirements. This text is both a learning resource and a field reference. In support of this, every data graph in the text has a MATLAB m-script in support of it and these m-scripts are kept simple, commented, and made available to readers for download from the CRC Press website for the book (http://www.crcpress.com/product/isbn/9781420043044). Taylor & Francis Group (CRC Press) acquired the rights to the first edition and have been relentless in encouraging me to update it in this second edition. There were also a surprising number of readers who found me online and encouraged me to make an updated second edition. Given the high cost of textbooks and engineering education, we are excited to cut the price significantly, make the book available electronically online, as well as for "rent" electronically which should be extremely helpful to students on a tight budget. Each chapter has a modest list of solved problems (answer book available from the publisher) and references for more information.

The second edition is organized into five parts, each of which could be used for a semester course in signal processing, or to supplement a more focused course textbook. The first two parts, "Fundamentals of Digital Signal Processing" and "Frequency Domain Processing," are appropriate for undergraduate courses in Electrical and/or Computer Engineering. Part III "Adaptive System Identification and Filtering" can work for senior-level undergraduate or a graduate-level course, as is Part IV on "Wave Number Sensor Systems" that applies the earlier techniques to beamforming, image processing, and signal detection systems. If you look carefully at the chapter titles, you will see these algorithm applications grouped differently from most texts. Rather than organizing these subjects strictly by application, we organize them by the algorithm, which naturally spans several applications. An example of this is the recursive least-squares algorithm, projection operator subspace decomposition, and Kalman filtering of state vectors, which all share the same basic recursive update algorithm. Another example is in Chapter 13 where we borrow the two-dimensional FFT usually reserved for image processing and compression and use it to explain available beam pattern responses for various array shapes.

Part V of the book covers advanced signal processing applications such as noise cancellation, transducers, features, pattern recognition, and modern sensor networking techniques using XML messaging and automation. It covers the critical subjects of noise, sensors, signal features, pattern matching, and automated logic association, and then creates generic data objects in XML so that all this information can be found. The situation recognition logic emerges as a cloud application in the network that automatically mines the sensor information organized in XML across the sensor nodes. This keeps the sensors as generic websites and information servers and allows very agile development of search engines to recognize situations, rather than just find documents. This is the current trend for sensor system networks in homeland security, business, and environmental and demographic information systems. It is a nervous system for the planet, and to that end I hope this contribution is useful.

MATLAB® is a registered trademark of The MathWorks, Inc. For product information, please contact:

The MathWorks, Inc.
3 Apple Hill Drive
Natick, MA 01760-2098 USA
Tel: 508 647 7000
Fax: 508-647-7001
E-mail: info@mathworks.com
Web: www.mathworks.com

Acknowledgments

I am professionally indebted to all the research sponsors who supported my colleagues, students, and me over the years on a broad range of sensor applications and network automation. It was through these experiences and by teaching that I obtained the knowledge behind this textbook. The Applied Research Laboratory at The Pennsylvania State University is one of the premier engineering laboratories in the world, and my colleagues there will likely never know how much I have learnt from them and respect them. A special thanks goes to Mr. Arnim Littek, a great engineer in the beautiful country of New Zealand, who thought enough of the first edition to send me a very detailed list of typographical errors and suggestions for this edition. There were others, too, who found me through the Internet, and I really loved the feedback which served as an inspiration to write the second edition. Finally to my wife Nadine, and children Drew, Anya, Erik, and Ava, your support means everything to me.

Author

David C. Swanson has over 30 years of experience with sensor electronics and signal processing algorithms and 15 years of experience with networking sensors. He has been a professor in the Graduate Program in Acoustics at The Pennsylvania State University since 1989 and has done extensive research in the areas of advanced signal processing for acoustic and vibration sensors including active noise and vibration control. In the late 1990s, his research shifted to rotating equipment monitoring and failure prognostics, and since 1999 has again shifted into the areas of chemical, biological, and nuclear detection. This broad range of sensor signal processing applications culminates in his book *Signal Processing for Intelligent Sensor Systems*, now in its second edition. Dr. Swanson has written over 100 articles for conferences and symposia, dozens of journal articles and patents, and three chapters in books other than his own. He has also worked in industry for Hewlett-Packard and Textron Defense Systems, and has had many sponsored industrial research projects. He is a fellow of the Acoustical Society of America, a board-certified member of the Institute of Noise Control engineers and a member of the IEEE. His current research is in the areas of advanced biomimetic sensing for chemicals and explosives, ion chemistry signal processing, and advanced materials for neutron detection. Dr. Swanson received a BEE (1981) from the University of Delaware, Newark, and an MS (1984) and PhD (1986) from The Pennsylvania State University, University Park, where he currently lives with his wife and four children. Dr. Swanson enjoys music, football, and home brewing.

Part I

Fundamentals of Digital Signal Processing

It was in the late 1970s that the author first learned about digital signal processing as a freshman electrical engineering student. Digital signals were a new technology and generally only existed inside computer programs and as hard disk files on cutting edge engineering projects. At the time, and reflected in the texts of that time, much of the emphasis was on the mathematics of a sampled signal, and how sampling made the signal different from the analog signal equivalent. Analog signal processing is very much a domain of applied mathematics, and looking back over 40 years later, it is quite remarkable how the equations we process easily today in a computer program were implemented eloquently in analog electronic circuits. Today there is little controversy about the equivalence of digital and analog signals except perhaps among audio extremists/purists. Our emphasis in this part is on explaining how signals are sampled, compressed, and reconstructed, how to filter signals, how to process signals creatively for images and audio, and how to process signal information "states" for engineering applications. We present how to manage the *nonlinearity* of converting a system defined mathematically in the analog s-plane to an equivalent system in the digital z-plane. These nonlinearities become small in a given low-frequency range as one increases the digital sample rate of the digital system, but numerical errors can become a problem if too much oversampling is done. There are also options for warping the frequency scale between digital and analog systems.

We present some interesting and useful applications of signal processing in the areas of audio signal processing, image processing, and tracking filters. This provides for a first semester course to cover the basics of digital signals and provide useful applications in audio and images in addition to the concept of signal kinematic states that are used to estimate and control the dynamics of a signal or system. Together these applications cover most of the signal processing people encounter in everyday life. This should help make the material interesting and accessible to students new to the field while avoiding too much theory and detailed mathematics. For example, we show frequency response functions for digital filters in this part, but we do not go into spectral processing of signals until Part II. This also allows some time for MATLAB® use to develop where students can get used to making m-scripts and plots of simple functions. The application of fixed-gain tracking filters on a rocket launch example will make detailed use of signal state estimation and prediction as well as computer graphics in plotting multiple functions correctly. Also, using a digital photograph and two-dimensional low- and high-pass filters provide an interesting introduction to image processing using simple digital filters. Over 40 years ago, one could not imagine teaching signal processing fundamentals while covering such a broad range of applications. However, any cell phone today has all of these applications built in, such as sampling, filtering, and compression of the audio signal, image capture and filtering, and even a global positioning system (GPS) for estimating location, speed, and direction.

1 Sampled Data Systems

Figure 1.1 shows a basic general architecture that can be seen to depict most adaptive signal processing systems. The number of inputs to the system can be very large, especially for image processing sensor systems. Since an adaptive signal processing system is constructed using a computer, the inputs generally fall into the categories of analog "sensor" inputs from the physical world and digital inputs from other computers or human communication. The outputs also can be categorized into digital information, such as identified patterns, and analog outputs that may drive actuators (active electrical, mechanical, and/or acoustical sources) to instigate physical *control* over some part of the outside world. In this chapter, we examine the basic constructs of signal input, processing using digital filters, and output. While these very basic operations may seem rather simple compared to the algorithms presented later in the text, careful consideration is needed to insure a high-fidelity adaptive processing system. Figure 1.1 also shows how the adaptive processing can extract the salient information from the signal and automatically arrange it into XML (eXtensible Markup Language) databases, which allows broad use by network processes. Later in the book we will discuss this from the perspective of pattern recognition and web services for sensor networks. The next chapter will focus on fundamental techniques for extracting information from the signals.

Consider a transducer system that produces a voltage in response to some electromagnetic or mechanical wave. In the case of a microphone, the transducer sensitivity would have units of volts/Pascal. For the case of a video camera pixel sensor, it would be volts per lumen/m^2, while for an infrared imaging system the sensitivity might be given as volts per degree Kelvin. In any case, the transducer voltage is conditioned by filtering and amplification in order to make the best use of the analog-to-digital converter (ADC) system. While most adaptive signal processing systems use floating-point numbers for computation, the ADC converters generally produce fixed-point (integer) digital samples. The integer samples from the ADC are further converted to floating-point format by the signal processor chip before subsequent processing. This relieves the algorithm developer from the problem of controlling numerical dynamic range to avoid underflow or overflow errors in fixed-point processing unless lesser expensive fixed-point processors are used. If the processed signals are to be output, then floating-point samples are simply reconverted to integer and an analog voltage is produced using a digital-to-analog converter (DAC) system and filtered and amplified.

1.1 A/D CONVERSION

Quite often, adaptive signal processing systems are used to dynamically calibrate and adjust input and output gains of their respective ADC and DAC devices. This extremely useful technique requires a clear understanding of how most data acquisition systems really work. Consider a generic successive approximation 8-bit ADC as seen in Figure 1.2. The operation of the ADC actually involves an internal DAC that produces an analog voltage for the "current" decoded digital output. A DAC simply sums the appropriate voltages corresponding to the bits set to 1. If the analog input to the ADC does not match the internal DAC output, the binary counter counts up or down to compensate. The actual voltage from the transducer must be sampled and held constant (on a capacitor) while the successive approximation completes. On completion, the least significant bit (LSB) of the digital output number will randomly toggle between 0 and 1 as the internal D/A analog output voltage converges about the analog input voltage. The "settling time" for this process increases with the number of bits quantized in the digital output. The shorter the settling time, the faster the digital

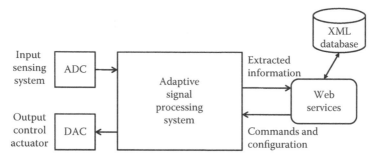

FIGURE 1.1 A generic architecture for an adaptive signal processing system, including sensor inputs, control outputs, and information formatting in XML databases for access through the Internet.

output sample rate may be. The toggling of the LSB as it approximates the analog input signal leads to a low level of uniformly distributed (between 0 and 1) random noise in the digitized signal. This is normal, expected, and not a problem as long as the sensor signal strengths are sufficient enough such that the quantization noise is small compared to signal levels. It is important to understand how transducer and data acquisition systems work so that the adaptive signal processing algorithms can exploit and control their operation.

While there are many digital coding schemes, the binary number produced by the ADC is usually *coded in either offset binary or in two's complement formats* [1]. Offset binary is used for either all-positive or all-negative data such as absolute temperature. The internal DAC in Figure 1.2 is set to produce a voltage V_{min} that corresponds to the number 0, and V_{max} for the biggest number or 255 (11111111), for the 8-bit ADC. The largest number produced by an M-bit ADC is therefore $2^M - 1$. The smallest number, or LSB, will actually be wrong about 50% of the time due to the approximation process. Most data acquisition systems are built around either 8-, 12-, 16-, or 24-bit ADCs giving maximum offset binary numbers of 255, 4095, 65535, and 16777215, respectively. If a "noise-less" signal corresponds to a number of, say 1000, on a 12-bit A/D, the signal-to-noise ratio (SNR) of the quantization is 1000:1, or approximately 60 dB.

Signed numbers are generally encoded in two's complement format where the most significant bit (MSB) is 1 for negative numbers and 0 for positive numbers. This is the normal "signed integer" format in programming languages such as "C." If the MSB is 1 indicating a negative number, the

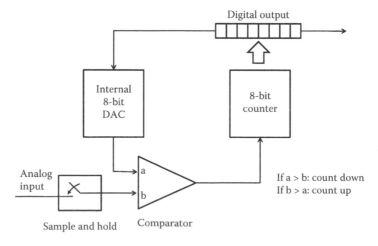

FIGURE 1.2 A generic successive approximation type 8-bit ADC showing the internal DAC converter to compare the counter result to the input voltage.

magnitude of the negative binary number is found by complementing (changing 0–1 or 1–0) all of the bits and adding 1. The reason for this apparently confusing coding scheme has to do with the binary requirements of logic-based addition and subtraction circuitry in all of today's computers [2,3]. The logical simplicity of two's complement arithmetic can be seen when considering that the sum of 2 two's complement numbers, N_1 and N_2, is done exactly the same as for offset binary numbers, except any carryover from the MSB is simply ignored. Subtraction of N_1 from N_2 is done simply by forming the two's complement of N_1 (complementing the bits and adding 1), and then adding the two numbers together ignoring any MSB carryover. An 8-, 12-, 16-, or 24-bit two's complement ADC with numbers over ranges of (+127, −128), (+2047, −2048), (+32767, −32768), and (+8388607, −8388608), respectively.

Table 1.1 shows two's complement binary for a 3-bit ±3.5 V A/D and shows the effect of subtracting the number +2 (010 or +2.5 V) from each of the possible 3-bit numbers. Note that the complement of +2 is (101) and adding 1 gives the "two's complement" of (110), which is equal to numerical −2 or −1.5 V in Table 1.1.

As can be seen in Table 1.1, the numbers and voltages with an asterisk are rather grossly in error. This type of numerical error is the single most reason to use floating-point rather than fixed-point signal processors. It is true that fixed-point signal processor chips are very inexpensive, lower power, and faster at fixed-point arithmetic. However, a great deal of attention must be paid to insuring that no numerical errors of the type in Table 1.1 occur in a fixed-point processor. Fixed-point processing severely limits the numerical dynamic range of the adaptive algorithms used. In particular, algorithms involving many divisions, matrix operations, or transcendental functions such as logarithms or trigonometric functions are generally not good candidates for fixed-point processing. All the subtractions are off by at least 0.5 V, or half the LSB. A final point worth noting from Table 1.1 is that while the analog voltages of the ADC are symmetric about 0 V, the coded binary numbers are not, giving a small numerical offset from the two's complement coding. In general, the design of analog circuits with nearly zero offset voltage is a difficult enough task that one should always assume some nonzero offset in all digitized sensor data.

The maximum M-bit two's complement positive number is $2^{M-1} - 1$ and the minimum negative number is -2^{M-1}. This is because one of the bits is used to represent the sign of the number and one number is reserved to correspond to zero. We want zero to be "digital zero" and we could just leave it at that but it would make addition and subtraction logically more complicated. That is why two's complement format is used for signed integers. Even though the ADC and analog circuitry offset is small, it is good practice in any signal processing system to numerically remove it. This is simply done by recursively computing the mean of the A/D samples and subtracting this time-averaged mean from each ADC sample.

TABLE 1.1
Effect of Subtracting 2 from the Range of Numbers from a 3-bit Two's Complement A/D

Voltage N	Binary N	Binary N_2	Voltage N_2
+3.5	011	001	+1.5
+2.5	010	000	+0.5
+1.5	001	111	−0.5
+0.5	000	110	−1.5
−0.5	111	101	−2.5
−1.5	110	100	−3.5
−2.5	101	011*	+1.5*
−3.5	100	010*	+0.5*

1.2   SAMPLING THEORY

We now consider the effect of the periodic rate of ADC relative to the frequency of the waveform of interest. There appear to be certain advantages to randomly spaced ADC conversions or "dithering" [1], but this separate issue will not be addressed here. According to Fourier's theorem, any waveform can be represented as a weighted sum of complex exponentials of the form $A_m e^{j\omega mt}$; $-\infty < m < +\infty$. A low-frequency waveform will have plenty of samples per wavelength and will be well represented in the digital domain. However, as one considers higher-frequency components of the waveform relative to the sampling rate, the number of samples per wavelength declines. As will be seen below for a real sinusoid, at least two equally spaced samples per wavelength are needed to adequately represent the waveform in the digital domain. Consider the arbitrary waveform in equation

$$x(t) = A\cos(\omega t) = \frac{A}{2}e^{j\omega t} + \frac{A}{2}e^{-j\omega t}. \tag{1.1}$$

We now sample $x(t)$ every T seconds giving a sampling *frequency* of f_s Hz (samples per second). The digital waveform is denoted as $x[n]$, where n refers to the nth sample in the digitized sequence in equation

$$x[n] = x(nT) = A\cos(\omega nT)$$
$$= A\cos\left(\frac{2\pi f}{f_s}n\right). \tag{1.2}$$

Equation 1.2 shows a "digital frequency" of $\Omega = 2\pi f/f_s$, which has the same period as an analog waveform of frequency f so long as f is less than $f_s/2$. Clearly, for the real sampled cosine waveform, a digital frequency of 1.1π is basically indistinguishable from 0.9π except that the period of the 1.1π waveform will actually be longer than the analog frequency f! Figures 1.3 and 1.4 graphically illustrate this phenomenon well-known as aliasing. Figure 1.3 shows a 100-Hz analog waveform sampled 1000 times/s. Figure 1.4 shows a 950-Hz analog signal with the same 1000 Hz sample rate. Since the periods of the sampled and analog signals match only when $f \leq f_s/2$, the frequency components of the analog waveform are said to be *unaliased*, and adequately represented in the digital domain [4].

Restricting real analog frequencies to be less than $f_s/2$ have become widely known as the Nyquist sampling criterion. This restriction is generally implemented by a low-pass filter (LPF) with -3 dB cutoff frequency in the range of $0.4 f_s$ to insure a wide margin of attenuation for frequencies above $f_s/2$. However, as will be discussed in the rest of this chapter, the "antialiasing" filters can have environment-dependent frequency responses which adaptive signal processing systems can intelligently compensate.

It will be very useful for us to explore the mathematics of aliasing to fully understand the phenomenon, and to take advantage of its properties in high-frequency bandlimited ADC systems. Consider a complex exponential representation of the digital waveform in Equation 1.3 showing both positive and negative frequencies

$$x[n] = A\cos(\Omega n)$$
$$= \frac{A}{2}e^{+j\Omega n} + \frac{A}{2}e^{-j\Omega n}. \tag{1.3}$$

While Equation 1.3 compares well with 1.1, there is a big difference due to the digital sampling. Assuming that no antialiasing filters are used, the digital frequency of $\Omega = 2\pi f/f_s$ (from the analog waveform sampled every T seconds) could represent a multiplicity of analog frequencies

$$A\cos(\Omega n) = A\cos(\Omega n \pm 2\pi m); \quad m = 0, 1, 2, \ldots. \tag{1.4}$$

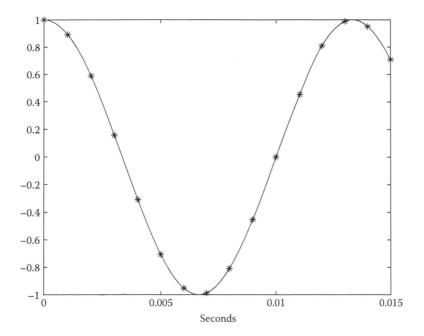

FIGURE 1.3   A 75-Hz sinusoid (solid line) is sampled at 1 kHz (1 ms per sample) as seen by each asterisk (*) showing that the digital signal accurately represents the frequency and amplitude of the analog signal.

or the real signal in Equation 1.3, both the positive and negative frequencies have images at $\pm 2\pi m$; $m = 0, 1, 2,\ldots$. Therefore, if the analog frequency f is outside the Nyquist bandwidth of $0 - f_s/2$ Hz, one of the images of $\pm f$ will appear within the Nyquist bandwidth, but at the wrong (aliased) frequency. Since we want the digital waveform to a linear approximation to the original analog waveform, the frequencies of the two must be equal. One must always suppress frequencies

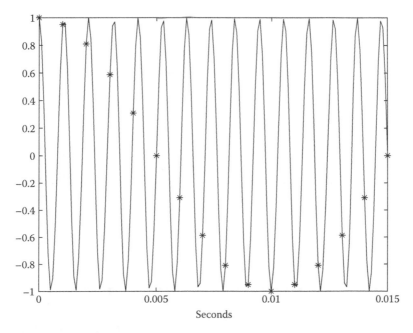

FIGURE 1.4   A 950-Hz sinusoid sampled at 1 kHz clearly show the aliasing effect as the digital samples (*) appear as a 50-Hz signal.

outside the Nyquist bandwidth to be sure that no aliasing occurs. In practice, it is not possible to make an analog signal filter that perfectly passes signals in the Nyquist band while completely suppressing all frequencies outside this range. One should expect a transition zone near the Nyquist band upper frequency where unaliased frequencies are attenuated and some aliased frequency "images" are detectable. Most spectral analysis equipment will implement an antialias filter with a −3 dB cutoff frequency of about 1/3 the sampling frequency. The frequency range from $1/3 f_s$ to $1/2 f_s$ is usually not displayed as part of the observed spectrum so the user does not notice the antialias filter's transition region and the filter very effectively suppresses frequencies above $f_s/2$.

Figure 1.5 shows a graphical representation of the digital frequencies and images for a sample rate of 1000 Hz and a range of analog frequencies including those of 100 and 950 Hz in Figures 1.3 and 1.4, respectively. When the analog frequency exceeds the Nyquist rate of $f_s/2$ (π on the Ω axis), one of the negative frequency images (dotted lines) appears in the Nyquist band with the wrong (aliased) frequency, violating assumptions of system linearity.

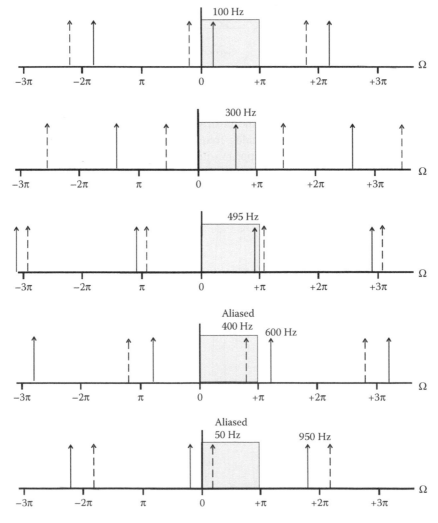

FIGURE 1.5 A graphical view of 100, 300, 495, 600, and 950 Hz analog signals sampled at 1 kHz in the frequency domain showing the aliased "images" of the positive and negative frequency components where the shaded box represents the digital signal bandwidth.

1.3 COMPLEX BANDPASS SAMPLING

Bandpass sampling systems are extremely useful to adaptive signal processing systems which use high-frequency sensor data but with a very narrow bandwidth of interest. Some excellent examples of these systems are active sonar, radar, and ultrasonic systems for medical imaging or nondestructive testing and evaluation of materials. These systems in general require highly directional transmit and receive transducers that physically means that the wavelengths used must be much smaller than the size of the transducers. The transmitted and received "beams" (comparable to a flashlight beam) can then be used to scan a volume for echoes from relatively big objects (relative to wavelength) with different impedances than the medium. The travel time from transmission to the received echo is related to the object's range by the wave speed.

Wave propagation speeds for active radar and sonar vary from a speedy 300 m/µs for electromagnetic waves, to 1500 m/s for sound waves in water, to a relatively slow 345 m/s for sound waves in air at room temperature. Also of interest is the relative motion of the object along the beam. If the object is approaching, the received echo will be shifted higher in frequency due to Doppler, and lower in frequency if the object is moving away. The use of Doppler, time of arrival, and bearing of arrival provide the basic target tracking inputs to active radar and sonar systems. Doppler radar has also become a standard meteorological tool for observing wind patterns. Doppler ultrasound has found important uses in monitoring fluid flow both in industrial processes and in the human cardiovascular system.

Given the sensor system's need for high-frequency operation and relatively narrow signal bandwidth, a digital data acquisition system can exploit the phenomenon of aliasing to drastically reduce the Nyquist rate from twice the highest frequency of interest down to the bandwidth of interest. For example, suppose a Doppler ultrasound system operates at 1 MHz to measure fluid flow of approximately ±0.15 m/s. If the speed of sound is approximately 1500 m/s, one might expect a Doppler shift of only ±100 Hz. Therefore, if the received ultrasound is bandpass filtered from 999.9 kHz to 1.0001 MHz, it should be possible to extract the information using a sample rate on the order of 1 kHz rather than the over 2 MHz required to sample the full frequency range. From an information point of view, bandpass sampling makes a lot of sense because only 0.01% of the 1.0001 MHz frequency range is actually required.

We can show a straightforward example using *real* aliased samples for the above case of a 1-MHz frequency with Doppler bandwidth of ±100 Hz. First, the analog signal is bandpass filtered attenuating all frequencies outside the 999.9 kHz to 1.0001 MHz frequency range of interest. By sampling at a rate commensurate with the signal bandwidth rather than absolute frequency, one of the aliased images will appear in the baseband between 0 Hz and the Nyquist rate. As seen in Figure 1.5, as the analog frequency increases to the right, the negative images all move to the left. Therefore, one of the positive images of the analog frequency is sought in the baseband. Figure 1.6 depicts the aliased bands in terms of the sample rate f_s.

Hence, if the 1 MHz, ±100 Hz signal is bandpass filtered from 999.9 kHz to 1.0001 MHz, we can sample at a rate of 1000.75 Hz putting the analog signal in the middle of the 999th positive image band. Therefore, one would expect to find a 1.0000 MHz signal aliased at 250.1875 Hz, 1.0001 MHz aliased at 350.1875 Hz, and 999.9 kHz at 150.1875 Hz in the digital domain. The extra 150 Hz at the

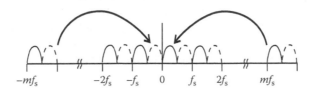

FIGURE 1.6 Analog frequencies bandpass filtered in the mth band will naturally appear in the baseband from 0 to $f_s/2$ Hz, just shifted in frequency.

top and bottom of the digital baseband allow for a transition zone of the antialiasing filters. Practical use of this technique requires precise bandpass filtering and selection of the sample rate. However, Figure 1.6 should also raise concerns about the effects of high-frequency analog noise "leaking" into digital signal processing systems at the point of ADC. The problem of aliased electronic noise is particularly acute in systems where many high-speed digital signal processors operate in close proximity to high-impedance analog circuits and the ADC subsystem has a large number of resolution bits.

For the case of a very narrow bandwidth at a high frequency it is obvious to see the numerical savings, and it is relatively easy to pick a sample rate where only a little bandwidth is left unused. However, for wider analog signal bandwidths a more general approach is needed where the bandwidth of interest is not required to lie within a multiple of the digital baseband. To accomplish this we must insure that the negative images of the sampled data do not mix with the positive images for some arbitrary bandwidth of interest. The best way to do this is to simply get rid of the negative frequency and its images entirely by using complex (real plus imaginary) samples.

How can one obtain complex samples from the real output of the ADC? Mathematically, one can describe a "cosine" waveform as the real part of a complex exponential. However, in the real world where we live (at least most of us some of the time), the sinusoidal waveform is generally observed and measured as a real quantity. Some exceptions to this are simultaneous measurement of spatially orthogonal (e.g., horizontal and vertical polarized) wave components such as polarization of electromagnetic waves, surface Rayleigh waves, or orbital vibrations of rotating equipment, all of which can directly generate complex digital samples. To generate a complex sample from a single real ADC convertor, we must tolerate a signal-phase delay which varies with frequency. However, since this phase response of the complex sampling process is known, one can easily remove the phase effect in the frequency domain.

The usual approach is to gather the real part as before and to subtract in the imaginary part using a T/4 delayed sample

$$x^R[n] = A\cos(2\pi f\, nT + \varphi),$$

$$j\,x^I[n] = -A\cos\left(2\pi f\left[nT + \frac{T}{4}\right] + \varphi\right). \tag{1.5}$$

The parameter φ in Equation 1.5 is just an arbitrary phase angle for generality. For the frequency $f = f_s$, Equation 1.5 reduces to

$$x^R[n] = A\cos(2\pi n + \varphi),$$
$$j\,x^I[n] = -A\cos\left(2\pi n + \varphi + \frac{\pi}{2}\right)$$
$$= A\sin(2\pi n + \varphi) \tag{1.6}$$

so that for this particular frequency, the phase of the imaginary part is actually correct. We now have a usable bandwidth f_s, rather than $f_s/2$ as with real samples. However, each complex sample is actually two real samples, keeping the total information rate (number of samples per second) constant! As the frequency decreases toward 0, a phase error bias will increase toward a phase lag of $\pi/2$. However, since we wish to apply complex sampling to high-frequency bandpass systems, the phase bias can be changing very rapidly with frequency, but it will be fixed for the given sample rate. The complex samples in terms of the digital frequency Ω and analog frequency f are

$$x^R[n] = A\cos(\Omega n + \varphi),$$
$$j\,x^I[n] = -A\cos(\Omega n + \varphi + \pi f/2f_s), \tag{1.7}$$

giving a sampling phase bias (in the imaginary part only) of

$$\Delta\theta = -\frac{\pi}{2}\left(1 - \frac{f}{f_s}\right). \tag{1.8}$$

For adaptive signal processing systems that require phase information, usually two or more channels have their relative phases measured. Since the phase bias caused by the complex sampling is identical for all channels, the phase bias can usually be ignored if relative channel phase is needed. The scheme for complex sampling presented here is sometimes referred to as "quadrature sampling" or even "Hilbert transform sampling" due to the mathematical relationship between the real and imaginary parts of the sampled signal in the frequency domain.

Figure 1.7 shows how any arbitrary bandwidth can be complex sampled at a rate equal to the bandwidth in Hertz, and then digitally "demodulated" into the Nyquist baseband. If the signal bandwidth of interest extends from f_1 to f_2 Hz, an analog bandpass filter is used to band limit the signal and complex samples are formed as seen in Figure 1.7 at a sample rate of $f_s = f_2 - f_1$ samples per second. To move the complex data with frequency f_1 down to 0 Hz and the data at f_2 down to f_s Hz, all one needs to do is multiply the complex samples by $e^{-j\Omega_1 n}$, where Ω_1 is simply $2\pi f_1/f_s$. Therefore, the complex samples in Equation 1.5 are demodulated as seen in equation

$$x^R[n] = A\cos(\Omega n + \varphi)e^{-j\Omega_1 n},$$

$$jx^I[n] = -A\cos\left(\Omega\left[n + \frac{1}{4}\right] + \varphi\right)e^{-j\Omega_1 n}. \tag{1.9}$$

Analog signal reconstruction can be done by remodulating the real and imaginary samples by f_1 in the analog domain. Two oscillators are needed, one for the $\cos(2\pi f_1 t)$ and the other for the $\sin(2\pi f_1 t)$. A real analog waveform can be reconstructed from the analog multiplication of the DAC real sample times the cosine minus the DAC imaginary sample times the sinusoid. As with the complex sample construction, some phase bias will occur. However, the technique of modulation and demodulation is well established in amplitude-modulated (AM) radio. In fact, one could have just as easily demodulated (i.e., via an analog heterodyne circuit) a high-frequency signal, band-limited it to a low-pass frequency range of half the sample rate, and ADC it as real samples. Reconstruction would simply involve DAC, low-pass filtering, and remodulation by a cosine

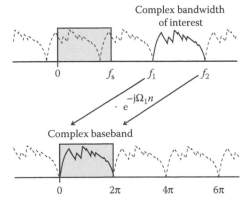

FIGURE 1.7 An arbitrary high-frequency signal may be bandpass filtered and complex sampled and demodulated to a meaningful baseband for digital processing.

waveform. In either case, the net signal information rate (number of total samples per second) is constant for the same signal bandwidth. It is merely a matter of algorithm convenience and desired analog circuitry complexity from which the system developer must decide how to handle high-frequency band-limited signals.

1.4 DELTA–SIGMA ANALOG CONVERSION

A relatively new type of ADC emerged during the 1990s called a "Delta–Sigma" ADC, or DSC (denoted here for brevity).* To the user, a DSC offers the profound advantage of not only eliminating the antialiasing filters but also having the cutoff frequency of the antialiasing filter track a programmable sampling rate. However, the DSC also carries a latency delay for each conversion and has a different effective number of bits (ENOB), both of which vary with the selected sample rate by increasing at lower sample rates. There is also a minimum sample rate, usually around 8 kHz, below which the DSC can operate, but an external antialiasing filter is required. In this section we explain the subtleties of the DSC in a general way, but the reader should note that the many different manufacturers of DSCs have slightly different algorithms and proprietary circuitry that may differ from our simplified presentation [5,6].

The first thing helpful to understand the DSC is that one can oversample and integrate to increase the ENOB and LPF at the same time. Suppose we have 8-bit signed (range +127 to −128) samples of a 50-Hz sine wave sampled at 10 kHz. We can add every two successive samples together to give 9-bit samples (range +254 to −256) at a rate of 5 kHz. Repeat the process of halving the sample rate for each bit added to the samples a few more times and you have 12-bit samples at a rate of 612.5 Hz, and so on. The process of adding successive samples together is essentially a LPF. Frequencies in the waveform near the Nyquist rate (near two-samples per wavelength) are nearly cancelled by the successive sample adding process, while low frequencies are amplified. Assuming the LSB noise is uniform and the analog signal electronic noise is zero mean Gaussian, the noise adds incoherently, so that the zero mean stays zero while the signal is added. It can be seen that simple oversampling and integrating samples gives a 6-dB improvement to the available SNR of the ADC for each having the output sample rate of the process. This can be seen by using a simple equation for the maximum possible SNR for an N-bit ADC based on the quantization noise being spread evenly over the signal Nyquist bandwidth defined by $f_s/2$, where f_s is the sample rate.

$$\text{SNR} = (6.02\,N + 1.76)\,\text{dB}. \tag{1.10}$$

The "6.02" is 20 times the base-10 logarithm of 2, and 1.76 is 10 times the base-10 logarithm of 1.5, which is apparently added into account for quantization noise in the LSB giving the correct bit setting 50% of the time. Hence, for a 16-bit sample, one might use Equation 1.10 to say that the SNR is over 97 dB, which is not correct. N should refer to the number of precision bits, which are 15 for a 16-bit sample because the LSB is wrong 50% of the time. Therefore, for a single-ended 16-bit sample the maximum SNR is approximately 92.06 dB. For signed integer (two's compliment) sample where the SNR is measured for sinusoids in white noise the maximum SNR is only 86.04 dB, because one bit is used to represent the sign. The ENOB is simply the SNR divided by 6.02.

The DSC actually gets theoretical 9 dB SNR improvement each halving of sample rate due to something called quantization noise shaping inherent in the delta modulator circuit, and by increasing the number of bits in the binary sample by 1 with each addition in the digital filtering. The integrator in the delta modulator and the feedback differencing operation have the effect of shifting the quantization noise to higher frequencies while enhancing the signal more at lower frequencies. Because of this, it makes sense to add a bit with each addition in the low-pass decimation filter,

---

* Also called a "Sigma–Delta" ADC.

giving three additional bits with each halving of the sample rate. Hence for a 6.4 MHz 1-bit sample bitstream (12.8 MHz modulation clock), one gets 12-bit samples at a rate of 400 kHz. However, the low-frequency signal enhancement means that the signal bandwidth is not flat, but rather rolls off significantly near the Nyquist rate. Hence, most DSC designs also employ a cascade of digital filters to correct this rolloff in the passband and enhance the filtering in the stopband. The additions in these filters add 2 bits per halving of the sample rate and provide an undistorted waveform (linear phase response) with a little added delay. The 12-bit samples at 400 kHz emerge delayed but with 16-bits at a 100 kHz sample rate and neatly filtered at a Nyquist cutoff frequency of 50 kHz. The DSC has a built in low-pass antialiasing filter, usually a simple R-C filter with a cutoff around 100 kHz, which attenuates by about 36 dB at 6.4 MHz, six octaves higher at the 1-bit delta modulator input. Any aliased signal images are therefore 72 dB attenuated back down at 100 kHz, and more as you go lower in frequency. At 25 kHz, aliased signals are 84 dB attenuated, so for audioband recording with 16-bit samples there is effectively no aliasing problem.

At the heart of a DSC is a device called a "delta modulator" that can be seen depicted in Figure 1.8. The delta modulator produces a 1-bit digital signal called a bitstream at a very high sample rate where one can convert a frame of N-bits to a $\log_2 N$-bit word. The analog voltage level at the end of the frame will be a filtered sum of the bits within the frame. Hence, if the analog input in Figure 1.8 was very close to V_{max}, the bitstream would be nearly all ones; if it were close to 0, the bitstream would be nearly all zeros; and if it were near $V_{max}/2$, about 50% of the bits within the frame would be 1's. The delta-modulated bitstream can be found today on "super audio DVD discs," which typically have 24-bit samples at sample rates of 96 kHz, and sometimes even 192 kHz, much higher resolution than the 16-bit 44.1 kHz samples of the standard compact disc.*

The DSC has some very interesting frequency response properties. The action of the integrator and latch give a transfer function which essentially filters out low-frequency quantization noise, improving the theoretical SNR about 3 dB each time the bandwidth is halved. The quantization noise attenuation allows one to keep additional bits from the filtering and summing, which yields a theoretical 9 dB improvement overall each time the frame rate is halved. This makes generating large sample words at lower frequencies very accurate. However, the noise-shaping effect also makes the upper end of the signal frequency response roll off well below the Nyquist rate. DSC manufacturers correct for this using a digital filter to restore the high-frequency response, but this also brings a time delay to the DSC output sample. This will be discussed in more detail in Chapter 3 in the sections on digital filtering with finite impulse response (FIR) filters. For some devices this delay can be on the order of 32 samples, and hence the designer must be careful with this detail for

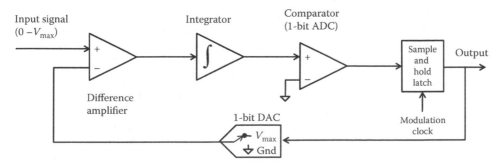

FIGURE 1.8 A delta modulator is used to convert an analog voltage to a 1-bit "bitstream" signal where the amplitude of the signal is proportional to the number of 0s and 1s in a given section of the bitstream.

---

* The author was very skeptical of this technology until he actually heard it. The oversampling and bigger bit-depth really does make a difference since most movie and music recordings are compilations of sounds with a wide range of loudness dynamics.

applications that require real-time signal inputs and outputs, such as control loop applications. The maximum theoretical SNR of a DSC can be estimated by considering the noise shaping of the delta modulator and the oversampling ratio (OSR).

$$\text{SNR} = (6.02N + 1.76) + 10\log_{10}(\text{OSR}), \tag{1.11}$$

where OSR is the ratio of the 1-bit sample rate f_s divided by the N-bit decimated sample rate f_{sN}. This SNR improvement is more of a marketing nuance than a useful engineering parameter because one only has a finite dynamic range available based on the number of bits in the output samples. For our 6.4 MHz sampled bitstream processed down to 16-bit samples at 100 kHz, the theoretical SNR from Equation 1.11 is 116.1 dB using $N = 16$ and 104.1 dB using $N = 14$ (1 bit for sign and ignoring the LSB). What does all this marketing rhetoric mean? It means that the DSC does not introduce quantization noise, and so the effective SNR is about 90 dB for 16-bit signed samples. However, by using more elaborate filters some DSC will produce more useful bits because of this higher theoretical limit. It is common to see 24-bit samples from a DSC which have SNRs in the range of 120 dB for audio bandwidths. The 24-bit sample word format conveniently maps to 3 bytes per sample, even though the actual SNR is not using all 24 bits. An SNR of 120 dB is ratio of about 1 million to 1. Since most signals are recorded with a maximum of ±10 V or less, and the analog electronic noise floor at room temperature is of the order of microvolts for audio bandwidths (unless one employs cooling to reduce thermal noise in electronic devices), an ENOB of around 20 can be seen as adequate to exceed the dynamic range of most sensors and electronics. As such, using a 24-bit DSC with effectively 20 bits of real SNR, one no longer needs to be concerned with setting the voltage gain to match the sensor signal to the ADC! For most applications where the signal is simply recorded and used, the DSC filter delay is not important either. As a result of the accuracy and convenience of the DSC, it is now the most common ADC in use.

1.5   MATLAB® EXAMPLES

Throughout the second edition of this text, we present in each chapter a summary of useful MATLAB® scripts for implementing the plots of the chapter figures and for further study. The book website at www.taylorandfrancis.com and/or www.crcpress.com will have downloadable m-scripts for the user to enjoy further study.* The main m-script used to make Figures 1.3 and 1.4 is called "demoa2d.m" and contains some very simple statements to generate simple plots in MATLAB. This is a good script to study if the user has no experience with MATLAB and needs to generate a simple plot from an equation. Table 1.2 contains the complete m-script.

The m-script in Table 1.2 can be seen as an overlay of two digital sinusoid plots, one in black to simulate the analog signal and the other using black asterisk "*" symbols to depict the digital signal. The "analog" is simply sampled at a much higher rate and drawn with smooth lines. The statement "Ta = 0:T_analog:Tstop;" creates a row vector with elements [0, T_analog, 2*T_analog, 3*T_analog, ..., Tstop]. The semicolon ";" at the end of each line in the script stops the MATLAB execution from dumping the result on the command line as the script executes. The vector "Ta" represents the sample times in seconds for each of the "analog waveform" samples to plot. We then define a row vector "ya" the same size as Ta, but filled with zeros. The parameter "w0" depicts the radian frequency where "pi" is by default set to 3.1415927... in MATLAB. The statement "ya = cos(w0.*Ta);" is both really convenient and very confusing to new MATLAB users. When you use ".*" for a multiply, MATLAB multiplies each element as you would in a "dot product." In our case the scalar "w0" multiplies each element of "Ta." The built-in math functions (and there are many of them) generally extend from scalar arguments to vectors, matrices, and even multidimensional matrices. This is very powerful and saves the user from the drudgery of writing many nested

---

* Provided the user pledges to only use the author's m-scripts for good, not evil.

TABLE 1.2
m-Script Example for Generating Simple Graphs of Sampled Sinusoids

```
% MATLAB m-file for Figures 1.2 and 1.3 A2D-Demo
fs = 1000;                        % sample rate
Ts = 1/fs;                        % sample time interval
fs_analog = 10000;                % "our" display sample rate (analog signal points)
npts_analog = 200;                % number of analog display points
T_analog = 1/fs_analog;           % "our" display sample interval
f0 = 950;                         % use 75 Hz for Fig 1.3 and 950 Hz for Fig 1.4
Tstop = 0.015;                    % show 15 msecs of data
Ta = 0:T_analog:Tstop;            % analog "samples"
Td = 0:Ts:Tstop;                  % digital samples
ya = zeros(size(Ta));             % zero out data vectors same length as time
yd = zeros(size(Td));
w0 = 2*pi*f0;
ya = cos(w0.*Ta);                 % note scalar by vector multiply (.*) gives vector in
                                  % the cosine argument and a vector in the output ya
yd = cos(w0.*Td);
figure(1);                        % initialize a new figure window for plotting
plot(Ta,ya,'k');                  % plot in black
hold on;                          % keep the current plot and add another layer
plot(Td,yd,'k*');                 % plot in black "*"
hold off;                         % return figure to normal state
xlabel('Seconds');
```

for-loops. It also executes substantially faster than a for-loop and leaves a script that is very easily read. The ".*" dot product extends to vectors and matrices. Conversely, one has to consider matrix algebra rules when multiplying and dividing matrices and vectors. If "x" and "y" are both row vectors, the statement "$x*y$" will generate an error. Using the transpose operator on "y" will do a Hermitian transpose (flip a row vector into a column and replace the elements with complex conjugates) so that "$x*y'$" will yield a scalar result. If you do not want a complex conjugate (it does not matter for real signals) the correct syntax is "$x*y'$". The "dot-transpose" means just transpose the vector without the conjugate operation. Once one masters this "vector concept" in the m-scripts generating plots of all the signal processing in this book presented in m-scripts will become very straightforward. The "plot" statement has to have the x and y components defined as identical-sized vectors to execute properly. The most common difficulty the author has seen is these vectors not matching (rows–columns need to be flipped or vectors of different lengths) in functions like "plot". The statement "hold on" allows one to overlay plots, which can also be done by adding multiple $x - y$ vector pairs to the plot argument. On the MATLAB command line one can enter "help plot" to get more details as well as through the help window. The reason MATLAB is part of this book is that it has emerged as one of the most effective ways to quickly visualize and test signal processing algorithms. The m-scripts are deliberately kept very simple for brevity and to expose the algorithm coding details, but many users will embed the algorithms into very sophisticated MATLAB-based graphical user interfaces (GUIs) or port the algorithms to other languages such as C, C++, C#, Visual Basic, and Web-based script languages such as Java script or Flash script.

1.6 SUMMARY, PROBLEMS, AND REFERENCES

This section has reviewed the basic process of analog waveform digitization and sampling. The binary numbers from an ADC can be coded into offset-binary or into two's complement formats for use with unsigned or signed integer arithmetic, respectively. Floating-point digital signal processors subsequently convert the integers from the ADC to their internal floating-point format for processing, and then back to the appropriate integer format for DAC conversion. Even though floating-point arithmetic has a huge numerical dynamic range, the limited dynamic range of the ADC and DAC convertors must always be considered. Adaptive signal processing systems can, and should, adaptively adjust input and output gains while maintaining floating-point data calibration. This is much less of an issue when using ADC and DAC with over 20 bits of precision. Adaptive signal calibration is straightforwardly based on known transducer sensitivities, signal conditioning gains, and the voltage sensitivity and number of bits in the ADC and DAC convertors. The LSB is considered to be a random noise source both numerically for the ADC convertor and electronically for the DAC convertor. Given a periodic rate for sampling analog data and reconstruction of analog data from digital samples, analog filters must be applied before ADC and after DAC conversion to avoid unwanted signal aliasing. The DSC has a built-in antialiasing filter, and one can alter the clock of the device over a fairly wide range and still have high-fidelity samples down to a frequency of approximately 8 kHz. Below that, an external antialias filter is needed. For real digital data, the sample rate must be at least twice the highest frequency which passes through the analog "antialiasing" filters. For complex samples, the complex-pair sample rate equals the bandwidth of interest, which may be demodulated to baseband if the bandwidth of interest was in a high-frequency range. The frequency response of DAC conversion as well as sophisticated techniques for analog signal reconstruction will be discussed in Section 4.6 later in the text.

PROBLEMS

1. An accelerometer with sensitivity 10 mV/G (1.0 G is 9.801 m/s²) is subjected to a ±25 G acceleration. The electrical output of the accelerometer is amplified by 11.5 dB before A/D conversion with a 14-bit two's complement encoder with an input sensitivity of 0.305 mV/bit.
   a. What is the numerical range of the digitized data?
   b. If the amplifier can be programmed in 1.5 dB steps, what would be the amplification for maximum SNR? What is the SNR?
2. An 8-bit two's complement A/D system is to have no detectable signal aliasing at a sample rate of 100,000 samples per second. An eighth-order (−48 dB/octave) programmable cutoff frequency LPF is available.
   a. What is a possible cutoff frequency fc?
   b. For a 16-bit signed A/D what would the cutoff frequency be?
   c. If you could tolerate some aliasing between fc and the Nyquist rate, what is the highest fc possible for the 16-bit system in part b?
3. An acceptable resolution for a medical ultrasonic image is declared to be 1 mm. Assume sound travels at 1500 m/s in the human body.
   a. What is the absolute minimum A/D sample rate for a receiver if it is to detect echoes from scatterers as close as 1 mm apart?
   b. If the velocity of blood flow is to be measured in the range of ±1 m/s (we do not need resolution here) using a 5 MHz ultrasonic sinusoidal burst, what is the minimum required bandwidth and sample rate for an A/D convertor? (Hint: a Doppler-shifted frequency f_d can be determined by $f_d = f(1 + v/c)$, $-c < v < +c$; where f is the transmitted frequency, c is the wave speed, and v is the velocity of the scatterer toward the receiver.)
4. A microphone has a voltage sensitivity of 12 mV/Pa (1 Pascal = 1 Nt/m²). If a sinusoidal sound of about 94 dB (approximately 1 Pa rms in the atmosphere) is to be digitally recorded, how much gain would be needed to insure a "clean" recording for a 10 V 16-bit signed A/D system?
5. A standard analog television in the United States has 525 vertical lines scanned in even and odd frames 30 times/s.
   a. If the vertical field of view covers a distance of 1.0 m, what is the size of the smallest horizontal line thickness which would appear unaliased?

    b.   A high-definition television provides 1080 vertical lines of resolution. What is the spatial resolution?

6. A new car television commercial is being produced where the wheels of the car have 12 stylish holes spaced every 30° around the rim. If the wheels are 0.7 m in diameter, how fast can the car move before the wheels start appearing to be rotating backward?

7. Suppose a very low-frequency high SNR is being sampled at a high rate by a limited dynamic range 8-bit signed A/D convertor. If one simply adds consecutive pairs of samples together one has 9-bit data at half the sample rate. Adding consecutive pairs of the 9-bit samples together gives 10-bit data at 1/4 the 8-bit sample rate, and so on.

    a.   If one continued on to get 16-bit data from the original 8-data sampled at 10,000 Hz, what would the data rate be for the 16-bit data?

    b.   Suppose we had a very fast device that samples data using only 1-bit, 0 for negative and 1 for positive. How fast would the 1-bit A/D have to sample to produce 16-bit data at the standard digital audio rate of 44,100 samples per second?

8. A DSC with noise shaping and FIR filters to correct for frequency response roll-off adds about two bits of precision for each halving of the internal sample rate.

    a.   If the delta modulator produces a bitstream at 6.4 MHz, what is the actual ENOB for this device at 50 kHz sample rate?

    b.   What is the theoretical limit for the number of DSC bits available at 50 kHz?

9. An electronic sensor system produces a full-scale voltage of ±1 V and has a flat spectral noise density specified at 50 nV/√Hz. If the bandwidth of the data acquisition is 20 kHz, what is the SNR and the required number of bits for the ADC?

REFERENCES

1. N. S. Jayant and P. Noll, *Digital Coding of Waveforms*. Englewood Cliffs, NJ: Prentice-Hall, 1984.
2. K. Hwang, *Computer Arithmetic*. New York, NY: Wiley, 1979, p. 71.
3. A. Gill, *Machine and Assembly Language Programming of the PDP-11*. Englewood Cliffs, NJ: Prentice-Hall, 1978.
4. A. V. Oppenheim and R. W. Schafer, *Discrete-Time Signal Processing*. Englewood Cliffs, NJ: Prentice-Hall, 1973.
5. P. M. Aziz et al., An overview of sigma-delta converters, *IEEE Sig Proc Mag*, Jan 1996, pp. 61–83.
6. S. Park, *Principles of Sigma–Delta Modulation for Analog-to-Digital Converters*, Motorola Application Notes. Schaumburg, IL: Motorola, Inc., 1999, /D, Rev 1.

2 z-Transform

Given a complete mathematical expression for a discrete time-domain signal, why transform it to another domain? The main reason for time–frequency transforms is that many mathematical reductions are much simpler in one domain than the other [1]. The z-transform in the digital domain is the counterpart to the Laplace transform in the analog domain. The z-transform is an extremely useful tool for analyzing the stability of digital sequences, designing stable digital filters, and relating digital signal processing operations to the equivalent mathematics in the analog domain. The Laplace transform provides a systematic method for solving analog systems described by differential equations. Both the z-transform and the Laplace transform map their respective finite-difference or differential systems of equations in the time or spatial domain to much simpler algebraic systems in the frequency or wavenumber domains, respectively. However, the relationship between the z-domain and the s-domain of the Laplace transform is not linear, meaning that the digital filter designer will have to decide whether to match the system poles, zeros, or impulse response. As will be seen later in this chapter, one can warp the frequency axis to control where and how well the digital system matches the analog system. We begin by that assuming time t increases as life progresses into the future, and a general signal of the form e^{st}, $s = \sigma + j\omega$, is *stable* for $\sigma \leq 0$. A plot of our general signal is shown in Figure 2.1.

The quantity $s = \sigma + j\omega$ is a complex frequency where the real part σ represents the damping of the signal ($\sigma = -10.0$ Nepers/s and $\omega = 50\pi$ rad/s, or 25 Hz, in Figure 2.1). All signals, both digital and analog, can be described in terms of sums of the general waveform shown in Figure 2.1. This includes transient characteristics governed by σ. For $\sigma = 0$, one has a steady-state sinusoid. For $\sigma < 0$ as shown in Figure 2.1, one has an exponentially decaying sinusoid. If $\sigma > 0$, the exponentially increasing sinusoid is seen as unstable, since eventually it will become infinite in magnitude. Signals which change levels over time can be mathematically described using piecewise sums of stable and unstable complex exponentials for various periods of time as needed.

The same process of generalized signal modeling is applied to the *signal responses* of systems such as mechanical or electrical filters, wave propagation "systems," and digital signal processing algorithms. We define a "linear system" as an operator which changes the amplitude and/or phase (time delay) of an input signal to give an output signal with the same frequencies as the input, independent of the input signal's amplitude, phase, or frequency content. Linear systems can be dispersive, where some frequencies travel through them faster than others, as long as the same system input–output response occurs independent of the input signal. Since there are an infinite number of input signal types, we focus on one very special input signal type called an impulse. An impulse waveform contains the same energy level at all frequencies including 0 Hz (direct current or constant voltage), and is exactly reproducible. For a digital waveform, a digital impulse simply has only one sample nonzero. The response of linear systems to the standard impulse input is called the *system impulse response*. The impulse response is simply the system's response to a Dirac delta function (or the unity amplitude digital domain equivalent), when the system has zero initial conditions. The impulse response for a linear system is unique and a great deal of useful information about the system can be extracted from its analog or digital domain transform [2].

2.1 COMPARISON OF LAPLACE AND z-TRANSFORMS

Equation 2.1 describes a general integral transform where $y(t)$ is transformed to $Y(s)$ using the kernel $K(s,t)$.

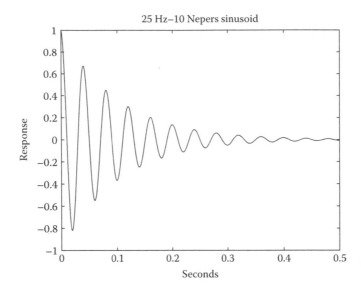

FIGURE 2.1   A "general" stable signal of the form $e^{(\sigma+j\omega)t}$ where $\sigma \leq 0$ indicates a stable waveform for positive time.

$$Y(s) = \int\limits_{\infty}^{+\infty} K(s,t)\, y(t)\, dt \qquad (2.1)$$

The Laplace transform makes use of the kernel $K(s,t) = e^{-st}$, which is also in the form of our "general" signal as shown in Figure 2.1. We present the Laplace transform $L\{\ \}$ as a pair of integral transforms in Equation 2.2 relating the time "t" and frequency "s" domains.

$$Y(s) = L\{y(t)\} = \int\limits_{0}^{+\infty} y(t)e^{-st}\, dt$$

$$y(t) = L^{-1}\{Y(s)\} = \frac{1}{2\pi j} \int\limits_{\sigma j\infty}^{\sigma+j\infty} Y(s)e^{+st}\, ds \qquad (2.2)$$

The corresponding z-transform pair for discrete signals is seen in Equation 2.3 where t is replaced with nT and denoted as $[n]$, and $z = e^{st}$.

$$Y[z] = Z\{y[n]\} = \sum_{n=0}^{+\infty} y[n]z^{-n}$$

$$y[n] = Z^{-11}\{Y[z]\} = \frac{1}{2\pi j} \oint\limits_{\Gamma} Y[z]z^{n-1}\, dz \qquad (2.3)$$

The closed contour Γ in Equation 2.3 must enclose all the poles of the function $Y[z]\, z^{n-1}$. Both $Y(s)$ and $Y[z]$ are, in the most general terms, ratios of polynomials where the zeros of the numerator are also zeros of the system. Since the system response tends to diverge if excited near a zero of the denominator polynomial, the zeros of the denominator are called the system poles. The transforms in Equations 2.2 and 2.3 are applied to signals, but if these "signals" represent system impulse or frequency responses, our subsequent analysis will refer to them as "systems," or "system responses."

There are two key points which must be discussed regarding the Laplace and z-transforms. First, we present what is called a "one-sided" or "causal" transform. This is seen in the time integral of Equation 2.2 starting at $t = 0$, and the sum in Equation 2.3 starting at $n = 0$. Physically, this means that the current system response is a result of the current and past inputs, *and specifically not future inputs*. Conversely, a current system input can have no effect on previous system outputs. Only time moves forward in the real physical world (at least as we know it in the twentieth century), and so a distinction must be made in our mathematical models to represent this fact. Our positive time movement mathematical convention has a critical role to play in designating stable and unstable signals and systems mathematically. Second, in the Laplace transform's s-plane ($s = \sigma + j\omega$), only signals and system responses with $\sigma \leq 0$ are mathematically stable in their causal response (time moving forward). This means e^{st} is either of constant amplitude ($\sigma = 0$), or decaying amplitude ($\sigma < 0$) as time increases. Therefore, system responses represented by values of s on the left-hand plane ($j\omega$ is the vertical Cartesian axis) are stable causal response systems. As will be seen below, the nonlinear mapping from the s-plane (analog signals and systems) to z-plane (digital signals and systems) maps the stable causal left-half s-plane to the region inside a unity radius circle on the z-plane, called the *unit circle*.

The comparison of the Laplace and z-transforms is most useful when considering the mapping between the complex s-plane and the complex z-plane, where $z = e^{sT}$, T being the time interval in seconds between digital samples of the analog signal. The structure of this mapping depends on the digital sample rate and whether real or complex samples are used. An understanding of this mapping will allow one to easily design digital systems which model (or control) real physical systems in the analog domain. Also, adaptive system modeling in the digital domain of real physical systems can be quantitatively interpreted and related to other information processing in the adaptive system. However, if we have an analytical expression for a signal or system in the frequency domain, it may or may not be realizable as a stable causal signal or system response in the time domain (digital or analog). Again, this is due to the obliviousness of time to positive or negative direction. If we are mostly concerned with the magnitude response, we can generally adjust the phase (by adding time delay) to realize any desired response as a stable causal system. Table 2.1 gives a partial listing of some useful Laplace transforms and the corresponding z-transforms assuming regularly sampled data every T seconds ($f_s = 1/T$ samples/s).

One of the subtler distinctions between the Laplace transforms and the corresponding z-transforms in Table 2.1 are how some of the z-transform magnitudes scale with the sample interval T. It can be seen that the result of the scaling is that the sampled impulse responses may not match the inverse z-transform if a simple direct s-to-z mapping is used. Since adaptive digital signal processing can be used to measure and model physical system responses, we must be diligent to eliminate digital system responses where the amplitude depends on the sample rate. However, in Section 2.3, it will be shown that careful consideration of the scaling for each system resonance or pole will yield a very close match between the digital system and its analog counterpart. At this point in our presentation of the z-transform, we compare the critical mathematical properties for linear time-invariant systems in both the analog Laplace transform and the digital z-transform.

The Laplace transform and the z-transform have many mathematical similarities, the most important of which are the properties of linearity and shift invariance. Linear shift-invariant system modeling is essential to adaptive signal processing since most optimizations are based on a quadratic squared output error minimization. But even more significantly, linear time-invariant physical systems allow a wide-range linear algebra to apply for the straightforward analysis of such systems. Most of the world around us is linear and time invariant, provided the responses we model are relatively small in amplitude and quick in time. For example, the vibration response of a beam slowly corroding due to weather and rust is linear and time invariant for small vibration amplitudes over a period of, say, days or weeks. But, over a period of years the beam's corrosion changes the vibration response, thereby making it time varying in the frequency domain. If the forces on the beam approach its yield strength, the stress–strain relationship is no longer linear and single-frequency vibration

Signal Processing for Intelligent Sensor Systems with MATLAB®

TABLE 2.1
Some Useful Signal Transforms

Time Domain	s Domain	z Domain
1 for $t \geq 0$ 0 for $t < 0$	$\dfrac{1}{s}$	$\dfrac{z}{(z-1)}$
$e^{s_0 t}$	$\dfrac{1}{s - s_0}$	$\dfrac{z}{z - e^{s_0 T}}$
$t\,e^{s_0 t}$	$\dfrac{1}{(s - s_0)^2}$	$\dfrac{T z\, e^{s_0 T}}{(z - e^{s_0 T})^2}$
$e^{at} \sin \omega_0 t$	$\dfrac{\omega_0}{s^2 + 2as + a^2 + \omega_0^2}$	$\dfrac{z\, e^{-aT} \sin \omega_0 T}{z^2 - 2z e^{-aT} \cos \omega_0 T + e^{-2aT}}$
$e^{-at} \cos(\omega_0 t - \theta)$	$\dfrac{(\cos\theta\,(s+a) + \omega_0 \sin\theta)}{(s+a)^2 + \omega_0^2}$	$\dfrac{z\cos\theta(z - \alpha) - z\beta\sin\theta}{(z-\alpha)^2 + \beta^2}$ $\alpha = e^{-aT}\cos\omega_0 T$ $\beta = e^{-aT}\sin\omega_0 T$
$\dfrac{1}{ab} + \dfrac{e^{-at}}{a\,(a-b)} + \dfrac{e^{-bt}}{b(b-a)}$	$\dfrac{1}{s(s+a)(s+b)}$	$\dfrac{(Az + B)z}{(z - e^{-aT})(z - e^{-bT})(z - 1)}$ $A = \dfrac{b(1 - e^{-aT}) - a(1 - e^{-bT})}{ab\,(b-a)}$ $B = \dfrac{ae^{-aT}(1 - e^{-bT}) - be^{-bT}(1 - e^{-aT})}{ab\,(b-1)}$

inputs into the beam will yield nonlinear multiple frequency outputs. Nonlinear signals are rich in physical information but require very complicated models. From a signal processing point of view, it is extremely valuable to respect the physics of the world around us, which is only linear and time invariant within specific physical constraints, and exploit linearity and time invariance wherever possible. Nonlinear signal processing is still something much to be developed in the future. Following is a summary of comparison of Laplace and z-transforms.

Linearity: Both the Laplace and z-transforms are linear operators. The inverse Laplace and z-transforms are also linear.

$$L\{af(t) + bg(t)\} = aF(s) + bG(s)$$
$$Z\{af[k] + bg[k]\} = aF[z] + bG[z]$$

(2.4)

Delay Shift Invariance: Assuming one-sided signals $f(t) = f[k] = 0$ for t, $k < 0$ (no initial conditions),

$$L\{f(t - \tau)\} = e^{-s\tau} F(s)$$
$$Z\{f[k - N]\} = z^{-N} F[z]$$

(2.5)

Convolution: Linear shift-invariant systems have the following property: a multiplication of two signals in one domain is equivalent to a convolution in the other domain.

$$L\{f(t)*g(t)\} = L\left\{\int_0^t f(\tau)g(t-\tau)\,d\tau\right\} = F(s)G(s) \tag{2.6}$$

A more detailed derivation of Equation 2.6 will be presented in the next section. In the digital domain, the convolution integral becomes a simple summation.

$$Z\{f[k]*g[k]\} = Z\left\{\sum_{k=0}^m f[k]g[m-k]\right\} = F[z]G[z] \tag{2.7}$$

If $f[k]$ is the impulse response of a system and $g[k]$ is an input signal to the system, the system output response to the input excitation $g[k]$ is found in the time domain by the convolution of $g[k]$ and $f[k]$. However, the system must be both linear and shift invariant (a shift of k samples in the input gives a shift of k samples in the output), for the convolution property to apply. Equation 2.7 is fundamental to digital systems theory and will be discussed in great detail later.

Initial Value: The initial value of a one-sided (causal) impulse response is found by taking the limit as s or z approaches infinity.

$$\lim_{t \to 0} f(t) = \lim_{s \to \infty} sF(s) \tag{2.8}$$

The initial value of the digital impulse response can be found in an analogous manner.

$$f[0] = \lim_{z \to \infty} F[z] \tag{2.9}$$

Final Value: The final value of a causal impulse response can be used as an indication of the stability of a system as well as to determine any static offsets.

$$\lim_{t \to \infty} f(t) = \lim_{s \to 0} sF(s) \tag{2.10}$$

Equation 2.10 holds so long as $sF(s)$ is analytic in the right-half of the s-plane (no poles on the jω-axis and for $\sigma \geq 0$). $F(s)$ is allowed to have one pole at the origin and still be stable at $t = \infty$. The final value in the digital domain is

$$\lim_{k \to \infty} f[k] = \lim_{z \to 1}(1 - z^{-1})F[z] \tag{2.11}$$

$(1 - z^{-1})F[z]$ must also be analytic in the region on and outside the unit circle on the z-plane. The region $|z| \geq 1$, on and outside the unit circle on the z-plane, corresponds to the region $\sigma \geq 0$, on the jω-axis and on the right-hand s-plane. The s-plane pole $F(s)$ is allowed to have $s = 0$ in equation maps to a z-plane pole for $F[z]$ at $z = 1$ since $z = e^{sT}$. The allowance of these poles is related to the restriction of causality for one-sided transforms. The mapping between the s and z planes will be discussed in some more detail in the following text.

Frequency Translation/Scaling: Multiplication of the analog time-domain signal by an exponential leads directly to a frequency shift.

$$L\{e^{-at}f(t)\} = F(s+a) \tag{2.12}$$

In the digital domain, multiplication of the sequence $f[k]$ by a geometric sequence α^k results in scaling the frequency range.

$$Z\{\alpha^k f[k]\} = \sum_{k=0}^{\infty} f[k]\left(\frac{z}{\alpha}\right)^{-k} = F\left[\frac{z}{\alpha}\right] \tag{2.13}$$

Differentiation: The Laplace transform of the derivative of the function $f(t)$ is found using integration by parts.

$$L\left\{\frac{\partial f}{\partial t}\right\} = sF(s) - f(0) \tag{2.14}$$

Carrying out integration by parts as in Equation 2.14 for higher-order derivatives yields the general formula

$$L\left\{\frac{\partial f^N}{\partial t^N}\right\} = s^N F(s) - \sum_{k=0}^{N-1} s^{N-1-k} \; f^{(k)}(0) \tag{2.15}$$

where $f^{(k)}(0)$ is the kth derivative of $f(t)$ at $t = 0$. The initial conditions for $f(t)$ are necessary to its Laplace transform just as they are necessary for the complete solution of an ordinary differential equation. For the digital case, we must first employ a formula for carrying forward initial conditions in the z-transform of a time-advanced signal.

$$Z\{x[n+N]\} = x^N X[z] - \sum_{k=0}^{N-1} z^{N-k} x[k] \tag{2.16}$$

For a causal sequence, Equation 2.16 can be easily proved from the definition of the z-transform. Using an approximation based on the definition of the derivatives, the first derivatives of a digital sequence is

$$\dot{x}[n+1] = \frac{1}{T}(x[n+1] - x[n]) \tag{2.17}$$

where T is the sample increment. Applying the time-advance formula in Equation 2.16 gives the z-transform of the first derivative.

$$Z\{\dot{x}[n+1]\} = \frac{1}{T}\{(z-1)X[z] - zx[0]\} \tag{2.18}$$

Delaying the sequence by one sample shows the *z*-transform of the first derivative of $x[n]$ at sample n.

$$Z\{\dot{x}[n]\} = \frac{1}{T}\{(1-z^{-1})X[z] - x[0]\} \tag{2.19}$$

The second derivative can be seen to be

$$Z\{\ddot{x}[n]\} = \frac{1}{T^2}\left\{(1-z^{-1})^2 X[z] - \left[(1-2z^{-1})x[0] + z^{-1}x[1]\right]\right\} \tag{2.20}$$

The pattern of how the initial samples enter into the derivatives can be more easily seen in the third derivative of $x[n]$, where the polynomial coefficients weighting the initial samples can be seen as fragments of the binomial polynomial created by the triple zero at $z = 1$.

$$Z\{\dddot{x}[n]\} = \frac{1}{T^3}\left\{(1-z^{-1})^3 X[z] - (1-3z^{-1}+3z^{-2})x[0] - (z^{-1}-3z^{-2})x[1] - z^{-2}x[2]\right\} \tag{2.21}$$

Putting aside the initial conditions on the digital domain definitive, it is straightforward to show that the *z*-transform of the Nth definitive of $x[n]$ simply has N zeros at $z = 1$ corresponding to the analogous N zeros at $s = 0$ in the analog domain.

$$Z\{x^{(N)}[n]\} = \frac{1}{T^N}\{(1-z^{-1})^N X[z] - \text{initial conditions}\} \tag{2.22}$$

Mapping between the s and z Planes: As with the aliased data in Section 1.1, the effect of sampling can be seen as a mapping between the series of analog frequency bands and the digital baseband defined by the sample rate and type (real or complex). To make sampling useful, one must band limit the analog frequency response to a bandwidth equal to the sample rate for complex samples, or LPF to half the sample rate (called the Nyquist rate) for real samples. Consider the effect of replacing the analog t in $z^n = e^{st}$ with nT, where n is the sample number and $T = 1/f_s$ is the sampling interval in seconds.

$$z^n = e^{(\sigma+j\omega)nT} = e^{\left(\frac{\sigma}{f_s}+j\frac{2\pi f}{f_s}\right)n} \tag{2.23}$$

As in Equation 2.23, the analog frequency repeats every multiple of f_s (a full f_s Hz bandwidth is available for complex samples). For real samples (represented by a phase-shifted sine or cosine rather than a complex exponential), a f_s Hz-wide frequency band will be centered about 0 Hz giving an effective signal bandwidth of only $f_s/2$ Hz for positive frequency. The real part of the complex spectrum is symmetric for positive and negative frequencies while the imaginary part is skew symmetric (negative frequency amplitude is opposite in sign from positive frequency amplitude). This follows directly from the imaginary part of $e^{j\theta}$ being $j\sin\theta$. The amplitudes of the real and imaginary parts of the signal spectrum are determined by the phase shift of the sine or cosine. For real time-domain signals sampled at f_s samples/s, the effective bandwidth of the digital signal is from 0 to $f_s/2$ Hz. For $\sigma \leq 0$, a strip within $\pm\omega_s/2$ for the left-half of the complex *s*-plane maps into a region inside a unit radius circle on the complex *z*-plane. For complex sampled systems, each multiple of f_s Hz on the *s*-plane corresponds to a complete trip around the unit circle on the *z*-plane. In other words, the left-half of the *s*-plane is subdivided into an infinite number of parallel strips, each

ω_s radians wide which all map into the unit circle of the z-plane. As described in Chapter 1, accurate digital representation requires that one limits the bandwidth of the analog signal to one of the s-plane strips before sampling [3].

For real sampled systems, the upper half of the unit circle from the angles from 0 to π is a "mirror image" of the lower half of the circle from the angles of 0 to $-\pi$. Figure 2.2 shows a series of s-values (depicted as the complex values "A" through "I") and their corresponding positions on the complex z-plane for a real sampled system. As long as the analog signals a filter appropriately before sampling (and after D/A conversion for output analog signal reconstruction), the digital representation on the complex z-plane will be accurate. The letters "A" through "I" depict mappings between the analog s-plane and the digital z-plane in Figure 2.2.

The mapping between the s and z planes can be seen as a straightforward implementation of periodic sampling. If one had a mathematical expression for the Laplace transform $H(s)$ of some system impulse response $h(t)$, the poles and zeros of $H(s)$ can be directly mapped to the z-plane as shown in Figure 2.2 giving a digital domain system response $H[z]$ with inverse z-transform $h[n]$. The sampled system impulse response $h[n]$ should simply be the analog domain impulse response $h(t)$ sampled every nT seconds, provided of course that the analog signal is first appropriately band limited for the sample rate.

While the mapping given here is similar to bilinear transformation, the mapping from the s-plane $j\omega$-axis onto the z-plane unit circle in Figure 2.2 is actually linear in frequency. Bilinear transformations include a nonlinear mapping to the "w-plane" where the frequency part of the mapping follows a tangent, rather than linear, relationship. Only at very low frequencies on the w-plane do the two frequency scales approximate one another. The w-plane is typically used for stability analysis using techniques such as the Routh–Hurwitz criterion or Jury's test. Software tools such as MATLAB, offer another warping technique where the user can select a frequency where the z and s domains match best. In our work, we simply map the poles and zeros back and forth as needed between the z-plane and a band-limited s-plane primary strip bounded by $\pm\omega_s/2$. The interior of the z-plane unit circle corresponds to the left-half ($\sigma < 0$) of the s-plane primary strip.

The approach of mapping between the s and z planes does have an area where some attention to physics is needed. The mapping of the s-plane and z-plane poles and zeros ensures that the frequency response of the two systems is relatively accurate at low frequencies. But, an accurate digital system representation of a physical analog domain system must have an impulse response that is identical in both the digital and analog domains. As will be seen below, the digital domain impulse

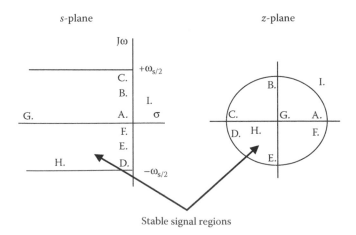

FIGURE 2.2 The region between $\pm\omega_s/2$ (and its images) on the left-hand s-plane maps to a region inside the unit circle on the z-plane.

response will differ from the analog impulse response by a scale factor for any system with more than a single pole mapped directly between the s and z planes [4]. The scale factor is found in a straightforward manner by comparing the analog and digital impulse responses from the inverse Laplace and inverse z-transforms, respectively. Each system "mode" can be isolated using a partial fraction expansion and scaled independently to make the analog and digital domain systems very nearly identical. Because $z = e^{sT}$ is not a linear function, the transition from analog to digital domain is not a perfect match. One can match the system modes which will give a nearly identical impulse response but will match the spectral zeros (or nulls in the frequency response), or one can match the frequencies of the poles (spectral peaks or resonances) and zeros, but the modal amplitudes and impulse response will differ between the digital and analog system responses. The next section illustrates the process for a simple mass–spring–damper (MSD) mechanical oscillator and its corresponding digital system model.

2.2 SYSTEM THEORY

Common systems such as the wave equation, electrical circuits made up of resistors, capacitors, and inductors, or mechanical systems made up of dampers, springs, and masses are described in terms of second-order linear differential equations of the form

$$M \frac{\partial^2 y(t)}{\partial t^2} + R \frac{\partial y(t)}{\partial t} + Ky(t) = Ax(t) \tag{2.24}$$

where $Ax(t)$ is the applied input (a force) to the system for $t > 0$ and $y(t)$ is the resulting output (displacement of the mass) assuming that all system initial conditions are zero. If Equation 2.24 describes a mechanical oscillator, $Ax(t)$ has units of force, $y(t)$ is displacement in meters m, M is mass in kg, K is stiffness in kg/s^2, and R is damping in kg/s. Note that each of the three terms on the left side of Equation 2.24 has units of force in N t, where 1 N t = 1 kg m/s^2. Assuming that the input force waveform $Ax(t)$ is a force impulse $f_0\delta(t)$, the displacement output $y(t)$ of Equation 2.24 can be seen as a system impulse response only if there are zero initial conditions on $y(t)$ and all its time derivatives. Taking Laplace transforms of both sides of Equation 2.24 reveals the general system response.

$$M\{s^2Y(s) - sy(0) - \dot{y}(0)\} + R\{sY(s) - y(0)\} + KY(s) = f_0 \tag{2.25}$$

$$Y(s) = \frac{F(s) + (Ms + R)y(0) + M\dot{y}(0)}{\{Ms^2 + Rs + K\}}$$

$$= H(s)G(s) \tag{2.26}$$

Since the Laplace transform of $f(t) = f_0\delta(t)$ is $F(s) = f_0$, where $H(s)$ is called the system function

$$H(s) = \frac{1}{\{Ms^2 + Rs + K\}} \tag{2.27}$$

and $G(s)$ is called the system excitation function.

$$G(s) = F(s) + (Ms + R)y(0) + M\dot{y}(0) \tag{2.28}$$

Usually, one separates the initial conditions from the system response

$$Y(s) = F(s)H(s) + \frac{(Ms + R)y(0) + M\dot{y}(0)}{\{Ms^2 + Rs + K\}} \tag{2.29}$$

since the effect of the initial conditions on the system eventually die out relative to a steady-state excitation $F(s)$. Figure 2.3 depicts an—mass-spring-damper mechanical oscillator and its Laplace transform system model equivalent.

Taking inverse Laplace transforms of Equation 2.29 gives the system displacement response $y(t)$ to the force input $f(t)$ and the initial displacement $y(0)$ and velocity $\dot{y}(0)$. The inverse Laplace transform of the s-domain product $F(s)H(s)$ brings us to a very important relationship called the *convolution integral*. We now examine this important relationship between the time and frequency domains. Applying the definition of the Laplace transform in Equation 2.2, to $F(s)H(s)$, one obtains

$$F(s)H(s) = \int_0^\infty f(\tau)e^{-s\tau}\,d\tau \int_0^\infty h(\beta)e^{-s\beta}\,d\beta \tag{2.30}$$

We can rearrange Equation 2.30 to give

$$F(s)H(s) = \int_0^\infty f(\tau)d\tau \int_0^\infty h(\beta)e^{-s(\beta+\tau)}\,d\beta \tag{2.31}$$

Equation 2.31 is now changed; so $\beta = t - \tau$

$$F(s)H(s) = \int_0^\infty f(\tau)d\tau \int_\tau^\infty h(t-\tau)e^{-st}\,dt \tag{2.32}$$

Finally, interchanging the order of integration, we obtain

$$F(s)H(s) = \int_0^\infty \left(\int_0^t f(\tau)h(t-\tau)\,d\tau \right)e^{-st}\,dt \tag{2.33}$$

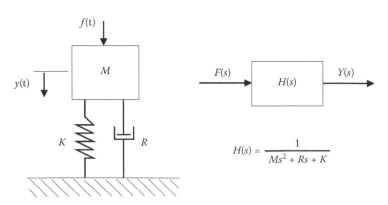

FIGURE 2.3 A mechanical oscillator and its equivalent Laplace transform system model with force input and displacement output.

where the middle integral in parenthesis in Equation 2.33 is the convolution of $f(t)$ and $h(t)$. It can be seen that the product of two system functions in the frequency domain (*s*- or *z*-plane) results in the convolution of the corresponding time-domain functions. Conversely, the product of two time-domain functions is equivalent to the convolution of their corresponding frequency-domain functions. Also note that the conjugate product in the frequency domain is equivalent to a cross-correlation of the two time-domain signals. The inverse Laplace transform of our mechanical oscillator system in Equation 2.29 is therefore

$$
y(t) = \int_0^t f(\tau)h(t-\tau)\,d\tau + y(0)\,L^{-1}\left\{\frac{Ms+R}{Ms^2+Rs+K}\right\}
$$

$$
+ \dot{y}(0)L^{-1}\left\{\frac{M}{Ms^2+Rs+K}\right\} \tag{2.34}
$$

where $h(t)$ is the inverse Laplace transform of the system function $H(s)$.

If the initial conditions are both zero, the second and third terms in Equation 2.34 are zero, and if the force input to the system $f(t)$ is a unity amplitude Dirac delta function $f(t) = f_0\delta(t); f_0 = 1$, the displacement output response $y(t)$ is exactly the same as the system impulse response $h(t)$. The system impulse response $h(t)$ can be seen as the solution to a *homogeneous* differential Equation 2.24 while the first term of the right-hand side of Equation 2.34 is seen as the forced response to a *nonhomogeneous* equation. The last two right-hand terms form the complete general solution to Equation 2.24.

In our chosen case of a unity impulse input force with zero initial conditions, we can write the system response $H(s)$ as

$$
H(s) = \frac{1}{Ms^2+Rs+K} \tag{2.35}
$$

which can be written as a partial fraction expansion as

$$
H(s) = \frac{1}{s_1-s_2}\left\{\frac{1}{s-s_1}\ \frac{1}{s-s_2}\right\} \tag{2.36}
$$

where the two system poles s_1 and s_2 are simply

$$
s_{1,2} = \frac{R}{2M} + s\sqrt{\frac{K}{M}\left(\frac{R}{2M}\right)^2}
$$

$$
= \zeta + j\omega_d \tag{2.37}
$$

The inverse Laplace transform of Equation 2.36 is therefore

$$
h(t) = \frac{e^{-\zeta t}}{j2\omega_d}\left[e^{+j\omega_d t} - e^{-j\omega_d t}\right] = \frac{e^{-\zeta t}}{\omega_d}\sin\omega_d t \tag{2.38}
$$

Equation 2.38 is the analog domain system impulse response $h(t)$. This simple physical example of a system and its mathematical model serves us to show the importance of linearity as well as initial conditions. Given this well-founded mathematical model, we can transfer, or map, the model to the digital domain using *z*-transforms in place of the Laplace transform. The physical relationship between physical system models in the digital and analog domains is extremely important to understand since we can use adaptive digital signal processing to identify physical models in the digital domain. Generating *information* about a physical system from the digital models requires a seamless transition from the *z*-plane back to the *s*-plane, independent of digital system sample rates, number of A/D bits, and transducer sensitivities. In other words, if we build a processing algorithm that adapts to the input and output signals of an unknown system, and we wish to represent physical quantities (such as mass, stiffness, damping, etc.) from their digital counterparts, we must master the transition between the digital and analog domains. This is slightly detailed and subtle because the mapping is nonlinear (warped) in frequency as presented in most of the literature where a bilinear transform is used. The bilinear transform maps the analog frequency of infinity Hz to the digital Nyquist frequency on the unit circle at $e^{j\pi}$. This is accurate only at low frequencies when the system is grossly oversampled. We present a more straightforward mapping here where the poles in the analog and digital domains are at the same frequencies. We then proceed to scale the amplitude of the system so that the impulse responses match as close as possible. The amplitude scaling is most sensitive at high frequencies in the digital domain.

2.3   MAPPING OF s-PLANE SYSTEMS TO THE DIGITAL DOMAIN

We now examine the implications of mapping the s-plane poles in Equation 2.37 to the z-plane using the transformation $z = e^{sT}$. This transformation alone places the poles at the correct frequencies, but not necessarily the same amplitudes as the analog domain counterpart.

$$
\begin{aligned}
z_1 &= e^{-\zeta T + j\omega_d T} \\
z_2 &= e^{-\zeta T - j\omega_d T}
\end{aligned}
\tag{2.39}
$$

The mapped z-transform for the system is depicted as $H^m[z]$ and written as

$$
H^m[z] = \frac{1}{(z - z_1)(z - z_2)} = z^{-2} \frac{1}{(1 - z_1 z^{-1})(1 - z_2 z^{-1})}
\tag{2.40}
$$

where the negative powers of z are preferred to give a causal z-transform using some additional time delay. Expanding Equation 2.40 using partial fractions gives a sum of two poles which will simplify the inverse z-transform.

$$
H^m[z] = \frac{z^{-1}}{(z_1 - z_2)} \left\{ \frac{1}{(1 - z_1 z^{-1})} - \frac{1}{(1 - z_2 z^{-1})} \right\}
\tag{2.41}
$$

The two terms in the braces in Equation 2.41 can be rewritten as an infinite geometric series.

$$
H^m[z] = \frac{z^{-1}}{(z_1 - z_2)} \sum_{k=0}^{\infty} (z_1^k - z_2^k) z^{-k}
\tag{2.42}
$$

By inserting the z-plane poles into Equation 2.42 and substituting $n = k + 1$ we have

$$
H^m[z] = \frac{1}{e^{-\zeta T} \sin(\omega_d T)} \sum_{n=1}^{\infty} e^{-\zeta T(n-1)} \sin(\omega_d T(n-1)) z^{-k}
\tag{2.43}
$$

From examination of Equations 2.3 and 2.43 the inverse z-transform of the mapped poles is

$$
h^m[n] = \frac{e^{-\zeta T(n-1)} \sin(\omega_d T(n-1))}{e^{-\zeta T} \sin(\omega_d T)}; \quad n > 0
\tag{2.44}
$$

Clearly, the mapped system impulse response of Equation 2.38 has a scale factor (in the denominator of Equation 2.44) which is dependent on the sample interval T and the damped resonance ω_d. As ω_d approaches $\omega_s/2$ or 0, $h^m[n]$ will have a large amplitude compared to the sampled analog domain impulse response $h(nT)$ given in Equation 2.38. The time delay of one sample is a direct result of causality constraints on digital systems. Figure 2.4 shows the impulse response of our system with a poles near ± 25 Hz and a damping factor of $\zeta = 10$ sampled at rates of 57, 125, and 300 Hz. At 57 samples/s, $\omega_d T$ is nearly π and the mapped impulse response of Equation 2.44 is larger in amplitude than the physical system. At 125 samples/s, the pole at $\omega_d T$ is nearly $\pi/2$, making the sin function nearly unity and the two impulse responses to be a pretty good match (except the factor of ω_d). However, as we sample at faster and faster rates, $\omega_d T$ approaches 0 and the unscaled digital impulse response becomes artificially huge. Note that we multiplied the "true impulse response" of Equation 2.38 by ω_d just to get the plots in each graph comparable.

Since we want the digital impulse response to match the true physical system's impulse response as perfectly as possible (this is the "impulse-invariant" case in the literature), we must scale the digital system to remove the sample rate dependence as well as the pole location dependence.

$$h([n-1]T) = \left\{ \frac{e^{-\zeta T} \sin(\omega_d T)}{\omega_d} \right\} h^m[n]; \quad n > 0 \tag{2.45}$$

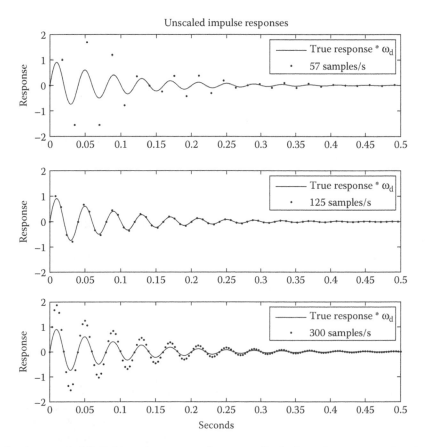

FIGURE 2.4 An unscaled digital impulse response has an amplitude dependence on sample rate.

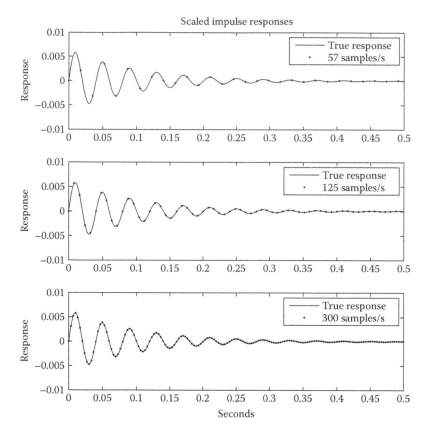

FIGURE 2.5   An accurate impulse response is obtained through proper modal scaling.

Figure 2.5 shows that this scaling allows any sample rate to be used for our mechanical oscillator system's digital model. The scaling required is actually a result of the partial fraction expansions in the two domains. The time delay in the scaled mapped digital impulse response is unavoidable due to causality constraints on real causal digital systems. Since the sample interval is usually quite small, the delay is generally of little consequence.

The scale factor in Equation 2.44 allows the digital impulse response to very nearly match the analog system. For very fast sample rates where ω_d is small compared to $2\pi f_s$, Equation 2.46 can be approximated by $T = 1/f_s$ as presented in Ref. [3].

The good news about the scaling between the z and s planes is that it is always a constant factor (linear) regardless of the number of poles and zeros being mapped. For the case of a pair of conjugate poles mapped from the s-plane to the z-plane, the scaling given in Equation 2.46 is exact, making the impulse responses match perfectly.

$$\frac{e^{-\zeta T}\sin(\omega_d T)}{\omega_d} = \frac{z_1 - z_2}{s_1 - s_2} \tag{2.46}$$

Comparing the frequency responses of the system Laplace transform $H(s)$ evaluated for $j0 \leq s \leq j\omega_s/2$, and the scaled and mapped digital system $H[z]$ evaluated on the unit circle provides another measure of the importance of scaling. Evaluating the frequency response of a digital system $H[z]$ requires an additional scale factor of T, the digital sampling time interval which can be seen to be the counterpart of the differential "dt" in the Laplace transform. The properly scaled frequency response of the digital domain $H[z]$ is seen to closely approximate that for the analog domain $H(s)$ for any nonaliased sample rate as

$$H(s)\big|_{s=j\omega} \approx H[z]\big|_{z=e^{j\omega T}} = T\left\{\frac{z_1 - z_2}{s_1 - s_2}\right\}\frac{1}{(z - z_1)(z - z_2)}\bigg|_{z=e^{j\omega T}} \tag{2.47}$$

Figure 2.6 compares the frequency responses of $H(s)$ and $H[z]$ in terms of dB-magnitude and phase in degrees. The case shown in Figure 2.6 is for two 25 Hz conjugate poles with a damping factor of $\zeta = 10$ Nepers and a sampling frequency of 300 samples/s. Except for the linear phase response due to the delay of one sample in $H[z]$, the magnitude and phase compare quite well, especially when considering the sensitivity of the log scale. For a high sampling frequency, or relatively low resonance, the net scale factor can be approximated by T^2. However, if the resonance is near the Nyquist frequency, the scaling following Equation 2.47 should be employed.

As in Figure 2.6, the match is very close at low frequencies. If we remove the sample delay from the digital system phase frequency response, the two phase curves almost exactly match. It can be seen that even with careful scaling of digital systems, the transference to the digital domain is not perfect. At the upper end of the spectrum where system features are observed through signals which are sparsely sampled, the system response errors are larger in general. From a system fidelity point of view, oversampling is the most common way to drive the dominant system features to the low

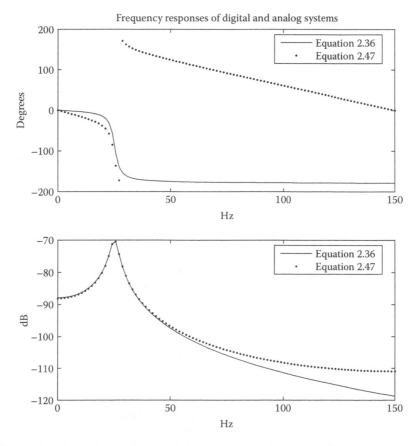

FIGURE 2.6 A comparison of the magnitude and phase of the analog system $H(s)$ and the digital approximation, with scaling $H(z)$ (seen as dots in the plot) shows a good magnitude match at the pole frequency and a slight phase shift due to the digital delay using a sample rate of 300 Hz.

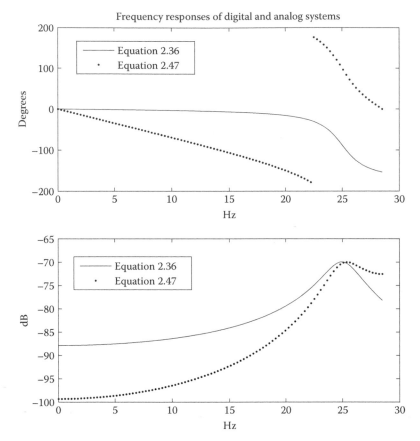

FIGURE 2.7 The same systems shown in Figure 2.6 but with a sampling frequency of 57 Hz showing a good amplitude match at the pole frequency but a much greater phase error due to the delay.

digital frequency range where very precise system modeling can occur. Figure 2.7 shows the same $H(s)$ system model but for a digital system $H[z]$ sampled at only 57 samples/s. Clearly, the scaling shown in Equation 2.47 succeeds in providing a close match near the system resonance, but high-frequency problems exist nonetheless. This is because we map a strip of the s-plane, from $-f_s/2$ to $+f_s/2$, $\sigma < 0$, to the region inside the unite circle on the z-plane. But because of the periodicity of frequency around the unit circle, we are also mapping the response of all the aliased strips from $2nf_s$ to $(4n + 1)f_s/2$, $n = 1,2,3, \ldots$, into the band from 0 to $f_s/2$, which can be seen as the "leakage" near the Nyquist rate in the digital system response. In practice, this is not a concern because we always force the analog counterpart to a digital system to be band limited by the antialiasing filter. However, it can also be seen from Figures 2.6 and 2.7 that it is a good idea to oversample the analog system by a factor of 2 or beyond that required for anti-aliasing to completely avoid any loss of precision.

Finding the precise scale constants becomes much more difficult for complicated systems with many system resonances, or modes. However, each mode (or system resonance) can be scaled in the same manner as the simple oscillator above to give a reasonable match for both the time-domain impulse responses, and the frequency-domain system responses. But for systems with both high-frequency poles and zeros (resonances and antiresonances, respectively), a design choice must be made. The choice is between either matching the frequencies of the poles and zeros with an error in system response magnitude or matching the impulse response and system resonances with an error in the zero (system antiresonance) frequencies. As will be seen below, the only recourse for precise analog and digital system and impulse response matching is to substantially "oversample" the digital system, where all the poles and zeros are at very low frequencies.

Consider the following two-zero, four-pole system with a real impulse response requiring that the poles and zeros be conjugate pairs. Let the two zeros z_1^s be at $\sigma = +5$, $j\omega/2\pi = \pm 130$ Hz, the first pair of poles p_1^s at $\sigma = -20$, $j\omega/2\pi = \pm 160$ Hz, and the second pair of poles p_2^s at $\sigma = -10$, $j\omega/2\pi = \pm 240$ Hz on the s-plane. Note that the positive damping for the zeros does not cause an unstable system, but rather a *nonminimum phase* system, which will be discussed in more detail in the next section. Our s-plane system response function is

$$H(s) = \frac{(s - z_1^s)(s - z_1^{s*})}{(s - p_1^s)(s - p_1^{s*})(s - p_2^s)(s - p_2^{s*}\ _2^s)} \tag{2.48}$$

where the impulse response is found using partial fraction expansions as seen in Equation 2.49.

$$h(t) = A_s e^{p_1^s t} + B_s e^{p_1^{s*} t} + C_s e^{p_2^s t} + D_s e^{p_2^{s*} t} \tag{2.49}$$

where

$$A_s = \frac{(p_1^s - z_1^s)(p_1^s - z_1^{s*})}{(p_1^s - p_1^{s*})(p_1^s - p_2^s)(p_1^s - p_2^{s*})}$$

$$B_s = \frac{(p_1^{s*} - z_1^s)(p_1^{s*} - z_1^{s*})}{(p_1^{s*} - p_1^s)(p_1^{s*} - p_2^s)(p_1^{s*} - p_2^{s*})}$$

$$\tag{2.50}$$

$$C_s = \frac{(p_2^s - z_1^s)(p_2^s - z_1^{s*})}{(p_2^s - p_1^s)(p_2^s - p_1^{s*})(p_2^s - p_2^{s*})}$$

$$D_s = \frac{(p_2^{s*} - z_1^s)(p_2^{s*} - z_1^{s*})}{(p_2^{s*} - p_1^s)(p_2^{s*} - p_1^{s*})(p_2^{s*} - p_2^s)}$$

Applying the mapping and modal scaling technique described previously in this section, the discrete impulse response which is seen to closely approximate Equation 2.49 is

$$h[n-1] = A_s (p_1^z)^n + B_s (p_1^{z*})^n + C_s (p_2^z)^n + D_s (p_2^{z*})^n \tag{2.51}$$

where p_1^z is the z-plane mapped pole corresponding to p_1^s, and so on. The discrete-system frequency response is found following Equation 2.47.

$$H[z] = T\left[\frac{A_s}{z - p_1^z} + \frac{B_s}{z - p_1^{z*}} + \frac{C_s}{z - p_2^z} + \frac{D_s}{z - p_2^{z*}} \right] \tag{2.52}$$

Note that the above method is essentially the same as the "impulse invariant method" outlined in Parks and Burrus [5], although we present more details on the scaling and frequency matching. The technique gives a good match between the impulse responses in the digital and analog domains, but the frequency responses are not possible to match perfectly. The spectral peaks in Equation 2.52 will be well matched to the analog system in Equation 2.48 due to the modal scaling. However, the change in relative modal amplitudes causes the zero locations to also change as will be seen graphically below.

There appears to be little one can do to match both the poles and zeros in general at high frequencies with an algorithm other than empirical means. However, a fairly close match (much closer than using the mapped poles and zeros alone) can be achieved by writing $H[z]$ as a product of an all-zero system and an all-pole system. The two systems are then scaled separately where the zeros are each divided by T and the poles are scaled according to the modal scaling described previously in this section. A compensated system response with separate "linear" scaling is depicted as $H^c[z]$ in the following equation.

$$H^c[z] = T\frac{(z - z_1^z)}{T}\frac{(z - z_1^{z*})}{T}\left[\frac{A_c}{z - p_1^z} + \frac{B_c}{z - p_1^{z*}} + \frac{C_c}{z - p_2^z} + \frac{D_c}{z - p_2^{z*}}\right] \tag{2.53}$$

$$A_c = \frac{1}{(p_1^s - p_1^{s*})(p_1^s - p_2^s)(p_1^s - p_2^{s*})}$$

$$B_c = \frac{1}{(p_1^{s*} - p_1^s)(p_1^{s*} - p_2^s)(p_1^{s*} - p_2^{s*})}$$

$$C_c = \frac{1}{(p_2^s - p_1^s)(p_2^s - p_1^{s*})(p_2^s - p_2^{s*})} \tag{2.54}$$

$$D_c = \frac{1}{(p_2^{s*} - p_1^s)(p_2^{s*} - p_1^{s*})(p_2^{s*} - p_2^s)}$$

The partial fraction expansion coefficients A_c, B_c, C_c, and D_c are seen in Equation 2.54 for the all-pole part of the system response.

The compensated "linear" scaling shown in Equations 2.53 and 2.54 is seen as a compromise between matching the peak levels and maintaining consistent pole–zero frequencies.

Another "linear scaling" technique seen in Equation 2.55 applies to the z-plane mapped impulse response

$$h^c[n-1] = A_z(p_1^z)^n + B_z(p_1^{z*})^n + C_z(p_2^z)^n + D_z(p_2^{z*})^n \tag{2.55}$$

where A_z, B_z, C_z, and D_z are the unscaled mapped z-plane counterparts to the s-plane coefficients given in Equation 2.50. Linear scaling gives an approximate impulse response match which degrades as the system resonances increase in frequency. Consider the impulse responses shown in Figure 2.8 for the system sampled at 600 Hz. As in Figure 2.8, the effect of modal scaling is to give a much

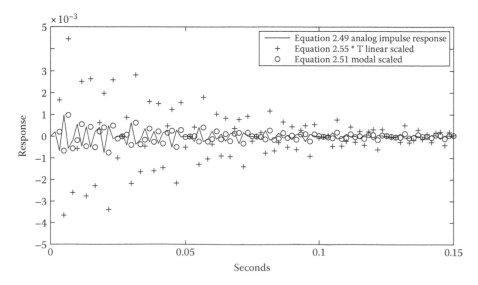

FIGURE 2.8 Comparison of the two-zero, four-pole analog (solid line) impulse response to an approximate linear-scaled (+) and more precise modal-scaled (o) digital impulse response sampled at 600 Hz.

closer match to the real system than simple linear scaling. Figure 2.8 compares the frequency responses for the same system and sampling rate.

As in Figures 2.8 and 2.9, the modal scaling gives an accurate impulse response and spectral peaks but a significant error in the shifted location of the zero. This could be an issue of concern if the zero is strong and used explicitly to filter out a particular frequency. The "linear-scaled" trade-off can place the zeros accurately, but at the expense of overall scaling. Note that we multiplied the linear-scaled impulse response by T to decrease its amplitude enough for all three plots to be examined in Figure 2.8. Clearly, it is up to the designer to decide whether an accurate match of the frequencies of the poles and zeros is needed more than an accurate impulse response. The important lesson here is that mapping from the s-plane to the z-plane (and z-plane back to the s-plane) is not without some trade-offs. For systems where accurate transient response is needed, one would prefer modal mapping. If accurate steady-state response is needed in terms of resonance and antiresonance frequencies, linear-scaled mapping is preferred. However, if high-precision matching between the z-plane and s-plane systems is required, it can best be achieved by "oversampling" the system. Oversampling can be seen as using a digital system bandwidth much greater than that required for the dominant system features (poles and zeros). As in Figures 2.10 and 2.11, where the former figure's system is oversampled 5:1 at 3 kHz, an excellent spectral and impulse response match occurs for both digital systems using either type of scaling [5].

The literature on mapping between the s-plane and z-place often cites the use of a "bilinear transform" which maps the entire left-hand s-plane inside the unit circle on the z-plane. While this is mathematically satisfying because the stable part of the s-plane maps to the inside of the unit circle, it is not very practical because the Nyquist sample rate maps to an infinite analog frequency. This is why our presentation above examines a linear frequency mapping using $z = e^{sT}$ allowing us to ignore the frequency response beyond the Nyquist rate (helped by a good antialiasing filter and a little oversampling). To be complete, we briefly present the bilinear transformation to explain how this popular technique significantly warps the frequency mapping at the higher end of the digital frequency spectrum. The bilinear mapping is defined by

$$z = \frac{1+(T/2)s}{1-T/2\,s} \tag{2.56}$$

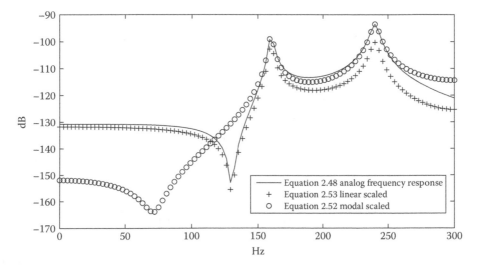

FIGURE 2.9 Comparison of the magnitude frequency response of the system in Figure 2.8.

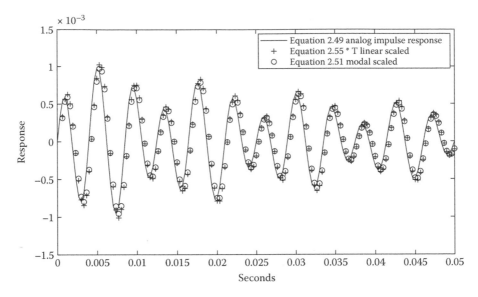

FIGURE 2.10   The impulse response systems in Figure 2.8 using a 3 kHz sample rate.

Rearranging Equation 2.56, we obtain

$$s = \sigma + j\omega = \frac{2}{T}\frac{1-z}{1+z} \tag{2.57}$$

If we examine Equation 2.57 along the jω-axis on the s-plane (on the unit circle on the z-plane), we can substitute $z = e^{j\Omega}$ where $\Omega = 2\pi f/f_s$ (see Equation 1.2). Equation 2.57 then reduces to

$$j\omega = \frac{2}{T}\frac{1-e^{j\Omega}}{1+e^{j\Omega}} = \frac{2}{T}j\tan\left(\frac{\Omega}{2}\right) \tag{2.58}$$

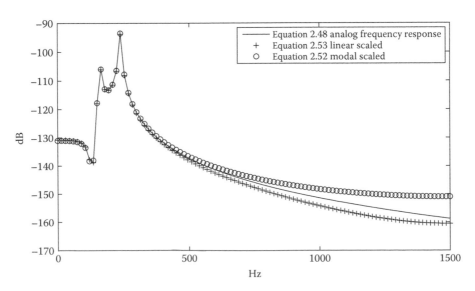

FIGURE 2.11   Comparison of the magnitude frequency response of the system in Figure 2.10 using a 3 kHz sample rate.

So that the "frequency warping" in the digital domain follows an inverse tangent.

$$\Omega = 2\tan^{-1}\left(\frac{\omega T}{2}\right) = 2\tan^{-1}\left(\frac{\omega}{2f_s}\right) \tag{2.59}$$

Equation 2.59 clearly shows the digital and analog frequencies being approximately equal only at very low frequencies and at the Nyquist rate ($f = f_s/2$) the analog frequency is infinite. MATLAB offers a way to make the digital and analog frequencies match at one frequency f_p.

$$\omega = 2\tan^{-1}\left(\frac{\Omega\tan\left(\pi\left(f_p/f_s\right)\right)}{2\pi f_p}\right) \tag{2.60}$$

This "prewarping" of the frequency axis is useful, albeit approximate for extending the range where the digital and analog domain mappings are accurate. It can be seen that Equations 2.58 and 2.60 are equal when $f_p = 0$.

2.4 MATLAB® EXAMPLES

Table 2.2 presents the MATLAB m-scripts used in making the various figures in this chapter. Readers can download these m-scripts directly from Taylor & Francis for further study.

There is nothing really fancy in Chapter 2 m-scripts, just an extensive use of ".*" and "./" to perform vector dot products and dot divides (element-by-element division). This technique is most foreign to new users of MATLAB and usually gives them the most trouble. But once mastered, it drastically simplifies the MATLAB code making it about as "readable" as possible for any software source code. For example, the impulse response coefficients in Equation 2.49 are scalars, but the impulse response is computed in one line as a vector, based on the "n-vector" defined as "$n = [0\ 1\ 2\ ...\ ntimepts-1]$." This allows Equation 2.49 to be written explicitly in one line as presented in Table 2.3.

The source for the code in Table 2.3 is "z2p4impl.m" and the nomenclature for "sp1" means "s-plane pole 1" and "sp2c" means "s-plane pole 2 conjugate." "Td" is the sample time interval in seconds so that "$n.*Td$" gives a vector of increasing sample times spaced Td seconds apart starting at zero second.

We also introduce some very useful graphics techniques in the figures of Chapter 2 as seen in the m-script code snippet in Table 2.4 (from "p2system.m"). Note that "ifs" is 201 frequencies starting at 0 Hz going up to the Nyquist rate ($f_s/2$). The vectors "s" and "z" are the same lengths as "ifs" as is the analog frequency response "hw" and its digital counterpart "hwzc." "T" is a scalar and is simple "$1/f_s$" or the time in seconds between digital samples.

TABLE 2.2
MATLAB m-Scripts for This Chapter

Script Name	Used in Figure(s)
z1se2sT.m	2.1
scaledimpulse.m	2.4, 2.5
p2system	2.6, 2.7
z2p4impl.m	2.8–2.11

TABLE 2.3
m-Script Snippets Showing Explicit Impulse Response

```
As = ((sp1-sz1)*(sp1-sz1c))/((sp1-sp1c)*(sp1-sp2)*(sp1-sp2c));

% analog system impulse response eqn (2.3.11)
n = 0:ntimepts-1;
h=As*exp(sp1.*n.*Td) + Cs*exp(sp2.*n.*Td) + Ds*exp(sp2c.*n.*Td);
```

The plots generated in the m-script snippet in Table 2.4 show two important techniques. First, the "subplot(2,1,1)" command creates a figure of two plot areas, one on top and one at the bottom, and sets the immediate plotting area to the top one. Second, the actual plot command overlays two separate plots, the first in a black line (defined by "*k*") and the second using black dots (defined by "*k.*"). See MATLAB help for more details on the use of "subplot" and "plot." This is a very useful feature for generating magnitude and phase plots for a system transfer function.

2.5 SUMMARY

The z-transform is a useful tool for describing sampled-data systems mathematically. Its counterpart in the continuous signal analog domain is the Laplace transform. By using linear frequency mapping between the analog s-plane of the Laplace transform and the digital z-plane of the z-transform, one can model any physical system band limited to a frequency strip on the s-plane using a digital system. In general, the guarantee of a band-limited signal in the analog domain requires electronic filtering before A/D conversion at a suitable rate high enough for the bandwidth to be represented without frequency aliasing. The parameters of the digital system model can be determined by mapping the s-plane analog system poles and zeros directly to the z-plane for the digital system. However, this mapping is not without accuracy limitations. It can be seen in this section that

TABLE 2.4
m-Script Snippet Showing a Magnitude and Phase Plot with Overlay

```
% system frequency responses

ifs = 0:fs/200:fs/2;
s = 0 + j.*2.*pi.*ifs;
z = exp(s.*T);

hw = ( 1./(sp1-sp1c) ).*( 1./(s-sp1) - 1./(s-sp1c) ); % eqn 2.2.13
hwzc= T.*( (zp1-zp1c)./(sp1-sp1c) ).*( 1./((z-zp1).*(z-zp1c)) ); % eqn 2.3.9
% note: multiply hwzc by z and they match, due to delay in digital system

figure(2);
subplot(2,1,1),plot(ifs,57.296.*(angle(hw)),'k',ifs,(180./pi).*(angle(hwzc)),'k.');
ylabel('Degrees');
xlabel('Hz');
legend('Equation 2.2.13','Equation 2.3.9');
title('Frequency Responses of Digital and Analog Systems');

subplot(2,1,2),plot(ifs,20.*log10(abs(hw)),'k',ifs,20.*log10(abs(hwzc)),'k.');
ylabel('dB');
xlabel('Hz');
legend('Equation 2.2.13','Equation 2.3.9');
```

the mapping is only precise at relatively low frequencies compared to the sampling rate f_s. In general, an accurate match is attainable in the $0 < f < f_s/4$ range, and only an approximate match is practical in the frequency range between $f_s/4$ and the Nyquist rate $f_s/2$ for real signals. For the frequency range approaching the Nyquist rate (f_s for complex signals), one must decide whether the impulse response match is more important than the match of the zeros in the steady-state frequency response. If so, each "mode" or system resonance for the digital system can be scaled independently to ensure a match to their analog system counterparts, giving a very accurate impulse response in the digital domain.

However, the modal scaling will affect the zero frequencies (or antiresonances) in the digital domain since one cannot control the phase of a digital pole–zero system at all frequencies. One could match both the magnitudes of the pole and zero frequencies in the digital system, but the phase at the system pole frequencies would have to be altered in order to have the zeros in the right place. Therefore, with the phases of the resonances adjusted, the digital impulse response would not match the true physical system. It should be clear that one cannot simultaneously match the impulse and frequency responses of a digital and corresponding analog pole–zero system unless one "over-samples" at a very high rate driving the spectral features of interest to low frequencies. Finally, MATLAB provides some tools for dealing with the frequency "warping" if a compromise is needed. However, we also show how to map the frequencies linearly and to properly scale the digital responses for a given analog system. This is a technique that is typically overlooked in the literature on digital filtering.

PROBLEMS

1. Prove the properties of linearity and shift invariance for both the Laplace transform and *z*-transform.
2. Show that for positive-moving time, the stable region for the *z*-transform is the region inside the unit circle and the stable region for the Laplace transform is the left-half *s*-plane.
3. Show that for the frequency scaling in Equation 2.13 where $|\alpha| = 1$, the digital domain scaling is actually the same as the analog domain frequency shifting in Equation 2.12 where "*a*" is imaginary.
4. Show that eliminating the digital delay in Equation 2.41, the digital and analog systems compared in Figures 2.6 and 2.7 become nearly identical.
5. Show that conjugate zeros and poles are needed in order to have real-valued coefficients and output given a real input to a system (digital or analog).
6. Assuming zero initial conditions, show the equivalence of differentiation in the analog and digital domains by mapping the Laplace transform in Equation 2.15 to the *z*-transform in Equation 2.22.
7. Build a digital system model for a mechanical oscillator as in Figure 2.3 where the system resonance is 175 Hz, with a damping of 40 Nepers/s, using a digital sample rate of 1000 real samples/s. Find the *z*-plane poles and scaled impulse response as described in Equations 2.41 through 2.45.

REFERENCES

1. W. E. Boyce and R. C. DiPrima, *Elementary Differential Equations and Boundary Value Problems*, 3rd ed., New York, NY: Wiley, 1977.
2. C. L. Phillips and H. T. Nagle, Jr., *Digital Control System Analysis and Design,* Englewood Cliffs, NJ: Prentice-Hall, 1984.
3. H. F. VanLandingham, *Introduction to Digital Control Systems,* New York, NY: MacMillan, 1985.
4. A. V. Oppenheim and R. W. Schafer, *Digital Signal Processing,* Englewood Cliffs, NJ: Prentice-Hall, 1975.
5. T. W. Parks and C. S. Burrus, *Digital Filter Design,* New York, NY: Wiley, 1987.

3 Digital Filtering

Digital filtering is a fundamental technique that allows digital computers to process sensor signals from the environment and generate output signals of almost any desired waveform. The digital computer in the 1990s has either replaced, or begun to replace, the analog electronics commonly found throughout the later part of the twentieth century. The twenty-first century begins with widespread digital filtering and compression of both audio and video information streamed across the Internet. The ¼″ phone plug and the "rca jack" are being replaced by the USB port, Firewire (IEEE 1394), and Ethernet RJ45 jacks for digital connectivity of devices and systems. For electronic control systems, the advent of the digital computer has meant the replacement of analog circuitry with digital filters. Microprocessor-based digital filters have the advantages of incredible versatility, repeatability, reliability, small size, and low power requirements. During the last quarter of the twentieth century, digital signal processors have replaced even low-technology electronics and mechanical systems such as thermostats, clocks/timers, scales, and even the mechanical control systems on the internal combustion engine. Digital filters are generally used to detect important information from electronic sensors, provide precision output signals to actuators in the environment, or to form a control loop between sensors and actuators. The most sophisticated kind of signal processing system makes use of additional information, human knowledge, sensor data, and algorithms to adaptively optimize the digital filter parameters. Before we engage the full potential of adaptive signal processing, we will first examine the fundamentals of digital filtering and its application in real physical systems.

In this section, we first describe the two most fundamental types of digital filter: the finite impulse response (FIR) and infinite impulse response (IIR) filters. Using results from system theory, FIR and IIR filter design techniques are presented with emphasis on physical applications to real-world filtering problems. Of particular importance are the techniques for insuring stable IIR filters and the effects of delay on the digital system parameters. Digital systems can also be designed based on state variables and this relationship to IIR filtering is presented. Generally, state variables are used when the information of interest is directly seen from the system state equations (position, velocity, acceleration, etc.). For processing two-dimensional (2D) data such as images or graphs, 2D FIR filters are presented. Image processing using 2D FIR filters (often referred to as convolution kernels) allows blurred or out-of-focus images to be sharpened. Finally, a set of popular applications of digital filters are given. Throughout the rest of this book, we will relate the more advanced adaptive processing techniques to some of these basic families of applications.

3.1 FIR DIGITAL FILTER DESIGN

There are many techniques for digital filter design which go far beyond the scope of this book [1]. In this section we will concentrate on the fundamental relationship between a digital filter and its frequency response. In subsequent chapters, we will also examine several more sophisticated design procedures. Consider the z-transform of Equation 2.3, but with z restricted to the unit circle on the z-plane ($z = e^{j2\pi f/f_s}$) as a means to obtain the frequency response of some digital signal $y[n]$.

$$Y[\Omega] = \frac{1}{N} \sum_{n=0}^{N-1} y[n] z^{-n}; \quad z = e^{j\Omega}; \quad \Omega = 2\pi \frac{f}{f_s} \tag{3.1}$$

43

The infinite sum in Equation 2.3 is made finite for our practical use where N is large. Equation 3.1 is called a *discrete time Fourier transform* (DTFT), the details of which will be explained in much greater detail in Chapter 6. If $y[n]$ is the output of a digital filter driven by a broadband input signal $x[n]$ containing every frequency of interest, the frequency response of the filter is found from system theory [2].

$$H[\Omega] = \frac{Y[\Omega]}{X[\Omega]} \tag{3.2}$$

The impulse response of our digital filter is found most conveniently by computing the *inverse discrete time Fourier transform* (IDTFT) using the sum of the responses at K frequencies of interest.

$$h_n = \frac{1}{2\pi} \sum_{k=0}^{K-1} H[\Omega_k] e^{+j\Omega_k n} \tag{3.3}$$

If $y[n]$ and $x[n]$ are both real (rather than complex-sampled data), $h[n]$ must also be real. The DTFT of a real signal gives both real and imaginary frequency response components in what are known as Hilbert transform pairs. The real part of the frequency response is symmetric for positive and negative frequency; $Re\{H[\Omega]\} = Re\{H[-\Omega]\}$. The imaginary part of the frequency response is skew-symmetric; where $Im\{H[\Omega]\} = -Im\{H[-\Omega]\}$. It follows from the fact that $e^{j\Omega}$ has the same symmetry of its real and imaginary components. In order to have $h[n]$ real, both positive, and their corresponding negative, frequencies must be included in the IDTFT in Equation 3.3. For complex time domain sampled data, the frequency range need only be from the lowest to highest positive frequency of interest. Any number of samples for $h[n]$ can be generated using the IDTFT in Equation 3.3. The most important aspect of using the IDTFT to design a digital filter is that *any* frequency response may be specified and inverted to give the filter impulse response.

FIR digital filters are usually designed by specifying a desired frequency response and computing an inverse Fourier transform to get the impulse response [3]. This highly convenient approach to filter design is limited only by the number of samples desired in the impulse response and the corresponding spectral resolution desired in the frequency domain. The finer the desired spectral resolution is, the greater the number of samples in the impulse response will need to be. For example, if the desired resolution in the frequency domain is 8 Hz, the impulse response will have to be at least 1/8th of a second long. At a sample rate of 1024 Hz, the impulse response would have at least 128 samples.

$$Y[z] = X[z]\{h_0 + h_1 z^{-1} + \cdots + h_{N-1} z^{-N+1}\} \tag{3.4}$$

The method for implementing a digital filter in the time domain is to use a difference equation based on the inverse z-transform of Equation 3.2.

Noting the delay property in Equation 2.5 where $z^{-1}X[z]$ transforms to $x[n-1]$, the inverse z-transform is seen to be

$$y[n] = h_0 x[n] + h_1 x[n-1] + \cdots + h_{N-1} x[n-N+1]$$
$$= \sum_{k=0}^{N-1} h_k x[n-k] \tag{3.5}$$

where the output $y[n]$ is the discrete convolution of the input $x[n]$ with the filter impulse response $h[n]$ as also seen in Equation 2.7. The impulse response is finite in Equation 3.5 due to the truncation at N samples. This type of filter is called a *finite impulse response*, or FIR, digital filter and has a convenient design implementation using the DTFT.

The question arises as to what penalty one pays for shortening the impulse response and simplifying Equation 3.5. In Figure 3.1 the response of a mass–spring–damper (MSD) system, with a damped resonance at 375 Hz and damping factor 10 Nepers, is modeled using an FIR filter and a sampling rate of 1024 samples/s. This system is similar to the one in Figure 2.3. The frequency responses for 1024-, 128-, 16-sample impulse responses are given in Figure 3.1 to graphically show the effect of truncation. The 8-sample FIR filter has only 128 Hz resolution, while the 32-sample FIR has 32 Hz, and the 1024-sample filter has 1 Hz resolution. Clearly, the resolution of the FIR filter in the frequency domain should be finer than the underlying structure of the system being modeled, as is the case for the 1024-sample FIR filter. All one needs to do to design an FIR filter is to specify the magnitude and phase and the desired frequency resolution and sample rate will determine the length of the FIR impulse response. Given a long enough FIR filter, *any frequency response can be realized.*

However, if the desired frequency response contains only antiresonances, or "dips," an FIR filter can model the response with absolute precision. Consider the case of multipath cancellation as shown in Figure 3.2. A distant transmitter radiates a plane wave to a receiver a distance "d" away from a reflecting boundary. The two different path lengths from source to receiver cause cancellation at frequencies where they differ by an odd multiple of half-wavelengths. From the geometry seen in Figure 3.2, it can be seen that the path difference for some angle θ measured from the normal to the reflecting boundary is

$$\Delta \ell = 2d \cos \theta \tag{3.6}$$

The geometry in Figure 3.2 is straightforward to solve. The segment AD is "d" meters long. Therefore, segments AC and AE are each $d/\cos \theta$ long. The angles in BAD, DAE, and BED are all θ, making the segment CE be $2d \tan \theta$ long. Therefore, BC is $2d \tan \theta \sin \theta$ long and the path difference along BAE is $(2d/\cos \theta)(1 - \sin^2 \theta)$, or simply $2d \cos \theta$. If we consider an acoustic wave where d is 0.3 m, θ is 30°, and the reflecting boundary has a reflection coefficient of +1, the direct plus reflected waves combine to give the following analog system frequency response

$$H(\omega) = 1 + e^{-jk\Delta\ell}; \quad k = \frac{2\pi f}{c} \tag{3.7}$$

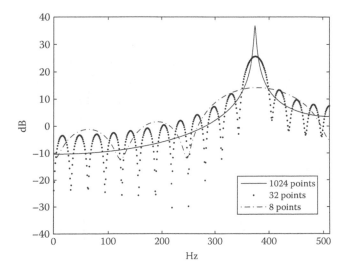

FIGURE 3.1 Comparison of 1024-, 32-, and 8-sample FIR filter models for a system with a damped resonance at 374 Hz.

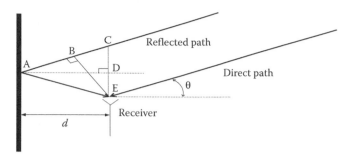

FIGURE 3.2 Multipath from a distant source gives a path difference seen along BAE which causes cancellation of some frequencies and enhancement of others.

where c is the speed of sound in air (about 343 m/s at 20°C). The path difference computes to be 0.5196 m giving a time delay between the direct and reflected paths of approximately 1.5 ms. If we are interested in frequencies up to 5 kHz, the digital sample rate is set to 10 kHz making the time delay approximately equal to 15 sample periods. The z-transform for a digital filter with a direct output plus a 15-sampled delayed output is simply

$$H[z] = 1 + z^{-15} \tag{3.8}$$

The digital filter model output $y[n]$ is found for the input $x[n]$ to be

$$y[n] = x[n] + x[n-15] \tag{3.9}$$

As can be seen in Equation 3.9, there's no need to include more samples in the FIR filter! Equation 3.8 is nearly an exact representation of the analog system and its modeling in Equation 3.7 and Figure 3.2. Figure 3.3 gives the frequency response of the system and also shows the effect of an imperfect reflection at the boundary. The weakness of the reflection changes the coefficient of $x[n-15]$ from 1.0 to 0.95 for 95% reflection, 0.5 for 50% reflection, and 0.10 for 10% reflection. Note that the acoustic pressure-squared increases (doubles for 100% reflection), or all multiples of

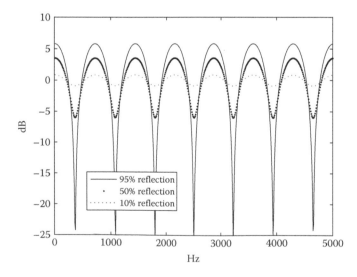

FIGURE 3.3 Imperfect reflection at the boundary causes only partial cancellation and less enhancement of some frequencies.

667 Hz and decreases (cancels for 100% reflection) for odd multiples of 333 Hz. This frequency response structure is generally referred to as a comb filter (affectionately named after the basic personal hygiene tool many signal processors used to be acquainted with). Comb filter effects are common ailments for sensors placed in multipath environments or near reflectors. It also occurs when multiple microphones near each other are mixed together, giving an unintended response like listening through a pipe. There is little one can do to correct comb filtering problems that is more effective than simply moving the sensor to a better location. However, the comb filter is the natural match for the FIR filter model since both systems contain only "dips" or "antiresonances" in their frequency responses. The FIR polynomial in the z-domain can be factored into complex zeros which are located near the unit circle at the appropriate frequency angle. FIR filters are often referred to as "all-zero" filters, or even Moving Average (MA) filters since their outputs appear as a weighted average of the inputs.

3.2   IIR FILTER DESIGN AND STABILITY

IIR filters are essentially the spectral inverse of FIR filters. IIR filters are described by a ratio of polynomials in z and can have poles and zeros or only poles. The all-pole form of an IIR filter is often referred to as an Auto-Regressive (AR) filter [4]. The pole-zero form of an IIR is called Auto-Regressive Moving Average (ARMA). We will first examine the structure of an AR filter and its difference equation to establish a straightforward rule for IIR stability. Consider the following Mth-order AR filter.

$$H[z] = \frac{1}{1 + a_1 z^{-1} + a_2 z^{-2} + \cdots + a_M z^{-M}} \tag{3.10}$$

For an IIR filter input $x[n]$ and output $y[n]$, taking z-transforms gives the relation

$$Y[z] = X[z]H[z] = X[z]\frac{1}{1 + a_1 z^{-1} + a_2 z^{-2} + \cdots + a_M z^{-M}} \tag{3.11}$$

Rearranging Equation 3.11 one has

$$Y[z]\{1 + a_1 z^{-1} + a_2 z^{-2} + \cdots + a_M z^{-M}\} = X[z] \tag{3.12}$$

where taking inverse z-transforms gives the AR difference equation

$$y[n] = x[n] - \sum_{i=1}^{M} a_i y[n-i] \tag{3.13}$$

Equation 3.13 clearly shows that an IIR is generated for an arbitrary input $x[n]$. This is because each output sample $y[n]$ contains past output samples, so the impulse response of the filter lingers on well beyond M samples. However, because of the feedback of past outputs back into the current output, we must be very careful not to create an unstable system. Instability occurs whenever a past output is regenerated at larger amplitudes, so that each "echo" gets louder. In addition, for the IIR filter to be realizable in the time domain, the current output $y[n]$ must not contain and future inputs or outputs. A straightforward way to examine stability is write $H[z]$ as a partial fraction expansion.

$$H[z] = \frac{A_1}{1 - z_1 z^{-1}} + \frac{A_2}{1 - z_2 z^{-1}} + \cdots + \frac{A_M}{1 - z_M z^{-1}} \tag{3.14}$$

Applying the inverse transform techniques described in Equations 2.41 and 2.42 it can be seen that the AR impulse response is a sum of M geometric series.

$$h[n] = A_1 \sum_{k=0}^{\infty} z_1^k z^{-k} + A_2 \sum_{k=0}^{\infty} z_2^k z^{-k} + \cdots + A_M \sum_{k=0}^{\infty} z_M^k z^{-k} \tag{3.15}$$

Clearly, for $h[n]$ to produce a stable output $y[n]$ from a stable input $x[n]$, the magnitude of the quotient z_i/z; $i = 0, 1, 2, \ldots, M$ must be less than unity (i.e., $|z_i/z| < 1$). Therefore, it can be seen that the zeros of the denominator polynomial of $H[z]$, or the poles of $H[z]$, must all have magnitude less than unity requiring them to all lie within the unit circle on the complex z-plane to guarantee stability.

Consider the MSD system modeled with FIR filters in Figure 3.1. We can derive an IIR filter with a very precise match to the actual system in both frequency response and impulse response using the s-plane to z-plane mapping techniques of the previous section. From examination of Equations 2.39 through 2.45, the mapped and modal-scaled IIR filter is

$$H[z] = \left\{ \frac{e^{-\zeta T} \sin(\omega_d T)}{\omega_d} \right\} \frac{z^{-2}}{1 + a_1 z^{-1} + a_2 z^{-2}} \tag{3.16}$$

where $\omega_d = 750\pi$, $T = 1/1024$ s, and $\zeta = 10$ Nepers. The IIR coefficient $a_1 = -(z_1 + z_2) = -2e^{-\zeta T}$ $\cos(\omega_d T)$ and $a_2 = z_1 z_2 = e^{-2\zeta T}$. Numerical calculations reveal that the pole magnitudes are approximately 0.99 giving a stable IIR filter. The scale factor, defined as b_0, works out to be 3.19×10^{-4} while $a_1 = -1.32$ and $a_2 = 0.98$.

$$H[z] = \frac{b_0 z^{-2}}{1 + a_1 z^{-1} + a_2 z^{-2}} \tag{3.17}$$

The IIR finite difference equation is simply

$$y[n] = b_0 x[n-2] - a_1 y[n-1] - a_2 y[n-2] \tag{3.18}$$

The three-term sum in Equation 3.18 is actually more accurate than the 1024-term sum shown in Figure 3.1! In fact, a surprisingly easy way to generate a sinusoid of frequency ω_0 with very low distortion is to implement Equation 3.18 with $a_1 = -(z_1 + z_2) = -2 \cos(\omega_0 T)$ and $a_2 = z_1 z_2 = 1$ where $x[n]$ is zero and $y[n-1]$ and $y[n-2]$ are two adjacent samples of the desired waveform, allowing the phase and amplitude to also be set. IIR filtering is very precise when the actual physical system being modeled is composed of spectral peaks which are easily modeled using poles in a digital filter. The same is true for pole-zero systems modeled by ARMA digital IIR filters. However, a serious question to be answered is how one can determine the ARMA numerator and denominator polynomial orders in the model. Most techniques for ARMA system identification provide a "best fit" for the model, given the number of poles and zeros to be modeled. Guessing the model orders wrong may still lead to a reasonable match between the desired and actual filter responses, but it may not be the best match possible. In general, a pole–zero difference equation for an ARMA model with Q zeros and P poles is seen in Equation 3.19. ARMA models derived from mapped s-plane poles and zeros are also subject to the limitations of matching both the impulse and frequency responses simultaneously at high frequencies as described in detail in Chapter 2. Figure 3.4 gives a block diagram for an ARMA IIR filter.

$$y[n] = \sum_{i=0}^{Q} b_i x[n-i]x - \sum_{j=1}^{P} a_j y[n-j] \tag{3.19}$$

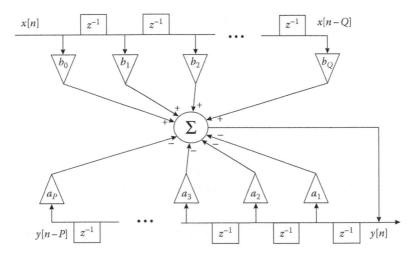

FIGURE 3.4 A block diagram of an ARMA IIR filter showing tapped delay lines for the input signal $x[n]$ and output signal $y[n]$.

3.3 WHITENING FILTERS, INVERTIBILITY, AND MINIMUM PHASE

The whitening filter for the ARMA IIR system $H[z] = B[z]/A[z]$ is its spectral inverse, $H^w[z] = A[z]/B[z]$. The spectral product of $H[z]$ and $H^w[z]$ is unity, or a spectrally "white" flat frequency response with unity gain. If both $H[z]$ and its inverse $H^w[z]$ exist as stable filters, both filters must have a property known as *minimum phase*. While IIR filters are far more flexible in modeling transfer functions of physical systems with spectral peaks than FIR filters, the major design concern is stability. This concern is due to the feedback of past IIR filter outputs into the current output as seen in the last term in Equation 3.19. This feedback must be *causal*, meaning that the current output $y[n]$ must be a function of only current and past inputs and past outputs. From Equation 3.15, it can be seen that the denominator polynomial $A[z]$ of an IIR system must have all of its zeros inside the unit circle on the z-plane for a stable impulse response. A polynomial with all its zeros inside the unit circle is said to be a *minimum phase* polynomial. For $H[z]$ to be stable, $A[z]$ must be minimum phase and for $H^w[z]$ to be stable $B[z]$ must also be minimum phase. For both $H[z]$ and $H^w[z]$ to be stable, both systems must be minimum phase in order to be invertible into a stable causal system.

Consider the effect of a delay in the form of $H[z] = z^{-d}B[z]/A[z]$ where both $B[z]$ and $A[z]$ are minimum phase. $H[z]$ is stable and causal, but $H^w[z] = z^{+d}A[z]/B[z]$ is not causal. Applying the ARMA difference equation 3.19, it can be seen that for output $x[n]$ and input $y[n]$, the whitening filter $H^w[z]$ is unrealizable because future inputs $y[n + d - j]$ are needed to compute the current output $x[n]$.

$$x[n] = \frac{1}{b_0}\left\{\sum_{j=0}^{P} a_j y[n+d-j] - \sum_{i=1}^{Q} b_j x[n-j]\right\} \tag{3.20}$$

Nonminimum phase numerator polynomials can be seen to contribute to the system delay. To examine this further, we consider a simple FIR system for real signals with a pair of conjugate-symmetric zeros corresponding to 60 Nepers at 300 Hz. With a digital sample rate of 1000 Hz, the two zeros map to a magnitude of $e^{+0.06}$, or 1.062, and angles $\pm 0.6\pi$, or $\pm 108°$, on the z-plane. Since the zeros are slightly outside the unit circle, the inverse filter (which is an AR IIR filter) is unstable. However, we can minimize the FIR filter's phase without affecting its magnitude response by adding an *all-pass* filter $H^{ap}[z]$ in series. As seen in Figure 3.5, the all-pass filter has a pole to cancel each

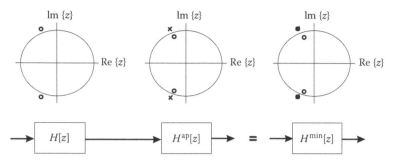

FIGURE 3.5 The all-pass filter $H^{ap}[z]$ replaces nonminimum phase zeros with their inverse conjugates maintaining the same magnitude response while minimizing the phase response.

zero outside the unit circle, and a zero in the inverse conjugate position of the pole to maintain a constant frequency response.

The all-pass filter is by itself unstable due to the poles outside the unit circle. However, its purpose is to replace the nonminimum phase zeros with their minimum phase counterparts. The unstable poles are canceled by the nonminimum phase zeros, and therefore are of no consequence. The resulting minimum phase system $H^{min}[z]$ will have the same magnitude frequency response, but a very different phase response. We should also expect the impulse response to be significantly affected by the imposition of a minimum phase condition. Figure 3.6 compares the magnitude and phase of the original 2-zero FIR system, the all-pass filter, and the resulting 2-zero FIR minimum phase filter. As the nonminimum phase zeros move toward infinity, the amount of phase lag in the

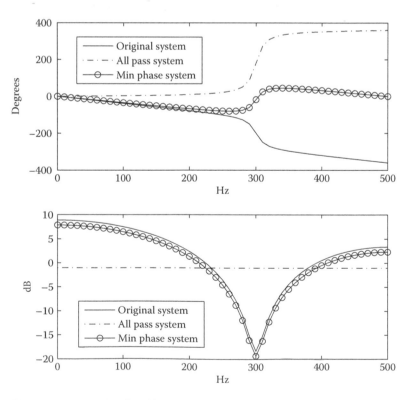

FIGURE 3.6 The all-pass filter has the effect of minimizing the phase of the original FIR system without causing a significant change in magnitude response.

FIR filter increases. *In the extreme case with a pair of conjugate zeros at infinity, the all-pass filter will leave a pair of zeros at the origin which correspond to a delay of two samples in the minimum phase system.* Minimum phase filters are also referred to as minimum delay filters since phase is also defined as radian frequency times delay. It follows then that constant delay filters have a linear phase response with negative slope.

As can be seen in Figures 3.5 and 3.6, the minimum phase system is invertible and differs only in phase from the nonminimum phase system (the less than 1 dB amplitude difference in Figure 3.6 is of little consequence to most people). But since we are relentless in our drive for signal processing omnipotence, we notice that the magnitude error gets worse with decreasing sample rate and better when the zero to be compensated moves toward the unit circle. Following the mapping between the s-plane and z-plane as seen in Chapter 2, we find that the scale factor for an all-pass filter compensating a pair of conjugate zeros z_1 and z_1^* is

$$H^{AP}[z] = |z_1|^2 \frac{\left(z - \left(1/z_1^*\right)\right)\left(z - \left(1/z_1\right)\right)}{(z - z_1)(z - z_1^*)} \tag{3.21}$$

Equation 3.21 has poles at z_1 and z_1^* to cancel the original filters nonminimum phase zeros, and replaces the canceled zeros with ones located at their inverse conjugate positions (inside the unit circle). The scale factor $|z_1|^2$ comes from the mapping of the all-pass system from the s-plane to the z-plane, assuming no scaling is needed for the s-plane all-pass system. Figure 3.7 shows the results of including the scale factor for the system in Figure 3.6.

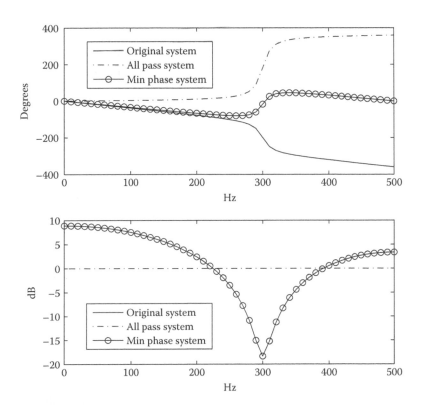

FIGURE 3.7 Applying the all-pass scale factor from Equation 3.21 gives a precise magnitude match between the original system and the minimum phase system.

If we wish to design an all-pole IIR filter to model an arbitrary frequency response where the spectral peaks are of primary interest, we could estimate an FIR model from the inverse Fourier transform of the spectral inverse of the IIR system. The FIR model is seen as a whitening filter because the product of its frequency response times the original frequency response of interest is a constant (spectrally flat or "white"). One would invert the spectrum of interest and compute the FIR whitening filter model from the inverse Fourier transform of the inverted spectrum. Because the phase of the original spectrum is arbitrary, the phase of the FIR whitening filter may or may not be minimum, giving one or more zeros outside the unit circle. This is typically the case when large numbers of zeros are modeled with long FIR filters. To invert the FIR whitening filter to get the desired stable IIR filter model, a minimum phase condition is imposed to insure stability of the IIR filter. One estimates the zeros of the FIR polynomial and "reflects" any nonminimum phase zeros into their respective inverse conjugate position to obtain an invertible minimum phase FIR filter. The phase of an FIR polynomial used as an FIR whitening filter can be arbitrarily defined in its Fourier transform, but the phase of an IIR is significantly restricted by the fact that its denominator polynomial must be minimum phase.

3.4 FILTER BASIS POLYNOMIALS

The advent of commercial radio in the 1930s created the need for electronic filters that adhered to precise specifications [5]. The frequency response of a filter can be modeled as a polynomial of the complex variable $s = \sigma + j\omega$ based on the impedances of a network of resistors, inductors, capacitors, and amplifiers. The section briefly outlines a number of these polynomials, their characteristics, and derivation. Using the techniques presented earlier in this chapter, one can map the s-plane poles and zeros to the z-plane for evaluation and use. MATLAB also offers many filter models in the signal processing toolkit which allow comparison of the frequency and phase responses, but using the bilinear or frequency warping techniques for the s-to-z plane mapping. Using these filter optimization features of MATLAB, it is recommended that the filter be designed specifically for the domain it is to be used, rather than mapping the poles and zeros. This is because the polynomials behave a little differently on the s-plane and z-plane. The various frequency responses of the polynomials trade off flatness of the loudness response in the pass band for steepness of the filter cutoff response. An ideal filter allows a constant (flat) response in the pass band and an infinitely steep cutoff transition to the stop band, where in theory zero signal passes. Real practical filters transition at roll-off slopes of −6 dB/octave (first-order filter) to −48 dB/octave (eighth-order filter) and may have some amount of amplitude "ripple" (deviation from flat response) in the pass band. Also, depending how steep the filter roll-off, the phase response can become quite distorted with increasing delay added to frequencies in the roll-off region.

3.4.1 BUTTERWORTH FILTERS

We begin with the oldest and simplest of the filter basis polynomials called the Butterworth filter, named after a British engineer named Stephan Butterworth who first published it in 1930. We begin by defining a general low-pass transfer function for an all-pole system of order n.

$$H(s) = \frac{K}{1 + c_1\left(s/\omega_c\right) + c_2\left(s/\omega_c\right)^2 + \cdots + c_n\left(s/\omega_c\right)^n} \tag{3.22}$$

Since the variable $s = \sigma + j\omega$ is complex, $H(s)$ has both a varying magnitude and phase which can be evaluated along the $j\omega$ axis of the s-plane. Since an ideal low-pass filter has a constant (flat) response in the pass band and zero response above the cutoff frequency ω_c we can solve for the coefficients c_i; $i = 1, 2, \ldots, n$ indirectly by first evaluating the derivatives of the magnitude-squared of $H(s)$ to make the function as flat as possible at $\omega = 0$.

$$|H(\omega)|^2 = \frac{K^2}{1 + C_1 (s/\omega_c)^2 + C_2 (s/\omega_c)^4 + \cdots + C_n (s/\omega_c)^{2n}}\Bigg|_{s=j\omega} \tag{3.23}$$

By inspection, for a derivative of Equation 3.23 to be zero, the derivative of the denominator must also be zero, so we can simplify our analysis by just inspecting the denominator. Clearly, for the second derivative to be zero, C_1 must be zero. For the fourth derivative to be zero, C_2 must be zero, and so on up to C_n which must be nonzero for the filter to reject high frequencies as desired. With $C_n = 1$ the "half power" response of our low-pass filter will occur at $\omega = \omega_c$. Equation 3.23 approximates an ideal low-pass filter by having nearly all its derivatives zero at $\omega = 0$ and its power response (squared magnitude) down by 1/2 at $\omega = \omega_c$.

$$|H(\omega)|^2 = \frac{K^2}{1 + (\omega/\omega_c)^{2n}} \tag{3.24}$$

Examining the poles of Equation 3.24 on the s-plane is facilitated by the following reduction:

$$|H(s)|^2 = H(s)H(-s) = \frac{K^2}{\left[1 + j(s/\omega_c)^n\right]\left[1 - j(s/\omega_c)^n\right]} \tag{3.25}$$

The poles of Equation 3.25 lie along a circle of radius ω_c on the s-plane, symmetric with the real σ-axis and imaginary $j\omega$ axis. There are two sets of poles as defined in Equation 3.26.

$$s_{1,m} = \omega_c \sqrt[n]{j} = \omega_c e^{j((1+4m)/2n)\pi}$$
$$s_{2,m} = \omega_c \sqrt[n]{-j} = \omega_c e^{j((3+4m)/2n)\pi} \tag{3.26}$$
$$m = 0, 1, 2, \ldots, n-1$$

Poles from either set in Equation 3.26 alternate around the circle on the s-plane defined by a radius of ω_c centered at the origin. For the case of $n = 1$ there are a total of two poles where the one at $-j\omega_c$ is associated with $H(s)$ and the one at $+j\omega_c$ is associated with $H(-s)$. All of the poles in the left-half s-plane are associated with $H(s)$ and the remaining poles in the right-half s-plane are associated with $H(-s)$. This allows the frequency response of $H(s)$ to be a stable system when evaluated for frequencies along the positive $j\omega$ axis on the s-plane. The circle of poles for the Butterworth filter also leads to some clever algorithms (not presented here) for calculating the filter coefficients c_i (not C_i used in the magnitude-squared response). Figure 3.8 compares the magnitude and phase response of several orders of Butterworth low-pass filters with cutoff frequency of 1 kHz.

As can be seen in Figure 3.8, even an eighth-order Butterworth filter has an undesirable transition into the stop band, although the slope is −48 dB/octave (−6 dB per model order per octave). The ideal low-pass filter would have unity gain in the pass band up to 1 kHz for our example and zero gain above 1 kHz. Why wouldn't one just make a much higher order polynomial filter? As it turns out, while the magnitude response gets better with higher order filters, the phase and time-delay

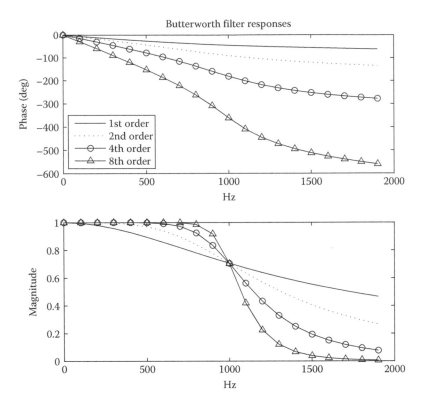

FIGURE 3.8 Magnitude and phase responses of first-, second-, fourth-, and eighth-order Butterworth filters showing nearly linear phase response and smooth magnitude response but with gentle roll-off.

distortion gets worse. To examine these phase distortions we introduce the *phase delay* τ_φ which measures the time delay as a function of frequency for the system.

$$\tau_\varphi = -\frac{\varphi(\omega)}{\omega} \tag{3.27}$$

The *group delay* is also useful for measuring where the phase response differs from *linear phase*.

$$\tau_g = -\frac{\partial\varphi(\omega)}{\partial\omega} \tag{3.28}$$

For a linear phase system, all frequencies arrive at the output with the same time delay relative to the input, which is the ideal case where $\tau_\varphi = \tau_g$. Figure 3.9 shows the group and phase delay for the Butterworth filters seen in Figure 3.8. As the filter order increases, the overall time delay of the filter increases along with the phase distortion in the filter transition region from the pass band to the stop band.

There are a number of common filter polynomials other than Butterworth which have useful response properties including Chebyshev (Type I and II), Elliptic, and Bessel responses. These are fairly common choice for filter design in either the analog of digital domains. Mapping the poles and zeros from the *s* to *z* planes can be done by using the bilinear transformation, frequency warping, or the linear frequency mapping presented earlier in Chapter 2. Amplitude scaling is a very simple matter because we generally want the pass band gain to be unity, so whatever scaling error that arises from the mapping and/or sample rate can be empirically adjusted for unity passband gain.

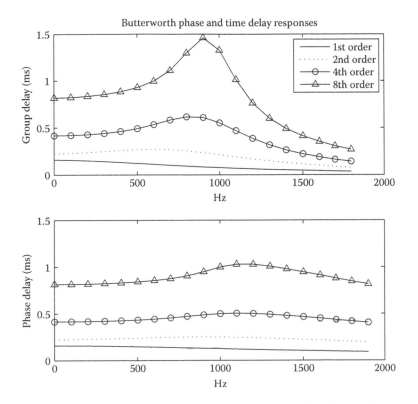

FIGURE 3.9 Phase and group delays for the Butterworth filters in Figure 3.8 showing distortions around the cutoff frequency of 1 kHz.

3.4.2 CHEBYSHEV TYPE I FILTERS

Chebyshev Type I filters have the desirable characteristic of a steeper roll-off into the stop band at the expense of a small ripple gain variation in the pass band. Chebyshev polynomials $T_n(x)$ have the special property of oscillating between ± 1 in the interval $-1 \leq x \leq +1$ and are solutions to a Chebyshev differential equation, which is a special case of a Sturm–Liouville equation. The generating functions are as follows:

$$
\begin{aligned}
T_0(x) &= 1 \\
T_1(x) &= x \\
T_{n+1}(x) &= 2x T_n(x) - T_{n-1}(x)
\end{aligned}
\tag{3.29}
$$

Note that for n odd, $T_n(0) = 0$ and for n even $T_n(0) = 1$. The magnitude-squared of the Chebyshev filter is

$$
\begin{aligned}
|H(s)|^2_{s=j\omega} = H(j\omega)H(-j\omega) &= \frac{K[1+j\varepsilon T_n(0)]}{\left[1+j\varepsilon T_n\left(\omega/\omega_c\right)\right]} \frac{K[1-j\varepsilon T_n(0)]}{\left[1-j\varepsilon T_n\left(\omega/\omega_c\right)\right]} \\
&= K^2 \frac{1+\varepsilon^2 T_n^2(0)}{1+\varepsilon^2 T_n^2\left(\omega/\omega_c\right)}
\end{aligned}
\tag{3.30}
$$

The parameter ω_c in Equation 3.30 is NOT the -3 dB cutoff frequency, but rather the frequency where the response leaves the "ripple box" of the pass band. The parameter K is just a scale factor to insure unity gain in the pass band. The actual −3 dB frequency can be found by direct solving ω in Equation 3.30 equal to 1/2 for the order chosen. It is generally a little higher in frequency for a low-pass filter. It can be shown that the ripple in the pass band is

$$\pm dB = 10\log_{10}(1+\varepsilon^2) \tag{3.31}$$

So, if $\varepsilon = 0.15$, the ripple is ± 0.1 dB. For $\varepsilon = 0.5$ the ripple is ± 1 dB and for $\varepsilon = 1.0$ the ripple is ± 3 dB. However, a smaller ε also means less roll-off in the stop band. Figures 3.10 through 3.12 compare the Chebyshev Type I to other filters of the same order. The poles of a Chebyshev Type I filter happen to lie in an ellipse centered on the s-plane origin and is not to be confused with an elliptical filter to be discussed below.

3.4.3 CHEBYSHEV TYPE II FILTERS

A second kind of Chebyshev filter is called Type II and uses an inverse Chebyshev polynomial so that the low-pass system actually has zeroes in the stop band and no poles (or ripple) in the pass band.

$$\left.|H(s)|^2\right|_{s=j\omega} = H(j\omega)H(-j\omega) = \frac{K}{\left[1+j\left(\dfrac{1}{\left(\varepsilon T_n\left(\omega/\omega_c\right)\right)}\right)\right]}\frac{K}{\left[1-j\left(\dfrac{1}{\varepsilon T_n\left(\omega/\omega_c\right)}\right)\right]}$$

$$= K^2 \frac{1}{1+\left(\dfrac{1}{\varepsilon^2 T_n^2\left(\omega/\omega_c\right)}\right)} \tag{3.32}$$

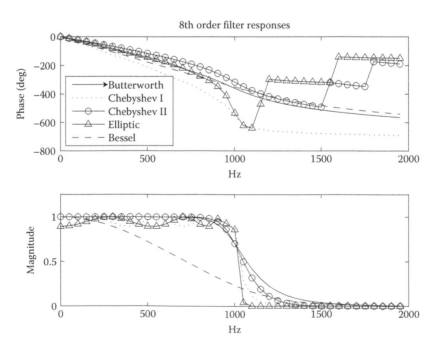

FIGURE 3.10 Magnitude and phase comparison of 8th order Butterworth, Chebyshev, Elliptic, and Bessel filters.

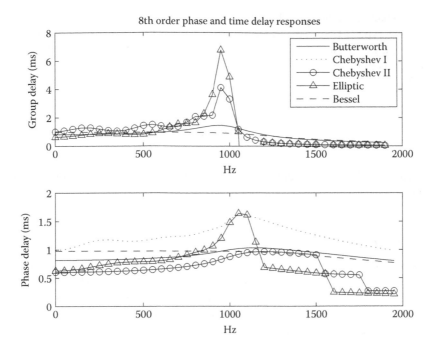

FIGURE 3.11 Phase and group delays for the eighth-order Butterworth, Chebyshev, Elliptic, and Bessel filters in Figure 3.10 showing distortions around the cutoff frequency of 1 kHz.

Now ε defines the gain in the stop band through maxima of the ripple. Usually, one defines the attenuation (worse case) in the stop band as Rs in dB. Then ε can be calculated by

$$\varepsilon = \sqrt{\frac{10^{-Rs/10}}{1-10^{-Rs/10}}} \tag{3.33}$$

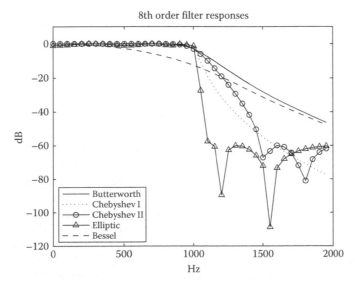

FIGURE 3.12 Log scale magnitude response comparisons of the Butterworth, Chebyshev, Elliptic, and Bessel filters showing the high stopband performance of the Elliptic filter.

So that a stop band attenuation of 10 dB leads to $\varepsilon = 0.333$. A 60 dB stop band attenuation gives an $\varepsilon = 0.001$. The Type II Chebyshev filter breaks out of the ripple box up in the stop band, so the 1/2 power frequency of the low-pass filter is substantially lower than ω_c. For the Type I filter the 1/2 power frequency is higher than ω_c. If we desire a 1/2 power frequency of $\omega_{1/2}$, the needed ω_c for Equation 3.32 is

$$\omega_c = \omega_{1/2} \cosh\left(\frac{1}{n} \cosh^{-1} \frac{1}{\varepsilon}\right) \tag{3.34}$$

Figures 3.10 through 3.12 compare the Chebyshev Type II (with ω_c compensated by Equation 3.34) with the other types of low-pass filters.

3.4.4 ELLIPTICAL FILTERS

Elliptical filters have both poles and zeros to maximize the filter magnitude response relative to the ideal low-pass filter. There is some ripple in the pass band (like the Chebyshev Type I) and zeros in the stop band (like the Chebyshev Type II) to give the steepest possible roll-off and solid attenuation in the stop band while giving good gain in the pass band. The pass band ripple and the stop band attenuation can be defined independently.

$$
\begin{aligned}
|H(s)|^2_{s=j\omega} = H(j\omega)H(-j\omega) &= \frac{K[1+ j\varepsilon R_n(\xi, 0)]}{\left[1+ j\varepsilon R_n\left(\xi, (\omega/\omega_c)\right)\right]} \frac{K[1- j\varepsilon R_n(\xi, 0)]}{\left[1- j\varepsilon R_n\left(\xi, (\omega/\omega_c)\right)\right]} \\
&= K^2 \frac{1+\varepsilon^2 R_n^2(\xi,0)}{1+\varepsilon^2 R_n^2\left(\xi, (\omega/\omega_c)\right)}
\end{aligned} \tag{3.35}
$$

The function $R_n(\xi, \omega)$ in Equation 3.35 is known as an elliptic rational function, or a Chebyshev rational function where ξ is called the selectivity factor. The mathematical details of the elliptic rational function are significantly beyond the scope of this book. The ripple in the pass band is set by ε (in dB using Equation 3.31), and the worst-case attenuation in the stop band is set by solving

$$
\begin{aligned}
dB &= 10 \log_{10} (1+\varepsilon^2 L_n^2) \\
L_n^2 &= R_n(\xi, \xi)
\end{aligned} \tag{3.36}
$$

The elliptic filter is seen as a general class of filter. In the limit as $\xi \to \infty$ the elliptic rational function becomes a Chebyshev polynomial and the elliptic filter becomes a Chebyshev Type I filter. If $\xi \to \infty$, $\omega_c \to 0$, and $\varepsilon \to 0$ such that $\varepsilon R_n(\xi, 1/\omega_c) = 1$, then the elliptic filter becomes a Butterworth filter. And if $\xi \to \infty$, $\omega_c \to 0$, and $\varepsilon \to 0$ such that $\xi\omega_c = 1$ and $\varepsilon L_n = \varepsilon'$, then the elliptic filter becomes a Chebyshev Type II were $\varepsilon = \varepsilon'$. The poles of an elliptic filter lie in an ellipse centered on the origin of the s-plane, and the zeros lie on the $j\omega$ axis above $+j\omega_c$ and below $-j\omega_c$. The frequency where the roll-off (transition region between ω_c and ξ) crosses the stop band attenuation level is ξ. The elliptic filter is the best approximation to the ideal "brick wall" low-pass filter response, but has the most phase and time delay distortion in the roll-off region. Figures 3.10 through 3.12 compare the elliptic filter response with other filter types.

3.4.5 Bessel Filters

The Bessel filter has the best possible phase response (nearly linear) but suffers from a rather poor roll-off into the stop band. Not to be confused with Bessel functions, the Bessel filter is based on reverse Bessel polynomials. The transfer function of a low-pass Bessel filter of order n is

$$H(s) = \frac{\theta_n(0)}{\theta_n(s/\omega_c)} \tag{3.37}$$

where $\theta_n(x)$ is the nth-order reverse Bessel polynomial defined by

$$\theta_n(x) = \sum_{k=0}^{n} \frac{(2n-k)!}{(n-k)!k!} \frac{x^k}{2^{n-k}} = \sqrt{\frac{2}{\pi}} x^{n+1/2} e^x K_{(n+1/2)}(x) \tag{3.38}$$

where $K_{(n+1/2)}(x)$ is a modified Bessel function of the second kind. Bessel filters are usually found in applications where linear phase (a simple time delay) is of high importance, such as loudspeaker crossover filter designs. The Bessel responses can be seen in Figures 3.10 through 3.12.

3.4.6 High-Pass, Band-Pass, and Band-Stop Filter Transformations

The various low-pass filter basis functions can be easily transformed by simple and well-known frequency transformations. Let the normalized frequency s/ω_c be s'.

Low-Pass to High-Pass: Replace s' by $1/s'$
Low-Pass to Band-Pass: Replace s' by $B(s' + 1/s')$, where B is the bandwidth in radians
Low-Pass to Band-Stop: Replace s' by $B^{-1}(s' + 1/s')^{-1}$

For the cases of the band-pass and band-stop filters, ω_c ends up in the middle of the pass band or stop band, respectively, and the filter order is doubled. As it appears, this transformation is algebraically straightforward and signal processing tools such as MATLAB's signal processing toolkit embed these transformations into their processing functions.

3.4.7 MA Digital Integration Filter

The MA filter gets its name from a simple process of making the output sample equal to the average of the last n input samples. An Nth order MA filter is depicted in Equation 3.39 and for $N = 10$ in Figure 3.13.

$$y[n] = \frac{1}{N} \sum_{k=0}^{N-1} x[n-k] \tag{3.39}$$

It's interesting to note that the $N - 1$ zeros of the filter end up evenly spaced around the unit circle on the z-plane except that one is missing at the point $z = 1.0$, or a 0 Hz. Clearly, the resulting low-pass filter has the shape of a $\sin(x)/x$ or $\text{sinc}(x)$ function, which is the expected Fourier transform of a rectangular impulse response. For a 10th order MA filter, the output sample $y[n]$ is the average of the current input sample $x[n]$ and the previous nine input samples. The first zero occurs exactly at 1/10 the sample rate as seen on the right in Figure 3.13. This operation acts as a rather poor low-pass

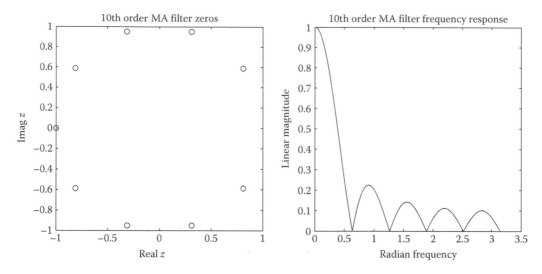

FIGURE 3.13 The zeros (left) and magnitude response (right) of a 10th-order MA averaging filter.

filter, which suffers from dropping pass-band response near the cutoff frequency (1/10th the sample rate for Figure 3.13), and also rather poor leakage in the stop band. However, "averaging" the MA filter in Equation 3.39 is most accurate at 0 Hz. But if this filtering is part of a delta sigma convertor, a better choice for the filter would be an ARMA Elliptical filter. To increase the effective number of bits in the fixed-point samples, the filter coefficients are simply scaled up, analogous to eliminating the "1/N" in the averaging MA filter. The output signal, with its smaller bandwidth (say 1/N) can then be decimated by a factor of 1/N, yielding more bits per sample but fewer samples per second. Section 1.4 explains this in the context of the sigma–delta convertor, but using a crude averaging filter rather than a more precise low-pass filter. In Section 4.4, we will apply this technology in a signal compression algorithm where a bank of filters produce an array of signal streams where the bit-depth of each stream can be set to reduce the total bit rate and associated file size. These compression techniques are the basis for a wide range of standards for digital file size compression, image compression, audio compression, and video compression.

3.5 MATLAB® EXAMPLES

The techniques used in this chapter are best simulated in MATLAB by also using the signal processing toolbox from the MathWorks, makers of MATLAB. One of the most important features of the filter design algorithms in the signal processing toolkit is that they are optimized for the z-plane and s-plane separately. While in theory, one could take the basis polynomial on the s-plane and map the poles and zeros to the z-plane using the bilinear transform, the frequency warping transform, or the "linear" mapping presented at the end of Chapter 2. In practice, there are numerical errors from the polynomial optimization which can be a problem when mapping from one domain to the other. *We recommend optimizing a digital filter using the z-plane and an analog filter using the s-plane.* Consider the eighth-order elliptical bandpass filter with passband from 1.2 to 1.4 kHz defined by −0.1 dB ripple in the pass band and −60 dB attenuation in the stop band seen in Figure 3.14. The solid curve in Figure 3.14 shows the outstanding performance of the elliptical filter response on the s-plane. However, neither the bilinear transformed response (dotted line, Equation 2.56), nor the warped bilinear response (dashed line, f_p = 1.3 kHz, Equation 2.60) match the corner frequencies at the pass band edges. Figure 3.15 compares the s-plane response with the "linear" mapping (dashed line) from the s-plane to z-plane following Equation 2.47, which matches corner frequencies but has

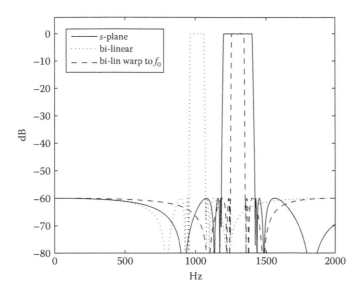

FIGURE 3.14 Numerical errors from the nonlinear mapping of *s*-plane poles and zeros to the *z*-plane lead to performance errors in practice. Direct design in the *z*-plane is recommended although mapping should work in theory.

an offset and tilt in the pass band. The actual scaling for the "linear mapped" *z*-plane response is found by

$$H(\omega) \approx H[z]\big|_{z=e^{j\omega T}} = T \prod_{n=1}^{N_p} \frac{(p^z_{2n} - p^z_{2n-1})}{(p^s_{2n} - p^s_{2n-1})} \prod_{k=1}^{K_z} \frac{(z^s_{2n} - z^s_{2n-1})}{(z^z_{2n} - z^z_{2n-1})} H^m[z] \tag{3.40}$$

where $H^m[z]$ is the *z*-plane transfer function with the mapped poles and zeros from the *s*-plane, the "$2n$" and "$2n-1$" pole or zero pair are complex conjugates, and T is the sample interval. This

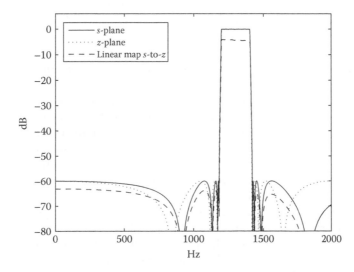

FIGURE 3.15 Linear mapping of the *s*-plane poles and zeros provides a better matching of the filter corner frequencies, but there is still a slight error in magnitude in the pass band due to numerical errors in the mapping.

TABLE 3.1
m-Scripts Used in Chapter 3

Script Name	Used in Figures
P1FFT.M	1
Multipath.m	3
Minphase.m	6
Filtersresp.m	8,9,10,11,12
MAFILT.M	13
bpellip.m	14,15

mapping get one fairly close for low-order filters with corner frequencies near the middle of the Nyquist band of the digital system.

For cases where the corner frequencies are near 0 Hz or the Nyquist rate in the digital system, there can be a range of additional numerical errors from optimizing the basis polynomials to the design parameters. This is due to the polynomial behavior on the s-plane as compared to the z-plane, but it also can arise from the polynomial optimization algorithm itself. The z-plane is periodic in frequency and also subject to frequency aliasing of the s-plane polynomial behavior outside the Nyquist band of the digital system. Depending on where one places the bass band, the filter order (which actually doubles for band pass and band stop filters), and the polynomial order, the linear mapped z-plane response may or may not be a good match with the original s-plane response. The MATLAB scripts used for the various figures of this chapter are summarized in Table 3.1.

3.6 SUMMARY

In this chapter, we presented two main linear digital filter structures in the form of FIR and IIR filter systems. There is far more to digital filter design than is presented here and the interested reader should refer to some of the excellent texts available on the subject. In order to achieve the best possible results in designing linear digital filters which model real physical systems, one must be driven by the underlying physical structure of the real physical system of interest. For example, if a relatively short impulse response is required where the system frequency response is composed mainly of spectral dips (antiresonances), an FIR filter design is generally the best choice. If the system of interest has a long impulse response, or has a frequency response characterized by dominant spectral peaks (resonances), a stabilized IIR filter should give the best results. However, the phase of the IIR filter will be very strictly constrained by the causality and minimum phase requirements for stability. If both the magnitude and phase of the physical system and its corresponding digital filter must be precisely matched with resolution Δf Hz, then one should specify an FIR filter with impulse response duration $1/\Delta f$ seconds. The main advantage for IIR digital filters is that they can represent spectral peaks accurately using a low number of coefficients. One can specify the physical parameters and map them to the z-plane (as done in the previous section), and obtain an accurate impulse response through modal scaling or oversampling techniques. Digital filters are the building blocks of real-time control systems and many signal processing operations.

This chapter discusses a lot of detail in mapping s-plane filters to the z-plane and the associated numerical errors. The relationship between the analog s-plane and the digital z-plane is not linear and the mapping from one domain to the other requires a few considerations because not all features in one domain will map properly to the other domain. One way to control these mapping errors is to significantly oversample the digital system so that the spectral features of the filter are confined to very low frequencies in the digital bandwidth. For oversampling, the classic bilinear transformation can be used to map the poles and zeros from the s-plane to the z-plane. If one only needs an approximate match but oversampling is not desired, one can map the z-plane pole or zero using $z = e^{sT}$ where T is the sample period. This will place the pole or zero at the correct frequency in the digital domain, but the spectral response of the filter will vary slightly at both the spectral peaks (poles) and dips (zeros). One can also use an overall scale factor for this case of $1/T$ as described in Chapter 2. For the case where the impulse response of the filter needs to be a close match, a more careful "modal scaling" can be used, also known as the impulse invariant method which matches the spectral peaks very accurately, not causes the zeros to shift in frequency. Given well know filter polynomials in the analog s domain, one can use these various mapping strategies to create the equivalent (albeit approximate) filter in the digital domain. However, the algorithms for creating Butterworth, Chebyshev, and Elliptical filters for the s-plane can also be directly applied to the z-plane. This produces a very accurate filter in the digital domain with all the filter properties intact. The filter gain is easily set for unity gain in the pass band and this is the preferred method for generating digital filters. All of these methods can be analyzed in detail using the Signal Processing Toolbox in MATLAB.

PROBLEMS

1. For a sampling frequency of 44,100 Hz and real digital signals, design a FIR "notch" filter to reject 8000 Hz. Write the difference equation for computing the filtered output sequence.
2. How many coefficients are needed for a FIR filter to precisely model a system with sharp resonances approximately 10 Hz wide around 10 kHz where the sample rate is 50 kHz?
3. For modeling real physical systems with both poles and zeros, what is the best strategy for matching the frequency response as closely as possible.
   a. Scaling each mode appropriately
   b. Scaling the digital system by the sample period T
   c. Significantly oversampling the digital system and scaling appropriately
4. To match the impulse response of a physical system using a minimum bandwidth digital filter, which of the listed strategies in Problem 3.3 makes the most sense?
5. Show that as a conjugate pair of zeros of an ARMA filter moves further outside the unit circle, the net system delay increases.

REFERENCES

1. T. Kailath, *Linear Systems,* Englewood Cliffs, NJ: Prentice-Hall, 1980.
2. K. Steiglitz, *An Introduction to Discrete Systems,* New York, NY: Wiley, 1974.
3. J. Vlach and K Singhal, *Computer Methods for Circuit Analysis and Design,* New York, NY: Van Nostrand Reinhold, 1983.
4. F. J. Taylor, *Digital Filter Design Handbook,* New York, NY: Marcel-Dekker, 1983, 4th printing.
5. S. Butterworth, On the theory of filter amplifiers, *Wireless Engineer,* 1930, 7, pp. 536–541.

4 Digital Audio Processing

Apart from the interest expected from musicians and audio enthusiasts, audio engineering has significantly affected human art and culture. This chapter presents some interesting applications of basic signal processing including a historical description of how electronic sound effects emerge in music and movie recordings. After World War II, improvements in amplifiers and loudspeakers, driven mostly by the movie theater industry, led to improved recording techniques, and by the late 1950s, high-fidelity (HiFi) sound systems for the home. Improvements in the phonograph transducer and the vinyl plastic "records" extended Edison's ingenuous invention for well over a century by improving the dynamic range, playing time, and bandwidth. Les Paul, a famous guitarist and audio engineer in the 1950s developed a multitrack magnetic tape recorder by literally stacking single-channel tape recorders so that the tape reels could share a common axel and thus remain time aligned. This allowed Les and his then wife, singer Mary Ford, to separately record themselves singing in harmony with the existing recordings of themselves playing different parts, later to mix all the "recording tracks" together for the public's amazement. By the 1960s, artists such as the Beatles and Jimi Hendrix raised multitrack recording and mixing to an art form that persists today.

Another important development from audio engineering came from the precise use of filters and compression–expansion processors to compensate for the losses and distortions from recording on magnetic tape and vinyl records. The long play (LP) record required this preequalization which became a standard by the Recording Industry Association of America (RIAA). This allows audio signals to be prefiltered using the so-called "pre-emphasis" with a nonconstant frequency response introduced so that playback of the recording would have higher SNR after equalization with the inverse filter. HiFi playback systems also included adjustments for "bass" and "treble" to further tailor the frequency response to suit the listener's needs and the loudspeakers of the day. These tools were also employed in the recording studio to unnaturally alter sounds, first in movie sound tracks and later in music recordings. *These so-called "special effects" became part of the art, rather than just a preserver of the art.* This is a significant step where the signal processing technology crosses over into the arts. It appears that this crossover is both permanent and accelerating. Today there are very sophisticated efforts to restore old recordings and movies but removing distortions and noise to more faithfully preserve the art for future generations. There are also very sophisticated digital music data compression techniques which we will explain briefly. These techniques exploit both the signal dynamics in some frequency bands as well as the limitations of audio processing in the human ear and brain.

4.1 BASIC ROOM ACOUSTICS

The performance space provides a necessary enhancement of audio signals meant to be heard by a group of people whether it is governmental communication, theater performing arts, music performance, or religious celebration. This is most true deep in the audience space where millions of echoes of acoustic waves from the speech or music source form a reverberation which can either enhance or degrade the audio signal. Recording audio signals in the reverberant field carries this acoustic information about the space along with the recording so that the listener to the recording can experience the event as if he/she was there. The recording industry began to produce reverberation in the 1930s using large empty buildings, microphones, and loudspeakers

("I am the great and powerful Oz!" from the movie *The Wizard of Oz*). Even today, Capitol Records in Hollywood, California maintains several long concrete tunnels for reverberation use that are causing controversy with a new planned subway system nearby. By the 1950s, electronic reverberation was introduced by the Hammond Organ Company [1]. The original device was based on a Bell Labs mechanical delay to simulate long-distance telephone calls. The Hammond Organ Company eventually made a small device using springs vibrated by a transducer to transmit and reverberate sound waves detected electronically on the opposite end. To recreate the sound of a pipe organ in a large cathedral this device was essential, but it later made its way to guitar amplifiers and recording studios. More sophisticated electromechanical reverberation later emerged using sheet metal plates suspended by springs to diversify the echo times, but they were basically using the same principle as the spring reverberation device. These were certainly much less expensive than a building or concrete tunnel with a microphone at one end and a loudspeaker at the other.

Consider a mechanical signal delay from a 0.1 kg spring, 0.5 m long, under 10 Nt tension, transversely excited by a transmitting transducer on one end, and producing an electrical signal from a receiving transducer on the other end. The transverse wave speed is

$$c_s = \sqrt{\frac{T_s}{\rho_s}} = \sqrt{\frac{10.0}{0.1/0.5}} = 7.071 \, \text{m/s} \tag{4.1}$$

where T_s is the spring tension and ρ_s is the mass/unit length. The delay τ through the spring is therefore 70.72 ms, which is the same delay as a sound wave moving through about 78 ft of air (the speed of sound in air at room temperature is about 1100 ft/s).

Our spring delay device also naturally produces a reverberation pattern of echoes known as a "flutter echo" because the wave bounces back and forth along the spring, losing a little energy with each reflection but eventually decaying exponentially to zero. One can hear a nice flutter echo between any two parallel walls, or floor and ceiling, when the walls are not sound absorbing, such as hard plaster or concrete. If the energy absorption (loss) at each reflection is α $(0.0 < \alpha < 1.0)$, then the energy of the reflected wave is $(1 - \alpha)$ times the incident wave energy. The amplitude of the reflection can be modeled as $\sqrt{1-\alpha}$. Let the input signal to the mechanical spring delay be $x(t)$. The output signal can be shown to be

$$y(t) = x(t-\tau) + \sum_{k=1}^{\infty} (1-\alpha)^{2k} x(t-2k\tau) \tag{4.2}$$

where one sees factor of "$2k$" because a round trip of the wave in the spring involves two reflections.

For most people, a flutter echo type of reverberation as described in Equation 4.2 is very annoying. If the delay is short, the echoes arrive in small regular intervals and have a tonal or buzzing sound quality to them. If the delay is long, then the distinct regular echoes are very distracting to understanding speech or music passages. To correct this, the Hammond Company used several springs under different tensions. This created a more life-like diffuse reverberation where the flutter echoes were no longer distinct. This is because the vibrations from one spring excite the other springs mixing up the echoes. The same principle is applied to plate reverberation where a piece of non-square sheet metal is used instead of springs and the waves bounce around the plate. More expensive plate reverberation devices usually use a number of different-sized plates to increase diffusion of the reverberation. Diffuse reverberation is generally found in large enclosed spaces with rough boundaries such as stone cathedrals, large caverns, and indoor stadiums.

For a real rectangular room, the reverberation from a simple hand clap very quickly becomes diffuse even with very simple plane walls and ceilings [2]. This is because the sound waves from the source are spherical, not plane waves, and thus are made up of an infinite number of plane waves that scatter in all directions. When the wall/ceiling/floor surfaces are parallel one can add a flutter echo on top of the diffuse room reverberation. Only a few objects are needed on the walls, such as sconce lights, statues, or balconies, to diffuse the flutter echoes. A detailed formal analysis of sound scattering in rooms can be seen in Kuttruff [3], where the average time between reflections in a rectangular room is derived to be

$$\bar{t} = \frac{4V}{cS} \tag{4.3}$$

where V is the room volume, c is the speed of sound, and S is the room surface area. It is fascinating (at least to some) that this result is independent of room shape. The mean time between reflections is dominated by the smallest dimension of the room due to the ratio of V over S. When the room becomes small in one dimension (pan like), or in two dimensions (tunnel like), the echoes repeat rapidly as the sound bounces back and forth between the closely spaced boundaries with a smooth diffuse reverberation in the background. Even a long tunnel can produce a smooth diffuse reverberation signal when the flutter echoes are controlled by absorption near the source and receiver.

To derive a physics-based model for room reverberation we will assume an average wall energy absorption coefficient $\bar{\alpha}$ and a diffuse sound energy density of $D(t)$. Following a wave through each reflection we have

$$\begin{aligned} D(\bar{t}) &= D(0)(1-\bar{\alpha}) \\ D(2\bar{t}) &= D(0)(1-\bar{\alpha})(1-\bar{\alpha}) \\ &\vdots \\ D(N\bar{t}) &= D(0)(1-\bar{\alpha})^N \end{aligned} \tag{4.4}$$

which can be rewritten in terms of time, rather than an echo number, as follows:

$$D(t) = D(0)(1-\bar{\alpha})^{(cS/4V)t} \tag{4.5}$$

We can write Equation 4.5 in terms of exponential decay as follows:

$$D(t) = D(0)e^{-(cS/4V)[-\ln(1-\bar{\alpha})]t} \tag{4.6}$$

Since measuring the sound pressure level on a decibel (dB) scale is much more convenient than energy density, we can multiply both sides of Equation 4.6 by ρc^2 (ρ is air density and c is the speed of sound) to get sound pressure squared.

$$p^2(t) = p^2(0)e^{-(cS/4V)[-\ln(1-\bar{\alpha})]t} \tag{4.7}$$

Pressure squared is converted to dB by taking 10 times the logarithm base 10.

$$L_p(t) = L_p(0) - 4.34 \frac{cS}{4V}\left[-\ln(1-\bar{\alpha})\right]t \tag{4.8}$$

Note that $10/\ln(10)$ is about 4.34 and the natural logarithm in 4.8 is always negative, since $\bar{\alpha}$ is between 0 and 1. This result may seem unusual, but on a dB scale, the sound energy decays linearly with time which makes it very easy to measure. Given the known room volume and surface area, one can estimate the average sound absorption from the dB drop in a known period of time. The reverberation time T_{60} for the room is defined as the time it takes the reverberation to decay 60 dB. This works out to be

$$T_{60} = \frac{0.161V}{-S\ln(1-\bar{\alpha})} \tag{4.9}$$

in metric units and

$$T_{60} = \frac{0.049V}{-S\ln(1-\bar{\alpha})} \tag{4.10}$$

in English units for volume and area. *The room reverberation time T_{60} is about 13.8 times the exponential decay time constant τ for the room.* Equations 4.9 and 4.10 are known as Eyring's reverberation time equations and are reasonably accurate. One estimates the average absorption by averaging the absorption coefficients for each surface in the room with distinct absorption, weighted by the area of each surface divided by the total surface area. This can be done even more accurately in octave frequency bands since the absorption usually varies by frequency. A more simple formula was originally developed empirically by Sabine [4].

$$T_{60} = 0.049 \frac{V}{\bar{\alpha}S} \tag{4.11}$$

Sabine's equation (English units) is very easy to use, but it is really only accurate for low-absorption (very reverberant) rooms where $\bar{\alpha} \leq 0.2$.

To complete our brief survey of room acoustics we need to introduce the room constant R and the critical distance r_c. The room constant is essentially the denominator of equations (4.9 through 4.11, depending on which model is used) and represents the effective area of perfect absorption. For example, if $S = 10,000$ and $\bar{\alpha} = 0.1$, the room constant is $R = 1000$. This is equivalent to a room with a subarea of 9000 with no absorption ($\bar{\alpha} = 0.0$) and a subarea of 1000 with total absorption ($\bar{\alpha} = 1.0$). Note that for two rooms with the same room constant, the larger room will have a longer reverberation time. Finally, the critical distance, r_c, is the distance from a sound source where the sound directly from the source is of the same loudness as that of the reverberant field and is defined as

$$r_c = \sqrt{\frac{QR}{16\pi}} \tag{4.12}$$

where Q is the directivity of the source (1 for monopole, 2 for dipole, etc.). If the listener is closer than the critical distance, he/she will hear the reverberation but it will not interfere with the sound directly from the source. At distances greater than the critical distance, the reverberation dominates the sound signal, and speech patterns are difficult to understand. The critical distance depends only on room absorption, not on room volume.

It is useful to be able to relate the physical properties of the room acoustics and the location of the listener relative to the sound source to design an appropriate reverberation generating filter. For example, a small-volume but low-absorption room such as a shower stall would have a very short critical distance and the corresponding reverberation generating filter would have the direct signal (dry or anechoic sound) at a lower volume than the reverberation and the T_{60} of the reverberation would be very short (due to small room volume) and "flutter like." For a large cathedral or indoor stadium, the reverberation would be very long and diffuse, but relatively low in volume compared to the direct signal due to the large critical distance. This understanding of the basic room acoustics allows one to design an artificial reverberation processor to suit the desired effect.

4.2 ARTIFICIAL REVERBERATION AND ECHO GENERATORS

In the previous section, we provided a physical basis for the decaying reverberation in a real room, which generally provides a diffuse echo response, as well as a simple spring delay which provides a flutter echo pattern. Suppose we were to make a digital filter simulate the spring reverberation device as described in Section 4.1 which has a 70.72 ms delay for the direct signal and a 141.44 ms delay for the round-trip echo back and forth in the spring. Figure 4.1 shows the mechanical spring and the equivalent digital filter model. If the sample rate for the digital system is 44,100 samples/s (standard for CD quality sound), the direct path delay Q in Figure 4.1 is 3129 samples and the reverberant path delay is 6258 samples. The difference equation for the output reverberation is

$$
\begin{aligned}
y[n] &= x[n-Q] + \beta^2 y[n-2Q] \\
&= x[n-Q] + \beta^2 x[n-3Q] + \beta^4 y[n-4Q] \\
&= x[n-Q] + \beta^2 x[n-3Q] + \beta^4 x[n-5Q] + \beta^6 y[n-6Q] \\
&\vdots \\
&= \sum_{k=0}^{\infty} \beta^{2k} x[n-(1+2k)Q]
\end{aligned}
\tag{4.13}
$$

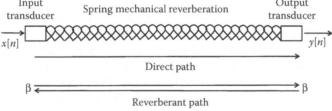

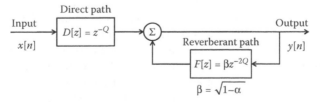

FIGURE 4.1 Mechanical spring reverberation and equivalent digital filter model showing the direct path and reverberant paths where the delay is depicted as Q samples.

From Figure 4.1, $\beta = \sqrt{1-\alpha}$, where α is the acoustic energy absorption for the reflection at each end of the spring. Clearly, when $0 < \alpha < 1$, it follows that $0 < \beta < 1$ and the output of Equation 4.13 is stable. However, there is an amplification factor of α^{-1} from the summation of many echoes.

Suppose we want our "flutter echo" to have an effective reverberation time of, say $T_{60} = 2.5$ s. This means that the acoustic energy output declines by a factor of 1/1000 (-60 dB) in about 2.5 s after the steady-state input is switched off. The amplitude of each round-trip echo in the spring declines by $\beta^2 = 1 - \alpha$ while the equivalent energy declines by $\beta^4 = (1 - \alpha)^2$ every 141.44 ms. To estimate the absorption required in the reverberant path, we first determine the number round-trip echoes ($T_{FB} = 141.44$ ms each) in the desired reverberation time of 2.5 s, which is about $N \approx T_{60}/T_{FB} \approx 18$. Solving for α one obtains

$$\alpha = 1 - e^{(\ln(0.001)/N)} = 0.1746 \tag{4.14}$$

so that about 17% energy absorption is needed at each end of the spring for a reverberation time of about 2.5 s and the reverberation would increase the overall signal level by a factor of about 5.7, or 15 dB. Another way to look at Equation 4.14 is that $(1 - \alpha)^{18} \approx 0.001$ or -60 dB.

A "flutter echo" that persists for 2.5 s would be very annoying and would interfere with speech, where syllables emerge 3–5 times/s (faster when angry or excited). To avoid interfering with speech the reverberation time would need to be less than about 200–300 ms. Shortening the reverberation for our flutter echo would give the impression of two people speaking (or singing) the same words. This is sometimes called a "slap-back" echo or voice-doubling echo and still is frequently used in music recording and movies to "fatten up" a voice or to provide an unusual special effect. The Beatles made extensive use of vocal double tracking throughout the 1960s, first by overdubbing a second or third vocal track, and later using a device they called the "auto double tracker," which was really just a reel-to-reel tape recorder where the output was mixed back into the overall recording. Separate recording and playback heads on the tape deck provided a delay, which could be lengthened by slowing the tape speed down. This technique was copied by many studios throughout the 1960s because it gave the impression of a "wall of sound" when listened on a small car radio [5,6]. If the reverberant path delay T_{FB} is shortened to something less than about 50 ms, the ear and brain can no longer distinguish separate echoes and begin to blend the signals together where the envelope and timbre of the sound is changed and speech intelligibility is less affected. These echoes are called "early reflections" in architectural acoustics because they have little negative impact on speech intelligibility.

Good-sounding acoustic spaces are devoid of flutter echoes in favor of a diffuse smooth reverberation with slightly different characteristics in each receiving channel if recorded in stereo of multichannel formats. The diffuse echoes occur naturally due to the scattering of sound waves and the coupling between various types of room modes. Consider the digital reverberation model in Figure 4.2. It shows the direct path, plus a number of "early reflections" and then a number of reverberant paths. Examples of early reflections are ray paths from the sound source bouncing off the floor to the listener, or ceiling, or walls. All these paths generally attenuate inversely with distance due to spherical spreading of the sound waves, but also the reflection surface takes some energy. There are only a limited number of early reflections that are important. In large auditoriums acousticians purposely design surfaces to provide early reflections to enhance the volume and clarity of sound. These are usually seen as reflecting panels near the podium or over the orchestral areas, relatively close to the sound source. The γs or gains of these reflections are less than unity relative to the direct path because they are longer paths and they involve loss at the surface reflection. Generally, the sound absorption mechanisms for an air-filled room are porous materials of finite thickness, which provide more absorption at higher frequencies than lower frequencies with longer wavelengths. A very realistic option for the γs in Figure 4.2 is to use a low-pass filter, so that the reflected amplitude is small as one goes higher in frequency. The same is true for the βs of the reverberant

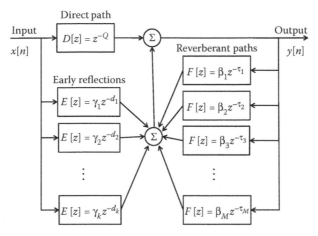

FIGURE 4.2 Digital filter model for typical room reverberation showing the direct path, separate early reflections (defined via ray tracing), and reverberant paths.

paths. However, having multiple reverberant paths with different delays is essential for eliminating the flutter echo and creating a diffuse reverberation filter. If there are M reverberant paths with unique (noninteger multiple) round trip delays, the mixing in the feedback path allows for M^2 combinations of delays with each group of echoes. In a short time there are literally millions of echoes involved in the output of the reverberation filter. In an actual room it is even more complicated than what we can do with a digital filter due to scattering and diffraction of the waves. However, the room reverberation filter structure in Equation 4.15 provides a physical basis to make a reasonable sounding reverberation.

$$y[n] = x[n-Q] + \sum_{k=1}^{K} \gamma_k x[n-d_k] + \sum_{m=1}^{M} \beta_m y[n-\tau_m] \qquad (4.15)$$

The room amplification factor relative to the direct path only is $1 + \sum_{k=1}^{K} \gamma_k + \bar{\alpha}^{-1}$, where $\bar{\alpha}$ is the average absorption coefficient for all the reverberant paths. Each early reflection path adds directly to the loudness at the listener location, which is why the stage end of a well-designed auditorium is very reflective.

Finally, we consider the reverberant path delays, τ_m samples in Equation 4.15. For a rectangular room of dimensions L_x, L_y, L_z, we can define three classes of room modes where plane waves bounce around in a repeating pattern. Each mode has a unique frequency if the room dimensions are distinct, but the round-trip travel time for all modes of a given type is the same. The first type is the *axial mode*, which like a flutter echo bounces back and forth between two parallel surfaces. The axial mode delay times in seconds are

$$t_x = \frac{2L_x}{c}; \quad t_y = \frac{2L_y}{c}; \quad t_z = \frac{2L_z}{c} \qquad (4.16)$$

and hence, for a digital sample rate of f_s, the sample offset for $\tau_x = f_s t_x$, and so on. The next type of rectangular room mode is called a *tangential mode* and bounces around in a rectangular pattern normal to a pair of parallel surfaces. The tangential mode delays are

$$t_{xy} = \frac{2}{c}\sqrt{L_x^2 + L_y^2}; \quad t_{yz} = \frac{2}{c}\sqrt{L_y^2 + L_z^2}; \quad t_{xz} = \frac{2}{c}\sqrt{L_x^2 + L_z^2} \qquad (4.17)$$

The last type of rectangular room mode is called an oblique mode and involves a plane wave reflecting off of all surfaces. The oblique mode delay time is

$$t_{xyz} = \frac{2}{c} \sqrt{L_x^2 + L_y^2 + L_z^2} \tag{4.18}$$

Clearly, the delay times in Equations 4.16 through 4.18 will result in a diverse pattern of echoes which very quickly emerge into diffuse reverberation. Since the tangential and oblique modes involve more than one reflection/round trip, the associated losses are greater than the axial modes. This is also due to the fact that absorption in air-filled rooms generally occurs from porous materials of finite thickness, such as ceiling tiles or fabric covered fiberglass wool panels mounted on walls. Since the tangential and oblique reflect off of these surfaces at a nonnormal angle, the effective depth of the absorption is greater relative to wavelength, giving a greater absorption.

4.3 FLANGING AND CHORUS EFFECTS

These types of "special effects" are usually only known to musicians and recording engineers, particularly electric guitar enthusiasts, because of their importance in modern recorded music and motion picture soundtracks. The term "flanger" or "flanging" is attributed to the late John Lennon of the Beatles [6], but a number of artists in the 1960s were experimenting with magnetic tape recorders to achieve long echoes, amazing constructive/destructive filtering effects, and when the delay could be varied at will, flanging and chorus effects. Since these effects produced sounds never heard before, their use was artistically sensational at the time, and their discussion here is just plain interesting because it ties together 50 years of electronics and acoustics.

The "musical" instrument here is the "open-reel magnetic tape recorder" where several were put together in tandem to allow a controllable time-delayed signal to be produced. Typical reel-to-reel tape recorders pulled the tape over a set of erase, playback, and recording "heads" that are electromagnet transducers, at a constant speed controlled by a heavy cylindrical capstan pinching the tape against a rubber "idling" wheel. The take-up (right in Figure 4.3) and supply reels (left in Figure 4.3) were motor driven and the tape was held tight using spring-loaded control arm guides that could take up any slack in the tape as the reel speeds synchronized. These were very sophisticated machines and great care was taken to ensure that the tape moved at precise constant speeds. However, if two tape machines (tape decks) were arranged side by side as shown in Figure 4.3, the

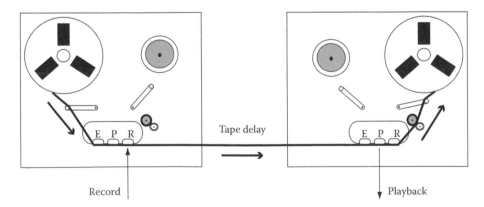

FIGURE 4.3 A pair of open-reel tape recorders showing the erase (E), playback (P), and record (R) heads and how the tape is cascaded to produce a controllable delay for producing artificial echoes, flanging, and chorusing effects.

tape could follow a path from the supply reel over the tape heads on the left-hand tape deck, then over the heads on the right-hand deck and up to the take-up reel there. One could record on the tape using the left-hand deck and playback the recorded signal from the right-hand deck with a pure delay of anywhere from a fraction of a second to even seconds later. The time delay between recording and playback could be adjusted by varying the tape speed or the distance the tape traveled between the two tape decks. The "tape-delayed" signal is a pure HiFi copy of the original recorded signal, unlike the rich and complex echoes of the reverberation devices described in the previous section. Mixing the delayed signal back into the original recorded signal is what produced all the interesting and artful sound effects.

Consider the frequency response of a signal delayed by τ seconds and added with the undelayed signal input. The frequency response of a constant time delay is a linear phase shift with negative slope. At 0 Hz and even multiples of $1/2\tau$ Hz the delayed signal will constructively add to the original signal input resulting in twice the signal amplitude at these frequencies. At odd multiples of $1/2\tau$ the output signal has destructive cancellation. So if we have a delay of 100 ms, the output will have cancellation dips at frequencies of 5, 15, 25 Hz, and so on giving rise to a "comb filter" effect, because the log-magnitude frequency response plot resembles a comb (the only use a geeky electrical engineer would have for one). Using a gain of α instead of unity would cause the constructive gain to be $1 + \alpha$ and the amplitude of the cancellation to be $1 - \alpha$. Instead of a feedforward FIR filter response, one could also feed back the delayed output back to the input at the recording tape deck leading to poles (peaks in the frequency response) at odd multiples of $1/2\tau$ rather than zeros (dips in the frequency response) with a stable output only if $\alpha < 1$. With $\alpha \geq 1$ the fed-back echoes very quickly would saturate the tape recording giving an interesting "alien space ship" sound. Using relatively short delays one would have to filter where the frequency response could be dramatically changed just by varying the tape speed slightly. Longer tape delays produced more of an audible echo effect along with a less than obvious frequency response effect, since the peaks or dips are much closer together. The human ear can perceive echoes easily when they are separated in time by around 50 ms or more and can detect frequency response changes in bandwidths greater than about 1/3 of an octave over the speech frequency range from around 300 Hz to 3 kHz [7]. Outside of these approximate ranges only very few people can detect and identify the effect, although most would sense something different such as a change in timbre when comparing the effect to the original signal only.

Consider the Beatles' "auto double tracker" (ADT) used to create two singers out of one performance. It was just the tape deck arrangement in Figure 4.3, but it saved them a tremendous amount of time since each vocal track needed to be closely identical to hide the process. In earlier recordings by the Beatles you can occasionally hear the second vocal track out of synchronization on a word here and there. By the mid-1960s the ADT they used produced perfect double vocal tracks giving them a very unique vocal sound. If the ADT was repeated on the recording with a slightly different delay, the original single vocal recording would result in four vocal recordings slightly offset in time, and so on. This is called a "chorus effect" for obvious reasons.

One can imagine varying the delay between the two tape decks in Figure 4.3 by physically moving them closer or farther apart as if the two decks were placed on movable carts. But the tape alignment needed to be very precise, and so the practical way to vary the tape delay was by pressing a pen or pencil body against the tape between the two decks and simply moving it up or down. Even at fast tape speeds this could produce considerable variation in delay. For small variations in delay, all one needed to do was touch the tape reel, or flange, of the supply reel (on the left) so that the tape speed would vary during recording differently than during playback. This not only produced pitch shifts, but also filtering effects that could be quite dramatic. Sometimes the tape itself could actually stretch due to the momentary stress permanently putting a pitch shift into the recording. This would drive audio purists crazy, but in the hands of the artist, these "flanging" machines became musical instruments still in use nearly 50 years later whenever a sonic "alternative reality" is needed. However, from a signal processing perspective, one should always consider that a *changing* filter is not time invariant and may even be nonlinear while it is changing.

To see mathematically how a change in tape delay can lead to a frequency shift, consider the frequency response of a delayed output added to the input in a feedforward arrangement.

$$H(\omega) = 1 + e^{-j\omega\tau} \tag{4.19}$$

We now let the delay τ be a function of time.

$$H(\omega) = 1 + e^{-j\omega\tau(t)} \tag{4.20}$$

Since frequency (rad/s) is the time derivative of phase (radians), the output frequency at any moment of time is equal to the input frequency plus the phase time derivative.

$$\begin{aligned}
\omega_{out} &= \omega_{in} + \frac{\partial}{\partial t}\{-\omega_{in}\tau(t)\} \\
&= \omega_{in}\left[1 - \frac{\partial\tau(t)}{\partial t}\right]
\end{aligned} \tag{4.21}$$

For the tape decks in Figure 4.3, consider a tape speed of 30 inches per second (ips) and the separation of the record head on the left from the playback head on the right of 15 in. This gives a delay of about 500 ms. If the length of tape between the tape decks (and delay) is increased by 3 in., or 20% over a 1-s time period, the tape has to move faster over the record head on the left than when it plays back 600 ms later on the deck on the right, and so the frequencies on playback will be lower. The amount of frequency shift depends on the *rate* of change in the time delay. For this example, the delay increased 100 ms over a 1-s interval, or a rate of 0.1 giving a 10% drop in frequency. This is about one whole step on the musical scale (two piano keys) or the amount of Doppler shift created by a sound source moving away from you at a speed of 34.5 m/s (about 77 miles/h). If the tape path were shortened again the frequency shift would go in the opposite direction. So by varying the tape length, an artist can generate the sonic illusion of sound sources moving toward or away from the listener. Since the rate of change of delay directly caused the Doppler frequency shift, one could alter the speed of the tape reel, or flange, using a finger if desired to create the effect, hence the term "flanging."

Eddie Kramer, a well-known recording engineer and engineer for most of Jimi Hendrix's recordings [8], took tape echoes and flanging to a significantly higher artistic level by flanging across the left and right channels of a stereo recording. This not only gave the listener a perception of motion, but also placement. In some cases the music is completely out of phase between the left and right channels, and so if one listened in mono, or just moved the left and right loudspeakers together, parts of the recording (such as vocals or a guitar solo) are cancelled. Although not much was technically known about how humans perceive the direction of a sound source at the time, these phase cancellations between the left and right channels of a stereo recording were deliberately done to fool the listener into thinking that the sound source was behind them, racing toward or away from them, or even spinning around them. Perhaps the best-recorded example of this is found on Jimi Hendrix's *Electric Ladyland* album [9] on a recording track called "... and the gods made love ..." Human hearing can perceive front and rear sources from the directional response of the outer ear (pina) and the diffraction differences of the head, which are symmetrical from left to right but not front to back. These same effects and phase cancellations are exploited in surround sound recordings played through stereo loudspeakers using special filters and are quite effective.

4.4 BASS, TREBLE, AND PARAMETRIC FILTERS

The basic tools of a recording engineer to control the sound signals are filters and faders (attenuators or volume controls). But to make these devices useful to nonelectrical engineers and artists, they have to be broken down into essential parameters or functions. Clearly, the output of a low-pass filter can be mixed with the output of a high-pass filter where the −3 dB corner frequencies are matched so that the relative loudness of the lower band to the higher band can control the "bass" (lower band) or "treble" (higher band) relatively. Sometimes a mid-frequency band is used as well to allow simple tailoring of the overall frequency response. These controls are designed to maximize simplicity and provide a gain boost or cut only in their designated frequency range with a unit frequency response everywhere else. They are sometimes called "shelf" filters because of their flat responses away from the transition band. Since these filters are by design with smooth frequency transition responses, they are usually composed of only first-order Butterworth functions to eliminate any phase cancellation issues in the transition bands.

A *Bass* filter is defined for corner frequency ω_0 and boost/cut gain G where $G > 1$ boosts the bass frequencies below ω_0 and $0 < G < 1$ cuts (attenuates) the bass frequencies below ω_0. For the analog s-plane ($s = \sigma + j\omega$) the first-order bass filter is

$$H(s) = \frac{s + \omega_0(1+\alpha)}{s + \omega_0(1-\alpha)}$$
$$\alpha = \frac{G-1}{G+1}$$

(4.22)

To make a digital version of the bass shelf filter, the zero at $s_z = -\omega_0(1 + \alpha)$ and the pole at $s_p = -\omega_0(1 - \alpha)$ are mapped to the z-plane using $z = e^{sT}$ where T is the sampling interval in seconds. However, because the transition from the bass band controlled by G to the unity response band above ω_0 is so gradual, the best way to scale the bass filter in the digital domain is to normalize the gain to unity at the Nyquist frequency (half the sample rate).

A *Treble* filter has a slightly different formulation.

$$H(s) = G\frac{s + \omega_0(1-\alpha)}{s + \omega_0(1+\alpha)}$$

(4.23)

Scaling the treble filter in the digital domain is best done by normalizing the gain to unity at 0 Hz. The pole–zero mapping is done using $z = e^{sT}$.

A *Mid* filter can be composed of a treble filter followed by a bass filter where the gains are tied together and the treble corner frequency is below the bass corner frequency. The simplicity of these "shelf" filters allows a very easy way to make broad tonal adjustments on audio signals using very simple concepts. Also, note how simple they are to implement in the digital domain.

The *Parametric* filter is an all-pass filter except for a narrowband where the boost/cut gain G, center frequency ω_0, and quality factor Q have been "parameterized" usually into three convenient knobs. A high Q makes a very narrow boost or cut frequency response at ω_0 while $0 < Q < 0.707$ makes a broader frequency range boost or cut.

$$H(s) = \frac{s^2 + \left(\dfrac{\alpha_z\omega_0}{Q}\right)s + \omega_0^2}{s^2 + \left(\dfrac{\alpha_p\omega_0}{Q}\right)s + \omega_0^2}$$

(4.24)

The parametric boost or cut gain is defined as $G = \alpha_z/\alpha_p$ but as can be seen in Equation 4.24, the effective Q factor for the filter is also changed with the boost/cut gain. We note that for a boost, the denominator polynomial in Equation 4.24 dominates. For a consistent frequency response for both boost and cut, we need to constrain the middle term in the numerator for a cut, and the denominator for a boost. For a boost, α_p should affect the gain and α_z should be unity. For a cut, the numerator polynomial dominates the response of Equation 4.24, and so α_p should be unity and α_z adjusted to affect the depth of the dip. This detail is much less problematic to implement in the digital domain than in the analog domain. The parametric filter has zeros

$$s_{z1,z2} = -\frac{\alpha_z \omega_0}{2Q} \pm j\omega_0 \sqrt{1 - \frac{\alpha_z^2}{4Q^2}} \tag{4.25}$$

and poles

$$s_{p1,p2} = -\frac{\alpha_p \omega_0}{2Q} \pm j\omega_0 \sqrt{1 - \frac{\alpha_p^2}{4Q^2}} \tag{4.26}$$

which are mapped to the z-plane using $z = e^{sT}$. To scale the digital parametric filter, one can either normalize the gain to unity at the Nyquist rate or 0 Hz, or use the following precise normalization as discussed in Chapter 2. At a gain of unity, the parametric filter should have a unity response at all frequencies.

$$H[z] = \left(\frac{z_{p1} - z_{2p}}{s_{p1} - s_{p2}}\right)\left(\frac{s_{z1} - s_{z2}}{z_{z1} - z_{z2}}\right)\frac{1 + (-z_{z1} - z_{z2})z^{-1} + z_{z1}z_{z2}z^{-2}}{1 + (-z_{p1} - z_{p2})z^{-1} + z_{p1}z_{p2}z^{-2}} \tag{4.27}$$

Figure 4.4 shows a set of parametric filter responses for a range of gains and a constant Q and center frequency. A high Q yields a very sharp peak for gain greater than unity or a very sharp dip for a gain less than unity. The filter Q, or quality factor is often defined as

$$Q = \frac{f_0}{f_2 - f_1} \tag{4.28}$$

where f_2 is the frequency above the center frequency where the response is 3 dB down from the peak (or 3 dB up from the dip) and f_1 is the frequency below the center frequency with this response. The closer f_2 is to f_1, the sharper the peak or dip and the higher the Q. For peak/dip gains near unity, these "half-power" frequencies can be approximated by finding the response where the amplitude is the geometric mean of unity and the peak or dip (the square root of the gain). Note that for a constant Q, the width of the peak or dip increases with increasing center frequency.

Parametric equalization is a very powerful and simple technique and can dramatically change the tonal character or fix problems such as resonances or cancellations in the observed frequency response. They are particularly useful where arrays of microphones are used to record large acoustic instruments such as drums, grand pianos, orchestras, board meetings, and choirs. The parametric equalizer is one of the most sonically powerful tools and with a group of them almost any filter response can be created with a few adjustments of the appropriate parameters. This is far more intuitive than trying to derive a filter polynomial or topology as discussed in Chapter 3.

4.5　AMPLIFIER AND COMPRESSION/EXPANSION PROCESSORS

A compressor or expander is an amplifier with a nonlinear gain that depends on the amplitude of the input signal. A compressor gradually limits the gain of the output when the input exceeds a prescribed threshold. The result is a higher gain for low-level signals and a lower gain for high-level signals. Therefore, the dynamics of the signal are compressed in amplitude by a nonlinear response of the compressor. An expander performs the opposite of a compressor by applying low gain for low-level

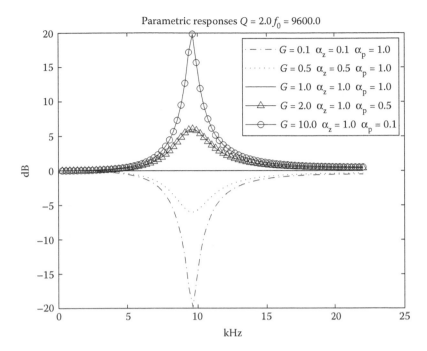

FIGURE 4.4 Frequency responses for a parametric equalizer filter with $Q = 2$ and $f_0 = 9.6$ kHz, varying the control band gains.

signals and higher gain for high-level signals. An expander can, in theory, exactly cancel the effects of a compressor, and vice versa. Compression of signals is very useful when the dynamic range of the signal must be kept within a specific limit. Good examples of this are terrestrial broadcast radio where overmodulation would cause interference between radio stations and analog magnetic tape recording. The original signal can be restored by employing the mirror expander to the signal compressor.

This kind of processing is sometimes called "companding" and was very useful as a means of noise reduction on magnetic tape recording. The moving magnetic tape produces a nearly white noise from the random orientation of the magnetic particles on the tape and their respective magnetic fields. This created a noticeable "tape hiss" background noise especially when the source signal was weak in amplitude. By compressing the source signal, the SNR is increased during low-level passages in the signal, and decreased during high-level signal passages due to the compression. When the compressed signal is played back from the tape deck and run through an expander that mirrors the compressor response, the original signal is heard undistorted but the tape hiss noise is significantly reduced during quiet passages. The tape hiss is usually unnoticeable during loud passages with or without companding. These tape-noise-reducing devices often used different companding in different frequency bands as well as prefiltering the signal to boost the high frequencies prior to tape recording, and restore the original frequency response during playback with another compensation filter. While analog tape recording is largely antiquated, there is a large body of recordings, both audio and video, which have been compressed deliberately by virtue of the signal distortions that occur in the recording process. For the future preservation and restoration of these recordings one must understand these processes to create digital processor that can correct for these distortions, or bring them back as part of the restoration.

Figure 4.5 compares the linear response of an ideal amplifier to that of a compressing amplifier and expanding amplifier. The voltage output for this example is

$$v_{out} = v_{in} + 0.0035 v_{in}^3 \qquad (4.29)$$

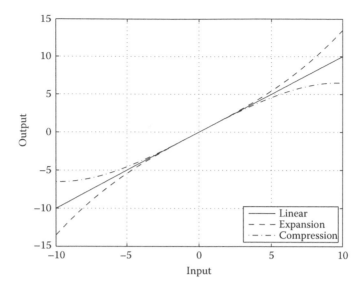

FIGURE 4.5 Amplifier nonlinear responses used for low levels of compression and expansion based on a third-order harmonic response.

for a mild expansion and

$$v_{\text{out}} = v_{\text{in}} - 0.0035v_{\text{in}}^3 \qquad (4.30)$$

for the mirror compression response. Note that introducing odd powers of the input signal as seen in Equations 4.29 and 4.30 leads to a signal distortion symmetrical with 0 V and yields odd numbered harmonics of a sinusoid input signal at the output. Even-order powers in the distortion are not symmetric with 0 input volts and would generate even-order harmonics which do not sound as pleasant due to the lopsided distortion in the output waveform and the critical bandwidth for the ear. The fundamental and third harmonics naturally form a major fifth musical interval. Applying compression or expansion to a signal is a form of "soft clipping" where the nonlinearity only affects the high-level portions of the signals. This is a very mild amount of compression or expansion.

There are several reasons as to why odd-order harmonic distortion is subjectively more pleasing than even-order distortion. The first is that things which vibrate back and forth to make sound naturally have odd-order-type nonlinearities when their spring stiffness is stretched or compressed beyond its linear range. When a vibration is asymmetric, the nonlinear motion is not balanced giving a less natural "buzzing" kind of sound. But third, the human ear can be modeled as an array of sensing cells which respond independently when the frequencies are separated by around 1/3 of an octave. For closely spaced frequencies, the sensed vibration tends to mix in the sensing tissue giving rise to amplitude beat frequencies and other perceptual effects.

Vacuum tube amplifiers and solid-state amplifiers produce slightly different kinds of distortion. The vacuum tube, or "valve" amplifier, is still popular with audio enthusiasts and especially electric guitar players because of a unique nonlinear response. The early transistor guitar amplifiers used bipolar transistors (switch type) which provided linear amplification to a fixed voltage level, and then a "hard" clip of the waveforms which sound quite bad. Later metal oxide field effect transistors (MOSFETs) amplifiers softened the clip and sound much more pleasant like the vacuum tube ancestors, but not quite as good. Why is this? Why do audio enthusiasts still debate this? Is amplification not just a linear multiplication? The answer is seen when considering the whole system including the loudspeakers. The loudspeaker impedance is around 6 Ω near 0 Hz, rises to as high as 90 Ω at the loudspeaker cone resonance, then drops back down to around 6 Ω until at higher frequencies (say over a few hundred Hz) the voice coil inductance raises the impedance gradually with increasing frequency. This would be a typical impedance curve for a 12″ (30 cm) heavy-duty guitar amplifier "8-Ω" loudspeaker. MOSFETs have no issue driving this impedance directly because their internal output impedance is very low. However, a vacuum tube has very high output impedance and requires a transformer to match its output to the low impedance of the loudspeaker. This transformer and the loudspeaker's nonuniform input impedance cause the vacuum tube to drive a very high impedance at high frequencies and the loudspeaker diaphragm resonance (40–80 Hz typically), but drive a much lower impedance in the range between the loudspeaker resonance and the high-frequency loud-speaker inductance-dominated impedance range (say over 1 kHz). It takes little current for the vacuum tube to drive a high-impedance load but much more to drive a lower impedance load. This is why vacuum tube amps provide 4, 8, and 16 Ω output transformer taps to best match the loudspeakers being used. The upshot of this is that the vacuum tube amplifier produces most of its harmonic distortion in the range of 100 Hz–1 kHz where the tube saturates and the loudspeaker impedance is low. The saturation causes a soft symmetric clip of the high-amplitude signals which in turn produce odd harmonics of a major 3rd, major 5th, and major 7th, and so on, right in the range of most musical instruments. The low-frequency fundamental notes of a chord can have lower distortion due to the loudspeaker diaphragm impedance resonance and the very high frequencies also can have lower distortion, sounding more like the strings natural harmonics. Audiophiles often swear by their vintage vacuum tube amplifiers sounding more natural, and they probably hear a very subtle nonlinear effect that cannot be reproduced in a solid-state amplifier. But, it might be closely reproducible in a digital processor "plug-in" using carefully designed filters and nonlinearity.

To produce a much stronger compression, one would not use amplifier nonlinearity, but a more brute force set of devices called a signal-level measurement (SLM) and a voltage-controlled amplifier (VCA). The SLM estimates the envelope of the signal amplitude dynamics by making a continuous estimate of the root-mean-square (RMS) signal level. The RMS estimate could have a fast time constant or slow and sophisticated compressors offered separate frequency bands to independently optimize the processing. As the RMS signal went up, the gain of the VCA is reduced and as the RMS signal decreased, the VCA gain is increased. The time constant of the RMS estimator and the proportional gain applied to the VCA must be precisely mirrored in the expander to accurately restore the original signal. Even a small mismatch between the two would leave artifacts in the signal such as soft clipping or artificial variations in signal loudness. This required precision electronics to achieve using analog circuitry of the day, but can be very precisely done in the digital domain. For example, the SLM-estimated envelope naturally has a slight delay relative to the sound signal waveform. This lag causes an audible effect at the beginning and end of sound transients called "compressor breathing," which is generally not restored precisely in the expander, which also suffers from a slight lag between its SLM and VCA. These audible effects of compressors and expanders were soon exploited to "improve" vocals, guitar, drums, and all sorts of sounds used in movies, as well as a preemphasis deemphasis technique to maximize SNR in recordings. In those days these artifacts were a small price to pay for enhanced SNR (reduced tape hiss in recordings). Today, audio restoration in the digital domain can remove even these nonlinearities because the processing can be noncausal (future signal inputs are always available).

While audio engineers in search of HiFi exploited the compressor and expander to maximize SNR in the limited dynamic range of recorded tape, vinyl records, or a radio broadcast channel, recording artists discovered that compressors made their guitar playing sound smooth, bass drums sound big and fat, while in general signal compression made it more difficult to overmodulate a broadcast or hard clip a recording. From the mid-1960s until digital recordings became more common in the mid-1980s, one finds compression signals almost everywhere in recorded music, television, and film. When these recordings are restored in a HiFi digital format, it can leave the listener wondering what the original sound actually was. Some would argue that these artifacts are part of the recording and should not be removed in the restoration. We would argue that unless it is an artistic effect, precise restoration of the signal is always desirable and also provides a means to recreate the artistic effects. Such is the case today with "modeled" amplifiers for guitars, basses, and keyboards where the frequency response and distortion of classic or vintage equipment is captured allowing a facsimile reproduction from modern sound sources. Today many vintage instruments and amplifiers are available as "plug-ins" or software modules that can be added to digital recording software.

4.6 DIGITAL-TO-ANALOG RECONSTRUCTION FILTERS

For signal processing and control systems where digital outputs are to be converted back to analog signals, the process of signal reconstruction is important to the fidelity of the total system. When an analog signal is sampled, its frequency range is band limited according to the well-known rules for avoiding aliased frequencies. The ADC process simply misses the signal information between samples (this is actually insignificant due to the antialiasing filter's attenuation of high frequencies). However, upon reconstruction of the output signal using DAC, no information is available to reconstruct the continuous signal in between digital samples. Therefore, one must interpolate the continuous signal between samples. The simplest approach electronically is to simply hold the output voltage constant until the next digital sample is converted to analog. This holding of the voltage between samples is known as a *zero-order-hold* DAC conversion and is the most straightforward way to produce analog signals from a sequence of digital numbers. In fact, if the frequency of the signal is very low compared to the sample rate, the reproduction is quite accurate.

One can visualize a DAC as a simple bank of analog current sources, each allocated to a bit in the binary number to be converted and with corresponding current strength, where the total output current across a resistor produces the desired DAC voltage. A simple digital latch register keeps the DAC output voltage constant until the next digital sample is loaded into the register. This practical electronic circuitry for DAC conversion results in a "staircase" reproduction of the digital waveform as shown in Figure 4.6 where the frequency is approximately fs/5.

The zero-order-hold circuitry has the effect of low-pass filtering the analog output. If one were to reproduce each sample of the digital sequence as a short pulse of appropriate amplitude (analogous to a weighted Dirac delta function), both the positive and negative frequency components of a real sampled signal will have aliased components with multiples of the sampling frequency. For example, if the sample rate is 1000 Hz and the signal of interest is 400 Hz, D/A reproduction with impulses would produce the additional frequencies of ±600 Hz (from ±400 to ±1000), ±1400 Hz,

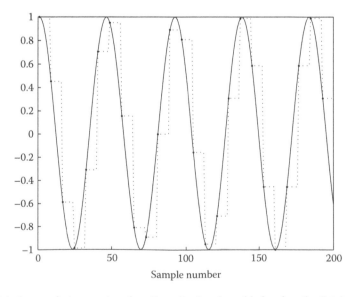

FIGURE 4.6 Digital-to-analog reconstruction (dotted) of a sinusoid showing the "staircase-"reconstructed signal which requires an analog filter to smooth out the edges and remove aliased images at high frequencies.

and so on. With the zero-order hold in place, the higher aliased frequencies will be attenuated somewhat, but generally speaking, additional analog low-pass filtering is required to faithfully reproduce the desired analog signal. In the early 1980s when digital music recording and playback equipment were in their commercial infancy, the quality of digital music reproduction came into controversy from an initial negative reaction by many Hi-Fi enthusiasts (many of whom still today swear by their vacuum tubes). There was little controversy about the elimination of tape hiss and phonograph rumble and the wear improvement for digital recordings. But many music enthusiasts could hear subtle effects of the aliased digital frequencies and low-pass filters in the D/A systems. By the mid-1980s, most digital audio players employed a technique known as *oversampling* on the output which is an extremely useful method for achieving HiFi without increasing the system sample rate.

Perhaps the most common example of sophisticated oversampling for analog signal reconstruction is in the modern audio compact disk (CD) player. Often these devices are marketed with features noted as "4× oversampling" or even "8× oversampling." Some ultra HiFi systems take a direct digital output from a CD or digital audio tape (DAT) and using about 100 million floating-point operations per second (100 MFLOPS), produce a 32× oversampled audio signal. For a typical CD with two channels sampled at 44,100 Hz, a 32× oversampled system produces over 2,822,400 samples/s! Where do the other 2,778,300 samples come from? To clearly illustrate what DAC oversampling is and how it works we will present below cases for 2×, 4×, and 8× oversampling.

The process of adding zeros to the digital sequence between the samples is the first step in upsampling using an interpolation filter. One zero is added if 2× oversampling is desired, three zeros if 4× is desired, and seven zeros if 8× oversampling is desired. Figure 4.7 shows the samples from Figure 4.6 with seven zeros added between each sample. The zero-filled digital sequence is then digitally processed by an *FIR interpolation filter*. Figure 4.8 shows interpolation filters for 1×, 2×, 4×, and 4× oversampling. The output sequence of the interpolation filter has the added zeros replaced by synthesized samples which "smooth out" the staircase effect seen with a zero-order hold. A first-order hold (connect the two actual samples with a straight line) is achieved by an FIR interpolation filter which moves the added zero sample to the average of the two adjacent actual samples for 2× oversampling. A third-order hold is achieved with an FIR filter which approximates a cubic function for 4× oversampling, and so on. Figure 4.9 shows the FIR impulse responses for the

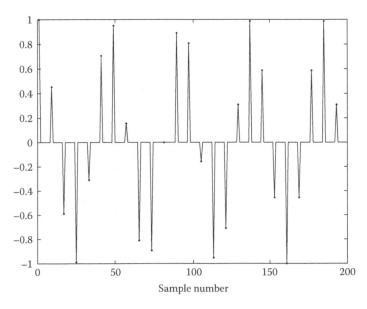

FIGURE 4.7   Adding zeros between output samples is the first step toward producing a digitally oversampled output. The dots are the original samples.

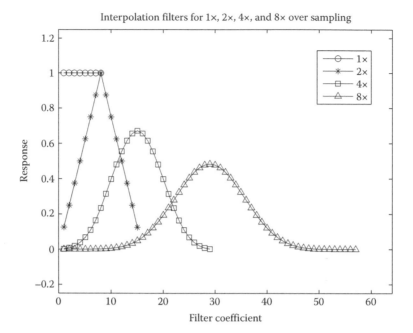

FIGURE 4.8 FIR interpolation filters for producing oversampled DAC waveforms from digital signals with inserted zeros as shown in Figure 4.7.

zero, first-, third-, and seventh-order holds corresponding to 1×, 2×, 4×, and 8× oversampling. However, we had to do some delay compensation and normalization for these waveforms to overlay. The 2× filter delays the signal by eight samples while the 4× filter delays 15 (twice 8 minus 1) and the 8× filter delays the signal 29 samples (twice 15 minus 1). Also, one can see that the maximum of the filter impulse responses in Figure 4.8 is not consistent. Normalization by the square root of the filter impulse response maximum aligns the interpolated signal amplitudes. This is a result of the convolution process. As will be discussed in Section 4.7 on waveform file compression, much higher-performing interpolation filters can be used.

4.7 AUDIO FILE COMPRESSION TECHNIQUES

Perhaps one of the most significant developments in audio engineering in terms of its impact on the music recording industry was the development of file compression techniques that permitted easy transfer of copyrighted music over the Internet. Prior to this development, there were significant technical barriers that made it difficult for widespread copyright infringement. The music business actually has a long history of legal battles over the rights (and profits) from recorded music. Back in the 1930s, musician's unions picked establishments that played recorded music and a fair compensation system was not yet established for the musicians performing on the recording. Songwriters and publishers were compensated, but not musicians. Throughout the 1940s through the 1960s, the record industry flourished due to its control of vinyl record manufacture. Only the record factories had the equipment to stamp out records. During the 1950s, there was a scandal of "payola" where radio stations were accepting payments from record companies to play their records creating an unfair market advantage for the large and powerful record companies. In the 1970s, when inexpensive cassette tape recording became popular, the copyright infringement issue emerged again, but this time against the customers who copied records or tapes and shared them with their friends. The issue was settled by charging a small royalty on every blank cassette tape sold and distributing them according to the popularity of various songs of the day, similar to the way radio stations pay royalties for the recorded music they broadcast. By the late 1980s, CDs essentially replaced the vinyl record and since it was

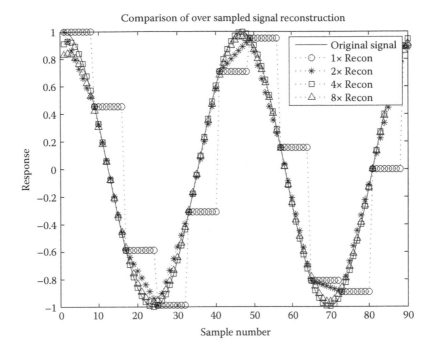

FIGURE 4.9 Comparison of oversampling levels on the original analog waveform where delays and amplitude have been compensated to align the digital waveforms.

the same CD device used as a data storage device for personal computers (PCs), users noticed they could make an exact copy of a CD with no loss of fidelity and share with their friends. However, a song in a wave (*.wav) format required two channels of 16-bit-sampled data at a rate of 44,100 samples/s. A typical 5-min song used 300 s times 44,100 samples/s, times two channels/sample, times 2 bytes/channel, or 52 megabytes (MB). The 5-min song as a data file was still too large for many PCs of the day to deal with, let alone send to someone else over a telephone line connection.

By the early 1990s, the motion picture experts group (MPEG) was developing standards for file compression of images and audio files. One such standard, technically known as MPEG layer III, or "MP3," in short, changes everything. The idea was to charge a royalty for using the encrypter, the program which converted the wave file to MP3 format, but give away the decoder, which is needed to generate the audio signals from the compressed format. The problem was that the source code for an MP3 encoder was widely available from many Internet sites. It did not take very long for music enthusiasts to set up Internet sites with huge libraries of MP3-formatted music where each file on average only used about 10% of the space of the original wave file. Note that the "bit-rate" of a 44,100 Hz-sampled stereo wave file is 1,411,200 bits per second (bps) and the bit rate of a "HiFi MP3" file is about 128 kbps, or less. If the listener could tolerate lower fidelity, much smaller file sizes in the MP3 format could be used. Much of the algorithm development for digital image and audio compression was driven by the broadcast and cell phone industries which intended to maximize the amount of information squeezed into their cables and broadcast channels. However, the musicians, publishers, and record companies had no control over these rapid developments. Within a few years CD sales were decimated and the profitable recording industry was near financial ruin. This is reflected in a comment by Eric Clapton in his autobiography [10] that the music recording industry may be completely gone in the near future, but that music will always find a way to us. Perhaps musicians will successfully self-publish via the Internet if a financial incentive can be preserved for them. There is an important lesson here for engineers and scientists: unintended consequences from advanced technology can result in bad things.

Today the most efficient method to distribute recordings is via the Internet and music download sites which account for and pay fair royalties to artists and publishers. The small amount of electricity and equipment it takes to complete the transaction can actually result in a larger royalty payment to the artist and a lower cost to the customer. What is wrong with that? The job gets done using fewer resources and is faster and potentially much more accountable. The problem is human nature in that if people can take advantage they will. People still share music files and convert them to whatever format they wish and move data anonymously on the Internet, which is structured in such a way making it nearly impossible to stop such activity. Publishers and their agents offer new improved file formats such as Windows Media Audio (*.wma) and Apple's Advanced Audio Coding (*.aac) (AAC), which offer improved fidelity at about the same bit rate plus limited free copying (typically up to 10 copies are allowed). At the maximum fidelity end of the encoding marketplace is digital theater system (DTS) encoding, which is about the same bps as the original stereo wave file, but actually has six channels (five surround sound channels plus a subwoofer channel—known as 5.1 format) where the decoded signal is 24 bits/sample, rather than 16 bits/sample. The DTS format delivers a *better sound quality* than the 16-bit CD but using about the same file size [11]. By keeping these encoding algorithms under strict control, offering customers convenience and other incentives, and managing the royalty stream, music files in these formats are pirated much less, although the skilled user can easily find ways to convert the recording to another format.

But how can one convert an audio stream of over 1.4 million bps (Mbps) to something only 1/10 of that with little audible difference? Music requires silence between the notes, beats, and phrasing of a song, as does speech. The silent blocks of data can be handled more efficiently than sending two 16-bit words of zeros 44,100 times/s. Given some signal processing power, the time record can be chopped up into small blocks (e.g., 1152 samples for MP3) and each block can be analyzed and further broken down into frequency bands, called *subbands*. These subbands are each analyzed in terms of loudness, transients, dynamic range over time, and dynamic range across frequencies. Then, one uses *perceptual coding* to apply data compression for components of the sound not very audible by the human ear. While back in the 1960s audiophiles argued about phonograph stylus materials, loudspeaker systems, and vacuum tube amplifier design, today, almost 50 years later the arguments are over data compression formats! Careful listening will generally reveal some artifacts of the compression algorithm, even at very high bit rates. However, for digital radios and cellular phones, it is amazing just how much compression can be applied and it still provides intelligible speech. Another aspect of music is that the low frequencies require much more loudness and dynamic range than the high frequencies. Low frequencies change more slowly than high frequencies, and so on. Other texts [12,13] can provide the reader with specific details of the various formats and compression algorithms, which are usually proprietary in the area of how they do the perceptual coding. However, for audio signals, virtually all these formats and compression schemes follow the same general framework, which we will discuss here.

Perceptual coding is a fascinating process that selectively removes data bits that correspond to sounds the human ear is unlikely to detect. There is a low sound-level limit called the threshold of hearing approximately equal to 20 micro-Pascals (μPa) RMS pressure at 1 kHz and is the 0 dB reference for sound pressure levels in air on the decibel scale. However, this threshold increases at lower and higher frequencies. There is also a masking effect from a sound which raises this hearing threshold at nearby frequencies. All perceptual coding systems require a calibrated ADC and so the absolute minimum audible bits can be determined for the threshold of human hearing over a range of frequency bands. These frequency bands can be defined by the ears response in terms of frequency resolution and are called *critical bands*. The ear's critical bands are roughly 1/3 octave in extent, meaning that they are of narrow bandwidth below 300 Hz and very broad bandwidth above 3 kHz. About 30 1/3 octave bands cover the human hearing frequency range from 20 Hz to 20 kHz. A critical band can be defined using two closely spaced sinusoids in frequency, but of the same amplitude. When the sinusoids are nearly exactly the same frequency, one hears a single frequency (the average of the two tones) but with an amplitude "beat" frequency of 1/2 (the difference between the two tones). One can clearly see this effect on a graph of two sinusoids equal in amplitude but slightly different in frequency. As the frequency separation widens, the beat frequency increases. As the beat frequency increases to 15–20 Hz, it no longer sounds like an amplitude-modulated signal, but rather sounds like a dissonant (bad sounding) tone. This is because the nerves in the ear are alternating between one frequency and the other while the amplitude is perceived as constant. Separating the two frequencies further, the ear begins to sense two distinct tones. This separation is the critical bandwidth at the average of the two frequencies. It represents the frequency resolution of the ear and the bandwidths where one sound can affect a frequency nearby. Given a tone in one critical band, it raises the hearing threshold for frequencies nearby, including the hearing threshold in nearby critical bands. By exploiting this property of human hearing, one can split the audio signal into a number of subbands, assign a separate gain to each subband, and eliminate the bits deemed inaudible by the perception algorithm's guess at what the hearing threshold mask should be. In addition, the encoding can slide the signal in small time blocks and dynamically apply the perception mask on each block in each subband. This can result in a very significant reduction in data bits with little audible degradation. For applications where some degradation is allowable (such as cell phone voice communications), one can achieve a much greater degree of data compression.

Figure 4.10 depicts the minimum sound pressure level in dB that a normal human ear can detect as the curve defined by the minimum audible field (MAF). The MAF at 1 kHz corresponds to 20 μPa, or the convention for 0 dB in air. The ear is not as sensitive at lower frequencies and very high frequencies. However, a sinusoid such as sinusoid B shown in Figure 4.10 generates its own masking curve due to vibration leakage in the inner ear between critical bands. This makes sinusoid C inaudible even though it is louder than the MAF at its frequency. The minimum number of bits needed to quantize sinusoid B is the number of bits required to cover the dynamic range from the peak of sinusoid B (60 dB) down to the lower of the masking levels at the edge of the critical band. This is about 30 dB in Figure 4.10. Each bit provides 6 dB of dynamics and we need 1 bit for the sign, and so we need only 6 bits and a scale factor to reproduce the signal in Figure 4.10. But the signal is confined to one critical band with a small fraction of the potential bandwidth defined by the Nyquist rate, and so we can filter and decimate the sample rate to get very significant data compression.

Quadrature mirror filter (QMF) banks are generally used for splitting the Nyquist bandwidth into subbands where the signal in each subband has a proportionally lower sample rate [14]. Most music data compression schemes use 32 subbands each with equal bandwidth. The 44,100 Hz

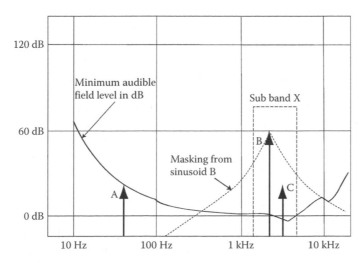

FIGURE 4.10 Graph showing the MAF of the ear and three sinusoids. Sinusoid A need not be quantized because it is below the MAF and inaudible. Sinusoid C would normally be audible but sinusoid B masks it out due to B's loudness, and so B need not be digitized.

sample rate of the original CD audio is reduced to 1378.125 Hz in each subband filter output. Recall that about 30 1/3 octave bands are a good approximation for the critical bands in a human ear, but we use 32 equal bandwidth QMF filters, or polyphase filters (polyphase refers to a different input sample rate than output sample rate). This means that the low-frequency QMFs are larger than the ear's critical bands and the high-frequency QMFs are smaller in bandwidth than the corresponding critical bands. To compensate for this, the lower subbands are sometimes allowed to use more bits than the higher subbands. This is one of the improvements of MPEG-2 layer III (widely referred to as MP3 files) over earlier versions.

A QMF process takes an input digital signal and splits it into two equal frequency bands which can then be decimated by 2. The upper half-band output can be seen as a heterodyned baseband signal which can be modulated back up in frequency when the signal is reconstructed. In principle, errors due to aliasing and aliased images cancel exactly during reconstruction in a QMF design. In practice, the errors introduced by QMF processing are quite small and manageable. Using five layers of cascaded QMF processes yields the needed 32 subbands.

The time samples are arranged into frames (1152 samples long for MP3 files) and the frames are input serially to the bank of QMF processors. Using a choice of 1152 samples in MPEG is the result of PAL video having 576 lines of resolution. In this way the audio frame can be interleaved with the video frame easily. The input audio frame is also split off into a side path and an FFT is performed to support perceptual coding analysis and estimating the masking for each subband. The QMF outputs are decimated to 1/32 of the original sample rate without loss of information, giving 36 samples at a rate of 1378.125 Hz if the original data were sampled at 44,100 Hz. Audio for DVD movies is sampled at 48 kHz giving a QMF filter output rate of 1.5 kHz/band. These 36 samples in a frame correspond to 26.12 ms (24 ms if the input is sampled at 48 kHz) of the original audio signal stream and may contain transients in a given subband. If the transient takes place in the beginning of the frame, the scale factors set for the frame will probably be okay as the transient decays and the ear recovers toward the end of the frame covering the low-quantization noise. But if the transient takes place near the end of a frame, the listener may hear a quantization noise burst prior to the transient called "pre-echo." To reduce this artifact, layer III of the MPEG 1 and 2 standards (and most subsequent compression algorithms) divide the frame into smaller packets on the order of 8 ms each and employ a modified discrete cosine transform (MDCT) to analyze each QMF outputs for transients. The MDCT coefficients in each frame packet are used instead of the time samples from the QMF output because for stationary signals the MDCT coefficients do not change over the frame time, offering a further bit rate reduction. Reconstruction from the MDCT uses an overlapped window scheme where the envelope window is selected by the encoder. There is a short window (for transients), a long window (for stationary signals), a long-short window (for prior to transients late in the frame), and a short-long window (for after transients early in the frame). For layer III, the scale factor and bit allocation also carry one of four window selections for the reconstruction MDCT in the decoder. We have used the MPEG-2 layer III as an example, but most of the features in MP3 are used by other data compression algorithms, just to different specific degrees. This is why most codecs (coder–decoders) keep their details proprietary and it is very easy for another codec to be developed with a unique file format and used without patent infringement. The encoders can get quite sophisticated while the decoders are essentially an open standard defined by the published data format. We discuss the distinguishing features of a few of these data compression formats below, while the general data compression scheme is shown in Figure 4.11.

MP3 (MPEG-1 layer III) is the widely used format for sharing music recording over the Internet. MP3 files have 32 equal bandwidth subbands where the low-frequency bands are allowed up to 15 bits, middle bands are allowed up to 7 bits, and high bands allowed up to 3 bits [12,13]. The number of bits used for scale factors is adjustable and the bit rate can be variable, or held within a fixed range using a buffering scheme with various resolutions to maintain the bit rate. Constant MP3 bit rates are easier for playback devices to synchronize with and maintain an uninterrupted audio output. There are also "granules" which are bit fields packed with various bit-length samples and

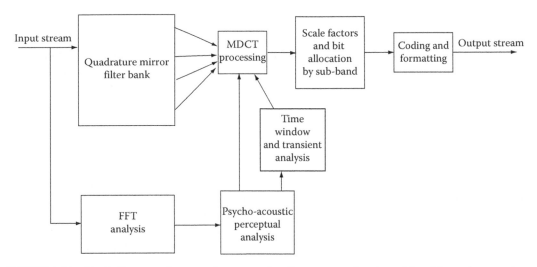

FIGURE 4.11 Block diagram of a general perceptual coding process showing subband formation, psychoacoustical modeling for masking estimation, time windowing, and bit allocation and coding.

"Huffman coding" where a symbol list is used to code the highest frequency bit patterns with fewer bit symbols, which also increases the data compression efficiency. Listening tests show that for most people, CD-quality sound can be heard with 192 kbps, although there are always audiophiles who "hear things" most people do not. At a bit rate of 128 kbps one sees about a 10:1 data compression factor with reasonable audio quality. MPEG-2 layer III adds more available sample rates and surround sound in support of digital television, but surround sound MP3 recordings are rare.

AAC is used by Apple, Inc., for their I-Tunes music store and the I-Pod series of players. AAC is not backward compatible with MP3 and offers finer time resolution (2.6 ms versus 8 ms for MP3), twice the MDCT frequency resolution, and most importantly, temporal noise shaping (TNS) [12,13]. TNS is a predictive windowing algorithm that exploits the higher MDCT resolution to provide better transient reproduction and better pitch-based reproduction. Like MP3, AAC supports coding of stereo and surround sound signals. The AAC format offer more scale factors and signal resolution but also requires more processing effort in the encoder. The result is a slightly better sound quality than MP3, where at a bit rate of 128 kbps the sound quality is close to a CD audio recording. AAC also supports surround sound at a bit rate of 384 kbps.

AC-3 (Dolby Digital) is the widely used surround sound encoder from Dolby Laboratories found on many DVD and DVD audio discs [12]. Often called "Dolby Digital," its encoding and decoding are actually very complex but achieve an outstanding data compression ratio of 13:1 while providing 18-bit (102 dB) SNR in surround sound. AC-3 has a nominal bit rate of 384 kbps and produces a surround sound signal with up to 102 dB of dynamic range. This high dynamics range is essential for DVD movies where one moment the actors are whispering and the next moment they are surrounded by explosions. However, AC-3 can carry some slight artifacts due to the massive compression carried out by analog companding, 50 subbands, and a hybrid forward/backward prediction adaptive bit allocation algorithm that operates in both the encoder and decoder. AC-3 also allows dynamic range control and features such as "dialog normalization" to aid in keeping speech around the same loudness level. These features are very well suited for movie theaters and home theater systems. AC-3 achieves its amazing data reduction by quantizing a frequency domain representation of the signal in a floating point. The frequency resolution (and also time resolution) can be adjusted on the fly to best suit the signal. It takes a moderately powerful (by today's standards) processor on the order of 50 million instructions/s to decode an AC-3 stream, which are found not only on DVDs, but digital cable and satellite television streams.

DTS is a perceptual coder with uncompromising audio fidelity and supports high-end audio and computer games as well as movie sound recordings [11,12]. DTS streams can support stereo recordings of 20 bit (120 dB) dynamic range up to 192 kHz sample rate, or over 4× the audible bandwidth of a human. Why so much bandwidth? Maybe the audiophile wants his/her dog or cat to enjoy the music as well. It certainly is more accurate to have the waveforms oversampled by eight times, rather than twice the highest frequency. As for the 120 dB dynamic range, this is about the SNR limit of room temperature electronics with 20 kHz bandwidth. It means that the signal can go from a few volts at the maximum amplitude down to a few microvolts (μV) with precision, with a background electronic noise in the microvolts range or less. This is also the limit for most acoustic transducers. A DTS bit rate for a 20-bit surround sound (six-channel) source sampled at 48 kHz typically requires a bit rate of 1.5 Mbps, but it supports bit rates from 8 to 512 kbps/channel with sampling frequencies ranging from 24 to 192 kHz and bit depths from 16 to 24 bits. The low-frequency effects (LFEs) channel is usually companded (dynamically compressed before recording and expanded during playback) in DTS giving it a very strong low-end response if the reproducing sound system can support it. The subwoofers to support the LFE channel can use well over 75% of the total amplifier power in the surround sound system.

Direct stream transfer (DST) is a *nonperceptual lossless* data compression algorithm used with direct stream digital (DSD) super audio CD (SACD) formats, which has a 2.8224 Mbps stream of 1-bit data/channel. Doing the math, this stream can produce 24-bit samples at over 96 kHz sample rate! The SACD supports sample rates of 32, 44.1, 48, 88.2, and 96 kHz. If this does not get your VU meter up into the red nothing will. However, a mono recording of 23-bit samples (24-bit ADC words) at 96 kHz would only allow about 222 min of recording on a 4.7 GB DVD. That is only about 37 min for six-channel surround sound and leaves no room for other information. The SACD standard has an area for stereo and an area for surround sound (2 + 6 = 8 channels total), and so the total maximum playing time is often quoted as 27 min, 45 s. However, by using DST compression on the bit stream, the SACD achieves a 74-min playing time with room left over for text and graphics. DST uses a constant frame of 37632 bits, 75 frames/s for DSD signals [15]. The DSD bit stream is called 64× oversampled (relative to a CD's 44,100 Hz sample rate) but it is only 1 bit. At 16 bits there is still 4× oversampling relative to a CD. With each halving of the oversampling rate one doubles the available bits, and so the SACD format actually provides a bit archive that can be reproduced over a range of sample rates and bit depths. This is no accident. The engineers who designed the first commercial digital recorders picked a format that exceeded the then proposed playback capabilities of the CD but appropriate for the foreseeable limitations of transducers and electronics. This allows recordings to be "remastered" for improved dynamic range as future consumer formats emerged. For each DST frame of 37,632 bits, a symbol table is produced grouping redundant bit patterns together. The most frequent bit patterns are assigned a symbol with the fewest bits while the rarest bit patterns get symbols with more bits. This is called entropy coding in the literature. The symbol table is unique to each frame, and so each frame of 37,632 bits gets converted to a symbol table followed by a frame of symbols during encoding and reversed during decoding. This achieves only about a factor of 50% reduction in bit rate for a mono channel, but it is lossless, meaning that the original bit frame is decoded exactly and the capacity of the storage media is effectively doubled. The encoded (data compressed) bit rate is variable but the decoded bit rate is constant and exactly restored using a buffering process. For multichannel stereo and surround sound data, there is redundancy across channels making the entropy coding more effective and increasing the SACD playing time from 27 to 74 min with zero data loss. The SACD format with its lossless data compression is a truly archival format.

4.8 MATLAB® EXAMPLES

Rather than work through the traditional m-scripts as with other chapters, audio signal processing provides an excellent opportunity to learn Simulink, a real-time simulation feature of the MATLAB package. We will use a simple 16-bit mono wave file sampled at the standard CD rate of 44.1 kHz as an input and connect the output to the PC sound board to listen to the results! This capability is pure fun but it does require a newer PC with enough RAM and processor speed to seamlessly process the audio signals. However, one can always use a wave file with a lower sample rate to ease the burden on the PC, but do not go below around 8 kHz sample rate, or you will hear artifacts of the aliased images of the output leaking through the DAC filters. From the MATLAB command window you can start Simulink by either typing "simulink" and hitting carriage return, or by clicking on the Simulink icon on the toolbar. The "Simulink Library Browser" window will appear where you can either open a file, as shown in Figure 4.12 for the file "audiosim.mdl" or create a new blank Simulink model. The current MATLAB directory or path has to be set first to the directory containing the wave file and the Simulink file. Also, under the "Simulation" menu item for the model window, select "Configuration Parameters" and make sure the edit box for "Type" is set to "Fixed Step," and the "Solver" is set to "Discrete." This is the typical setup for audio signal processing.

The lower part of Figure 4.12 is just a simple oscillator (around 0.5 Hz) and an offset to wiggle around the time delay in the block called "variable integer delay (VID)" and can also be viewed by clicking on the "time scope." One can set autoscale by right clicking on the axis and save the plot parameters using the toolbar icons. The "source from wave file" lists the filename, sample rate, channels, and bits for the file. All of the connected devices in the Simulink model must have the same sample rate and frame size (in this case 1 sample/frame). Patching the output of the VID to the PC loudspeaker is just amazing for lots of fun experiments. This output is also delayed and attenuated before being fed back to the VID input via a summing junction. The frequency response of the VID and these feedback paths will have sharp peaks which move around in frequency due to the oscillating delay to make a strange sound. There are of course many variations of this circuit—which is the whole point of playing in Simulink!

Figure 4.13 shows another Simulink model (racketballrev.mdl), but this time using delays so long that they exceed the author's PC capability for real-time audition. So we simply write the output to

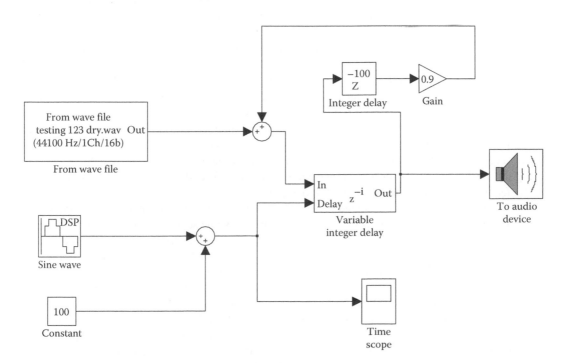

FIGURE 4.12 Simulink screen showing a simple digital flanger with and oscillating and static feedback delay using a wave file as input and the PC soundcard as output.

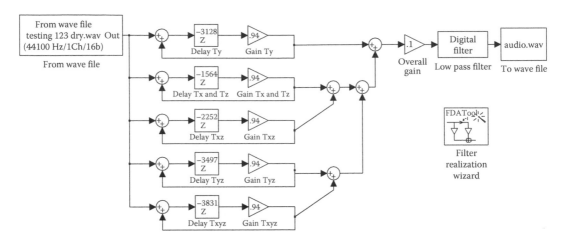

FIGURE 4.13 Simulink screen showing a simple reverberation generator for auralization of a racquetball court with a 10-s reverberation time.

a file called "audio.wav" and listen to it with a standard wave file player. Note that the delay, gain, and summing devices are standard Simulink objects and the sample rates all match the input wave file. Since wave files clip at a level of around ±1.0 in MATLAB (it is unspecified but usually about 1.228 V rms, or +4 dBu for most sound boards), we have an "overall gain" of 0.10 applied to the output to avoid clipping the output wave file. Note that we added an output low-pass filter which makes the reverberation sound a little more realistic by attenuating the higher frequencies more. This is typical for real rooms where the sound absorption increases with increasing frequency. Also shown in Figure 4.13 is the "Filter Realization Wizard" which with a few mouse clicks one can drop just about any type of digital filter into the Simulink model. Given the background earlier in this and the previous chapters on audio filters you have earned the right to click-and-drop these into the model without sweating all the m-script code. The sample delays are chosen to simulate the reverberation in a racquetball court, which is 20 ft wide × 20 ft tall × 40 ft long. We chose a 10-s reverberation time, not unusual for a well-painted court this size (16,000 ft³ volume and 4000 ft² surface area). Working back through the reverberation Equation 4.12 we obtain the average wall energy absorption of 0.02. Since the axial modes have two reflections this suggests a feedback gain of 0.98, which for the tangential modes its 0.96 and the oblique modes its 0.94. However, listening to the reverberation, it sounds a bit too flutter like. Another way of designing the reverberation is to consider the longest round-trip echo mode for the room, which is always the oblique mode. In our room this is about 87 ms and about 115 round-trip echoes in the 10-s reverberation time where the sound decays by 60 dB, or a factor of 1000. Dividing the natural logarithm of 0.001 × 115 and taking the antilog gives 0.94 for the average energy remaining/round-trip echo, and so we just use it for the feedback gain and it seems to work fairly well.

Real acoustic reverberation is considerable more complex than what we model with plane waves in a rectangular room. When the wave hit a surface there is some scattering, more so if the surface is not flat or has roughness or objects. This causes the modes to leak energy into other modes and individual echoes and flutter echoes to become indistinct. Smooth diffuse reverberation is generally very pleasant and typically found in large stone structures with carvings, statues, columns, and balconies such as cathedrals. Figure 4.14 shows a nice trick to add diffusion to the reverberation by feeding back the reverberation output back into the input (racketballrev2.mdl). Only a small amount is needed to smooth out the reverberation, but it does significantly increase the output, and so the overall gain must be further reduced. This technique is useful when it is literally impossible to account for all the modes, scattering, and diffusion in the space using physical analysis.

The book website does provide the m-scripts "treble.m," "parametriceval.m," "DTOA.m," and "amplineasity.m" used in making Figures 4.4 through 4.9. The essential elements of these m-scripts

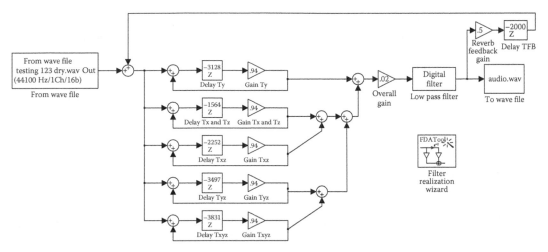

FIGURE 4.14 Simulink screen showing the same racquetball court but with a more diffuse reverberation typical of structures like cathedrals with lots of sound scattering.

could easily be put into Simulink using the common objects or other block sets if it is desired to work with a more graphical interface and easily listen to these filtering and nonlinear effects.

4.9 SUMMARY

This chapter presents some interesting and fun applications of basic signal processing to digital audio. It also discusses audio signal processing in the music and performing arts starting with the first man-made electronic sound affects in the 1950s. Many students and enthusiasts of signal processing initially became interested in the discipline through audio engineering and sound effects heard in music and motion pictures while as a child. This makes a quick exploration great fun and certainly a useful application of basic filtering and processing. However, in many electrical engineering texts there is really no tie between the signal processing and the physics of the sound produced or the perception of the sound by the human ear and brain. These aspects are quite important and form the basis for modern perceptive coding of waveforms as a means of signal bandwidth compression or file size compression.

We presented a very brief discussion on room acoustics and reverberation based on the well-established physical models and approximations. We then discussed the history of artificial electro-mechanical reverberation signal devices and how those responses relate to the physics of real room reverberation. We then show how one can generate reverberation using simple digital filters and delays. This led to a discussion of flanging and chorusing filters derived from the Beatles automatic double tracking (ADT) system using multiple tape recorders. This is an area of audio engineering where the artists led the engineers into a very creative activity. We also showed some very simple filters for bass, treble, and midrange frequencies and how to realize them in the digital domain. We also presented the "parametric filter" and explained its use in audio engineering. For recording and other applications, the concept of signal compression and expansion (companding), distortion, and restoration techniques are discussed. Today these nonlinear effects are developed as software "plug-ins" for computer-based audio workstations and are growing in sophistication. We also introduced interpolation filters for upsampling waveforms which has the benefit of simple antialiasing filters on the DAC outputs for better fidelity.

Finally, we present a very brief discussion on audio file compression techniques and perceptive coding of audio waveforms. These techniques are now nearly standard in digital broadcasting, recorded music, and movie soundtracks. The key elements of this technology are multirate filtering using QMFs or poly-phase filters since the output sample rate is not equal to the input sample rate, and perceptual coding, where bits are discarded that the ear will not likely hear. In music files, these compression techniques can result in a file size 1/10 of the original, but with some small audible effects. For cellular telephones, the perceptual coding can be pushed much farther allowing many digital channels to fit in a limited bandwidth and for each cell phone to use a minimal amount of battery power during transmission. We discuss the general differences between several of the more popular techniques, including HiFi surround sound coding found on most DVD movies. We also discuss the SACD and the lossless compression coding it uses to store ultra HiFi recordings on a DVD.

PROBLEMS

1. A home theater is desired with a reverberation time of 0.6 s. The room is 8 ft tall, 18 ft wide, and 26 ft long.
   a. If the floor is stone with an absorption coefficient $\alpha = 0.01$, what must the average absorption of the walls be for the desired reverberation time?
   b. What is the critical distance for an omnidirectional ($Q = 1$) source?
2. An artificial spring reverberation device is limited to fit in a 30-cm-long space. We would like to use three springs, each with mass 100 g and unstretched length 10 cm to get a minimum delay of 35 ms when the spring is stretched to 30 cm.
   a. What stiffness is needed for the springs?
   b. If the other two springs were stretched to 20 and 12 cm, what would their respective delays be?
3. Plot the frequency response extremes of a flanger with a feedback delay that oscillates between 10 and 2 ms with 0.9 feedback gain.
4. Plot the frequency response of a chorus ADT with summed delays of 50 and 75 ms.
5. Design a bass shelving filter with 20 dB gain at low frequencies, a crossover frequency of 400 Hz in the digital domain for a sample rate of 44.1 kHz, and plot its response from 0 Hz to 5 kHz.
6. Design a parametric filter with a Q of 50, a center frequency of 800 Hz, and a cut of -20 dB at the center frequency.
7. Show that if a compressor has a nonlinear response where the output is the arctangent of the input, then it only generates odd-numbered harmonics.
8. A sinusoid of high SNR and virtually no distortion goes into a compressor and the output is plotted as abscissa relative to the input as ordinate. How could you make a model for the nonlinearity of the amplifier?
9. Which would sound better at a bit rate of 128 kbps, MP3 or AC3?
10. Explain the difference between MPEG-1 layer III and MPEG-2 layer III.
11. A DVD movie offers two surround sound choices of AC-3 (Dolby Digital) and DTS. Which should sound better and which is more sophisticated (complex in processing)? Which has the higher bit rate (and recording file size)?
12. Are the data stored on an archival SACD also compressed?

REFERENCES

1. Sound Enhancement Products, Inc., Cary, IL, http://www.accutronicsreverb.com/history.htm. Accessed July 2009.
2. M. Long, *Architectural Acoustics*, New York, NY: Elsevier Academic Press, 2006, pp. 298–303.
3. H. Kuttruff, *Room Acoustics*, New York, NY: Taylor & Francis, 1999, pp. 122–130.
4. W. C. Sabine, *Collected Papers on Acoustics*, Los Altos, CA: Peninsula Publishing, 1992.
5. W. Richard. *Phil Spector: Out of His Head*, Abacus, London: Omnibus Press, 2003.
6. L. Mark. *The Complete Beatles Recording Sessions (First Hardback Edition)*, New York: Random House, 1988.
7. H. Newby, *Audiology*, Englewood Cliffs, NJ: Prentice-Hall, 1979, pp. 5–61.
8. J. McDermott and E. Kramer, *Hendrix*, M. Lewisohn ed., New York, NY: Grand Central Publishing, 1992, Chapter 7, pp. 94–95.

9. The Jimi Hendrix Experience. *Electric Ladyland*, New York: Reprise Records, div of Warner Communications, 1968.
10. E. Clapton, *Clapton: The Autobiography*, New York, NY: Broadway Books, 2007, pp. 327–328.
11. 2002. *DTS Coherent Acoustics; Core and Extensions.* European Broadcasting Union, ETSI TS 102 114 v1.2.1, downloadable from http://www.etsi.org. Accessed July 2009.
12. K. C. Pohlmann, *Principles of Digital Audio*, 4th ed., New York, NY: McGraw-Hill, 2000, pp. 303–362.
13. J. Watkinson,*The Art of Digital Audio*, 3rd edition, New York, NY: Elsevier, 2001, pp. 275–326.
14. P. P. Vaidyanathan, Quadrature mirror filter banks, M-band extensions and perfect reconstruction techniques, *IEEE Signal Process Mag*, 4(3), 1987, pp. 4–20.
15. Digital Audio Industrial Supply (DAISy), http://www.daisy-laser.com/technology/techsacd/techsacd15. htm Accessed July 2009.

5 Linear Filter Applications

Linear digital filters have as a typical application, frequency filtering for attenuation of certain frequency ranges such as high-pass, low-pass, band-pass, and band-stop filters. In this chapter, we explore some important applications of digital filtering which will be used later in this book and are also widely used in the signal processing community. State variable theory is presented for applications where the system state (position, velocity, acceleration, etc.) is of interest. Any system modeled as a mathematical function can be completely described provided sufficient samples of the function are known, or if the function output and all its derivatives are known for a particular input. The mass–spring–damper oscillator is formulated in a state variable model for comparison with the IIR models developed in the previous chapters. It will be shown that for discrete state variable filter, oversampling must be employed for an accurate system impulse response. Tracking filters are introduced for smoothing observed system output data and for examining unobservable system states using the nonadaptive α–β tracker. Tracking filters can also be used for detecting events based on the derivatives of the signal, rather than just the signal level.

We present 2D FIR filters [1] which are widely used in image and video processing. Almost all image filtering involves FIR rather than IIR filters because the spatial causality constraints must be relaxed in order for a small 2D FIR convolution filter to process a much larger image by scanning and filtering. Perhaps the most common 2D filter in most households is an autofocus system in a video camera. A high-pass 2D filter produces a maximum output signal when the image is in focus and giving many sharp edges. By constantly adjusting the lens system using a servo-motor to maximize the 2D high-pass filter output, the image stays in focus even as the camera moves about. In poor light or for very low contrast images, the high-pass filter approach to autofocus has difficulty finding the maximum output point for the lens. Many other applications of 2D filters to image processing can be done including nonlinear operations such as brightness normalization, edge detection, and texture filtering.

The chapter closes with a brief discussion of 2D filters for "upsampling" image data. Much like the audio oversampling reconstruction filters in Chapter 4; these 2D oversampling filters provide an easy way to interpolate pixel samples in two dimensions. Such image filtering can remove pattern aliasing which can occur when the spatial pattern of the original image is close in spatial sampling to the pixel grid of the sampled image. For example, a black and white checkered pattern in the image will likely align with the sampled image pixels poorly such that some pixels have all black, some have all white, but many have a mix of black and white giving a wavy-aliased pattern not part of the original image. By applying a 2D filter that extends over several of the original pixels, these wavy patterns can be smoothed out in an image with higher pixel densities (upsampled or oversampled). While this cannot increase the real resolution of the image, it can make an image look more natural and even make text easier to read.

5.1 STATE VARIABLE THEORY

A popular formulation for digital systems is state variable formulation where the state variables completely describe the dynamics of the system. State variables can represent almost any physical quantity, so long as the set of state variables represent the minimum amount of information which is necessary to determine both future states and system outputs for the given inputs. State variable formulations in digital systems can be seen as a carryover from analog electronic control systems. Linear time-invariant systems can be described as parametric models such as ARMA systems, or as linear differential equations such as a state variable formulation. A function polynomial of order

N can be completely described by $N + 1$ samples of its input and output, or by the value at one sample point plus the values of N derivatives at that point. Early analog control systems used state variables in the form of derivatives because the individual state elements could be linked together using integrators (single-pole low-pass filter).

The design of analog control systems is still a precise art requiring a great deal of skill and inventiveness from the designer. However, the digital age has almost completely eclipsed analog controllers due to the high reliability and consistent operation of digital controllers. The state variable in a digital control system can be some intermediate signal to the system, rather than specifically a representation of the system derivatives. The exceptions to this are the so-called alpha–beta (α–β) nonadaptive position tracking filter and the adaptive Kalman–Bucy tracking filter whose states are the derivatives of the underlying system model. Tracking filters will be covered in some detail later in the book with particular attention given to position tracking.

We now consider a general form of a state variable digital filter [2] at iteration k with "r" inputs $\bar{u}(k)$, "m" outputs $\bar{y}(k)$, and "n" system states $\bar{x}(k)$. The system state at time $k + 1$ has the following functional relationship:

$$\bar{x}(k+1) = f[\bar{x}(k), \bar{u}(k)]$$
$$= \mathbf{A}\bar{x}(k) + \mathbf{B}\bar{u}(k) \tag{5.1}$$

The system output also has a functional relationship with the state and input signals.

$$\bar{y}(k) = g[\bar{x}(k), \bar{u}(k)]$$
$$= \mathbf{C}\bar{x}(k) + \mathbf{D}\bar{u}(k) \tag{5.2}$$

The dimensions of $\mathbf{A}$ is $n \times n$, $\mathbf{B}$ is $n \times r$, $\mathbf{C}$ is $m \times n$, and $\mathbf{D}$ is $m \times r$. For single-input single-output systems, $\mathbf{B}$ is a $n \times 1$ vector and $\mathbf{C}$ is a $1 \times n$ vector, while $\mathbf{D}$ becomes a scalar.

Consider the following ARMA filter system to be represented in a state variable model.

$$G[z] = \frac{b_0 + b_1 z^1 + \cdots + b_Q z^Q}{1 + a_1 z^1 + \cdots + a_P z^P} = \frac{Y[z]}{U[z]} \tag{5.3}$$

The ARMA system in Equation 5.3 will need P states making $\mathbf{A}$ $P \times P$, $\mathbf{B}$ $P \times 1$, $\mathbf{C}$ $1 \times P$, and $\mathbf{D}$ 1×1 in size. An expanded Equation 5.1 is

$$\begin{bmatrix} x_1[k+1] \\ x_2[k+1] \\ \vdots \\ x_P[k+1] \end{bmatrix} = \begin{bmatrix} 0 & 1 & 0 & 0 & \cdots & 0 \\ 0 & 0 & 1 & 0 & \cdots & 0 \\ & & & \vdots & & \\ -a_P & -a_{P-1} & & \cdots & & -a_1 \end{bmatrix} \begin{bmatrix} x_1[k] \\ x_2[k] \\ \vdots \\ x_P[k] \end{bmatrix} + \begin{bmatrix} 0 \\ 0 \\ \vdots \\ 1 \end{bmatrix} u[k] \tag{5.4}$$

and the system output is

$$y[k] = \begin{bmatrix} b_Q & b_{Q-1} & \cdots & b_0 \end{bmatrix} \begin{bmatrix} x_1[k] \\ x_2[k] \\ \vdots \\ x_P[k] \end{bmatrix} \tag{5.5}$$

Equation 5.5 is valid for $P = Q$ as written. If $P > Q$, $P - Q$ zeros follow to the right of b_0 in Equation 5.5. If $Q > P$, $Q - P$ columns of zeros are added to the left of $-a_P$ in Equation 5.4. State variable formulations are usually depicted in a signal flow diagram as shown in Figure 5.1 for the ARMA system described in Equations 5.4 and 5.5. The state variable implementation of the ARMA filter in Figure 5.1 is numerically identical to the IIR digital filter presented in the previous section [1]. However, when the state vector elements represent derivatives rather than delay samples, the state variable formulation takes on an entirely different meaning.

5.1.1 CONTINUOUS STATE VARIABLE FORMULATION

It is straightforward and useful for us to derive a continuous state variable system for the mass–spring–damper oscillator of Figure 2.3. Letting the force input $f(t)$ be denoted as $u(t)$ (so our notation is consistent with most publications on state variable filters), we have the second-order differential equation

$$u(t) = M\ddot{y}(t) + R\dot{y}(t) + Ky(t) \tag{5.6}$$

where $\ddot{y} = \partial^2 y/\partial t^2$ and $\dot{y} = \partial y/\partial t$. We can write the second-order system in Equation 5.6, as well as much higher order systems, as a first-order system by defining the necessary states in a vector which is linearly proportional to the observed output $y(t)$. The updates for the state vector are defined by the differential equation in $y(t)$, but the state vector itself is only required to be linearly proportional to $y(t)$ [2].

$$\overline{x}(t) = \begin{bmatrix} x_1(t) \\ x_2(t) \end{bmatrix} = \begin{bmatrix} y(t) \\ \dot{y}(t) \end{bmatrix} \tag{5.7}$$

The first-order differential equation for the state vector is simply

$$\dot{\overline{x}}(t) = \begin{bmatrix} 0 & 1 \\ -\dfrac{K}{M} & -\dfrac{R}{M} \end{bmatrix} \begin{bmatrix} x_1(t) \\ x_2(t) \end{bmatrix} + \begin{bmatrix} 0 \\ \dfrac{1}{M} \end{bmatrix} u(t) \tag{5.8}$$

$$= \mathbf{A}^c \overline{x}(t) + \mathbf{B}^c \overline{u}(t)$$

The state variable system's frequency and impulse responses are derived using Laplace transforms

$$s\overline{X}(s) - \overline{x}(0^+) = \mathbf{A}^c \overline{X}(s) + \mathbf{B}^c U(s) \tag{5.9}$$

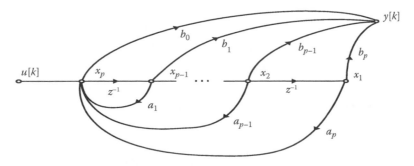

FIGURE 5.1 A flow diagram for the state variable formulation of an ARMA system showing a simple tapped delay line propagating the state variables $x_i[k]$.

giving the state vector *s*-plane response as

$$\bar{X}(s) = [sI - \mathbf{A}^c]^{-1}\bar{x}(0^+) + [sI - \mathbf{A}]^{-1}\mathbf{B}^c U(s) \tag{5.10}$$

The time-domain system response for arbitrary initial conditions and input forcing function $u(t)$ is

$$\bar{x}(t) = \bar{\phi}^{A^c}(t)\,\bar{x}\,(0^+) + \int_0^t \bar{\phi}^{A^c}(t-\tau)\,B^c\,u\,(\tau)\,d\tau \tag{5.11}$$

where $\bar{\phi}^{A^c}(t) = L^{-1}\{[sI - \mathbf{A}^c]^{-1}\}$. For the system impulse response, the initial position and velocity are both zero and $u(t)$ is a Dirac delta function. The system impulse response is therefore

$$\bar{h}(t) = \mathbf{B}^c \int_0^t \bar{\phi}^{A^c}(t-\tau)\delta(\tau)\,d\tau = \bar{\phi}^{A^c}(t)\mathbf{B}^c \tag{5.12}$$

Note that the impulse response signal is a 2×1 vector with the first element linearly proportional to the output $y(t)$ and the second element linearly proportional to the output velocity. The proportionality constants are elements of the $\mathbf{C}$ matrix in Equation 5.2 and are derived by equating the state vector impulse response to the actual response for the physical system.

$$\bar{h}(t) = L^{-1}\left\{ \frac{\left[\left(s+\dfrac{R}{M}\right) \quad -\dfrac{K}{M} \right]\left[\begin{matrix} 0 \\ \dfrac{1}{M} \end{matrix} \right]}{\left(s^2+\dfrac{R}{M}+\dfrac{K}{M}\right)} \right\} = L^{-1}\left\{ \frac{\left[\begin{matrix} -\dfrac{K}{M^2} \\[2mm] \dfrac{s}{M} \end{matrix} \right]}{\left(s^2+\dfrac{R}{M}+\dfrac{K}{M}\right)} \right\} \tag{5.13}$$

The mass position given in Equation 2.38 turns out to be $-M^2/K$ times $x_1(t)$ and the velocity of the mass is simply M times $x_2(t)$ where ω_d and ζ are defined in Equation 2.37.

$$\bar{h}(t) = \begin{bmatrix} x_1(t) \\ x_2(t) \end{bmatrix} = \begin{bmatrix} -\dfrac{k}{M^2\omega_d}\,e^{-\zeta t}\sin\omega_d t \\[4mm] \dfrac{1}{M}\,e^{-\zeta t}\cos\omega_d t - \dfrac{\zeta}{M\omega_d}\,e^{-\zeta t}\sin\omega_d t \end{bmatrix} \tag{5.14}$$

Both an output position and velocity are available from the state variables using $\bar{y}(t) = \mathbf{C}^c\bar{x}(t)$.

$$\bar{y}(t) = \begin{bmatrix} y(t) \\ \dot{y}(t) \end{bmatrix} = \begin{bmatrix} -\dfrac{M^2}{K} & 0 \\[2mm] 0 & M \end{bmatrix}\bar{x}(t) \tag{5.15}$$

5.1.2 Discrete State Variable Formulation

To implement a discrete state variable digital filter and show the effects of sampling we start with the continuous system sampled every T seconds. The state vector elements are defined the same as for the continuous case.

$$\dot{\bar{x}}(t) = \begin{bmatrix} 0 & 1 \\ -\dfrac{K}{M} & -\dfrac{R}{M} \end{bmatrix} \begin{bmatrix} x_1(kT) \\ x_2(kT) \end{bmatrix} + \begin{bmatrix} 0 \\ \dfrac{1}{M} \end{bmatrix} u(kT)$$
$$= \mathbf{A}\bar{x}_k + \mathbf{B}\bar{x}_k \tag{5.16}$$

The discrete estimate for the derivative given in Equation 2.17 is used to obtain

$$\bar{x}_{k+1} = \left[T\mathbf{A} + I \right]\bar{x}_k + T\mathbf{B}u_k$$
$$= \tilde{\mathbf{A}}\bar{x}_k + \tilde{\mathbf{B}}u_k \tag{5.17}$$

where $\tilde{\mathbf{A}}$ and $\tilde{\mathbf{B}}$ are the state transition and control input matrices, respectively, for the digital system. To examine the digital frequency response, the z-transform is used to yield

$$\bar{X}[z] = \frac{z}{[zI - \tilde{\mathbf{A}}]}\bar{x}(0^+) + \frac{1}{[zI - \tilde{\mathbf{A}}]}\tilde{\mathbf{B}}U[z] \tag{5.18}$$

The digital time-domain state response to the input u_k is therefore

$$\bar{x}_k = \tilde{\mathbf{A}}^k \tilde{x}(0^+) + \sum_{i=0}^{k-1} \tilde{\mathbf{A}}^{k-i-1} \tilde{\mathbf{B}} u_{k-i} \tag{5.19}$$

and the system impulse response is

$$\bar{h}_k = \tilde{\mathbf{A}}^{k-1} \tilde{\mathbf{B}} \tag{5.20}$$

A plot of the true position impulse response for the case of $M = 1$ kg, $R = 4$ kg/s, and $K = 314$ kg/s^2 where the sample rate f_s is 500 Hz is shown in Figure 5.2. For 500 samples/s, or $T = 2$ ms, there is clearly a reasonable, but not perfect, agreement between the continuous and digital state variable systems.

The reason for the difference between the two systems is that $\tilde{\mathbf{A}}$ scales with the sample rate. As the sample rate decreases (T increases), the error between the two systems becomes much larger as shown in Figure 5.3 for a sample rate of only 75 Hz.

The instability in the example shown in Figure 5.3 illustrates an important design criteria for discrete state variable filters: *significant oversampling is required for an accurate and stable impulse response*. It can be shown that for the impulse response to be stable, the determinant of the discrete state transition matrix must be less than unity.

$$|\tilde{\mathbf{A}}| = \begin{vmatrix} 1 & T \\ -\dfrac{K}{M}T & -\dfrac{R}{M}T + 1 \end{vmatrix} = -T\frac{R}{M} + 1 + T^2\frac{K}{M} < 1 \tag{5.21}$$

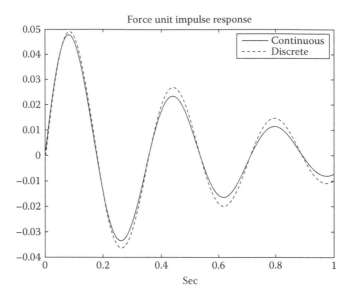

FIGURE 5.2 Comparison of state variable impulse response in the analog (solid) and digital (dotted) domain sampled at 500 Hz.

After some algebra, $0 < T < R/K$, or since $T > 0$ always, $T < 2\zeta/\omega_0^2$, where $f_0 = \omega_0/2\pi$ is the undamped frequency of resonance in Hertz. Clearly, the sampling frequency $f_s > \omega_0^2/2\zeta$ for stability. For the example shown in Figure 5.3, $f_0 = 27.65$ Hz and the minimum stable f_s is 78.75 Hz. It can be seen that while $f_s = 75$ Hz is easily high enough to represent the signal frequency properly, it is not high enough for a stable state variable system. Note that the sample rate must be even higher as the real physical system's damping decreases, and/or undamped frequency of resonance increases. An approximate guideline for *accurate* impulse responses can be found to be $f_s > \omega_0^2/20\zeta$ or higher. The cause of the error is seen to be the finite-difference approximation for the derivative of the state vector in Equation 5.17. Contrast the state variable filter oversampling requirement to a properly

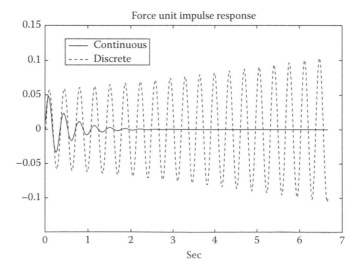

FIGURE 5.3 At a sample rate of only 75 Hz the digital state variable system is actually unstable due to poor approximations for the states leading to state error growth.

scaled IIR digital filter (which matches the impulse response for any unaliased sample rate) and one might wonder why one would use a digital state variable system. However, the reason as to why state variable systems are important can be seen in the simple fact that in some adaptive signal processing applications, the state variables themselves are the main point of interest.

5.2   FIXED-GAIN TRACKING FILTERS

Perhaps the most important and popular use of digital state variable systems is in tracking systems. Common examples of tracking filter use include air-traffic control systems, stock and futures market-programmed trading, autopilot controllers, and even some sophisticated heating, ventilation, and air conditioning (HVAC) temperature and humidity controllers. What makes a tracking filter unique from other digital filters is the underlying kinematic model [3]. Rather than a state transition matrix based on the specific parameters of the physical system, such as the mass, stiffness, and damping in Equation 5.16, a tracking filter's state transition matrix is based on the Newtonian relationship between position, velocity, acceleration, jerk, and so on. "Jerk" is the term used for the rate of change of acceleration with time. However, our discussion here will be mainly limited to "position-velocity" states with no input signal u_k to simplify presentation of the tracking filter concept.

A tracking filter can have input signals u_k, but the more usual case involves just passive observations to be smoothed and predicted in the future. The position state is predicted based on the last time's position and velocity and the Newtonian kinematic model. The prediction is then compared with an actual measurement of the position, and the resulting prediction error is weighted and used to correct or "update" the state variables. The weights for the state updates are α for the position, β for the velocity, and γ for the acceleration. The effect of the tracking filter is to "follow" the measurements and to maintain a kinematic model (position, velocity, acceleration, etc.) for predicting future positions. If the true kinematic system for the target has position, velocity, and acceleration components, and the tracking filter only has position and velocity, the α–β tracker will eventually "lose" a target under constant acceleration. However, if the target stops accelerating, an α–β tracker will fairly quickly converge to the true target track. Clearly, the usefulness of tracking filters can be seen in the air-traffic control system where collision avoidance requires estimates of target velocities and future positions.

Another benefit of tracking filters is to reduce the noise in the measurements and estimated velocities, accelerations, and so on. All measurement systems have inherent errors due to SNR and unmodeled physics in the measurement environment. The underlying assumption for tracking filters of all types is that the measurement errors are zero-mean Gaussian (ZMG) with a known standard deviation of σ_w. This is a fairly broad and occasionally problematic assumption particularly when unmodeled environment dynamics produce occasional biases or periodic (chaotic), rather than ZMG random measurement errors. However, if the measurement is affected by a large number of random processes (such as from signal propagation through atmospheric turbulence), the measurement noise statistics will tend to be Gaussian, following the central limit theorem.

State variable tracking is also useful for the detection of an event, which manifests itself by a pattern in the position, velocity, and/or acceleration states. For example, in a nuclear reactor, a change in the core temperature can happen extremely fast as a result of a control rod malfunction, and so detection of the rate of change (velocity) of the temperature is as important as the actual temperature (position) exceeding a threshold. Since the position, velocity, and acceleration of the state variable vector also have associated state error variances, a detection threshold can be set in terms of statistical metrics to balance the probability of detection verses the probability of false alarm.

If one has an unbiased measurement system, the ZMG assumption is very practical for most applications. The amount of position noise reduction is equal to α, the weighting factor for the position state updates. Therefore, if $\alpha = 0.10$, one would expect a 90% reduction in noise for the predicted position state as compared with the raw measurements. For $\alpha > 1$, one would expect noise amplification. Let the raw measurements be depicted as z_k

$$z_k = Hx_k + w_k \tag{5.22}$$

where H is a matrix (analogous to C above) which relates the components of the state vector x_k to the measurements, while w_k is the ZMG measurement noise. Using the Newtonian kinematic model one can predict the state vector one time step in advance (assuming no "process" noise in the state vector).

$$x_{k+1|k} = Fx_{k|k} = \begin{bmatrix} 1 & T & \frac{T^2}{2} \\ 0 & 1 & T \\ 0 & 0 & 1 \end{bmatrix} \begin{bmatrix} x_{k|k}^p \\ x_{k|k}^v \\ x_{k|k}^a \end{bmatrix} \tag{5.23}$$

where $x_{k|k}$ is a three-row, one-column position–velocity–acceleration state vector updated at time step k. Examining the top row of the F matrix (the state transition matrix), it can be clearly seen how the new position state is predicted for step $k+1$, given updated data at step k from Newton's laws of motion.

$$x_{k+1|k}^p = x_{k|k}^p + x_{k|k}^v T + \frac{1}{2} x_{k|k}^a T^2 \tag{5.24}$$

The updated state vector elements from time step k are $x_{k|k}^p$ for position, $x_{k|k}^v$ for velocity, and $x_{k|k}^a$ for acceleration, and the time interval between steps is T s. The state transition in Equation 5.23 amounts to a *noiseless* perpetual velocity and acceleration. That is, the current velocity and acceleration are assumed constant for a prediction N iterations into the future. If the "1's" in the 2,2 and 3,3 position of the matrix in Equation 5.23 were, say, 0.99, then the position would eventually come to rest for N iterations into the future. If all three 1's on the main diagonal were less than unity, the position would eventually return to the origin after many iterations into the future. Future predictions of the state vector are a useful application of tracking filters and the state transition matrix F in Equation 5.23 can be used to bias these predictions if desired. For some applications, measurements can be sporadically missing from the update cycle, so one would use a prediction instead of an actual measurement. A "bias" such as the leaky (main diagonal of F less than unity) state transition is usually done on the acceleration state at least to avoid a possible large track divergence during the period where measurements are missing. The normal operation of a tracking filter involves making a state prediction, using an actual measurement to calculate the prediction error, and then adjusting the predicted state through corrections proportional to the measured error in a separate state vector called the "updated state." The distinction between predicted and updated states is important. This way the position, velocity, and acceleration states "track" the actual measurements and the predicted states offer a "smoothed" estimate with less measurement noise than the actual measurements or updated states.

The error between the predicted state vector and the actual measurement is

$$\varepsilon_{k+1} = z_{k+1} - Hx_{k+1|k} \tag{5.25}$$

One then produces an "updated" state vector, separate from the "predicted" state vector, using the α–β–γ weights on the error.

$$x_{x+1|k+1} = x_{k+1|k} + W\varepsilon_{k+1}; \quad W = \left[\alpha, \frac{\beta}{T}, \frac{\gamma}{2T^2} \right] \tag{5.26}$$

The updated state vector will tend to follow the measurements more closely than the predicted state vector, but will also have measurement noise in it. Hence, the predicted state $x_{k+1|k}$ is often referred to as the "smoothed" state estimate, although both state estimates will be less reactive to measurement noise (as well as target maneuvers) as α decreases. We use the term "target" to refer to the position state estimate. The "target" is the information being tracked, be it an actual moving target or a simple parameter such as temperature or financial stock price. The task of optimally setting the tracking filter gains α, β, and γ is presented in Section 11.1, since it involves least-squared error processing. The derivation at the end of Section 11.1 is fairly tedious, but it allows one to set α, β, and γ optimally with respect to each other, and to determine the amount of smoothing the tracking filter is doing on the raw target measurements.

Choosing the tracking filter gains requires additional information in the form of the tolerable state vector *process noise*. Unlike the measurement noise, the process noise has nothing to do with the environment or measurement system. It is determined by the underlying assumptions in the target kinematic model where we have state elements backed up by measurements, state elements with no measurements, and unmodeled dynamics of the target such as higher derivatives. The later collection of unmodeled dynamics is what we lump together into the process noise for the target. For example, for an α–β tracker there is an implicit assumption of a ZMG acceleration process with standard deviation σ_v. A reasonable guide for an α–β tracker would be to set σ_v to the maximum expected acceleration of the target. For an α–β–γ tracker, σ_v would be set to the biggest expected jerk. The underlying assumption of an unpredictable acceleration in an α–β tracker from step to step allows the tracking filter to follow changes in velocity from target maneuvers (e.g., changes in target velocity and/or acceleration). Therefore, the state prediction equation is actually

$$x_{k+1|k} = Fx_{k|k} + v_k \qquad (5.27)$$

where v_k is a ZMG random *process noise* with standard deviation σ_v substituting for the unmodeled dynamics of the target.

For the α–β tracker, the process noise would nominally be set to be on the order of the maximum expected acceleration of the target. For the α–β–γ tracker, one assumes an unpredictable white jerk (not to be confused with an obnoxious Caucasian), to allow for ZMG changes in the rate of change in acceleration. Similar to the α–β tracker, the process noise for the α–β–γ tracker would be set on the order of the biggest jerk (no comment this time) expected from the target track. Consider how the state transition scales the process noise in Equation 5.27. If the RMS jerk noise is σ_v, the acceleration RMS-predicted state error can be no smaller than σ_v, the velocity RMS-predicted state error can be no smaller than $T \sigma_v$, and the position RMS-predicted state error can be no smaller than $1/2 \, T^2 \, \sigma_v$. This is the case if the measurement noise is zero. But the measurement noise is greater than zero and presumably the reason for using the tracking filter to reduce the state noise in the first place. The updated state error under steady-state conditions (process noise is zero) can be no smaller than the measurement noise, as in Equation 5.22.

We have to deal with the measurement noise as it is but we have to select the process noise to suit our assumptions about the target's unmodeled state dynamics. Why not select a process noise of zero? One could certainly do this and the associated underlying assumption would be that the target kinematic states do not change (the target never maneuvers). The smaller one makes the process noise, the more the measurements are smoothed by the tracking filter, and also, the tracking filter will be very sluggish to track real changes in the target kinematics. Making the process noise larger make the tracker more responsive to the data measurements, and also more measurement noise will be present in the smoothed (predicted) state estimates.

The process noise assumptions lead us to a very important parameter in fixed-gain tracking filters called the target maneuvering index λ_M.

$$\lambda_M = \frac{\sigma_v}{\sigma_w} T^2 \tag{5.28}$$

The RMS-predicted position noise lower bound (if the measurement noise is zero) is found to be $\sigma_v T^2/2$. Choosing a smaller target-maneuvering index (reducing the tolerated process noise for the available measurement noise) for the tracker will give a "sluggish" target track which greatly reduces track noise but is slow to respond to target maneuvers. Low target-maneuvering indices are appropriate for targets such as large aircraft, ships, or very stable, slow moving, systems. Larger maneuverability indices may be appropriate for cases where one is not as interested in smoothing the states, but rather having low bias for real-time dynamic state tracking.

In Chapter 11, we will present a derivation of the optimum least-squared error tracking gains based on the underlying statistics and kinematic model. Below we simply provide the solution for the tracking filter weights based on either of two simple design criteria: constant maneuvering index; and nonconstant measurement noise. For both cases the target process noise is assumed to be constant and set based on the target kinematics. The constant maneuvering index case is therefore most appropriate when the measurement noise variance is assumed constant. One might then simply choose an α based on the amount of noise reduction desired. The smoothed (predicted) position state will have a variance approaching $\alpha\sigma_w^2$ for small α and steady-state conditions. For $\alpha > 0.1$ the predicted state variance is $\alpha\sigma_w^2/(1-\alpha)$. The β-gain can be derived directly from α.

$$\beta = 2(2-\alpha) - 4\sqrt{1-\alpha} \tag{5.29}$$

The gain for γ is then set as β^2/α if an acceleration state exists. The target maneuvering index in terms of α and β is therefore

$$\lambda_M = \frac{\beta}{\sqrt{1-\alpha}} = \frac{\sigma_v}{\sigma_w} T^2 \tag{5.30}$$

One can evaluate the resulting process noise σ_v from the choice of α, but the more typical way to design a fixed gain tracking filter is to choose σ_v based on the known target kinematics, determine σ_w objectively for the particular measurement sensor system in use, and compute the optimal α, β, and γ gains from the resulting target maneuvering index.

$$\beta = \frac{1}{4}\left(\lambda_M^2 + 4\lambda_M - \lambda_M\sqrt{\lambda_M^2 + 8\lambda_M}\right) \tag{5.31}$$

The α gain can then be determined directly from λ_M, or more conveniently from β.

$$\alpha = -\frac{1}{8}\left\{\lambda_M^2 + 8\lambda_M - (\lambda_M + 4)\sqrt{\lambda_M^2 + 8\lambda_M}\right\} = \sqrt{2\beta} - \frac{\beta}{2} \tag{5.32}$$

If an acceleration state is used, $\gamma = \beta^2/\alpha$ as previously noted. The optimal tracking filter fixed gains are determined from the algebraic solution of several nonlinear equations derived from the least-squared error solution. The solutions for α, β, and γ presented here assume a piecewise constant ZMG acceleration for the α–β tracker and piecewise constant ZMG acceleration increment (the jerk) for the α–β–γ tracker.

Consider the following example of an elevation tracking system for a small rocket which launches vertically from the ground, burns for 5 s, and then falls back to earth. Neglecting changes in the mass of the rocket, the thrust would produce an acceleration of 15 m/s² in a zero gravity vacuum. The rocket is also assumed to be subject to a drag deceleration of 0.25 times the velocity. During lift-off, the rocket's acceleration slowly decreases as the drag forces build up with velocity. At burn-out, the maximum deceleration is imposed on the rocket from both gravity and drag. But as the rocket falls back to earth the drag forces again build up in the other direction, slowing the rocket's acceleration back toward the ground. Our measurement system provides altitude data 10 times/s with a standard deviation of 3 m. It is estimated that the maximum deceleration is around 13 m/s² which occurs at burnout. Therefore, we will assume $\sigma_y = 13$, $\sigma_w = 3$, and $T = 0.1$ giving a target maneuvering index of $\lambda_M = 0.0433$, and tracking filter gains of $\alpha = 0.2548$, $\beta = 0.0374$, and $\gamma = 0.0055$. Figure 5.4 shows the results of the α–β–γ tracking filter.

Figure 5.4 clearly shows the benefits of tracking filters when one needs a good estimate of the target velocity. The velocity measurements shown in Figure 5.4 are computed by a simple finite difference and show significant errors as compared to the tracking filter's velocity estimate. The measurement errors for acceleration based on finite difference are so great that they are omitted from the acceleration graph. Once the acceleration begins to settle into a nearly constant range, the acceleration states start to converge. The underlying assumption for the α–β–γ tracker is that the acceleration increment, or jerk, is ZMG. For our real physical problem this is indeed not the case, and so it is not surprising that the acceleration losses track during changes in the target's acceleration

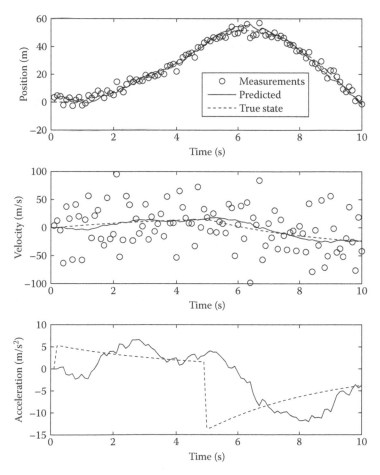

FIGURE 5.4 Typical α–β–γ results for a constant measurement error standard deviation of 3 m and a process noise set to be on the order of the maximum acceleration, of $\sigma_y = 13$. Time steps are 100 ms, dotted lines represent true states, circles represent measurements, and solid lines represent tracking filter states are.

(maneuvers). If one were to choose a process noise too small for the expected target maneuvers, such as $\sigma_v = 3$ rather than 13, the sluggish track ($\lambda_M = 0.01$) contains significant errors in position velocity and acceleration as shown in Figure 5.5. Choosing too large a process noise produces a very reactive track to target maneuvers, but the value of the estimated velocity and acceleration is greatly diminished by the huge increase in unnecessary tracking noise as shown in Figure 5.6 ($\sigma_v = 100$, $\lambda_M = 0.333$). The acceleration and velocity estimates from the tracking filter are almost useless with high track maneuverability, but they are still much better than the finite-difference estimates.

The technique of determining the maneuvering index based on the target kinematics and measurement system noise can be very effective if the measurement error variance is not constant. Computing λ_M, α, β, and γ for each new estimated measurement noise σ_w allows the tracking filter to "ignore" noisy measurements and pay close attention to accurate measurements. In Figure 5.7, a burst of extra measurement noise occurs between 7 and 8 s during the simulation where σ_w goes from 3 to 40 m and then back to 3 m. With the tracking filter assuming a constant measurement noise standard deviation of 3 m, a significant "glitch" appears in the state estimates. Figure 5.8 shows the performance possible when the target maneuvering index and filter gains are updated at every step to maintain optimality during changes in measurement noise.

It can be seen that the varying maneuvering index which follows the varying measurement noise is similar to the fully adaptive gain Kalman tracking filter. However, the big difference in the

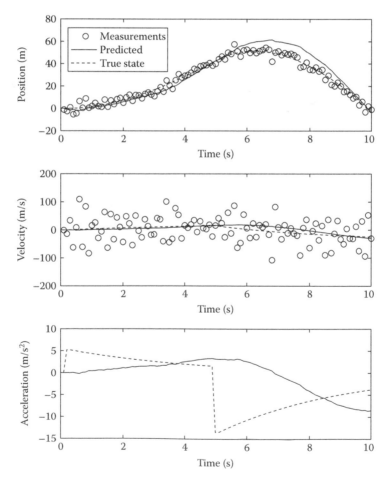

FIGURE 5.5 Tracking filter response using a sluggish process noise of $\sigma_v = 3$; all other parameters are the same as in Figure 5.4.

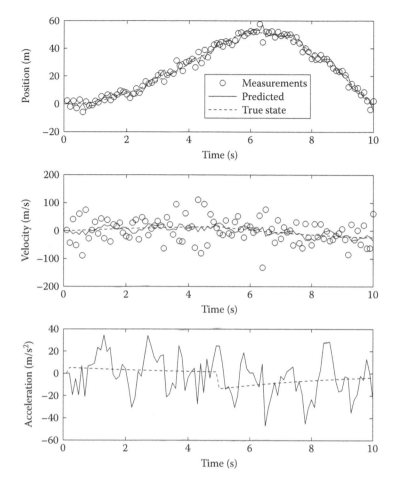

FIGURE 5.6 Tracking filter response using a hyperactive $\sigma_y = 100$ showing close tracking of the position state but at the expense of very noisy velocity and acceleration states.

Kalman filter (as presented in Chapter 11) is that the adaptive gains are determined from the state vector error variances as well as the measurement and process noises. In other words, the sophisticated Kalman filter considers its own estimates of the state error uncertainty when adapting to a new measurement. If the new measurement noise error variance is bigger than the estimated state error variance, the Kalman filter will place less emphasis on the new measurement than if the state error were larger than the measurement error. The much simpler fixed-gain tracker presented above simply determines the tracking filter gains based on the assigned target maneuvering index.

5.3 2D FIR FILTERS

The 2D FIR filters are most often used to process image data from camera systems [4]. However, the techniques of sharpening, smoothing, edge detection, and contrast enhancement can be applied to any 2D data such as level versus frequency versus time, for example. Image processing techniques applied to nonimage data such as acoustics, radar, or even business spreadsheets can be very useful in assisting the human eye in extracting important information from large complicated and/or noisy data sets. Data visualization is probably the most important benefit of modern computing in science. The amount of information passed through the human optic nerve is incredible but, the human brain's ability to rapidly process visual information completely eclipses all man-made computing

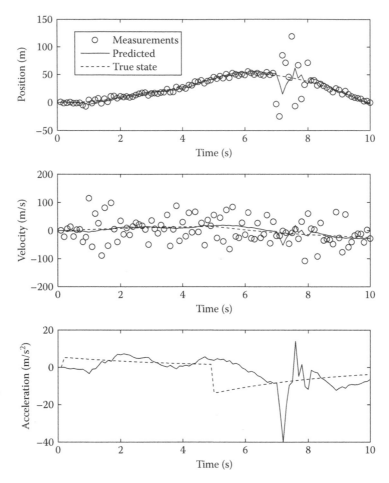

FIGURE 5.7 Tracking response for the same data shown in Figure 5.4 except a measurement noise burst of $\sigma_w = 40$ occurs between 7 and 8 s where the tracking filter assumes a constant $\sigma_w = 3$.

machines. A robot can select, assemble, and manipulate nuts and bolts with efficiency well beyond that of a human. But if the robot drops a bolt into a pile of screws and has to recover it, the required computing complexity increases by an almost immeasurable amount. Even a child only a few years old could find the bolt with ease. One should never underestimate the value of even unskilled human labor in automated industries, particularly in the area of visual inspection.

Ultimately, one would like to develop algorithms constructed of relatively simple 2D processes which can enhance, detect, and hopefully recognize patterns in the data indicative of useful information. This is an extremely daunting task for an automatic computing system even for the simplest tasks. However, we can define a few simple operations using 2D FIR filters and show their usefulness by examining the effects on a digital 256-level gray-scale image of a house (the author's), as shown in Figure 5.9. For those of you who bought the first edition, you will notice the addition to the house (royalties much appreciated) and the removal of some trees.

Each picture element (pixel) in Figure 5.9 is an 8-bit number representing the brightness of the image at the corresponding location on the camera focal plane. It is straightforward to see that if one were to combine adjacent pixels together, say in a weighted average, the sharpness, resolution, contrast, and even texture of the image can be altered. Even more interesting is the idea that the signal processing system can apply a filter to the image in the form of a template detector to search for features such as windows, roof lines, walls, trees, and so on. The template detector filter will have

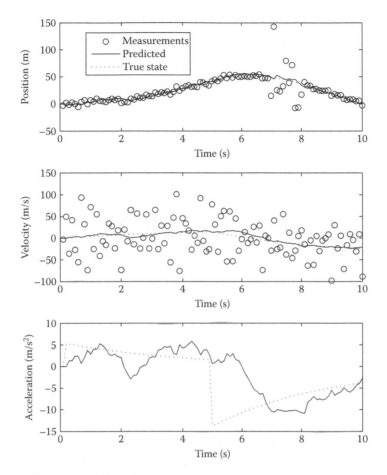

FIGURE 5.8 Tracking response for the same data shown in Figure 5.4 except a measurement noise burst of $\sigma_w = 40$ occurs between 7 and 8 s where the tracking filter the changes in σ_w. If measurement noise is variable and can be estimated, it can significantly improve fixed gain tracking filter performance.

FIGURE 5.9 Test image for 2D filtering examples 1168×1760 pixels wide and 256 gray-scale levels.

high output for the regions where there is a good match between the template and the local pixel distribution. These "vision" features can be used by a computer to recognize one house from another (depending on the diversity of house designs in the neighborhood). The process of developing computer algorithms which enable a signal processing system to have visual recognition capability is awesome in scope. The fundamental processing element for computer vision is the 2D digital filter or convolver. We denote the gray-scale image brightness as $B(x,y)$, where x represents the row and y the column of the pixel. A 2D filter which combines regional pixels using weights $w_{i,j}$ to produce the filtered output image brightness $B'(x,y)$ can be expressed as follows:.

$$B'(x,y) = \sum_{i=-N}^{+N} \sum_{j=-M}^{+M} w_{i,j} B(x+i, y+j) \qquad (5.33)$$

The process of computing a brightness output based on a weighted average of the adjacent pixels is likely where the term "moving average" for FIR digital filters originates. A wide range of visual effects can be implemented using simple 2D filters to alter the image focus, contrast, sharpness, and even color. 2D FIR filters can be designed to enhance or reduce various types of textures in an image. However, rarely would one try to implement an IIR filter since causality constraints would limit severely the direction of movement for some types of image patterns. For example, if the 2D filter moves in the $+x$ direction, "future-scanned" inputs $B(x+i,y)$ are used to compute the current output $B'(x,y)$. This is acceptable for an FIR filter. However, an IIR filter must only use past outputs to avoid a noncausal instability. In general, one uses only FIR filters for image processing since the direction of scan for the filter should not be a factor in the processed output image. To simply illustrate the 2D FIR filtering process, Figure 5.10 shows a process often referred to as "pixilation" where all the pixels in a block are replaced with the average of the higher-resolution original image pixels. Essentially, the image sampling rate is reduced or *decimated*. This process can often be seen in television broadcast to obscure from view offensive or libelous material in a section of an image while leaving the rest of the image unchanged. Figure 5.10 shows a 16×16 pixilation of the original image in Figure 5.9 reducing the original 1168×1760 pixel image to only 73×110 pixels. This reduces the file size of the image from 2,055,680 bytes to only 8030 bytes, but with considerable loss of resolution. We present pixilation because it is easy to conceptualize. But, why not skip the pixel decimation and replace each pixel in the original 1168×1760 pixel image with the average of the 16×16 pixels around it. For this "boxcar" FIR filter, the weights $w_{i,j}$ in Equation 5.33 are each 1/1024. Figure 5.11 shows the result of this equal-weighted pixel averaging, which is similar to

FIGURE 5.10 Pixelated test image where blocks of 16×16 pixels are replaced by their average giving an effective image resolution of 73×110 pixels.

FIGURE 5.11 A low-pass image filter replacing each output pixel by the average of the 16×16 block of neighboring pixels.

low-pass filtering except that the averaging is spatially greater along the diagonals (about 22 pixels averaged) and shorter in the vertical and horizontal directions.

Another interesting application of 2D filters to image processing is edge enhancement and detection. Edges in images can provide useful features for determining object size and dimension, and therefore are very useful elements in computer recognition algorithms. The most basic edge detector filter is a simple gradient which is shown in the positive x-direction (to the right) below. Since it is usually undesirable for a filtering operation to give an output image offset by 1/2 a pixel in the direction of the gradient, one typically estimates the gradient using a symmetric finite-difference approximation.

$$\begin{aligned} B'(x, y) &= -\frac{\partial}{\partial x} B(x, y) \\ &= -\left[B(x+1, y) - B(x-1, y) \right] \end{aligned} \tag{5.34}$$

The negative gradient in the x-direction in Equation 5.34 can be seen in Figure 5.12 giving the digital image an "embossed" look. Examining the area between the tree trunk and house, one can clearly see that the transition from the black shadow to the white wall in the original image produces a dark vertical line along the edge of the house wall. The image is normalized to a mid-level gray which allows the transition from bright to dark (seen in the left edges of the windows) to be represented as a bright vertical line. Note how the horizontal features of the image are almost completely suppressed. Figure 5.13 shows the test image with a negative gradient in the positive y-direction (upward).

Derivatives based on a finite-difference operation are inherently noisy. The process of computing the difference between two pixels tends to amplify any random noise in the image while averaging pixels tends to "smooth" or suppress image noise. Generally, one can suppress the noise by including more pixels in the filter. An easy way to accomplish this is to simply average the derivatives in adjacent rows when the derivative is in the x-direction, and adjacent columns when the derivative is in the y-direction. A 2D FIR filter results with $N = M = 1$ for Equation 4.19 where the filter weights for the negative gradient in the x-direction are given in Equation 5.35.

$$-\nabla_x = \begin{bmatrix} w_{-1,+1} & w_{0,+1} & w_{+1,+1} \\ w_{-1,0} & w_{0,0} & w_{+1,0} \\ w_{-1,-1} & w_{0,-1} & w_{+1,-1} \end{bmatrix} = \begin{bmatrix} +1 & 0 & -1 \\ +2 & 0 & -2 \\ +1 & 0 & -1 \end{bmatrix} \tag{5.35}$$

+*x*-dir derivative

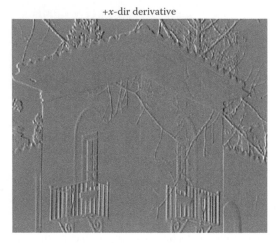

FIGURE 5.12 An *x*-direction gradient 2D filter output image showing enhancement of horizontal changes in brightness (vertical features are emphasized).

The 2D filter weights in Equation 5.35 are known as a kernel because the weights can be rotated along with the direction of the desired gradient. For example, a negative gradient in the positive *y*-direction (upward) can be realized by rotating the weights in Equation 5.35 counterclockwise 90° as seen in

$$-\nabla_y = \begin{bmatrix} -1 & -2 & -1 \\ 0 & 0 & 0 \\ +1 & +2 & +1 \end{bmatrix} \tag{5.36}$$

The kernel as written above can be conveniently rotated in 45° increments since the pixel in the center has eight neighbors. There are many possible kernels for a 2D gradient including larger, more complex, filters based on higher-order finite-difference approximations. However, it is also useful to take the eight neighboring pixels and *estimate* the vector gradient for the local area. Gradient information from the 2D image data is very useful for simplifying the task of automated detection of

+*y*-dir derivative

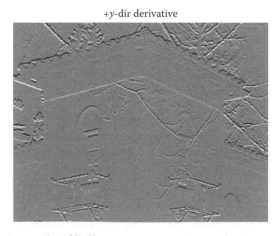

FIGURE 5.13 A *y*-direction gradient 2D filter output image shows enhancement of the vertical changes in brightness emphasizing horizontal features in the image.

geometrical features. These features of the image can subsequently be compared with a computer database of features as a means of automated detection of patterns leading to computer image recognition algorithms.

While directional derivatives can be very useful in detecting the orientation of image features such as edges, sometimes it is desirable to detect all edges simultaneously. The geometry of the detected edges can then be used to identify important information in the image such as shape, relative size, and orientation. A straightforward edge detection method, known as Sobel edge detection, computes the spatial derivatives in the x- and y-directions, sum their squares, and compute the square-root of the sum as the output of the filter. A less complex operator, known as the Kirsh operator, accomplishes a more economical result without the need for squares and square roots by estimating all eight gradients and taking the maximum absolute value as the edge detection output. The application of Sobel edge detection, with the Kirsh approximation to our test image in Figure 5.9 can be seen in Figure 5.14, where we have inverted the image to see the edges as black in a white field. This line-art appearance to the image is much easier to let the computer calculate than manually drawing it or even using photographic techniques! While edge detection is useful for extracting various features from the image for use in pattern recognition algorithms, it can also be used to enhance the visual quality of the image. The edge detector operator can easily be seen as a type of high-pass filter allowing only abrupt changes in spatial brightness to pass through to the output. If one could amplify the high frequencies in an image, or attenuate the low frequencies, one could increase the sharpness and apparent visual acuity. Typically, sharpness control filtering is done using a rotationally invariant Laplacian operator as follows:

$$\nabla^2 B = \frac{\partial^2 B}{\partial x^2} + \frac{\partial^2 B}{\partial y^2} \tag{5.37}$$

The Laplacian, like the gradient, is approximated using finite differences. The sum of the x- and y-direction second (negative) derivatives is expressed in the following equation:

$$-\nabla^2 B = \begin{bmatrix} 0 & -1 & 0 \\ -1 & +4 & -1 \\ 0 & -1 & 0 \end{bmatrix} \tag{5.38}$$

Inverted Sobel edges approx by Kirsh operator

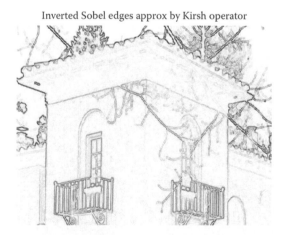

FIGURE 5.14 A Sobel edge detector approximated by a Kirsh operator (inverted here to save black ink) which enhances the squared magnitude of vertical and horizontal gradients, giving a line-art appearance to the original image.

Since we prefer to suppress noise by including all eight neighboring pixels in the edge detection operator, we simply add in the diagonal components to the negative Laplacian as follows:

$$-\nabla^2 B = \begin{bmatrix} -1 & -1 & -1 \\ -1 & +8 & -1 \\ -1 & -1 & -1 \end{bmatrix} \tag{5.39}$$

Note that both Equations 5.38 and 5.39 are normalized operators that approximate a negative Laplacian operator based on finite differences. They are called "normalized" because they do not cause a shift in the average brightness of the image. However, the operator will cause a slowly varying brightness to be nearly canceled since the sum of the eight neighbors will be nearly the same value as eight times the central pixel. But, because we calculate derivatives, random variations from pixel to pixel are amplified by the Laplacian operator, making the image appear more noisy. Figure 5.15 shows the house image with the Laplacian 2D FIR filter in Equation 5.39 and a mean image brightness of 127 (the 8-bit gray scale has a range of 0–255) to allow the negative edges to be seen as black and the positive edges to be seen as white. Note how the image compression by the camera (Joint Photographics Expert Group (JPEG)) artifacts is amplified by the Laplacian operator. By adding some percent of the negative Laplacian image back to the original, one can get some enhancement of the edges and "sharpen" the image features, but also this approach adds noise from the derivative estimates. It is a naming convention in the image processing literature that the image filter that enhances the spatial high frequencies is called a "sharpening" filter, but the image it produces looks nothing like the original image, such as what is shown in Figure 5.15. To improve the original image's "sharpness" we need to add or subtract some percentage the image in Figure 5.15 back to the original. This operation, by naming convention in the literature, is called an *unsharp filter* [5].

The unsharp filter mask actually dates back to the 1930s and is still used today in large format photography. One takes the glass negative (typically 8″ × 10″ in size) and creates a slightly blurry glass positive image by passing light from a point source through the negative emulsion and its glass plate, then the positive glass plate with its photo emulsion on the underside. By varying the aperture of the source light one can control the amount of diffusion, or blurriness, on the positive plate. Then the positive plate is placed above the negative plate and a regular paper photograph is made, but since the blurry positive plate cancels some of the negative features, in particular the low spatial frequency features, the resulting photograph on the paper is sharpened. The process is very useful

Laplacian

FIGURE 5.15 The Laplacian operator approximated using a 2D filter enhances all the high frequencies in the image, but also amplified noise artifacts (in this case, JPEG compression residuals).

Original LOG σ = 0.66

FIGURE 5.16 Comparison of the original to an "unsharp" filter using a LOG filter to enhance the edges without an increase in image noise.

in industrial photography and photolithography where the geometry of the photo features must be very precise and the contrast and dynamic range of the image must also be tightly controlled. This very technical process was employed by the most skillful photographers to enhance or salvage important images. The digital unsharp filter can be done by adding/subtracting the Laplacian filtered image to the original, or by subtracting the low-pass-filtered image from the original and using that as the high-frequency image that is added/subtracted to the original in place of the Laplacian image. The sign used in adding/subtracting the high-frequency or Laplacian image to the original is also a matter of convention, depending on the unsharp effect desired.

A less noisy approach to image sharpening is to first low-pass filter the image to reduce the noise and then apply the Laplacian. The 2D convolution operation filtering the image is associative, that is we can take the Laplacian of a Gaussian smoothing filter and use the result as the 2D FIR filter. This is called a "Laplacian of Gaussian" (LOG) filter in the literature because of the Gaussian smoothing done as well, and to not confuse it with the Laplacian. The Laplacian of a 2D Gaussian function is

$$\tilde{N}^2 G = \frac{-1}{\pi\sigma^4}\left(1 - \frac{x^2 + y^2}{2\sigma^2}\right)e^{-(x^2+y^2)/(2\sigma^2)} \tag{5.40}$$

where the range of x and y are from -1 to $+1$ pixels. Taking $\sigma = 0.66$ the unsharpen operator is

$$\nabla^2 G \approx \begin{bmatrix} -0.2189 & -0.0787 & -0.2189 \\ -0.0787 & +1.677 & -0.0787 \\ -0.2189 & -0.0787 & -0.2189 \end{bmatrix} \tag{5.41}$$

Figure 5.16 compares the original image with the unsharp LOG-filtered image on the right. Close examination shows a slight increased brightness around the dark objects such as tree branches and wrought iron. The "ringing" of the image edge signals aids visual contrast giving the appearance of sharpening the image. The noise from the Laplacian sharpening is suppressed with the LOG operator which also gives visual benefits.

5.4 IMAGE UPSAMPLING RECONSTRUCTION FILTERS

Sometimes we need to increase the pixel density of an image and smooth the pixilated approximation as shown in Figure 5.10, which is a 99.61% file size reduction (8030 bytes) of the original 2,055,680 byte image. The block averaging in each pixel happens in every digital image, in particular if one zooms in on a particular area. The effect of pixilation is seen in a rather extreme case in Figure 5.10, comparing with the original image in Figure 5.9. However, we can take a somewhat less extreme down sampling of 8×8 rather than 16×16 to work with a 146×220 pixel image which has 32,120 bytes, or about 1.5625% of the file size of the original image. How does one "upsample" this image back to the original resolution? Can one use 2D filters to help restore some of the original image? The image details within the 8×8 pixelation block are lost forever as a result of the averaging process. This is true in the camera as well. The projected image on the sensor may have higher optical resolution than the pixel resolution of the sensor, but each pixel can only estimate the average brightness by integrating all the photons striking the pixel surface in the sensor chip. So, there is not much we can do about the center of the downsampled pixels, but there is a lot we can do about the edges of those pixels in the upsampled image.

The checkered pattern of the pixels in the upsampled image can be seen as a zeroth-order sample and hold type of filter, where the downsampled pixel is just copied to the 63 surrounding pixels with an 8×8 pixel block. This is for our consumption only as a textbook example. Our approach follows with that in Figures 4.8 and 4.9 for an oversampled audio signal. For the 2D image oversampling, each of the downsampled pixels is surrounded by 63 zeros such that an 8×8 filter will replicate the downsampled pixel in all the surrounding "zero-padded" pixels in the upsampled image. Following the one-dimensional (1D) case of the audio oversample signal, we can derive higher-order filters to process the zero-padded image through the use of convolutions. Figure 5.17 shows the 8×8 zeroth-order filter in the upper left, and the 15×15 first-order 2D filter in the upper right. Convolving the first-order filter with itself yields a third-order filter of size 29×29 pixels, shown in the lower left corner of Figure 5.17. With each convolution operation the filter grows in size, making it more of a low-pass filter. But, the jagged edges of the pixilated image soften symmetrically for all directions in the image. This is a very nice property especially for lines that are not parallel to the horizontal or vertical, as well as any type of curve along a contrasting boundary in the image.

Figure 5.18 compares the upsampled image after the 2D filters have been applied. The upper left is the 8×8 pixelated version of the original image in Figure 5.9. Note that as the higher-order oversampling filters are applied, the image appears softer and the jagged edges around the arched window become smooth, as do the telephone wires seen on the upper right edge of the image. It appears that the first- and third-order filters do the best overall job of restoration at the higher resolution, while the seventh-order filter is doing a bit too much low-pass filtering. It really depends on the particular image and on ones perception of what looks best.

Given the zero-padded oversample-filtered image, we can restore some of the sharp edges and still keep the jagged pixilation under control by high-pass filtering. We can compute a Laplacian filtered image and scale and subtract it from the seventh-order-filtered image to restore the edges somewhat, as shown in Figure 5.19. In this image, the curves of the arch windows remain smooth while the edges are enhanced. A close examination of the telephone lines in the upper right show some zigzags, but they have been softened considerably. Doing these types of operations, one has to carefully scale the result, and so the pixels fit in the 0–255 range of an unsigned byte/pixel. Typically, one does the image processing in a floating point, then rescales so that the overall brightness is comparable to the original. One can see the enhanced contrast near the vertical edges of the front part of the house where the wall gets a little brighter just before a dark edge. This is how the Laplacian operator captures the spatial derivatives to enhance the sharpness. Many of the small tree branches are gone, but the window mullions can be seen more clearly. Overall, the upsampled and filtered image in Figure 5.19 is more cartoon like relative to the original in Figure 5.19. Textures in the original are replaced by averages of brightness. But, the brightness follows a smoothed contour of the image, not the downsampled pixel grid, and this is why it is useful to do this type of filtering in an attempt to restore the image. Similar filtering operations, sometimes called "image antialiasing filters" are applied to standard definition digital television displayed on a high-definition screen, although the specifics vary from manufacturer to manufacturer.

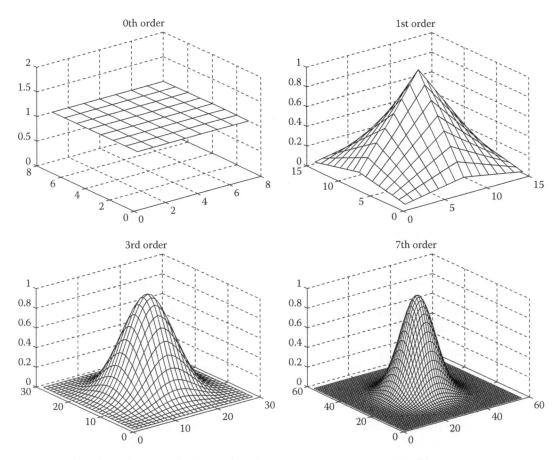

FIGURE 5.17 2D FIR filters for interpolating zero-padded image pixels showing the zeroth-order sample and hold (upper left), linear interpolator (upper right), and higher-order convolutions in the lower plots for a greater degree of smoothing.

5.5 MATLAB® EXAMPLES

Table 5.1 lists the particular MATLAB m-scripts used and the associated figure numbers for this chapter. These m-scripts should be available for download from the book website and also have comments in them to aid the reader in understanding the examples.

This is a good place to discuss/revisit matrix math in MATLAB. As a teacher, the subtle syntax MATLAB used to support matrix algebra has been the prime challenge for new users of MATLAB who may have extensive computer programming skills in C, Fortran, Basic, Java, and so on. Suppose we define a row vector in MATLAB as $v1 = [1\ 2\ 3]$ and a column vector as $v2 = [4;\ 5;\ 6]$. Note the semicolons which tell the script compiler "new row." Only a space is needed to define a new column. The square brackets define a nonscalar variable. The expression $v1*v2$ results in a scalar output of 32. Using the transpose operator " ' " we can get a 3×3 matrix result from $v1'*v2'$, since a 3×1 vector times a 1×3 vector results in a 3×3 matrix. The transpose operator automatically conjugates complex numbers for you, and so if you do not want a complex conjugate transpose (Hermitian transpose, named after Charles Hermite), you can use " .' " instead, but this step is not necessary for real numbers in matrix or vector elements. Dot products can be done only on identical dimensioned vectors and matrices. For example, $v1.*v2' = [4\ 10\ 18]$. The " .* " signifies a vector dot product, but it can extend to matrices too. The multiply or divide " . " operation happens element by element. These operations are extremely convenient, but only once you understand the subtlety.

0th order   1st order

3rd order   7th order

FIGURE 5.18 Reconstruction upsampling of an 8× zero-padded image (8 × 8 decimated original image) using sample and hold (zeroth-order upper left), linear (upper right), and higher-order convolution filters to smooth the jagged edges.

To make matters a little more challenging, the m-script interpreter lets you get away with sloppy syntax, such as dimensioning vectors or matrices on the fly, rather than explicitly at the beginning of a program. Again, this is very convenient, but can frustrate a new user trying to debug an m-script. Using dot products can save a great deal of computation time in MATLAB while also keeping your m-script free of for-loops and counters. The bright side is that once you have mastered the concepts of embedded matrix variables and the dot-operators, you can produce very fast running m-scripts that read like the equations in a textbook. It is well worth the effort caused by the flexibility to let you program the scripts the old fashioned way.

7th order minus 80*Laplacian

FIGURE 5.19 Subtracting a Laplacian image from the seventh-order upsampled and filtered image restores much of the edges and contrast but still cannot recover the texture detail in the original.

Table 5.1
m-Scripts Used in This Chapter

MATLAB m-Script Name	Figures Used
discrete.m	5.2, 5.3
abtrk2.m	5.4–5.8
imreader.m	5.9–5.11
imreaderfilter.m	5.12–5.16
imupsampled.m	5.17–5.19

The plots in Figures 5.1 and 5.3 were generated using "discrete.m" which happens to be a great example of the convenience of matrix algebra in an m-script. As presented in Table 5.2, a matrix "**A**" is a 2×2 matrix and "**B**" is a 2×1 vector. The system state "X" is stored as a 2×500 matrix where each column is a time sample of the 2×1 state vector. Note how simply Equation 5.20 is implemented in the first line in the for-loop. The colon in the parentheses after "X" mean "fill both rows" at the specified nth column. This expression allows the entire state history to be saved as a matrix and later plotted relative to the analog system being modeled in "$y(n)$" shown in Figures 5.2 and 5.3. Note the " .*" used in the second line in the for-loop for the discrete output "$yd(n)$." This is actually not necessary (call it old school), but used to explicitly show that $(1/T)$ is a scalar multiplying the "row 1," "column n" element of the matrix "X."

Using the appropriate matrix math, the for-loop in Table 5.2 could be eliminated and the code would run faster, it just would not be very readable from a computer programming point of view. We will opt for clarity, rather than the most elegant m-script to help keep all readers onboard. Since MATLAB now support matrices with three or more dimensions, one could in theory never use a for-loop again. May the Geek in you prosper!

The m-script snippet in Table 5.3 is for the α–β–γ tracker shown in Figures 5.4 through 5.8. Here the m-script might have well been written in BASIC or FORTRAN. Arrays defined are zeroed out first and all variables are essentially operated on as scalars. This makes the m-script easily read and manipulated, such as is done with the noise burst in the measurements between 7 and 8 s.

5.6 SUMMARY

Linear digital filters have many applications beyond simple frequency suppression of the time-domain signals. We have shown formulations of the digital state vector and the difference between a tapped delay line state vector and a differential state vector which can be used to model many physical systems. Conspicuously, one must have either a firm grasp of the underlying physics

---

Table 5.2
m-Script Snippet Evaluating Equation 5.20

```
A = [1  T;  (-T.*km)  (-Trm+1)] ;
B = [0;  Tm] ;
for  n = 1:500,
     X(:,n) = (A^(n-1))*B;
     yd(n) = (1/T).*X(1,n) ;
     y(n) = (1/wd)*exp(-sig.*n.*T).*sin(wd.*n.*T) ;
end;
```

Table 5.3
m-Script Snippet for the α–β–γ Tracker

```
xpred = zeros(size([1,npts]));
vpred = zeros(size([1,npts]));
apred = zeros(size([1:npts]));
xupdat = zeros(size([1,npts]));
vupdat = zeros(size([1,npts]));
aupdat = zeros(size([1:npts]));
for k = 2:npts,
    l = sigmav*dt^2/sigmaw;
    b = .25*(l^2+4*l-l*sqrt(l^2+8*l));
    a = sqrt(2*b)-b/2;
    g = b^2/a;
    xpred(k) = xupdat(k-1) + dt*vupdat(k-1) + (.5*dt^2)*aupdat(k-1);
    vpred(k) = vupdat(k-1) + dt*aupdat(k-1);
    apred(k) = aupdat(k-1);
    error = xmeas(k)-xpred(k);
    xupdat(k) = xpred(k) + a*error;
    vupdat(k) = vpred(k) + b*error/dt;
    aupdat(k) = apred(k) + g*error/(2*dt^2);
end
```

behind the modeling problem, or a firm grasp of a physicist to make sure that the model inputs and outputs are reasonable. However, what is often overlooked in applications of digital filtering to real physical problems is the physics of the digital system's operation. Converting analog signals to digital data for computer processing, while extremely powerful and versatile, does carry with it some undesirable artifacts. For example, sampling without band-limiting the analog signal leads to aliased frequencies. Mapping analog domain system poles and zeros to the digital domain requires modal scaling to obtain a reasonable, but not perfect, match between analog and digital impulse and frequency responses. Unless one significantly oversamples the signals, one cannot simultaneously match both poles and zeros for the same frequency response in the analog and digital domains. A similar problem arises in the digital state variable problem where the finite-difference errors become significant as the frequencies of interest approach the Nyquist sampling rate. For all digital filters, we must maintain a stable causal response giving rise to the strict requirement for all system poles to be interior to the unit circle on the digital z-plane. In 2D filters we generally limit the structure of operators to be of the FIR type so that the stability of the filtering operation is not dependent on which direction the 2D filter is moved over the 2D data. These physical attributes of digital signal processing are perhaps the most important fundamental concepts to understand before moving on to frequency transforms and the adaptive processes presented in the next few chapters.

PROBLEMS

1. What is the discrete state variable flow diagram for the system $(b_0 + b_1 z^{-1})/(1 + a_1 z^{-1} + a_2 z^{-2})$?
2. For a series resonant tank circuit with $R = 10\ \Omega$, $C = 1\ \mu F$, and $L = 2\ mH$, what is the minimum sample period in seconds for a stable state variable model? What sample period is recommended?

3. A volatile stock is being smoothed by an α–β–γ-type fixed-gain tracking filter to help remove random hourly fluctuations and improve the observability of market trends. If we wish to reduce the observed fluctuations to 10% of the original data, what are the values of α, β, and γ?

4. We are interested in detecting lines in an image that run along a −45° line (from upper left to lower right). Define a 2D FIR filter to enhance detection of these lines.

5. Show that the first-order (2×) interpolation filter is the convolution of the zeroth-order filter with itself, the third order is the convolution of the second order with itself, and so on. That being the case, it proves that the frequency response of the first-order interpolation filter is the square of the zeroth-order (taking into account the doubled sample rate), third-order square of the second order, and so on.

REFERENCES

1. A. V. Oppenheim and R. W. Schafer, *Discrete-Time Signal Processing*, Englewood Cliffs, NJ: Prentice-Hall, 1989.
2. T. Kailath, *Linear Systems*, Englewood Cliffs, NJ: Prentice-Hall, 1980.
3. Y. Bar-Shalom and X. Li, *Estimation and Tracking: Principles, Techniques, and Software*, Boston, MA: Artech House, 1993.
4. J. C. Russ, *The Image Processing Handbook*, Boca Raton, FL: CRC Press, 1992.
5. J. Phillips, *Unsharp Masking—A Beginners Primer*, http://www.largeformatphotography.info/unsharp/. Accessed May 2010.

Part II

Frequency Domain Processing

Frequency domain processing of signals is an essential technique for extracting signal information with physical meaning as well as a filtering technique for enhancing detection of periodic signal components. The genesis of frequency domain techniques dates back to the later half of the nineteenth century when Fourier (pronounced "4-E-A") published a theory which suggested that any waveform could be represented by an infinite series of sinusoids of appropriate amplitudes and phases. This revolutionary thought led to the mathematical basis for many fundamental areas in physics such as diffraction theory and optics, field theory, structural vibrations, and acoustics, just to name a few. However, it was the development of the digital computer in the 1950s and 1960s which allowed the widespread use of digital Fourier transformations to be applied to recorded signals. Now, in the last decade of the twentieth century, real-time digital frequency transformations are commonplace using ever more astonishing rates of numerical computation for applications in almost every area of modern technology. The topic of frequency domain processing is of such extreme importance to modern adaptive signal processing that an entire chapter is dedicated to it here. One can do Fourier processing in the time, space, frequency, and wavenumber domains for steady-state signals. However, one must adhere to the underlying physics and mathematical assumptions behind the particular frequency transformation of interest to be sure that the correct signal information is being extracted.

The Fourier transform mathematically is an integral of the product of the waveform of interest and a complex sinusoid with the frequency for which one would like to know the amplitude and phase of the waveform's sinusoidal component at that frequency. If one has an analytic function of the waveform in the time domain, $x(t)$, then one could analytically integrate $x(t)e^{j\omega t}\, dt$ over infinite time to obtain an equation for the frequency domain representation of the waveform $X(\omega)$. This has obvious mathematical utility because many differential equations are more easily solved algebraically in the frequency domain. However, in the straightforward applied mathematics of the signal processing world, one has a finite-length digital recording, $x[n]$, representing the waveform of interest. The analytic indefinite integral becomes a finite (truncated from t_1 to t_2) discrete sum which takes on the characteristics of spectral leakage, finite resolution, and the possibility of frequency aliasing. The limitations on the Fourier transform imposed by real-world characteristics of digital systems are manageable by controlling the size, or number of samples, of the discrete-time Fourier transform, the use of data envelope windows to control resolution and leakage, and by controlling and optimizing frequency resolution as needed for the application of interest. Our goal in this presentation of Fourier transforms is to demonstrate the effect of finite time (space) integration on frequency (wavenumber) resolution and spectral leakage.

Given the underlying physics of the waves of interest, the frequency domain data can be used to observe many important aspects of the field such as potential and kinetic energy densities, power flow, directivity, and wave coupling effects between media of differing wave impedance. When the sampled waveform represents spatial data, rather than time-sampled data, the frequency response represents the wavenumber spectrum which is a very important field parameter describing the wavelengths in the data. One can describe the physical manmade device of a pin-hole or lens-system camera as an inverse wavenumber Fourier transform system in two dimensions where the optical wavelengths are physically filtered in direction to reconstruct the spatial image data on a screen. The importance of wavenumber spectra will become evident when we use them to control focus in images as well as show radiation directivities (beamforming) by various transmitting arrays of sources. Perhaps what is most useful to keep in mind about Fourier transforms is the orthogonality of $e^{j\omega t}$ for time (t) and frequency (ω) transforms and e^{jkx} for space (x) and wavenumber ($k = 2\pi/\lambda$). Note that if t has units of seconds, ω has units of radians/second and if x has units of meters, k has units of radians/meter. Since the kernel sinusoids are orthogonal to each other, one obtains a discrete number of frequencies with amplitudes and phases independent of one another as the output of the "forward" (i.e., time to frequency or space to wavenumber) Fourier transform. The independence of the frequencies (or wavenumbers if space is the dimension rather than time) is what allows differential equations in the time domain to be solved algebraically in the frequency domain.

It is interesting to note the many natural frequency transforms occurring all around us such as rainbows and cameras, musical harmony, speech and hearing, and the manner in which materials and structures radiate, transmit, and reflect mechanical, acoustic, and electromagnetic waves. In the most simple terms, a frequency transform can be seen as a wave filter, not unlike an audio graphic equalizer in many high-fidelity audio systems. The frequency band volume slide controls on an audio equalizer can approximately represent the various frequency bands of the input signal for which the output volume is to be controlled by the slider position. Imagine for a moment thousands of audio equalizer slide controls and signal level meters for filtering and monitoring a given sound with very high-frequency precision. While the analog electronic circuitry to build such a device would be extremely complicated, it can be achieved easily in a digital signal processing system in real-time (instant response with no missed or skipped data) using current technology.

Rainbows are formed from the white light of the sun passing through water droplets which naturally have a chromatic aberration (slightly different wave speed for each color in white light). The angle of a ray of white light entering a droplet changes inside the droplet differently for each color wavelength. Upon reaching the other side of the droplet, the curvature of the surface causes the rays to be transmitted back into the air at slightly different angles, allowing one to see a rainbow after a rain storm. Chromatic and geometric distortions can be minimized in lens systems by elaborate design and corrective optics, but it is also possible to do so using signal processing on the digital image. In modern astronomy, imperfections in mirrors and lens systems are usually corrected in the digital wavenumber domain by characterization of the two-dimensional (2-D) wavenumber response of the telescope (usually on a distant star near the object of interest) and "normalization" of the received image by the inverse of the wavenumber response of the telescope. The process is often referred to as "de-speckling" of the image because before the process a distant star appears as a group of dots due to the telescope distortions and atmospheric multipath due to turbulence. The "system" 2-D wavenumber transfer function which restores the distant star clarifies the entire image.

For musical harmony, the modes of vibration in a string, or acoustic resonances in a horn or woodwind instrument form a Fourier series of overtone frequencies, each nearly an exact integer multiple of a fundamental frequency. Western musical scales are based on the first 12 natural harmonics, where the frequency difference between the 11th and 12th harmonic is the smallest musical interval, or semitone, at the frequency of the 12th harmonic. In other words, there are 12 semitones in a musical octave, but the frequency difference between a note and the next semitone up on the scale is $\sqrt[12]{2}$ (or approximately 1.059 times the lower note), higher in frequency. This limits the frequency complexity of music by insuring a large number of shared overtones in musical chords. Indeed, an octave chord sounds "rock solid" because all overtones are shared, while a minor 4th (5 semitones) and minor 5th (7 semitones) intervals form the basis for blues, and most rock and roll music. In many ancient eastern cultures, musical scales are based on 15 semitones per octave, and some are even based on 17-note octaves giving very interesting musical patterns and harmony. Chords using adjacent semitones have a very complex overtone structure as do many percussive instruments such as cymbals, snare drums, and so on. In practical musical instruments, only the first few overtones are accurate harmonic multiples while the upper harmonics have slight "mis-tunings" because of the acoustic properties of the instrument, such as structural modes, excitation nonlinearities, and even control by the musician. The same is certainly true for the human voice due to nasal cavities and nonlinearities in the vocal chords.

In speech and hearing, vocal chords in most animal life vibrate harmonically to increase sound power output, but in human speech, it is largely the time rate of change of the frequency response of the vocal tract which determines the information content. For example, one can easily understand a (clearly heard) spoken whisper as well as much louder voiced speech. However, the frequency content of voiced speech also provides informational clues about the speaker's ages, health, gender, emotional state, and so on. Speech recognition by computers is a classic example of adaptive pattern recognition addressed in later chapters of this book. We are already talking to electronic telephone operators, office computers, and even remote controls for home videotape players! It is likely in the near future that acoustic recognition technology will be applied to monitoring insect populations, fish and wildlife, and perhaps even stress in endangered species and animals in captivity or under study.

The operation of the human ear (as well as many animals) can be seen as a frequency transformer/detector. As sound enters the inner ear it excites the Basilar membrane which has tiny hair-like cells attached to it, which when stretched, emit electrical signals to the brain. It is believed that some of the hair cells actually operate as little loudspeakers, receiving electrical signals from the brain and responding mechanically like high-frequency muscles. The structure of the membrane wrapped up in the snail-shell-shaped cochlea (about the size of a wooden pencil eraser), along with the active feedback from the brain, cause certain areas of the membrane to resonate with a very high Q, brain-controllable sensitivity, and a more distinct and adaptive frequency selectivity. The sensor hair cell outputs represent the various frequency bands (called critical bands) of hearing acuity, and are nearly analogous to one-third octave filters at most frequencies. At higher frequencies in the speech range, things become much more complex and the hair cells fire in response vibrations in very complex spatial patterns. Evolution has made our speech understanding abilities absolutely remarkable where the majority of the neural signal processing is done within the brain, not the ear. It is fascinating to point out that the "background ringing," or mild tinnitus, everyone notices in their hearing is actually a real sound measurable using laser vibrometry on the ear drum, and can even be canceled using a carefully phased sound source in the ear canal. Tinnitus is thought to be from the "active" hair cells, where the corresponding sensor hair cells have been damaged by disease or excessive sound levels, driving the acoustic feedback loop through the brain to instability. It is most often in the 4 kHz to 16 kHz frequency region because this area of the Basilar membrane is closest to where the sound enters the cochlea. Severe tinnitus often accompanies deafness, and in extreme cases can lead to a loss of sanity for the sufferer. Future intelligent hearing aid adaptive signal processors will likely address issues of tinnitus and outside noise cancellation while also providing hearing condition information to the doctor. Based on current trends in hearing loss and increases in life expectancy, intelligent hearing aids will likely be an application of adaptive processing with enormous benefits to society.

Essential to the development of intelligent adaptive signal processing and control systems is a thorough understanding of the underlying physics of the system being monitored and/or controlled. For example, the mass loading of fluids or gases in a reactor vessel will change the structural vibration resonances providing the opportunity to monitor the chemical reaction and control product quality using inexpensive vibration sensors, provided one has identified detectable vibration features which are causal to the reaction of interest. Given a good physical model for sound propagation in the sea or in the human body, one can adaptively optimize the pulse transmission to maximize the detection of the target backscatter of interest. Analogous optimizations can be done for radar, lidar (light detection and ranging), and optical corrective processing as mentioned earlier. The characterization of wave propagation media and structures is most often done in the frequency-wavenumber domain. An intelligent adaptive processor will encounter many waves and signals where useful information on situation awareness will be detected using practical frequency-wavenumber transforms. On a final note, it is interesting to point out that Fourier's general idea that any waveform, including sound waveforms, could be represented by a weighted sum of sinusoids was a very radical idea at the time. His application of the Fourier transform to heat transfer is considered one of the great scientific contributions of all time.

6 Fourier Transform

The French scientist Joseph B. Fourier (1768–1830) developed the first important development in the theory of heat conduction presented to The Academy of Sciences in Paris in 1807. To induce Fourier to extend and improve his theory, the Academy of Sciences in Paris assigned the problem of heat propagation as its prize competition in 1812. The judges were Laplace, Lagrange, and Legendre. Fourier became a member of the Academy of Sciences in 1817. He continued to develop his ideas and eventually authored the applied mathematics classic *Théorie Analytique de la Chaleur* (or *Analytical Theory of Heat*) in 1822. Fourier got much due credit, for his techniques revolutionized methods for the solution of partial differential equations and have led us to perhaps the most prominent signal processing operation in use today.

6.1 MATHEMATICAL BASIS FOR THE FOURIER TRANSFORM

Consider the Fourier transform pair [1]

$$Y(\omega) = \int_{-\infty}^{+\infty} y(t)\,e^{-j\omega t}\,dt \tag{6.1}$$

$$y(t) = \frac{1}{2\pi}\int_{-\infty}^{+\infty} Y(\omega)e^{+j\omega t}\,d\omega$$

where $y(t)$ is a time-domain waveform and $Y(\omega)$ is the frequency-domain Fourier transform. Note the similarity with the Laplace transform pair in Equation 2.2 where the factor of "j" in the Laplace transform is simply from the change of variable from "$j\omega$" to "s." We can eliminate the factor of "$1/2\pi$" by switching from "ω" to "$2\pi f$" as seen in Equation 6.2, but this is not the historically preferred notation.

$$Y(f) = \int_{-\infty}^{+\infty} y(t)e^{-j2\pi ft}\,dt \tag{6.2}$$

$$y(t) = \int_{-\infty}^{+\infty} Y(f)e^{+j2\pi ft}\,df$$

The Fourier transform pair notation in Equation 6.2, although slightly longer, may actually be slightly more useful to us in this development because it is easier to physically relate to frequencies in Hertz (rather than rad/s), and in the digital domain, all frequencies are relative to the sampling frequency. Consider below a short practical example of the Fourier transform for the case $y(t) = \sin(2\pi f_0 t)$.

$$Y(f) = \int\limits_{-\infty}^{+\infty} \sin(2\pi f_0 t) e^{-j2\pi ft}\, dt$$

$$= \int\limits_{-\infty}^{+\infty} \frac{e^{+j2\pi[f_0 - f]t} - e^{-j2\pi[f_0 + f]t}}{2j}\, dt$$

$$= \lim_{T\to\infty} \int\limits_{-T/2}^{+T/2} \frac{e^{+j2\pi[f_0 - f]t}}{2j}\, dt - \lim_{T\to\infty} \int\limits_{-T/2}^{+T/2} \frac{e^{-j2\pi[f_0 + f]t}}{2j}\, dt$$

(6.3)

The limit operations are needed to examine the details of the Fourier transform by first evaluating the definite integral. It is straightforward to show the results of the definite integral as seen in Equation 6.4.

$$Y(f) = \frac{1}{2j}\left\{ \lim_{T\to\infty} \frac{e^{+j\pi[f_0 - f]T}}{j2\pi[f_0 - f]} - \lim_{T\to\infty} \frac{e^{-j\pi[f_0 - f]T}}{j2\pi[f_0 - f]} \right\}$$
$$- \frac{1}{2j}\left\{ \lim_{T\to\infty} \frac{e^{-j\pi[f_0 + f]T}}{-j2\pi[f_0 + f]} - \lim_{T\to\infty} \frac{e^{+j\pi[f_0 + f]T}}{-j2\pi[f_0 + f]} \right\}$$

(6.4)

Combining the complex exponentials yields a more familiar result for analysis in Equation 6.5.

$$Y(f) = \frac{1}{2j}\left\{ \lim_{T\to\infty} \frac{\sin(\pi[f_0 - f]T)}{\pi[f_0 - f]} - \lim_{T\to\infty} \frac{\sin(\pi[f_0 + f]T)}{\pi[f_0 + f]} \right\}$$

(6.5)

For the cases where f is not equal to $\mp f_0$, $Y(f)$ is zero because the oscillations of the sinusoid in Equation 6.5 integrate to zero in the limit as T approaches infinity. This very important property is the result of *orthogonality* of sinusoids where a residue is generated in the Fourier transform for the corresponding frequency components in $y(t)$. In other words, if $y(t)$ contains one sinusoid only at f_0, $Y(f)$ is zero for all $f \ne f_0$. For the case where $f = +f_0$ and $f = -f_0$, we have an indeterminate condition of the form 0/0 which can be evaluated in the limit as f approaches $\mp f_0$ using L'Hôpital's rule.

$$\lim_{x\to a} \frac{g(x)}{f(x)} = \lim_{x\to a} \frac{\left(\dfrac{\partial}{\partial x}\right)g(x)}{\left(\dfrac{\partial}{\partial x}\right)f(x)}$$

(6.6)

Taking the partial derivative with respect to f of the numerators and denominators in Equation 6.5 separately and then applying L'Hôpital's rule we find that the magnitudes of the "peaks" at $\mp f_0$ are both equal to $T/2$.

$$Y(f) = \frac{1}{2j} \lim_{T\to\infty}\left\{ \lim_{f\to f_0} \frac{-\pi T \cos(\pi[f_0 - f]T)}{-\pi} - \lim_{f\to f_0} \frac{+\pi T \cos(\pi[f_0 + f]T)}{+\pi} \right\}$$

(6.7)

Clearly, the peak at $f = f_0$ has a value of $+T/2j$, or $-jT/2$, and the peak at $f = -f_0$ is $-T/2j$ or $+jT/2$. The imaginary peak values and the skew sign symmetry with frequency are due to the phase of the sine wave. A cosine wave would have two real peaks at $\mp f_0$ with value $+T/2$. A complex sinusoid ($y(t) = e^{j\pi f_0 t}$) would have a single peak at $f = f_0$ of value $+T$. For real signals $y(t)$, one should always expect the Fourier transform to have a symmetric real part where the value at some positive frequency

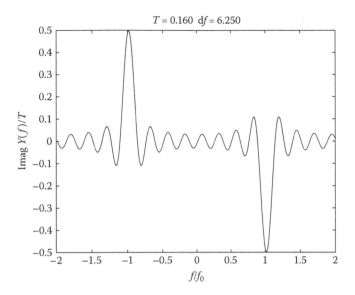

FIGURE 6.1 Fourier transform of $\sin(2\pi f_0 t)$ limiting the integration to $\pm T/2$ to evaluate the characteristics of the Dirac delta function.

is the same as that at the "mirror image" negative frequency. The imaginary parts will be opposite in sign at positive and negative frequencies. The real and imaginary components of the Fourier transform of a real signal are said to be Hilbert transform pairs. Figure 6.1 shows the frequency response $Y(f)$ for finite T where the amplitude in the figure has been normalized by $2T$. The width of the peaks are found by noting that the first zero crossing of the sine function in Equation 6.5 occurs when $f - f_0 = 1/T$, giving an approximate width of $1/T$. Therefore, as T approaches infinity, the amplitude of the peak becomes infinite, while the area remains constant at one. This is one way to define the Dirac delta function $\delta(x)$, which is zero for all $x \neq 0$, and infinite amplitude, but with unity area at $x = 0$. Therefore, for $y(t) = \sin(2\pi f_0 t)$, $Y(f) = -j\delta(f_0 - f)/2 + j\delta(f_0 + f)/2$, or if the radian notation is used, one can simply multiply by 2π to obtain $Y(\omega) = -j\pi\delta(\omega_0 - \omega) + j\pi\delta(\omega_0 + \omega)$, as listed in most texts for the sine function. The cosine function is $Y(f) = \delta(f_0 - f)/2 + \delta(f_0 + f)/2$, or if the radian notation is used, $Y(\omega) = \pi\delta(\omega_0 - \omega) + \pi\delta(\omega_0 + \omega)$. For the complex sinusoid $e^{j2\pi f_0 t}$, $Y(f) = \delta(f_0 - f)$, or $Y(\omega) = 2\pi\delta(\omega_0 - \omega)$, which can be verified by inspection using the sifting property of the Dirac delta function. To complete the example for the case of $y(t) = \sin(2\pi f_0 t)$, we simply evaluate the inverse Fourier transform integral.

$$y(t) = \int_{-\infty}^{+\infty} \frac{\delta(f_0 - f) - \delta(f_0 + f)}{2j} e^{+j2\pi ft} \, df$$
$$= \frac{e^{+j2\pi f_0 t} - e^{-j2\pi f_0 t}}{2j}$$
$$= \sin(2\pi f_0 t) \tag{6.8}$$

Clearly, one can see the symmetry properties of the Fourier transforms of real signals giving a Hilbert transform pair relationship to the symmetric real and skew-symmetric imaginary parts. As T approaches infinity, the resolution of the Fourier transform increases until all frequencies are completely independent, or orthogonal, from each other. As will be seen in Section 6.2, the discrete Fourier transform (DFT) has limited resolution due to the finite number of data samples in the digital transform sum [2].

6.2 SPECTRAL RESOLUTION

We consider a regularly sampled digital signal (as described in Section 1.2) $y[n]$, sampled every T seconds and converted to a signed integer [3]. The sampling frequency $f_s = 1/T$ Hz, or samples/s. The Fourier transform integral in Equation 6.2 becomes the N-point DFT sum

$$Y[f_m] = \sum_{n=0}^{N-1} y[n] e^{-j2\pi f_m nT}, \quad m = 0, 1, 2, \ldots, M-1$$

$$y[n] = \frac{1}{N} \sum_{m=0}^{M-1} Y[f_m] e^{+j2\pi f_m nT}, \quad n = 0, 1, 2, \ldots, N-1$$

(6.9)

where $y[n]$ is a digital sample of $y(nT)$, nT is t, and f_m is a discrete frequency in the integral. The factor of $1/N$ is necessary to restore the correct amplitude to the time-domain sequence in the inverse DFT (IDFT) of the second line in Equation 6.9. Note that we do not require any specific number of frequencies M or sample points N or that the frequencies and sample points are equally spaced. It follows algebraically that the DFT and IDFT can only be exact for any sampled signal when $N = M$ and the sample points and frequencies are equally spaced. When this is not the case, the DFT and IDFT are approximations, meaning that some of the waveform information may be lost during the transform. The larger the N and M are, the more the resolution to the DFT and IDFT.

We will depart from the literature slightly by including the factor of $1/N$ in the DFT to make an "N-normalized" DFT denoted as NDFT. Most texts avoid the $1/N$ factor in the forward transform (including it in the inverse transform) since the "height" of Dirac delta function approaches infinity as the time integral gets longer. Therefore, the "height" in the DFT should also increase with increasing N. However, from a physical signal processing point of view, *we prefer here to have the amplitude of the frequency-domain sinusoidal signal independent of N*, the number of points in the transform. This normalization is also applied to the power spectrum in Chapter 7. We must be careful to distinguish the NDFT presented here from the standard DFT to avoid confusion with the literature. Any random noise in $y[n]$ will now be suppressed in the NDFT relative to the amplitude of a sinusoid(s) as the number of points in the NDFT increases. This will also be discussed subsequently in Chapter 7 on spectral density. Consider the NDFT pair where $0 \leq f \leq 0.5/T$ (only analog frequencies less than the Nyquist rate $f_s/2 = 0.5/T$ are present in $y[n]$, which is real).

$$Y[f_m] = \frac{1}{N} \sum_{n=0}^{N-1} y[n] e^{-j2\pi f_m nT}$$

$$y[n] = \sum_{m=0}^{M-1} Y[m\Delta f] e^{+j2\pi (m\Delta f)nT}, \quad m\Delta f = f_m$$

(6.10)

We have assumed a set of M evenly spaced frequencies in the inverse NDFT in Equation 6.10 where Δf is f_s/M Hz. However, the resolution Δf and number of frequencies can actually be varied considerably according to the needs of the application, but the underlying resolution available is a function of the integration limits in time or space. In general, the spectral resolution available in a forward NDFT can be simply found using the result from the Dirac delta function shown in Figure 6.1. With N samples spanning a time interval of NT seconds in Equation 6.10, the available resolution in the DFT is approximately $1/(NT)$ Hz. Therefore, it is reasonable to set $\Delta f = 1/(NT)$ giving N frequencies between 0 and f_s, or $N = M$ in the NDFT pair. Evaluating the NDFT at closer-spaced frequencies than Δf will simply result in more frequency points on the "$\sin(x)/x$," or $\text{sinc}(x)$, envelope function for a sinusoidal signal. This envelope is clearly shown in Figure 6.1 and in Equation 6.5, where the Fourier transform is evaluated over a finite time interval. The only way to increase the spectral resolution (narrowing Δf) is to increase NT, the physical span of the DFT sum for a given f_s. Increasing the number of samples N for a fixed sample interval T (fixed sample rate or bandwidth) simply increases the length of time for the Fourier transform and allows a finer frequency resolution to be computed. Therefore, one can easily see that a 1-s forward Fourier transform can yield a 1 Hz resolution in the frequency domain while a 100 ms transform would have only 10 Hz resolution available, and so on. This relationship between the length of time record and frequency resolution is essential to know.

Consider the following short proof of the NDFT for an arbitrary $y[n]$, N time samples and $M = N$ frequency samples giving a $\Delta f = f_s/N$, or $\Delta f\, nT = n/N$, to simplify the algebra and represent a typical DFT operation.

$$y[p] = \sum_{k=0}^{N-1} \left\{ \frac{1}{N} \sum_{n=0}^{N-1} y[n]\, e^{-j2\pi k\, n/N} \right\} e^{+j2\pi k\, p/N}, \quad p = 0, 1, \ldots, N-1 \tag{6.11}$$

The forward NDFT in the braces of Equation 6.11 can be seen as $Y(k\Delta f) = Y[k]$. Rearranging the summations provides a clearer presentation of the equality.

$$y[p] = \frac{1}{N} \sum_{n=0}^{N-1} y[n] \left\{ \sum_{k=0}^{N-1} e^{+j2\pi k(p-n)/N} \right\}, \quad p = 0, 1, \ldots, N-1 \tag{6.12}$$

Clearly, the expression in the braces of Equation 6.12 is equal to N, when $p = n$. We now argue that the expression in the braces is zero for $p \neq n$. Note that the expression in the braces in Equation 6.12 is of the form of a geometric series.

$$\sum_{k=0}^{N-1} a^k = \frac{1 - a^N}{1 - a}, \quad a = e^{j2\pi(p-n)/N} \tag{6.13}$$

Since for our argument $p \neq n$ and $(p-n)$ goes from $-(N-1)$ to $+(N-1)$, the denominator term "a" cannot equal 1 thus making the series convergent. However, since the numerator term $a^N = e^{j2\pi(p-n)}$, a^N is always 1 since the angle is a multiple of 2π. Therefore, the numerator of Equation 6.13 is zero and Equation 6.12 is nonzero only for the case where $p = n$, the sum over n reduces to the $p = n$ case, the factors of N and $1/N$ cancel (as they also do for an analogous proof for the DFT), and $y[n] = y[p]$; $p = n$.

We now examine the N-point NDFT resolution for the case of a sine wave. Consider the case where $y[n] = \sin(2\pi f_0 nT) = \sin(2\pi f_0 n/f_s)$ and f_0 is an integer multiple of Δf, or $f_0 = m_0 \Delta f$. We will consider an N-point NDFT and N frequency points $\Delta f = f_s/N$ Hz apart. In other words, the frequency of our sine wave will be exactly at one of the discrete frequencies in the DFT. Since $f_0 = m_0 f_s/N$, we can write our sine wave as a simple function of m_0, n, and N by $y[n] = \sin(2\pi m_0 n/N)$. Equation 6.14 shows the N-point NDFT expressed in terms of $Y[m]$, for the mth frequency bin, by again writing the sine function as the sum of two complex exponentials.

$$Y[m] = \frac{1}{N} \sum_{n=0}^{N-1} \sin\left(\frac{2\pi m_0 n}{N} \right) e^{-j2\pi(m\, n/N)}$$

$$= \frac{1}{2jN} \sum_{n=0}^{N-1} e^{+j2\pi(m_0-m)(n/N)} - \frac{1}{2jN} \sum_{n=0}^{N-1} e^{-j2\pi(m_0+m)\, n/N} \tag{6.14}$$

Applying the finite geometric series formula of Equation 6.13 gives

$$
\begin{aligned}
Y[m] &= \frac{1}{2jN}\left[\frac{1-e^{+j2\pi(m_0-m)}}{1-e^{+j2\pi(m_0-m)/N}}-\frac{1-e^{-j2\pi(m_0+m)}}{1-e^{-j2\pi(m_0+m)/N}}\right] \\
&= \frac{e^{+j\pi(m_0-m)\frac{N-1}{N}}}{2jN}\frac{\sin\left[\pi(m_0-m)\right]}{\sin\left[\pi(m_0-m)\frac{1}{N}\right]}-\frac{e^{-j\pi(m_0-m)\frac{N-1}{N}}}{2jN}\frac{\sin\left[\pi(m_0+m)\right]}{\sin\left[\pi(m_0+m)\frac{1}{N}\right]}
\end{aligned}
\tag{6.15}
$$

As m approaches $\pm m_0$, one can clearly see the indeterminate condition 0/0 as evaluated using L'Hôpital's rule in Equation 6.6. The result is a peak at $+m_0$ with amplitude $-j/2$, and a peak at $-m_0$ with amplitude $+j/2$. This is consistent with the continuous Fourier transform result in Equations 6.6 and 6.7 keeping in mind that we divided our forward DFT by N (as we have defined as the NDFT). Note that the indeterminate condition 0/0 repeats when $\pi(m_0-m)/N=k\pi;=0,\pm1,\pm2,\ldots$, since the numerator sine angle is simply N times the denominator sine angle. The actual frequency of m in Hz is $m\Delta f$, where $\Delta f = f_s/N$, N being the number of frequency points in the NDFT. Therefore, it can be seen that the peak at $+m_0\Delta f$ Hz repeats every f_s Hz up and down the frequency axis to $\pm\infty$, as also does the peak at $-m_0$. Figure 1.5 clearly illustrates the frequency aliasing for a cosine wave showing how a wave frequency greater than $f_s/2$ will appear "aliased" as an incorrect frequency within the Nyquist band. If the original continuous waveform $y(t)$ is band limited to a frequency band less than $f_s/2$, all frequencies observed after performing an NDFT will appear in the correct place. Mathematically, it can be seen that the NDFT (and DFT) *assumes* the waveform within its finite-length input buffer repeats periodically outside that buffer for both time- and frequency-domain signals.

Consider the example of a 16-point NDFT on a unity amplitude sine wave of 25 Hz where the sample rate is 100 Hz. With 16 frequency "bins" in the NDFT, we find $\Delta f = 100/16$, or 6.25 Hz. The sampling time interval is 10 ms, and so the time window of data is 160 ms long, also indicating a possible resolution of 6.25 Hz for the NDFT. Figure 6.2 shows the NDFT of the sine wave in the Nyquist band (between ±50 Hz) where the "-o-" symbols indicate the 16 discrete integer "m" values and the curve depicts $Y[m]$ as a continuous function of m showing the resolution limitations for the

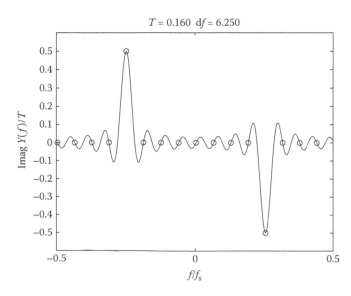

FIGURE 6.2 The imaginary part of the 16-point NDFT of a 25 Hz sine wave sampled at 100 Hz showing the NDFT bins (O) and a zero-padded 1024-point NDFT as a continuous curve to show the available resolution.

160 ms wave. Note that if a 16-point DFT were used instead of the NDFT, the peak magnitudes would be 8, or $N/2$. The real part of the 16-point NDFT is zero as shown in Figure 6.3. But, it is more customary to show the magnitude of the NDFT as shown in Figure 6.4.

The discrete NDFT bin samples in Figures 6.2 through 6.4 depicted by the "-o-" symbols resemble the ideal Dirac delta function (normalized by N) in the context of the 16-point NDFT output because the 25 Hz sine wave has exactly four wavelengths in the 160 ms-long digital input signal. Therefore, the assumptions of periodicity outside the finite-NDFT input buffer are valid. However, if the input sine wave has a frequency of, say, 28.125 Hz, the wave has exactly 4.5 wavelengths in the NDFT input buffer sampled at 100 Hz. For 28.125 Hz, the periodicity assumption is violated because of the half-cycle discontinuity and the resulting magnitude of the NDFT is shown in Figure 6.5.

As Figure 6.5 clearly shows, when the sine wave frequency does not match up with one of the 16 discrete frequency bins of the NDFT, the spectral energy is "smeared" into all other NDFT bins to some degree. This is called *spectral leakage*. There are three things one can do to eliminate the smeared NDFT (or DFT) resolution. First, one could synchronize the sample frequency to be an integer multiple of the sine wave frequency, which in the case of 28.125 Hz would require a sample rate of 112.5, 140.612 Hz, and so on. Synchronized Fourier transforms are most often found in "order-tracking" for vibration analysis of rotating equipment such as engines, motor/generators, turbines, fans, and so on. Synchronizing the sample rate to the frequencies naturally occurring in the data ensures that the vibration fundamental and its harmonics are all well matched to the NDFT bin frequencies giving reliable amplitude levels independent of machinery rotation speed.

The second technique to minimize spectral leakage is to increase the input buffer size, or increase N. Increasing N decreases Δf and the number of output NDFT bins. Eventually, with enough increase in resolution, there will be so many bins in the NDFT that the sine wave of interest will lie on or very close to one of the discrete bins. For the case of the 16-point NDFT data in Figure 6.5, doubling the NDFT size to 32 (320 ms input time buffer), narrows the resolution to 3.125 Hz allowing the 28.125 Hz wave to lie exactly on a bin. All the other NDFT bins will be at a zero crossing of the underlying sinc function giving the appearance of a "normalized Dirac-like" spectrum. Another way to look at doubling the input buffer size is to consider that 28.125 Hz will have exactly nine full wavelengths in the input buffer, making the periodicity assumptions of the Fourier transform on a finite interval correct. However, if one kept the 16-point input buffer and simply analyzed the NDFT output at many more frequencies closer together than the original 16 NDFT bins, one would simply be computing points on the continuous curves in Figures 6.1 through 6.5. These continuous curves represent the underlying resolution available for the 160 ms input buffer.

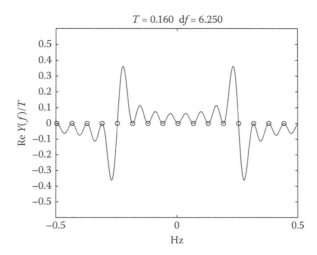

FIGURE 6.3 The real part of the 16-point NDFT is zero at all 16 bin frequencies.

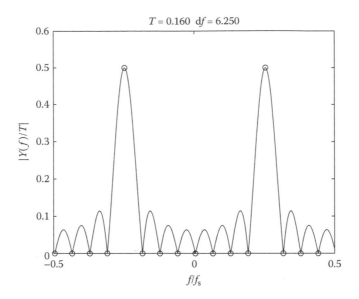

FIGURE 6.4 The magnitude of the 16-point NDFT of a 25 Hz sine wave sampled at 100 Hz showing the discrete NDFTS bins (O) and a 1024-point NDFT with the 16 input samples zero-padded to show the available resolution as a continuous curve.

The third technique is to apply a data envelope window to the DFT or NDFT input. This technique will be discussed in detail in Section 6.4. Windowing the data envelope makes the data "appear" more periodic from one input buffer to the next by attenuating the data amplitude near the beginning and end of the input buffer. This of course changes the spectral amplitude as well as causes some spectral leakage to occur even if the input frequency is exactly on one of the discrete DFT frequency bins.

Generally speaking for stationary signals, larger and/or synchronized Fourier transforms are desirable to have a very high-resolution, well-defined spectrum. Also, if there is any random

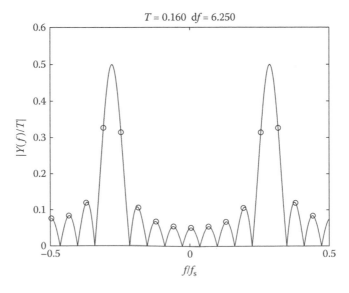

FIGURE 6.5 The magnitude of a 16-point NDFT of a 28.128 Hz sine wave sampled at 100 Hz showing the NDFT bins (O) and the spectral leakage in the zero-padded continuous curve.

broadband noise in the input data, the more the number of NDFT bins one has, the less the random noise in each bin since the noise is divided over all the bins. Large NDFTs will tend to focus on periodic signals and suppress random noise which is a very desirable property. However, even if the sines and cosines of the NDFT (or DFT) are computed in advance and stored in memory, the number of complex multiplies in a straightforward N-point DFT is N^2. Therefore, very large broadband DFTs have historically been prohibitively expensive in terms of the required computing resources. To address this need in the 1960s, Cooley and Tukey [4] developed a signal processing algorithm.

6.3 FAST FOURIER TRANSFORM

Early on in the history of digital signal processing it was recognized that many of the multiplies in a DFT are actually redundant, making a fast Fourier transform (FFT) possible through computer algorithm optimization. We will not present a detailed derivation or discussion on the FFT as there are many signal processing books that do so with eloquence not likely here. It should suffice to say that the FFT is an efficient implementation of the DFT where the number of input samples and output frequency bins are the same, and in general, a power of 2 in length (this is called a radix-2 FFT). An N-point FFT where N is a power of 2 (256, 512, 1024, etc.) requires $N \log_2 N$ complex multiplies, as opposed to N^2 multiplies for the exact same N-point output result from the DFT. For a 1024-point transform, that is a reduction from 1,048,576 complex multiplies for the DFT to 105,240 complex multiplies for the FFT. Clearly, to call the FFT efficient is an understatement for large N! Now, by the year 2010, your average PC is capable of over 10 billion floating-point multiplies per second (10 GFLOPS); so why should we care about the FFT? The thought process of careful examination of a general algorithm to eliminate redundant operations not only saves processing time, but also energy, while allowing one to do more processing with a given amount of computing power. The FFT process leads to algorithms such as wavelet and discrete cosine transform algorithms which are crucial to the data compression used today in cell phones, digital television, and portable digital music players.

Both the FFT and DFT give exactly the same result when the DFT has N equally spaced output frequency bins over the Nyquist band. In the early days of digital signal processing with vacuum tube flip-flops and small magnetic core memories, the FFT was an absolutely essential algorithm. It is (and likely always will be) the standard way to compute digital Fourier transforms of evenly spaced data primarily for economical reasons. What is to be emphasized here is that the resolution of the DFT or FFT is always limited by the length of the input buffer. The spectral leakage within the DFT or FFT is caused by a breakdown of the periodicity assumptions in the Fourier transform when a noninteger number of full wavelengths appear in the input buffer. One can use an FFT to evaluate more frequency bins than input samples by "zero padding" the input buffer to fill out the larger-sized FFT buffer. This is sometimes a useful alternative to synchronized sampling as a means of finding an accurate peak level even with the underlying sinc function envelope. One can also perform a "zoom-FFT" by band limiting the input signal of interest centered at some high center frequency f_c rather than zero Hz, and essentially performing a complex demodulation down to 0 Hz as described at the end of Section 1.3 and Figure 1.7 before computing the FFT.

Consider briefly the case of an 8-point DFT which we will convert to an 8-point FFT through a process known as time decomposition. First, we split the unnormalized (we will drop the $1/N$ normalization here for simplicity of presentation) DFT into two $N/2$-point DFTs as in

$$Y[m] = \sum_{n=0}^{(N/2)-1} y[2n] e^{-j2\pi(2n)m/N} + \sum_{n=0}^{(N/2)-1} y[2n+1] e^{-j2\pi(2n+1)m/N} \qquad (6.16)$$

We can make our notation even more compact by introducing $W_N = \underline{e}^{-j2\pi/N}$, commonly referred to in most texts on the FFT as the "twiddle factor," or the time/frequency invariant part of the complex exponentials. Equation 5.22 is then seen as a combination of 2 $N/2$-point DFTs on the even and odd samples of $y[n]$.

$$Y[m] = \sum_{n=0}^{(N/2)-1} y[2n] W_N^{2nm} + W_N^m \sum_{n=0}^{(N/2)-1} y[2n+1] W_N^{2nm} \tag{6.17}$$

Note that by doing two $N/2$-point DFTs rather than one N-point DFTs we have reduced the number of complex multiplies from N^2 to $N^2/4 + N/2$, which in itself is a significant saving. Each of the $N/2$-point DFTs in Equation 6.17 can be split into even and odd samples to give four $N/4$-point DFTs, as in

$$Y[m] = W_N^0 \sum_{n=0}^{(N/4)-1} y[4n] W_N^{4nm} + W_N^{2m} \sum_{n=0}^{(N/4)-1} y[4n+2] W_N^{4nm}$$

$$+ W_N^m \sum_{n=0}^{(N/4)-1} y[4n+1] W_N^{4nm} + W_N^{3m} \sum_{n=0}^{(N/4)-1} y[4n+3] W_N^{4nm} \tag{6.18}$$

For the case $N = 8$, Equation 6.18 can be seen as a series of four 2-point DFTs and the time decomposition need not go further. However, with N some power of 2, the time decomposition continues q times where $N = 2^q$ until we have a series of 2-point DFTs on the data. This type of FFT algorithms is known as a radix-2 FFT using time decomposition. Many other forms of the FFT can be found in the literature including radices 3, 4, 5, and so on, as well as frequency decomposition formulations. Radix-2 FFTs are by far the most common used. For blocks of data which are not exactly the length of the particular FFT input buffer, one simply "pads" the input buffer with zeros. The underlying spectral resolution in Hz (wavenumber for spatial FFTs) is still the inverse of the actual data record in seconds (meters) excluding the zero-padded elements.

Figure 6.6 shows a flow diagram for the 8-point FFT derived above in Equations 6.16 through 6.18. The 2-point DFTs are arranged in a convenient order for a recursive algorithm which computes the FFT in-place (the original input data are overwritten) during the computation. The 2-point DFTs

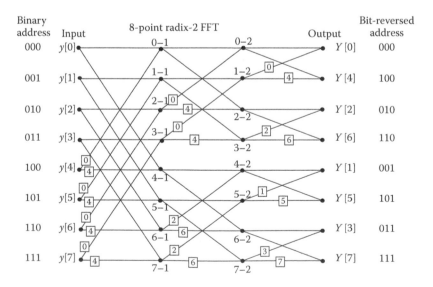

FIGURE 6.6 An 8-point radix-2 FFT showing twiddle factors in the square boxes for time decomposition and bit-reversed outputs.

in the various sections are typically called "butterflies" because of their X-like signal flow patterns in the flow diagram. However, the convenient ordering of the butterflies leads curiously to the outputs being addressed in binary bit-reversed order. This is probably the main reason the radix-2 butterfly is so popular. "Unscrambling" of the bit reversed output addresses is actually quite straightforward and most of today's digital signal processing processors usually offer a hardware-based bit-reversed addressing instruction for speedy FFT programs.

The twiddle factor structure in the flow diagram in Figure 6.6 also has a much more subtle design feature which is not widely discussed and is extremely important to algorithm efficiency. Note that for the input column only $y[4]$ through $y[7]$ are multiplied by twiddle factors while $y[0]$ through $y[3]$ pass straight through to the second column, where again, only half the nodes are multiplied by twiddle factors, and so on. As will be seen, the ordering of the twiddle factors at each node is critical, *because at any one node, the two twiddle factors differ only in sign!* It seems the ingenious algorithm design has also allowed $W_8^0 = -W_8^4$, $W_8^2 = -W_8^6$, $W_8^1 = -W_8^5$, and $W_8^3 = -W_8^7$. Figure 6.7 graphically shows how the twiddle factors differ in sign.

Obviously, one would do the complex multiply just once at each node, and either add or subtract the result in the computation for the nodes in the next column. For example, $y[4]$ would be multiplied by W_8^0 (which is unity by the way) and saved into complex temporary storage "ctemp1," $y[0]$ is copied into "ctemp2," then is overwritten by $y[0] + \text{temp1}$ (shown in node 0–1 in Figure 6.6). Then, ctemp1–ctemp2 is seen to be $y[0] + y[4]W_8^4$ and overwrites $y[4]$ in node 4–1. This type of FFT algorithm computes the Fourier transform *in-place* using only two complex temporary storage locations, meaning that the input data are overwritten and no memory storage space is wasted—a critical design feature in the 1960s and an economical algorithm feature today. To complete the 8-point FFT, one would proceed to compute the unity twiddle factor (W_8^0) butterflies filling out all the nodes in column 1 in Figure 6.6, then nodes 0–2 through 3–2 in column 2, and finally the outputs $Y[0]$ and $Y[4]$. Then using W_8^2, nodes 4–2 through 7–2 are computed followed by the outputs $Y[2]$ and $Y[6]$. Then W_8^1 can be used to compute the outputs $Y[1]$ and $Y[5]$, followed by W_8^3 to compute the outputs $Y[3]$ and $Y[7]$. A hardware assembly code instruction or a straightforward sorting algorithm in software reorders the outputs in non-bit-reversed address form for use.

It can be seen that for a radix-2 N-point FFT algorithm there will be $\log_2 N$ columns of in-place computations where each column actually requires only $N/2$ complex multiplies and N additions.

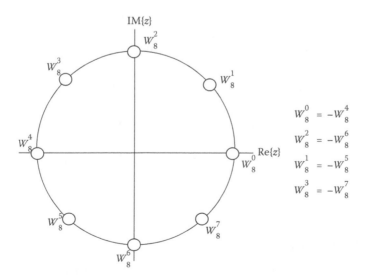

FIGURE 6.7 Twiddle factors for the 8-point FFT can be grouped into pairs which differ only in sign to speed computations.

Since multiplications are generally more intensive than additions, we can focus the computational cost on only multiplications. The radix-2 N-point FFT, where N is a power of 2 ($N = 2^q$) requires $(N/2)\log_2 N$ complex multiplies. Since a significant number of those multiplies are using the twiddle factor W_N^0, which is unity, the required number of complex multiplies is actually less than $(N/2)$ $\log_2 N$. However, a complex multiply actually involves four numerical multiplies. Therefore, for complex input and output, the FFT is seen to require $2N\log_2 N$ numerical multiplies. Once again, most FFT applications start with real sampled input data, and so the widely accepted computational estimate for the FFT is $N\log_2 N$ multiplies, but the actual number will depend on the specific application and how well the designer has optimized the algorithm. Note that the inverse FFT simply requires the opposite sign to the complex exponential which is most easily achieved by reversing the order of the input data from 0 to $N-1$ to $N-1$ to 0. The scale factor of N, or $1/N$ depending on one's definition of the DFT/FFT, is generally handled outside the main FFT operations.

Usually we work with real sampled data as described in Section 1.2, yet most FFT algorithms allow for a fully complex input buffer and provide a complex output buffer. For real input samples, one simply uses zeros for the imaginary component and a significant number of multiplies are wasted unless the user optimizes the FFT algorithm further for real input data. This can be done in different ways including packing the even samples into the real part of the input and odd samples into the imaginary part of the input. Another approach is to simply eliminate all unnecessary multiplications involving imaginary components of the input data. Another technique would be to bit-reverse the input data and neglect all computations which lead to the upper half of the in-order output data, and so on. But, perhaps an even more intriguing technique is simultaneously computing the FFTs of two real signals assembled into a single complex array using one complex FFT operation. It was discussed earlier that real input data (imaginary part is zero) yields real frequency-domain data with even symmetry (positive and negative frequency components equal), or $\mathrm{Re}\{Y[m]\} = \mathrm{Re}\{Y[N-m]\}$, $m = 0, 1, \ldots, N-1$. The imaginary part will have skew symmetry, or $\mathrm{Im}\{Y[m]\} = -\mathrm{Im}\{Y[N-m]\}$, $m = 0, 1, \ldots, N-1$. The amplitudes of the real and imaginary parts are really only a function of the phases of each frequency in the FFT input buffer. If we put our real input data into the imaginary part of the input leaving the real part of the FFT input zero, the opposite is true where $\mathrm{Re}\{Y[m]\} = -\mathrm{Re}\{Y[N-m]\}$ and $\mathrm{Im}\{Y[m]\} = \mathrm{Im}\{Y[N-m]\}$.

We can exploit the Hilbert transform pair symmetry [5] by packing two real data channels into the complex FFT input buffer and actually recovering two separate FFTs from the output. For "channel 1" in the real part of our complex input data, and "channel 2" in the imaginary part of our complex input data, the following relationships in Equation 5.25 can be used to completely recover the two separate frequency spectra $Y_1[m]$ and $Y_2[m]$ for positive frequency. Note that the "0 Hz" components are recovered as $\mathrm{Re}\{Y_1[0]\} = \mathrm{Re}\{Y[0]\}$, $\mathrm{Re}\{Y_2[0]\} = \mathrm{Im}\{Y[0]\}$, and $\mathrm{Im}\{Y_1[0]\} = \mathrm{Im}\{Y_2[0]\} = 0$.

$$\mathrm{Re}\{Y_1[m]\} = \frac{1}{2}[\{\mathrm{Re}\{Y[m]\} + \mathrm{Re}\{Y[N-m]\}\}]$$

$$\mathrm{Im}\{Y_1[m]\} = \frac{1}{2}[\{\mathrm{Im}\{Y[m]\} - \mathrm{Im}\{Y[N-m]\}\}]$$

$$\mathrm{Re}\{Y_2[m]\} = \frac{1}{2}[\{\mathrm{Im}\{Y[N-m]\} + \mathrm{Im}\{Y[m]\}\}] \tag{6.19}$$

$$\mathrm{Im}\{Y_2[m]\} = \frac{1}{2}[\{\mathrm{Re}\{Y[N-m]\} - \mathrm{Re}\{Y[m]\}\}]$$

6.4   DATA WINDOWING

Controlling the resolution of the Fourier transform using NDFT, DFT, or FFT algorithms is an important design consideration when dealing with signal frequencies which may lie in between the discrete frequency bins. The reason why a peak's signal level in the frequency domain is reduced, and leakage into other frequency bins increases, when the frequency is between bins, can be seen as a result of a nonperiodic waveform in the input buffer. Therefore, it is possible to reduce the leakage effects along with the peak-level sensitivity to frequency alignment by forcing the input wave to appear periodic. The most common way this is done is to multiply the input buffer by a raised cosine wave which gradually attenuates the amplitude of the input at either end of the input buffer. This "data window" is known as the Hanning window, after its developer, and has the effect of reducing the signal leakage into adjacent bins when the frequency of the input signal is between bins. It also has the effect of making the peak signal level in the frequency domain less sensitive to the alignment of the input frequency to the frequency bins. However, when the input signal is aligned with a frequency bin and the Hanning window is applied, some spectral leakage, which otherwise would not have occurred, will appear in the bins adjacent to the peak frequency bin. This trade-off is generally worth the price in most Fourier processing applications. As we will see below, a great deal of artifice has gone into data window design and controlling the amount of spectral leakage. We present a wide range of data windows for the FFT (the NDFT, DFT, and FFT behave the same) and discuss their individual attributes. First, we consider the Hanning window, which is implemented with the following equation:

$$W^{Hn}[n] = \frac{1}{2}\left\{1 - \cos\left(\frac{2\pi n}{N-1}\right)\right\}, \quad n = 0, 1, 2, \ldots, N-1 \tag{6.20}$$

The input buffer of the Fourier transform can be seen as the product of the data window times a "infinitely long" waveform of the input signal. The finite sum required by the FFT can be seen as having an input signal which is the product of a rectangular window N samples long and the infinitely long input waveform. A product in the time domain is a convolution in the frequency domain, as explained in detail in Chapter 2, and the sinc-function response of the FFT due to the finite sum can be seen as the convolution of the Fourier transform of the rectangular data window with the Dirac delta function of the infinitely long sinusoid. The Hanning window can be written as the sum of three terms, each of which has a "sinc-like" Fourier transform for finite N.

$$W^{Hn}[n] = \frac{1}{2} - \frac{1}{4}e^{(+j2\pi n/N-1)} - \frac{1}{4}e^{(-j2\pi n/N-1)} \tag{6.21}$$

The effect of the three sinc functions convolved with the delta function for the sinusoid is a "fatter," or broader, main lobe of the resulting resolution envelope and lower leakage levels in the side lobes. This is depicted in Figure 6.8 which compares the resolution for the rectangular data window (no data window) to that for the Hanning "raised cosine" data window. Note that zero padding is used to generate the line plot.

Narrowband Calibration Factors: The resolutions shown in Figure 6.8 are normalized by N (as we defined in the NDFT), but an additional narrowband correction factor of 2.0317 for $N = 64$ (it approaches 2.00 for large N) is needed to boost the Hanning window peak level to match the rectangular window peak level. For any given window type, one can "calibrate" a bin-aligned sine wave to determine the narrowband correction factor. However, normalizing the integral of the window to match the integral of the rectangular window is a stricter definition of narrowband normalization. Therefore, one simply sums the window function and divides by N, the rectangular window sum, to get the narrowband correction factor. In other words, *the narrowband correction factor is the ratio of the window integral to the integral of a rectangular window of the same length*. The narrowband normalization constant multiplied by the window function is important for signal calibration purposes in signal processing systems, as we would like the peak levels of the spectra to match when the peak is bin aligned for whichever data window is used. But still, the amplitude of the spectral peak will vary depending on its location relative to the discrete frequency bins of the FFT.

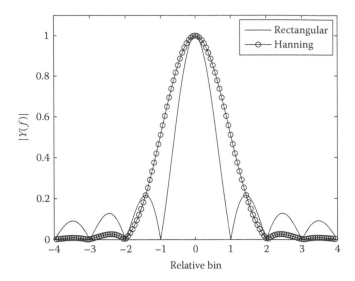

FIGURE 6.8 Comparison of the detailed spectral resolution of the rectangular (no data window) and Hanning window where $N = 64$ points (1024-point zero-padded FFT's are used to show the underlying available resolution).

Broadband Calibration Factors: There will be a slightly different correction factor for matching broadband Hanning-windowed signals to the rectangular window levels. Broadband normalization is important for determining the total signal power in the frequency domain by summing all the NDFT bins. Application of a window on the time-domain data before the Fourier transform introduces a small level error in the total spectral power due to the controlled spectral leakage of the window. Since power is measured as a magnitude-squared spectrum, *the broadband correction factor is determined by the square root of the integral of the window function squared, then divided by N (the integral of the squared rectangular window function).* The broadband correction factor for an $N = 64$ point Hanning window is 1.6459, and 1.6338 for $N = 1024$. For N large, the broadband correction factor works out to be exactly $\sqrt{8/3}$.

Why are broadband and narrowband signals affected differently in the frequency domain by a data window? The reason has to do with how the time-domain signal's energy is distributed in the spectrum. A narrowband sinusoid is concentrated into a few frequency bins while a broadband signal such as Gaussian noise is distributed over every frequency bin. This is because a random variable when multiplied by a sinusoid simply gives another random variable with the same statistical metrics (mean, variance, etc.). A sinusoid is not coherent with the noise signal, and so it does nothing to even change the statistics. Another way to view this is that 50% of the time the sinusoid is multiplying the random signal positively, and 50% of the time the sinusoid is multiplying the random signal negatively in exactly the same way. One could argue that the net amplification of a sinusoid of unity amplitude on the random signal is $1/\sqrt{2}$, but there is also an imaginary part to the Fourier transform sinusoid adding another factor of $1/\sqrt{2}$. The net effect is that the random signal with variance σ_t^2 in the time domain produces a new random variable in each frequency where the real and imaginary components have variance $\sigma_f^2 = (1/2)\sigma_t^2$ assuming the input buffer to the Fourier transform algorithm is real. If the broadband input is fully complex, the factor of 1/2 is not needed. Since for broadband signals it makes sense to measure the signal in terms of variance, we "calibrate" the window for broadband data windows by comparing the variances in the time and frequency domains to remove the effects of the data window on these metrics.

Careful definition of narrowband and broadband normalization constants is consistent with our reasoning for the NDFT, where we would like periodic signal levels to be independent of N, the size of the DFT, as well as independent of the type of data window used. By keeping track of time- and frequency-domain scale factors we will make applications of frequency-domain signal processing techniques to adaptive systems, pattern recognition algorithms, and control applications much clearer. *Narrowband and broadband correction factors are critically important to power spectrum amplitude calibration in the frequency domain.* A top-quality spectral analyzer or software analysis package will provide one with the precise type of data window used as well as broadband and narrowband correction factors for calibration purposes, but one generally has to look for these details since the average user has no idea of the effect of data windows on spectral calibration.

There are several other noteworthy data windows which are presented below ranging from highest-resolution and side-lobe levels to lowest-resolution and side-lobe levels. For the NDFT, these windows need not be normalized by N (as seen in most texts for DFT applications). The calibration factors are presented for $N = 64$ for comparison.

The Welch window is essentially a concave down parabola centered in the middle of the data input buffer. The Welch window requires a narrowband scale factor of 1.5242 to boost its peak level up to the rectangular window's level.

$$W^{Wc}[n] = 1 - \left(\frac{n - \left(\frac{1}{2}\right)(N-1)}{\left(\frac{1}{2}\right)(N-1)} \right)^2 \tag{6.22}$$

Another simple but very useful data window is the Parzen, or triangle window, described in Equation 5.36, which requires a narrowband scale factor of 2.0323.

$$W^{Pz}[n] = 1 - \left| \frac{n - \left(\frac{1}{2}\right)(N-1)}{\left(\frac{1}{2}\right)(N-1)} \right| \tag{6.23}$$

Raising the Hanning window by a small amount (so the end points are nonzero) produces a significant improvement by narrowing resolution and lowering side-lobe leakage as seen in the Hamming window.

$$W^{Hm}[n] = \frac{1}{2}\left\{ 1.08 - 0.92\cos\left(\frac{2\pi n}{N-1} \right) \right\} \tag{6.24}$$

The small change between the Hanning and Hamming windows (small spelling change too) is indicative of the art that has gone into data window design. The Hamming window requires a narrowband scale factor of 1.8768 for its peak levels to match the rectangular window. One can also define a generic exponential window which can be modified to have almost any spectral width with very smooth low side-lobe levels.

$$W^{Ex}(n) = \sqrt{\frac{\pi}{3}}\, e^{-(1/2)(n-(N/2)-1)^2\ (3.43\times10^{-k})} \tag{6.25}$$

When $k \geq 4$, the exponential window very closely matches the rectangular window and the narrowband scale factor is 1.0347. When $k = 3$, the window very closely matches the Welch window and the narrowband scale factor is 1.5562. However, k also depends on N, and so a wide variety of window shapes and resolution can be realized. We show the exponential window below with $N = 64$,

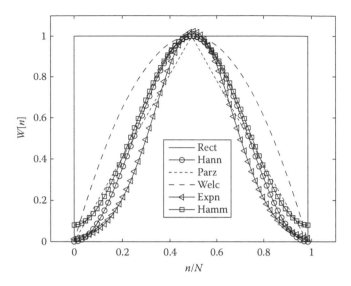

FIGURE 6.9 Comparison of the rectangular, Welch, Parzen, Hamming, Hanning, and exponential data windows in the time domain.

$k = 2.6$, and a narrowband scale factor of 2.3229, giving a broader resolution than the other windows but with the lowest side-lobe levels.

Figure 6.9 shows the six data windows together in the time domain unnormalized. The narrowband correction scale factors noted above for $N = 64$ and in Table 6.1 for the windows should be applied (multiply the windows by them) to ensure consistent peak or levels for all windows on frequency bin-aligned signals. For FFT sizes not equal to $N = 64$, calculate the narrowband scale factor by the ratio of the window integral to the rectangular window integral and the broadband scale factor by the ratio of the window-squared integral to the integral of the rectangular window. Note that if the input signals are not bin aligned, or a broadband correction factor is used, their peak levels will differ slightly depending on the particular data window used. The best we can do is to calibrate each data window so that the peak levels for each window match when the input signal frequency matches the FFT bin. The spectral resolutions shown in Figure 6.10 include the narrowband correction factors (window calibration) constants named with each window above and in Table 6.1. Broadband correction factors are applied to ensure that the total signal power is consistent from the (unwindowed) time domain to the frequency domain. Table 6.1 also presents the broadband corrections for $N = 64$-point windows. All correction factors given will vary slightly with N, and so a simple calculation is in order if precise signal level calibrations are needed in the frequency domain.

TABLE 6.1
Narrowband and Broadband Spectral Level Correction Factors for Data Windows

Correction Factor	Rectangular	Hanning	Welch	Parzen	Hamming	Exponential ($N = 64, k = 2.6$) ($N = 1024, k = 5.2$)	N
Narrowband	1.000	2.0317	1.5242	2.0323	1.8768	2.3229	64
Broadband	1.000	1.6459	1.3801	1.7460	1.5986	1.7890	64
Narrowband	1.000	2.0020	1.5015	2.0020	1.8534	2.8897	1024
Broadband	1.000	1.6338	1.3700	1.7329	1.5871	1.6026	1024

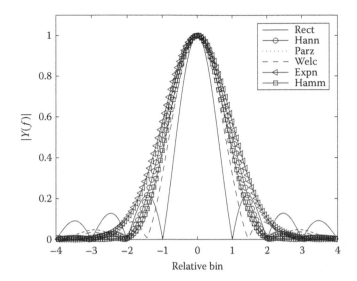

FIGURE 6.10 Comparison of the rectangular, Welch, Parzen, Hamming, Hanning, and exponential data windows in the frequency domain showing the slight differences in spectral resolution.

Comparing the nonrectangular data windows in Figure 6.9 we see only very slight differences. However, we can make some interesting observations which hold in general for data window applications. First, the broader rectangular and Welch data windows are the ones with better resolution, as shown in Figure 6.10. Second, the data windows with nearly zero slope at either end (near $n = 0$ and $n = N - 1$) tend to have the lowest side-lobe levels, as shown in Figure 6.10 for the exponential, Hamming, and Hanning window's spectral resolution. It is hard to say which data window is the best because it depends on the particular applications emphasis on main-lobe resolution and side-lobe suppression.

The reason why one desires low side lobes is that often one has strong and weak signal peaks in the same frequency range. With high leakage, the side lobes of the strong peak can cover up the main lobe of the weak peak. For applications where a wide dynamic range of sinusoidal peak levels are expected, application of a data window is essential. However, other applications require very precise frequency and phase measurement of spectral peaks. In this case, higher resolution is of primary importance and the rectangular window may be best suited. As always, a thorough understanding of the underlying physics of the signal processing application is needed to be sure of an optimal design approach and implementation. Perhaps the most important reason to have a variety of data windows available is for sensor array shading, which is presented in more detail in Sections 8.1 and 13.2, and Chapter 14. Array shading using data windows is the reason why the spectral leakage is referred to as "side lobes" while the resolution response of the FFT bin is called the "main lobe." These terms have a perfect analogy in array beamforming directivity responses.

6.5 CIRCULAR CONVOLUTION ISSUES

In this section we briefly discuss some very important and often unappreciated issues of circular convolution in the proper use of the FFT (or DFT) in real-world signal processing systems. Often one can define a desired system response in the frequency domain using FFTs of various input and output signals or perhaps a particular frequency-response function. If the desired system response is found from spectral products and/or quotients, and the system is to be implemented by an inverse FFT to give an FIR digital filter, one has to be very careful about the implications of the finite Fourier integral. The mathematics of the "analog-domain" Fourier integral imply that the signals of interest are linear and stationary, and that the finite sum (finite resolution) of the FFT provides an accurate signal representation (no spectral leakage). Since we often have leakage if the signal frequencies are mismatched to the FFT bins, the product of two spectral functions leads an FIR filter impulse response based on *circular convolution*, rather than linear convolution. Even if it has a broadband input signal to a linear system, spectral leakage in the output signal can occur if the system has a sharp (very high Q or low damping) resonance with a narrower bandwidth than the FFT bins. But in general broadband signals will have a circular convolution which very closely matches the linear convolution. One should also note that the inverse FFT of a conjugate multiply in the frequency domain (a cross spectrum) is equivalent to a cross-correlation in the time domain, which can also be affected by circular correlation errors when narrowband signals or responses are involved.

The Classic "Optimal Detection Filter" Problem: Consider the classic optimal filtering problem as depicted in Figure 6.11. This general filtering problem gives a result known as an Eckhart filter [6] and will be used in Chapters 12 and 13 for optimal signal detection. The least-squares technique will also be seen in a more general form in Chapter 9. The problem in Figure 6.11 is to recover the system input $U(f)$, given the output $C(f)$ which contains both signal $S(f)$ from the known system $R(f)$, and a noise waveform $N(f)$. With no noise, the answer is almost trivial because $C(f) = S(f) = U(f)R(f)$. Therefore, the "optimal" filter $A(f)$ to recover $U(f)$ from $C(f)$ is simply $A(f) = 1/R(f)$, and the recovered input is $U(f) = C(f)A(f)$. However, in the presence of noise $N(f)$ uncorrelated to the input $U(f)$, we have one of the classic adaptive signal processing problems known as the Weiner filtering problem. The beauty of solving the problem in the (analytic) frequency domain is that with infinite spectral resolution and no spectral leakage, all frequencies are perfectly orthogonal. Therefore, the equations can be solved frequency by frequency as a simple algebra equation. We examine this theoretical solution first and then address the issues of circular convolution which can arise when discrete FFTs are used.

We start by defining our model output $U'(f)$ where $U'(f) = C(f)A'(f)/R(f)$. We wish our model output to match with the actual $U(f)$ as close as possible for all frequencies. Therefore, we seek to choose $A'(f)$ such that the total squared spectral error defined in Equation 6.26 is minimized.

$$\int_{-\infty}^{+\infty} E(f)\,df = \int_{-\infty}^{+\infty} \left|U'(f) - U(f)\right|^2 df$$

$$= \int_{-\infty}^{+\infty} \left| \frac{[S(f) + N(f)]A'(f)}{R(f)} - \frac{S(f)}{R(f)} \right|^2 df \tag{6.26}$$

We can let $0 \geq A'(f) \geq 1$ and evaluate the integrand algebraically, dropping the (f) notation for brevity.

$$E = \left[|S|^2 + 2\,\mathrm{Re}\{NS\} + |N|^2\right]A'^2 + |S|^2 - 2\,\mathrm{Re}\{SN^*A'^*\} - 2A'|S|^2 + A'^2|N|^2 \tag{6.27}$$

Since the signal S and noise N are assumed uncorrelated, any spectral products involving SN are assumed to be zero.

$$E = |S|^2\left[A'^2 - 2A' + 1\right] + A'^2|N|^2 \tag{6.28}$$

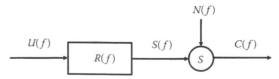

FIGURE 6.11 The optimal Weiner filtering problem described in the frequency domain using spectral products to derive a filter which recovers $U(f)$, given $R(f)$, $S(f)$, and $N(f)$.

We can solve Equation 6.28 for an $A'(f)$ which minimizes the squared spectral error $E(f)$ at each frequency by examining the first and second derivatives of E with respect to A'.

$$\frac{\partial E}{\partial A'} = |S|^2 (2A' - 2) + 2|N|^2 A'$$

$$\frac{\partial^2 E}{\partial A'^2} = 2 (|S|^2 + |N|^2) \tag{6.29}$$

Clearly, the parabolic structure of the squared spectral error shows a positive second derivative in Equation 6.29, indicating that the squared error function (a parabola) is concave up. Therefore, a solution for the first derivative equal to zero is a minimum of the squared error. We will repeatedly use variations of this basic approach to adaptively solving for optimal systems in many places later on in this text. The optimal solution for $A'(f)$ is clearly bounded between zero and unity and is known as an Eckhart filter, which maximizes the SNR of the output signal.

$$A'(f) = \frac{|S|^2}{|S|^2 + |N|^2} \tag{6.30}$$

The optimal filter for recovering $U(f)$ from the noise-corrupted $C(f)$ given that we know $R(f)$ is therefore $H(f) = A'(f)/R(f)$, or the spectral product of $A'(f)$ and $B(f) = 1/R(f)$. For the analog-domain indefinite-integral Fourier transform representation of $A'(f)$ and $B(f)$ there is no mathematical problem with designating the linear convolution of $a'(t)$ and $b(t)$ in the time-domain as the inverse Fourier transform of the spectral product $A'(f)$ and $B(f)$. Note that one could express the cross-correlation $R^{ab}(\tau) = E\{a'(t)b(t-\tau)\}$ as the inverse Fourier transform of the spectral conjugate product $A'(f)B^*(f)$. These relations are very important for the process of using measured signals from a system of interest to design or optimize the performance of signal filters for control or detection of useful information. Therefore, it is critical that we recognize the effect of the finite sum in the DFT and FFT on the convolution–correlation relations. *It is only for the case where all narrowband signal frequency components are bin aligned that the spectral product of two functions results in the equivalent linear correlation or convolution in the time domain.* For the more common case where leakage occurs between the discrete spectral bins in the frequency domain, the discrete spectral product results in a *circular convolution* or *circular correlation* (for the conjugate spectral product) in the digital time domain [7]. Unexpected circular convolution or correlation effects in an FIR from an inverse Fourier transform of the desired frequency response of the FIR filter can be extremely detrimental to the filter's performance in the system.

Consider the following simple example of a system with a short delay modeled in the frequency domain using a single frequency. If the frequency matched one of the DFT bin frequencies (for an exact integer number of wavelengths in the DFT input buffer as shown in the DFT spectra in Figure 6.3), there would be no problem determining the input–output cross-correlation from the IDFT of the spectral product of the input and conjugate of the output signal spectra. However, as shown in Figure 6.12, the frequency happens to lie exactly in between two discrete bins as can be seen by the odd number of half-wavelengths in the DFT input buffer and by the spectral leakage in Figure 6.4. The true linear correlation is shown in the solid curve in Figure 6.12, while the IDFT of the conjugate spectral product is shown as the curve denoted with (-o-) in Figure 6.12. Clearly, the additional frequencies from the product of the spectral leakage terms leads to significant amplitude and phase errors in the cross-correlation.

One can avoid circular convolution errors only by strictly limiting input and output signals to only have frequency components bin aligned to the discrete DFT bins or by the following technique for spilling over the circular convolution errors into a range where they can be discarded [5]. Note

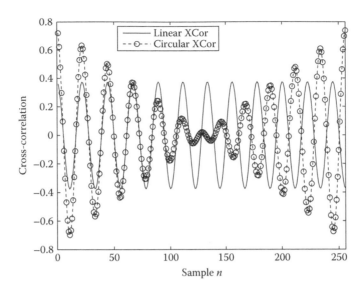

FIGURE 6.12 Comparison of the frequency domain estimated circular cross-correlation with spectral leakage (o) with the true linear cross-correlation.

that only bin-aligned sinusoids will work for the former case. Random noise sequences are found to suffer from the leakage problem. To implement the latter case, we simply double the number of data points in the DFT or FFT where one of the signals has the latter half of its input buffer *zero padded*, or filled with zeros. Suppose we only really need 128 cross-correlation data points in Figure 6.12. We would fill one signal buffer with 256 points and compute a 256-point DFT. The other signal buffer would have the most recent 128 points of its 256-point input buffer filled with zeros, the latter 128 points filled with the corresponding input time data, and its 256-point DFT is computed. The zero-padded DFT actually has a bit too much spectral resolution for the input data, but this added frequency sampling of the "sinc" function offers a special trick for the user. When the IDFT is computed after the spectral product is computed, all of the circular correlation errors are shifted to the upper half of the system impulse response and the lower half matches the true linear correlation result. Figure 6.13 shows the results of this operation graphically where the solid curve is again the true linear cross-correlation and the (-o-) curve is the result using spectral products.

Doubling the number of input data points and zero padding the input for one of the spectral products are the accepted methods for controlling circular convolution and/or circular cross-correlation errors from spectral products using the DFT or FFT. As shown in Figure 6.13, the circular correlation errors are completely corrected in the first 128 points, and the latter 128 points are simply discarded. This technique is well documented in the literature, and yet it is still surprising that many system designers fail to embrace its importance. It is likely that the widespread use of the analytical Fourier transforms and the corresponding convolution and correlation integral relations make it straightforward to assume that the same relations hold in the digital domain. The fact that these relations do hold but for only a very special case of exact signal frequency bin alignment for the DFT is the technical point which has ruined many a "robust" adaptive system design! Truly robust adaptive processes need to work with precision for any input and output signals.

6.6 UNEVEN-SAMPLED FOURIER TRANSFORMS

In our discrete summations for the NDFT, DFT, and FFT, we have always assumed regular equally spaced signal samples. There is a very good reason for this in the use of the trapezoidal rule for integration. However, in real-world signal processing applications, one may be faced with missing or corrupted data samples in the input to an FFT, or irregularly spaced samples from asynchronous

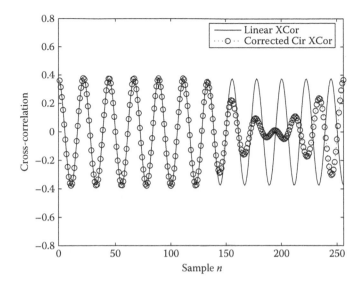

FIGURE 6.13 Comparison of the circular correlation correction (o) with the true linear cross-correlation showing the spectral leakage confined to the upper half of the output buffer from zero padding one signal before the Fourier transforms and spectral product.

communications between the sensor and the processor. Perhaps the most common occurrence of uneven samples is with spatial Fourier transforms of sensor array systems. As described in wave-number Fourier transforms, the spatial sampling of a group of sensors may not be regular, requiring special consideration when calculating the Fourier transform. These types of data problems are not at all unusual in real applications of signal processing, and are actually quite common in astrophysics and radio astronomy, where a considerable amount of attention has been given to the subject in the literature.

An area which can frustrate the use of Fourier transforms is the so-called "missing-data" or "unevenly sampled" data cases [8]. We have generally assumed regular-timed data samples in a continuous numerical stream as this allows simple application of discrete integration to replace the Fourier integral with a discrete summation. The discrete summation approximation can be seen as a finite "trapezoidal rule" summation where the average value of the integrand at two adjacent time sample points is taken as a "height" allowing the area under the integrand curve to be approximated by a rectangle with the width equal to the sampling interval. If the sampling interval is very small and regular, the errors in the approximation are quite small. However, if the sample points are not evenly spaced, or if various data points are missing from the series of input waveform numbers, the errors in estimating the DFT become very significant. A good example of how unevenly spaced input data samples to a Fourier transform can occur in practice can be observed in the application of sensor arrays for spatial Fourier transforms (wavenumber transforms) where the sensors are physically spaced in irregular positions. The irregular spacing can be due to practical limitations of the particular application or it can be purposely done as a simple means of controlling the wavenumber, or spatial (directional), response of the sensor array.

An example of missing or irregular time-sampled data can be seen in global networks of radio astronomy receiver networks. By linking the receiver dishes across the planet, a directional high-gain signal sensitivity response can be obtained. However, due to varying atmospheric conditions and the earth's magnetic field fluctuations, the signal pulses from a distant transmitter such as the Voyager I and II probes do not arrive at each receiver site at precisely the same time. Furthermore, some receiver site may experience momentary "drop-out" of the received signal due to multipath-induced destructive interference or other technical problems. The central question (among many) is: how should one deal with a section of missing or corrupted data? Should one "clamp" the data at the last known level until a new valid sample comes along? Is it better to simply set the missing or bad data to zero? Should one interpolate or spline fit the missing data to fit the rest of the sequence? A good point to consider is the relationship of the wavelength of interest with respect to the gap in the data sequence. For really long wavelengths, the relatively small irregularity is in general not a problem. However, if the data irregularity is on the order of a quarter wavelength or bigger, clearly a substantial error will occur if the problem is ignored.

In the following examples, we will simulate the missing or corrupted input data problem by imposing a Gaussian randomization on the sample interval T making the input data sequence to be actually sampled randomly in time (random spatial samples are the same mathematically). We will then examine the FFT of a low and high frequency with varying degrees of sample time randomization. For small variations in sample period T with respect to the wave period, simply applying the FFT as if there was no variation appears to be the best strategy. However, for relatively high frequencies with a shorter wave period, the variation in sample time is much more significant and the FFT completely breaks down. In the astrophysics literature, a transform developed by N. R. Lomb allows the spectrum to be accurately computed even when significant missing data or sample time variations occur in the input data. The Lomb-normalized periodogram, as referred to in the literature [8], starts out with N samples $y[n]$ $n = 0, 1, 2, \ldots, N - 1$, sampled at time intervals $t[n]$ $n = 0, 1, 2, \ldots, N - 1$ where the mean and variance of $y[n]$ are approximated by

$$\bar{y} = \frac{1}{N} \sum_{n=0}^{N-1} y[n]$$

$$\sigma^2 = \frac{1}{N} \sum_{n=0}^{N-1} (y[n] - \bar{y})^2 \tag{6.31}$$

and used in the normalized periodogram

$$Y^{\text{Lomb}}(\omega) = \frac{1}{2\sigma^2} \left\{ \frac{\left[\sum_{n=0}^{N-1} (y[n] - \bar{y}) \cos \omega (t[n] - \tau) \right]^2}{\sum_{n=0}^{N-1} \cos^2 \omega (t[n] - \tau)} \right. $$
$$\left. + \frac{\left[\sum_{n=0}^{N-1} (y[n] - \bar{y}) \sin \omega (t[n] - \tau) \right]^2}{\sum_{n=0}^{N-1} \sin^2 \omega (t[n] - \tau)} \right\} \tag{6.32}$$

where τ is determined by

$$\tau = \frac{1}{2\omega} \tan^{-1} \left\{ \frac{\sum_{n=0}^{N-1} \sin 2\omega t[n]}{\sum_{n=0}^{N-1} \cos 2\omega t[n]} \right\} \tag{6.33}$$

The setting of τ in the Lomb transform according to Equation 6.33 not only makes the periodogram independent of time shift, but is actually essential for a least-squared error model of fit of sinusoids to the input data. The model fit is particularly important if the $t[n]$ span is somewhat short. It should be noted that *the Lomb transform is a very slow algorithm* compared to even a DFT since the time shift τ must be estimated for each frequency of the periodogram, which has the equivalent magnitude of the magnitude-squared DFT, normalized by N.

A much simpler transform which is essentially equivalent to the Lomb transform for reasonably large data sets will be referred to here as simply the uneven Fourier transform (UFT). The UFT

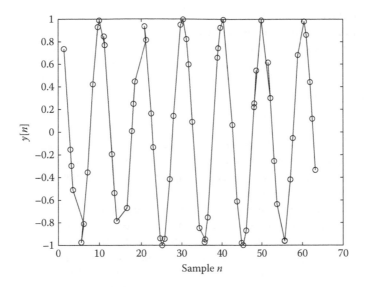

FIGURE 6.14 A 64-point 0.1 fs sine wave sampled with 0.5 T standard deviation of sample time.

cannot be implemented as an FFT because it requires the precise time sample value for each data value which mathematically requires a unique sample set of complex sinusoids.

$$Y^{\text{UFT}}(\omega) = \sum_{n=0}^{N-1} y[n] e^{j\omega t[n]} \tag{6.34}$$

We can compare the normalized magnitude-squared spectra of Equation 6.34, $|Y^{\text{UFT}}(\omega)|^2/N = Y^{\text{lomb}}(\omega)$ to examine performance of the two approaches. Consider the following data set of a sinusoid 0.1 times the mean sample frequency ($f/fs = 0.1$) where the sample error standard deviation is 0.5 times the sample interval T as shown in Figure 6.14. The FFT of the signal in Figure 6.14 is shown in Figure 6.15.

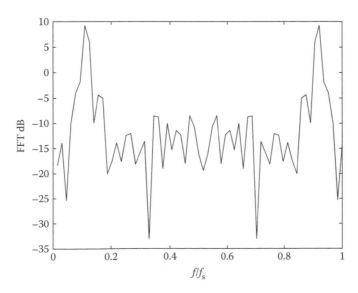

FIGURE 6.15 An FFT of the 0.1 fs sine wave in Figure 6.14 showing reasonable results and the expected "mirror image" peak at 0.9 fs.

The FFT of the 0.1 fs frequency sine wave in Figure 6.15 shows a "mirror image" of the spectral peak at 0.9 fs on the right side of the figure. This peak is expected for real data and can be seen as the aliased image of the negative frequency component of the sine wave which would also be seen at −0.1 fs. The Lomb and UFTs are shown in Figure 6.16 and *do not show this "mirror-image"* due to the randomization of the sample times. In other words, the time randomization of the input samples causes some very short sample time intervals, effectively raising the Nyquist rate for the data sequence. However, we must know the precise sample times for each input sample.

Note how the UFT and Lomb transforms are nearly identical. As presented, the two algorithms are not identical and the Lomb approach appears to be more robust on very short data records. However, the UFT is so much more efficient to implement and does not appear to be significantly less accurate than the Lomb algorithm.

Now if we examine a case where the frequency is increased to 0.45 fs with the same sampling error, the Lomb and UFT transforms outperform the FFT as shown in Figures 6.17 through 6.19. Clearly, the results shown in Figure 6.18 are unacceptable. It can be seen that the sine wave at 0.45 fs has a wave period nearly equal to twice the sample period *T*. With a standard deviation on the FFT input data samples of 0.5 T, or approximately a quarter wave period for the sine wave, the integration errors in the FFT are so great as to render the algorithm useless. It is in this realm where the Lomb and UFT algorithms show their value, as shown in Figure 6.19.

While Figures 6.14 through 6.19 are convincing, the next example will be somewhat stunning in its ability to show the usefulness of the UFT or Lomb transforms on extremely erratic sampled input data. Again, all we require is an accurate sample level and the precise time of the sample. Consider the case where the standard deviation of the data samples is 10 T (yes, 10 sample periods)! Figure 6.20 shows the input data for the Fourier transform. Even though it looks like a 2-year-old baby's refrigerator artwork, the UFT or Lomb algorithms can easily unscramble the mess and produce a useful Fourier transform! We will also violate the typical even-sampled Nyquist rate by letting the frequency rise to 0.7 fs, just to demonstrate some positive effects of sample randomization. Figure 6.21 shows the Fourier transform using the Lomb and UFT algorithms actually recovering the sinusoid!

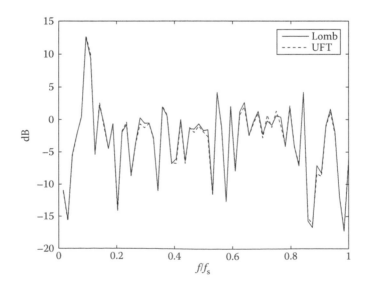

FIGURE 6.16 The UFT and Lomb transforms show no mirror image peak at 0.9 fs but actually have more noise than the FFT for the 0.1 fs sine wave with 0.5 T sampling deviation.

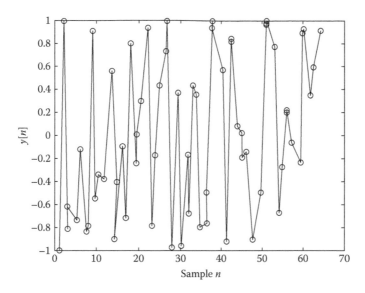

FIGURE 6.17   A 0.45 fs sine wave with 0.5 T sample deviation shows much less structure in the time domain than lower frequencies.

Why do the Lomb and UFT algorithms work so well for large randomizations of the input data, and yet not so well for the case where a low randomization was given for random sample times with a standard deviation of 0.5 T? The answer can be seen in the fact that small randomizations of the time samples of 0.5 T as shown in Figures 6.14 through 6.16 are of little consequence to the low-frequency wave with frequency 0.1 fs, which has a wave period of 10 T, or 20 times the standard deviation. The random signal levels in the FFT and UFT bins are due to errors in the assumptions in the trapezoidal rule behind the numerical integration carried out by the discrete sum in the DFT. It can be postulated, but will not be proven here that the regular sample assumptions in the FFT have a better averaging effect on the integration errors than do the UFT and Lomb algorithms. Clearly, the discrete sum in the DFT does not care in what order the products are summed up. *All that really*

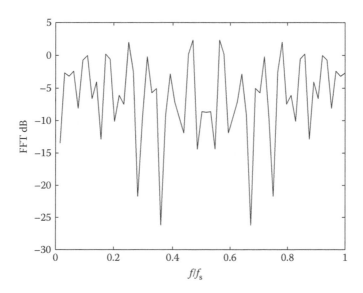

FIGURE 6.18   An FFT of the 0.45 fs sine wave with 0.5 T sample deviation shows no sign of a spectral peak.

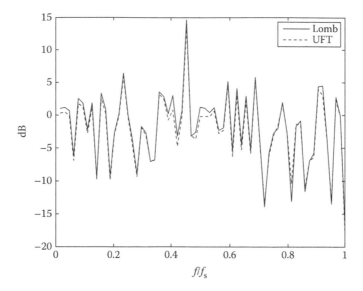

FIGURE 6.19 Both the Lomb and UFT algorithms perform quite well at recovering the 0.45 fs sine wave peak, even though these sample deviations are approaching one quarter of a wavelength.

matters is that the correct sample times are used in the Fourier integral. As the sample time randomizations increase with respect to the wave period, the regular sample assumptions in the FFT simply become a huge source of numerical error. The random noise in the UFT and Lomb algorithms tends to stay at a constant level, depending on the number of data points in the transform. Since the normalized Lomb periodogram is approximately the magnitude-squared of the DFT divided by N, it can be seen that the narrowband signal-to-noise gain the Lomb periodogram is $10 \log_{10} N$, or about 21 dB for a complex sinusoid. Therefore, one would expect a unity amplitude sine wave to be approximately +18 dB in the periodogram and the correspondingly scaled UFT and FFT responses with some small losses due to bin misalignment. It would appear from these

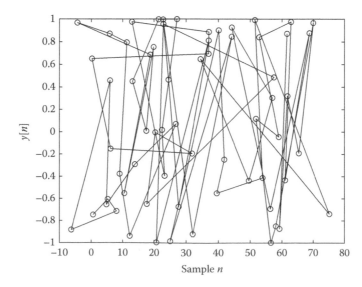

FIGURE 6.20 This is actually a plot of a 0.7 fs sine wave where the sample deviation is 10 times the sample period.

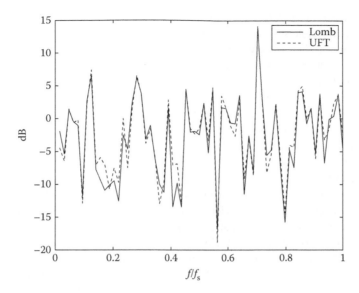

FIGURE 6.21 Both the UFT and Lomb algorithms show a significant peak at the correct frequency even for the extreme case of a sine wave frequency near above the Nyquist rate and a sample time deviation of 10 times the sample interval.

simulations that the Lomb Fourier transform periodogram may require more computation without significant improvement in performance when compared with the simpler UFT approach to the random Fourier integral. It should be clear that as long as the precise time and/or sensor position (for spatial transforms) is known for the Fourier input data, an accurate Fourier transform can be computed but at the expense of giving up the efficient FFT algorithm and some added noise in the spectral result.

6.7 WAVELET AND CHIRPLET TRANSFORMS

We have shown that the frequency resolution of a DFT or FFT is inversely proportional to the length of time of the input buffer such that the resolution Δf in Hz cannot exceed $1/T$, where T is the time length in seconds of the input buffer. For a spatial sampled DFT or FFT, the resolution in wavenumber Δk (rad/m) is inversely proportional to D, the extent in meters of the spatial input samples. Resolution has nothing to do with the number of samples in T or D, or whether the samples are evenly spaced, such as in the UFT. The sample rate (or sample density in spatial transforms) determines the range of unaliased frequencies (or wavenumbers) that can be represented. We have also shown that when one has a sinusoid as the input with an integer number of wavelengths in the input buffer, the resulting FFT spectral peak will be exactly on a frequency bin, where all other bins will be zero. If the frequency is altered so that a noninteger number of wavelengths exist in the input buffer, there still will be a spectral peak at the closest frequency bin, but the level will be less and the adjacent bins will have "spectral leakage" throughout the range of FFT bins. By applying a data window to the input buffer, this leakage is significantly attenuated in the bins away from the peak area, but there is always some leakage around the peak when a window is used whether the input frequency is bin aligned or not.

One of the interesting aspects of spectral analysis is the idea of time–bandwidth relations. To measure spectral changes in very narrow bandwidths (high spectral resolution) it requires very long input data buffers. Even if the input data buffers are heavily overlapped to allow update rates faster than the buffer time length T, each buffer represents a spectral average over the length T. This makes it difficult to see rapid changes in the spectrum from impulsive events. If we shorten T, the spectra will be more responsive to sudden changes in the input data since there is less of a time interval for the average, but the available spectral resolution must decrease, meaning the spectral bandwidth of each bin must get larger. There is no way to control bandwidth/resolution independent of the analysis time window length. But we can offer more flexibility in the display of the spectral estimate in terms of time and bandwidth considerations for analysis, as will be seen later in this section. The full details of wavelet analysis are very mathematically interesting, but beyond the scope of this book.

Our brief discussion of wavelets will be a simplified description and their operational (practical) use in the analysis of data, and in particular breaking a waveform down into constitutive frequency bands. As discussed in Section 4.4, MPEG 2 layer 3 (MP3) coding of audio waveforms broke the signal down into multiple frequency bands, each signal band being decimated to the appropriate lower sample rate, and each band being assigned a scale factor and bit depth for the samples to allow a substantial data rate compression. This compression concept so widely used today in audio, video, and telephone really originated from wavelet analysis. The MP3 signal is broken down into multiple frequency bands where each band is about 1/3 of an octave, similar to the critical bands in human hearing. In wavelet analysis the bands are related by octaves starting with the band from $fs/4$ to $fs/2$, where fs is the sample rate of the original waveform. All discrete wavelet transforms use an input buffer with a length that is a power of 2, split the signal into a high-pass and low-pass band, and decimate the two bands each by a factor of 2. The high-pass filter, called the wavelet function, produces an output signal that retains the original signal details with the average signal properties attenuated. The low-pass filter, called the scaling function, produces an output signal that retains only the average signal properties and attenuates the details. The decimated low-pass signal is then wavelet processed again, dividing its bandwidth into wavelet function (high-pass detailed) and scaling function (low-pass average) signals, which again can be decimated by 2 and the low-pass signal processed again, and so on. If the original signal buffer contained 2^N samples with a bandwidth $fs/2$ and sample rate fs, the wavelet analysis yields N signals: the top (Nth) octave with bandwidth $fs/4$, sample rate $fs/2$; $N - 1$st octave with bandwidth $fs/8$, sample rate $fs/4$, and so on. The first (lowest) octave scaling function output would have bandwidth $fs/(2N)$ and sample rate fs/N.

So what is all the wavelet hype about? Some of these octave bands may have little signal at all, or have noise only, allowing one to represent the original signal with the empty or noise octave band removed. With each iteration of the wavelet processing, the initial (or the so-called Mother wavelet) is reused but a half the sample rate, dilating the wavelet by a factor of two relative to the original sample rate. As more of the input data are processed the wavelet is translated (shifted) across the samples, although both dilation and translation are a bit more interesting properties for the continuous wavelet transform. If the wavelet and scaling functions (high and low pass, respectively) are orthonormal, the original signal input can be perfectly reconstructed from the base octave scaling function (low-pass) output and the upper octave wavelet function (high-pass) outputs. Orthogonal wavelets allow a signal to be broken down into octaves, analyzed for compression and noise reduction opportunities, and reconstructed. In many cases, noise can be removed, or the signal can be compressed by removing one or more of these octaves. These data compression and denoising scheme applies to two-dimensional image data as well. Also, because the higher-frequency octave bands are so wide, the time resolution is very high for locating waveform discontinuities compared to Fourier transforms.

The Haar Wavelet Transform: One of the most straightforward transforms in terms of linear algebra wavelet transforms is called the Haar transform [9], developed for generalized orthogonal expansion of functions long before it was called wavelet analysis. The Haar wavelet is the oldest (first published in 1910), and like all wavelet algorithms, breaks down the signal, y_t, in a hierarchy of high-pass "details" called the wavelet coefficients c_t, and low-pass-filtered signal "averages" called the wavelet scaling coefficients a_t.

$$c_t = \frac{y_t - y_{t+1}}{2}$$

$$a_t = \frac{y_t + y_{t+1}}{2} \tag{6.35}$$

One can view the expressions in Equation 6.35 as FIR filters with coefficients

$$\begin{aligned}
\bar{g} &= \begin{bmatrix} g_0 & g_1 \end{bmatrix} = \begin{bmatrix} 0.5 & -0.5 \end{bmatrix} \\
\bar{h} &= \begin{bmatrix} h_0 & h_1 \end{bmatrix} = \begin{bmatrix} 0.5 & 0.5 \end{bmatrix}
\end{aligned} \tag{6.36}$$

where $\bar{g}$ is clearly the high-pass filter impulse response (wavelet function coefficients) and $\bar{h}$ is the low-pass filter impulse response (scaling function coefficients). The even and odd samples of the input signal can be reconstructed from $y_t = a_t + c_t$ and $y_{t+1} = a_t - c_t$, and so it can be seen that both a_t and c_t are half the sample rate of y_t and also have half the bandwidth. So, the next set of inputs needs to be shifted by two samples instead of one. Following this logic, the FIR filter equations in Equation 6.35 become

$$\begin{aligned}
a_t &= h_0 y_{2t} + h_1 y_{2t+1} \\
c_t &= g_0 y_{2t} + g_1 y_{2t+1}
\end{aligned} \tag{6.37}$$

which is a bit more illustrative of the wavelet process. The low-pass signal a_t becomes the input signal for the next level of octave band division, and so on.

The Daubechies D4 Wavelet Transform: The Daubechies Wavelet transform [10] uses four coefficients for the wavelet and scaling functions and is one of the mathematically more interesting wavelets.

$$\begin{aligned}
\bar{h} &= \begin{bmatrix} \dfrac{1+\sqrt{3}}{4\sqrt{2}} & \dfrac{3+\sqrt{3}}{4\sqrt{2}} & \dfrac{3-\sqrt{3}}{4\sqrt{2}} & \dfrac{1-\sqrt{3}}{4\sqrt{2}} \end{bmatrix} \\
\bar{g} &= \begin{bmatrix} h_3 & -h_2 & h_1 & -h_0 \end{bmatrix}
\end{aligned} \tag{6.38}$$

Even though there are four coefficients to the wavelet and scaling functions, the shift is still two samples between iterations so that the average and detail outputs have one-half the same rate and bandwidth.

$$\begin{aligned}
a_t &= \sum_{i=0}^{3} y_{2t+i} h_i \\
c_t &= \sum_{i=0}^{3} y_{2t+i} g_i
\end{aligned} \tag{6.39}$$

There are many other wavelet functions, including many more by Daubechies, each with its particular properties, some orthogonal, and some not. The discussion is perhaps more interesting mathematically for continuous wavelet functions and analysis than the discrete case as some of the wavelets have fractal properties and produce beautiful analysis displays called "scaleograms." The general principle is to decompose a signal for analysis, compression, or detection of otherwise hidden features and have the ability to reconstruct the original signal from the salient wavelet components. This raises a more general question for spectral analysis: "Can we have each spectral bin have any bandwidth we wish?" The answer is yes so long as we start out with an appropriately bandlimited sampled signal.

General Multiresolution Spectral Analysis: Consider that in a standard DFT, the input data buffer is multiplied by the data window and a complex sinusoid and summed as if it were a row vector of the input data times a column vector of the window and complex sinusoid for the DFT bin. We can think of the window and complex sinusoid as a wavelet, but instead of filtering the signal as a convolution in the time domain, we can just multiply them as a vector inner product and the result is a complex number for the real and imaginary part of the DFT at the particular bin frequency. As it turns out, a typical DFT or FFT has all the frequency bins equally spaced and with the same resolution, since the same data window is used for each frequency bin. What if we used a different data window for each bin? There is nothing wrong with that. But the effective bandwidth for each frequency bin would depend on the time extent of the data window and so it would make sense to space the frequency bins nonuniformly so that they do not overlap. This is just a practical matter to minimize cross-talk between adjacent bins. We are not trying to carry out an orthogonal Fourier transform to manipulate the signal and restore it; we are just flexing the resolution in the frequency domain. For example, examining transducer frequency responses is often done on a logarithmic frequency scale and acoustical responses as they pertain to human hearing are often done using 1/3 octave frequency bands.

Consider a general wavelet formula for constructing a bandpass filter in the frequency domain with lower frequency f_1 and upper frequency f_2 and impulse response $h(t)$.

$$h(t) = \int_{f_1}^{f_2} e^{j2\pi ft} df$$

$$= \frac{1}{j2\pi t} \left\{ e^{j2\pi f_2 t} - e^{-j2\pi f_1 t} \right\} \tag{6.40}$$

Equation 6.40 is just the inverse Fourier transform written as an analytic integral. Let the arithmetic mean of f_2 and f_1 be $f_a = (1/2)(f_1 + f_2)$ and $\Delta f = (1/2)(f_2 - f_1)$ allowing us to write the impulse response in the form of a sinusoid times an envelope.

$$h(t) = 2\Delta f \cdot e^{j2\pi f_a t} \cdot \left\{ \frac{\sin(2\pi \Delta f t)}{2\pi \Delta f t} \right\} \tag{6.41}$$

Equation 6.41 has a complex sinusoid because we specified the transfer function of the bandpass filter as real in the frequency domain. One can also see that the term in the braces is the familiar $\sin(x)/x = \text{sinc}(x)$ function, which is expected because the corresponding function in the frequency domain is a "boxcar" function (it is constant amplitude from f_1 to f_2 as prescribed).

Let us consider the digital counterpart for our wavelet filter where the signal is sampled at a rate of f_s samples/s, or $T = 1/f_s$ seconds between each sample. We have a buffer of N samples and shift the maximum of the sinc() function to the center of the buffer by

$$h[n] = h(nT) = 2\Delta f \cdot \text{sinc}\left(2\pi \Delta f [n - N/2]T\right) \cdot e^{j2\pi f_a [n-N/2]T} \tag{6.42}$$

which will cause the peak of the envelope to occur in the middle of the buffer. The final step is to control spectral leakage by also applying a Hanning window from Equation 6.20 to ensure that the impulse response is zero at each end of the buffer ($n = 0$ and $n = N - 1$). Windowing this way does little to the frequency response except for reduce "ringing" or oscillations of the frequency response near the ends of the boxcar (f_1 and f_2). Windowing is mostly necessary for narrowband wavelet filters where the impulse response is fairly long compared to the buffer length. Our "mother wavelet" is now seen in Equation 6.43.

$$h[n] = \Delta f \cdot \left\{ 1 - \cos\left(\frac{2\pi n}{N-1}\right)\right\} \cdot \mathrm{sinc}\left(2\pi\Delta f\left[n - N/2\right]T\right) \cdot e^{j2\pi f_a[n-N/2]T} \qquad (6.43)$$

We note here that an additional scale factor is needed which depends on the specific calibration required for the spectral analysis which will vary for broadband signals versus narrowband sinusoids, the size of the buffer N, and the center frequency of the wavelet filter band. As it turns out, a scale factor of $1/f_s$ works well for the wavelet transfer function to be 0 dB in the passband for complex input buffer signals, independent of N or other frequency-domain scaling. If the input is real only, then an additional factor of 2 will be needed to recover the input signal amplitude. The factor of $1/f_s$ puts $\Delta f/f_s$ in dimensionless units. Equation 6.44 eliminates T to give the digital wavelet.

$$h[n] = \frac{\Delta f}{f_s} \cdot \left\{ 1 - \cos\left(\frac{2\pi n}{N-1}\right)\right\} \cdot \mathrm{sinc}\left(2\pi\frac{\Delta f}{f_s}\left[n - N/2\right]\right) \cdot e^{j2\pi(f_a/f)[n-N/2]} \qquad (6.44)$$

Proportional frequency bands are ones where the ratio of f_2/f_1 is the same for every band and the ratio of the center frequency of two adjacent bands k and $k + 1$ is $f_c(k + 1)/f_c(k) = f_2/f_1$. The center frequency for a given band is defined as the geometric mean $f_c = (f_2 f_1)^{1/2}$, which is not the arithmetic mean defined by f_a in Equation 6.43. Proportional bands are typically octave bands, $f_c(k + 1) = 2f_c(k)$, 1/3 octave bands, $f_c(k + 1) = 2^{1/3} f_c(k)$, or even 1/10 octave bands, $f_c(k + 1) = 2^{1/10} f_c(k)$. We could even make an array of wavelet filters aligned with the frequencies of the keys of a piano by using $f_c(k + 1) = 2^{1/12} f_c(k)$! Given the center frequency of a given band, the upper and lower frequencies can be conveniently found using

$$f_1 = \frac{f_c}{\sqrt[2Q]{2}}, \quad f_2 = \sqrt[2Q]{2}\, f_c \qquad (6.45)$$

where $Q = 1$ for octave bands, $Q = 3$ for 1/3 octave bands, and so on. So for a 1/3 octave band filter centered at 1000 Hz, f_1 is 890 Hz, and f_2 is 1122 Hz, approximately. For an octave band filter centered at 1000 Hz, f_1 is 707 Hz, and f_2 is 1414 Hz, approximately. Figure 6.22 shows a 64 Hz center band 1/3 octave wavelet (top) and filter frequency response (bottom) and Figure 6.23 shows the same for a 4096 Hz 1/3 octave spectral wavelet filter. Note that in both figures the frequency response is shown on a logarithmic frequency axis. Clearly, the use of wavelet technology is a very interesting way of designing FIR filters as well as does multiband spectral analysis. All one needs to do is to replace $e^{j2\pi(f_a/f_s)n}$ in the DFT with $h[n]$ in Equation 6.44 and the result of the inner product of the complex input buffer with the wavelet filter impulse response summed over the buffer length gives the signal amplitude in the band. If the input buffer is real only (the imaginary part is zero), which is usually the case for acoustical analysis, an additional factor of 2 is needed to recover the real input signal amplitude.

This type of multiband filtering for spectral analysis is not as efficient as the FFT, but usually fewer bands are used than FFT bins. For example, only 30 1/3 octave bands cover the range from 20 Hz to 20 kHz. There are N multiplies and adds (multiply–accumulate or MAC) per band, and so the algorithm requires $30N$ MACs. The FFT requires $N\log_2 N$ MACs, and if we want to exceed the resolution of a 1/3 octave filter at 20 Hz (4 Hz resolution required) with a sample rate of 44,100 Hz, a 16,384-point FFT is required (2^{14}). For this resolution example, the FFT requires $14*16,384 = 229,376$ MACs and the 30 1/3 octave wavelet filters require 491,520 MACs, or just over twice as many MAC operations. Note that a DFT of the same size requires $N^2 = 268,435,456$

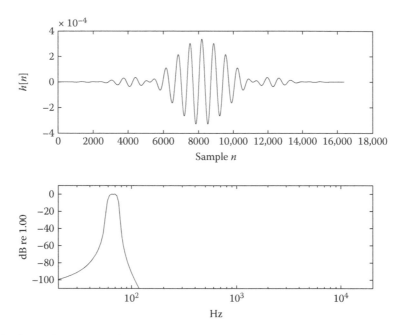

FIGURE 6.22 A wavelet based FIR filter (16,384 coefficients) for a precision 1/3 octave 64 Hz center frequency filter where the amplitude of the output is estimated from an inner product of the wavelet and the input data.

MACs or over 546 times the computation of the wavelet 1/3 octave bands! If you implemented 30 FIR 1/3 octave digital filters and estimated the amplitude of each filter output, it would require over $30N^2$ MACs since each filter requires a convolution to produce an output waveform—clearly the wrong choice is all you need to do is spectral analysis. With today's desktop and laptop computers capable of well over 1 giga-FLOP (1 billion floating-point operations per second), the computational

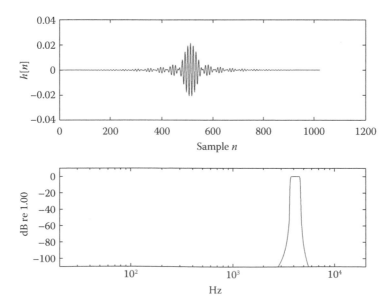

FIGURE 6.23 A wavelet based FIR filter (1024 coefficients) for a precision 1/3 octave 4096 Hz center frequency filter where the amplitude of the output is estimated from an inner product of the wavelet and the input data.

aspect is fast becoming a historical note. The lesson here is to not limit one's spectral analysis to fixed resolution FFT bins and to creatively use a spectral resolution appropriate for the application. For audio or acoustical analysis of sounds that affect human hearing, 1/3 octave bands are a great choice. For musical analysis, 1/12 octave bands can be used to correspond to the notes of an even-tempered musical tuning.

Chirplet Analysis: When the wavelet frequency changes over time we call it a "chirplet" named after common sounds found in bird calls. In a crowded forest with many different species of birds, it would be very difficult for mates to find each other or communicate warnings of danger or to establish a territory without defining a unique pattern of "chirps." One can think of the auditory system of a given species of bird as a "matched filter" to the transmitted chirp pattern, allowing that species of bird to single out a transmission by a mate, or foe, in the vicinity even in the presence of chirps from many other species of birds or other animals making acoustical calls. The matched filter concept is used today in many spread spectrum communication schemes based on the idea of correlating a known frequency chirping or hopping pattern with the received signal to detect and read a given message. But these kinds of chirplets do not scale and translate as we have shown with wavelets and proportional filters.

A very common and natural example of chirplets can be found in acoustic Doppler when listening to sounds from moving objects. Doppler shift (named after Austrian physicist Christian Doppler who proposed it in 1842 for application to astronomy), or the Doppler effect, is the shift in frequency upward of a wave whose source is moving toward the receiver, or down in frequency when the source is moving away from the receiver. Where the speeds of the source and the receiver in the medium are slow compared to the wave speed c, the received frequency f is

$$f = \left(\frac{c + v_R}{c + v_S} \right) f_0 \tag{6.46}$$

where v_R is the speed of the receiver relative to the medium, v_S is the speed of the source relative to the medium, and f_0 is the transmitted frequency at the source. It is interesting how the velocities of the source and the receiver relative to the medium affect the Doppler slightly differently. However, when the velocities are small relative to the wave speed in the medium, Equation 6.46 can be rewritten in terms of the difference in velocity between the source and the receiver $\Delta v_{SR} = v_S - v_R$.

$$f = \left(\frac{[c + v_S] - \Delta v_{SR}}{[c + v_S]} \right) f_0 = \left(1 - \frac{\Delta v_{SR}}{c + v_S} \right) f_0 \approx \left(1 - \frac{\Delta v_{SR}}{c} \right) f_0 \tag{6.47}$$

so that when the *source* and the receiver are moving toward each other Δv_{SR} is negative and the frequency is shifted upward. This is a *bistatic* source and receiver setup meaning that the source and the receiver are separated and moving independently.

Consider the problem of a collision avoidance sensor based on Doppler chirplets. Let us say we have a robot that transmits and receives an acoustic wave omnidirectionally. The velocity of the robot and the transmitted frequency f_0 are known at all times. Since the transmitter and receiver are colocated, their relative velocities are zero. However, the sound transmitted from the robot in the forward direction is Doppler shifted upward at a reflection from an obstacle, and since the robot is closing in, Doppler shifted upward again at the receiver. This is called a *monostatic* configuration of source and receiver, and is common for a sonar platform on a vehicle. For a monostatic sonar, the Doppler equation becomes

$$f = \left(1 - \frac{2v}{c} \right) f_0 \tag{6.48}$$

where v is the vehicle speed relative to the reflector (v is negative when the reflector approaches the vehicle) and the factor of 2 comes from the transmitted Doppler to the reflector and the received Doppler due to the closing speed at the receiver. Since our robot vehicle speed is known, we can tell if the object is directly in front, behind, or anywhere off to the sides, based on the received frequency.

The Doppler pattern of a passing reflector depends on the closest point of approach (CPA), the transmitted frequency, and the ratio of vehicle velocity to wave speed in the medium. If the CPA is small the frequency will change very abruptly from $f_0 + 2vf_0 / c$ to $f_0 - 2vf_0 / c$ as the reflector passes by the moving vehicle. If the CPA is large the rate of frequency change will be much more gradual. If the receiving array of sensors could be split into a right and left semicircular array, one could associate a bearing angle based on the Doppler frequency and a CPA based on the rate of frequency change.

To see how all this could be estimated, one just has to examine the geometry. If the reflector is an angle θ relative to the vehicle direction so that $\theta = 0$ is the direction of travel, $\theta = 90°$ is the direction at CPA, and $\theta = 180°$ if the reflector is directly behind the vehicle, then the effective velocity to the reflector is $v\cos\theta$. Another way of looking at this is the distance from the vehicle to the reflector is $r = \sqrt{CPA^2 + (vt)^2}$ where $t = 0$ at CPA and both v and t are negative as the reflector approaches CPA. The velocity between the reflector and vehicle is therefore

$$\frac{\partial r}{\partial t} = \frac{1}{2}(CPA^2 + v^2t^2)^{-1/2}(2v^2t)$$

$$= \frac{v^2t}{\sqrt{CPA^2 + v^2t^2}} = v\cos\theta \tag{6.49}$$

Combining Equations 6.48 and 6.49 gives the modeled Doppler frequency as a function of time.

$$f(t) = f_0\left(1 - \frac{2}{c} \cdot \frac{v^2t}{\sqrt{CPA^2 + v^2t^2}}\right) \tag{6.50}$$

Now let us look at the rate of change in Doppler frequency.

$$\frac{\partial f}{\partial t} = \frac{\partial}{\partial t}\left\{f_0\left(1 - \frac{2}{c} \cdot \frac{v^2t}{\sqrt{CPA^2 + v^2t^2}}\right)\right\}$$

$$= -\frac{\partial}{\partial t}\left\{\frac{2f_0v^2}{c}(t)(CPA^2 + v^2t^2)^{-1/2}\right\}$$

$$= -\frac{2f_0v^2}{c} \cdot \left\{(CPA^2 + v^2t^2)^{-1/2} + t\left(-\frac{1}{2}\right)(CPA^2 + v^2t^2)^{-3/2}(2v^2t)\right\} \tag{6.51}$$

Note that the maximum rate of frequency change occurs at CPA ($t = 0$) and the rate tends to zero at $\pm\infty$ for CPA > 0. If CPA = 0 (collision case), the rate of frequency change is zero up to the point of impact, then $-\infty$ at CPA, then zero after CPA such that the Doppler frequency function in Equation 6.50 is a downward step function at $t = 0$ and CPA. The maximum Doppler rate is at CPA and $t = 0$.

$$\left.\frac{\partial f}{\partial t}\right|_{t=0} = -\frac{2f_0v}{c}\frac{v}{CPA} = \Delta f_{max} \tag{6.52}$$

CPA of the reflector is easily solved by the expression in Equation 6.53.

$$\text{CPA} = -\frac{2f_0 v^2}{c\Delta f_{\text{max}}} \tag{6.53}$$

Equations 6.50 and 6.51 are scientifically interesting because the Doppler frequency and the Doppler rate of change are unique for a given CPA on the right half-plane or left half-plane defined by the vehicle direction of motion. This suggests a way to spatially map all the reflectors in a half-plane based on the Doppler and Doppler rate, but we would have to know the time until CPA for each reflector. If we waited until CPA for each reflector (when the Doppler rate is a maximum downward) we could solve for CPA, as shown in Equation 6.53 but that is not much help for collision avoidance since we would like to know where reflecting objects are *before* we collide with them. We also want to know which way to turn, and so it is helpful to map the location of all reflectors in each half-plane.

The Doppler chirplet is completed by also including a magnitude function based on the range to the reflector and back to the vehicle and the fact that for an omnidirectional source the amplitude of the wave will decay from spherical wave spreading. We do not know how well the object reflects the sound wave. If the object is large compared to wavelength, the reflection may vary with the aspect angle to the vehicle. If the object is small compared to wavelength, or happens to absorb sound, the reflection will be very weak. So, in general, we cannot reliably use the amplitude of the reflected sound to determine the distance to the reflector, and the wave attenuation may not follow spherical spreading precisely. Variations in amplitude will define the scattering strength of the reflecting object. Equation 6.54 provides the Doppler chirplet for CPA happening at $t = 0$.

$$h(\text{CPA}, t) = \frac{e^{j2\pi f_0 \left(1 - (2/c)\left(v^2 t/\text{CPA}^2 + v^2 t^2\right)\right)t}}{\sqrt{\text{CPA}^2 + v^2 t^2}} \tag{6.54}$$

Figure 6.24 compares two Doppler chirplets, one for CPA = 200 m (top) and the other for 20 m using very low frequency (2 Hz) and high velocity (100 m/s) so that the Doppler effect can be readily seen in the chirplet plot.

So how can one map all the reflectors in a half-plane with the Doppler chirplet in Equation 6.54? This is where chirplet analysis (and fast computers) provides a completely different way of solving a practical problem. We can lay out a grid in the half-plane of interest. The coordinate of a grid cell has both a time to CPA and a CPA distance. Therefore, each grid cell has a chirplet of the form in Equation 6.54. In fact, each grid cell will have a unique Doppler frequency and Doppler rate of change in frequency. We also note that each grid cell will have a rate of change of amplitude, but we do not need to, and should not depend upon the amplitude due to acoustical complexities of sound scattering and propagation. Given a received data buffer of length dT seconds, we can do a wavelet inner product using the chirplet in Equation 6.54 to recover the amplitude of the reflector in a given grid cell, if there is one there. The orthogonality of sinusoids, including ones with nonstationary frequencies, means that we need a match in both frequency and chirp rate for a given cell to produce any amplitude in the chirplet analysis.

Considerations for the Doppler chirplet are that for low velocities (a few m/s) one would need to use a high frequency to get an appreciable Doppler shift. The spatial resolution is a function of the transmitted frequency, vehicle velocity, and received buffer length dT, which in turn determines the maximum resolution in the frequency domain. The received buffer processing can be overlapped in time for finer resolution grid updates. The immediate collision avoidance is obviously the case with little downward Doppler shift and a rapidly rising amplitude. Knowing the location of other objects either side of the vehicle path helps in choosing the best course change. The key signal processing

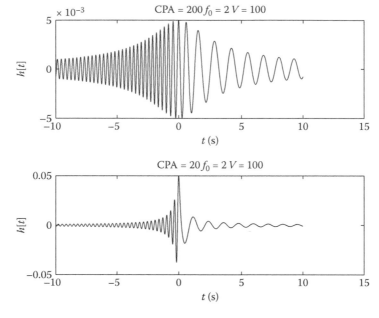

FIGURE 6.24 A Doppler chirplet filter showing one for CPA = 200 m (top) and 20 m (bottom) using an exaggerated low vehicle velocity (100 m/s) and low frequency (2 Hz) to make the wave structure more visible. A more practical case would be 40 kHz and 1 m/s for most robots.

concept here is to define a chirplet based on the geometry and physics of the problem, then apply the available computing power to process the received into useful information. One of the major reasons as to why this Doppler chirplet scheme works is that we assume that we know the vehicle velocity, we know the transmitted frequency, we know the medium wave speed, and we assume that the reflectors are not moving. The solution gets a lot more complicated if any of these assumptions are unknown.

6.8 MATLAB® EXAMPLES

While this is a fairly extensive chapter, there are only 6 m-scripts used to make all the figure plots. Table 6.2 summarizes each m-script and the associated figures. These m-scripts are available for download from the website for this book and are deliberately kept unsophisticated to emphasize the algorithm details in the most straightforward manner.

For the m-script "FourierRes.m," our goal is to examine the underlying resolution in Hertz of the Fourier transform on a simple sine wave in terms of the length in seconds of the input buffer.

TABLE 6.2
MATLAB m-Scripts Used in Various Figures

m-Script	Figures Used
FourierRes.m	Figures 6.1–6.5
dwindow.m	Figures 6.8–6.10
Circonv.m	Figures 6.12–6.13
uft.m	Figures 6.14–6.21
octbwavlet.m	Figures 6.22–6.23
DopChirp.m	Figure 6.24

Table 6.3 gives a code snippet for our discussion. This code allows us to compare the Fourier transform of a sinusoid with the theoretical result in Equation 6.5. However, because we chose $f_0 = f_s/4$, we have to be very careful to start the sine wave with $n = 0$ making the first sample in the input buffer zero. Otherwise, there would be a small phase shift giving both real and imaginary parts to the transform which do not match the theoretical result in Equation 6.5. Why? Consider the fact that for $f_0 = f_s/4$, there are only four samples per wavelength, and so an offset of only one sample is a phase shift of $\pi/4$. The variable "dzp" is the zero-padded buffer length. So we have a 256-point input buffer where the first 16 samples have the sine wave and the rest are zero. The FFT of the zero-padded input buffer produces the continuous curves seen in Figures 6.1 through 6.5. This shows the underlying available resolution of the 16-sample input buffer. Even though the input buffer is 256 samples, the effective time span of the data is only those 16 samples. Figure 6.2 shows the imaginary part normalized by T, the effective length of the input data matching the theoretical result in Figure 6.1 and Equation 6.5. Note how in Figures 6.2 through 6.4 the imaginary part, real part, and magnitude of 16-point FFT show zero leakage and "perfect" digital Dirac delta functions. This is because the frequency of the sine wave exactly matches the frequency of the fifth bin above 0 Hz. Also, note how we use "fftshift" and manipulated the frequency axis so that the 0 Hz bin appears in the middle of the spectral plots.

The m-script "dwindow.m" computes all the windows and does a normalized zero-padded FFT on each, applying "fftshift()" so that we can have a symmetric plot of the detailed spectral resolution. The plots in Figures 6.8 and 6.10 used 1024-point FFTs with the window length of only 64 points followed by 960 zeros for the input buffer. The observation here is clear. There is a big difference between a rectangular window (really no data window) and any of the nonrectangular windows, but the difference between the Hanning, Hamming, Parzen, or exponential windows is slight in terms of spectral resolution. The Welch window has a bit higher resolution than the others as seen in Figure 6.10 (it is closest to the rectangular), and is also a bit broader as a window as shown in Figure 6.9. All of the windows exhibit a trade-off between side-lobe spectral leakage and frequency resolution. Since resolution can be improved by a larger FFT (longer input buffer in seconds) at the same sample rate, the side-lobe leakage is generally best controlled by using a data window and a longer FFT. Why are the bumps on either side of the main peak called "side lobes?" It is because a line array of sensors summed together will have a directivity response where the "main lobe" is in the normal direction to the line axis. So attenuating the sensors near either end of the line array

---

TABLE 6.3
Code Snippet from "FourierRes.m"

```
y = sin((2*pi*f0/fs).*(n-1));
% must start @ 0 for real-imag to match theory when f0 = fs/4
dzp = 16; % play with this to see different resolutions

T = dzp./fs; % total seconds of integration time
df = 1/T;
titbuf = sprintf('T = %.3f df = %.3f',T,df);

for k = dzp + 1:npts, % zero pad
  y(k) = 0;
end;

NT = dzp;
Yf = (1/NT).*fftshift(fft(y,npts));
Yfdzp = (1/NT).*fftshift(fft(y(1:dzp),dzp));

F = (n-npts/2).*fs./npts;
Fdzp = ((1:(npts/dzp):npts)-npts/2).*fs./npts;
```

following the response of a "data window" reduced the leakage of sound into the array from directions away from the normal direction. This was a very simple way of improving the beamforming response of line arrays just by how you added the sensor signals together which is still useful today for long line arrays such as towed sonar arrays or geophysical arrays used in oil exploration. Much more on this topic will be presented in Chapters 8, 13, and 14. The most important aspect of data windows is that when used, they cause a known amount of spectral leakage whether the frequency is bin aligned or not. Furthermore, use of a data window also changes the signal level in the frequency domain, such that calibration adjustments are needed that are slightly different for narrowband and broadband signals.

While system transfer functions are routinely estimated in the frequency domain (see Chapter 7), estimating the system impulse response from the transfer function frequency response appears to be a straightforward inverse Fourier transform. It is in theory at least. But we must consider the effects of finite integration in the Fourier transform in that the input waveform is assumed periodic leading to spectral leakage unless we have the signals as sinusoids exactly aligned on the FFT bins. So the problem is that even small amounts of spectral leakage can lead significant errors in the inverse transform. A bin-by-bin conjugate product of two frequency-domain spectra is the equivalent of a cross-correlation of the two corresponding time-domain signals. But in theory these are infinite integrals. For the finite integrals of FFTs, the spectral multiply is the equivalent of *a circular convolution* and the conjugate spectral multiply is equivalent to a *circular cross-correlation*. So solving a transfer function problem such as the Eckhart filter example can lead to very poor results which do not improve by increasing the size of the FFT. However, as seen in the code snippet from "Circonv.m" in Table 6.4, a simple trick of zero padding on one of the signals in the product can shift the errors to the upper half of the inverse Fourier transform buffer with the cross-correlation result. This technique is useful for broadband as well as sinusoidal signals in the system response.

The uneven FFT and Lomb algorithm provide a way to analyze the spectral content of data where the samples are not uniformly spaced in time, but are sampled at known time intervals. While well off the beaten path, this is a very interesting concept that is very useful in some unusual cases. The Lomb periodogram described in Equations 6.31 through 6.33 is a fairly elaborate statistical

---

TABLE 6.4
Code Snippet from "Circonv.m"

```
% our "system is a simple delay with amplitude h0
for n = 1:N, % oldest to newest in old-school loop
    XN(n) = cos(w*n*T);
    YN(n) = h0*cos(w*T*(n-delay));
    YNZP(n) = YN(n);
    % RN is the correct cross correlation
    RN(n) = h0.*cos(w.*T.*(n+delay-1))./2; % n-1 here to include 0!
end
for n = N./2:N,
    YNZP(n) = 0.; % zero n/2 most recent (zero padding)
end
FXN = fft(XN);              % compute FFT's for spectral product
FYN = fft(YN);
FYNZP = fft(YNZP);
FZN = FXN.*conj(FYN);      % FZ = FX * Conj(FY)
FZNZP = FXN.*conj(FYNZP);
ZN = ifft(FZN)./(.5.*N); % ZN is the circular correlation
ZNZP = ifft(FZNZP)./(.5.*N); % ZNZP is the corrected circular correlation
```

estimation given the specific sample times of each input sample. It should be noted that once the input samples are "dithered" or shifted randomly from regular sample intervals, the spectral estimate will have spectral leakage. But without this dithering of sample times, there is no difference between the UFT and FFT (The Lomb Periodogram differs at the Nyquist rate). However, a normally distributed dither with a standard deviation approaching 1 sample period extends the bandwidth past the Nyquist rate at the expense of increased spectral leakage. A random sequence can be generated repeatedly using an algorithm known as a maximal length sequence (see Section 14.3) so that if the sampling follows a known random sequence it can be precisely recovered in the UFT or Lomb periodogram algorithms. Table 6.5 shows a snippet from "uft.m" showing the UFT algorithm computed as a DFT where the sample times are stored in the array $t(n)$.

The m-script "octbwavlet.m" is a really useful demonstration of how one can design a high-performance FIR filter for Qth-octave filtering ($Q = 1$ for octave, 3 for 1/3 octave, etc.) by designing a simple wavelet and using it in place of the complex sinusoid in a DFT. Table 6.6 shows a snippet of the key parts of the m-script where the wavelet is generated in the array "h" and a testing complex sinusoid is generated in the array "y" allowing the amplitude of y to be recovered in the simple inner product output scalar "outbin." Because the bandwidth of the wavelet filter is quite large compared with a regular FFT or DFT bin, the phase estimate is not very accurate, but phase estimation is generally not why one would use a wider bandwidth filter. One generally wants to know the signal energy in the band. However, the wavelet inner product avoids the added computation of first convolving the wavelet with the input signal and then estimating the mean-square signal energy.

6.9   SUMMARY

The Fourier transform is a fundamental signal processing algorithm which allows time-domain signals to be expressed in the frequency domain as a spectrum of a system transfer function. Frequency-domain representation of signals is generally more physically intuitive for periodic (sinusoidal) signals. Conversely, impulsive signals such as sonic booms are not very intuitive in the frequency domain, although the impulse response of a physical system is a special case where its Fourier transform is the system frequency response. For discrete signals and systems, the infinite time integration of the Fourier transform is replaced by a finite discrete summation. The result of this practical implementation issue is that the frequency separating resolution of the DFT is limited by the length of time spanned by the summed discrete signal samples. For example, the maximum resolution in Hertz is the inverse of the time span in seconds. For a DFT with 10 Hz bin resolution, the time span of the input data must be at least 100 ms. For 0.1 Hz resolution, the time span must be at least 10 s. The resolution of a particular DFT depends on the discrete sample rate and the number of data points used, which both translate physically into the time span of the input data to the DFT.

TABLE 6.5
The UFT Algorithm Based on a DFT

```
j = sqrt(-1);
for m = 1:npts,
  f(m) = m*fs/npts;
  ang = -j*2*pi*f(m)/fs;
  ytemp = 0.;
  for n = 1:npts,
      ytemp = ytemp + y(n)*exp(ang*t(n));
  end
  yf(m) = ytemp;
end
```

TABLE 6.6
The Wavelet Inner Product as a General Filter

```
% wavelet parameters
df = .5*(f2-f1);
fa = .5*(f2+f1);
n = 1.01:N; % adding .01 avoid a log of zero
ampscale = 1/fs; % this makes the passband 0 dB with a regular fft
j = sqrt(-1);
h = ampscale.*df.*(1-cos(2.*pi.*n./N)).*exp(j.*2.*pi.*(fa/fs).*(n-N/2)).
    *sin(2.*pi.*(df/fs).*...
(n-N/2))./(2.*pi.*(df/fs).*(n-N/2));
H = fft(h,N);   % for ploting filter frequency response
F = fs.*n./N;
% lets do a check - note phase not very accurate due to df
Amp = 4.2
phz = 0.5
y = Amp.*(cos(2.*pi.*(fc/fs).*n+phz) + j.*sin(2.*pi.*(fc/fs).*n+phz));
outbin = h*y';
outmag = abs(outbin)
outphz = angle(outbin)
```

The FFT is an engineered DFT such that the number of multiplications and additions are minimized. This is done by cleverly arranging the multiplies in the summation so that nothing is repeated. The FFT frequency bins are orthogonal for input frequencies which are exactly aligned with the bin frequencies, and thus, produce no leakage into adjacent FFT bins. The sine and cosine components of the FFT and DFT can be precomputed into tables to minimize redundant computation. Computationally, the FFT is significantly more efficient than the more explicit DFT requiring only $N\log_2 N$ multiplies as compared with the DFT's N^2. For a 1024-point FFT this difference is 10,240–1,048,576 or a reduction of 102.4:1. However, in order to use the FFT's marvelous efficiency, the input samples must be regularly spaced and the number of input samples must equal the total number of FFT bins. While an FFT typically has N bins where N is a power of 2, any length of input buffer can be processed by adding zeros (zero padding) to bring the input buffer up to the next power of 2, but the effective frequency resolution is determined by the inverse of the length in time of the nonzero-padded input samples.

For cases where spectral leakage is unavoidable (nonstationary input frequencies), one may weigh the input data (by multiplying it by a data "window") to keep the spectral leakage approximately the same for both bin-aligned and nonaligned input data frequencies. Window design and use is an approximate art and, like shoes, fast-food, music, and breakfast cereal, there is a wide variety of both old and new designs which do the same basic job slightly differently. For most applications the Hanning window gives excellent performance. However, one of the most overlooked aspects of the use of data windows is their effect on signal levels in the frequency domain. Using a bin-aligned narrowband reference for the rectangular data window (no window applied), we provide "narrowband correction" scale factors in Table 6.1) which if multiplied by the window samples, allows bin-aligned narrowband signal levels to be consistent, independently of which window is used. The narrowband correction factor is simply N divided by the sum of the window's N samples. For broadband signals the sum of the energy in the time domain should be consistent with the sum of the spectral energy. The broadband correction factor is found by N divided by the square root of the sum of the squared window samples. These values can also be found in Table 6.1 for a wide range of data windows. Both narrowband and broadband signals cannot be simultaneously calibrated in the frequency domain if a data window is used to control spectral leakage, a nonlinear phenomenon.

Another area of Fourier processing of data often overlooked is the effect of circular convolution when spectral products or divisions are inverse Fourier transformed back into the time domain to give impulse responses, cross-correlations, or other physical results. If there is spectral leakage in the frequency domain, the effect of the leakage is clearly seen in a plot of the spectral operation but can be far more devastating in the time domain. Spectral products (or divisions) bin by bin are only valid when the spectra accurately represent the data as an orthogonal frequency transform. The assumptions of signal/system linearity and time invariance carry forward to imply long-time records transformed into high-resolution Fourier transforms. In the physical world, these assumptions all have limits of applicability which can lead to serious sources of error in frequency-domain processing.

It is possible to transform nonregularly spaced samples into the frequency domain provided that the precise sample times are known for each sample. The UFT cannot use the efficient form of the FFT, but does offer the benefit of no frequency aliasing when significantly random sample time dithering is present. The UFT may be most useful for spatial Fourier transforms where sensor positions rather than sample times may be randomized for various reasons. The UFT and its more formal brother, the Lomb transform, allow for unevenly sampled data to be transformed to the frequency domain for subsequent analysis and processing.

Wavelet processing is important for orthogonally decomposing a signal into octave bands, allowing bands with little signal or noisy signal to be discarded, and the original signal reconstructed without the noise and from a smaller data file. Wavelet transforms are quite effective at data compression in this regard. However, the wavelet concept can also be used for multiband spectral analysis and detection of nonstationary frequencies using "chirplets." These techniques are not necessarily designed for efficiency like the FFT, but are rather designed for performance where ample computing resources are available.

PROBLEMS

1. A digital signal is sampled 100,000 times/s and contains white noise and two sinusoids. One sinusoid is 20,000 Hz and the other has the same amplitude at 20,072 Hz. Assuming that a DFT can resolve the two sinusoids if at least one bin lies between the two spectral peaks to be resolved, what size DFT (i.e., number of samples) is required to just resolve the two sinusoids? What is the associated integration time in seconds?

2. Given a 1990s era digital signal processor (DSP) chip capable of 25 million floating-point operations per second (25 MFLOPS), what is the real-time (no data missed) N-point FFTs processing rate assuming N operations are needed to move the real input data into place and N operations are needed to move the $N/2$ complex FFT output away when $N = 128$, 1024, 8192? What are the computational times for the FFTs?

3. By 2010, a typical desktop PC is capable of over 5 GFLOPS. How many 8192-point FFTs can be done per second? How many 65,536-point FFTs can be done per second? If a PC in 1990 could do 1 MFLOP what is the average annual processing speed increase between 1990 and 2010?

4. Show how two real signals can be loaded into a complex input, one signal in the real part and the other in the imaginary part, a normal FFT computed, and the two signal spectra can be recovered from the real and imaginary parts of the resulting FFT output.

5. Given a real-time data sequence, the total real spectrum is found by summing the positive and negative frequency bins and the total imaginary spectrum is found by subtracting the negative frequency bin values from the positive frequency bins. This is due to the Hilbert transform pair relationship for the real and imaginary spectra, given a real input time data sequence. Suppose the time data sequence is placed in the imaginary part of the input leaving the real part zero. How does one recover the proper real and imaginary spectra?

6. Given a real (imaginary part zero) digital waveform consisting of 1024 points of a sinusoid of amplitude 100 and white Gaussian noise of standard deviation 5, what would the

spectral levels be for a standard (unnormalized) FFT assuming the sinusoid is aligned to an FFT bin for a 1024-point FFT with a rectangular window? How about a normalized NFFT? What is the SNR gain of either FFT or NFFT?

7. Suppose we have a data window that is a straight line starting 1.00 at $n = 1$ and decreasing to 0.00 at $n = N$, the window size. Show that the narrowband correction factor is 2.00 and the broadband correction factor is $3^{1/2}$ or 1.732.

8. A real sampled signal can be converted to a complex signal using a Hilbert transform. Given an N-point FFT of the real signal, the positive frequency Hilbert transform for each bin is the sum of the positive frequency FFT bin with the complex conjugate its corresponding negative frequency FFT bin. Show that for a real sinusoid, the Hilbert transform can be used to recover the amplitude as a function of time and the frequency as a function of time.

9. Design a narrowband filter that passes frequencies from 39 to 41 kHz for real signals sampled at 200 kHz using a wavelet-based FIR filter. Show that the length of the FIR filter needs to be at least 1000 points at this sample rate.

10. A house fan 1 m in diameter has four blades with total area half the total fan disc area, tilted 5 cm front to back and turns at a rate of 1800 rotations per minute (RPM). Design a chirplet to model the sound reflected from the fan assuming a sound speed of 344 m/s and a point source and point receiver.

REFERENCES

1. A. Populus, *Signal Analysis,* New York, NY: McGraw-Hill, 1977.
2. S. D. Sterns, *Digital Signal Analysis,* Rochelle Park, NJ: Hayden, 1975.
3. K. Steiglitz, *An Introduction to Discrete Systems,* New York, NY: Wiley, 1974.
4. J. W. Cooley and J. W. Tukey, An algorithm for machine calculation of complex Fourier series, *Math Comput*, 1965, 19, pp. 291–301.
5. A. V. Oppenheim and R. W. Schafer, *Discrete-Time Signal Processing,* Englewood Cliffs, NJ: Prentice-Hall, NJ, 1989.
6. E. K. Al-Hussaini and S. A. Kassam, Robust Eckhart filters for time-delay estimation, *IEEE Trans ASSP*, 1984, 32, pp. 1052–1063.
7. J. J. Shynk, Frequency-domain multirate adaptive filtering, *IEEE Signal Process Mag*, 1992, 9(1), pp. 14–37.
8. W. H. Press, B. P. Flannery, S. A. Teukolsky, and W. T. Vetterling, *Numerical Recipies: The Art of Scientific Computing,* New York, NY: Cambridge University Press, 1986.
9. H. A. Zur, Theorie der orthogonalen Funktionensysteme, *Mathematische Annalen*, 1910, 69, pp. 331–371.
10. I. Daubechies, Ten lectures on wavelets, CBMS-NSF Reg. Conf Series in Applied Math. SIAM volume 61, Philadelphia, 1992.

7 Spectral Density

The spectral density of a signal is a statistical quantity very useful in determining the mean-square value, or power spectral density (PSD), for frequency-domain signals. Strictly speaking, the spectral density, as its name implies, is an expected power density per Hertz. As the time integral in the forward Fourier transform increases, so does the amplitude of the spectral density, while the resolutions of the Fourier transform narrows. Therefore, the expected signal power per Hertz in the frequency domain stays the same, matching the expected signal power at that frequency in the time domain. However, many current signal processing texts define the spectral density with some subtle differences which depend on whether the underlying Fourier transform integral limits are $-T$ to $+T$ or $-T/2$ to $+T/2$, and whether the original time signal is real or complex. Unfortunately, for the uninitiated student, both of these subtle parameters can either lead to a "factor of 2" difference in the spectral density definition or actually cancel one another deep in the algebra leading to significant confusion. In the derivation below (which is thankfully consistent with all texts), we will allude to the origins of the potential "factor of 2's" so that the student can easily see the consistency between the many available texts with derivations of spectral density.

7.1 SPECTRAL DENSITY DERIVATION

We start the derivation of spectral density with the assumption of an analytic time waveform $x(t)$ over the interval $t = -T/2$ to $+T/2$ and zero everywhere else on the range of t. Using the $\pm T/2$ integration interval for the Fourier transform, we define the spectral density as

$$S_X(\omega) = \lim_{T \to \infty} \frac{E\{|X(\omega)|^2\}}{T} \quad X(\omega) = \int_{-T/2}^{+T/2} x(t)\, \mathrm{e}^{-\mathrm{j}\omega t}\, \mathrm{d}t \tag{7.1}$$

Note that if the Fourier transform $X(\omega)$ is defined over the interval $-T$ to $+T$, a factor of $2T$ (rather than T) would be required in the denominator for the spectral density $S_X(\omega)$, as seen in some texts.

A good starting point for the derivation of spectral density is Parseval's theorem for two time functions $g(t)$ and $f(t)$ Fourier-transformable into $G(\omega)$ and $F(\omega)$, respectively.

$$\int_{-\infty}^{+\infty} f(t)g(t)\,\mathrm{d}t = \frac{1}{2\pi} \int_{-\infty}^{+\infty} F(\omega)G(-\omega)\,\mathrm{d}\omega = \int_{-\infty}^{+\infty} F(f)G(-f)\,\mathrm{d}f \tag{7.2}$$

One can easily see that the only difference between working in radian frequency ω in radians/s and the more familiar physical frequency f in Hertz (cycles/s) is the simple factor of $1/2\pi$ from the change of variables. We prefer to use f instead of ω because it will be more straightforward to compare spectral densities in the analog and digital domains. For our real signal $x(t)$ which is nonzero only in the range $-T/2 < t < T/2$, we have

$$\int_{-T/2}^{+T/2} x^2(t)\,\mathrm{d}t = \int_{-\infty}^{+\infty} X(f)X(-f)\,\mathrm{d}\tilde{f} \tag{7.3}$$

Dividing Equation 7.3 by T and taking the limit as T goes to infinity provides an equation for the expected mean-square signal (or signal power), in the time and frequency domains.

$$\lim_{T \to \infty} \frac{1}{T} \int_{-T/2}^{+T/2} x^2(t) \, dt = \lim_{T \to \infty} \frac{1}{T} \int_{-\infty}^{+\infty} X(f) X(-f) \, df \tag{7.4}$$

For the special (and most common) case where $x(t)$ is purely a real signal (the imaginary component is zero), we note the Hilbert transform pair relationship which allows $X(-f)$ to be seen as the complex conjugate of $X(+f)$. Taking the expected value of both sides of Equation 7.4 gives

$$E\left\{ \lim_{T \to \infty} \frac{1}{T} \int_{-T/2}^{+T/2} x^2(t) \, dt \right\} = \lim_{T \to \infty} \frac{1}{T} \int_{-\infty}^{+\infty} E\left\{ |X(f)|^2 \right\} df \tag{7.5}$$

which for stationary signals and random processes, the expected value of the time average of the signal-squared is just the mean-square signal.

$$\bar{x}^2 = E\{x^2(t)\} = \int_{-\infty}^{+\infty} \lim_{T \to \infty} \frac{E\left\{ |X(f)|^2 \right\}}{T} df = \int_{-\infty}^{+\infty} S_X(f) \, df \tag{7.6}$$

Equation 7.6 provides the very important result of the integral of the spectral density over the entire frequency domain is the mean-square time-domain signal, or signal power. The integral of the spectral density $S_X(f)$ in Equation 7.6 is referred to as a *two-sided PSD* because both positive and negative frequencies are integrated. Since $x(t)$ is real, the two-sided PSD is exactly the same as twice the integral of the spectral density over only positive frequencies.

$$\bar{x}^2 = E\{x^2(t)\} = 2 \int_0^{+\infty} S_X(f) \, df \tag{7.7}$$

The integral of the spectral density in Equation 7.7 is known as a *one-sided PSD*. Clearly, if one is not careful about the definition of the time interval ($\pm T$ or $\pm T/2$) for the Fourier transform and the definition of whether a one-sided or two-sided PSD is being used, some confusion can arise. And, just when things appear to be clear, the *power spectrum* estimate, $G_{XX}(f)$, of a real signal, $x(t)$, is defined as twice the spectral density at that frequency as a simple means of including positive and negative frequency power in the estimate.

Consider the following simple example. Let $x(t)$ be equal to a sine wave plus a constant.

$$x(t) = A + B \sin(\omega_1 t) \tag{7.8}$$

We can write the mean-square signal by inspection as $A^2 + B^2/2$. The Fourier transform of $x(t)$ is

$$X(\omega) = A T \frac{\sin(\omega T/2)}{\omega T/2} + \frac{BT}{2j} \frac{\sin([\omega_1 - \omega]T/2)}{[\omega_1 - \omega]T/2} - \frac{BT}{2j} \frac{\sin([\omega_1 + \omega]T/2)}{[\omega_1 + \omega]T/2} \tag{7.9}$$

The spectral density $S_X(\omega)$ is found by taking expected values and noting that the three "$\sin(x)/x$"-type functions in Equation 7.9 are orthogonal in the limit as T approaches infinity.

$$S_X(\omega) = \lim_{T \to \infty} \left\{ A^2 T \left| \frac{\sin \omega T/2}{\omega T/2} \right|^2 + \frac{B^2}{2} \left(\frac{T}{2} \right) \left| \frac{\sin\left([\omega_1 - \omega]T/2\right)}{[\omega_1 - \omega]T/2} \right|^2 \right.$$
$$\left. + \frac{B^2}{2} \left(\frac{T}{2} \right) \left| \frac{\sin\left([\omega_1 + \omega]T/2\right)}{[\omega_1 + \omega]T/2} \right|^2 \right\} \tag{7.10}$$

To find the signal power, we simply integrate the two-sided spectral density over positive and negative frequencies noting the following important definition of the Dirac delta function:

$$\lim_{T' \to \infty} \int_{-\infty}^{+\infty} T' \left| \frac{\sin(\omega T')}{\omega T'} \right|^2 \, d\omega = \pi \delta(\omega) = \frac{\delta(f)}{2} \tag{7.11}$$

Then, by a straightforward change of variables where $T' = T/2$ and $df = d\omega/2\pi$, it can be seen that the integral of the two-sided PSD is simply

$$\int_{-\infty}^{+\infty} S_X(f) \, df = 2A^2 \left(\frac{\delta(f)}{2} \right) + \frac{B^2}{2} \left(\frac{\delta(f_1 - f)}{2} \right) + \frac{B^2}{2} \left(\frac{\delta(f_1 + f)}{2} \right) \tag{7.12}$$

Since the integral of a Dirac delta function always gives unity area, the signal power is easily verified in Equation 7.12 as $A^2 + B^2/2$.

We now briefly consider the discrete sampled data case of $x[n] = x(nTs)$, where Ts is the sampling interval in seconds. For an N-point DFT or FFT of $x[n]$, we can compare the digital signal equivalent of Parseval's theorem.

$$\sum_{n=0}^{N-1} x^2[n] = \frac{1}{N} \sum_{m=0}^{N-1} |X[m]|^2 \tag{7.13}$$

The factor of $1/N$ in Equation 7.13 is very interesting. It is required to scale the transform appropriately as is the case for the IDFT formula. It essentially takes the place of "df" in the analog-domain inverse Fourier transform. For N equal spaced frequency samples of $X[m]$, the factor of $1/N$ represents the spectral width of a bin in terms of a fraction of the sample rate. If we consider the mean-square digital signal, or the expected value of the digital signal power, we have

$$\frac{1}{N} \sum_{n=0}^{N-1} x^2[n] = \frac{1}{N^2} \sum_{m=0}^{N-1} |X[m]|^2 \tag{7.14}$$

The equivalent digital domain spectral density can be seen as the magnitude-squared of the normalized DFT (NDFT) presented in the previous section. The mean-square value (or average power) of the digital signal is the sum of the magnitude-squared of the NDFT bins. But more importantly, the power at some frequency f_k is the sum of the squares of the NDFT bins for $\pm f_k$, or twice the value at real f_k for $x[n]$. The RMS value is simply the square root of the power, or more simply, $1/2^{1/2}$ times the magnitude of the NDFT bin. This is why the NDFT is very often convenient for real physical system applications.

Sensor calibration signals are usually specified in rms units at some frequency. Electronic component background noise spectra are usually specified in the standard Volts per square-root Hertz, which is the square root of the voltage PSD. Obviously, one would not want the size of the DFT or FFT to be part of a component noise standard. For intelligent adaptive systems, we generally want to change NDFT size according to some optimization scheme and do not want signal levels in the normalized PSD (NPSD) to depend on the number of points in the transform. Throughout Section 7.2, we will examine the statistics of the data in the NDFT bins and many useful signal functions well represented by physical applications in the frequency domain. To avoid confusion, we will refer to the mean-square values of the NDFT magnitude as the NPSD, while the PSD as defined in the literature is N times the NPSD (i.e., NPSD = PSD/N).

7.2 STATISTICAL METRICS OF SPECTRAL BINS

One often processes broadband signals in the frequency domain which can be seen as the sum of a large number of sinusoids of randomly distributed amplitudes. There are many naturally occurring random processes which generate broadband signals such as turbulence in the atmosphere, intrinsic atomic vibrations in materials above absolute zero temperature, electron position uncertainty and scattering in conductors, and the LSB error in a successive approximation ADC as described in Chapter 1. All physical signals harvested in an ADC and processed in an adaptive signal processing system contain some level of "background" random signals referred to as noise. However, with the prevalence 24-bit ADC in audio systems and 16-bit ADC in video systems, the background noise is usually from the sensing electronics or the medium being sensed. Fidelity is great, but we still have background noise. Even numerical errors in computing can be modeled as random noise, although this kind of noise is (hopefully) quite small. The NPSD (NPSD = $1/N$ times the PSD) in a given bin can also be seen as a random variable where we would like to know the expected value as well as the likelihood of the bin value falling within a particular range. These metrics from statistics provide a means to estimate the SNR as well as to measure the effect of spectral processing on the SNR in each frequency bin [1].

The *central limit theorem* states that regardless of the probability distribution of an independent random variable X_n, the sum of a large number of identical but statistically independent random variables (each of the same probability distribution) gives a random variable Y which tends to have a Gaussian probability distribution. This may seem like a remarkable assumption, but it is not from a detailed mathematical perspective.

$$Y = \frac{1}{\sqrt{N}}\left[X_1 + X_2 + X_3 + \cdots + X_N\right] \tag{7.15}$$

Equation 7.15 shows a peculiar normalization in the factor of $N^{-1/2}$ which averages the square of the random variables. This is a very typical metric for statistical signals since it allows the mean and variance (first and second moments) to be used as the salient descriptors. We will have to briefly summarize some basic statistics in this section in order to fully describe statistical signals.

Before we get into the mathematical details of statistics, let us do a verbal walk-through of the central limit theorem to see why Gaussian statistics are so important. When we have a random variable, perhaps the most basic measurement we can do is called a histogram, which is plot counting the number of sample occurrences (y-axis) versus the sample values (x-axis). If there is an equal likelihood of sample values between say ±1 and zero elsewhere, we can say that the distribution is uniform between ±1. The histogram shape would look like a square centered over zero on the x-axis. Now let us sum two of these uniform distributions together. The range on the x-axis extends now to ±2 and the likelihood increases near the origin. The shape of this new distribution is a triangle with the peak at $x = 0$ and the base extending from -2 to $+2$. It is the convolution of the two square distributions. If we continue this process to sum N uniform distributions, the extreme case of all the random variables lining up at $+1$ or -1 becomes very rare compared to the cases where the sum of the random variables is near zero, which is the mean of each random variable. As such, one ends up with the familiar "bell curve" of a Gaussian function to describe the histogram of likely sample values. When we normalize this distribution so that the integral of it is equal to unity, it is called a *probability density function* (PDF). Integrating the PDF between two limits of x conveniently gives a measure of the probability of a sample of x falling within the prescribed range. It does not matter what the shape of the original random variable distributions is. If you sum a large number of identical but independent random variables, one ends up with a Gaussian distribution. A common example of this is the electronic noise on a wire stimulated by the thermal energy in the wire atoms. Since there are an extremely large number of individual atoms involved with essentially the same independent electronic statistics individually, one gets a Gaussian distribution of electronic "thermal noise" which is indeed observed. The central limit theorem will be discussed again in greater detail in Chapter 12.

7.2.1 PROBABILITY DISTRIBUTIONS AND PDFs

A probability distribution $P_Y(y)$ is defined as the probability that the observed random variable Y is less than or equal to the value y. For example, a six-sided die cube has an equal likelihood for showing each of its faces upward when it comes to rest after being rolled vigorously. With sides numbered 1 through 6, the probability of getting a number less than or equal to 6 is 1.0 (or 100%), less than or equal to 3 is 0.5 (50%), equal to 1 is 0.1667 (16.67% or a 1 in 6 chance), and so on. A probability distribution has the following functional properties:

$$
\begin{aligned}
&(a) \quad P_Y(y) = \Pr(y \leq y) \\
&(b) \quad 0 \leq P_Y(y) \leq 1 \quad -\infty < y < +\infty \\
&(c) \quad P_Y(-\infty) = 0 \quad P_Y(-\infty) = 1.00 \\
&(d) \quad \frac{dP_Y(y)}{dy} \geq 0 \\
&(e) \quad \Pr(y_1 < Y \leq y_2) = P_Y(y_2) - P_Y(y_1)
\end{aligned}
\tag{7.16}
$$

The term "Pr" means "the probability of" in Equation 7.16. Also note that the probability curve never has a negative slope since it is the integral of the PDF. The PDF function $p_Y(y)$ has the following functional properties:

$$
\begin{aligned}
&(a) \quad p_Y(y) = \lim_{\varepsilon \to 0} \frac{P_Y(y+\varepsilon) - P_Y(y)}{\varepsilon} = \frac{dP_Y(y)}{dy} \\
&(b) \quad p_Y(y)\,dy = \Pr(y < Y \leq y+dy) \\
&(c) \quad p_Y(y) \geq 0 \quad -\infty < y \leq +\infty \\
&(d) \quad \int_{-\infty}^{\infty} p_Y(y)\,dy = 1 \\
&(e) \quad P_Y(y) = \int_{-\infty}^{y} p_Y(u)\,du \\
&(f) \quad \int_{y_1}^{y_2} p_Y(y)\,dy = \Pr(y_1 < Y \leq y_2)
\end{aligned}
\tag{7.17}
$$

For our die example above, the probability distribution is a straight line from 1/6 for a value of 1–1.00 for a value less than or equal to 6. The PDF is the slope of the probability distribution, or simply 1/6. Since the PDF for the die is the same for all values between 1 and 6 and zero elsewhere, it is said to be a *uniform probability distribution*. The expected value (or average value) of a random variable can be found by computing the first moment, which is defined as the mean.

$$m_Y = \bar{Y} = E\{Y\} = \int_{-\infty}^{+\infty} y p_Y(y) \, dy \tag{7.18}$$

The mean-square value is the second moment of the PDF.

$$\bar{Y}^2 = E\{Y^2\} = \int_{-\infty}^{+\infty} y^2 p_Y(y) \, dy \tag{7.19}$$

The variance is defined as the second central moment, or the mean-square value minus the mean value-squared. The standard deviation for the random variable, σ_Y, is the square root of the variance.

$$\sigma_Y^2 = \bar{Y}^2 - m_Y^2 = \int_{-\infty}^{+\infty} (y - m_Y)^2 p_Y(y) \, dy \tag{7.20}$$

For our die example, the mean is 1/6 times the integral from 0 to 6 of y, or 1/12 times y^2 evaluated at 6 and zero, which gives a mean of 3. The mean-square value is 1/6 times the integral from 0 to 6 of y^2, or 1/24 times y^3 evaluated at 6 and 0, which gives a mean-square value of 12. The variance is simply $12 - 9$ or 3 as expected.

Consider the statistics of rolling 100 dice and noting the central limit theorem in Equation 7.15 and the probability distribution for the sum (giving possible numbers between 100 and 600). Clearly, the likelihood of having all 100 dice roll up as 1's or 6's is extremely low compared to the many possibilities for dice sums in the range of 300. This "bell curve" shape to the probability density is well known to approach a Gaussian PDF in the limit as n approaches infinity. The Gaussian PDF is seen in Equation 7.21 and also in Figure 7.1 for $m_Y = 0$ and $\sigma_Y = 1$.

$$p_Y(y) = \frac{1}{\sigma_Y \sqrt{2\pi}} e^{-(y - m_Y)^2 / 2\sigma_Y^2} \tag{7.21}$$

For real-world sensor systems where there are nearly an infinite number of random noise sources ranging from molecular vibrations to turbulence, it is not only reasonable, but prudent to assume Gaussian noise statistics. However, when one deals with a low number of samples of Gaussian random data in a signal processing system (say < 1000 samples), the computed mean and variance will not likely exactly match the expected values and a histogram of the observed data arranged to make a digital PDF will have a shape significantly different from the expected bell curve. This problem can be particularly acute where one has scarce data for adaptive algorithm training. There are various intricate tests, such as Student's *t*-test and others, that have been developed (but are not presented here) to determine whether one data set is statistically different from another. Results from statistical tests with small data sample sets can be misleading and should only be interpreted in

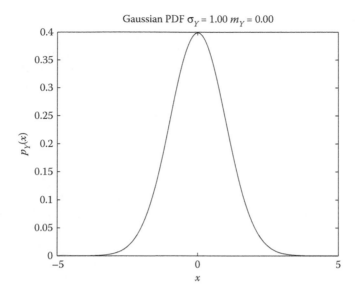

FIGURE 7.1 A Zero-Mean Gaussian (ZMG) PDF with variance equal to unity.

general as a cue, or a possible indicator for significant information. The general public very often is presented with statistical results from small sample size of clinical trials in the health sciences (human volunteers are rare, expensive, and potentially in harm's way). These statistical sample set tests can also be used in adaptive pattern recognition to insure low bias in small data sets. However, the best advice is to use an enormous data set for statistical training of adaptive algorithms wherever possible and to always check the data set distribution parameters to ensure that the proper assumptions are being applied.

7.2.2 STATISTICS OF THE NPSD BIN

Consider the Fourier transform of a real Zero-Mean Gaussian (ZMG) signal with a time-domain standard deviation σ_t. The Fourier transform (as well as the DFT, FFT, and NDFT defined above) is a linear transform where each transformed frequency component can be seen as the output of a narrowband filter with the ZMG input signal. Since the ZMG signal is spectrally white (it can be represented as the sum of an infinite number of sinusoids with random amplitudes), each bin in a DFT, FFT, or NDFT can be seen as a complex random variable where the real and imaginary components are ZMG random variables with standard deviation $\sigma_R = \sigma_I = \sigma_f$. If we use the NPSD to compare mean-square values in the time and frequency domains (DFT computed using a rectangular data window), we can easily determine σ_f^2 in terms of σ_t^2 using Equation 7.14 as

$$\begin{aligned}
\sigma_t^2 &= \frac{1}{N} \sum_{n=0}^{N-1} x^2[n] \\
&= \sum_{m=0}^{N-1} \frac{X_R[m]^2 + X_I[m]^2}{N^2} \\
&= N\left(\sigma_R^2 + \sigma_I^2\right) = 2N\sigma_f^2
\end{aligned} \tag{7.22}$$

Clearly, $\sigma_f^2 = \sigma_t^2/(2N)$, and the variance of the real and imaginary parts of an N-point NPSD bin decreases by a factor of $1/(2N)$ relative to the time-domain variance for the NPSD. For the standard

PSD, half the variance of the real time domain signal ends up in the real and the other half ends up in the imaginary PSD bin while a sinusoid in that bin would increase by a factor of N. If the time-domain input was complex with variance σ_t^2, the factor of 1/2 goes away as now the imaginary part of the time-domain signal is not zero. If a Hanning or other nonrectangular data window is used in the NDFT calculation, a broadband correction factor (1.2247 for the Hanning window, see Table 6.1 for other window types) must be applied for the broadband variances to match the rectangular window levels. For the PSD, the variance in the real and imaginary bins is simply 1/2 σ_t^2 while a sinusoid is coherently amplified by a factor of N. If the input data were complex with real and imaginary components each with variance σ_t^2, the NPSD drop in variance is $1/N$. It can be seen that the spectrally white input noise variance is equally divided into each of the N DFT bins. Note that the NDFT scaling (and NPSD definition) used allows the mean-square value for a sinusoid in the time domain to be matched to the value in the corresponding NPSD bin. Therefore, the Fourier transform is seen to provide a narrowband SNR enhancement of $10 \log_{10}N$ dB. The larger the N is, the greater the SNR enhancement in the power spectrum will be. A more physical interpretation of N is the ratio of the sample rate f_s (in Hz) over the available resolution (in Hz), which is inversely proportional to the total integration time in seconds. Therefore, the SNR enhancement for stationary sinusoids in white noise is $10 \log_{10}N$, where N can be derived from the product of the sample rate and the total integration time. A 1024-point PSD or NPSD provides about 30 dB SNR improvement while a 128-point transform gives only 21 dB. PSD measurement is an extraordinary tool for enhancing the SNR and associated observability of stationary sinusoids in white noise whether the normalization is used or not.

7.2.3 SNR Enhancement and the Zoom FFT

The SNR enhancement for sinusoids in Gaussian noise is $10 \log_{10}N$ for an N-point FFT, DFT, NDFT, PSD, and NPSD. This useful property of the spectral transform is simply due to the way signal energy for sinusoids is concentrated into a single-bin white noise is divided over the bins covering the signal bandwidth. The factor of N we throw around (as well as narrowband correction factors for data windows) is to conveniently put the peak in the frequency domain into a calibrated level relative to the time-domain amplitude. The SNR effect of the FFT is simply a matter of how the noise is divided across the frequency bins, which enhances detection of sinusoidal signals. The same is true for the so-called "zoom FFT" where the spectral bandwidth goes from f_1 to f_2 rather than from 0 Hz to $f_s/2$, f_s being the sample rate. In a zoom FFT, like zooming in on an image, one is interested only in the bandwidth between f_1 and f_2. This is done using demodulation in either the analog domain or the digital domain. In the analog domain, the real input signal is multiplied by $\cos(2\pi f_1 t)$ to make a real input to the zoom FFT and the real input signal is multiplied by $\sin(2\pi f_1 t)$ to make an imaginary input. The signal multiplications in the analog domain will result in the sum and difference of frequencies with f_1, so that what would appear to be 0 Hz in the modulated signal is actually f_1. There will also be a high-frequency component at $2f_1$, but our scheme will filter the high-frequency images before digital processing. The sample rate needs to be at least $\Delta f = (f_2 - f_1)$ and the low pass antialiasing filter needs to have a corner frequency of less than the sample rate since we have a complex modulated signal. Since we have a complex input signal to the FFT we can use the entire FFT spectral result rather than just the lower half of the bins. The SNR enhancement and FFT resolution is a function of the size of the FFT and the modulated sample rate Δf. However, the low-pass filtering reduces the noise of the input by a factor of at least $\Delta f/2f_2$ since we have reduced the overall bandwidth of the input signal.

The same zoom FFT can be done in the digital domain given a proper low-pass signal sampled at a rate over $f_s > 2f_2$ so that it can be properly represented digitally. The input sequence $x[n]$ is then multiplied by $\cos(2\pi n f_1/f_s)$ for the real part and $\sin(2\pi n f_1/f_s)$ for the imaginary part. We now have a complex modulated signal sampled at a rate much higher than the bandwidth that we are interested in. By low-pass filtering the real and imaginary modulated signals we can decimate the sample rate to something higher than Δf samples/s. This filtering reduces the noise by a factor of about $\Delta f/2f_2$ as with the analog case. Then, an additional SNR gain of $10\log_{10}N$ is seen from the FFT itself.

7.2.4 CONVERSION OF RANDOM VARIABLES

A very useful technique for converting one ZMG random variable to another when there is a known algebraic relationship between the two is presented below. To observe an approximation to the expected value of the power spectrum we are clearly interested in the average magnitude-squared value and its statistics for a given Fourier transform bin (with no overlap of input buffers). These statistics for the NPSD will be based on a limited number of spectral averages of the magnitude-squared data in each bin. We start by assuming real and imaginary ZMG processes each with variance σ^2. The PDF of the square of a ZMG random variable is found by employing a straightforward change of variables. Deriving the new PDF (for the squared ZMG variable) is accomplished by letting $y = x^2$ be the new random variable. The probability that y is less than some value Y must be the same as the probability that x is between $\pm Y^{1/2}$.

$$P_Y(y \leq Y) = P_X(-\sqrt{Y} \leq x \leq +\sqrt{Y}) = \int_{-\sqrt{Y}}^{+\sqrt{Y}} p(x)\, \mathrm{d}x \qquad (7.23)$$

Differentiating Equation 7.23 with respect to Y gives the PDF for the new random variable $y = x^2$.

$$\begin{aligned} p(Y) &= p(x = +\sqrt{Y})\frac{\mathrm{d}(+\sqrt{Y})}{\mathrm{d}Y} - p(x = -\sqrt{Y})\frac{\mathrm{d}(-\sqrt{Y})}{\mathrm{d}Y} \\ &= \frac{1}{2\sqrt{Y}}\left\{ p(x = +\sqrt{Y}) - p(x = -\sqrt{Y}) \right\} \quad Y \geq 0 \end{aligned} \qquad (7.24)$$

Since the Gaussian PDF is symmetric (it is an even function), we can write the new PDF simply as

$$p(y) = \frac{1}{\sigma\sqrt{2\pi y}}\, \mathrm{e}^{\left(-y/2\sigma^2\right)} \qquad (7.25)$$

The PDF in Equation 7.25 is known as a Chi-Square PDF of order $v = 1$ and is denoted here as $p(x^2|v = 1)$. The expected value for y (the mean of y) is now the variance for x since $E\{y\} = E\{x^2\} = \sigma^2$. The variance for the Chi-Square process is

$$\begin{aligned} \sigma_y^2 &= E\{y^2\} - (E\{y\})^2 \\ &= E\{x^4\} - (E\{x^2\})^2 \\ &= 3\sigma^4 - \sigma^4 = 2\sigma^4 \end{aligned} \qquad (7.26)$$

The fourth moment on the ZMG variable x in Equation 7.26 is conveniently found using the even central moments relationship for Gaussian PDFs (the odd central moments are all zero). For the nth central moment (n even),

$$E\{(x - \bar{x})^n\} = 1 \cdot 3 \cdot 5 \ldots (n-1)\, \sigma^n \qquad (7.27)$$

where $\bar{x}$ is the mean of x. So for $n = 4$, Equation 7.27 equals $3\sigma^4$.

We now consider the sum of two squared ZMG variables as is done to compute the magnitude-squared in a bin for the PSD. This PDF is a Chi-Square distribution of order $v = 2$. Equation 7.28 gives the general Chi-Square for v degrees of freedom.

$$p(x^2 \mid v) = \frac{\left(y/\sigma^2\right)^{v/2-1}}{\sigma^2 2^{v/2} \Gamma\left(v/2\right)} e^{-y/2\sigma^2}, \quad y = \sum_{n=1}^{v} x_n^2 \tag{7.28}$$

The means and variances for the two degrees of freedom Chi-Square process simply add up to twice that for the $v = 1$ case. The gamma function $\Gamma(v/2)$ equals $\pi^{1/2}$ for the $v = 1$ case and equals $(M - 1)!$ for $M = 2v$. Since we are interested in the average of pairs of squared ZMG variables, we introduce a PDF for the average of M pairs of NPSD bins.

$$p(z \mid M) = \frac{M}{\sigma^2} \frac{\left(Mz/\sigma^2\right)^{M-1}}{2^M (M-1)!} e^{-Mz/2\sigma^2} \quad z = \frac{1}{M} \sum_{n=1}^{M} (x_{R_n}^2 + x_{I_n}^2) \tag{7.29}$$

Figure 7.2 shows a number of the Chi-Square family PDF plots for an underlying ZMG process with unity variance. The $v = 1$ and $v = 2$ ($M = 1$) cases are classic Chi-Square. However, the $M = 2$, 4, and 32 cases use the averaging PDF in Equation 7.29. Clearly, as one averages more statistical samples, the PDF tends toward a Gaussian distribution as the central limit theorem predicts. Also for the averaging cases ($M > 1$), note how the mean stays the same (2 since $\sigma^2 = 1$) while the variance decreases with increasing M. In the limit as M approaches infinity, the PDF of Equation 7.29 takes on the appearance of a Dirac delta function.

It is important to note that the $v = 1$ case has a mean of $\sigma^2 = 1$, while all the other means are $2\sigma^2 = 2$. This $2\sigma^2$ mean for the NDFT bin relates to the time-domain variance by a factor of $1/(2N)$ for real data as mentioned above. It may not at first appear to be true that the mean values are all the same for the spectral averaging cases ($M = 1$, 2, 4, and 32). However, even though the

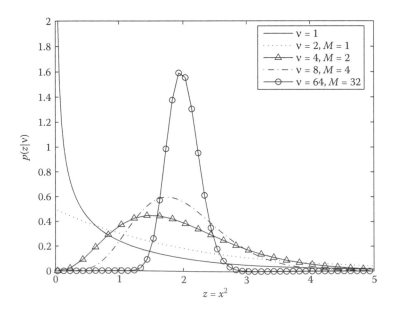

FIGURE 7.2 The family of Chi-Square PDFs useful for determining spectral averaging ($M > 1$) statistics.

skewness (asymmetry) of the spectral averaging PDFs is quite significant for small M as shown in Figure 7.2, the curves really flatten out as x approaches infinity adding a significant area to the probability integral. The variance for the $v = 1$ case is $2\sigma^4$ while the variances for the averaged cases ($M = 1, 2, 4,$ and 32) are found to be $4\sigma^4/M$ which can be clearly seen in Figure 7.2. Figure 7.3 shows the probability distribution curves corresponding to the PDFs in Figure 7.2. Equation 7.30 gives the probability of finding a value less than x^2. However, this solution is somewhat cumbersome to evaluate because the approximate convergence of the series depends on the magnitude of x^2.

$$P(x^2 \mid v) = \frac{1}{\Gamma(v/2)} \sum_{n=0}^{\infty} \frac{(-1)^n (x^2/2)^{(v/2+n)}}{n!(v/2+n)} \tag{7.30}$$

A forensic application of this analysis can be seen when one examines the apparent statistics of random noise amplitudes in the frequency domain. One can compare the mean and standard deviation of a group of frequency bins and determine the approximate number of spectral averages used. Suppose about 100 bins in the DFT or PSD (or normalized versions) are apparently from a ZMG "white" noise source. This means that the energy in each frequency bin is statistically the same. One could estimate the PDF by making a histogram of the 100 white-noise bin amplitudes. The mean of the 100 bins would be $2\sigma^2$ as expected, but the variance would be $4\sigma^4/M$, where M is the number of spectral averages. Note that increasing the number of spectral averages does not increase the SNR; it just smoothens out the noise bins. To increase the SNR one would need to increase the FFT size.

7.2.5 Confidence Intervals For Averaged NPSD Bins

What we are really interested in for spectrally averaged Fourier transforms is the probability of having the value of a particular NPSD bin to fall within a certain range of the mean. For example, for the $M = 32$ averages case in Figure 7.3, the probability of the value in the bin lying below 2.5 is

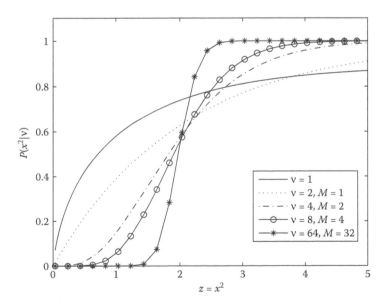

FIGURE 7.3   Probability distributions numerically integrated from the PDFs in Figure 7.2 where $\sigma = 1$.

about 0.99. The probability of the value being less than 1.5 is about 0.02. Therefore, the probability of the averaged value for the bin being between 1.5 and 2.5 is about 97%. With only four averages, the probability of the bin value being between 1.5 and 2.5 drops to 0.80–0.30, or about 50%, and with only one NPSD bin ($M = 1$), the probability is less than 20%. Clearly, spectral averaging of random data significantly reduces the variance, but not the mean, of the spectral bin data. The only way to reduce the mean of the noise (relative to a sinusoid) is to increase the resolution of the underlying FFT in the NPSD.

One statistical measure that is often very useful is a *confidence interval* for the spectral bin. A confidence interval is simply a range of levels for the spectral bin and the associated probability of having a value in that range. The 99% confidence interval for the $M = 2$ case could be from 0.04 to 4.9, while for $M = 4$, the interval narrows to 0.08–3.7, and so on. It can be seen as even more convenient to express the interval in dB about the mean value. Figure 7.4 shows the probability curves on dB value x-axis scale where 0 dB is the mean of $2\sigma^2 = 2.0$. Clearly, using Figure 7.4 as a graphical aid, one can easily determine the probability for a wide range of dB. For example, the ±3 dB confidence for the $M = 4$ case is approximately 98%, while for the $M = 1$ case it is only about 50%. This metric is useful for determining the number of spectral averages needed to achieve a desired probabilistic confidence for each spectral bin.

7.2.6 SYNCHRONOUS TIME AVERAGING

We can see that spectral averaging in the frequency domain reduces the variance of the random bin data as the number of complex magnitude-squared M averages increases. This approach is very useful for "cleaning up" spectral data, but it does not allow an increase in narrowband SNR beyond what the Fourier transform offers. However, if we know that the frequency to be detected has period T_p, we can synchronously average the time data such that only frequencies synchronous with $f_p = 1/T_p$ (i.e., f_p, $2f_p$, $3f_p$, ...) will remain. All noncoherent frequencies including ZMG random noise will average to zero. This is an extremely effective technique for enhancing known signal detectability. Examples are often found in active sonar and radar, vibration monitoring of rotating equipment such as bearings in machinery, and measurements of known periodic signals in high noise levels. For the case of

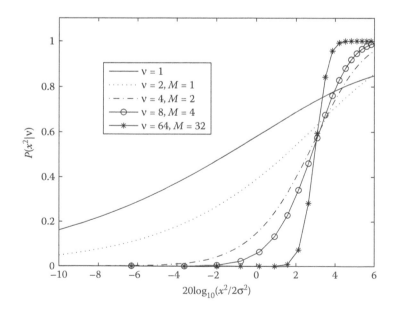

FIGURE 7.4 Probability distributions for the Chi-Square-averaged PDFs but with x^2 on a $10\log_{10}$ scale relative to the mean of 2σ.

rotating equipment, a tachometer signal can provide a trigger to synchronize all the averages such that shaft-rate-related vibrations would all be greatly enhanced while other periodic vibrations from other shafts would be suppressed. The averaging process has the effect of reducing the noncoherent signals by a factor of $1/N$ for N averages. Therefore, to increase the detectability by 40 dB, one would need at least 100 coherent time averages of the signal of interest. Synchronous averaging is often used where a repeatable signal can be precisely measured over and over again, such as response to an active transmission in sonar, radar, or medical imaging. With the background noise independent of the transmissions, the signal is enhanced significantly by the synchronous averaging process.

7.2.7 HIGHER-ORDER MOMENTS

Beyond the mean and mean-square (first and second general moments) of a PDF, there are several additional moments and moment sequences which are of distinct values to intelligent signal processing systems. In general, these "higher-order" statistics and their corresponding spectra each provide insightful features for statistically describing the data set of interest. We will start by considering the nth general moment of the probability distribution $p(x)$.

$$\overline{(X)^n} = E\left\{X^n\right\} = \int_{-\infty}^{+\infty} x^n p(x)\, dx \tag{7.31}$$

The central moment is defined as the moment of the difference of a random variable and its mean as seen in Equation 7.32. The central moments of a random variable are preferred because in many cases a zero-mean random variable will result in much simpler mathematical derivations.

$$\overline{(X - \overline{X})^n} = E\left\{(X - \overline{X})^n\right\} = \int_{-\infty}^{+\infty} (x - \overline{X})^n p(x)\, dx \tag{7.32}$$

As noted earlier, the second central moment is the variance σ^2. The third central moment leads one to the *skewness* which is typically defined with the difference between the random variable and its mean normalized by the standard deviation to give a dimensionless quantity.

$$S = E\left\{\left(\frac{X - \overline{X}}{\sigma}\right)^3\right\} = \int_{-\infty}^{+\infty} \left(\frac{x - \overline{X}}{\sigma}\right)^3 p(x)\, dx \tag{7.33}$$

A positive skewness means that the distribution tail extends out more in the positive x-direction than in the negative direction. Hence, a positively skewed distribution tends to lean "toward the right," while a negatively skewed PDF leans toward the left. The normalized fourth central moment is known as the *kurtosis* and is given in Equation 7.34 below.

$$K = E\left\{\left(\frac{X - \overline{X}}{\sigma}\right)^4\right\} - 3 \tag{7.34}$$

The kurtosis is a measure of how "peaky" the PDF is around its mean. A strongly positive kurtosis indicates a PDF with a very sharp peak at the mean and is called (by very few people) *leptokurtic*. A negative kurtosis indicates that the PDF is very flat in the vicinity of the mean and is called (by even fewer people) *platykurtic*. The measures of skewness and kurtosis are essentially relative to the Gaussian PDF because it has zero skewness and a kurtosis of –1. The mean, variance, skewness, and kurtosis form a set of features which provide a reasonable description of the shape of a Gaussian-like unimodal curve (one which has just one bump). PDFs can be bimodal (or even more complicated on a bad day) such that other measures of the PDF must be taken for adequate algorithmic description.

7.2.8 CHARACTERISTIC FUNCTION

Given an analytic function for the PDF or a numerical histogram, the general moments can be calculated through a numerical or analytic integration of Equation 7.31. However, analytic integration of the form in Equation 7.31 can be quite difficult. An alternative analytic method for computing the central moments of a PDF is the *characteristic function* $\varphi(u)$. The characteristic function of a random variable X is simply $\varphi(u) = E\{e^{juX}\}$.

$$\varphi(u) = E\left\{e^{juX}\right\} = \int_{-\infty}^{+\infty} p(x)\, e^{+juX}\, dx \tag{7.35}$$

It can be seen that, except for the positive exponent, the characteristic function is a Fourier transform of the PDF. Conversely, if we are given a characteristic function, the "negative exponent" inverse transform gives back the PDF.

$$p(x) = \frac{1}{2\pi} \int_{-\infty}^{+\infty} \varphi(u)\, e^{-juX}\, du \tag{7.36}$$

The sign in the exponent is really more of a historical convention than a requirement for the concept to work. To see the "trick" of the characteristic function, simply differentiate Equation 7.35 with respect to u and evaluate at $u = 0$.

$$\left.\frac{d\phi(u)}{du}\right|_{u=0} = j \int_{-\infty}^{+\infty} x p(x)\, dx = j\bar{X} \tag{7.37}$$

As one might expect, the nth general moment is found simply by

$$\overline{X^n} = E\left\{X^n\right\} = \frac{1}{j^n} \left.\frac{d^n\phi(u)}{du^n}\right|_{u=0} \tag{7.38}$$

Joint characteristic functions can be found by doing two-dimensional Fourier transforms of the respective joint PDF and the joint nth general moment $(n = i + k)$ can be found using the simple formula in Equation 7.39. It is not the first time a problem is much easier solved in the frequency domain.

$$\overline{X^i Y^k} = E\left\{X^i Y^k\right\} = \frac{1}{j^{i+k}} \left.\frac{\partial^{i+k}\varphi_{XY}(u,v)}{\partial u^i \partial v^k}\right|_{u=v=0} \tag{7.39}$$

7.2.9 CUMULANTS AND POLYSPECTRA

The general moments described above are really just the zeroth lag of a moment time sequence defined by

$$m_n^x(\tau_1, \tau_2, \ldots \tau_{n-1}) = E\left\{x(k)\, x(k+\tau_1)\, \ldots\, x(k+\tau_{n-1})\right\} \tag{7.40}$$

Therefore, one can see that $m_1^x = E\{x\}$, or just the mean, while $m_2^x(0)$ is the mean-square value for the random variable $x(k)$. The sequence $m_2^x(\tau)$ is defined as the autocorrelation of $x(k)$. The *second-order cumulant* is seen as the *covariance sequence* since the mean is subtracted from the mean-square value [2].

$$c_2^x(\tau_1) = m_2^x(\tau_1) - (m_1^x)^2 \tag{7.41}$$

The zeroth lag of the covariance sequence is simply the variance. The third-order cumulant is

$$c_3^x(\tau_1, \tau_2) = m_3^x(\tau_1, \tau_2) - m_1^x\left[m_2^x(\tau_1) + m_2^x(\tau_2) + m_2^x(\tau_1 - \tau_2)\right] + 2\,(m_1^x)^3 \tag{7.42}$$

The (0,0) lag of the third-order cumulant is actually the "unnormalized" skewness, or simply the skewness times the standard deviation cubed. As one might guess, the fourth-order cumulant (0,0,0) lag is the unnormalized kurtosis. The fourth-order cumulant can be written more compactly if we can assume a zero mean $m_1^x = 0$.

$$\begin{aligned}
c_4^x(\tau_1, \tau_2, \tau_3) = {}& m_4^x(\tau_1, \tau_2, \tau_3) - m_2^x(\tau_1)\,m_2^x(\tau_3 - \tau_2) \\
& - m_2^x(\tau_2)\,m_2^x(\tau_3 - \tau_1) - m_2^x(\tau_3)\,m_2^x(\tau_2 - \tau_1)
\end{aligned} \tag{7.43}$$

These cumulants are very useful for examining just how the "Gaussian" random noise process is, as well as, for discriminating linear processes from nonlinear processes. To see this we will examine the spectra of the cumulants known as the *power spectrum* for the second cumulant, *bispectrum* for the third cumulant, *trispectrum* for the fourth cumulant, and so on. While the investigation of *polyspectra* is a relatively new area in signal processing, the result of the power spectrum being the Fourier transform of the second cumulant is well-known as the Weiner–Khintchine theorem. Consider the bispectrum which is a two-dimensional Fourier transform of the third cumulant.

$$C_3^x(\omega_1, \omega_2) = \sum_{\tau_1=-\infty}^{+\infty}\sum_{\tau_2=-\infty}^{+\infty} c_3^x(\tau_1, \tau_2)\, e^{-j(\omega_1\tau_1 + \omega_2\tau_2)} \tag{7.44}$$

$$|\omega_1|, |\omega_2|, |\omega_1 + \omega_2| \le \pi$$

It can be shown that the bispectrum in the first octant of the ω_1, ω_2 plane ($\omega_2 > 0$, $\omega_1 \ge \omega_2$), bounded by $\omega_1 + \omega_2 \le \pi$ to ensure no aliasing, is actually all that is needed because of a high degree of symmetry. The trispectrum is a three-dimensional Fourier transform of the fourth cumulant.

$$C_4^x(\omega_1, \omega_2, \omega_3) = \sum_{\tau_1=-\infty}^{+\infty}\sum_{\tau_2=-\infty}^{+\infty}\sum_{\tau_3=-\infty}^{+\infty} c_4^x(\tau_1, \tau_2, \tau_3)\, e^{-j(\omega_1\tau_1 + \omega_2\tau_2 + \omega_3\tau_3)} \tag{7.45}$$

$$|\omega_1|, |\omega_2|, |\omega_3|, |\omega_1 + \omega_2 + \omega_3| \le \pi$$

The trispectrum covers a three-dimensional volume in ω and is reported to have 96 symmetry regions. Do not try to implement Equations 7.44 and 7.45 as written on a computer. There is a far more computationally efficient way to compute polyspectra without any penalty, given the N-point

DFT of $x(k)$, $X(\omega)$. While the power spectrum (or PSD) is well known to be $C_2^x(\omega) = X(\omega)X^*(\omega)/N$ the bispectrum and trispectrum can be computed as

$$C_3^x(\omega_1, \omega_2) = \frac{1}{N} X(\omega_1) X(\omega_2) X^*(\omega_1 + \omega_2)$$

$$C_4^x(\omega_1, \omega_2, \omega_3) = \frac{1}{N} X(\omega_1) X(\omega_2) X(\omega_3) X^*(\omega_1 + \omega_2 + \omega_3)$$

(7.46)

which is a rather trivial calculation compared to Equations 7.42 through 7.45. For purposes of polyspectral analysis, the factor of $1/N$ is not really critical and is included here for compatibility with power spectra of periodic signals.

Let us take an example of a pair of sinusoids filtered through a nonlinear process and use the bispectrum to analyze the frequencies that are the result of the nonlinearity. The bispectrum allows frequency components that are phase-coupled to coherently integrate, making them easy to identify. Let $x(k)$ be sampled at $f_s = 256$ Hz and

$$x(k) = A_1 \cos(2\pi \frac{f_1}{f_s} k + \theta_1) + A_2 \cos(2\pi \frac{f_2}{f_s} k + \theta_2) + w(k)$$

(7.47)

where $A_1 = 12$, $f_1 = 10$, $\theta_1 = 45°$, $A_2 = 10$, $f_2 = 14$, $\theta_1 = 170°$, and $w(k)$ is a ZMG noise process with standard deviation 0.01. Our nonlinear process for this example is

$$z(k) = x(k) + \varepsilon x(k)^2$$

(7.48)

where $\varepsilon = 0.035$, or a small 3.5% quadratic distortion of $x(k)$. The nonlinearity should produce a harmonic for each of the two frequencies (20 and 28 Hz) and sum and difference tones at 24 and 4 Hz. Figure 7.5 shows the original undistorted time waveform of $x(k)$ in the solid curve and the

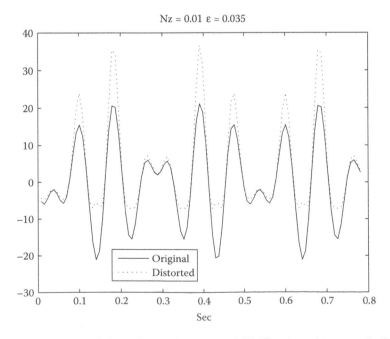

FIGURE 7.5 Input signal consisting of two sinusoids and ZMG noise with a small (3.5%) quadratic nonlinearity.

distorted waveform from Equation 7.48 in the dotted curve. Figure 7.6 shows the NPSD for each signal where a 256-point FFT was used. Note that all one needs to do is multiply the magnitude-squared time average of the FFT by $(2/N)^2$ to get the one-sided NPSD (except for the 0 Hz bin which gets an additional factor of 1/2). The NPSD in Figure 7.6 shows the 10 Hz peak at 21.58 dB which corresponds to $10\log_{10}(144)$ and the 14 Hz peak at exactly 20 dB, the correct NPSD values for amplitudes of 12 and 10, respectively. The sample rate is 128 Hz, so both sinusoids are perfectly bin aligned in the example, and thus no data window is used. The noise level is about −60 dB which is about right for $10\log_{10}(2(0.01)^2/256) = -61.07$ dB. The generated harmonics are clearly shown in Figure 7.6 from the small 3.5% quadratic nonlinearity.

If one calculates the bispectrum in the top line of Equation 7.46 one gets a visually interesting two-dimensional plot showing high levels of correlations between the various frequencies. But perhaps even more interesting from an analysis perspective is the bispectrum along the baseline of one of the frequency dimensions as shown in Figure 7.7. What can be seen is an overall shift upward of the distorted spectrum relative to the undistorted. Also one can see a peak near 0 Hz for the distorted spectrum. What does this mean? The peak near 0 Hz comes from the even-power of the distortion. In Equation 7.48 if the signal is cubed instead of squaring it the 0 Hz peak would go away, but the level shift remains. The amount of shift between the undistorted and distorted spectra increases with increased SNR as well as increased distortion. By comparing the bispectrum baseline in Figure 7.7 to the NPSD in Figure 7.6 we can tell that we have an expansion type of nonlinearity (from the elevated levels in Figure 7.7) and that the nonlinearity contains an even-order (from the peak near 0 Hz in the distorted bispectrum).

Consider what is going on with the nonlinearity. A multiplication in the time domain is equivalent to a convolution in the frequency domain. So, when we add the nonlinearity as in Equation 7.48, the spectrum is convolved with itself and added back to the original spectrum in the frequency domain. This means that the frequency bins are no longer independent and orthogonal. Since the distortion adds to the signal amplitude, the bispectral shift is upward and the nonlinearity is dynamically *expansive* of the signal. When one computes the bispectrum, there is now a marked increase in the cross-correlation between frequency bins adding to the original spectra. By examining the baseline where one of the frequencies is 0 Hz, the effect is a simple offset in level. It is a bit more complex, but gives two-dimensional plots, when one examines the bispectrum for all combinations

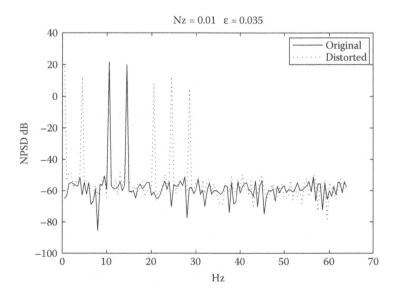

FIGURE 7.6 NPSD of the original input signal and the input with the 3.5% quadratic distortion.

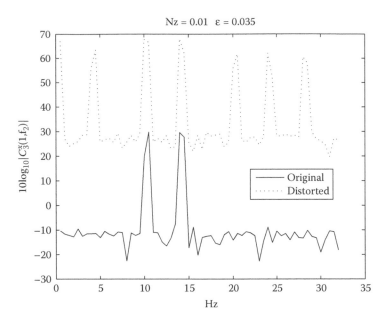

FIGURE 7.7 Bispectrum of the original and distorted input signal examined along the baseline $f_1 = 0$ and showing the effect of the expansive quadratic nonlinearity.

of frequencies. The added complexity comes from the relative levels at each frequency, and so it is difficult to determine if one is looking at a collection of harmonically related frequencies or a nonlinear effect.

Now let us consider a much more complicated but very common distortion called *saturation y distortion*. Using the same input signal we will introduce a distortion model using the formula

$$z(k) = x(k) + \alpha \tan^{-1} \left\{ \beta x(k) \right\} \tag{7.49}$$

where $\alpha = 10$ and $\beta = 0.05$. This type of distortion is typical of overdriving an amplifier or loudspeaker where the output signal is "soft clipped" by the limited amplifier power or loudspeaker excursion. The inverse tangent, or arctangent, can be approximated using a Taylor series, and so like Equation 7.48, Equation 7.49 now has a more complicated series of nonlinearities added into the original spectrum in the frequency domain. However, this time the nonlinearity is dynamically *compressive*, not expansive, in that it limits the signal amplitude and spectral levels. Figure 7.8 shows the original and distorted waveforms. Figure 7.9 shows the baseline bispectrum. Now we see the distorted spectrum lower than the undistorted. This is because of the compressive distortion shown in Figure 7.8. We also see intermodulation of the two frequencies, that is, harmonics that are spaced by the difference in frequency. This can be seen as a type of frequency modulation since the rate of change of phase is limited at the signal peaks. Clearly, the bispectrum and other higher-order statistical moments have uses in signal and system analysis.

Bispectral analysis of nonlinear systems in not just limited to audio systems. Recent articles in the literature show applications to electric motor faults [3], electroencephalograms (EEGs) [4], and even signal forensics [5]. EEG is an electrical measurement of neural activity in the brain and neural pathways. Evoked potentials (ones stimulated by an external stimulus) and spontaneous potentials are relatively slow-moving electrochemical reactions through the body, some taking several hundred milliseconds to reach to or come from the brain. At the cellular level, these electrochemical

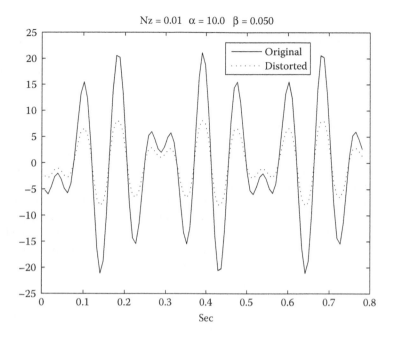

FIGURE 7.8 The original input of two sinusoids in ZMG noise and a compressive distortion typical of an over-driven amplifier or loudspeaker system.

transmissions from neuron to neutron depend on supply of potassium and other materials manufactured by the cells or supplied by the bloodstream. When there is overuse (from over stimulation), the neurotransmitter supplies are diminished and the neurons take longer to transmit, or fail to transmit altogether. As such, changes in transfer time and amplitude are measurable in the EEG and are quite nonlinear. The bispectrum can provide a means to monitor anesthesia, measure nerve damage or

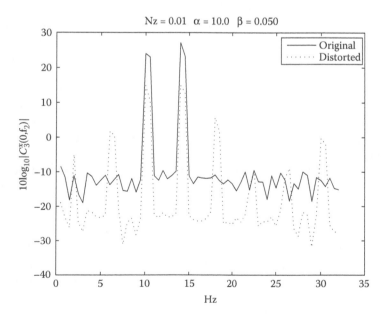

FIGURE 7.9 Baseline bispectrum of the original and compressive distortion signal showing the lowering of the signal level as well as the intermodulation products from the distortion.

recovery, or perhaps even measure pain. In instances such as signal forensics, in particular tampering with speech and photographic evidence, the bispectrum can be used to compare sections of the signal or detect a sudden break in the signal statistics, such as when one substitutes or deletes a portion of the signal. Clearly, these are very important higher-order statistical techniques for the future of signal processing with intelligent sensor systems.

7.3   TRANSFER FUNCTIONS AND SPECTRAL COHERENCE

Perhaps the most common method to measure the frequency response of a transducer or electrical/mechanical system is to compute the transfer function from the simultaneous ratio of the output signal spectrum to the input signal spectrum. For modal analysis of structures, the classic approach is to strike the structure with an instrumented hammer (it has an embedded force-measuring accelerometer), and record the input force impulse signal simultaneously with the surface-acceleration response output signal(s) at various points of interest on the structure. The ratio of the Fourier transforms of an output signal of interest to the force impulse input provides a measure of the mechanical force-to-acceleration transfer function between the two particular points. Noting that velocity is the integral of acceleration, one can divide the acceleration spectrum by "jω" to obtain a velocity spectrum except at 0 Hz). The ratio of velocity over force is the mechanical mobility (the inverse of mechanical impedance). An even older approach to measuring transfer functions involves the use of a swept sinusoid as the input signal and an RMS level for the output magnitude and a phase meter level for the system phase. The advantage of the swept sinusoid is a high-controllable SNR and the ability to synchronously average the signals in the time domain to increase the SNR. The disadvantage is that the sweep rate should be very slow in the regions of sharp peaks (resonances) and dips (antiresonances) in the response function and tracking filters must be used to eliminate harmonic distortion. More recently, random noise is used as the input and the steady-state response is estimated using spectral averaging.

Once the overall sample propagation time delay is separated from the transfer function to give a minimum phase response, the frequencies where the phase is 0 or π (the imaginary part is zero) indicate the system modal resonances and antiresonances, depending on how the system is defined. For mechanical input impedance (force over velocity) at the hammer input force location, a spectral peak indicates that a large input force is needed to have a significant structural velocity response at the output location on the structure. A spectral "dip" in the impedance response means that a very small force at that frequency will provide a large velocity response at the output location. It can therefore be seen that the frequencies of the peaks of the mechanical mobility transfer function are the structural resonances and the dips are the antiresonances. Recalling from system partial fraction expansions in Chapter 2, the frequencies of resonance are the system modes (poles) while the phases between modes determine the frequencies of the spectral dips, or zeros (antiresonances). Measurement of the mechanical system's modal response is essential to design optimization for vibration isolation (where mobility zeros are desired at force-input locations and frequencies) or vibration-communication optimization for transduction systems. For example, one needs a flat (no peaks or dips) response for good broadband transduction, but using a structural resonance as a narrowband "mechanical amplifier" is also often done to improve transducer efficiency and sensitivity.

It is anticipated that adaptive signal processing systems will be used well into the twenty-first century for monitoring structural modes for changes which may indicate a likely failure due to structural fatigue or damage, as well as to build in a sensor-health "self-awareness," or sentient capability for the monitoring system. It is therefore imperative that we establish the foundations of transfer function measurement especially in the area of measurement errors and error characterization. Consider the case of a general input spectrum $X(f)$, system response $H(f)$, and a coherent output spectrum $Y(f)$ as shown in Figure 7.10.

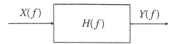

FIGURE 7.10   A linear time-invariant system $H(f)$ to be measured using the spectrum of the input $X(f)$ and the spectrum of the corresponding output $Y(f)$.

Clearly, $H(f) = Y(f)/X(f)$, but for several important reasons, we need to express $H(f)$ in terms of expected values of the ratio of the short-time Fourier transforms $X_n(f,T)$, and $Y_n(f,T)$ integrated over the time interval T where each block "n" will be averaged.

$$X_n(f,T) = \int_0^T x_n(t)\, e^{-j2\pi ft}\, dt \tag{7.50}$$

The data $x_n(t)$ in Equation 7.50 can be seen as being sliced up into blocks, T seconds long which will then be Fourier transformed and subsequently averaged [6]. This approach for spectral averaging is based on the property of *ergodic random processes*, where the expected values can be estimated from a finite number of averages. This requires that the signals be stationary. For example, the blows of the modal analysis hammer will in general not be exactly reproducible. However, the average of, say 32 blows of the hammer should provide the same average force impulse input as the average of thousands of hammer blows if the random variability (human mechanics, background noise, initial conditions, etc.) is ergodic. The same is true for the PSD Gaussian random noise where each DFT bin is a complex Gaussian random variable. However, after averaging the NDFT bins over a finite period of time (say $n = 32$ blocks or $32T$ s), the measured level in the bin will tend toward the expected mean as shown in Figure 7.2 for the cases of $M = 1$ and $M = 32$ and the autospectrum of a ZMG noise signal (Figure 7.11).

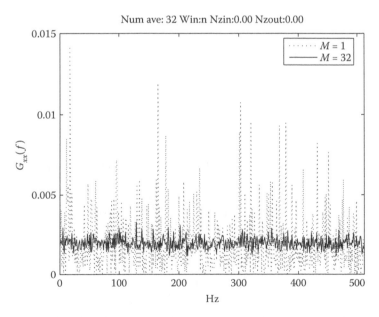

FIGURE 7.11   Comparison of the 1 Hz resolution autospectrum of unity variance ZMG noise using 32 averages (solid curve) and no averages (dotted curve).

Given the resolution and frequency span in Figure 7.11, it can be seen that the underlying NDFT is 1024-point. How? The frequency span is 512 Hz and the resolution is 1 Hz, indicating that 1 s of data is in the FFT input buffer and the sample rate is 1024 Hz, thus making the FFT size 1024 point. For a unity variance ZMG real time-domain signal, we expect the normalized power spectrum to have a mean $2\sigma^2 = 2/N$, or approximately 0.002. The bin variance of the $M = 32$ power spectrum is approximately 1/32 of the $M = 1$ case. See how easy it is to really read into a properly labeled spectral plot? It can be alarming how often spectral plots are poorly labeled and how common it is for signal processors to ignore these important details. For the transfer function measurement, $H_{xy}(f)$, we can exploit the ergodic random processes by estimating our transfer function as the expected value of the ratio of Fourier transforms.

$$H_{xy}(f) = E\left\{\frac{Y_n(f,T)}{X_n(f,T)}\right\} \tag{7.51}$$

The expected value in Equation 7.51 is the average of the ratio of the output spectrum over the input spectrum. It is not the ratio of the average output spectrum over the average input spectrum which does not preserve the input–output phase response. However there is an unnecessary computational burden with Equation 7.51 in that complex divides are very computationally demanding. But, if we premultiply the numerator and the denominator by $2X_n^*(f,T)$ we can define the transfer function as the ratio of the input–output cross-spectrum $G_{xy}(f)$ over the input autospectrum (referred to earlier as the one-sided power spectrum) $G_{xx}(f)$.

$$\begin{aligned}
H_{xy}(f) &= E\left\{\frac{2\,X_n^*(f,T)\,Y_n(f,T)}{2\,X_n^*(f,T)\,X_n(f,T)}\right\} \\
&= \frac{E\{2\,X_n^*(f,T)\,Y_n(f,T)\}}{E\{2\,X_n^*(f,T)\,X_n(f,T)\}} \\
&= \frac{G_{xy}(f)}{G_{xx}(f)}
\end{aligned} \tag{7.52}$$

Why the factor of 2? So that it is consistent with a one-sided power spectrum for real signals. Recall that the two-sided power spectrum $S_X(f)$ for real $x(t)$ is defined as

$$S_X(f) = \lim_{T\to\infty}\left\{\frac{|X(f,T)|^2}{T}\right\} + \lim_{T\to\infty}\left\{\frac{|X(-f,T)|^2}{T}\right\} \tag{7.53}$$

and the one-sided power spectrum is $G_{xx}(f)$, also known as the autospectrum of $x(t)$, is

$$\begin{aligned}
G_{xx}(f) &= 2\lim_{T\to\infty}\left\{\frac{|X(f,T)|^2}{T}\right\}; \quad f > 0 \\
&= \lim_{T\to\infty}\left\{\frac{|X(f,T)|^2}{T}\right\}; \quad f = 0 \\
&= 0; \quad f < 0
\end{aligned} \tag{7.54}$$

so that the factor of 2 only really applies to the nonzero Hz FFT bins, but it divides out anyway.

The autospectrum gets its name from the fact that *a conjugate multiply $X^*(f)X(f)$ in the frequency domain is an autocorrelation, not convolution, in the time domain*. The autospectrum (or power spectrum), $G_{xx}(f)$, is the Fourier transform of the autocorrelation of $x(t)$ in the time domain.

$$G_{xx}(f) = 2 \int_{-\infty}^{+\infty} R^x(\tau) e^{-j2\pi f\tau} d\tau; \quad f > 0 \tag{7.55}$$

$$R^x(\tau) = E\{x(t)\, x(t+\tau)\}$$

Likewise, the input–output cross-spectrum $G_{xy}(f)$ is the Fourier transform of the cross-correlation of x and y in the time domain.

$$G_{xx}(f) = 2 \int_{-\infty}^{+\infty} R^{xy}(\tau) e^{-j2\pi f\tau} d\tau; \quad f > 0 \tag{7.56}$$

$$R^{xy}(\tau) = E\{x(t)\, y(t+\tau)\}$$

Equations 7.52, 7.55, and 7.56 allow for a computationally efficient methodology for computing the transfer function $H_{xy}(f)$. One simply maintains a running average of the cross-spectrum $G_{xy}(f)$ and autospectrum $G_{xx}(f)$. After the averaging process is complete, the two are divided only once, and bin by bin, to give the transfer function estimate $H_{xy}(f)$.

However, due to numerical errors which can be caused by spectral leakage, environmental noise, or even the noise from the LSB of the ADC, we will introduce a measurement parameter known as an *ordinary coherence function* for characterization of transfer function accuracy.

$$\gamma_{xy}^2(f) = \frac{|G_{xy}(f)|^2}{G_{xx}(f)\, G_{yy}(f)} \tag{7.57}$$

The coherence function will be unity if no extraneous noise is detected, leading to errors in the transfer function measurement. The exception to this is when only one average is computed giving a coherence estimate which is algebraically unity and therefore not a proper statistical estimate. For a reasonably large number of averages, the cross-spectrum will differ slightly from the input and output autospectra, if only from the noise in the A/D system. However, the system impulse response is longer than the FFT buffers, some response from the previous input buffer will still be reverberating into the time frame of the subsequent buffer, thus appearing as noise incoherent with the present input. Since the input is ZMG noise in most cases for system identification, the residual reverberation is uncorrelated with the current input, and as such, appears as noise to the transfer function measurement. A simple coherence measurement will identify this effect. Correction requires that the FFT buffers be at least as long as the system impulse response, giving corresponding frequency-domain resolution higher than what the system peaks require. Significant differences will be seen at frequencies where an estimation problem occurs. We will use the coherence function along with the number of nonoverlapping spectral averages to estimate error bounds for the magnitude and phase of the measured transfer function.

Consider the example of a simple digital filter with two resonances at 60.5 and 374.5 Hz, and an antiresonance at 240.5 Hz, where the sample rate is 1024 Hz and the input is ZMG with unity variance. Figure 7.12 shows the measured transfer function (dotted line), true transfer function (solid line), and measured coherence for a single buffer in the spectral average ($M = 1$).

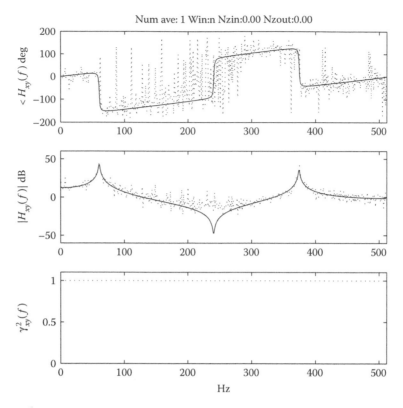

FIGURE 7.12 Transfer function phase (top), magnitude (middle), and coherence (bottom) calculated from the cross and autospectra but using only one buffer (no averaging).

Obviously, with only one pair of input and output buffers, the transfer function estimate is rather feeble due primarily to spectral leakage and nonuniform input noise. The coherence with only one average is algebraically unity (erroneously) since

$$
\begin{aligned}
\gamma_{xy}^2(f) &= \frac{|G_{xy}(f)|^2}{G_{xx}(f)\,G_{yy}(f)} \\
&\approx \frac{X_1^*(f)\,Y_1(f)\,Y_1^*(f)\,X_1(f)}{X_1^*(f)\,X_1(f)\,Y_1^*(f)\,Y_1(f)} = 1
\end{aligned}
\tag{7.58}
$$

Figure 7.13 gives the estimated result with only four averages. Considerable improvement can be seen in the frequency ranges where the output signal level is high. With only four averages, the cross-spectrum and autospectrum estimates begin to differ slightly where the output signal levels are high, and differ significantly where the spectral leakage dominates around the frequency range of the 240.5 Hz dip. The expected values can be approximated by sums.

$$
\gamma_{xy}^2(f) = \frac{\left|G_{xy}(f)\right|^2}{G_{xx}(f)\,G_{yy}(f)} \approx \frac{\left|\sum_{n=1}^{4} X_n^*(f)\,Y_n(f)\right|^2}{\left[\sum_{n=1}^{4} X_n^*(f)\,X_n(f)\right]\left[\sum_{n=1}^{4} Y_n^*(f)\,Y_n(f)\right]}
\tag{7.59}
$$

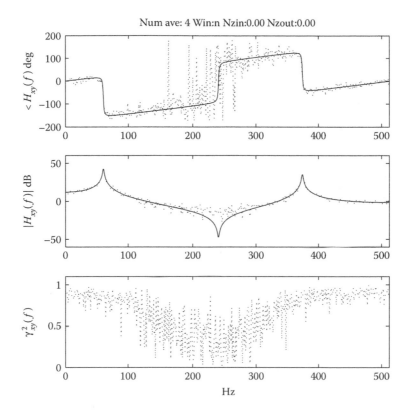

FIGURE 7.13 Transfer function phase (top), magnitude (middle), and coherence (bottom) calculated from the cross and autospectra but using four averaged buffers showing a more credible coherence.

Figure 7.14 clearly shows how the application of a Hanning data window significantly reduces the spectral leakage allowing the transfer function estimate to be accurately measured in the frequency range of the dip at 240.5 Hz. Note that the coherence drops slightly at the peak and dip frequencies due to the rapid system phase change there and the loss of resolution from the Hanning window. Clearly, the utility of a data window for transfer function measurements is seen in the improvement in the dip area and throughout the frequency response of the measured transfer function.

We now consider the real-world effects of uncorrelated input and output noise leaking into our sensors and causing errors in our transfer function measurement. Sources of uncorrelated noise are, at the very least, the random error in the LSB of the A/D process, but typically involve things like electromagnetic noise in amplifiers and intrinsic noise in transducers. A block diagram of the transfer function method including measurement noise is shown in Figure 7.15.

Consider the effect of the measurement noise on the estimated transfer function. Substituting $U(f) + Nx(f)$ for $X(f)$ and $V(f) + Ny(f)$ for $Y(f)$ in Equation 7.52 we have

$$H_{xy}(f) = \frac{E\{[U(f) + Nx(f)]^* [V(f) + Ny(f)]\}}{E\{[U(f) + Nx(f)]^* [U(f) + Nx(f)]\}} \tag{7.60}$$

We simplify the result if we can assume that the measurement noises $Nx(f)$ and $Ny(f)$ are uncorrelated with each other and uncorrelated with the input signal $U(f)$ and the output signal $V(f)$.

$$H_{xy}(f) = \frac{G_{UV}(f)}{G_{UU}(f) + G_{NxNx}(f)} \tag{7.61}$$

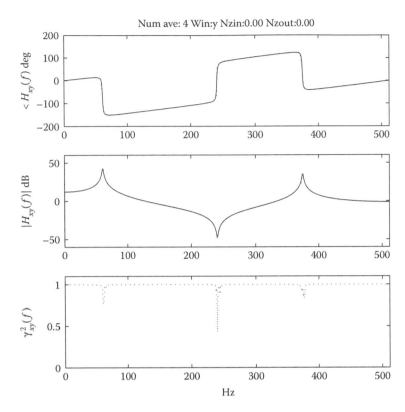

FIGURE 7.14 Same signals as in Figure 7.13 except that a Hanning window has been applied to the input and output signals removing spectral leakage in the low amplitude ranges, but introducing leakage at the peaks.

Equation 7.61 clearly shows a bias whenever the input SNR is low which causes the transfer function amplitude to be low. An overall random error is seen in Equation 7.60 when the output SNR is low. The ordinary coherence function also shows a random effect which reduces the coherence to <1 with uncorrelated input and output measurement noise.

$$\gamma_{xy}^2(f) = \frac{|G_{UV}(f)|^2}{[G_{UU}(f) + G_{NxNx}(f)][G_{VV}(f) + G_{NyNy}(f)]} \qquad (7.62)$$

Equation 7.62 clearly shows a breakdown in coherence whenever either the input or the output, or both SNRs become small. Figure 7.16 shows the effect of low-input SNR (both $U(f)$ and $Nx(f)$ are ZMG signals with unity variance) on the measured transfer function and coherence. The 0 dB

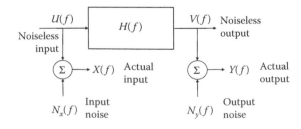

FIGURE 7.15 A general signal flow diagram showing a transfer function corrupted by input noise N_x and output noise N_y that do not pass through the measured system leading to estimation errors.

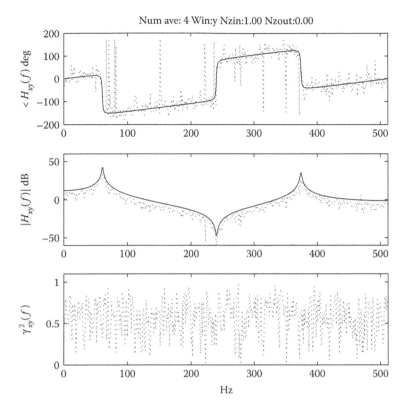

FIGURE 7.16 Transfer function phase (top), magnitude (middle), and coherence (bottom) calculated from the cross and autospectra using four averaged buffers, and a Hanning window, but with unity variance input noise added.

input SNR gives about a 3 dB bias to the transfer function magnitude, and an average coherence of about 0.5. The variances of the magnitude, phase, and coherence will be reduced if the number of averages in increased, but not the bias offsets.

While the low input SNR in Figure 7.16 is arguably a nearly worse case, and can easily be avoided with good data acquisition practice, it serves us to show the bias effect on the transfer-function magnitude. If the output measurement noise is ZMG, we should only expect problems in frequency ranges where the output of $H_{xy}(f)$ is low. Figure 7.17 shows the effect of unity variance ZMG output noise only for four averages when a Hanning window is used on both the input and the output data. Clearly, one can easily see the loss of coherence and the corresponding transfer function errors in the frequency region where the output of our system $H_{xy}(f)$ is low compared to the output measurement noise with unity variance. Since the output noise for our example is uncorrelated with the excitation input $U(f)$, it is possible to partially circumvent the low-output SNR by doing a large number of averages. Figure 7.18 shows the result for 32 averages which shows a mild improvement for the regions around 250 and 500 Hz, and an overall reduced spectral variance.

The situation depicted in Figure 7.18 is more typical of a real-world measurement scenario where the dynamic range of the system being measured exceeds the available SNR of the measurement system and environment. The errors in the transfer function magnitude and phase are due primarily to the breakdown in coherence as will be seen below. While one could arduously average the cross and input autospectra for quite some time to reduce the errors in the low-output frequency ranges, the best approach is usually to eliminate the source of the coherence losses.

The variance of the transfer function error is usually modeled by considering the case of output measurement noise (since the input measurement noise case is more easily controlled), and

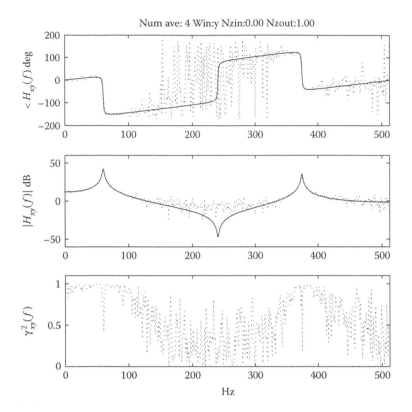

FIGURE 7.17 Transfer function phase (top), magnitude (middle), and coherence (bottom) calculated from the cross and autospectra using four averaged buffers, and a Hanning window, but with an added output noise with unity variance.

considering the output SNRs impact on the real and imaginary parts of the transfer function using the diagram depicted in Figure 7.19.

The circle in Figure 7.19 represents the standard deviation for the transfer function measurement which is a function of the number of averages n_d, the inverse of the output SNR, and the true amplitude of the transfer function represented by the "H" in the figure. Recall that for output noise only, the transfer function can be written as an expected value.

$$E\left\{H_{xy}(f)\right\} = H(f)\left(1 + E\left\{\frac{Ny(f)}{V(f)}\right\}\right) \tag{7.63}$$

The expected value $E\{Ny(f)/V(f)\}$ is zero because for a ZMG input $U(f)$ into the system, which is uncorrelated with the output noise $Ny(f)$, the real and imaginary bins of both $Ny(f)$ and $V(f)$ are ZMG processes. This centers the error circle around the actual $H(f)$ in Figure 7.19 where the radius of the circle is the standard deviation for the measurement. The mean magnitude-squared estimate for the transfer function is

$$E\left\{|H_{xy}(f)|^2\right\} = |H(f)|^2\left(1 + E\left\{\frac{Ny(f)}{V(f)}\right\} + E\left\{\frac{Ny^*(f)}{V^*(f)}\right\} + E\left\{\frac{Ny^*(f)Ny(f)}{V^*(f)V(f)}\right\}\right)$$

$$= |H(f)|^2\left(1 + \frac{G_{NyNy}(f)}{G_{VV}(f)}\right) \tag{7.64}$$

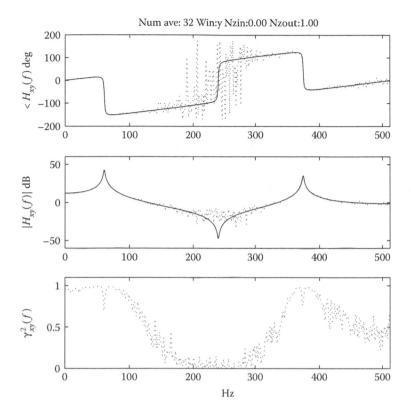

FIGURE 7.18 Same scenario as Figure 7.17 but with 32 averaged buffers.

The variance of the amplitude error for $H_{xy}(f)$ is therefore,

$$\sigma^2_{H_{xy}} = E\left\{\left|H_{xy}\right|^2\right\} - \left|H(f)\right|^2 = \frac{1}{2}\frac{G_{N_yN_y}(f)}{G_{VV}(f)}\left|H_{xy}(f)\right|^2\frac{1}{n_d} \tag{7.65}$$

where the factor of $1/(2n_d)$ takes into account the estimation error for the division of two autospectra averaged over n_d buffers. Clearly, $G_{N_yN_y}(f)/G_{VV}(f)$ is the magnitude-squared of the noise-to-signal ratio (NSR). We can write the NSR in terms of ordinary coherence by observing

$$G_{VV}(f) = |H_{xy}(f)|^2\ G_{xx}(f)$$

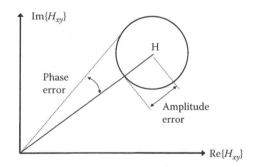

FIGURE 7.19 Phase diagram showing the magnitude and phase errors of the transfer function estimate due to added output noise interference.

$$
= \frac{|G_{xy}(f)|^2}{G_{xx}(f)G_{yy}(f)} \, G_{yy}(f)
$$

$$
= \gamma_{xy}^2(f) \, G_{yy}(f) \tag{7.66}
$$

and

$$
G_{N_yN_y}(f) = G_{yy}(f) - G_{VV}(f) = [1 - \gamma_{xy}^2(f)]G_{yy}(f) \tag{7.67}
$$

allowing the magnitude-squared of the NSR to be written as

$$
\frac{G_{N_yN_y}(f)}{G_{VV}(f)} = \frac{G_{N_yN_y}(f)}{G_{yy}(f)} \, \frac{G_{yy}(f)}{G_{VV}(f)} = \frac{[1 - \gamma_{xy}^2(f)]}{\gamma_{xy}^2(f)} \tag{7.68}
$$

We can also write the ordinary coherence function in terms of the output SNR magnitude squared

$$
\gamma_{xy}^2(f) = \frac{|SNR(f)|^2}{1 + |SNR(f)|^2} \tag{7.69}
$$

The transfer function amplitude error standard deviation (the circle radius in Figure 7.19), assuming there is no input measurement noise, is therefore

$$
\sigma_{H_{xy}} = \sqrt{\frac{1}{2\,n_d} \frac{[1 - \gamma_{xy}^2(f)]}{\gamma_{xy}^2(f)}} \, |H_{xy}(f)| \tag{7.70}
$$

The transfer function phase error standard deviation (the angle error in Figure 7.19) is found by an arctangent.

$$
\sigma_{\theta_{xy}} = \tan^{-1}\left\{ \sqrt{\frac{1}{2\,n_d} \frac{[1 - \gamma_{xy}^2(f)]}{\gamma_{xy}^2(f)}} \right\} \tag{7.71}
$$

Clearly, as the coherence drops, the accuracy expected in magnitude and phase also decreases. Spectral averaging improves the variance of the coherence, but does not restore coherence. As a consequence, compensating poor coherence in the transfer function input–output signals with increased spectral averaging has only a limited benefit. For example, where the coherence is below 0.1, the model estimated magnitude standard deviation is about 0.375 times the transfer function magnitude, or about −20 to −40 dB, as confirmed in Figure 7.18. The phase-error standard deviation is only about 20° and approaches 90° as the coherence approaches zero. This too is confirmed in Figure 7.18 although some of the more significant phase jumps are due to the wrapping of the phase angle to fit in the graph.

For systems where the impulse response is effectively longer than the FFT buffers, a loss of coherence occurs because some of the transfer function output is due to input signal before the current input buffer, and therefore is incoherent due to the random nature of the noise. It can be seen in these cases that the spectral resolution of the FFTs is inadequate to precisely model the sharp system resonances. However, a more problematic feature is that the measured magnitude and phase of a randomly excited transfer function will be inconsistent when the system impulse response is longer than the FFT buffers. If the FFT buffers cannot be increased, one solution is to use a chirped sinusoid as input, where the frequency sweep covers the entire measurement frequency range within a single FFT input buffer. Input–output coherence will be very high allowing a consistent measured transfer function, even if the spectral resolution of the FFT is less than optimal. The swept-sine technique also requires a tracking filter to insure that only the fundamental of the sinusoid is used for the transfer function measurement. However, tracking harmonics of the input fundamental is the preferred technique to measuring harmonic distortion. Again, the bandwidth of the moving tracking filter determines the amount of time delay measurable for a particular frequency. Accurate measurements of real systems can require very slow frequency sweeps to observe all the transfer function dynamics.

For applications where input and output noise are a serious problem, time-synchronous averaging of a periodic input–output signal can be used to increase SNR before the FFTs are computed. The time buffers are synchronized so that the input buffer recording starts at exactly the same place on the input signal waveform with each repetition of the input excitation. The simultaneously triggered and recorded input–output buffers may then be averaged in the time domain greatly enhancing the input signal and its response in the output signal relative to the outside nonsynchronous noise waveforms which average to zero. This technique is very common in active sonar, radar, and ultrasonic imaging to suppress clutter and noise. However, the time–bandwidth relationship for linear time-invariant system transfer functions still requires that the time buffer length must be long enough to allow the desired frequency resolution (i.e., $\Delta f = 1/T$). If the system represents a medium (such as in sonar) one must carefully consider time–bandwidth of any Doppler frequency shifts from moving objects or changing system dynamics. Optimal detection of nonstationary systems may require a wide-band processing technique using wavelet analysis, rather than narrowband Fourier analysis.

7.4 INTENSITY FIELD THEORY

There are many useful applications of sensor technology in mapping the power flow from a field source, be it electrical, magnetic, acoustic, or vibration. Given the spatial power flow density, or *intensity*, which has units of Watts per meter-squared (W/m^2), one can determine the total radiated power from a source by scanning and summing the intensity vectors over a closed surface [7]. This is true even for a field region where there are many interfering sources so long as the surface used for the integration encloses only the power source of interest. Recalling Gauss's law, one obtains the total charge in a volume either by integrating the charge density over the volume or by integrating the flux density over the surface enclosing the volume. Since it is generally much easier to scan a sensor system over a surface than throughout a volume, the flux field over the surface and the sensor technologies necessary to measure it are of considerable interest. In electromagnetic theory, the power flux is known as the *Poynting vector*, or $\mathbf{S} = \mathbf{E} \times \mathbf{H}$. The cross-product of the electric field vector $\mathbf{E}$ and the magnetic field vector $\mathbf{H}$ has units of W/m^2. A substantial amount of attention has been paid in the literature to *acoustic* intensity where the product of the acoustic pressure and particle velocity vector provides the power flux vector which can be used with great utility in noise control engineering. The same is true in the use of vibration intensity in structural acoustics; however, the structural intensity technique is somewhat difficult to measure in practice due to the many interfering vibration components (shear, torsion, compression waves, etc.). The Poynting vector in electromagnetic theory is of considerable interest in antenna design. Our presentation of intensity will be most detailed for acoustic intensity and will also show continuity to structural and electromagnetic intensity field measurements.

7.4.1 Point Sources and Plane Waves

When the size of the radiating source is much smaller than the radiating wavelength, it is said to be a *point source*. Point sources have the distinct characteristic of radiating power equally in all directions giving a spherical radiation pattern. At great distances with respect to wavelength from the source, the spherically spreading wave in an area occupied by sensors can be thought of as nearly a plane wave. Intensity calculation for a plane wave is greatly simplified since the product of the potential field (pressure, force, electric fields) and the flux field (particle velocity, mechanical velocity, magnetic fields) gives the intensity magnitude. The vector direction of intensity depends only on the particle velocity for acoustic waves in fluids (gasses and liquids), but depends on the vector cross-product for vibration waves in solid and electromagnetic waves. One can get an idea of the phase error for a plane wave assumption given a measurement region approximately the size of a wavelength at a distance R from the source as shown in Figure 7.20.

The phase error is due to the distance error $R - R \cos \theta$ between the chord of the plane wave and the spherical wave. The phase error $\Delta \phi$ is simply

$$\Delta \phi = 2\pi \frac{R}{\lambda} \left[1 - \cos \left(\tan^{-1} \left\{ \frac{\lambda}{2R} \right\} \right) \right] \qquad (7.72)$$

where λ is the wavelength. A plot of this phase error in degrees is shown in Figure 7.21 which clearly shows that for the measurement plane to have less than one degree phase error, the distance to the source should be more than about 100 wavelengths. For sound, the wavelength of 1 kHz is approximately 0.34 m and one would need to be over 30 m away from the source for the plane wave assumption to be accurate. Even at 40 kHz where many ultrasonic sonars for industrial robotics operate the plane wave region only exists beyond a meter. The phase error scales with the size of the measurement area shown in Figure 7.21 in terms of wavelength. Clearly, the plane wave assumption is met for small apertures (measurement patch) far from the source. Larger sensor array sizes must be farther away than smaller arrays for the plane wave assumption to be accurate. For multiple point sources, the radiation pattern is no longer spherical but the distance/aperture relationship for the plane wave assumption still holds. This issue of plane wave propagation will be revisited again in Sections 13.3 and 13.4 for field reconstruction and propagation modeling techniques such as acoustic holography.

The point of discussing the plane wave assumption is to make clear the fact that one simply cannot say that the power is proportional to the square of the pressure, force, electric fields, and so on, unless the measurement is captured with a small array located many wavelengths from the source. This is a very important distinction between field theory and circuit theory where the wave-propagation geometry is much less than a wavelength. Since we need to understand the field in more detail to see the signal-processing application of field intensity measurement, we develop the wave equation for acoustic waves as this is the most straightforward approach compared to electric or vibration fields.

7.4.2 Acoustic Field Theory

To understand acoustic intensity one must first review the relationship between acoustic pressure and particle velocity. Derivation of the acoustic wave equation is fairly straightforward and highly

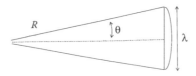

FIGURE 7.20 Over a wavelength-sized patch a distance R from the source, a plane-wave assumption introduces a spatial phase error which increases closer to the point source.

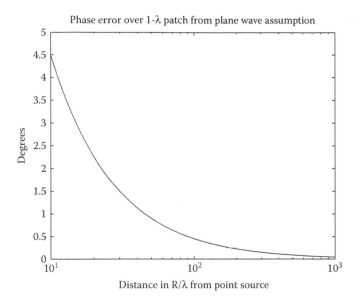

FIGURE 7.21 Phase error for a one-wavelength patch as a function of distance in wavelengths from the point source.

useful in gaining insight into acoustic intensity and the physical information which can be extracted. The starting point for the wave equation derivation is usually the *equation of State*, which simply put states that the change in pressure relative to density is a constant. In general, the pressure is a monic, but a nonlinear function of fluid density (gasses will also be referred to here as "fluids"). However, the static pressure of the atmosphere at sea level is about 101,325 Pa where 1 Pa = 1 Nt/m². An RMS acoustic signal of 1 Pa is about 94 dB, or about 100 times louder than normal speech, or about as loud as an obnoxious old lawnmower. So even for fairly loud sounds, the ratio of acoustic pressures to the static pressure is about 1:10^5 making the linearization assumption quite valid. However, for sound levels above around 155 dB linear acoustics assumptions are generally considered invalid and linear wave propagation is replaced by shock wave propagation theory. We can derive the equation of state for an ideal gas using Boyle's law

$$pv^\gamma = nRT \tag{7.73}$$

where n is the number of moles of an ideal gas (assumed 1), v is the volume of 1 mole of fluid, R is 8310 J/kmol K, and T is the temperature in degrees Kelvin. The thermodynamics are assumed to be *adiabatic*, meaning that no heat flow occurs as a result of the acoustic pressure wave. This is reasonable for low heat conduction media such as air and the relatively long wavelengths and rapid pressure/temperature changes of audio-band acoustics. Taking the natural logarithm of Equation 7.73 we have (assume $n = 1$)

$$\gamma \ln v + \ln p = \ln(RT) \tag{7.74}$$

We examine the differential of the pressure with respect to volume by computing

$$\gamma \frac{\partial v}{v} + \frac{\partial p}{p} = 0 \tag{7.75}$$

Since the volume v of 1 mole of gas is the molecular mass M divided by the density ρ, $dv/d\rho = -M/\rho^2$ and we can simplify the derivative of pressure with respect to volume as seen in Equation 7.76.

$$\frac{dp}{dv} = -\gamma \frac{p}{v} = -\frac{\rho^2}{M} \frac{dp}{d\rho} \tag{7.76}$$

The *linearized acoustic equation of state* is simply

$$\frac{dp}{d\rho} = \gamma \frac{p}{\rho} = c^2 \tag{7.77}$$

The parameter c^2 has the units of meter-squared per second-squared; so c is taken as the speed of sound in the fluid. For small perturbations $p = RT/v$, we can write the speed of sound in terms of the fluid's physical parameters.

$$c = \sqrt{\frac{\gamma RT}{M}} \tag{7.78}$$

At a balmy temperature of 20°C (293°K) we find nitrogen ($M = 28$) to have a speed of 349 m/s, and oxygen ($M = 32$) to have a speed of 326 m/s. Assuming that the atmosphere contains about 18% oxygen and the remaining nitrogen, we should expect the speed of sound in the atmosphere at 20°C to be around 343 m/s, which is indeed the case. Note how the speed of sound in an ideal gas does not depend on static pressure since the density also changes with pressure. The main contributors to the acoustic wave speed (which can be measured with a little signal processing) are temperature, mass flow along the direction of sound propagation, and fluid properties. Therefore, for intelligent sensor applications where temperature, fluid chemistry, and mass flow are properties of interest, measurement of sound speed can be an extremely important noninvasive observation technique. However, for extremely intense sound waves from explosions, sonic booms, or devices such as sirens where the sound pressure level exceeds approximately 150 dB, many other loss mechanisms become important and the "shock wave" speed significantly exceeds the linear wave speed of sound given by Equation 7.78.

The second step in deriving the acoustic wave equation is to apply Newton's second law ($f = ma$) to the fluid in what has become known as *Euler's equation*. Consider a cylindrical "slug" of fluid with circular cross-sectional area S and length dx. The force acting on the slug from the left is $S p(x)$, whereas the net force acting on the slug from the right is $S p(x + dx)$. With the positive x-direction pointing to the right, the net force on the slug resulting in an acceleration du/dt to the right of the mass $S dx \rho$ is $S[p(x) - p(x + dx)]$ or

$$S dx \frac{p(x) - p(x + dx)}{dx} = -S dx \frac{dp}{dx} = S dx \rho \frac{du}{dt} \tag{7.79}$$

which simplifies to

$$\frac{dp}{dx} = -\rho \left(\frac{du}{dt} + u \frac{du}{dx} \right) \tag{7.80}$$

The term du/dt in Equation 7.80 is simply the acoustic particle acceleration, whereas $u\,du/dx$ is known as the convective acceleration due to flow in and out of the cylinder of fluid. When the flow u is zero, or the spatial rate of change of flow du/dx is zero, Euler's equation is reduced to

$$\frac{dp}{dx} = -\rho\,\frac{du}{dt} \tag{7.81}$$

The continuity equation expresses the effect of the spatial rate of change of velocity to the rate of change of density. Consider again our slug of fluid where we have a "snapshot" dt s long where the matter flowing into the volume from the left is $S\,\rho(x)\,u(x)\,dt$ and the matter flowing out to the right is $S\,\rho(x+dx)\,u(x+dx)\,dt$. If the matter entering our cylindrical volume is greater than the matter leaving, the mass of fluid will increase in volume by $\partial\rho\,S\,dx$.

$$-S\,dx\,\frac{\partial(\rho u)}{\partial x}\,dt = -S\,dx\left[\rho_0\,\frac{\partial u}{\partial x} + \frac{\partial\rho}{\partial x}\,u_0\right]dt \tag{7.82}$$

The change in density due to sound relative to the static density ρ_0 is assumed small in Equation 7.82 allowing a linearization in Equation 7.83

$$\partial\rho_0\,S\,dx \approx -S\,dx\,\frac{\partial u}{\partial x}\,\rho_0\,dt \tag{7.83}$$

which is rearranged to give the standard form of the continuity equation.

$$\frac{\partial u}{\partial x} = -\frac{1}{\rho_0}\,\frac{\partial\rho}{\partial t} = -\frac{1}{\rho_0 c^2}\,\frac{\partial p}{\partial t} \tag{7.84}$$

The acoustic wave equation is derived by time differentiating the continuity equation in (7.84), taking a spatial derivative of Euler's equation in (7.81), and equating the terms $(d^2u/dx)\,dt$.

$$\frac{\partial^2 p}{\partial x^2} = \frac{1}{c^2}\,\frac{\partial^2 p}{\partial t^2} \tag{7.85}$$

7.4.3 Acoustic Intensity

For a plane wave, the solution of the one-dimensional wave equation in Cartesian coordinates is well known to be of the form

$$p(x,t) = C_1\,e^{j(\omega t - kx)} + C_2\,e^{j(\omega t + kx)} \tag{7.86}$$

where for positive moving time t (the causal world), the first term in Equation 7.86, is an outgoing wave moving in the positive x-direction and the second term is incoming. The parameter k is the radian frequency over wave speed ω/c, which can also be expressed as an inverse wavelength $2\pi/\lambda$. For an outgoing plane wave in free space many wavelengths from the source, we can write the velocity using Euler's equation.

$$u(x,t) = \frac{-1}{\rho}\int\frac{\partial p(x,t)}{\partial x}\,dt = \frac{1}{\rho c}\,p(x,t) \tag{7.87}$$

Since the pressure and velocity are in-phase and only differ by a constant, one can say under these strict *farfield* conditions that the time-averaged acoustic intensity is

$$< I(x) >_t = < p(x,t)u^*(x,t) >_t = \frac{1}{2\rho c} \mid p(x,t) \mid^2 \tag{7.88}$$

The specific acoustic impedance p/u of a plane wave is simply ρc. However, for cases where the intensity sensor is closer to the source, the pressure and velocity are not in-phase and accurate intensity measurement requires both a velocity and a pressure measurement.

Consider a finite-sized spherical source with radius r_0 pulsating with velocity $u_0 e^{j\omega t}$ to produce a spherically spreading sinusoidal wave in an infinite fluid medium. From the Laplacian in spherical coordinates it is well known that the solution of the wave equation for an outgoing wave in the radial direction r is

$$p(r,t) = \frac{A}{r} e^{j(\omega t - kr)} \tag{7.89}$$

where A is some amplitude constant jet to be determined. Again, using Euler's equation the particle velocity is found.

$$u(r,t) = \frac{A}{r} \frac{1}{\rho c} \left(\frac{1}{jkr} + 1 \right) e^{j(\omega t - kr)} \tag{7.90}$$

At the surface of the pulsating sphere $r = r_0$, the velocity is $u_0 e^{j\omega t}$ and it can be shown that the amplitude A is complex.

$$A = \frac{jkr_0^2 u_0 \rho c}{1 + jkr_0} e^{jkr_0} \tag{7.91}$$

The time-averaged acoustic intensity is simply the product of the pressure and velocity conjugate replacing the complex exponentials by their expected value of 1/2 for real waveforms. If the amplitude A is given as an RMS value, the factor of 1/2 is dropped.

$$< I(r)>_t = \frac{|A|^2}{r^2} \frac{1}{2\rho c} \left(1 + \frac{j}{kr} \right) \tag{7.92}$$

Equation 7.92 shows that for distances many wavelengths from the source ($kr \gg 1$), the intensity is approximately the pressure amplitude-squared divided by $2\rho c$. However, in regions close to the source ($kr \ll 1$), the intensity is quite different and more complicated. This complication is particularly important when one is attempting to scan a surface enclosing the source(s) to determine the radiated power. One must compute the total field intensity and use the real part to determine the total radiated power by a source. Figure 7.22 shows the magnitude and phase of the intensity as a function of source distance. It can be seen that inside a few 10s of wavelengths we have an acoustic *nearfield* with nonplane wave propagation and beyond an acoustic *farfield* where one has mainly plane-wave propagation. The consequence of the nonzero phase of the intensity in the nearfield is stored energy or *reactive intensity* in the field immediately around the source. The propagating part of the field is known as the *active intensity* and contains the real part of the full intensity field.

Figure 7.23 plots the radiation impedance for the source and shows a "mass load" in the nearfield consistent with the reactive intensity as well as the spherical wavefront curvature near the source.

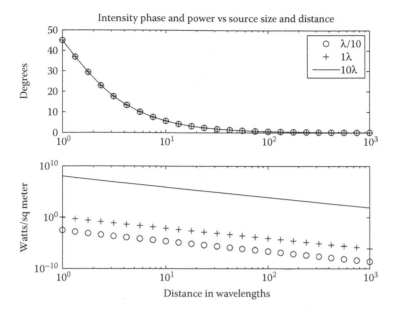

FIGURE 7.22 Intensity magnitude and phase as a function of source size and distance for a point source with 10 cm/s peak surface velocity.

Interestingly, the range where the nearfield and farfield meet, say around 20 wavelengths, does not depend on the size of the source so long as the source is radiating as a monopole (pulsating sphere). Real-world noise sources such as internal combustion engines or industrial ventilation systems have many "monopole" sources with varying amplitudes and phases all radiating at the same frequencies to make a very complex radiation pattern. Scanning the farfield radiated pressure (hundreds of wavelengths away) to integrate $p^2/2\rho c$ for the total radiated power is technically correct but very

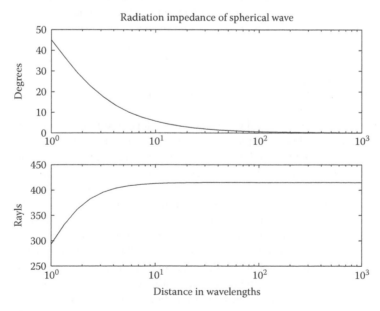

FIGURE 7.23 Radiation impedance magnitude and phase for a spherical wave as a function of distance from the point source.

impractical. Scanning closer to the noise source and computing the full complex acoustic intensity provides tremendous insight into the physical noise mechanisms (including machinery health) as well as allowing the real part to be integrated for the total radiated power.

A great deal of consistency can be seen in comparing Figures 7.21 through 7.23 where the acoustic nearfield and acoustic farfield can be seen to meet in the range of 10–100 wavelengths from the source. When the wave front has a spherical curvature the intensity indicates energy storage in the field and the wave impedance clearly shows a "mass-like" component. Note how the phase of the intensity field is also the phase of the wave impedance. If a sensor design places one conveniently in the farfield, great simplifications can be made. But, if nearfield measurements must be made or are desirable, techniques such as intensity are invaluable. The conservation of energy corollary governing all systems is seen "verbally" in Equation 7.93.

$$
\begin{pmatrix} \text{Power} \\ \text{from} \\ \text{sources} \end{pmatrix} + \begin{pmatrix} \text{Power} \\ \text{out of} \\ \text{surface} \end{pmatrix} + \begin{pmatrix} \text{Rate of} \\ \text{energy storage} \\ \text{in field} \end{pmatrix} = 0 \tag{7.93}
$$

If the radiation is steady state and medium passive, the rate of energy storage eventually settles to zero and the energy radiating out from the surface enclosing the sources is equal to the radiated energy of the sources. This is also well known in field theory as Gauss's Law. Intelligent adaptive sensor systems can exploit the physical aspects of the waves they are measuring to determine the accuracy and/or validity of the measurement being made.

7.4.4 STRUCTURAL INTENSITY

Structural acoustics is a very complicated discipline associated with the generation and propagation of force waves in solids, beams, shells, and so on. A complete treatment of structural intensity is well beyond the scope of this and many other books. However, from the perspective of intelligent sensor and control system design, it is important for us to contrast the differences between structural and fluid waves toward the goal of understanding the sensor system signal processing requirements. There are three main types of waves in solids: compression, or "p-waves" where the force and propagation directions are aligned together; transverse shear, or "s-waves" where a moment force and shear force are orthogonal to the propagation direction; and torsion shear, where a twisting motion has a shear force aligned with the rotation angle θ and is orthogonal to the propagation of the wave. There are also numerous composite forces such as combinations of p, s, and torsion waves in shells and other complex structural elements.

There are some simplified structural elements often referred to in textbooks which tend to have one dominant type of wave. For example, "rod" vibration refers to a long stiff element with compression waves propagating end-to-end like in the valves of a standard internal combustion engine. "String" vibration refers to mass–tension transverse vibration such as the strings of a guitar or the cables of a suspension bridge. "Shaft" vibration refers to torsion twisting of a drive shaft connecting gears, motors, and so on. Rod, string, and shaft vibrations all have a second-order wave equation much like the acoustic wave equation where the wave speed is constant for all frequencies and the solution is in the form of sines and cosines. "Beam" vibration refers to a long element with transverse bending shear vibrations such as a cantilever. "Plate" vibration refers to bending vibrations in two dimensions. Beams and plates have both shear and moment forces requiring a fourth-order wave equation, giving a wave speed which increases with increasing frequency, and a solution in the form of sines, cosines, and hyperbolic sines and hyperbolic cosines. "Shell" vibration is even more complex where the bending s-waves are coupled with compression p-waves. But here is a valuable secret about acoustics in general: it is almost impossible in nature to excite or isolate one type of wave. All types of waves are generated at the source and wave energy couples together between wave types at every discontinuity in impedance along the propagation path. One has to always be aware of wave mixture and coupling. It creates remarkably complicated fields, which is why measurement of parameters such as intensity is so useful.

These "textbook" structural models can be analytically solved (or numerically using finite-element software) to give natural frequencies, or modes, of vibration. The modes can be seen as the "energy storage" part of the conservation of energy corollary in Equation 7.93 making intensity measurement essential to the measurement of the vibration strength of the internal sources. When dealing with fluid-loaded structures such as pipes and vessels, strong coupling between the "supersonic" vibration modes (vibrations with wave speeds faster than the fluid wave speed) will transmit most of their energy to the fluid while the nonradiating "subsonic" remains in the structure dominating the remaining vibrations there. One can observe the net radiated power using a combination of modal filtering and structural intensity integration. Real-world structures are far more complex than the simplified "textbook" structural elements due to effect of holes, bolts, welds, ribs, and so on. These discontinuities in structural impedance tend to scatter modes. For example, a low-frequency bending mode (which is subsonic nonradiating) can spillover its stored energy into a supersonic mode at any discontinuity in the structure. Not surprisingly, required structural elements such as ribs, bolts, rivets, and weld seams tend to radiate noise "hot spots" on the structure. The modal scattering at discontinuities also contributes to high dynamic stress making the hot spots susceptible to fatigue cracks, corrosion, and eventual structural failure. The vibration and intensity response of the structure will change as factors such as fatigue and corrosion reduce the stiffness of the structural elements. Therefore, in theory at least, an intelligent sensor monitoring system can detect signs of impending structural failure from the vibration signature in time to save lives and huge sums of money.

P-waves in solids are much the same as acoustic waves except that they can have three components in three dimensions. Each component has the force and propagation direction on the same axis. In Cartesian coordinates, each p-wave component is orthogonal with respect to the other components. For the direction unit vectors a_x, a_y, and a_z, the p-wave time-averaged intensity for a sinusoidal p-wave is simply

$$< \vec{I}^P(\omega) >_t = \frac{1}{2} \left[F_x^p \, u_x^{p*} a_x + F_y^p \, u_y^{p*} a_y + F_z^p \, u_z^{p*} a_z \right] \qquad (7.94)$$

where the factor of 1/2 can be dropped if RMS values are used for the compression forces and velocity conjugates. Generally, p-waves in solids are extremely fast compared to s-waves and most acoustic waves. This is because solids are generally quite stiff (hard) allowing the compression force to be transmitted to great depths into the solid almost instantly. Torsion intensity is found in much the same manner except the shear force velocity in the θ-direction produce an intensity vector in the orthogonal direction.

Bending waves on lossless beams, or s-waves, there are two kinds of restoring forces, shear and moment, and two kinds of inertial forces, transverse and rotary mass acceleration. Combining all forces (neglecting damping), we have a fourth-order wave equation

$$EI \frac{\partial^4 y}{\partial x^4} + \rho S \frac{\partial^2 y}{\partial t^2} = f(x, t) \qquad (7.95)$$

where E is Young's modulus, I is the moment of inertia, ρ is the mass density, and S is the cross-sectional area. The transverse vibration is in the y-direction and the wave propagation is along the beam in the x-direction. To compute the intensity, one needs to measure the shear force $F^s = EI(\partial^3 y/\partial x^3)$, moment force $M^s = EI \, (\partial^2 y/\partial x^2)$, transverse velocity $u^s = (\partial y/\partial t)$, and rotary velocity $\Omega^s = (\partial^2 y/\partial x \partial t)$. Again, assuming peak values for force and velocity and a sinusoidal s-wave, the bending wave time-averaged intensity is

$$< I^s(\omega) >_t = \frac{1}{2}\left(F^s u^{s*} + M^s \Omega^{s*}\right) = \frac{EI}{2}\left[\frac{\partial^3 y}{\partial x^3}\left(\frac{\partial y}{\partial t}\right)^* + \frac{\partial^2 y}{\partial x^2}\left(\frac{\partial^2 y}{\partial x \partial t}\right)^*\right] \qquad (7.96)$$

where * indicates a complex conjugate of the time-varying part of the bending wave only. The spatial derivatives needed for the shear and moment forces are generally measured assuming EI and using a small array of accelerometers and finite-difference approximations, where the displacement is found by integrating the acceleration twice with respect to time (multiply by $-1/\omega^2$). However, other techniques exist such as using strain gauges or lasers to measure the bending action. Bending wave intensity, while quite useful for separating propagating power from the stored energy in standing wave fields, is generally quite difficult to measure accurately due to transducer response errors and the mixture of various waves always present in structures. Intensity measurements in plates and shells are even more tedious, but likely to be an important advanced sensor technique to be developed in the future.

7.4.5 ELECTROMAGNETIC INTENSITY

Electromagnetic intensity is quite similar to acoustic intensity except that the power flux is defined as the curl of the electric and magnetic fields, $\mathbf{S} = \mathbf{E} \times \mathbf{M}$ [8]. While the acoustic pressure is a scalar, the electric field is a full vector in three dimensions as well as the magnetic field. In simplest terms, if the electric field is pointing upward and the magnetic field is pointing to the right, the power flux is straight ahead. Measurement of the electric field simply requires emplacement of unshielded conductors in the field and measuring the voltage differences relative to a ground point. The E field is expressed in Volts per meter (V/m) and can be decomposed into orthogonal a_x, a_y, and a_z components by arranging three antennae separated along Cartesian axes, a finite distance with the ground point at the origin. The *dynamic* magnetic field $\mathbf{H}$ can be measured using a simple solenoid coil of wire on each axis of interest. Note that static magnetic fields require a flux-gate magnetometer which measures a static offset in the coil's hysteresis curve to determine the net static field. Flux-gate magnetometers are quite commonly used as electronic compasses. Three simple coils arranged orthogonally can be used to produce a voltage proportional to the time derivative of the magnetic flux Φ along each of three orthogonal axes. If the number of turns in the coil is N, the electromotive force or emf voltage will be $v = -N \, d\Phi/dt$, where $\Phi = \mu HS$, S being the cross-sectional area and μ the permeability of the medium inside the coil (μ for air is about $4\pi \times 10^{-7}$ Henrys/m). Unlike an electric field sensor, the coil will not be sensitive to static magnetic fields. For Cartesian coordinates, the curl of the electromagnetic field can be written as a determinant.

$$E \times H = \begin{vmatrix} a_x & a_y & a_z \\ E_x & E_y & E_z \\ H_x & H_y & H_z \end{vmatrix} \qquad (7.97)$$

which can be written out in detail for the time-averaged intensity assuming a sinusoidal wave.

$$< E \times H >_t = \frac{a_x}{2}\left(E_y H_z^* - E_z H_y^*\right) + \frac{a_y}{2}\left(E_x H_z^* - E_z H_x^*\right) + \frac{a_z}{2}\left(E_x H_y^* - E_y H_x^*\right) \qquad (7.98)$$

The electromagnetic intensity measurement provides the power flux in W/m^2 at the measurement point, and if scanned, the closed surface integral can provide the net power radiation in Watts. The real part of the intensity can be used to avoid the nearfield effects demonstrated for the acoustic case which still hold geometrically for electromagnetic waves. Clearly, the nearfield for electromagnetic waves can extend for enormous distances depending on frequency. The speed of light is 3×10^8 m/s giving a 100 MHz signal a wavelength of 3 m. For a modern computer or other digital consumer electronic devices, electromagnetic radiation can interfere with radio communications, aircraft navigation signals, and the operation of digital electronics. For hand-held cellular telephones, there can be potential health problems with the transmitter being in very close proximity to the head and brain, although these problems are extremely difficult to get objective research data due to the varying reaction of animal tissues to electromagnetic fields. However, for most frequencies of concern, a simple field voltage reading is not sufficient to produce an accurate measure of net power radiation in the farfield.

7.5 INTENSITY DISPLAY AND MEASUREMENT TECHNIQUES

The measurement of intensity can involve separate measurement of the potential and kinetic field components with specialized sensors, provided that sensor technologies are available which provide good field-component separation. In acoustics, velocity measurements in air are particularly difficult. Typically one indirectly measures the acoustic particle velocity by estimating the gradient of the acoustic pressure from a finite-difference approximation using two pressure measurements. The gradient estimation and intensity computation when done in the frequency domain can be approximated using a simple cross-spectrum measurement, which is why this technique is presented here. While we will concentrate on the derivation of the acoustic intensity using cross-spectral techniques here, one could also apply the technique to estimating the Poynting vector using only electric field sensors. We also present the structural intensity technique for basic compressional and shear waves in beams. Structural intensity measurement in shells and complicated structures are the subject of current acoustics research and are beyond the scope of this book.

We begin by deriving acoustic intensity and showing its relation to the cross-spectrum when only pressure sensors are used. An example is provided using an acoustic dipole to show the complete field reconstruction possible from an intensity scan. Following the well-established theory of acoustic intensity in air, we briefly show the intensity technique in the frequency domain for simple vibrations in beams. Finally, we will very briefly explore the Poynting vector field measurement and why intensity field measurements are important to adaptive signal processing.

7.5.1 GRAPHICAL DISPLAY OF THE ACOUSTIC DIPOLE

A dipole is defined as a source which is composed of two distinct closely spaced sources, or "monopoles," of differing phases. When the phases are identical and the spacing is much less than a wavelength, the two sources couple together into a single monopole. For simplicity, we consider the two sources to be of equal strength, opposite phase, and of small size with respect to wavelength ($r_0 \ll \lambda$, therefore $kr_0 \ll 1$). The pressure amplitude factor A in Equation 7.91 can be simplified to $A = jk\rho cQ/4\pi$, where $Q = 4\pi r_0^2 u_0$ is the source strength or volume velocity in m³/s. Figure 7.24 shows the pressure response in Pascals for two opposite-phased 0.1-m-diameter spherical sources separated by 1 m and radiating 171.5 Hz with 1 mm peak surface displacement. The sources are located at $y = 0$ and $x = \pm 0.5$ m and the net pressure field is found from summing the individual point source fields.

The classic dipole "figure-8" directivity pattern can be seen in Figure 7.24 by observing the zero pressure line along the $x = 0$ plane (the y-axis) where one expects the pressures from the two opposite-phased sources to sum to zero. For the assumed sound speed of 343 m/s, the sources are separated by exactly 1/2 λ so that no cancellation occurs along the $y = 0$ plane (the x-axis). Figure 7.25 shows the pressure response at 343 Hz giving a "4-leaf clover" directivity pattern. At 343 Hz and a 1 m spacing (1 m is the wavelength at 343 Hz if $c = 343$ m/s), cancellation also occurs along

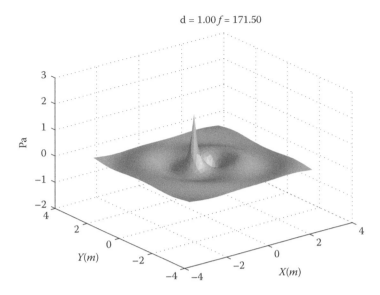

FIGURE 7.24 Sound pressure field for an opposite-phased dipole with 1 m separation at 171.5 Hz.

the $y = 0$ axis of the dipole. This frequency and phasing gives cancellation along the x- and y- axis, making four "lobes" to the directivity pattern in the x–y plane, symmetric about the axis of the dipole (the x-axis). At 786 Hz there would be eight lobes and if the two sources had the same phase at 1 m spacing at 343 Hz, there would only be two bobes. The directivity pattern of a dipole, or any multipole source, depends on phasing, spacing, and frequency.

The velocity field of a dipole is also interesting. Using Equation 7.90 we can calculate the velocity field of each source as a function of distance. In the x–y plane of the dipole, the velocity field is decomposed into x and y components for each source and then summed according to the phase between the sources. The resulting field is displayed using a vector field plot as shown in Figure 7.26

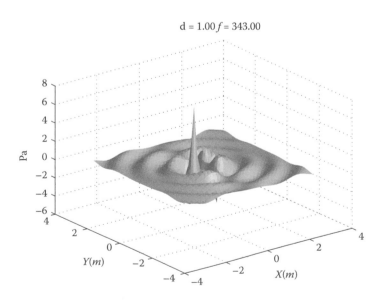

FIGURE 7.25 At 343 Hz, one can see a four-lobe directivity pattern due to the cancellation along the dipole axis as well as normal to the dipole axis.

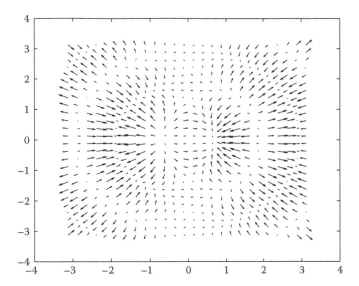

FIGURE 7.26 The velocity field for 171.5 Hz where $1/R$ spreading has been removed from the plot to show fluid motion easily throughout the plot.

for 171.5 Hz. For each of the measured field points, the direction of the vectors indicates the particle velocity direction, and the length of the vectors indicates the relative magnitude of the particle velocity, normalized by the distance from the center of the plot. This normalization was done so that one could see the arrows near the perimeter of the plot. Without this artificial normalization, only the arrows near the sources can be seen, since the velocity is so high near the sources. For a 1 mm peak displacement of the 10-cm-diameter spherical sources at 171.5 Hz, the surface velocity is 1.078 m/s and the Q (source strength) is 0.0338 m³/s. The vector field in Figure 7.26 is scaled for the best display and, like the pressure field in Figure 7.24, represents a "snapshot" of the velocity field at one time instant. The MATLAB scripts for these figures actually produce a movie of the fields through a cycle of the radiation.

One can compute, point by point, the instantaneous intensity $I(x,y,t) = p(x,y,t)\,u(x,y,t)$. However, the direction of real power flow is not always the same as the velocity because both pressure and velocity are complex, and because the product of a negative velocity and negative pressure yields a positive instantaneous intensity. This is mathematically correct but not physically correct nor intuitive. Intensity is really a time-average and spatial average power flow estimate where the real part is the *active intensity* representing the power flow and the imaginary part is the *reactive intensity* representing the power stored in the field. Figure 7.28 shows another representation of the field using time-averaged intensity which is also problematic technically.

Neither Figure 7.27 nor Figure 7.28 is technically correct representation of the actual intensity field. However, they can be seen as quite useful in describing the power flow between the two sources. It is expected that the instantaneous intensity field in Figure 7.27 would be zero where either the pressure or velocity is zero, but this can be misleading since the waves are propagating outward from the sources. The time-averaged intensity is found for sinusoidal excitation using $< I(x,y) >_t = 1/2\,p(x,y,t)\,u^*(x,y,t)$, where the real part represents the active intensity. Computing the time-averaged intensity for each source in the dipole and then superpositioning them in opposite phase yields the vector plot in Figure 7.28, which is physically intuitive since one can visualize the arrows oscillating back and forth with the phases of the two sources. However, Figure 7.28, like Figure 7.27, is misleading since it is neither a true time-average field nor does power actually flow into the source at $x = < +.5, 0 >$. To graphically depict the dipole intensity field, we first need a model for the total pressure and velocity fields in the radial direction for the dipole. It is

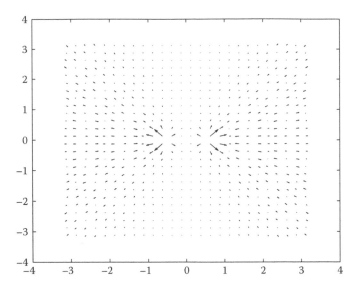

FIGURE 7.27 The real part of the instantaneous intensity with $1/R^2$ spreading removed to enhance detail at the perimeter, but also showing erroneous power flow for the out-of-phase source.

straightforward to show that for a separation distance d on the y-axis symmetric about the origin, the total pressure field is

$$p(r,\theta,t) = j\frac{k\rho cQe^{j(\omega t - kr)}}{4\pi r}\left[e^{+j(kd/2)\cos\theta} - e^{-j(kd/2)\cos\theta}\right] \tag{7.99}$$

where $r = (x^2 + y^2)^{1/2}$ and θ is measured counterclockwise from the positive x-axis. The radial component of velocity for the complete dipole field is

$$u(r,\theta,t) = j\frac{kQe^{j(\omega t - kr)}}{4\pi r}\left(1 - j\frac{1}{kr}\right)\left[e^{+j(kd/2)\cos\theta} - e^{-j(kd/2)\cos\theta}\right] \tag{7.100}$$

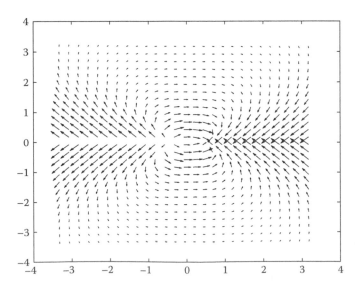

FIGURE 7.28 Superposition of the two-point source intensities showing (falsely) the circulation of power between the dipoles (arrow lengths on dB scale referenced to 1×10^{-12} W/m²).

Finally, the time-averaged intensity for the dipole field is found using $< I(r, \theta) >_t = 1/2 \, p(r, \theta, t)$ $u^*(r, \theta, t)$ as seen in Equation 7.101. We would drop the factor of 1/2 for RMS values of $p(r\, \theta, t)$ and $u(r\, \theta, t)$ rather than peak values.

$$< I(r, \theta) >_t = \frac{k^2 \rho c Q^2}{8\pi^2 r} \left(1 - j\frac{1}{kr} \right) \sin^2 \left(\frac{kd}{2} \cos \theta \right) \tag{7.101}$$

A vector plot of the correct time-averaged dipole intensity along the radial direction at 171.5 Hz is shown in Figure 7.29. We do not include the circumferential direction intensity component since its contribution to radiated power on a spherical surface is zero. Figure 7.29 correctly shows radiated power from both sources and a null axis along $x = 0$ corresponding to the zero pressure axis shown in Figure 7.24 and elliptical particle velocities in Figure 7.26.

It is important to understand the graphical display techniques in order to obtain correct physical insight into the radiated fields being measured. The vector plots in Figures 7.26 through 7.29 are typically generated from very large data sets which require significant resources to gather engineering-wise. Even though one is well inside the nearfield, the net radiated power can be estimated by integrating the intensity over a surface enclosing the sources of interest. Any other noise sources outside the enclosing scan surface will not contribute to the radiated power measurements. This surface is conveniently chosen to be a constant coordinate surface such as a sphere (for radial intensity measurements) or a rectangular box (for x–y–z intensity component measurements). The intensity field itself can be displayed creatively (such as in Figures 7.27 and 7.29) to gain insight into the physical processes at work. Intensity field measurements can also be used as a diagnostic and preventive maintenance tool to detect changes in systems and structures due to damage.

7.5.2 Calculation of Acoustic Intensity from Normalized Spectral Density

Given spectral measurements of the field at the points of interest, the time-averaged intensity can be calculated directly from the measured spectra. However, we need to be careful to handle spectral density properly. Recall that twice the bin magnitude for the NDFT (NDFT or DFT divided by the

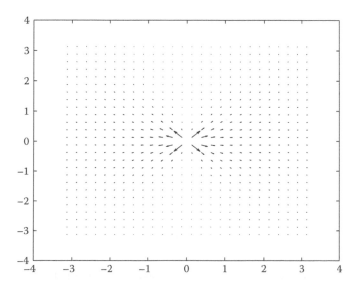

FIGURE 7.29 A true representation of the intensity field of a dipole shows radiation from both point sources and the cancellation along the $x = 0$ line normal to the axis of the dipole (arrow lengths on linear scale).

size of the transform N) with real data input gives the peak amplitude at the corresponding bin frequency. Therefore, the square root of 2 times the NDFT bin amplitude gives an RMS value at the corresponding bin frequency. When the necessary field components are measurable directly, the calibrated RMS bin values can be used directly in the field calculations resulting in vector displays such as that in Figure 7.29. When spatial derivatives are required to compute field components only indirectly observable (say via pressure or structural acceleration sensors only), some interesting spectral calculations can be done. For the acoustic intensity case, two pressure sensors signals $p_1(\omega)$ and $p_2(\omega)$ are separated by a distance Δr, where $p_2(\omega)$ is in the more positive position. From Euler's equation given here in Equation 7.80, the velocity can be estimated at a position in between the two sensors from the pressure by using a spatial derivative finite difference approximation.

$$u(\omega) = \frac{-1}{\rho} \int \frac{\partial p(\omega)}{\partial r} \, dt \approx j \frac{p_2(\omega) - p_1(\omega)}{\omega \rho \Delta r} \tag{7.102}$$

The pressure sensors need to be close together for the finite-difference approximation to be accurate, but if they are too close, the finite difference will be dominated by residual noise since the sensor signals will be nearly identical. Somewhere between about 1/16 and 1/4 of a wavelength spacing can be seen as a near optimum range. The pressure used in the intensity calculation is estimated at the center of the sensors by a simple average. The RMS intensity is simply

$$I(\omega) = \frac{1}{2}\left\langle p(\omega)u^*(\omega)\right\rangle_t = \frac{1}{2}\left\langle p^*(\omega)u(\omega)\right\rangle_t$$

$$= \frac{-j}{4\omega\rho\Delta r}\left\langle p_1(\omega) + p_2(\omega)\right\rangle\left\langle p_1^*(\omega) - p_2^*(\omega)\right\rangle_t$$

$$= \frac{-j}{4\omega\rho\Delta r}\left\langle p_1(\omega)p_2^*(\omega) + |p_2(\omega)|^2 - |p_1(\omega)|^2 - p_2(\omega)p_1^*(\omega)\right\rangle_t \tag{7.103}$$

Noting that both $p_1(\omega)$ and $p_2(\omega)$ have real and imaginary components, that is, $p_2(\omega) = p_2^R(\omega) + jp_2^I(\omega)$, the real part of the intensity, or *active intensity*, is

$$I^R(\omega) = \frac{1}{2\omega\rho\Delta r}\left\langle p_1^I(\omega)\, p_2^R(\omega) - p_1^R(\omega)\, p_2^I(\omega)\right\rangle_t \tag{7.104}$$

which represents the propagating power in the field in the units of W/m². The active intensity can be very conveniently calculated using the negative of the imaginary part of the cross-spectrum $G^{12}(\omega) = <p_1^*(\omega)p_2(\omega)>_t$.

$$I^R(\omega) \approx \frac{-\operatorname{Im}\{G_{12}(\omega)\}}{2\omega\rho\Delta r} = \frac{\left\langle p_1^I(\omega)\, p_2^R(\omega) - p_1^R(\omega)\, p_2^I(\omega)\right\rangle}{2\omega\rho\Delta r} \tag{7.105}$$

The imaginary part of the intensity or *reactive intensity* depicts the stored energy in the field and is given in Equation 7.106.

$$I^I(\omega) = \frac{1}{4\omega\rho\Delta r}\left\langle |p_1(\omega)|^2 + |p_2(\omega)|^2\right\rangle_t \tag{7.106}$$

The reactive intensity provides a measure of the stored energy in the field, such as the standing wave field in a duct or the nearfield of a dipole. Since the energy stored in the field eventually is coupled to the radiating part of the field, the transient response of a noise source can be significantly affected by the reactive intensity. For example, a loudspeaker in a rigid-walled enclosure with a small opening radiating sound to a free space would reproduce a steady-state sinusoid well, but will be very poor for speech intelligibility. The standing-waves inside the enclosure would store a significant part of the loudspeaker sound and reradiate it as the modes decay causing words to be reverberated over time. The nearfields of active sources for sonar and radar can also have their impulse responses affected by the stored energy in the reactive intensity field. The term nearfield generally refers to any component of the field in the vicinity of the source which does not propagate to the farfield. This could be the *evanescent* (nonpropagating) part of the velocity field as well as the "circulating" components of the pressure field which propagate around the source rather than away from it. One can consider the nearfield as an energy storage mechanism which can affect the farfield transient response of a transmitter source.

7.5.3 CALCULATION OF STRUCTURAL INTENSITY FOR COMPRESSIONAL AND BENDING WAVES

Calculation of structural intensity most commonly uses finite differences to estimate the force and velocity from spatial acceleration measurements and known physical parameters of the structure material, such as Young's modulus E and the moment of inertia I. Structural vibrations are enormously complex due to the presence of compression, shear, and torsion vibrations in three dimensions. At discontinuities (edges, holes, joints, welds, etc.), each kind of wave can couple energy into any of the others. The finite size of a structure leads to spatial modes of vibration, which with low damping are dominated by one frequency each, and again can couple to any other modes at an impedance discontinuity. Measurement of structural vibration power is not only important from a noise/vibration control point of view, but also from a fatigue monitoring and prognosis view. Controlling the dissipation of fatigue causing vibrations in a structure will have important economical implications for everything from vehicles to high-performance industrial equipment. The technical issues with structural intensity measurement are analogous to the acoustic case in that one wants the sensors as close together as possible for accurate finite-difference approximations, yet far enough apart to yield high SNRs in the finite-difference approximations. For simplification, one generally assumes plane waves, and separation of, say, shear and compression waves involves complicated sensor arrays for wavenumber separation through the use of wavenumber filtering and "beamforming," discussed later in this text. We note that newer sensor technologies in the area of laser processing and micromachined strain gauges offer the potential to eliminate the finite-difference approximations, and the associated systematic errors.

The type of sensor preferred for a particular vibration measurement depends on the frequency range of interest, wave speed, and effect of the sensor mass on the vibration field. At very low frequencies (say below the audio range of around 20 Hz), displacement sensing is preferred because velocity is quite small and acceleration even smaller (they scale as $j\omega$ and $-\omega^2$, respectively). Typical very low-frequency/static displacement sensors would be capacitive or magnetic proximity transducers, strain gauges, or optical sensors. At medium to low frequencies, velocity sensors such as geophones are preferred since a moving coil in a magnetic field can be made to have very high voltage sensitivity to a velocity excitation in the range of a few Hertz to several hundred Hertz. Above a few hundred Hertz, accelerometers are preferred due to their high sensitivities and low cost. An accelerometer is essentially a strain gauge with a known "proof" mass attached to one side. The known mass produces a traceable stress on the accelerometer material when accelerated along the common axis of the accelerometer material and mass. The accelerometer material is usually a piezoelectric material such as natural quartz or the man-made lead–zirconate–titanate (PZT) ceramic. These materials produce a small voltage potential when mechanical stress in applied. In recent years, a new class of accelerometer has become widely available making use of micromachined strain gauge technologies. Micromachining (accomplished chemically rather than with micromilling ...) allows very robust and low-cost manufacture of sensors with precise properties and integrated signal-conditioning electronics. For lightweight structures, the mass of an attached sensor is very problematic due to its effect of the vibration field. For lightweight structural vibration measurement, noncontact laser vibrometers are the most common sensors.

Fundamental to all intensity measurements is the need to simultaneously measure force and velocity at a field position. The preferred approach is to employ a sensor which directly measures force and another sensor which directly measures velocity. In heavy structures, velocity is easily observed directly using a geophone, by a simple time integration (low-pass filter with 6 dB/octave roll off) of acceleration, or by a time differentiation (high-pass filter with 6 dB/octave roll up) of measured displacement. Force measurement can be done with a direct strain gauge measurement which is usually a thin metal film attached under tension so that its electrical resistance changes with the resulting strain in the structure. However, an easier approach to force measurement is to use the known material properties of the structure to relate the spatial difference in, say, acceleration, to the force. For example, Young's modulus E for a particular material is known (measured *a priori*) and has units of Nt/m^2 (often listed as pounds per square-inch (psi)) or the ratio of longitudinal stress (Nt/m^2) over longitudinal strain (change in length over the initial length). Given two accelerometers spaced by Δ meters, the difference in acceleration spectra, $A_2(\omega) - A_1(\omega)$, where point 2 is further along the positive x-axis than point 1, can be multiplied by $-1/(\Delta\omega^2)$ to give the longitudinal strain spectrum. The structure cross-sectional area S times Young's modulus times the longitudinal strain spectrum gives the compression force spectrum, or p-force wave.

$$F^{\mathrm{p}}(\omega) = \frac{ES}{\Delta\omega^2}\left\{A_1(\omega) - A_2(\omega)\right\} \tag{7.107}$$

The compression force spectrum in Equation 7.107 represents the force between the two accelerometers oriented along the structure (usually referred to as a rod for this type of structural vibration). The velocity is simply the average of the two estimated velocities from the acceleration signals.

$$u^{\mathrm{p}}(\omega) = \frac{1}{2\mathrm{j}\omega}\left\{A_1(\omega) + A_2(\omega)\right\} \tag{7.108}$$

The RMS compression wave intensity in a rod is therefore

$$I^{\mathrm{p}}(\omega) = \frac{ES}{4\Delta\mathrm{j}\omega^3}\left\{ |A_1|^2 - A_2\,A_1^* + A_2^*\,A_1 - |A_2|^2\right\} \tag{7.109}$$

where the factor of 1/2 can be dropped if RMS-calibrated acceleration spectra are used (multiply by 2). Equation 7.109 provides a one-dimensional intensity p-wave which can be integrated into the three-dimensional compression time-averaged intensity for solids in Equation 7.94.

The largest source of error in Equation 7.109 comes from the acceleration differences in Equation 7.107, which can be near zero for long wavelengths and short accelerometer separations. The compression wave speed can be quite fast and is calculated by the square root of Young's modulus divided by density. For a given steel, if $E = 50 \times 10^{10}$ and density $\rho = 500$ kg/m^3, the compression wave speed is 31,623 m/s, or nearly 100 times the speed of sound in air. Therefore, the accelerometers need to be significantly separated to provide good SNR for the intensity measurement.

Bending wave intensity follows from Equation 7.96 where the various components of shear are calculated using finite-difference approximations for the required spatial derivatives. The time-averaged bending wave intensity in a structural beam is seen to be

$$< I^{\mathrm{s}}(\omega) >_t = \frac{1}{2}\left(F^s u^{s*} + M^s \Omega^{s*}\right) = \frac{EI}{2}\left[\frac{\partial^3 y}{\partial x^3}\left(\frac{\partial y}{\partial t}\right)^* + \frac{\partial^2 y}{\partial x^2}\left(\frac{\partial^2 y}{\partial x\,\partial t}\right)^*\right] \tag{7.110}$$

where again, the factor of 1/2 may be dropped if RMS spectra are used Typically one would use accelerometers to measure the beam response to bending waves where the displacement $y(x)$ is $-1/\omega^2$ times the acceleration spectra. One could estimate the spatial derivatives using a five-element linear array of accelerometers each spaced at a distance Δ meters apart, to give symmetric spatial derivative estimates about the middle accelerometer, $A_3(\omega)$. Consider the accelerometers to be spaced such that $A_5(\omega)$ is more positive on the x-axis than $A_1(\omega)$, and so on. Using standard finite-difference approximations for the spatial derivatives, the bending (shear wave) RMS intensity is

$$I^s(\omega) = j\,\frac{EI}{4\omega^3\Delta^3}\left\{(A_5 - 2A_4 + 2A_2 - A_1)\,A_3^* - (A_4 - 2A_3 + A_1)(A_4^* - A_2^*)\right\} \qquad (7.111)$$

where the accelerations $A_i(\omega)$ are written as A_i, $i = 1, 2, 3, 4, 5$; to save space. Clearly, the bending wave intensity error can be caused by finite-difference approximation errors which happen if the sensors are widely spaced (Δ must always be less than $\lambda/2$), and by SNR if Δ is too small. The bending wave wavelength for a beam with moment of inertia I, density ρ, cross-sectional area S, Young's modulus E, and frequency ω, is

$$\lambda_B = \frac{2\pi\sqrt[4]{EI}}{\sqrt[4]{\rho S \omega^2}} \qquad (7.112)$$

Clearly, the bending wavelength is much slower than the compression wave. But even more interesting is that the bending wave speed can be seen as a function of frequency to be slower at low frequencies than at higher frequencies. This is because a given beam of fixed dimensions (moment of inertia) is stiffer to shorter wavelengths that it is to longer wavelengths. While measurement of vibration power flow (bending or compression) is of itself valuable, changes in vibration power flow and wavelength due to material changes could be of paramount importance for monitoring structural integrity and prediction of material fatigue.

7.5.4 CALCULATION OF THE POYNTING VECTOR

Electromagnetic sensors can rather easily detect the electric field with a simple voltage probe and the alternating magnetic field with a simple wire coil. The static magnetic field requires a device known as a flux-gate magnetometer, where filtering a measured hysteresis response of a coil provides an indirect measurement of the static magnetic field. For strong magnetic fields in very close proximity to magnetic sources, a Hall effect transistor can be used. We are interested here in the process of measuring electrical power flow in fields using the intensity technique, or the Poynting vector. While one could use finite-difference approximations with either electric or magnetic field sensors, it is much simpler to directly observe the field using a directional voltage probe and search coil. The electric field is measured in volts per meter and can be done using a pair of probes, each probe exposing a small unshielded detector to the field at a known position. Probe voltage differences along each axis of a Cartesian coordinate system normalized by the separation distances provide the three-dimensional electric field vector, expressed as a spectrum where each bin has three directional components. This information alone is sufficient to resolve the direction of wave propagation with surprising accuracy. The time derivative of the magnetic field can be indirectly estimated from the curl of the electric field as seen in Equation 7.98, allowing the magnetic field to be estimated from spatial derivatives of the electric field. However, when the presence of a solenoid coil is not problematic for the field measurement (back emf from currents in the coil may affect the field to be measured), a direct magnetic field measurement can be made

$$H(\omega) = j\,\frac{\xi(\omega)}{NA\mu\omega} \qquad (7.113)$$

where the solenoid has N circular loops with average cross-sectional area A. The parameter $\xi(\omega)$ in Equation 7.113 is the voltage produced by the coil due to Faraday's law. The simple and effective use of coils to measure magnetic field is preferable because of the ability to detect very weak fields, while finite-difference approximations to the electric field would likely produce very low SNR. Also, for a given level of magnetic field strength, one can see from Equation 7.113 that a large coil with a large number of turns could produce substantial voltage for easy detection by a sensor system. However, too many turns in the coil will produce too much inductance for easy detection of high frequencies. Direct measurement of electric and magnetic fields allows one to compute the electromagnetic intensity using Equation 7.98 with spectral products in the frequency domain.

7.6 MATLAB® EXAMPLES

As noted at the end of each chapter, a collection of MATLAB m-scripts are being made available through the publisher for free downloading to enhance the lessons in this book. Table 7.1 associates each m-script with the various figures in this chapter. Readers are strongly encouraged to take advantage of this, even if one does not use MATLAB for graphing because the algorithms can easily be ported to most languages.

The first m-script "gausspdf.m" generates and plots the PDF for Gaussian and Chi-Square distributions used for modeling the magnitude-squared of an FFT bin. What we see in Figure 7.2 is that a few spectral averages (i.e., more than 8) lead to a Gaussian shape for the magnitude-squared PDF of a spectral bin where the mean is based on the noise variance, or the amplitude of the signal in the narrowband of the spectral bin, whichever is large. The signal and noise do not "add" because they are incoherent, but the noise PDF is used with the signal mean to model the statistical fluctuations in the signal bin. So, the noise PDF "floats" with the signal mean in the FFT bin when the SNR is >1.00. Given a time-averaged magnitude-squared spectrum (M averages), one can determine the 90% confidence interval by going to the integral in Figure 7.4 (or making a plot for the number of M averages), and finding the level that corresponds to 0.95, meaning that there is a 95% probability that the signal is less than the level, or a 5% chance that the signal is higher than the corresponding level. Since the PDF is fairly symmetric for $M > 8$, we can use this as the $\pm$ interval that the signal is within 90% of the time, centered at the FFT bin level. It is an important detail to understand that the PDF integrals in Figures 7.3 and 7.4 represent the probability that the signal is less than the level on the x-axis, where the noise-only mean is $2\sigma^2$ and is replaced by the signal level in the bin when the SNR is >1.

The m-script "bispectrum.m" generates a signal consisting of a couple of sinusoids in white noise and then processes the signal through nonlinearity for analysis using the bispectrum. Comparing the linear and nonlinear signals in the time and frequency domains, one can see some marked differences. One of the interesting aspects of nonlinearity is that the nonlinear change to the signal only happens at specific points in time when the signal amplitude extends into the nonlinear range.

TABLE 7.1
MATLAB Examples used in this Chapter

gausspdf.m	Figures 7.1–7.4
bispectrum.m	Figures 7.5–7.9
transcoher.m	Figures 7.11–7.18
nearfieldintensity.m	Figures 7.21–7.23
pressdipolemovie.m	Figures 7.24,7.25
DPVMOVIE.m	Figure 7.26
instintmovie.m	Figure 7.27
timeaveint.m	Figures 7.28,7.29

However, any periodic signal, nonlinear or not, can be expressed as a Fourier sum of sinusoids, and so the bispectrum is not a tool to detect nonlinearity, but is merely another tool to analyze nonlinearity from a statistical point of view.

The script "transcoher.m" is a very important set of measurements relating to transfer functions and the precision possible given some source of interfering noise. Transfer functions are measured quite commonly in mechanical systems and in audio systems and transducers. Unfortunately, coherence is often overlooked in these measurements. This section and the script hopes to point out the usefulness of the coherence function, both for estimating SNR and magnitude and phase error bounds on the estimated transfer function. The coherence can also be used to diagnose instrumentation problems and/or validate measurements, and so it is one of the most important and overlooked measurements.

The remaining two sections of the chapter and m-scripts in Table 7.1 deal with intensity measurements and display. These are great applications of spectral estimates and field calculations from measurements. While most readers may not have any experience with acoustical measurements, these techniques bring together field theory and spectral estimation together with advanced graphical display techniques. The script "nearfieldintensity.m" examines the errors associated with a "plane wave" assumption of the field from a point source. When you are far away from the point source the plane wave assumption is true, but how far? Examination of the field equations shows that the plane wave assumption is valid beyond 20–100 wavelengths, depending on the level of precision desired. MATLAB really makes this type of field modeling very convenient, such as with the various "movie" m-scripts, which are fun to watch and to play around with. Use of the "quiver" function for displays is a bit more tedious, but provides a very illustrative display of how the fluids move. This is because each field point needs an "X" and a "Y" component, which must be calculated separately. The result gives snapshots of power flow and particle motion that are fascinating. These techniques can be used to investigate fields from all sorts of products such as noise from engines, radiation from home theater loudspeaker systems, to the RF fields radiated from a cell phone.

7.7 SUMMARY

Most applications of the Fourier transform are for the purposes of steady-state sinusoidal signal detection and analysis, for which, the Fourier transform provides an enhancement relative to random noise. This is because of the orthogonality of sinusoids of different frequencies in the frequency domain. Provided that the time-domain signal recording is long enough, conversion to the frequency domain allows the levels of different frequencies to be easily resolved as peaks in the magnitude spectrum output. Peaks in the time domain are best detected in the time domain because the broad spectral energy would be spread over the entire spectrum in the frequency domain. For nonstationary frequencies, the length of time for the Fourier transform should be optimized so that the signal is approximately stationary (within the transforms frequency resolution). For most signal detection applications, we are interested in the statistics of the Fourier spectrum bin mean and standard deviation, given that the spectrum is estimated by averaging many Fourier transforms together (one assumes an ergodic input signal where each averaged buffer has the same underlying statistics). If the input signal is ZMG noise, both the real and imaginary bins are ZMG processes. The magnitude-squared of the spectral bin is found by summing the real part squared plus the imaginary part squared and results in a second-order Chi-Square PDF. For a time-domain ZMG signal with σ_t^2 variance, the N-point-normalized Fourier transform produces a magnitude-squared frequency bin with mean σ_t^2/N and variance σ_t^4/N^2. As one averages M Fourier magnitude-squared spectra, the mean for each bin stays the same, but the variance decreases as σ_t^4/MN^2 as M increases. The PDF for the averaged spectral bin is Chi-Square of order 2M. Given the underlying statistics of averaged spectra, one can assign a confidence interval where the bin value is say, 90% of the time. Examination of higher-order spectral statistics is also very useful in extracting more information about the spectrum, such as nonlinearities as seen in the bispectrum.

The transfer function of a linear time-invariant system can be measured by estimating the ratio of the time-averaged input–output cross-spectrum divided by the input autospectrum. Again, the statistics of the spectral bins plays a crucial role in determining the precision of measured system magnitude and phase. The spectral coherence is a very useful function in determining the precision of transfer function measurements as it is very sensitive to showing weak SNR as well as interfering signals in the input–output signal path. Using the observed coherence and the known number of averages, one can estimate a variance for the transfer function magnitude and phase responses. This error modeling is important because it allows one to determine the required number of spectral averages to achieve a desired precision in the estimated transfer function.

If the transfer function has very sharp modal features, the input–output buffers may have to be large for consistent ergodic averages to be possible. This is because the buffer must be long enough to measure the complete system response to the input. With random input signals in each buffer, some of the "reverberations" from the previous input buffers is not correlated with the current input, giving transfer function results which do not seem to improve with more averaging. This requires switching to a broadband periodic input signal such as an impulse or sinusoidal chirp, which is exactly reproduced in each input buffer. The magnitude response for a synchronous periodic input will quickly approach the spectral resolution limit defined by the buffer length, but some phase errors may still be present due to the "reverberation" from the previous buffers. Synchronous input signals also allow one to average in the time domain, or average the real and imaginary spectral bins separately, before computing the transfer function and magnitude and phase. This "time-synchronous averaging" virtually eliminates all interfering signals leaving only the input and output to be processed and is an extremely effective technique to maximize SNR on a difficult measurement.

While the extent to which we have explored acoustic, structural, and electromagnetic intensity here is somewhat involved, we will make use of these developments later in the text. The intensity techniques generally involve spectral products, cross-spectra, or various forms of finite-difference approximations using measured spectra. The underlying statistics of the spectral measurements translate directly into the precision of the various power flow measurements. Detailed measurements of the full field through active and reactive intensity scans can provide valuable information for controlling radiated power, optimizing sensor system performance, or for diagnostic/prognostic purposes in evaluating system integrity. Clearly, the most sophisticated adaptive sensor and control systems can benefit from the inclusion of the physics of full-field measurements.

PROBLEMS

1. Broadband electrical noise is known to have white spectral characteristics, zero mean, and a variance of $2.37v^2$. A spectrum analyzer (rectangular window) is used to average the magnitude-squared of 15 FFTs each of 1024 points in size. What is the mean and standard deviation of the spectral magnitude-squared bins? If the sample rate is 44.1 kHz, what is the noise level in $\mu v^2/Hz$?

2. Suppose a Hanning data window was used by mistake for the data in Question 1. What would the measured spectral density be for the noise?

3. A zoom FFT is done for the bandwidth from 200 to 210 kHz on digital data sampled a 2 MHz with a 1 MHz antialiasing filter. The signal levels are approximately 1 V rms in the frequency range from 200 to 210 kHz. The noise level for the sampled data in the 1 MHz band is approximately 1 mV rms. If a 1024-point FFT is used, what is the SNR in the zoom band in the estimated spectrum?

4. A transfer function of a transducer is known to have a very flat response. Would you expect a reasonable measurement of the transfer function using a rectangular window?

5. A transfer function is measured with poor results in a frequency range where the coherence is only 0.75. How many spectral averages are needed to bring the standard deviation of the transfer magnitude to with 5% of the actual value?

6. Derive an equation for estimating the SNR, given the coherence.

7. Given the measured intensity in three dimensions at random points on a closed surface around a source of interest, how does one compute the total radiated power? If the source is outside the closed surface, what is the measured power?

8. A 30 cm (12 in.)-diameter loudspeaker in a sealed cabinet is radiating 100 Hz. How far away does one need to be in order to estimate the intensity from the sound pressure-squared? Explain how this is done?

REFERENCES

1. W. S. Burdic, *Underwater Acoustic System Analysis*, 2nd ed., Chapters 9 and 13, Englewood Cliffs, NJ: Prentice-Hall, 1984.

2. W. H. Press, B. P. Flannery, S. A. Teukolsky, and W. T. Vetterling, *Numerical Recipes: The Art of Scientific Computing*, New York, NY: Cambridge University Press, 1986.

3. M. Boltezar and J. Slavic, Fault detection of DC electric motors using the bispectral analysis, *Mechanica*, 2006, 41, pp. 283–297.

4. S. Hagihira, M. Takashina, T. Mori, T. Mashimo, and I. Yoshiya, Practical issues in bispectral analysis of electroencephalographic signals, *Anesth Analg*, 2001, 93, pp. 966–70.

5. J. W. A. Fackrell and S. McLaughlin, Detecting nonlinearities in speech sounds using the bicoherence, *Proc Inst Acoust*, 1996, 18(9), pp. 123–130.

6. J. S. Bendat and A. G. Piersol, *Engineering Applications of Correlation and Spectral Analysis*, New York, NY: Wiley, 1993.

7. F. J. Fahy, *Sound Intensity*, New York, NY: Elsevier Science, 1989.

8. F. W. Sears, M. W. Zamansky, and H. D. Young, *University Physics*, Reading, MA: Addison-Wesley, 1977.

8 Wavenumber Transforms

The wavenumber k is physically the radian frequency divided by wave speed (ω/c), giving it dimensions of radians/m. Wavenumbers can be seen as a measure of wavelength relative to 2π. Actually, k also equals $2\pi/\lambda$ and can be decomposed into k_x, k_y, and k_z components to describe the wave for 3D spaces. The wavenumber k is the central parameter of *spatial signal processing* which is widely used in radar, sonar, astronomy, and in digital medical imaging [1]. Just as the phase response in the frequency domain relates to time delay in the time domain, the phase response in the wavenumber domain relates to the spatial position of the waves. The wavenumber transform is a Fourier transform, where time is replaced by space. Wavenumber processing allows spatial and even directional information to be extracted from waves sampled with an array of sensors at precisely known spatial positions. The array of sensors could be a linear array of hydrophones towed behind a submarine for long-range surveillance [2], a line of geophones used for oil exploration, or a 2D phased-array radar such as those used in missile defense systems [3]. The most common 2D signal data are video which can benefit enormously from wavenumber processing in the frequency domain. Wavenumber processing can be used to enhance imagery from telescopes, digitized photographs, and video where focus or camera steadiness is poor.

8.1 SPATIAL TRANSFORMS

We begin our analysis of wavenumber transforms by defining a plane wave moving with speed c (in m/s) in the positive r-direction as being of the form $e^{j(\omega t - kr)}$. With respect to the origin and time t, the wave is outgoing because the phase of the wave further out from the origin corresponds to a time in the past. As t increases a constant phase point on the wave propagates outward. If $r = (x^2 + y^2 + z^2)^{1/2}$ represents a distance relative to the origin, the respective wavenumber components k_x, k_y, and k_z in the x, y, and z directions can be used to determine the direction of wave propagation. Recall that to measure temporal frequency, a signal is measured at one field point and sampled at a series of known times allowing a time Fourier transform on the time series to produce a frequency spectrum. To measure the wavenumber spectrum, the waveform is sampled at one time "snapshot" over a series of known field positions allowing a spatial Fourier transform to be applied producing a wavenumber spectrum. Since the wavenumber is $2\pi/\lambda$, long wavelengths correspond to small wavenumbers and short wavelengths correspond to high wavenumbers.

Consider the pinhole camera in Figure 8.1. On any given spot on the outside of the camera box, the light scattered from the field produces a "brightness" wavenumber spectrum. Another way to look at the physics is that light rays from every object in the field can be found on any given spot on the outside of the camera box. By making a pinhole, the camera box allows reconstruction of the object field as an upside-down and left–right reversed image on the opposite side of the pinhole on what we call the focal plane.

One can describe the magic of the pinhole producing an image inside the camera as a 2D inverse wavenumber transform where the wavenumber spectrum at the pinhole spot is $I(k_x, k_y)$.

$$i(x, y) = \int\limits_{-\infty}^{+\infty} \int\limits_{-\infty}^{+\infty} I(k_x, k_y) e^{+j(k_x x + k_y y)} \, dx \, dy \tag{8.1}$$

The ray geometry which allows reconstruction of the image is represented by the wavenumber components k_x and k_y for the image coordinates x and y. Small wavenumbers near zero correspond to

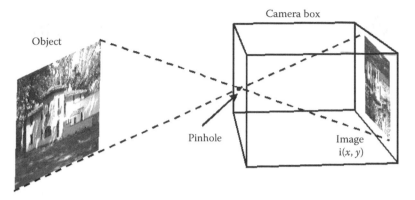

FIGURE 8.1   A pinhole camera produces a clear upside down and left–right reversed image due to the ray traces from the object to the image, but little light energy reaches the image plane because of the small aperture.

light with slowly varying brightness (long wavelengths) across the pinhole side of the camera while the large wavenumbers correspond to sharply varying brightness such as edges from shadows.

Some of the earliest cameras used a simple pinhole and extremely long exposures on the photographic plate due to the low light levels. Lens-type cameras are far more efficient at gathering and focusing light. The shape and index of refraction for the lens bends the light rays creating a focal point and an inverted image. The focal point is geometrically analogous to the pinhole, but light is highly concentrated there by the lens. This allows faster exposure times. The amount of light is also controlled by the aperture which is also referred to as the entrance pupil in most optics texts. A border usually forming a rectangular boundary around the photographic film is the exit pupil. Dilating or constricting the aperture does not affect the exit pupil boundary, but rather just impacts the brightness of the image since a large number of photons are collected on the focal plane in a given amount of time exposure.

As shown in Figure 8.2, the aperture can have an effect on focus of objects near and far. Note that for the far object parallel rays enter the lens and are refracted depending on the angle of incidence. A near object produces a focal point closer than a far object. Therefore, far objects in the background

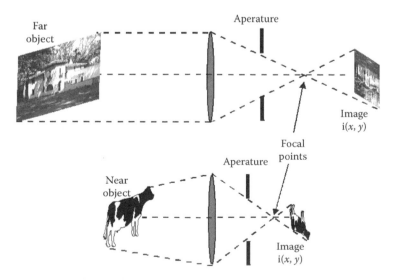

FIGURE 8.2   A lens system increases the amount of light for the photograph, but also introduces a finite depth of field where objects are in focus on the image plane.

can appear out of focus when a wide aperture (small F-stop on a camera lens) is used. Narrowing the aperture will tend to allow only the nearly parallel rays into the camera thus improving the focus of both near and far objects. The entire field of view is in focus for the pinhole camera. Focus can be seen as the situation where the wavenumbers are coherently "filtered" by the camera to faithfully reproduce a sharply defined image. The "fuzziness" of out-of-focus objects is really due to "leakage" of wavenumber components into other areas of the image surrounding the "true" intended spot. The physics to be noted here is that "focus" for a camera system is definable as a wavenumber filter and that out-of-focus objects can (in theory) be digitally recovered if the lens system wavenumber filtering response is known. The transfer function of the lens system is quite complicated and will change depending on aperture, focus, and also for objects near and far.

Before the corrective optics was installed in the orbiting Hubble Space Telescope, the focus of initial images was controlled using Fourier image processing. We will examine this technique in detail below. In short, the "fuzziness" in an astronomical image can be dealt with by picking a distant star in the field of view. Under ideal conditions, this distant star might be detected in only 1 pixel. However, the reality in the early days of Hubble was that the distant star's light would smear over all the given area of the image in what some old-school photographers sometimes call "circles of confusion" of poor focus. From a signal processing point of view, the circles of confusion can be seen as a spatial impulse response of a wavenumber filter. Performing a 2D Fourier transform of the spatial impulse response, or *point spread function* (PSF) gives the wavenumber frequency response. The optimal sharpening filter for the distant star is found from inverting the wavenumber frequency response for the fuzzy distant star and multiplying the entire image's wavenumber frequency response by this corrective wavenumber filter. The inverse 2D Fourier transform of the corrected wavenumber frequency response gives an image where the formerly fuzzy distant star appears sharp along with the rest of the image (assuming that the fuzziness was homogeneously distributed). The technique is analogous to measuring and inverting the frequency response of a microphone to calculate the optimal FIR filter via inverse 1D Fourier transform which will filter the microphone signals such that the microphone's frequency response appears perfectly uniform. Astronomers have been applying these very powerful image restoration techniques for some time to deal with atmospheric distortions, ray multipaths, and wind buffeting vibrations of ground-based telescopes.

8.2 SPATIAL FILTERING AND BEAMFORMING

A very common coherent wavenumber system is the parabolic reflector used widely in radar, solar heating, and for acoustic sensors. Consider an axially symmetric paraboloid around the y-axis with the equation $y = ar^2 + b$, where r is a radial measure from the y-axis. To find the focal point, or focus, one simply notes where the slope of the paraboloid is unity. Since the angle of incidence equals the angle of the reflected wave, the rays parallel to the y-axis reflect horizontally and intersect the y-axis at the focus as seen in the lightly dotted line in Figure 8.3. This unity slope ring on the parabola will be at $r = 1/(2a)$ giving a focus height of $y = 1/(4a) + b$. For rays parallel to the y-axis, all reflected rays will have exactly the same path length from a distant source to the focus. Physically, this has the effect of integrating the rays over the dish cross section of diameter d into the sensor located at the focus. This gives a wavelength λ dependent gain of $1 + 2d/\lambda$, which will be presented in more detail in Section 13.2. For high frequencies where the gain is quite large, the dish provides a very efficient directional antenna which allows waves from a particular direction to be received with high signal-to-noise ratio (SNR). At low frequencies, the directional effect and gain are much smaller. This gain defines a "beam" where the parabolic reflector can be aimed toward a source of interest for a very high-gain communication channel. The analogy to an arc-lamp with a parabolic reflector producing a spotlight beam is quite accurate for the high-frequency sound case.

The "beam" created by the parabolic reflector is the result of coherent wavenumber filtering. All waves along the "look" of the beam add direction coherently at the focus F in Figure 8.3. Waves

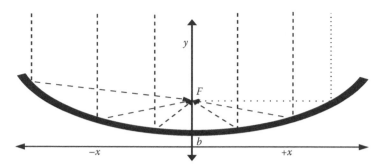

FIGURE 8.3 Parabolic reflectors focus plane waves from a normal direction onto a sensor at the focus of the parabola to greatly increase the amplitude especially at high frequencies.

from other directions will appear as well at the focus, but much weaker relative to the look direction (LD) beam. For example, some waves will scatter in all directions at the reflector's edges (edge diffraction) but the area for this scattering is extremely small compared to the area of the reflector as a whole which produces the coherent reflections for the LD beam.

The material of the parabolic reflector plays an important role in defining the frequency range of useful operation. For example, a mirror finish is required for visible light frequencies (due to their extremely short wavelengths), a conductive wire mesh will suffice for radar frequencies, and a solid dense material such as plastic will provide an atmospheric acoustic reflection. For underwater applications, one might provide an air volume in the shape of a parabolic dish to achieve a high reflection. The parabolic shape is somewhat problematic to manufacture, and often a spherical shape is used as an approximation. The parabolic reflector can be very useful as a movable "search beam" to scan a volume or plane for wave-emitting sources or reflections from an adjacent active transmitter allowing the locations of the reflecting sources to be determined. As will be discussed below, applications to direction-finding and location of sources or wave scatterers is a major part of wavenumber filtering applications. Moving of the parabolic dish mechanically is sometimes inconvenient and slow, and so many modern radar systems electronically "steer" the search beam. These operations are generally referred to as "array processing" because an array of sensors is used to gather the raw wavenumber data. The amount of computation is well suited for parallel processing architectures. Current microprocessors provide a very economical means to steer multiple search beams electronically as well as design beams which also have complete cancellation of waves from particular directions.

Consider a "line array" of acoustic sensors (radio frequency (RF) antennae could also be used) as shown in Figure 8.4, where three distant sources "A," "B," and "C" over 100 wavelengths away provide plane waves at the line array. The outputs of each of the array elements are filtered to control relative amplitude, time delay, and phase response and combined to give the line array the desired spatial response. If the array processing filters are all set to unity gain and zero phase for all frequencies and the sensor frequency responses are all identical in the array, the processor output for source A will be coherently summed while the waves from sources B and C will be incoherently summed. This is because the plane wave from source A has the same phase across the array while the phases of the waves from sources B and C will have a varying phase depending on the wavelength and the angle of incidence with respect to the array axis. A simple diagram for the first two array elements in Figure 8.5 illustrates the relative phase across the array from a distance source's plane waves. Obviously, when θ is 90°, the distance R to the source is the same, and thus, the plane wave phase is the same across the array and the summed array output (with unity weights) enhances the array response in the $\theta = 90°$ direction. The phase of the wave at the first element $y_{1,t}$ relative to the phase of the source is simply kR, where k is the wavenumber. Remembering that the wavenumber has units of rad/m, k can be written as either $2\pi/\lambda$, or $2\pi f/c = \omega/c$, where λ is wavelength,

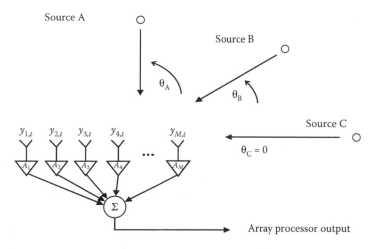

FIGURE 8.4   A line array of sensors can determine the direction of arrival of a plane wave from a distant source.

$2\pi f = \omega$ is the radian frequency, and c is the speed of the wave in m/s. With all the array weights $A_i = 1$; $i = 1, 2, \ldots, M$, the summed array output s_t is simply

$$s_t = P_0 e^{j(\omega t - kR)} \sum_{m=1}^{M} A_m e^{+jk(m-1)d\cos\theta} \tag{8.2}$$

where R is the distance from the source and d is the array element spacing. Note the opposite sign of the ωt and the kR in the exponent. As time increases, the distance R must increase for a given phase point on the plane wave. When $\theta = 90°$, s_t is M times louder than if a single receiver was used to detect the incoming plane wave. The mathematical structure of Equation 8.2 is identical to that of a Fourier transform. For other angles of arrival, the array output level depends on the ratio of d/λ and the resulting relative phases of the array elements determined by the exponent $j2\pi(m-1) \cos\theta/\lambda$ in Equation 8.2. When d/λ is quite small (low frequencies and small array spacings') the spatial phase changes across the array will be small and the array output will be nearly the same for any angle of arrival (AOA).

The net phase from the wave propagation time delay is really of no concern. The relative phases across the array are of interest because this information determines the direction of arrival within the half-plane above the line array. Given that we can measure frequency and array element positions with great accuracy, a reasonable assumption for the speed of sound or even wavelength for a given

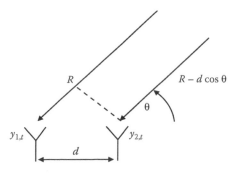

FIGURE 8.5   A distant source's plane wave will have a sinusoidal varying phase across the line array depending on the frequency and direction of arrival.

plane wave frequency is all that is actually needed to turn the measured relative phase data into source bearing information. Clearly, the phase difference, $\Delta\varphi_{21}$, found by subtracting the measured phase of sensor 1 from sensor 2 is simply $kd\cos\theta$. Therefore, the bearing to the distant source can be measured from the phase difference by

$$\theta = \cos^{-1}\left(\frac{\Delta\varphi_{21}}{kd}\right) \tag{8.3}$$

The simple formula in Equation 8.3 is sometimes referred to as a "direction cosine" or "phase interferometric" bearing estimation algorithm. Early aircraft navigation systems essentially used this direction-finding approach with a dipole antenna to determine the direction to a radio transmitter beacon. This technique is extremely simple and effective for many applications, but it is neither beamforming nor array processing, just a direct measurement of the AOA of a single plane wave frequency. If there are multiple angles of arrival for the same frequency then the direction cosine technique will not work. However, the relationship between measured spatial phase, the estimated wavenumber, and the AOA provides the necessary physics for the development of direction-finding sensor systems.

Returning now to the array output equation in Equation 8.2, we see that the array output depends on the chosen number of array elements M, weights A_i, source angle, and the ratio of element spacing to wavelength. Figure 8.6 displays the response of a 16-element line array with unity weights for ratios of element spacing to wavelength (d/λ) of 0.05, 0.10, and 0.50. As the frequency increases toward a wavelength of $2d$, the beam at 90° becomes highly focused. If the element spacing is greater than half a wavelength, there will be multiple angles of arrival where the array response will have high gain at a focused angle. These additional beams are called "grating lobes" after light diffraction gratings which separate light into its component colors using a periodic series of finely spaced slits. The same physics are at work for the operation of the line array beamforming algorithm and the optical spectroscopy device using diffraction gratings to allow detection of the various color intensities in light. Figure 8.7 shows the beam response as the frequency is increased to give d/λ ratios of 0.85, 1.00, and 1.50. In general, the array element spacing should be at most a half-wavelength to ensure no grating lobes for any AOA. The presence of grating lobes can be seen as

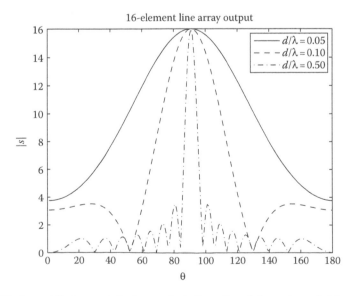

FIGURE 8.6 A 16-element line array output magnitude for ratios of element spacing to wavelength (d/λ) of 0.05, 0.10, and 0.50.

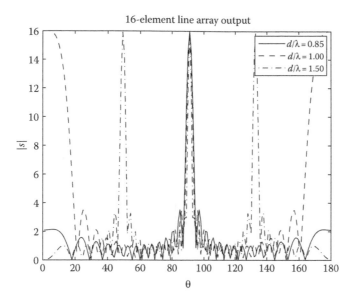

FIGURE 8.7 Beam response of a 16-element line array at high frequencies where the ratio d/λ is 0.85, 1.0, and 1.5 showing the emergence of "grating lobes" to the beam pattern.

aliasing in spatial sampling. Adding more elements to the array while keeping the same aperture will eliminate the grating lobes. Also, the narrowness of the main LD beam depends on the total array aperture (M times d). It can be shown that for large arrays ($M \gg 16$), the beam width in degrees is approximately $180\lambda/M\, d\pi$; so when the array aperture is many wavelengths in size, the resulting beam will narrow and highly desirable for detecting plane waves from specific directions.

One of the more interesting aspects of the weights A_m in Figure 8.4 and Equation 8.2 is that the magnitudes of the weights can be "shaded" using the spectral data windows described in Chapter 6 to suppress the side-lobe leakage. The expense of this is a slight widening of the main lobe, but this effect is relatively minor compared to the advantage of having a single main beam with smooth response. Figure 8.8 compares a Hanning window to a rectangular window for the weights A_m. Recall that the Hanning window is $A_m = 1/2\,[1 - \cos\,(2\pi m/M)]$ for an M-point data window; that is, $m = 1, 2, \ldots, M$. For large M, the narrowband data normalization factor is 2, allowing "on-the-bin" spectral peaks to have the same amplitude whether a rectangular or Hanning window is used. For small arrays, we have to be a little more careful and integrate the window and normalize it to M, the integral of the rectangular window. For the $M = 16$ Hanning window, the integral is 8.5 making the narrowband normalization factor 16/8.5 or 1.8823. As M becomes large the normalization factor approaches 2.0 as expected.

The phase of the array weights A_m can also be varied to "steer" the main lobe in a desired LD other than 90°. Electronic beam steering is one of the main advantages of the array processing technique because it avoids the requirement for mechanical steering of devices such as the parabolic dish to move the beam. Furthermore, given the signals from each array element, parallel processing can be used to build several simultaneous beams looking in different directions. A snapshot of array data can be recorded and then scanned in all directions of interest to search for sources of plane waves. To build a steering weight vector one simply subtracts the appropriate phase from each array element to make a plane wave from the desired LD θ_d, sum coherently in phase. The steering vector including the Hanning window weights is therefore

$$A_m = \frac{A_0}{2}\left[1 - \cos\left(\frac{2\pi m}{M}\right)\right] \mathrm{e}^{-\mathrm{j}2\pi(d(m-1)/\lambda)\cos\theta_d} \tag{8.4}$$

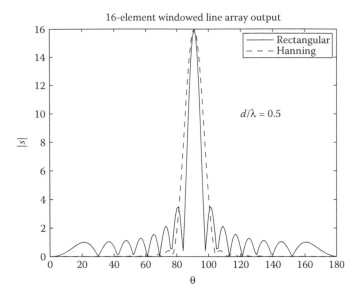

FIGURE 8.8 Use of windows to "shade" the line array elements greatly reduces side-lobe leakage, but broadens the main lobe giving less directional resolution to the beam.

where A_0 is the narrowband normalization scale factor for the Hanning window. Recalling Equation 8.2, the application of the weights in Equation 8.4 would allow a plane wave from a direction θ_d to pass to the array output in phase coherency giving an output M times that of a single sensor. Figure 8.9 shows the beam response for steering angles of 90°, 60°, and 30° with a Hanning window and a sensor spacing to aperture ratio of 0.5.

Some rather important physics are revealed in Figure 8.9. As the beam is steered away from the broadside direction ($\theta_d = 90°$) toward the endfire directions ($\theta_d = 0°$ or 180°), the beam get wider because the effective aperture of the line array decreases toward the endfire directions. This decrease

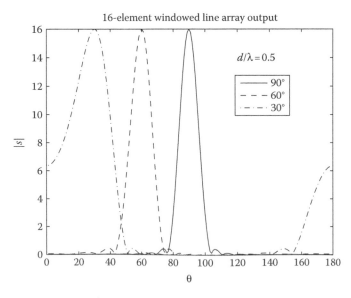

FIGURE 8.9 Adjusting the phase of the array weights allows the main-lobe LD beam to be electronically steered to a desired direction.

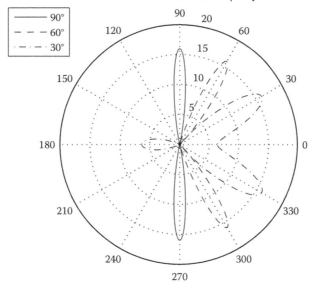

FIGURE 8.10 The complete 360° response of the line array shows the symmetry of the beam pattern about 0° for steering angles of 90° (—), 60° (– –), and 30° (· – · –).

in effective aperture can be expressed approximately as $Md \sin \theta_d$ for steering angles near 90°. But, as one steers the beam near the endfire directions, the beam response depends mainly on the spacing to wavelength ratio, as will be demonstrated shortly. Since the line array is 1D, its beam patterns are symmetric about the axis of the array. Figure 8.10 shows the complete 360° response of the line array for the three steering angles given in Figure 8.9. Note the appearance of a "backward" beam near 180° for the beam steered to 30°. This "forward–backward" array gain is often overlooked in poor array processing designs, but tends to go away at lower frequencies as shown in Figure 8.11 for

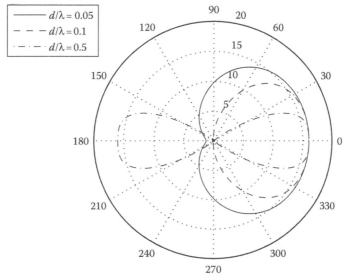

FIGURE 8.11 The complete beam pattern for steering angles of 0° at frequencies where $d/\lambda = 0.05$ (—), 0.10 (– –), and 0.50 (· – · –) showing the forward–backward gain loss of performance at 180° and high frequencies.

several frequencies steered to 0°. The overall size and symmetry of the array determines the maximum beam as a function of frequency and angle as well as the forward–backward gain performance. There is really no optimal sensor layout for an array, but some geometries offer various advantages over others. Perhaps more often overlooked are the effects of objects/structures near each array element such as sensor frames and supporting structures which can significantly change array performance at high frequencies.

The way one makes use of the steered beams in an electronic scanning system is to record a finite block of time data from the array elements and scan the data block *algorithmically* in all directions of interest. The directions where the steered beam has a high output level correspond to the directions of distant sources, provided the beam pattern is acceptable. Figure 8.12 shows a scanned output for the 32-element line array where the four sources at 45°, 85°, 105°, and 145° with levels of 10, 20, 25, and 15, respectively. The frequency to spacing ratio is 0.4 and a Hanning window is used to shade the array and suppress the side-lobe leakage compared to the unshaded rectangular window.

Figure 8.12 clearly shows four distinct peaks with angles of arrival and levels consistent with the sources (the array output magnitude is divided by $M = 32$). However, at lower frequencies the resolution of the scanning beam is not sufficient to separate the individual sources. Limited resolution is due to the finite aperture Md which must be populated with sufficient elements so that a reasonable high-frequency limit is determined by the intraelement spacing to wavelength ratio d/λ. Increasing the number of elements M increases the line array aperture Md and narrows the peaks in Figure 8.12. Obviously, one must also limit the number of elements to keep computational complexity and cost within reason. It is extremely important to understand the fundamental physics governing array size, resolution, element spacing, and upper frequency limits.

While a 1D line array can determine a plane wave AOA within a half-plane space, a 2D planar array can determine the azimuth and elevation angles of arrival in a half-space. We define the angle ψ as the angle from the normal to the plane of the 2D array. If the array plane is horizontal, a plane wave from the direction $\psi = 0$ would be coming from a direction straight up, or normal to the array plane. Waves from $\psi = 90°$ would be coming from a source on the horizon with an azimuth direction defined by θ, as defined earlier for the 1D line array. The elevation angle of the incoming plane wave is very important because it affects the measured wavelength in the array plane by making it

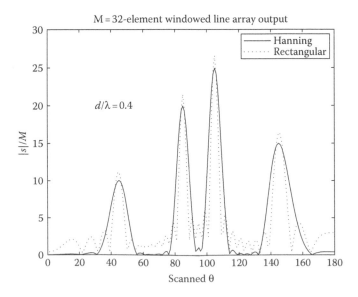

FIGURE 8.12 A scanned 32-element line array output showing sources at 45°, 85°, 105°, and 145° for both a Hanning-shaded beam and an unshaded (rectangular window) beam.

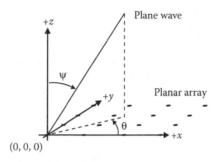

FIGURE 8.13 A 2D planar array showing the azimuth angle θ and elevation angle ψ relative to the Cartesian coordinate system.

appear longer as the elevation angle approaches zero. Figure 8.13 graphically depicts a 16-element planar array arranged in a 4 × 4 equally spaced sensor matrix.

Since the effective wave speed in the x–y plane of the 2D array is $c_{xy} = c/\sin \psi$, where c is the wave speed along the direction of propagation, we define the wavenumber k_{xy} for the x–y plane as $2\pi f/c_{xy}$, or more explicitly, $2\pi f \sin \psi/c$. Using the sensor layout in Figure 8.13 where the spacing along the x-axis d is the same as the spacing along the y-axis, we can define a beam steering weight $A_{m,n}$, by simply subtracting the appropriate phase for a given sensor to make the array output phase coherent for the desired LD in θ and ψ.

$$A_{m,n}(f) = A_{m,n}^{w} e^{-j\left[(2\pi f d \sin\varphi/c)\{(m-1)\cos\theta + (n-1)\sin\theta\}\right]} \tag{8.5}$$

The frequency-independent parameter $A_{m,n}^{w}$ in Equation 8.5 represents the spatial data window for controlling side-lobe leakage. The 2D Hanning window is defined as

$$A_{m,n}^{w} = \frac{A_0}{4}\left[1 - \cos\left(\frac{2\pi m}{M}\right)\right]\left[1 - \cos\left(\frac{2\pi n}{N}\right)\right] \tag{8.6}$$

where A_0 is the narrowband normalization factor to equalize the array output peak levels with what one would have using a rectangular window where all $A_{m,n}^{w}$ equal unity. For a desired beam LD defined by θ_d and ψ_d, one calculates the array beam steering weights $A_{m,n}(f)$ for each frequency of interest. A convenient way to process the array data is to compute the Fourier spectra for each array element and multiply each Fourier frequency bin by the corresponding frequency steering weight in a vector dot product to produce the array output. One can then produce an array output spectrum for each desired LD to scan for potential source detections. The steering vectors for each frequency and direction are independent of the signals from the array elements. Therefore, one typically "precomputes" a set of useful steering vectors and methodically scans a block of Fourier spectra from the array elements for directions and frequencies of high signal levels. Many of these applications can be vectorized as well as executed in parallel by multiple signal processing systems.

8.3 IMAGE ENHANCEMENT TECHNIQUES

One of the more astute signal processing techniques makes use of 2D Fourier transforms of images to control focus, reduce noise, and enhance features [4]. The first widespread use of digital image enhancement is found among the world's astronomers, where the ability to remove noise from imagery allows observations of new celestial bodies as well as increased information. Obviously, satellite imagery of earth, has for some time now, been fastidiously examined for state surveillance purposes.

The details of particular image enhancement techniques and how well they work for satellite surveillance are probably some of the tightest held state technology secrets. However, we can examine the basic processing operations through the use of simple examples which hopefully will solidify the concepts of wavenumber response and filtering. As presented in Section 5.3, image processing techniques can be applied to any two-, or even higher, dimensional data to extract useful information.

The PSF for an imaging system is analogous to the impulse response of a filter in the time domain. The input "impulse" to a lens system can be simply a black dot on a white page. If the lens system is perfectly focused, a sharp but not perfect dot will appear on the focal plane. The dot cannot be perfect because the index of refraction for a glass or plastic lens is not exactly constant as a function of light wavelength. Blue light will refract slightly different from red, and so on. Our nearly perfect dot on the focal plane will upon close examination have a "rainbow halo" around it from chromatic aberrations as well as other lens system imperfections. The 2D fast Fourier transform (2DFFT) of the PSF gives the wavenumber response of the camera. If our dot were perfect, the 2DFFT would be constant for all wavenumbers. With the small halo around the dot, the 2DFFT would reveal a gradual attenuation for the higher wavenumbers (shorter wavelengths) indicating a small loss of sharpness in the camera's imaging capability. As the lens is moved slightly out-of-focus, the high wavenumber attenuation increases. Focus can be restored, or at least enhanced, by normalizing the 2DFFT of the image by the 2DFFT of the PSF for the out-of-focus lens producing the image. This is analogous to inverse filtering the output of some system to obtain a signal nearly unchanged by that system. For example, one could amplify the bass frequency range of music before it passes through a loudspeaker with weak bass response to reproduce the music faithfully. However, for imagery, one's ability to restore focus is quite limited due to limited SNRs and that fact that the PSF is attenuated rapidly into the noise when spread circularly in two dimensions. Optical aberrations can cause the PSF to be a small cluster of dots, each representing a lens system path. For example, a pair of well-defined dots would represent "double vision" superimposing two complete images. For multiple lens systems, there can be multiple point PSFs which even vary across the focal plane. In theory, these imperfections can be removed in postprocessing. In practice, there exists a lot of work and best to be avoided by simply having the best possible imaging system to start with.

Astronomers can use distant stars in their images to try to measure the PSF, which includes the light refracting and scattering effect of the atmosphere as well as the telescope. The "twinkling" of stars actually happens due to atmospheric turbulence and scattering from dust. The PSF for a distant star can sometimes look like a cluster of stars which fluctuate in position and intensity. Inverting the 2DFFT for the distant star section of the image, one can use the inverse filter to clarify the entire image. The technique usually works well when the same refractive multipath seen in the distant star section of the image applies throughout the image, which is reasonable physically since the angle of the telescope's field of view is extremely small. Recently, real-time feedback controllers have been used to bend flexible mirrors using a "point source" PSF criteria on a distant star to effectively keep the image sharp while integrating.

Consider the house image shown earlier in the book, but this time clipped into a 256×256 pixel image where each pixel is an 8-bit gray scale in Figure 8.14. The house image reveals some camera noise as well as pixilation and Joint Photographics Expert Group (JPEG) data compression inherent in modern digital images. Figure 8.15 shows a 2DFFT of the house image where a log base 10 amplitude scale is used to make the details more visible. Otherwise, on a linear scale most of the signals would be concentrated in the center of the figure. We have applied an "fftshift" to move the 0 wavenumber point to the center of the 2DFFT image. The 2DFFT reveals some very interesting image characteristics. The equation for computing the 2D Fourier transform of an image $i(x, y)$ is

$$I(k_x, k_y) = \int_{-\infty}^{+\infty} \int_{-\infty}^{+\infty} i(x, y) e^{-j(k_x x + k_y y)} \, dx \, dy \qquad (8.7)$$

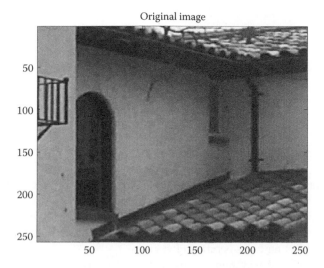

FIGURE 8.14   The house picture example shown in Chapter 5 clipped to 256 × 256 pixels and an 8-bit gray scale.

The equation for the inverse 2D Fourier transform is expressed in Equation 8.1.

One can easily see some dominant horizontal bands near $ky = 0$ (across the middle of Figure 8.15) indicating large horizontal bands of light and dark across the original image in Figure 8.14 (the dark tile roof, white stucco, and dark iron balcony). Horizontally, the arched windows and many different vertical band widths give a "sinusoidal" pattern in the region along $k_x = 0$ (the vertical band in the middle of Figure 8.15). The phase information in the 2DFFT places the various bright/dark regions on the correct spot on the original image. One can see a fair amount of noise throughout the 2DFFT as well as the original image, which we will show can be suppressed without much loss of visual information.

In the 2D wavenumber domain, one can do filtering to control sharpness just like one can filter audio signals to control "brightness" or "treble." The high wavenumber components which correspond to the sharp edges in the image are found in the outskirts of the 2DFFT in Figure 8.15. We can suppress the sharp edges by constructing a 2D wavenumber filter for the $M \times M$ discrete wavenumber spectrum.

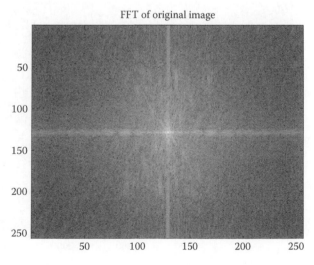

FIGURE 8.15   A 2DFFT of the house with the 0,0 wavenumber shifted to the center and the amplitude displayed on a $\log_{10}$ scale (base-10 logarithm) to show details at higher wavenumbers.

$$w^{lp}(m_x, m_y) = \left[1 - \cos\left(\frac{2\pi m_x}{M}\right)\right]^{12}\left[1 - \cos\left(\frac{2\pi m_y}{M}\right)\right]^{12} \tag{8.8}$$

For an $N \times N$ image, the discrete 2D Fourier transform is

$$I(m_x, m_y) = \sum_{n_x=1}^{N}\sum_{n_y=1}^{N} i(n_x, n_y)\, e^{-j2\pi\left(\frac{m_x n_x}{N} + \frac{m_y n_y}{N}\right)} \tag{8.9}$$

Multiplying the wavenumber transform in Equation 8.9 by the LPF in Equation 8.8, one obtains a "low-pass-filtered" wavenumber transform of the house image as seen in Figure 8.16 on a log scale. This is not a matrix multiply, but rather a matrix "dot product," where each wavenumber is attenuated according to the filter function in Equation 8.8. The filter function is rather steep in its "roll-off" so that the effect on focus is rather obvious.

Given the filtered $M \times M$ wavenumber spectrum (M is usually equal to N) $I^{lp}(m_x, m_y) = I(m_x, m_y)\, w^{lp}(m_x, m_y)$, the low-pass-filtered $N \times N$ image $i^{lp}(n_x, n_y)$ can be computed using a 2D inverse Fourier transform.

$$i^{lp}(n_x, n_y) = \sum_{m_x=1}^{M}\sum_{m_y=1}^{M} I^{lp}(m_x, m_y)\, e^{+j2\pi\left(\frac{m_x n_x}{M} + \frac{m_y n_y}{M}\right)} \tag{8.10}$$

Figure 8.17 shows the house image after an inverse 2DFFT in Equation 8.5. The softening of all the edges in the image is a direct result of the attenuation of high wavenumbers (high frequencies) in the original image.

Can we enhance all the sharp edges in Figure 8.14? We can use a high-pass filter somewhat similar to the LPF. The high-pass filter is unity at the highest wavenumbers and declines to zero for zero wavenumber (at the center). The inverse Fourier transform of the removed high-frequency

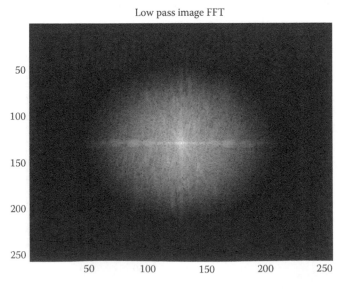

Low pass image FFT

FIGURE 8.16 Low-pass filtering in the frequency domain suppresses the high wavenumbers while passing the low wavenumbers near the center of the 2DFFT.

Low-pass-filtered image

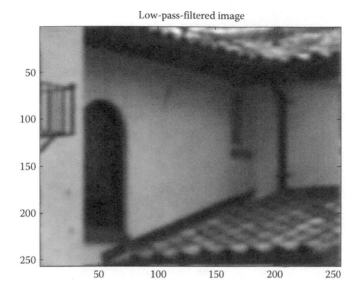

FIGURE 8.17 The inverse 2DFFT of the low-pass-filtered spectrum showing the "softening" or "defocusing" of the original image edges by attenuation of the high wavenumbers (short wavelengths).

components is shown in Figure 8.18 and the corresponding 2DFFT of the high-pass-filtered image is shown in Figure 8.19.

From an engineering point of view, one must ask how much visual information is actually in a given image, how can one determine what information is important, and how one can manage visual information to optimize cost. A reasonable approach is to threshold detect the dominant wavenumbers and attempt to reconstruct the image using information compression based on the strongest wavenumber components. But, the practical situation turns out to be even a bit more complicated than that since important textures and features can be lost by discarding all but the most

High-pass-filtered image

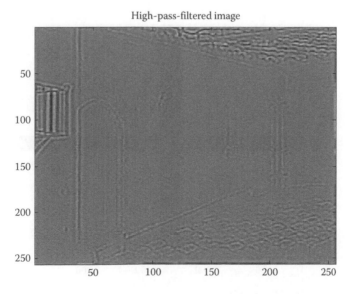

FIGURE 8.18 High wavenumber components of the image are enhanced by high-pass filtering in the frequency domain and then inverse Fourier transforming the filtered wavenumber spectrum.

High-pass image FFT

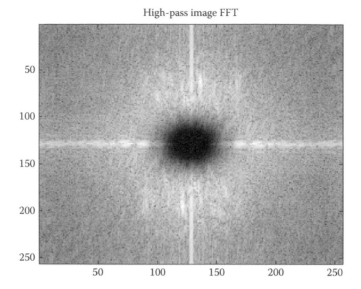

FIGURE 8.19 The spectrum of the high-pass-filtered image in Figure 8.18 showing the attenuation of the low wavenumbers near the 0,0 wavenumber bin at the center.

dominant wavenumbers. The short answer is that the optimal image data compression generally depends on the type of image data being compressed. The dominant wavenumber approach would work best on images with periodic patterns. Using a simple threshold in the image domain would work well for line art, or images with large areas on one solid shade. One of the dominant strategies in Motion Picture Experts Group (MPEG) and MPEG2 image/movie compression algorithms is run length encoding, where large areas of constant shade can be reduced to only a few bytes of data without loss of any information.

Just for fun, let us set the pixels brightness to zero on a vertical strip of the 2DFFT through to 0 wavenumber center as well as a horizontal strip as shown in Figure 8.20. The corresponding image

Just for fun FFT

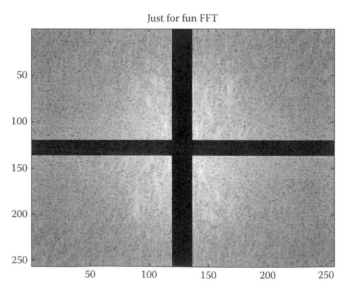

FIGURE 8.20 "Just for fun" we zero out the wavenumbers around the $k_x = 0$ (vertical strip) and $k_y = 0$ (horizontal strip).

Just to fun image

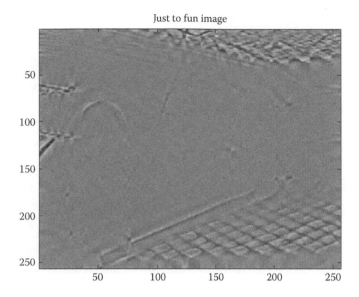

FIGURE 8.21 The inverse 2DFFT of the spectrum in Figure 8.20 has nearly all vertical and horizontal features removed, leaving curved and diagonal lines only as a result of the "Just for fun" filtering.

is given in Figure 8.21 and shows, amazingly, only the image features not associated with vertical or horizontal lines. Note how the diagonal support is prevalent in the image while the vertical bars of the wrought iron balcony are suppressed. Now for even more fun, let us do the opposite by throwing away all of the images' 2DFFT except for the vertical and horizontal strips through the origin as shown in Figure 8.22. These strips are only 7×256 pixels in the frequency domain; $1792 \times 2 = 3584$ wavenumber pixels. This is from the original image of 65536 pixels, or about 5.5% of the original image. The inverse 2DFFT is given in Figure 8.23 and amazingly, it shows

More fun FFT

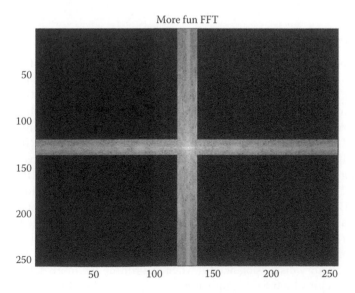

FIGURE 8.22 "More Fun" is seen by doing the opposite of Figures 8.20 and 8.21 and now zeroing out all the wavenumbers except those along the $k_x = 0$ (vertical strip) and $k_y = 0$ (horizontal strip).

More fun image

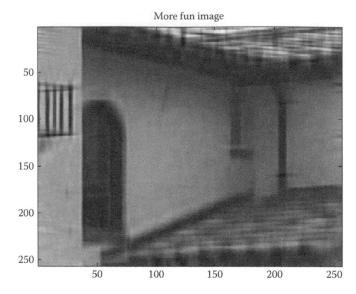

FIGURE 8.23 Here we see the vertical and horizontal image features restored from the inverse 2DFFT of the spectrum in Figure 8.22 and the curved and diagonal image features attenuated.

much of the original image, less some details in the roof tile and other minor areas. There are two important lessons to be remembered from this little exercise in image wavenumber filtering. First, the largest wavenumbers are near the origin and represent the overall brightness, or large areas of a given shade. Second, the high frequency (high wavenumber) details are relative low amplitude. These details would not be visible in the FFT plots of Figures 8.15, 8.19, 8.20, and 8.22 if we did not use a logarithmic scale in the plot. Therefore, much of the detailed information in a typical picture is not significant from an information point of view, as it is only visible in textures and the sharpness of boundaries. This is a solid basis for image compression algorithms discussed in Section 8.4.

8.4   JPEG AND MPEG COMPRESSION TECHNIQUES

The JPEG developed a number of international standards and file formats for lossy compression of image data files [5]. Lossy means that some of the image information is lost unrecoverably in the process, but this information is deemed unimportant to the image quality, or a viable trade-off for having a smaller image file to store, transmit, or compile into a web page. We cannot go into all details here, but a tremendous amount of up-to-date descriptions and example are available on the World Wide Web. Image compression has played a huge role in the success of the Internet since it allows web pages to download quickly with detailed photos of products, places, and services using a minimum of available Internet bandwidth. As with MP3 audio, the file formats and "players" are open public technologies, while many of the techniques for encoding (or generating the JPEG image from the raw digital file) are proprietary. This encourages innovation in products such as cameras and scanners, but avoids the hassle of formats that few people can read without proprietary software. Our description here is very cursory, but follows conveniently after our discussion in the last section of wavenumber processing of images. Hopefully, the reader will find this introductory discussion helpful when examining the more detailed online sources.

JPEG encoding begins by breaking up the picture into smaller blocks, usually 32×32 pixels. Each block is transformed to the wavenumber domain using a discrete cosine transform (DCT). A DCT of size N is essentially the same as an FFT of size $4N$ where all the imaginary inputs are zero and all the even-numbered input samples are zero. Equation 8.14 defines the DCT mathematically for an image matrix B_{mn} and DCT coefficients C_{pq}.

$$C_{pq} = \alpha_p \alpha_q \sum_{m=0}^{M-1} \sum_{n=0}^{N-1} B_{mn} \cos \frac{\pi(2m+1)p}{2M} \cos \frac{\pi(2n+1)q}{2N}, \quad \begin{array}{l} 0 \le p \le M-1 \\ 0 \le q \le N-1 \end{array} \tag{8.11}$$

Note that the samples of the cosines are all odd numbers and that for the $p = 0$ and $q = 0$ cases the normalization of the sum is twice that for all the other wavenumbers. Because of this, separate scaling factors are needed for the 0 wavenumber cases.

$$\alpha_p = \begin{cases} \dfrac{1}{\sqrt{M}}, & p = 0 \\ \sqrt{\dfrac{2}{M}}, & 1 \le p \le M-1 \end{cases} \qquad \alpha_q \begin{cases} \dfrac{1}{\sqrt{N}}, & q = 0 \\ \sqrt{\dfrac{2}{N}}, & 1 \le q \le N-1 \end{cases} \tag{8.12}$$

The inverse of the DCT is simply

$$B_{mn} = \sum_{p=0}^{M-1} \sum_{q=0}^{N-1} \alpha_p \alpha_q C_{pq} \cos \frac{\pi(2m+1)p}{2M} \cos \frac{\pi(2n+1)q}{2N}, \quad \begin{array}{l} 0 \le m \le M-1 \\ 0 \le n \le N-1 \end{array} \tag{8.13}$$

and that all the usual Fourier properties hold except that both the input and the output of the DCT are real.

Where are the computational savings? The DCT eliminates all the redundant and zero multiplies. The resulting spectrum has even symmetry. The DCT in two dimensions is essentially the same process as a 2DFFT, except that the spectrum is real with even symmetry which simplifies the mathematics and encoding software somewhat. Note that the 2DFFT in Figure 8.15 is highly symmetric. We really need only one quadrant of the plot to derive the other three due to this symmetry. Once the 32×32 point DCT is computed, the wavenumbers are arranged in a matrix so that the 0 wavenumber is in the upper left corner of the matrix. This is like the lower right quadrant of Figure 8.15. The low wavenumber elements generally have the largest amplitude. As one moves toward the lower right corner of the DCT matrix, the amplitudes become smaller and smaller. So, to compress the image data, one simply discards the low-amplitude high–wavenumber bins of the 32×32 DCT output matrix. Figure 8.24 shows a technique for ordering an 8×8 DCT output, and discarding the DCT wavenumbers of little importance to the image. The 2D DCT output is arranged into a 1D

1			k_x			8	
1	2	6	7	15	16	28	29
3	5	8	14	17	27	30	43
4	9	13	18	26	31	42	44
10	12	19	25	32	41	45	54
11	20	24	33	40	46	53	55
21	23	34	39	47	52	56	61
22	35	38	48	51	57	60	62
36	37	49	50	58	59	63	64

FIGURE 8.24 JPEG compression bin ordering (follow the numbers in the cells from 1 to 64) for arranging the 2DFFT bins in likely order of dominance in terms of amplitude and significance to the visual image balancing both the k_x and k_y as equally as possible.

array starting with the 0 wavenumber bin (overall brightness) and then wraps in a serpentine way across the low wavenumber bins in k_x and k_y equally in a simple algorithm. The result sorts the bins by wavenumber going from two dimensions to a one-dimension list, so that the array can be cut off at some point to save bytes in the file size. How does one decide where to cut the list off? It depends on the quality desired in the compressed image. In some schemes, the lower-level bins are quantized with fewer bits and the numbers are packed into bytes as "granules" less than 8 bits in size, again to save file space and maximize image quality.

Many of these blocks in our discussion are 32×32, but they could be 8×8, or 16×16, and are very similar to each other. When that is the case, one block can be used to "predict" the other, so that only the difference between the two blocks needs to be digitized for the subsequent occurrence. This scheme can eliminate a huge amount of redundant data in the compressed file, especially for images with large areas of nearly the same shade. In addition, Huffman coding can be used to create a reduced table of intensities to represent the image. When the block represents a part of the image with a "texture" only a few DCT bins are needed to make a reasonable representation of the texture in the original image. Examining JPEG-encoded images in detail can often reveal these subtle patterns in the coded blocks. Sometimes vertical or horizontal lines or patterns can be seen in the coded blocks, indicating that only a few dominant DCT bins were used to represent the whole block. When one steps back and looks at the whole image, these subtle artifacts are barely noticeable.

MPEG compression is a standard developed for motion pictures, and like JPEG, is a well-documented and very dynamic international standard that can be found via the World Wide Web. MPEG compression is an extension of JPEG compression. But now we have a stream of pictures with many nearly identical coded blocks. The general idea is include the significant DCT bins in each block as needed, use one block in a picture frame to "predict" other similar blocks by tagging them and encoding the differences in the blocks. For the stream of picture frames, there are periodic "key frames" where the image is completely represented. The frames in between the key frames use forward or backward prediction to point to blocks which can be substituted of to use for block differences. Again this saves a tremendous amount of file space and stream bandwidth. It also allows for a wide range of decoding to different target image resolutions, from high definition television (HDTV) down to small thumbnail videos on a web page displayed on a cellular telephone. Besides all this compatibility with networked receivers and computers, a typical HDTV broadcast transmission sends the HDTV signal, six surround sound super audio channels, plus several side video channels and other digital information in less bandwidth and power than a single analog TV channel from the 1990s. The broadcaster can vary the bandwidth depending on the source material. The receivers extract the bandwidth they can display. This is a great solution made possible by very inexpensive signal processors for video.

MPEG artifacts are really interesting to look for in broadcast HDTV. Many cable and satellite companies save bandwidth by varying the amount of compression in their transmissions. In many cases, an "HDTV" channel is simply not carrying an HDTV quality signal, even though the technical number of pixels displayed meets the standard. This is because the level of MPEG compression employed at the source end of the transmission can be very high. For example, the frame rate on some movies can be so slow that an actor's eyes and mouth appear to "float" around in their face. The received image looks as though it were underwater at times. This is because the key frames are too far apart. Other times one can see the JPEG like coded blocks with their patterns from compression, especially if the transmission is blocked temporarily. Sporting events such as baseball, football, golf, hockey, and tennis tend to have the very best source streams. This is because the backgrounds tend to be pretty stationary relative to the ball or puck. Instant replay affords HDTV the ability to really shine, showing viewers even more detail than they could see if they were right there in the action. For this to work, the key frames have to be frequent, the compression has to be very high quality, and the processors along the stream have to be very fast. These little artifacts may only be temporary as our consumer technology becomes faster and more commonplace, but these little artifacts make watching TV a little more interesting to us geeks. MPEG is continually improving as are the processors used in digital cameras, recorders, and receivers, but a 50-year-old mold has been broken. This author can remember as a child what an achievement it was for his father to "pull in" a distant TV station, especially an important baseball game by aiming the antenna just right. All that fussing inspired many a child to learn about electronics and signal processing. Maybe MPEG artifacts will catch the attention of the next generation of young signal processors only to be antiquated by the time their children become curious of signals and systems.

8.5 COMPUTER-AIDED TOMOGRAPHY

Computer-aided tomography, or "CAT-scans" are well known to the general public through its popular use in medical diagnosis. While the medical CAT-scan is synonymous with an x-ray generated "slice" of the human body, without actually cutting any tissue, it also has many industrial uses for inspection of pipes and vessels, structures, and in the manufacture of high-tech materials and chemicals. Webster's definition of tomography reads simply as "roentgenography of a selected plane in the body." After looking up "roentgenography," we find that it is named after Wilhelm Konrad Roentgen 1845–1923, the German physicist who discovered x-rays around the turn of the last century in 1895. X-rays have extremely short wavelengths (approximately 10^{-10} m) and are produced when electrons accelerated through a potential difference of perhaps 1 kV–1 MV strike a metal target. The electrons in the inner atomic electron shells jump into higher energy states in the surrounding electron shells and produce the x-rays when they return to their equilibrium state back in the original shell. This electron-to-photon "pumping" also occurs between the outermost electron shells at much lower voltages producing only visible light. The obvious utility of x-rays is their ability to penetrate many solid objects. Continued animal exposure to high levels of x-rays has been shown to cause radiation poisoning and some cancers. But, modern engineering has nearly perfected low-level x-ray systems for safe periodic use by the medical community, although all levels of x-rays can be ionizing radiation.

An x-ray image is typically created using high-resolution photographic film placed on the opposite side of the subject illuminated by the x-ray source. The extremely short wavelength of x-rays means that plane wave radiation is achieved easily within a short distance of the source. The x-ray image is composed of 3D translucent visual information on a 2D medium. Examination of x-ray images requires a specially trained medical doctor called a *radiologist* to interpret the many subtle variations due to flesh, organs, injury, or pathology. It was not until 1917 when Radon published what is now referred to as the Radon transform [6] that the concept of combining multiple 2D images to produce a volumetric cross-slice was available. It took another 50 years of computing and signal processing technology advancement before G. N. Hounsfield at EMI Ltd. in England and A. M. Cormack at Tufts University in the United States developed a practical mathematical approach and electronic hardware making the CAT-scan a practical piece of technology in 1972. In 1979, Hounsfield and Cormack shared the Nobel Prize for their enormous contribution to humanity and science.

A straightforward walk-through tour of the Radon transform and its application to the CAT-scan follows. The reader should refer to an outstanding book by J. C. Russ [4] for more details on the reconstruction of images. We start by selecting some artwork representative of a CAT-scan of a human brain as shown in Figure 8.25. The top part of the drawing is the area behind the forehead while the bottom part shows the cerebellum and brain stem.

Figure 8.26 shows a sketch of a single scan along an image angle of +40° above the horizontal. The scan line could be simply a line sampled from a 2D x-ray image or simply an array of photovoltaic cells sensitive to x-ray energy.

The bright and dark areas along the detection array are the result of the integration of the x-ray absorption along the path from the source through the brain to the detectors. If the x-ray propagation were along the horizontal path from left to right, the spatial Fourier transform of the detector array output corresponds exactly to a $k_x = 0$, k_y = whatever, 2D Fourier transform where the result would be graphed along the k_y vertical axis passing through the $k_x = 0$ origin. At the 40° angle in Figure 8.26,

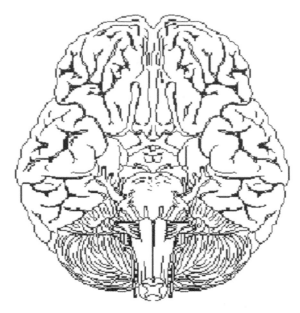

FIGURE 8.25 Cross-sectional drawing of the human brain used for our tomographic image reconstruction demonstration.

the wavenumber-domain line would be along 130° (the normal to this line is 40°), as shown in Figure 8.27.

For an $N \times N$ Fourier transform, we can write

$$I(m_{x'}, m_{y'}) = \sum_{n_{x'}=1}^{N} \sum_{n_{y'}=1}^{N} i(n_{x'}, n_{y'}) e^{-j(2\pi/N)(m_{x'} n_{x'} + m_{y'} n_{y'})} \tag{8.14}$$

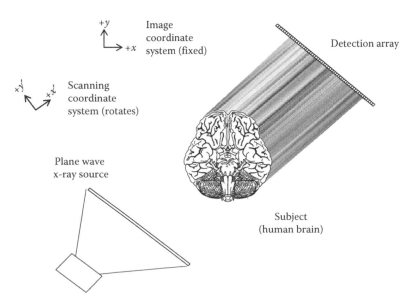

FIGURE 8.26 Layout of a single scan line produced by x-rays along a +40° angle with respect to the horizontal.

FIGURE 8.27 The spatial Fourier transform line corresponding integration of the brain cross section along a +40° angle with respect to the horizontal.

where $m_{x'} = m_x \cos \theta + m_y \sin \theta$, $m_{y'} = m_y \cos \theta - m_x \sin \theta$, $n_{x'} = n_x \cos \theta + n_y \sin \theta$, and $n_{y'} = n_y \cos \theta - n_x \sin \theta$. One simply rotates the x'-axis to align with the x-ray propagation path so that the rotated y'-axis aligns with the sensor array. The Fourier transform of the y' data gives the wavenumber spectra for $m_{y'}$ with $m_{x'} = 0$. With the integration of the absorption along the propagation path of the x-rays being done physically, rather than mathematically, only one of the summations in Equation 8.14 is necessary ($m_{x'}$ is zero). The inverse Fourier transform of the data in Figure 8.27 is shown in Figure 8.28. Equation 8.14 is known as a discrete Radon transform. For the simple case where $\theta = 0$, we are given a response where the only variation is along the vertical y-axis in the spatial image domain. The corresponding wavenumber response only has data along the vertical y-axis where $k_x = 0$. For $k_x = 0$, we have an infinite wavelength ($k = 2\pi/\lambda$), hence the straight integration along the x-axis.

Increasing the number of scan directions will begin to add additional image information which is combined in the wavenumber domain and then inverse Fourier transformed to present a reconstruction of the original image. Figure 8.29 shows the wavenumber response for eight scans, each at 22.5° spacing. The corresponding inverse Fourier transform is shown in Figure 8.30. Only the basic

FIGURE 8.28 Reconstruction inverse Fourier transform of the single scan wavenumber response in Figure 8.27.

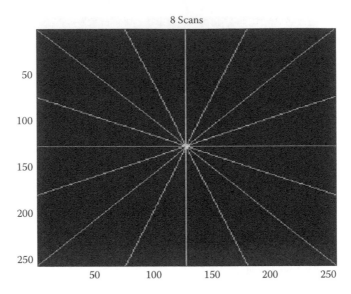

FIGURE 8.29   A total of eight scans as seen in the wavenumber domain for the brain image.

structures of the brain are visible and many lines are present as artifacts of the 65k pixel image being represented by only 256 × 8, or 2048 pixels in wavenumber space. Figures 8.31 and 8.32 show the wavenumber response and image reconstruction when using 32 scans. Figures 8.33 and 8.34 show the wavenumber response and image reconstruction when using 128 scans. A typical high-resolution CAT-scan would use several hundred scan angles over a range from 0° to 180°. It is not necessary to rescan between 180° and 360°, and even if one did, the chances of misregistration between the upper and lower scans are great in many applications using human or animal subjects due to possible movement.

It is truly amazing that a "slice" through the cross section of the subject can be reconstructed from outside 1D scans. The utility of being able to extract this kind of detailed information has been

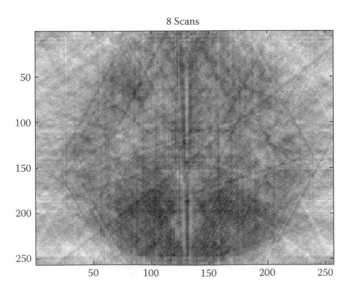

FIGURE 8.30   Reconstruction inverse Fourier transform of the eight scans wavenumber response in Figure 8.29.

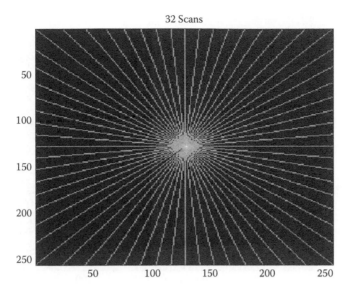

FIGURE 8.31   A total of 32 scans as seen in the wavenumber domain for the brain image.

invaluable to the lives of the millions of people who have benefited by the medical CAT-scan. However, what is even more fascinating is that the Fourier transforms can be completely eliminated by using the filtered backprojection technique. The largest computational burden is not all the individual scans, but the inverse 2D Fourier transform. For a 256×256 pixel crude image, each 256-point FFT requires only 2048 multiply operations, but the 256×256 inverse Fourier transform to reconstruct the image requires over 4 million multiplies. A typical medical CAT-scan image will have millions of pixels (over 1024×1024) requiring hundreds of millions of multiplies for the 2D inverse FFT, as well as consist of dozens of "slices" of the patient to be processed.

The filtered backprojection technique is extremely simple, thereby making CAT-scan equipment even more affordable and robust. In backpropagation, one simply "stacks" or adds up all the scan

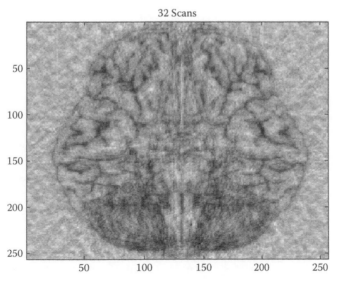

FIGURE 8.32   Reconstruction inverse Fourier transform of the 32 scans wavenumber response in Figure 8.31.

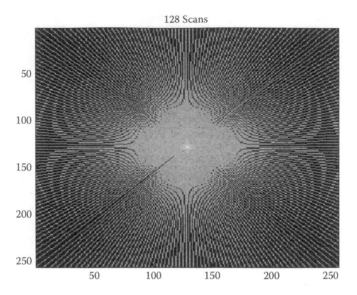

FIGURE 8.33   A total of 128 scans as seen in the wavenumber domain for the brain image.

data (one scan is shown in Figure 8.28) on a single image without doing any Fourier transforms. Adding all these scan data up pixel by pixel gives a sharp, yet foggy, image as shown in Figure 8.35 for 128 scans. The effect is due to the fact that the pixels in the center of the image are "summed" and "resummed" with essentially the same low wavenumber scan information with each added scan. The high wavenumber data near the center of the image tends to sum to zero for the different scan angles. In the wavenumber domain, this means that the low wavenumbers near the origin are being overemphasized relative to the high wavenumbers which correspond to the sharp edges in the image. A simple high-pass wavenumber filter corrects the fogginess as shown in Figure 8.36 yielding a clear image without a single Fourier transform. The wavenumber filtering can be accomplished very simply in the time domain using an FIR filter on the raw scan data. The filtered backprojection

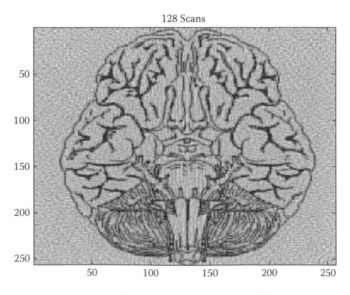

FIGURE 8.34   Reconstruction inverse Fourier transform of the 128 scans wavenumber response in Figure 8.33.

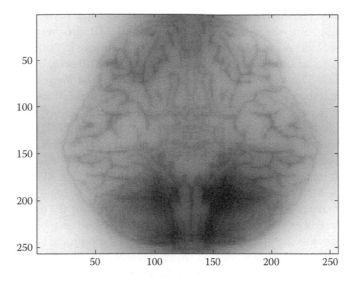

FIGURE 8.35 The unfiltered backprojection reconstruction of the brain cross section from 128 scans showing "fogginess" from redundant low wavenumber addition.

technique also allows filtering to enhance various image features to help medical diagnosis. Newer techniques such as the backpropagation technique even allow for propagation multipath and wave scattering effects to be (at least in theory) controlled for tomographic applications in ultrasonic imaging.

8.6 MAGNETIC RESONANCE IMAGING

A magnetic resonance image (MRI) is actually a nonlinear adaptation of the Radon transform [7]. After the discovery of nuclear magnetic resonance (NMR) in 1946 by Bloch [8], the phenomenon remained a basic science technique for investigating magnetic properties of the atom through much of the 1950s and 1960s. It was not until the sentinel paper by Lauterbur [9] in 1973 showing how

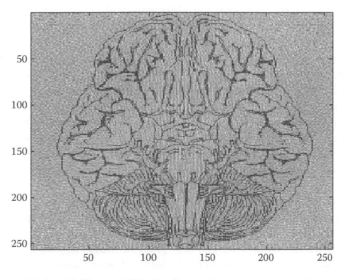

FIGURE 8.36 Simple high-pass filtering of the backprojection reconstruction balances the high and low wavenumber contributions and restores the sharpness to the reconstructed image.

magnetic gradients could be used to measure imaging projections that NMR became what we know today as MRI. Bloch and others received a Nobel prize for the NMR discovery in 1952 and Lauterbur and others received a Nobel prize for the gradient-imaging technique in 2003, which illustrates just how important these techniques have been to mankind. NMR still remains a potent atomic and chemical analysis tool. The term "nuclear" was dropped from the medical imaging application name to avoid potential fears by the public that the technique may involve nuclear radiation. A more efficient technique by Kumar [10] combines multiple field gradients for faster measurements and is what is widely in use today. There are two very significant aspects of the MRI compared to the CAT-scan. First, the MRI does not use ionizing radiation like that found in x-rays. This means an MRI is far less of a cancer risk to a patient than a CAT-scan. Second, the MRI offers high contrast to soft tissue types allowing imaging of nonbone structures in great detail. The main downside of MRI is that the room with the MRI machine must be heavily shielded and free of all magnetic objects since they would become dangerous projectiles once the field current is turned on. Items with magnetic information such as credit cards and computer hard disk drives must also be out of the area. It can cost millions of dollars just the properly construct and shield the room housing and MRI machine. There is far more to the NMR/MRI story than we can present here. But from the signal processing perspective, we can provide a salient discussion of the main attributes of the process. We will begin with a discussion of the imaging process, then go into the details of NMR, and then finish with a discussion of resolution in MRI systems.

Certain elements, those with odd number of protons or neutrons, have an inherent angular momentum, or "spin" in the nucleus. This is coupled with a magnetic dipole moment and allows the atom to respond to an external magnetic field (more details will be discussed later). In a strong static magnetic field B_z these atoms will magnetically align themselves with the field and will also have a precession resonance (think of a wobbling spinning top) that depends on the particular element and the strength of the magnetic field. This resonance frequency is called the *Larmor frequency* ω_L and for the hydrogen atoms in water, it is about 42.577 MHz in a 1 Telsa (T) static magnetic field.

$$\omega_L = \gamma B_z \tag{8.15}$$

Each NMR-capable atomic element has a slightly different Larmor frequency in the same static magnetic field making NMR a useful tool for atomic and chemical analysis. One can stimulate the atomic resonance by transmitting an alternating magnetic field, normal to the direction of the strong static magnetic field for maximum detection sensitivity, with a frequency near the Larmor frequency. We refer to this as the RF stimulation pulse and it is a relatively small magnetic amplitude B_1 compared to the large static magnetic flux B_z. Figure 8.37 depicts the RF pulse in an MRI machine where the static magnetic field is into the page inside the cross section of a solenoid-shaped electromagnet. After the RF pulse ends, one measures the Larmor frequency of all the stimulated elements in the static magnetic field zone using the same Faraday coil used to transmit the RF pulse. It takes the atoms anywhere from 30 ms to 1 s for the Larmor frequency resonance to "ring down" or perhaps more specifically to "wobble up" to alignment with the static magnetic field again. Only those elements with Larmor frequencies close to the RF pulse frequency will be significantly excited and measurable as NMR when a long RF pulse is used. For application to MRI of humans, hydrogen is the atom of choice due to its abundance throughout the body, and its density variation for different tissue structures such as muscles, ligaments, fat, and organ tissues.

Lauterbur's imaging breakthrough comes from exploiting the fact that the Larmor frequency depends on the amplitude of the static magnetic field. Immediately after the RF pulse is turned off, the static magnetic field is changed to a static magnetic field gradient along some axis in the measurement zone as shown in Figure 8.38. The field gradient means that the Larmor frequency will vary for elements along the axis of the gradient. By doing a Fourier transform of the received atomic response to the RF pulse we can locate particular frequencies along the axis of the gradient. This is like the single scan of the CAT-scan described in Figures 8.26 through 8.28. By repeating the RF

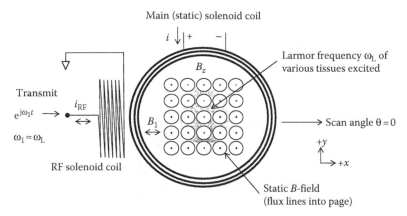

FIGURE 8.37 Typical MRI configuration where a strong static magnetic field B_0 (into page) aligns atomic spins while a weaker RF pulse tilts these spins off normal to allow detection by the same RF coil.

pulse stimulation and measuring the received spectrum for a number of scan angles where the added field gradient moves with each scan angle, the multiple scans can be combined in the Radon transform to produce a CAT-scan-like image slice as shown in Figure 8.39. For the case of hydrogen in water (humans are 62% water by weight), a 42.577 MHz RF pulse in a 1 T static magnetic field will stimulate predominately hydrogen atoms in water, and not other elements. Since the nuclear resonance can take nearly 100 ms to decay, the spectral resolution of the NMR is on the order of 10 Hz. This means a modest linear magnetic field gradient that results in a Larmor frequency shift of 500 Hz/cm can separate the elements in space with a resolution 50 pixels/cm in the received spectrum for a particular scan angle. The intensity of the spectrum at each frequency bin is proportional to the amount of hydrogen (via the water content) present along the ray path of the scan angle, offset

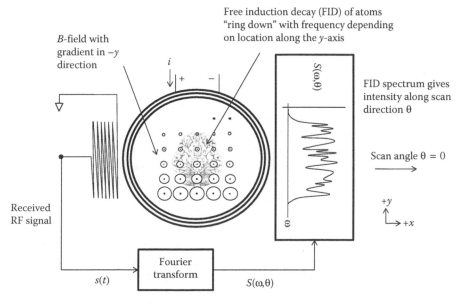

FIGURE 8.38 Immediately after the RF pulse a gradient is applied to the static magnetic field causing the received frequency to be associated with a spatial offset along the gradient direction. The received amplitude at a given frequency is proportional to the number of atoms excited by the RF pulse.

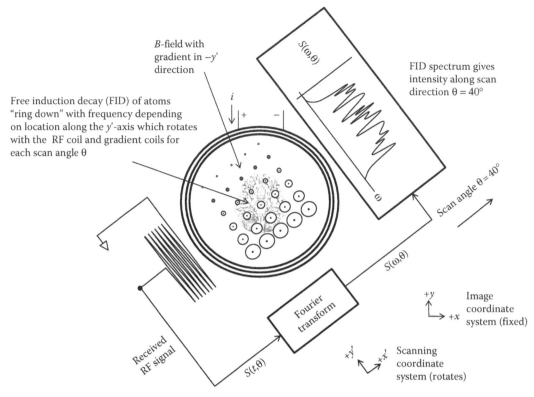

FIGURE 8.39 By rotating the RF transmit/receive apparatus a CT-like scan can be captured over a range of scan angles allowing a Radon transform and backprojection to produce an image of the tissue scanned.

spatially by the shift in Larmor frequency due to the field gradient. Typical MRI machines in use today divide the received signal into multiple segments and apply field gradients in many directions to speed the overall process and support SNR enhancement in the image.

Now let us discuss the details of what is happening at the atomic level. Most texts explaining MRI usually start at this point, but as signal processors we expect you to really want to know how the imaging works initially. We skipped over some very interesting details explained a bit more here, but the reader should explore other resources to get all the details. Figure 8.40 depicts the atom aligned with the static magnetic field on the left and immediately after the RF pulse on the right. The precession of the atomic spin (magnetization M of the atom) back to near alignment with the static external field B_0 has an important detail in the damping of the tilt angle, which is independent of the damping of the RF-induced Larmor precession and is called the free induction decay (FID). This is caused by the interaction of the atomic magnetization with neighboring atoms and the material atomic lattice as a whole. The exponential decay time constant T_1 of the tilt angle α (the amplitude of α proportional to RF pulse length) in Figure 8.40 is temperature dependent and is a result of the atom losing energy to the material lattice and is called *spin–lattice* relaxation (damping). As material temperature increases, T_1 decreases perhaps because there are more interactions with other atoms in the material lattice due to the higher overall vibration from increased temperature. This can be used to measure temperature in an MRI as a means to identify areas in inflammation or poor blood circulation. The damping of the Larmor oscillation is also due to interactions with the spin of neighboring atoms, but in a way which does not affect the tilt angle, and hence this damping is called *spin-spin* relaxation. The spin-lattice time constant T_1 is always larger than T_2. For example, in human tissue T_1 may be on the order of 100 ms or more while T_2 is less than 30 ms. For MRI, the RF pulse is generally long compared to these time constants which also narrows the bandwidth of

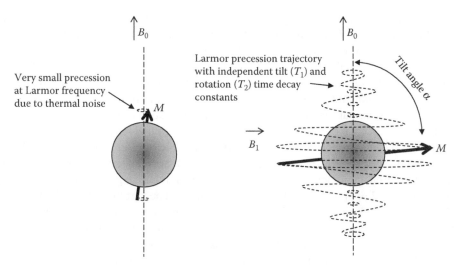

FIGURE 8.40 When an NMR active atom is placed in a strong static magnetic field (B_0) its nuclear spin axis aligns with the external field but a small oscillatory precession about this axis occurs at the Larmor frequency due to thermal in the material (left). After the RF pulse (B_1) is applied the spin axis of the atom realigns with B0 but with separate time constants for the tilt and Larmor amplitude decay.

RF excitation which also helps to ensure that only one type pf atom is excited and imaged (such as hydrogen for human MRI).

The interactions of the nuclei spins can also be excited acoustically [11] and measured as a change in NMR response when the material is saturated with ultrasound or as an FID stimulated by ultrasound. This process is called nuclear acoustic resonance (NAR) and provides a good means to measure the properties of materials where the RF magnetic field has difficulty propagating, such as conductors or magnetic materials. The mechanism at work here is primarily the nuclear quadrupole resonance, which is from the electric field of the nucleus not being spherical. The effect shows up as a sharp acoustic absorption increase near the nuclear resonance—the same frequency as the NMR stimulated by an RF pulse in NMR. But it can be seen also as a change in NMR amplitude and shortened T_2 decay when the same is saturated with ultrasound prior to the RF pulse. The ultrasound increases the nuclear interactions in both spin-lattice and quadrupole resonance interactions which dampens the NMR response. NAR mainly applies to solids and works for solid conductors because the acoustic wave can propagate well where the external RF pulse cannot. The NMR response, however, can be detected externally as a small change in acoustic attenuation. What is most fascinating here is how these wave–lattice interactions speak to the physical nature of energy absorption in materials at the atomic level.

When the RF pulse is short the bandwidth of the RF excitation increases, which extends the excitation to include more than one atomic element and measure chemical interactions (chemical shift) with the atomic spin. This is the basis for chemical NMR, a standard technique for validating specific chemicals and/or chemical mixtures for purity. Clearly, a very short RF pulse will excite a broader range of atom types, but how can chemicals be seen? Chemicals are of course made up of atoms bonded together based on the energy of the individual electron shells of each atom. Each electron shell has a unique electron cloud shape and energy and corresponds to a column in the chemical periodic table of elements. Atoms bond together when their respective open electron shells can "share" electrons, which in turn arrange the atoms in the molecule at specific angles. These electron shells can "shield" the nucleus of external magnetic fields from the spin of neighboring atoms in the molecules, or even other molecules. Electrons themselves have a magnetic field which is antiparallel to the externally applied static magnetic field. The measured spectrum of the NMR signal can have separate peaks for each type of atom excited, and also each type of molecule containing those excitable atoms. This gives rise to a repeatable spectral signature for a given static magnetic field B_0, RF pulse B_1 of duration t, and chemical sample, be it a pure "neat" chemical or mixture of chemicals. To make these chemical signatures standardized, the sample is typically mixed with a chemical standard such as tetramethylsilane (TMS). TMS has four methyl groups (CH_3) single bonded to a silicon atom. Chemical NMR spectra reference the peaks relative to this (or a stated) chemical standard in units of "parts per million" (ppm). In this case, ppm has nothing to do with concentration, but rather is a measure of Hertz relative to the chemical standard peak. Since the absolute frequency of all the peaks depend on the applied magnetic field B_0, dividing the frequency of the measured spectrum by the frequency of the RF pulse, which is in the MHz generally, gives an NMR spectrum in "ppm" for the ordinate. The normalized NMR spectrum is further standardized by referencing 0 ppm to the named chemical standard (e.g., TMS). Most chemicals lie in a range from 0 to 12 ppm on the NMR-normalized scale. The significance of chemical NMR is that it will eventually make its way into MRI given that the future will bring faster signal processing and inspired engineers. Just imagine picking a pixel on an image of a 3D MRI slice (really a volume element or "voxel"), and doing a chemical or pathological analysis!

The SNR in an NMR or MRI machine is a function of the coupling of the RF pulse into the atoms of interest, the number of NMR-excited atoms present, and the electronic background noise in the bandwidth of the instrument. The SNR improves with a larger static field since more atoms will be aligned and the Larmor frequency will be higher and overall instrument bandwidth wider. For a fixed RF pulse length and fixed FFT buffer size the frequency resolution of the instrument is determined. With a higher Larmor frequency one can use a steeper magnetic gradient and therefore observe a finer pixel resolution. But, this divides the signal over more pixels, and so in general higher pixel resolution means lower SNR. Other factors are the temperature of the coils which have resistance and thus thermal noise. Live subjects must be scanned at an appropriate temperature but a higher SNR can be obtained by lowering the temperature, for example, examining frozen tissues. Another factor is how much the subject fills the RF coil which directly affects the number of atoms excited. NMR is not considered a trace detection method since a bulk of material is needed. However, by averaging N spectra, one can improve the SNR by a factor of $N^{-1/2}$ which can allow small amounts of material to be identified.

Other MRI errors come from the practical limitation that the magnetic field in the main coil is not perfectly uniform, nor is the gradient field perfectly linear. These errors are not related to SNR, but are biases. In some cases combining pulses from opposite directions can cancel the bias. Another bias-canceling technique is to use two pulses, the first resulting in a 90° tilt and the second in a 180° tilt to create a *spin echo*. There also exists a *stimulated echo* where two 90° RF pulses are used, while a third 90° pulse is used to cause a shift of the result of the first two pulses, and so on. Highly linear magnetic fields can be created using arrays of magnets or arrays of controlled electromagnets with great precision. One such array used in efficient motors and electromechanical batteries is called a Hallbach array [12]. This and new magnetic materials will lead to very precise control of MRI bias errors for future machines.

8.7 MATLAB® EXAMPLES

Table 8.1 summarizes the m-scripts used in various figures in this chapter. These scripts are downloadable from the website for this book or available from the author. The script source is meant to illustrate the signal processing techniques without a lot of complexity. They are commented to further help the user. Users are free to download the scripts from the book publisher and use them for nonevil purposes (such as learning). We discuss certain small code "snippets" here to explain the implementation details of the underlying algorithm.

There are a number of line array scripts, each tailored to a specific figure to keep the software as simple as possible. The script "LineArray.m" has two useful code features. The first is just graphics. In the "legend()" function call, we use "\lambda" to invoke the LaTex interpreter in MATLAB to put

TABLE 8.1
m-Scripts Used in This Chapter

LineArray.m	Figures 8.6, 8.7
LarrayShd.m	Figure 8.8
LarrayStr.m	Figures 8.9, 8.10
LarrayEnd.m	Figure 8.11
LarrayOut.m	Figure 8.12
GreyImage.m	Figures 8.14–8.23
House Dec 2004 gs.jpg	Figures 8.14–8.23
CatScan.m	Figures 8.27–8.34
Headf.mat	Figures 8.27–8.36
BackProp.m	Figures 8.35, 8.36

the Greek symbol λ in the legend box where we want it. If we used a "sprintf()" to first print to a string, we could put "\\lambda" in the string allowing the second "/" to persist in the string and thus be interpreted as LaTex by MATLAB. This is very useful because you can automatically pass variable values into a string containing LaTex for Greek symbols and equations. The "LineArray.m" has some simple lines commented out allowing you to experiment with this. It will also work with the "text()" function which allows you to put a text string which can contain LaTex anywhere on the figure. The second technique is the way we compute the line array beam pattern, as shown in Table 8.2.

The snippet in Table 8.2 is very "old school," dare I say "FORTRAN-like" except for the for-loop syntax and semicolons, but as clear as possible. Table 8.3 shows a few simple MATLAB-centric vectorizations of the same code in Table 8.2. We show this detail because this is consistently the main hurdle new users of MATLAB have difficulty with.

Clearly, the vectorized code in Table 8.3 is much more compact yet still very easy to read once you are familiar with the MATLAB syntax for vectors and matrices. Variables such as "j" and "pi" are automatically initialized to $j = \sqrt{-1}$ and $\pi = 3.1415927 \ldots$ and the parameter "doverl" is d/λ. The syntax ".*" is a "dot product" where each vector element is multiplied, not a vector–vector multiply. Vector dimensions have to be the same for ".*" to be used, and so you may have to transpose a vector once in a while when the MATLAB interpreter flags the mismatch. Be mindful that a MATLAB transpose is a Hermitian transpose which will conjugate complex elements. Using "$X'.$" instead of "X'" will transpose the vector without conjugating a complex element. Note that the scalar product

TABLE 8.2
MATLAB Code Snippet Showing Array Beam Response

```
% n is the range over theta, m is the array element
% resp is a simple DFT sum
for n = 0:180,
    thetad(n) = n;
    theta = pi*thetad/180;
end;
resp = zeros(size(theta));
for n = 1:181,
    for m = 1:16
        resp(n) = resp(n) + exp(j*2*pi*(m-1)*doverl*cos(theta(n)));
    end;
end;
```

TABLE 8.3
Vectorized MATLAB Code Snippet of Table 8.2 Code

```
% here's another way to sum the array elements using the vector
% features of MATLAB
theta = pi.*(0:180)./180;
resp = zeros(size(theta));
for m = 1:16
    resp = resp + exp((j*2*pi*(m-1)*doverl).*cos(theta));
end;
```

"($j*2*pi*(m − 1)*doverl$)" is in parentheses in Table 8.3 to force the interpreter to calculate this first, and then do the dot products. A function with a vector as its argument such as "cos(theta)" produces a vector as its output. The code in Table 8.3 will run much faster than the code in Table 8.2 due to vectorization. One could further vectorize by arranging a matrix of the exponent phases and doing a one-line vector–matrix multiply to eliminate the for-loop in Table 8.3, but then we would lose quite a bit of clarity in the code. Some programmers take to this with great enthusiasm to make highly efficient m-scripts that few people can read or translate into other languages like C. Our goal here is to be as clear as possible about the algorithm and make the m-script produce a nice useable plot.

To "steer" the array main beam to a given LD it is a simple matter of adjusting the phase of the line array so that a plane wave from the LD is in-phase across all the line array elements, thus adding up to the maximum array gain for the LD. Waves from other directions will tend to cancel in this sum. Table 8.4 shows a code snippet for beam steering from the m-script "LarrayStr.m." The steering phase increment is added as a multiple to each array element output, with zero phase added to the $m = 1$ element. The phase is negative, meaning that the $M > 1$ elements will be delayed by the phase increment according to the LD angle. A 90° beam would result in no phase added, since the array elements are already in-phase for that direction. Given a spectrum from each array element, one can precalculate a steering vector for each frequency bin, array element, and LD. The array data can then be processed to provide spectra as a function of LD for detection of signals. The array processing is completely parallel in both frequency and LD.

TABLE 8.4
Beam Steering is a Simple Matter of Phasing the Array

```
% these phase increments are added to each element phase for steering
% the beam to 60 degrees (90 degrees is broadside, 0,180 degrees is endfire
steer60 = -j*2*pi*doverl*cos(pi*60/180);
for m = 1:16,
    Aw60(m) = Aw60(m).*exp(steer60.*(m-1));
end; % steering vectors
resp60 = zeros(size(theta));
% simple DFT sum for beam patterns over 360 degrees
for n = 1:360,
    for m = 1:16
        resp60(n) = resp60(n) + Aw60(m)*exp(j*2*pi*(m-1)*doverl*cos(theta(n)));
    end;
end;
```

Next, we need to make example waves from specified directions for testing our line array. Again, considering a plane wave from some AOA, we need to simulate the amplitude and phase across our line array. Let us assume that the line array is along the x-axis where the first element is on the left and the Mth element is on the right, with each element space d/λ apart for a given frequency. The broadside direction is 90° (along the y-axis) and the endfire directions are along the x-axis (0° in the $+x$ direction). We will ignore the symmetry from 180° ($-x$ direction) through 360° ($+x$ direction). For a plane wave, we implicitly focus the array to the farfield to an object many array apertures (line array length is M times d) away. For an AOA less than 90°, the Mth element "sees" the wave first and the first element last. So we need to add more phases to the signals from the elements farther away from the first element. This way, when we steer the array to the LD that matches the AOA for a given source, all the elements will add in phase to give a large output. Table 8.5 shows how to synthesize plane wave sources and then scans the array across all the LDs summing the output.

The source amplitudes in Table 8.5 are in the variable "$S(n)$" and the AOA is in "ts(n)" for each source. A Hanning window is used to shade the array element amplitudes to reduce side-lobe leakage. Note that the phase in "steer" is positive for the sources and negative for the steering vector for the LD of the array. The code is for one frequency but could be expanded for a whole spectrum. Figure 8.12 shows both the Hanning window and no window (rectangular where $Ah = 1$ for all elements) for 32 array elements, which shows the window performance better than 16 elements due to its smaller aperture.

The image processing section of this chapter mainly uses the m-script "GreyImage.m" which loads a JPEG encoded image and performs a number of frequency domain filtering operations. This script is amazingly simple. Once we have the image in the frequency domain and shifted so that the center is the wavenumber origin, it is extremely straightforward to do filtering in the frequency domain using MATLAB's "dot product" to multiply the images, pixel by pixel, as matrix dot products. Table 8.6 shows a snippet of the frequency-domain image filtering.

The matrix "lpwfilt" is a 256×256 matrix determined by a vector outer product of a Hanning window raised to the eighth power to make a nice big bump at the wavenumber origin. The house

---

TABLE 8.5

Source Synthesis and Scanning Using the Line Array

```
% here we sum up sources at the given directions of arrival (DOA)
y = zeros([1,M]); % array data vector
% 4 sources, M = 16 array elements to sum sources
for n = 1:4,
    steer = +j*2*pi*doverl*cos(pi*ts(n)/180); % phase offset for source
    for m = 1:M,
        y(m) = y(m) + S(n)*exp(steer*(m-1)); % phase for mth element
    end;
end;
Ah = hanning(M); % scaled Hanning shading
resp = zeros(size(theta));
% now we make a steering vector for each degree and "scan" 1 to 180
for n = 1:180,
    steer = -j*2*pi*doverl*cos(pi*n/180);
    for m = 1:M,
        Aw(m) = Ah(m)*exp(steer*(m-1)); % make steering vector
    end;
    resp(n) = y*Aw./M; % this is the array output for the nth look direction
end;
```

TABLE 8.6
Frequency Domain Filtering as a Matrix Dot Product

```
% this is a crude low-pass filter (easy to calculate)
han = hanning(256);
% note that raising to the 8th power increases the emphasis
lpwfilt = (han.^8)*(han.^8)';
% simple point by point low pass filter in wavenumber domain!
lpgifft = gifft.*lpwfilt;
% this is the resulting image - un-shift first, then ifft, the real part
% the imaginary part can have residual junk in it and we don't need it
lpgi = real(ifft2(fftshift(lpgifft)));
```

image 2DFFT is in the matrix "gifft" and the frequency-domain matrix dot product to get the frequency-domain low-pass-filtered image is in the matrix "lpgifft" in Table 8.6. It is that simple to code. To get the low-pass-filtered image you need to "unshift" the matrix, inverse FFT it, and then display the real part (the imaginary part just has some residual noise from round off errors). The filtering in Figures 8.17 through 8.23 is done in the frequency (2D wavenumber) domain using this simple type of operation.

The Radon transform is described in the m-script "CatScan.m" where we load an already 2DFFT image of a slice for processing. Here, we cheat shamelessly in the m-script using the wavenumber data to save an enormous amount of coding. But we are not really cheating, we are just not obliviously crunching data when we do not need to. To simulate a CAT-scan given an image of a brain slice, one would sum all the pixel values along the scan angle for each bin in the detection array. In a real CAT-scan, the x-rays are collimated and traverse a volume of tissue onto a 2D image. The brightness/darkness of each pixel is proportional to the amount of x-ray energy absorbed along the ray path from the transmitter to the receiver. The transmitted intensities are assumed to be the same within a collimated beam. If we take a slice out of the collimated volume represented by our brain slice image, and sum the pixels along each ray path to a receiving line array normal to the scan angle, we can simulate a "projection." This needs a lot of coding and work, especially for large number of scan angles. So what we do is pull a single trace, normal to the scan angle, from the 2DFFT of the slice image. The inverse 2DFFT of this line gives the projection for the scan angle chosen. It is the same data. So we gather (or sum for the backpropagation algorithm) the 2DFFT scan traces into the matrix "catf" in Table 8.7, and then inverse transform to show the equivalent image. It still represents a Radon transform, but perhaps explains the underlying mathematics even better.

To do the backpropagation algorithm, one could generate the complete one-scan image for each scan angle, and sum over all the scans. This would give the image in Figure 8.35. Since the low frequencies are added more often than the high frequencies, a simple high-pass filter corrects the problem. So one would do a 2DFFT, high-pass filter, and then inverse 2DFFT to get the image. But why work hard? We can just sum the frequency-domain traces, filter with a simple dot product, and inverse transform. That is the only difference in the m-script "BackProp.m" used for Figures 8.35 through 8.36.

8.8 SUMMARY

Waves can be represented in either time or space as a Fourier series of sinusoids which can be filtered in the time-frequency or spatial wavenumber domain depending on the chosen wave representation. We have shown that for sensor arrays spaced fractions of a wavelength apart (up to half of a wavelength), spatial filtering can be used to produce a single array output having very high spatial

TABLE 8.7
Simulation of a CAT-Scan is Done by Gathering Scan Traces in the Frequency Domain

```
for n = 1:numscan,
    th = (n-1)*dth;
    tanth = tan(th);
    tanthc = tan(.5*pi-th); % this is the complimentary angle
    if th < pi/4
      for m = 1:256,
        r = m;
        c = round(128 + (m-128)*tanth);
        catf(r,c) = headf(r,c); % here low frequencies can be overwritten
      end;
    elseif th < (3*pi/4)
      for m = 1:256,
        c = m;
        r = round(128 + (m-128)*tanthc);
        catf(r,c) = headf(r,c);
       end;
    elseif th < pi
      for m = 1:256,
        r = m;
        c = round(128 - (129-m)*tanth);
        catf(r,c) = headf(r,c);
      end;
    end;
end;
% now just do the inverse 2DFFT to see what the scans produced
headcat = ifft2(fftshift(catf));
figure(3);
colormap(gray);
imagesc(abs(headcat));
```

gain in a desired look direction. The LD "beam" can be electronically steered using signal process-ing to give a similar search beam to that for a parabolic reflector which would be mechanically steered. Using parallel signal processing, one could have several simultaneous beams independently steerable, where each beam produces a high-gain output on a signal from a particular AOA. Later in this book, Chapters 12 and 13 will discuss in detail adaptive beamforming techniques for sonar and radar which optimization of the search beam by steering beam nulls in the directions of unwanted signals.

Examination of images from optical sensors as wavenumber systems requires consideration of a whole different set of physics. Visible light wavelengths range from around 400×10^{-9} m, or 400 nm to around 700 nm, and frequencies around 1×10^{14} Hz, making spatial filtering using electronic beamforming impossible with today's technology—these are not the wavenumbers that we discuss here. However, when optical wave information is projected on a screen using either a pinhole aperture or lens system, we can decompose the 2D image data into spatial waves using Fourier's theorem. The image element samples, or pixels, form a 2D set of data for Fourier series analysis. We can control the sharpness of the image by filtering the wavenumbers in an analogous manner to the way one can control the treble in an audio signal by filtering frequencies. The short wavelengths in an audio time-domain signal correspond to the high frequencies while the short wavelengths in image data correspond to the high wavenumbers making up the sharp edges in the picture. If the pixels have enough bits of resolution, an out-of-focus image can be partially recovered by inverting the "PSF" for the lens in its out-of-focus position. The success of focus recover is sensitive to the number of bits of pixel brightness resolution because the out-of-focus lens spreads the light information out in two dimensions into adjacent pixels, which attenuates the brightness substantially in the surrounding pixels.

We can also use the 2D Fourier transform to detect the strongest wavenumber components, rejecting weaker (and noisier) wavenumbers, and thereby compress the total image data while also reducing noise. This is the basis for JPEG and MPEG compression of pictures and motion pictures, respectively. For both of these compression algorithms, a typical picture is broken up into smaller blocks, say 32 × 32 pixels each, and a 2D Fourier transform is done, usually as a DCT since only one quadrant of the wavenumber space is needed to represent all the wavenumbers. Then the block is arranged as a vector starting from the lowest wavenumber pair (0,0) to the highest, mapping the 1D vector to the 2D block in a serpentine manner. This way the pixels of the block are already arranged where the largest wavenumbers with the largest corresponding amplitudes tend to come first and the more detailed high-frequency wavenumber which tend to have lower amplitude come last in the vector. This scheme offers a range of choices on how to remove redundant and/or unimportant data for the image and give a "compressed" image file. One can simply cut off the vector below a threshold, or map frequently used amplitude levels to a symbol table look-up (Huffman coding), or simply "predict" the block using a similar already coded block. The later gets used extensively in most images since there can be large areas of similar color or texture. For MPEG video, this prediction of the block can extend from the "key frame" forward or backward in time to blocks from other frames. In the stream of image frames that make up a video signal, the only self-contained frames are the key frames which are sent at a rate much less than the frames/s rate of the video signal. This is quite sensible since successive image frames will share much of the same detail. MPEG video compression is employed in all digital broadcast video signals.

For scanning systems such as x-rays, the massive 3D data on an x-ray image can be resolved into a synthetic thin "slice" crosswise by combining many rays from a wide range of angles in a Radon transform. The technique of CAT solves an important problem in x-ray diagnosis by simplifying the amount of information on the translucent x-ray image into a more organized virtual slice. Radon recognized the need and solved the problem mathematically in the early part of the twentieth century. But, it was not until the arrival of computer's and the genius of Cormack and Hounsfield in the early 1970s that the CAT-scan became practical. The understanding of the Radon transform and wavenumber domain signal processing allows an even more simple approach to be used called the filtered backpropagation tomographic reconstruction. Filtered backprojection produces a sharp image without the computational demands of Fourier transforms by filtering each scan to better balance the wavenumbers in the final image. The CAT-scan is considered one of the premier technological achievements of the twentieth century because of the technical complexies which had to be solved and the enormous positive impact it has achieved for the millions of people who have benefited from its use in medical diagnosis.

Like the CAT-scan, and an MRI is also generated using a Radon transform, but the projections are generated by exploiting a nonlinear principle of the interactions of an external magnetic field with the atomic spin of some atoms. About 1/3 of the atoms in the periodic table have an imbalance due to the number of protons and neutrons in the nucleus. This is called "spin" in short since it results in a net torque on the nucleus when the atom is placed in a static magnetic field. Like the gyroscopic precession of a top the magnetization of the atom will wobble at a frequency proportional to the static magnetic field called the Larmor frequency. By transmitting an RF pulse near this frequency the wobble can be easily excited and subsequently measured as the FID of the atom, making a nice way to identify various material chemistries. But for an MRI, the static magnetic field is switched to a gradient magnetic field just after the RF pulse so that the elements along the gradient produce a Larmor frequency associative with their position. An FFT of the received FID in the gradient magnetic field gives a projection scan for the Radon transform, and thus allows 3D image reconstruction and slice analysis. The MRI images soft tissue much better than an x-ray image and does not use ionizing radiation which is harmful to the body. However, the MRI machine must be in a well-shielded facility with no magnetic materials or objects nearby.

PROBLEMS

1. For an arbitrary three-element planer (but not linear) sensor array with sensor spacings 1-2 and 1-3 each less than 1/2 wavelength, derive a general formula to determine the angle of arrival of a plane wave, given the phases of the three sensors at a particular frequency.

2. A linear array of 16 1-m spaced hydrophones receives a 50 Hz acoustic signal from a distant ship screw. If the angle of arrival is 80° relative to the array axis (the line of sensors), what is the steering vector which best detects the ship's signal? Assume a sound speed of 1530 m/s.

3. Consider a circular array of 16 evenly spaced microphones with diameter 2 m (sound speed in air is 345 m/s). Derive an equation for the array response as a function of k and θ when the array is steered to θ'.

4. Show that the Chapter 5 Laplacian operator in Equation 5.39 is a high-pass filter, by examining the operator's response in the wavenumber domain.

5. Explain how an out-of-focus image can have focus restored if the number of bits per pixel is high and there is a known "point object" in the field of view.

6. Show that the backpropagation algorithm works whether the scans are summed in the wavenumber domain or in the spatial domain.

7. Show mathematically how a DCT of order N is the same as the sum of two phase-modulated DFTs of order N, assuming the input is real.

8. Explain why a heavily compressed JPEG image seems to be made up of blocks with horizontal, vertical, or checked textures.

9. How does the gradient magnetic field and acquisition time affect the resolution in an MRI image?

10. Under what conditions could two or more materials (chemicals, elements, etc.) be imaged simultaneously in an MRI, given that you could control the RF pulse and gradient magnetic fields?

REFERENCES

1. R. E. Colin, *Antennas and Radio Wave Propagation*, New York, NY: McGraw-Hill, 1985.
2. W. S. Burdic, *Underwater Acoustic System Analysis*, Englewood Cliffs, NJ: Prentice-Hall, 1991.
3. R. A. Monzingo and T. A. Miller, *An Introduction to Adaptive Arrays*, New York, NY: Scitech, 2004.
4. JPEG 2000: ITU-T T.800, ISO/IEC IS 15444–1.
5. J. Radon, Über die Bestimmung von Funktionen durch ihre Integralwerte längs gewisser Mannigfaltigkeiten, *Berlin Sächsische Akad Wissen*, 1917, 29, pp. 262–279.
6. J. C. Russ, *The Image Processing Handbook*, Boca Raton, FL: CRC Press, 1992.
7. S. Stergiopoulos, ed., *Advanced Signal Processing Handbook*, Chapter 17, Boca Raton, FL: CRC Press, 2001.
8. F. Bloch, W. W. Hansen, and M. Packard, The nuclear induction experiment, *Phys Rev*, 1946, 70, p. 474.
9. P. C. Lauterbur, Image formation by induced local interactions: Examples employing nuclear magnetic resonance, *Nature*, 1973, 242, pp. 190–191.
10. A. Kumar, D. Welti, and R. Ernst, MNR Zeugmatography, *J Mag Res*, 1975, 18, pp. 69–83.
11. D. I. Bolef and R. K. Sundfors, *Nuclear Acoustic Resonance*, New York, NY: Academic Press, 1993.
12. K. Halbach, Design of permanent multipole magnets with oriented rare earth cobalt material, *Nucl Instr Methods*, 1980, 169, pp. 1–10.

Part III

Adaptive System Identification and Filtering

Adaptive signal processing is a fundamental technique for intelligent sensor and control systems which uses the computing resources to optimize digital system parameters as well as process digital signals. Currently, we generally use microprocessors as a stable, flexible, and robust way to filter sensor signals precisely for information detection, pattern recognition, and even closed-loop control of physical systems. The digital signal processing engine allows for precise design and consistent filter frequency response in a wide range of environments with little possibility for response drift characteristic of analog electronics in varying temperatures, and so on. However, the microprocessor is also a computer which while filtering digital signals can also simultaneously execute algorithms for the analysis of the input–output signal waveforms and update the digital filter coefficients to maintain a desired filter performance. An adaptive filter computes the optimal filter coefficients based on an analytic cost minimization function and adapts the filter continuously to maintain the desired optimal response. This self-correction feature requires some sort of simple cost function to be minimized as a means of deciding what adjustments must be made to the filter. In general, a positive (concave upward) quadratic cost function is sought since it has only one point of zero slope, which is a minimum. For linear time-invariant filters, the cost function is usually expressed as the square of the error between the adaptive filter output and the desired output. Since the filter output is a linear (multidimensional) function of the filter coefficients, the squared-error is a positive quadratic function of the filter coefficients. The minimum of the squared-error cost function is known as a least-squared-error solution.

In Chapter 9, we present a very concise development of a least-squared-error solution for an FIR filter basis function using simple matrix algebra. A projection operator is also presented as a more general framework which will be referred to in the development of adaptive lattice filter structures later in Chapter 10. The recursive adaptive algorithms in Chapters 10 and 11 allow the implementation of real-time adaptive processing where the filter coefficients are updated along with the filter computations for the output digital signal. All adaptive filtering algorithms presented are linked together through a common recursive update formula. The most simple and robust form of the recursive update is the least-mean-square (LMS) error adaptive filter algorithm. We present some important convergence properties of the LMS algorithm and contrast the convergence speed of the LMS to other more complicated but faster adaptive algorithms. Chapter 11 details a wide range of adaptive filtering applications including Kalman filtering for state vector adaptive updates and frequency-domain adaptive filtering. Recursive system identification is also explored using the adaptive filter to model an "unknown" system. The issues of mapping between the digital and analog domains are again revisited from Chapter 2 for physical system modeling.

One powerful adaptive algorithm, called the "Genetic" algorithm, is not presented in this book. The reason for excluding this technique is the space limit of an already broad book. The genetic algorithm is neither optimal nor least-squared error. It is used with great joy if all you need to do is find "a pretty good solution" rather than the optimal solution and you do not have computing resources to evaluate every solution. For problems where the squared-error surface can have multiple minima, a genetic recursive algorithm can be a really interesting way to get a reasonable solution, but without a guarantee that it is optimal. It works by "spawning" a group of directed guesses, evaluating the result, killing off the worst guesses, and repeating. It is a Darwinian approach to adaptive signal processing, which is made even more fascinating by today's enormous computing power. There are many flavors and techniques to genetic algorithms, and although really tempting to explore, we are just not doing it in this book.

9 Linear Least-Squared Error Modeling

We owe the method of least squares to the genius of Carl Friedrich Gauss (1777–1855), who at the age of only 24 made the first widely accepted application of least-squared error modeling to astronomy in the prediction of the orbit of the asteroid Ceres from only a few position measurements before it was lost from view [1]. This was an amazing calculation even by today's standards. When the asteroid reappeared months later, it was very close to the position Gauss predicted it would be in, a fact which stunned the astronomical community around the world. Gauss made many other even more significant contributions to astronomy and mathematics. But, without much doubt, least-squared error modeling is one of the most important algorithms to the art and science of engineering, and will likely remain so well into the twenty-first century [2].

9.1 BLOCK LEAST SQUARES

We begin by considering the problem of adaptively identifying a linear time-invariant causal system by processing only the input and output signals. Figure 9.1 depicts the process with a block diagram sketch showing the "unknown" system, its digital input $x[n] = x_n$, its digital output y_n, our FIR filter model $H[z]$, and the model output y'_n. The difference between the unknown system output y_n and the model output y'_n gives an error signal ε_n which is a linear function of the model filter coefficients. If we can find the model filter coefficients which give an error signal of zero, we can say that our model exactly matches the unknown system's response to the given input signal x_n. If x_n is spectrally white (an impulse, ZMG noise, sinusoidal chirp, etc.), or even nearly white (some signal energy at all frequencies), then our model's response should match the unknown system. However, if the unknown system is not well represented by an FIR filter, or if our model has fewer coefficients, the error signal cannot possibly be made exactly zero, but can only be minimized. The *least-squared error* solution represents the best possible match for our model to the unknown system given the constraints of the chosen FIR and number of coefficients [3].

Real-world adaptive system identification often suffers from incoherent noise interference as described in Section 7.3 for frequency-domain transfer functions. In general, only in the sterile world of computer simulation will the error signal actually converge to zero. The issue of incoherent noise interference will be addressed in Section 11.2 on Weiner filtering applications of adaptive system identification [4]. The FIR filter model $H[z]$ has $M + 1$ coefficients as depicted in Equation 9.1.

$$H[z] = h_0 + h_1 z^{-1} + h_2 z^{-1} + \cdots + h_M z^{-M} \tag{9.1}$$

We define a *basis function* for the least-squared error model φ_n as a 1 by $M + 1$ row vector

$$\varphi_n = \begin{bmatrix} x_n & x_{n-1} & x_{n-2} & \cdots & x_{n-M} \end{bmatrix} \tag{9.2}$$

and a model impulse response $M + 1$ by 1 column vector as

$$H = [h_0 \quad h_1 \quad h_2 \quad \cdots \quad h_M]^{\mathrm{T}} \tag{9.3}$$

265

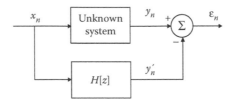

FIGURE 9.1 Block diagram depicting adaptive system identification using least-squared error system modeling.

where superscript T depicts a vector transpose. The FIR filter model output y_n' can now be written as a simple vector dot product.

$$y_n' = \varphi_n H \tag{9.4}$$

For this application of FIR system identification, the basis function is a vector of the sampled digital inputs from sample n back to sample $n - M$. Other forms of basis functions are possible such as a power series, function series (exponents, sinusoids, etc.), or other functions defined by the structure of the model which one wishes to obtain a least-squared error fit to the actual unknown system of interest. Least-squares modeling using other forms of basis function will be addressed in Section 9.3. The model error is simply the unknown system output minus the modeled output signal.

$$\begin{aligned}\varepsilon_n &= y_n - y_n' \\ &= y_n - \varphi_n H\end{aligned} \tag{9.5}$$

The least-squared error solution is computed over an observation window starting at sample n and ranging backward to sample $n - N + 1$, or a total of N samples. The error signal over the observation window is written as an N by 1 column vector.

$$\begin{bmatrix} \varepsilon_n \\ \varepsilon_{n-1} \\ \vdots \\ \varepsilon_{n-N+1} \end{bmatrix} = \begin{bmatrix} y_n \\ y_{n-1} \\ \vdots \\ y_{n-N+1} \end{bmatrix} - \begin{bmatrix} x_n & x_{n-1} & \cdots & x_{n-M} \\ x_{n-1} & x_{n-2} & \cdots & x_{n-M-1} \\ \vdots & \vdots & \cdots & \vdots \\ x_{n-N+1} & x_{n-N} & \cdots & x_{n-N+1-M} \end{bmatrix} \begin{bmatrix} h_0 \\ h_1 \\ \vdots \\ h_M \end{bmatrix} \tag{9.6}$$

Equation 9.6 is written more compactly in matrix form

$$\overline{\varepsilon} = \overline{y} - \overline{X} H \tag{9.7}$$

where the rows of $\overline{X}$ are the basis functions for the individual input–output signal samples.

We now write a compact matrix expression for the sum of the squared error over the observation window by computing a complex inner product for the error signal vector (superscript H denotes Hermitian transpose or transpose and complex conjugate for the most general notation).

$$\begin{aligned}\overline{\varepsilon}^H \overline{\varepsilon} &= (\overline{y} - \overline{X} H)^H (\overline{y} - \overline{X} H) \\ &= \overline{y}^H \overline{y} - \overline{y}^H \overline{X} H - H^H \overline{X}^H \overline{y} + H^H \overline{X}^H \overline{X} H\end{aligned} \tag{9.8}$$

If H is a scalar ($M = 0$), the sum of the squared error in Equation 9.8 is clearly a quadratic function of the FIR filter coefficient h_0. If $M = 1$, one could visualize a bowl-shaped error "surface" which is a function of h_0 and h_1. For $M > 1$, the error surface is multidimensional and not very practical to visualize. However, from a mathematical point of view, Equation 9.8 is quadratic with respect to the coefficients in H. The desirable aspect of a quadratic matrix equation is that there is only one solution where the slope of the error surface is zero. The value of H where the slope of the error surface is zero represents a minimum of the cost function in Equation 9.8, provided that the second derivatives with respect to H is positive definite (it is), indicating that the error surface is concave up. If the second derivative with respect to H is not positive definite, Equation 9.8 represents a "profit" function, rather than a cost function, which is maximized at the value of H where the error slope is zero.

Calculating the derivative of a matrix equation is straightforward with the exception that we must include the components of the derivative with respect to H and H^H in a single matrix-dimensioned result. To accomplish this with persnickety (paying too much attention to detail), we simply compute a partial derivative with respect to H (treating H^H as a constant) and adding this result to the Hermitian transpose of the derivative with respect to H^H (treating H as a constant). Note that this approach of summing the two partial derivatives provides a consistent result with the scalar case.

$$\frac{\partial \bar{\varepsilon}^H \bar{\varepsilon}}{\partial H} + \left\{\frac{\partial \bar{\varepsilon}^H \bar{\varepsilon}}{\partial H^H}\right\}^H = -\bar{y}^H \bar{X} + H^H \bar{X}^H \bar{X} + \left\{-\bar{X}^H \bar{y} + \bar{X}^H \bar{X} H\right\}^H$$

$$= -2\bar{y}^H \bar{X} + 2H^H \bar{X}^H \bar{X} \tag{9.9}$$

The solution for the value of H^H which gives a zero error surface slope in Equation 9.9 is

$$H^H = \bar{y}^H \bar{X} (\bar{X}^H \bar{X})^{-1} \tag{9.10}$$

The Hermitian transpose of Equation 9.10 gives the value of H for zero slope in the error surface.

$$H = (\bar{X}^H \bar{X})^{-1} \bar{X}^H \bar{y} \tag{9.11}$$

Given that we now have a value for H where the error surface is flat, the second derivative of the squared error is calculated to verify that the surface is concave up making the solution for H in Equation 9.11 a least-squared error solution.

$$\frac{\partial^2 \bar{\varepsilon}^H \bar{\varepsilon}}{\partial H^2} = 2\bar{X}^H \bar{X} = 2\sum_{k=0}^{N-1} \varphi_{n-k}^H \varphi_{n-k} \geq 0 \tag{9.12}$$

Closer examination of the complex inner product of the basis function reveals that the second derivative is simply the autocorrelation matrix times a scalar [5].

$$2\sum_{k=0}^{N-1} \varphi_{n-k}^H \varphi_{n-k} \approx 2N \begin{bmatrix} R_0^x & R_1^x & R_2^x & \cdots & R_M^x \\ R_{-1}^x & R_0^x & R_1^x & \cdots & R_{M-1}^x \\ \vdots & \vdots & \ddots & & \vdots \\ R_{-M}^x & R_{-M+1}^x & \cdots & R_{-1}^x & R_0^x \end{bmatrix} \tag{9.13}$$

where $R_j^x = E\{x_n^* x_{n-k}\}$ is the jth autocorrelation lag of x_n. If x_n is spectrally white, the matrix in Equation 9.13 is diagonal and positive. If x_n is nearly white, then the matrix is approximately diagonal (or can be made diagonal by Gaussian elimination or QR decomposition). In either case, the autocorrelation matrix is positive definite even for complex signals, the error surface is concave up, and the solution in Equation 9.11 is the least-squared error solution for the system identification problem. As the observation window grows large, the least-squared error solution can be seen to asymptotically approach

$$
\lim_{N \to \infty} H =
\begin{bmatrix}
R_0^x & R_1^x & R_2^x & \cdots & R_M^x \\
R_{-1}^x & R_0^x & R_1^x & \cdots & R_{M-1}^x \\
\vdots & \vdots & \ddots & & \vdots \\
R_{-M}^x & R_{-M+1}^x & \cdots & R_{-1}^x & R_0^x
\end{bmatrix}^{-1}
\begin{bmatrix}
R_0^{xy} \\
R_{-1}^{xy} \\
\vdots \\
R_{-M}^{xy}
\end{bmatrix}
\tag{9.14}
$$

where $R_j^{xy} = E\{x_n^* y_{n-j}\}$ is the cross correlation of the input and output data for the unknown system and the scale factor N divides out of the result.

Equation 9.14 has an exact analogy in the frequency domain where the frequency response of the filter $H[z]$ can be defined as the cross-spectrum of the input–output signals divided by the autospectrum of the input signal as described in Section 7.3. This should come as no surprise, since the same information (the input and output signals) is used in both the frequency domain and adaptive system identification cases. It makes no difference to the least-squares solution, so long as the Fourier transform resolution is comparable to the number of coefficients chosen for the filer model (i.e., M-point FFTs should be used). However, as we will see in Chapters 10 and 11, the adaptive filter approach can be made to work very efficiently and can be used to track nonstationary systems.

Figure 9.2 presents an example where a two-parameter FIR filter ($M = 1$) with $h_0 = -1.5$ and $h_1 = -2.5$ showing the squared-error surface for a range of model parameters. The least-squared error solution can be seen with the coordinates of H matching the actual "unknown" system where the squared error is a minimum.

The term "block least squares" is used to depict the idea that a least-squared error solution is calculated on a block of input and output data defined by the N samples of the observation window. Obviously, the larger the observation window, the better the model results will be, assuming that

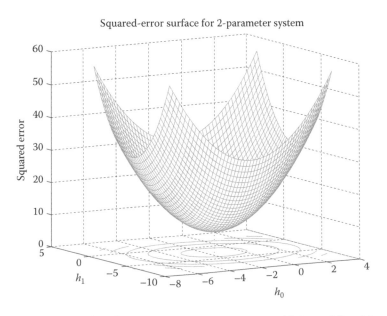

Squared-error surface for 2-parameter system

FIGURE 9.2 Squared-error surface for a two-parameter FIR system with $h_0 = -1.5$ and $h_1 = -2.5$ showing the minimum error for the optimum model.

some level of noise interference is unavoidable. However, if the unknown system being modeled is not a linear function of the filter coefficients H, the error surface depicted in Figure 9.2 would have more than one minimum. Nonlinear system identification is beyond the scope of this book. In addition, it is not clear from the above development what affect the chosen model order M has on the results of the least-squared error fit. If the unknown system is linear and FIR in structure, the squared error will decline as model order is increased approaching the correct model order. Choosing a model order higher than the unknown FIR system will not reduce the squared error further (the higher-order coefficients are computed at or near zero). If the unknown system is linear and IIR in structure, the squared error will continue to improve as model order is increased.

9.2 PROJECTION-BASED LEAST SQUARES

In this section, we present the least-squared error solution for modeling a linear time-invariant system in the most general mathematical terms. The reasons for this development approach may become of more interest later when we develop the fast-converging adaptive least-squares lattice filter and Schur recursions. The systems presented in this book can all be modeled as a weighted sum of linearly independent functions. For example, Chapter 2 showed how any pole–zero filter could be expressed as a weighted sum of resonances, defined by conjugate pole pairs. For a stable causal system with real input and output signals, the conjugate pole pairs each constitute a subsystem with the impulse response represented by a simple damped sinusoid, or mode. The total system response to any set of initial conditions and excitation signal can be completely described by the proper weighted sum of modes. The system response can be seen as a finite subset of the infinite number of possibilities of signals.

An abstract Hilbert space $\hat{H}$ is an *infinite-dimensional linear inner product space* [6]. A linear time-invariant system can be seen as a linear manifold spanned by the subspace $\hat{G}$. The difference between a Hilbert space and a subspace is that the Hilbert space has an infinite number of linearly independent elements and the subspace has only a finite number of linearly independent elements. Casting our linear time-invariant system into a subspace allows us to write the least-squared error signal as an orthogonal projection of the system output signal onto the subspace. Why do we care? The subspace can be expanded using mathematically defined orthogonal projections, which represent by definition, the minimum distance between the old subspace and the new expanded subspace. For new observations added to the signal subspace, the orthogonal expansion represents the least-squared error between the model output prediction and the unknown system output! The mathematical equations for this orthogonal expression give exactly the same least-squared error solution derived in the previous section. However, since subspaces can be defined in both time and order, we can define least-squared error updates in either time, order, or both, which leads to the fastest possible convergence of an adaptive algorithm.

The signal subspace can also be expanded in model order using orthogonal projections, allowing a least-squared error $m = M + 1$ system model to be derived from the $m = M$ order model, and so on. Given a new observation, one would calculate the $m = 0$ solution, then calculate the $m = 1$ solution using the $m = 0$'s information, and so on up to the $m = M$ order model. Since the response of linear time-invariant systems can be represented by a weighted sum of linearly independent functions, it is straightforward to consider the orthogonal projection order updates a well matched-framework for identification of linear systems. The projection operator framework allows the subspace to be decomposed in time and order as an effective means to achieve very rapid convergence to a least-squared error solution using very few observations. The block least-squares approach will obtain exactly the same result for a particular model order, but the solution contains only that model order. The projection operator framework actually allows the model order to be evaluated and updated along with the model parameters. Under the conditions of noiseless signals and too high a chosen model order, the required matrix inverse in block least squares is ill-conditioned due to linear dependence. Updating the model order in a projection operator framework allows one to avoid overdetermining the model order and the associated linear dependence which prohibits the matrix inversion in the least-squares solution. The projection operator framework can also be seen as a more general representation of the eigenvalue problem and singular value decomposition. The recursive update of both the block least squares and projection operator framework will be dealt with in Chapter 10.

Without going into finer details of linear operators and Hilbert space theory, we can simply state that $\hat{G}$ is a subspace of the infinite Hilbert space $\hat{H}$. Given a new observation y_n in the vector $\bar{y}$ but not in the subspace spanned by $\hat{G}$, we define a vector $\mathbf{g}$ as the projection of $\bar{y}$ onto the subspace $\hat{G}$, and $\mathbf{f}$ as a vector orthogonal to the subspace $\hat{G}$ which connects $\mathbf{g}$ to $\bar{y}$. Simply put, $\bar{y} = \mathbf{g} + \mathbf{f}$. One can think of $\mathbf{g}$ as the component of $\bar{y}$ predictable in the subspace $\hat{G}$ and $\mathbf{f}$ is the least-squared error (shortest distance between the new observation and the prediction as defined by an orthogonal vector) of the prediction for $\bar{y}$. The prediction error is $\bar{\varepsilon} = \mathbf{f} = \bar{y} - \mathbf{g}$. Substituting the least-squared error solution for our model H, we can examine the implications of casting our solution in a projection operator framework.

$$\bar{\varepsilon} = \bar{y} - \bar{X}H$$
$$= \bar{y} - \bar{X}(\bar{X}^H \bar{X})^{-1} \bar{X}^H \bar{y} \tag{9.15}$$

The projection of $\bar{y}$ onto the subspace $\hat{G}$ denoted as the vector $\mathbf{g}$ above is simply

$$P_X \bar{y} = \bar{X}(\bar{X}^H \bar{X})^{-1} \bar{X}^H \bar{y} \tag{9.16}$$

where P_X is a projection operator for the subspace $\hat{G}$ spanned by the elements of $\hat{H}$. The projection operator outlined in Equations 9.15 and 9.16 has the interesting properties of being *bounded, having unity norm*, and being *self-adjoint*. Consider the square of a projection operator.

$$P_X^2 = \bar{X}(\bar{X}^H \bar{X})^{-1} \bar{X}^H \bar{X} \left(\bar{X}^H \bar{X}\right)^{-1} \bar{X}^H$$
$$= \bar{X} \left(\bar{X}^H \bar{X}\right)^{-1} \bar{X}^H = P_X \tag{9.17}$$

The projection operator orthogonal to the subspace $\hat{G}$ is simply

$$I - P_X = I - \bar{X}(\bar{X}^H \bar{X})^{-1} \bar{X}^H \tag{9.18}$$

and thus, it is easy to show that $(I - P_X)P_X = 0$, where I is the identity matrix (all elements zero except the main diagonals which are all unity). Therefore, the error vector can be written as the orthogonal projection of the observations to the subspace as shown in Figure 9.3 and Equation 9.19.

$$\bar{\varepsilon} = (I - P_X)\bar{y} \tag{9.19}$$

This seems like a roundabout way to reinvent least squares, but consider updating the subspace spanned by the elements of $\bar{X}$ to the space spanned by $\bar{X} + \bar{S}$. We can update the subspace by again applying an orthogonal projection.

$$\bar{X} + \bar{S} = \bar{X} + \bar{S}(I - P_X) \tag{9.20}$$

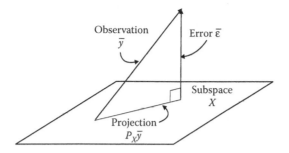

FIGURE 9.3 Graphical depiction of projection-based least-squared error prediction showing the error vector as an orthogonal projection.

The projection operator for the updated subspace is

$$P_{\{X+S\}} = P_X + P_{\{S(I-P_X)\}}$$
$$= P_X + (I - P_X)\,\overline{S}\,\left[\overline{S}^{\mathrm{H}}\,(I - P_X)\,\overline{S}\right]^{-1}\overline{S}^{\mathrm{H}}\,(I - P_X) \tag{9.21}$$

The orthogonal projection operator to the updated subspace $S + X$ is simply

$$I - P_{\{X+S\}} = I - P_X + (I - P_X)\,\overline{S}\,\left[\overline{S}^{\mathrm{H}}\,(I - P_X)\,\overline{S}\right]^{-1}\overline{S}^{\mathrm{H}}\,(I - P_X) \tag{9.22}$$

Equation 9.22 can be used to generate a wide range of recursive updates for least-squared error algorithms. The recursion in Equation 9.22 is remarkably close to the recursion derived from the matrix inversion lemma presented in Chapter 10. But, this difference is only due to the different applications of matrix inversion and orthogonal projection and will be revisited in more detail in Chapter 10. However, it is enlightening to observe the mathematical uniformity of least-squares recursions. Casting an error signal as an orthogonal projection by definition guarantees the minimum error as the shortest distance between a plane and a point in space is a line normal to the plane and intersecting the point.

9.3 GENERAL BASIS SYSTEM IDENTIFICATION

Consider a more general framework for least-squared error system modeling. We have shown a straightforward application for system identification using a digital FIR filter model and the input–output signals. The basis function for the FIR filter model was a digital filter's tapped delay line, or input sequence

$$\left[x_n\ x_{n-1}\ x_{n-2}\cdots x_{n-M}\right] \tag{9.23}$$

where x_n is the input signal to the filter. Gauss's technique, developed nearly two centuries ago, was used for many things long before digital filters even existed. By applying several other basis functions, we will see the great power of least-squared error system modeling, even for nonlinear systems that can be written to produce a linear error function. The reader should understand that any basis function can be used to produce a linear least-squared error model. But the choice of basis function(s) will determine the *optimality* of the least-squared error model.

Consider a very simple task of fitting a polynomial curve to four pairs of observations $\bar{y} = [1.1\ 6.0\ 8.0\ 26.5]$ at the ordinates $x = [1\ 2\ 3\ 4]$. For a linear curve fit, the basis function is simply $\varphi_n = [x_n\ 1]$ and the subspace X is spanned by the elements of $\bar{X}$ defined by

$$\bar{X} = \begin{bmatrix} 1 & 1 \\ 2 & 1 \\ 3 & 1 \\ 4 & 1 \end{bmatrix} \tag{9.24}$$

The least-squared error solution for a linear (straight line) fit is

$$H = \left(\bar{X}^H \bar{X}\right)^{-1} \bar{X}^H \bar{y}$$

$$= \begin{bmatrix} 7.82 \\ -9.15 \end{bmatrix} = \begin{bmatrix} 0.2 & -0.5 \\ -0.5 & 1.5 \end{bmatrix} \begin{bmatrix} 1 & 2 & 3 & 4 \\ 1 & 1 & 1 & 1 \end{bmatrix} \begin{bmatrix} 1.1 \\ 6 \\ 8 \\ 21.5 \end{bmatrix} \tag{9.25}$$

If we were to write the equation of the line in slope–intercept form, $y = 7.82x - 9.15$, where the slope is 7.82 and the intercept of the y-axis is -9.15. The projection operator for the linear fit is symmetric about both diagonals, but the main diagonal does not dominate the magnitudes indicating a rather poor fit as seen in the error vector.

$$P_X = \begin{bmatrix} 0.7 & 0.4 & 0.1 & -0.2 \\ 0.4 & 0.3 & 0.2 & 0.1 \\ 0.1 & 0.2 & 0.3 & 0.4 \\ -0.2 & 0.1 & 0.4 & 0.7 \end{bmatrix} \quad \bar{\varepsilon} = \left(I - P_X\right)\bar{y} = \begin{bmatrix} 2.43 \\ -0.49 \\ -6.31 \\ 4.37 \end{bmatrix} \tag{9.26}$$

The mean-squared error for the linear fit in Equations 9.26 through 9.28 is 16.26.

Fitting a quadratic function to the data simply involves expanding the basis function as $\varphi_n = \left[x_n^2\ x_n\ 1\right]$.

$$H = \left(\bar{X}^H \bar{X}\right)^{-1} \bar{X}^H \bar{y}$$

$$= \begin{bmatrix} 3.40 \\ -9.18 \\ 7.85 \end{bmatrix} = \begin{bmatrix} 0.25 & -1.25 & 1.25 \\ -1.25 & 6.45 & -6.75 \\ 1.25 & -6.75 & 7.75 \end{bmatrix} \begin{bmatrix} 1 & 4 & 9 & 16 \\ 1 & 2 & 3 & 4 \\ 1 & 1 & 1 & 1 \end{bmatrix} \begin{bmatrix} 1.1 \\ 6 \\ 8 \\ 21.5 \end{bmatrix} \tag{9.27}$$

The projection operator and the error vector for the quadratic fit are

$$P_X = \begin{bmatrix} 0.95 & 0.15 & -0.15 & -0.05 \\ 0.15 & 0.55 & 0.45 & -0.15 \\ -0.15 & 0.45 & 0.55 & 0.15 \\ 0.05 & -0.15 & 0.15 & 0.95 \end{bmatrix}, \quad \bar{\varepsilon} = \left(I - P_X\right)\bar{y} = \begin{bmatrix} -0.97 \\ 2.91 \\ -2.91 \\ 0.99 \end{bmatrix} \tag{9.28}$$

The mean-squared error for the quadratic fit is 4.71, significantly less than the linear fit. Clearly, the projection operator for the quadratic fit, while still symmetric about both diagonals, is dominated by the main diagonal element magnitudes, indicating a better overall fit for the data. Fitting a cubic basic function to the data we get the model parameters $H = [3.233\ -20.85\ 44.82\ -26.10]$, P_X is the identity matrix, and the error is essentially zero. Figure 9.4 shows the linear, quadratic, and cubic modeling results graphically.

It is expected that the least-squared error fit for the cubic basis function would be zero (within numerical error limits—the result in this case was 1×10^{-23}). This is because there are only four data observations and four elements to the basis function, or four equations and four unknowns. When the data are noisy (as is the case here), having too few observations can cause a misrepresentation of the data. In this case these data are really from a quadratic function, but with added noise. If we had many more data points it would be clear that the quadratic model really provides the best overall fit, especially when comparing the cubic model for the four data observation to a more densely populated distribution of observations. In any modeling problem with noisy data it is extremely important to overdetermine the model by using many more observations than the basis model order. This is why distribution measures such as the Student's t test, F-test, and so on are used extensively in clinical trials where only a small number of subjects are available. In most adaptive signal processing models, large data sets are readily available, and this should be exploited wherever possible.

9.3.1 Mechanics of the Human Ear

Consider the highly nonlinear frequency–loudness response of the human ear as an example. Audiologists and noise control engineers often use a weighted decibel value to describe sound in terms of human perception. According to Newby [7], the A-weighted curve describes a decibel correction factor for the human ear for sound levels below 55 dB relative to 20 µPa, the B curve for levels between 55 and 85 dB, and the C curve for levels over 85 dB. The A and C curves are predominantly used. Most hearing conservation legislation uses the A curve since occupational hearing impairment occurs in the middle frequency range where the A curve is most sensitive. For loud sounds, the brain restricts the levels entering the ear via tendons on the bones of the middle ear and neural control of the mechanical response of the cochlea in the organ of Corti. This action is

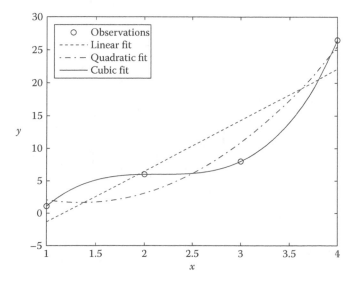

FIGURE 9.4 Linear, quadratic, and cubic least-squared error models fitting only four observations points to the model.

analogous to the action of the iris in the eye closing down in response to bright light. In the age of signal processing tools such as spectrum analyzers and MATLAB, it is useful to have audio band curves to quickly convert a narrowband sound spectrum to an A-weighted level. To obtain a straightforward algorithm for curve generation, the method of least-squared error is employed to fit a curve accurately through a series of well-published decibel–frequency coordinates. The ear is an absolutely remarkable intelligent sensor system which uses adaptive response to optimize hearing in a wide range of environments. It is very useful from a sensor technology point of view, to analyze the physics and physiology of the ear in the context of the least-squared error modeling of its frequency–loudness response to acoustic signals.

The human ear is a remarkable transducer, which by either evolution or divine design has mechanisms to adapt to protect itself while providing us detection of a wide range of sounds. We define 0 dB, or 20 µPa RMS, as the "minimum" audible sound detectable above the internal noise from breathing and blood flow for an average ear, and 130 dB as about the loudest sound tolerable, our ears have typically a 10^6 dynamic pressure range. This is greater than most microphones and certainly greater than most recording systems capable of the ear's frequency response from 20 to 20 kHz. We know that a large part of human hearing occurs in the brain, and that speech is processed differently from other sounds. We also know that the brain controls the sensitivity of the ear through the auditory nerve. Figure 9.5 shows the anatomy of the right ear looking from front to rear.

Like the retina of the eye, overstimulation of the vibration-sensing "hair cells" in the cochlea can certainly lead to pain and even permanent neural cell damage. The brain actually has two mechanisms to control the stimulation levels in the cochlea. First, the bones of the middle ear (mallus, incus, and stapes) can be restricted from motion by the *tendon of Stapedious muscle* as shown in Figure 9.5 (it attaches to the stapes bone) in much the same way the iris in the eye restricts light. As this muscle tightens, the amplitude of the vibration diminishes and the frequency response of the middle ear actually flattens out so that low, middle, and high frequencies have nearly the same sensitivity. For very low-level sounds, the tendon relaxes allowing the stapes to move more freely. The relaxed state tends to significantly enhance hearing sensitivity in the middle speech band (from 300 to 6000 Hz approximately). Hence the A weighting reflects a significant drop in sensitivity at very low and very high frequencies. The second way the brain suppresses overstimulation in the cochlea involves the response of the hair cells directly. It is not known whether the neural response for vibration control affects the frequency response of the ear, or whether the tendon of *Stapedious* in conjunction with the neural response together cause the change in the ear's frequency response as a

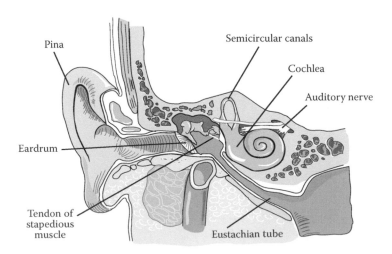

FIGURE 9.5 Sketch of the anatomy of the right human ear looking from front to rear showing the tendons used by the brain to control the loudness of sound reaching the inner ear cochlea and auditory nerve.

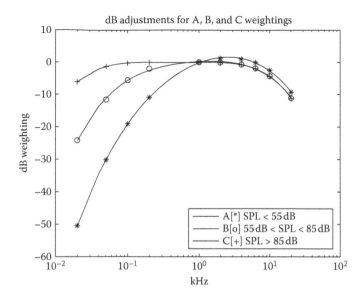

FIGURE 9.6 A, B, and C spectral weighting curves for modeling the human ear's relative frequency response to sound at various levels of loudness.

function of loudness level. However, the average frequency response's of healthy ears have been objectively measured for quite some time. The "loudness" button on most high fidelity music playback systems inverts the ear's frequency response in conjunction with the "volume" control so that music can have a rich sounding bass response at low listening levels. Figure 9.6 shows the relative frequency correction weightings for the A, B, and C curves.

It is useful to have the ability to precisely generate the A, B, or C weighting curves for the human ear's response for any desired frequency in the audible range. This capability allows one to easily convert a power spectrum in decibels relative to $20\,\mu Pa$ of some arbitrary frequency resolution directly to an A-, B-, or C-weighted decibel reading. The American National Standards Institute (ANSI) standard for sound level meter (SLM) specifications provides tables of relative decibel weightings as a function of frequency for the human ear [8]. Additional tables are given specifying the accuracy required in ±dB versus frequency for a SLM to be considered Type 0 (most accurate), Type 1 (roughly ±1 dB in the 50 Hz to 4000 Hz range), or Type 2 (economical accuracy). The ANSI standard does not specify A, B, and C curve equations, just the Tables representing the accepted "normal" human response.

9.3.2 LEAST-SQUARES CURVE FITTING

To generate a model for a continuous curve to be used to map narrowband FFT spectra calibrated in Pascals to A-, B-, or C- weighted sound pressure levels (SPLs), the least-squared error technique is employed. We begin with 7 dB observations at seven frequencies which are simply read from

TABLE 9.1
Decibel Correction Factors for A, B, and C Weightings

Frequency (Hz)	19.95	50.12	100	199.5	1000	1995	3981	6310	10000	20000
A	−50.5	−30.2	−19.1	−10.9	0	1.2	+1.0	−0.1	−3.0	−9.3
B	−24.2	−11.6	−5.6	−2	0	−0.1	−0.7	−1.9	−4.3	−11.1
C	−6.2	−1.3	−0.3	0	0	−0.2	−0.8	−2.0	−4.0	−11.2

existing A, B, and C curves in the literature given in Table 9.1. These data observations can be seen in Figure 9.6 as the symbols are plotted on the corresponding curves. As noted earlier, the B curve is generally only used in audiology, but is useful to illustrate the sensitivity changes in the ear between 55 and 85 dB.

To improve the conditioning of the curve-fitting problem, we use the base-10 logarithm of the frequency in kiloHertz in the basis function and a curve model of the form

$$dB_H^x(f) = \sum_{m=1}^{M} H_m^x f_\ell^{m-1} \tag{9.29}$$

where f_1 is the log base 10 of the frequency in kiloHertz, the superscript x refers to either A, B, or C weighting, and the subscript H means that the decibel value is predicted using the weights H_m. The use of the base-10 logarithm of the frequency in kiloHertz may seem unnecessary, but it helps considerably in conditioning the matrices in the least-squares fitting problem over the nine octave frequency range from 20 Hz to 20 kHz. The error between our model in Equation 9.29 and the ANSI table data are

$$\varepsilon(f) = dB_T^x(f) - dB_H^x(f) = dB_T^x(f) - \begin{bmatrix} 1 & f_\ell & f_\ell^2 \cdots f_\ell^{M-1} \end{bmatrix} \begin{bmatrix} H_1^x \\ H_2^x \\ \vdots \\ H_M^x \end{bmatrix} \tag{9.30}$$

where the subscript T means that the decibel data are from the ANSI tables and H is from the model. We want our model, defined by the weights H_m and basis function $[1\; f_1\; f_1^2\; ...\; f_1^{M-1}]$, to provide minimum error over the frequency range of interest. Therefore, we define Equation 9.30 in matrix form for a range of N frequencies.

$$\begin{bmatrix} \varepsilon(f_1) \\ \varepsilon(f_2) \\ \vdots \\ \varepsilon(f_N) \end{bmatrix} = \begin{bmatrix} dB_T^x(f_1) \\ dB_T^x(f_2) \\ \vdots \\ dB_T^x(f_N) \end{bmatrix} - \begin{bmatrix} 1 & f_{l,1} & f_{l,1}^2 & \cdots & f_{l,1}^{M-1} \\ 1 & f_{l,2} & f_{l,2}^2 & \cdots & f_{l,2}^{M-1} \\ & & \vdots & & \\ 1 & f_{l,N} & f_{l,N}^2 & \cdots & f_{l,N}^{M-1} \end{bmatrix} \begin{bmatrix} H_1^x \\ H_2^x \\ \vdots \\ H_M^x \end{bmatrix} \tag{9.31}$$

Equation 9.31 is simply written in compact matrix form as

$$\bar{E} = \bar{D} - \bar{F}H \tag{9.32}$$

Therefore, the least-squared error solution for the weights which best fit a curve through the ANSI table decibel values is seen in Equation 9.33. The resulting fifth-order model coefficients for the A-, B-, and C-weighted curves in Figure 9.6 are given in Table 9.2.

$$\bar{H} = \left(\bar{F}'\bar{F} \right)^{-1} \bar{F}' \bar{D} \tag{9.33}$$

9.3.3 Pole–Zero Filter Models

However, historically, the functions for the A, B, and C curves are given in terms of pole–zero transfer functions, which can be implemented as an analog circuit directly in the SLM. The method for

TABLE 9.2
Fifth-Order Least-Squares Fit Coefficients for A, B, and C Curves

M	H^A	H^B	H^C
1	−0.1940747	+0.1807204	−0.07478983
2	+8.387643	+1.257416	+0.3047574
3	−9.616735	−3.327762	−0.2513878
4	−0.2017488	−0.7022932	−2.416345
5	−1.111944	−1.945072	−2.006099

generating the curves is wisely not part of the ANSI standard, since there is no closed form model defined from physics. One such pole–zero model is given in Appendix C of ANSI S1.4-1983 (with the disclaimer "for informational purposes only"). The C-weighting curve in dB is defined as

$$\mathrm{dB}^C(f) = 10\log_{10}\left(\frac{K_1 f^4}{(f^2 + f_1^2)^2 (f^2 + f_4^2)^2}\right) \tag{9.34}$$

where K_1 is 2.242881×10^{16}, $f_1 = 20.598997$, $f_4 = 12194.22$, and f is the frequency in Hertz. The B-weighting is defined as

$$\mathrm{dB}^B(f) = 10\log_{10}\left(\frac{K_2 f^2}{f^2 + f_5^2}\right) + \mathrm{dB}^C(f) \tag{9.35}$$

where K_2 is 1.025119 and f_5 is 158.48932. The A-weighting curve is

$$\mathrm{dB}^A(f) = 10\log_{10}\left(\frac{K_3 f^4}{(f^2 + f_2^2)(f^2 + f_3^2)}\right) + \mathrm{dB}^C(f) \tag{9.36}$$

where K_3 is 1.562339, f_2 is 107.65265, and f_3 is 737.86223. The curves generated by Equations 9.34 and 9.35 are virtually identical to those generated using least squares. A comparison between the least-squares fit and the "ANSI" pole–zero models is shown in Figure 9.7.

Since the highest required precision for an SLM is Type 0 where the highest precision in any frequency range is ±0.7 dB, the variations in Figure 9.7 for the least-squares fit are within a Type 0 specification. The European Norm EN60651 (IEC651) has slightly different formulation for a pole–zero model for the A, B, and C curves, but the listed numbers tabulated are identical to the ANSI curve responses. Using either curve definition, one can access the A, B, and C weightings for conversion of narrowband spectra to broadband weighted readings for modeling the human ear's response to sound. Accuracy of the least-squared error model can be improved by using more frequency samples in between the samples given in Table 9.1. In general, the least-squares fit will best match the model *at the basis function sample points*. In between these input values, the model can be significantly off the expected trend most notably when the number of input samples in the basis function N is close to the model order M. N must be greater than M for a least-squared error solution to exist, but $N \gg M$ for a very accurate model to be found from potentially noisy data.

As our final example basis function in this section, we consider the Fourier transform cast in the form of a least-squares error modeling problem. The basis function is now a Fourier series of

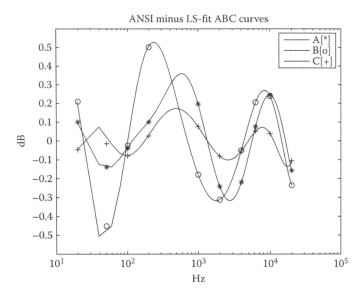

FIGURE 9.7 Decibel error as a function of log-frequency between the models found using a least-squares fit and those published in Appendix C of ANSI S1.41983.

complex exponentials of chosen particular frequencies and our N observations $\bar{y} = [y_n \; y_{n-1} \; \cdots \; y_{n-N+1}]$ are of some time series of interest.

$$\bar{\varepsilon} = \bar{y} - \begin{bmatrix} e^{j\omega_1 nT} & e^{j\omega_2 nT} & \cdots & e^{j\omega_M nT} \\ e^{j\omega_1 (n-1)T} & e^{j\omega_2 (n-1)T} & \cdots & e^{j\omega_M (n-1)T} \\ \vdots & \vdots & \cdots & \vdots \\ e^{j\omega_1 (n-N+1)T} & e^{j\omega_2 (n-N+1)T} & \cdots & e^{j\omega_M (n-N+1)T} \end{bmatrix} \begin{bmatrix} A_1 \\ A_2 \\ \vdots \\ A_M \end{bmatrix} \tag{9.37}$$

The least-squared error solution for the chosen basis functions (choice of ω's and M) is simply

$$A_m = \frac{1}{N} \sum_{n' = n}^{n'=n-N+1} y_{n'} \; e^{-j\omega_m n'T} \tag{9.38}$$

which is consistent with the normalize DFT presented in Chapter 6, Equation 6.2. The solution is greatly simplified by the orthogonality of the complex sinusoids (provided that the span of N samples constitutes an integer number of wavelengths in the observation window). Orthogonality of the basis functions simplifies the matrix inverse $(X^H X)^{-1}$ by making it essentially a scalar N^{-1} times the identity matrix I. Complex sinusoids are also orthonormal when the factor of N is removed by scaling. Nonorthogonal basis functions can always be used in a least-squared error model, but the matrix inverse must be completely calculated.

If the basis function is a spectrum of the input to an unknown system $X(\omega)$, and the observations are a spectrum of the output signal from an unknown system $Y(\omega)$, then the model is the frequency response of the unknown system $H(\omega)$. It is useful to compare the least-squared error spectrum solution to the transfer function measurement previously developed in Section 6.2. The transfer function solution using least-squares and time-domain input–output data are seen in Equation 9.39 for $X(\omega)$ white.

$$
\begin{bmatrix} h_0 \\ h_1 \\ \vdots \\ h_{M-1} \end{bmatrix} = \begin{bmatrix} R_0^x & R_1^x & \cdots & R_{M-1}^x \\ R_{-1}^x & R_0^x & \cdots & R_{M-2} \\ \vdots & \vdots & \vdots & \vdots \\ R_{-M+1}^x & R_{-M+2}^x & \cdots & R_0^x \end{bmatrix}^{-1} \begin{bmatrix} R_0^{xy} \\ R_{-1}^{xy} \\ \vdots \\ R_{-M+1}^{xy} \end{bmatrix} = \frac{1}{R_0^x} \begin{bmatrix} R_0^{xy} \\ R_{-1}^{xy} \\ \vdots \\ R_{-M+1}^{xy} \end{bmatrix}
\tag{9.39}
$$

Equation 9.39 is the same result presented in Equation 9.14 in the limit as N approaches infinity. Since $X(\omega)$ is spectrally white, the basis functions are all "digital" Dirac delta functions making the matrix inverse in Equation 9.39 a simple scalar inverse. Taking Fourier transforms of both sides of Equation 9.39 we have the transfer function expression from Section 7.3.

$$
H(\omega) = \frac{G^{xy}(\omega)}{G^{xx}(\omega)}
\tag{9.40}
$$

The projection operator is the identity matrix when the input signal is orthogonal for the observation window N. For large N and reasonably white $X(\omega)$ the characteristics of Equations 9.39 through 9.40 generally hold as stated. However, for nonorthogonal N (spectral leakage cases), the frequency-domain solution for the frequency response $H(\omega)$ in Equation 9.40 may have significant error compared to the time-domain solution including the full matrix inverse. The reason for this is that in the time-domain solution, the model is free to move its zeros anywhere to best match the dominant peaks in the spectral response. But with a frequency-domain basis function, the specific frequencies are prechosen and can result in a bias error in the model. Obviously, for very large N and high model orders, the two solutions are essentially identical.

9.4 MATLAB® EXAMPLES

We used only four simple MATLAB m-scripts for this chapter, listed in Table 9.3. The script "lserrorbowl.m" is just what the name implies, generation of a 3D error surface for a two-parameter least-squared error model. However, rather than going through the details of implementing Equation 9.8, we just look at a bowl-shaped error surface for a range of H values in X and Y dimensions that contain the optimal H_0. We added an offset to the error so that it would make a visually nice plot using the mesh-contour function in MATLAB which projects a contour plot on the bottom surface boundary of the 3D surface plot shown in Figure 9.2.

The script "lspolynomial.m" uses a simple input data vector "xdata" and output data vector "ydata" and sets up the basis matrices for linear, quadratic, and cubic curve fitting and compares the results in Figure 9.4. This makes the script very illustrative. It also show how MATLAB's matrix-math

---

TABLE 9.3
MATLAB m-Scripts Used in This Chapter

Figure	m-Script
9.2	lserrorbowl.m
9.4	lspolynomial.m
9.6	ABCsplwts.m
9.7	compareABC.m

makes programming extremely simple since the matrix operations can all be written in one line as if it was a scalar algebra equation!

The m-scripts "ABCsplwts.m" and "compareABC.m" are quite a bit more detailed because they produce a model for the A, B, and C weighting curves for SPLs from the ANSI standard. "ACSsplwts.m" produces the A, B, and C weighting curves one sees in Figure 9.6 using a log-frequency basis function (Vandermode matrix) and writes out the fitted function coefficients to a text file. The script "compareABC.m" regenerates these curves and then compares them to ANSI standard functions for A, B, and C, weighting given by Equations 9.34 through 9.36. The small differences in the least-squared error curve fits and the ANSI standards are shown in Figure 9.7.

9.5 SUMMARY

The manual computations required in Gauss's day severely limited the applications of least squares to only a few well-defined problems. It is humbling, to say the least, that an individual at the turn of the nineteenth century could manually make the numerical calculations necessary to predict the position of an asteroid nine months in advance. Remarking once that the least-squares technique would be useful on a much wider scale if a suitable machine could be built to automate some of the calculations, Gauss made probably the greatest understatement of the past two centuries! The advent of the digital computer has made application of least-squared error modeling as pervasive in business law, sociology, psychology, medicine, and politics as it is in engineering and the hard sciences.

The least-squared error problem can be cast into a projection operator inner product space which is very useful for derivation of fast recursive adaptive processing algorithms. The observation data for a particular problem can be seen as a subspace where the system of interest is a linear manifold, the response of which can be modeled as a weighted linear combination of orthogonal eigenvectors. The subspace spanned by the observations of the system input signal form the basis of a projection operator. The orthogonal projection of the observed system output signal vector onto the input signal subspace determines the least-squared error solution. The value of the projection operator framework is that the subspace can be expanded using orthogonal projections for model order as well as observation window size. The orthogonal decomposition of the subspace allows for very efficient (fast converging) recursive adaptive algorithms to be developed such as the least-squares lattice adaptive algorithm.

We have shown a very straightforward application of least-squared error modeling on an FIR filter where the basis function is a simple delay line of filter input samples. However, a much wider range of basis functions can be used, including nonlinear functions. What is critical is that the error response, the difference between the actual and predicted outputs, be a linear function of the model coefficients. The linear error model allows a quadratic squared error surface which is minimized through the choice of model coefficients which gives a zero gradient on the error surface. The fact that the zero-gradient solution is a minimum of the error surface is verified by examination of the second derivative for a positive definite condition.

PROBLEMS

1. Given four temperatures and times, find β and T_0 for the model $T(t) = T_0 e^{-t/\beta}$. $T(60) = 63.8$, $T(120) = 19.2$, $T(180) = 5.8$, and $T(240) = 1.7$. Assume that t is in seconds and T is in degree Celsius.
2. Do basis functions have to be orthogonal for the least-squared error technique to work?
3. Does the underlying error model have to linear for the least-squared error technique to work?
4. A stock market index has the following values for Monday through Thursday:{1257, 1189, 1205, 1200}. Using a four-day moving average linear model fit, what do you predict the index will be by Friday close?
5. Show that the optimal mean-squared error can be written as $\overline{\varepsilon}^H \overline{\varepsilon} = \frac{1}{N} \overline{y}^H (I - P_X) \overline{y}$.

6. Show that the projection operator is self-adjoint and has unity norm.
7. Show that as the projection operator P_X approaches the identity matrix, the mean-square error must go to zero.
8. Show that $P_X \bar{y} = \bar{X} H_0$, where H_0 is the least-squared error set of coefficients.
9. You buy a stock at \$105/share. Monday's closing price is \$110. Tuesday's closing price drops to \$100. Wednesday's closing price is back up to \$108. But then Thursday's closing price only drops to \$105. Based on a linear model, should you sell on Friday morning?

REFERENCES

1. C. B. Boyer, *A History of Mathematics*, 2nd ed. New York, NY: Wiley, 1991, pp. 496–508.
2. T. Kailath, *Linear Systems,* Englewood Cliffs, NJ: Prentice-Hall, 1980.
3. S. J. Orfandis, *Optimum Signal Processing*, 2nd ed. New York, NY: McGraw-Hill, 1988.
4. B. Widrow and S. D. Sterns, *Adaptive Signal Processing*, Englewood Cliffs, NJ: Prentice-Hall, 1985.
5. M. Bellanger, *Adaptive Digital Filters and Signal Analysis*, New York, NY: Marcel-Dekker, 1988.
6. N. I. Akheizer and I. M. Glazman, *Theory of Linear Operators in Hilbert Space*, Mineola: Dover, 1993.
7. H. A. Newby, *Audiology*, 4th ed. Englewood Cliffs, NJ: Prentice-Hall, 1979.
8. ANSI S1.4–1983, Specification for sound level meters, ASA Catalog No. 47–1983, American Institute of Physics, New York, NY.

10 Recursive Least-Squares Techniques

The powerful technique of fitting a linear system model to the input–output response data with a least-squared error can be made even more useful by developing a recursive form with limited signal data memory to adapt with nonstationary systems and signals. One could successfully implement the block least-squares in Section 9.1 on a sliding record of N input–output signal samples. However, even with today's inexpensive, speedy, and nimble computing resources, a sliding-block approach is ill-advised. It can be seen that the previous $N-1$ input–output samples and their correlations are simply being recomputed over and over again with each new sample as time marches on. By writing a recursive matrix equation update for the input data autocorrelations and input–output data cross-correlations, we can simply add on the necessary terms for the current input–output signal data to the previous correlation estimates, thereby saving a very significant amount of redundant computations. Furthermore, the most precise and efficient adaptive modeling algorithms will generally require the least amount of overall computation. Every arithmetic operation is a potential source of numerical error, and so fewer the redundant computations the better. Later, we will show simplifications to the recursive least-squares (RLS) algorithm which require very few operations but converge more slowly. However, these simplified algorithms can actually require more operations over many more iterations to reach a least-squared error solution and may actually not produce a precise result. This is particularly true for nonstationary signals and systems.

To make the RLS algorithm adaptive, we define an exponentially decaying memory window, which weights the most recent data the strongest and slowly "forgets" older signal data which are less interesting and likely from an "out-of-date" system relative to the current model. Exponential memory weighting is very simple and only requires the previous correlation estimate. However, a number of other shapes of recursive data memory windows can be found [1], but are rarely called for. Consider a simple integrator for a physical parameter w_t for the wind speed measured by a cup-vane-type anemometer of the form $\bar{W}_t = \alpha \bar{W}_{t-1} + \beta w_t$, where the averaging weights are $\alpha = (N-1)/N$, $\beta = 1/N$. The parameter N is the length of a linear memory window in terms of number of wind samples. If the linear memory window starts at $N = 1$, $\alpha = 0$, and $\beta = 1$, at $N = 2$, $\alpha = 1/2$, and $\beta = 1/2$, at $N = 3$, $\alpha = 2/3$ and $\beta = 1/3$, $N = 4$, $\alpha = 1/4$ and $\beta = 3/4$, and so on. For any given linear data window length N, the computed α and β provide an *unbiased* estimate of the mean wind speed based on the last N samples. If at sample 100 we fix $\alpha = 0.99$ and $\beta = 0.01$, we have essentially created an exponentially weighted data memory window effectively 100 samples long. In other words, a wind speed sample 100 samples old is discounted by $1/e$ in the current estimate of the mean wind speed. The "forgetting factor" of 0.99 on the old data can be seen as a low-pass moving average filter. The wind speed data are made more useful by averaging the short-term turbulence while still providing dynamic wind speeds in the changing environment. If the anemometer is sampled at a 1 Hz rate, the estimated mean applies for the previous 2 min (approximately), but is still adaptive enough to provide measurements of wind changes and turbulence. The simple integrator with an exponential memory window is fundamental in its simplicity and importance. All sensor information has a value, time, and/or spatial context and extent, as well as a measurement confidence. To make the most use of raw sensor information signals in an intelligent signal processing system, one must optimize the information confidence and context that are typically done through various forms of integration and filtering.

10.1 RLS ALGORITHM AND MATRIX INVERSION LEMMA

The RLS algorithm simply applies the recursive mean estimation of the simple integrator to the autocorrelation and cross-correlation data outlined in Section 9.1 for the block least-squares algorithm. Using the formulation given in Equations 9.1 through 9.7, a recursive estimate for the input autocorrelation matrix data is

$$\left(\bar{X}^H \bar{X}\right)_{n+1} = \alpha\left(\bar{X}^H \bar{X}\right)_n + \varphi_{n+1}^H \varphi_{n+1} \tag{10.1}$$

where we do not need a β term on the right-hand side since

$$\lim_{n \to N} E\left\{\left(\bar{X}^H \bar{X}\right)_n\right\} = N \lim_{n \to N} E\left\{\varphi_n^H \varphi_n\right\} \tag{10.2}$$

where N is the size of the data memory window. A similar relation is expressed for the cross-correlation of the input data x_t and output data y_t.

$$\lim_{n \to N} E\left\{\bar{X}^H \bar{y}\right\} = N \lim_{n \to N} E\left\{\varphi_n^H y_n\right\} \tag{10.3}$$

A recursive update for the cross-correlation is simply

$$\left(\bar{X}^H \bar{y}\right)_{n+1} = \alpha\left(\bar{X}^H \bar{y}\right)_n + \varphi_{n+1}^H y_{n+1} \tag{10.4}$$

However, the optimal filter H requires the inverse of Equation 10.1 times 10.4. Therefore, what we really need is a recursive matrix inverse algorithm.

10.1.1 MATRIX INVERSION LEMMA

The matrix inversion lemma provides a means to recursively compute a matrix inverse when the matrix itself is recursively updated in the form of "$A_{new} = A_{old} + BCD$." The matrix inversion lemma states [2]

$$\left(A + BCD\right)^{-1} = A^{-1} - A^{-1} B\left(C^{-1} + DA^{-1}B\right)^{-1} DA^{-1} \tag{10.5}$$

where A, C, $DA^{-1}B$, and BCD are all invertible matrices or invertible matrix products. It is straightforward to prove the lemma in Equation 10.5 by simply multiplying both sides by $(A + BCD)$. Rewriting Equation 10.1 in the $A + BCD$ form, we have

$$\alpha^{-1}\left(\bar{X}^H \bar{X}\right)_{n+1} = \left(\bar{X}^H \bar{X}\right)_n + \varphi_{n+1}^H \alpha^{-1} \varphi_{n+1} \tag{10.6}$$

where $A = (\bar{X}^H \bar{X})_n$ and $BCD = \varphi_{n+1}^H \alpha^{-1} \varphi_{n+1}$. Taking the inverse by directly applying the matrix inversion lemma in Equation 10.5 gives a recursive equation for the autocorrelation matrix

$$\alpha\left(\bar{X}^H \bar{X}\right)_{n+1}^{-1} = \left(\bar{X}^H \bar{X}\right)_n^{-1} - \frac{\left(\bar{X}^H \bar{X}\right)_n^{-1} \varphi_{n+1}^H \varphi_{n+1} \left(\bar{X}^H \bar{X}\right)_n^{-1}}{\alpha + \varphi_{n+1} \left(\bar{X}^H \bar{X}\right)_n^{-1} \varphi_{n+1}^H} \tag{10.7}$$

which is more compactly written in terms of a Kalman gain vector K_{n+1} in

$$\left(\overline{X}^H \overline{X}\right)^{-1}_{n+1} = \alpha^{-1}\left[I - K_{n+1}\,\varphi_{n+1}\right]\left(\overline{X}^H \overline{X}\right)^{-1}_{n} \tag{10.8}$$

The Kalman gain vector K_{n+1} has significance well beyond a notational convenience. Equations 10.7 and 10.9 show that the denominator term in the Kalman gain is a simple scalar when the basis function vector φ_{n+1} is a row vector. This computationally replaces a matrix inversion with several matrix products and a scalar division and can be seen as a great gift from applied mathematics to signal processing.

$$K_{n+1} = \frac{\left(\overline{X}^H \overline{X}\right)^{-1}_{n}\varphi^H_{n+1}}{\alpha + \varphi_{n+1}\left(\overline{X}^H \overline{X}\right)^{-1}_{n}\varphi^H_{n+1}} \tag{10.9}$$

Clearly, inversion of a scalar is much less a computational burden than inversion of a matrix. When x_n is stationary, one can see that the Kalman gain simply decreases with an increasing size of the data memory window N as would be expected for an unbiased estimate. However, if the statistics (autocorrelation) of the most recent data is different from the autocorrelation matrix, the Kalman gain will automatically increase causing the recursion to "quickly forget" the old outdated data. One of the more fascinating aspects of recursive adaptive algorithms is the mathematical ability to maintain least-squared error for nonstationary input–output data. Several approximations to Equation 10.9 will be shown later which still converge to the same result for stationary data, but more slowly due to the approximations.

Our task at the moment is to derive a recursion for the optimal (least-squared error) filter H_{n+1}, given the previous filter estimate H_n and the most recent data. Combining Equations 10.4 and 10.8 in recursive form gives

$$H_{n+1} = \left[\alpha^{-1}\left(\overline{X}^H \overline{X}\right)^{-1}_{n} - \alpha^{-1}K_{n+1}\varphi_{n+1}\left(\overline{X}^H \overline{X}\right)^{-1}_{n}\right]\left[\alpha\left(\overline{X}^H \overline{X}\right)_{n} + \varphi^H_{n+1}y_{n+1}\right] \tag{10.10}$$

which can be shown to reduce the result in

$$H_{n+1} = H_n + K_{n+1}\left(y_{n+1} - y'_{n+1}\right) \tag{10.11}$$

The prediction of y_{n+1} (denoted as y'_{n+1}) is derived from $\varphi_{n+1}H_n$, and is sometimes denoted as $y_{n+1|n}$, which literally means that "the prediction of y_{n+1} given a model last updated at time n." Again, the significance of the Kalman gain vector can be seen intuitively in Equation 10.11. The filter does not change if the prediction error is zero. But, there is always some residual error, if not due to extraneous noise or model error is due to the approximation of the least-significant bit in the A/D converters in a real system. The impact of the residual noise on the filter coefficients is determined by the Kalman gain K_{n+1}, which is based on the data memory window length and the match between the most recent basis vector statistics and the long-term average in the correlation data. It is humanistic to note that the algorithm has to make an error in order to learn. The memory window and variations on the Kalman gain in the form of approximations determine the speed of convergence to the least-squared error solution.

The RLS algorithm is summarized in Table 10.1 at step "$n+1$" given a new input sample x_{n+1} and an output sample y_{n+1}, and the previous estimates for the optimal filter H_n and inverse autocorrelation matrix $P_n = \left(\overline{X}^H \overline{X}\right)^{-1}_{n}$.

TABLE 10.1
RLS Algorithm

Description	Equation
Basis function containing x_{n+1}	$\phi_{n+1} = \left[x_{n+1} \, x_n \, x_{n-1} \cdots x_{n-M+1} \right]$
Inverse autocorrelation matrix; output prediction	$P_n = \left(\bar{X}^H \bar{X} \right)_n^{-1}; \quad y'_{n+1} = \phi_{n+1} H_n$
Kalman gain using RLS and exponential memory window N samples long. $\alpha = (N-1)/N$	$K_{n+1} = \left(P_n \, \phi_{n+1}^H \right) \Big/ \left(\alpha + \phi_{n+1} \, P_n \, \phi_{n+1}^H \right)$
Update for autocorrelation matrix inverse	$P_{n+1} = \alpha^{-1} \left[I - K_{n+1} \phi_{n+1} \right] P_n$
Optimal filter update using the prediction error	$H_{n+1} = H_n + K_{n+1} \left[y_{n+1} - y'_{n+1} \right]$

10.1.2 Approximations to RLS

Approximations to RLS can offer significant computational savings, but at the expense of slower convergence to the least-squared error solution for the optimal filter. So long as the convergence is faster than the underlying model changes being tracked by the adaptive algorithm through its input and output signals, one can expect same optimal solution. However, as one shortens the data memory window, the adaptive algorithm becomes more reactive to the input–output data resulting in more noise in the filter coefficient estimates. These design trades are conveniently exploited to produce adaptive filtering systems well matched to the application of interest. The largest reduction in complexity comes from eliminating P_n from the algorithm. This is widely known as the projection algorithm.

$$K_{n+1}^{PA} = \frac{\gamma \, \phi_{n+1}^H}{\alpha + \phi_{n+1} \, \phi_{n+1}^H}; \quad 0 < \gamma < 2, \quad 0 < \alpha < 1 \tag{10.12}$$

It will be shown later in the section on the convergence properties of the LMS algorithm why γ must be less than 2 for stable convergence. Also, when α is positive, K_{n+1} remains stable even if the input data becomes zero for a time.

Another useful approximation comes from knowledge that the input data will always have some noise (making $\phi\phi^H$ nonzero) allowing the elimination of α which makes the memory window essentially of length M. This is known as the stochastic approximation to RLS. Of course, choosing a value of $\gamma < 2$ effectively increases the data memory window as desired.

$$K_{n+1}^{SA} = \frac{\gamma \, \phi_{n+1}^H}{\phi_{n+1} \, \phi_{n+1}^H}; \quad 0 < \gamma < 2 \tag{10.13}$$

By far, the most popular approximation to RLS is the least-mean-squared error, or LMS algorithm [3] defined in

$$K_{n+1}^{LMS} = 2\mu \, \phi_{n+1}^H; \quad \mu \leq E \left\{ \phi_{n+1} \, \phi_{n+1}^H \right\}^{-1} \tag{10.14}$$

The LMS algorithm is extremely simple and robust. Since the adaptive step size μ is determined by the inverse of the upper bound of the model order $M+1$ times the expected mean-square of the input data, it is often referred to as the normalized LMS algorithm as presented here. The nice feature of the LMS adaptive filtering algorithm is that only one equation is needed as seen in Equation 10.15, and no divide operations are needed except to estimate μ. In the early days of fixed-point-embedded adaptive filters, the omission of a division operation was a key advantage. Even with today's very powerful DSP processors, the simple LMS adaptive filter implementation is important because of the need for processing wider bandwidth signals with faster sample rates. However, the slowdown in convergence can be a problem for large model-order filters. An additional factor $\mu_{rel} < 1$ as seen in Equation 10.15 is generally needed in the LMS algorithm as a margin of safety to insure stability. A small μ_{rel} has the effect of an increased memory window approximately as $N = 1/\mu_{rel}$, which is often beneficial in reducing noise in the parameter estimates at the expense of even slow convergence. We will see in Section 10.2 that the LMS algorithm convergence depends on the eigenvalues of the input data.

$$H_{n+1} = H_n + 2\mu_{max}\mu_{rel}\,\varphi_{n+1}^{H}\left(y_{n+1} - y'_{n+1}\right) \tag{10.15}$$

The parameter $\mu_{max} = 1/\{\sigma_x^2\}$, and σ_x^2 is the variance of the input data φ_{n+1} and μ_{rel} is the step size component which creates an effective "data memory window" with an exponential forgetting property of approximate length $N = 1/\mu_{rel}$. It can be seen that the shortest possible memory window length for a stable LMS algorithm is the filter length $M + 1$ and occurs if $\mu_{rel} = 1/(M + 1)$ such that $\mu_{max}\mu_{rel} < 1/\{\varphi_{n+1}\varphi_{n+1}^{H}\}$. Separating the LMS step size in this way may seem to be adding unneeded complexity, but it actually adds clarity and eases parameter setting in a running LMS algorithm. A simple integrator can be used to estimate μ_{max} allowing the memory window length to be independently adjusted as desired while the LMS algorithm is running with little concern for causing an unstable divergence of the adaptive filter. While almost all other texts on adaptive filtering develop the LMS algorithm on its own, doing so here in the context of the complete recursive least-squares algorithm based on the matrix inversion lemma creates a very clear view of the nature of the LMS approximation and how to best exploit its simplicity in a robust adaptive filter algorithm.

10.2 LMS CONVERGENCE PROPERTIES

The least-mean-squared error adaptive filter algorithm is a staple technique in realm of signal processing where one is tasked with adaptive modeling of signals or systems. In general, modeling of a system (system of linear dynamical responses) requires both the input signal exciting the system and the system's response in the form of the corresponding output signal. A signal model is obtainable by assuming that the signal is the output of a system driven by ZMG, or spectrally white, random noise. The signal model is computed by "whitening" the available output signal using an adaptive filter. The inverse of the converged filter can then be used to generate a model of the signal of interest from a white-noise input, called the *innovation*. We will first discuss the dynamical response of the LMS algorithm by comparing it to the more complicated RLS algorithm for system modeling, also known as Wiener filtering, where both input and output signals are available to the "unknown system," as shown in Figure 10.1.

10.2.1 System Modeling Using Adaptive System Identification

From the input–output signals is a powerful technique made easy with the LMS or RLS algorithms. Clearly, when the input–output response (or frequency response) of the filter model $H[z]$ in Figure 10.1 closely matches the response of the unknown system, the matching error $\varepsilon_n = y_n - y'_n$ will be quite small. The error signal is therefore used to drive the adaptive algorithm symbolized by the system box with the angled arrow across it. This symbolism has origins in the standard electrical circuitry symbols for variable capacitance, variable inductance, or the widely used variable resistance (potentiometer). It is a reasonable symbolic analogy except that when the error signal approaches

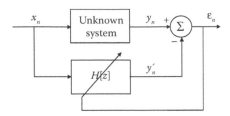

FIGURE 10.1 Block diagram depicting adaptive system identification using an adaptive filter with input x_n, output y_n, predicted output y'_n, and error signal ε_n.

zero, the output of $H[z]$ is not zero, but rather $H[z]$ stops adapting and should in theory be closely matched to the unknown system.

The LMS algorithm offers significant savings in computational complexity over the RLS algorithm for adaptive filtering. Examination of the convergence properties of the LMS algorithm in detail will show where the effects of the simplification of the algorithm are most significant. We will show both mathematically and through numerical experiment that the performance penalty for the LMS simplification can be seen not only for the modeling of more complex systems, but also for system input excitation signals with a wide eigenvalue spread. System identification with a ZMG "white noise" input excitation and the LMS algorithm will converge nearly identically to the more complex RLS algorithm, except for systems with large model orders (requiring a smaller, slower, step size in the LMS algorithm). When the input signal has sinusoids at the low- and/or high-frequency extremes of the Nyquist band, the resulting wide eigenvalue spread of the input signal correlation matrix results in very slow convergence of the LMS algorithm.

Consider the LMS coefficient update in Equation 10.15 where we set $\mu_{rel} = 1$ for the moment and write the predicted output signal y'_{n+1} as $\varphi_{n+1} H_n$.

$$H_{n+1} = H_n + 2\mu \varphi_{n+1}^H \left(y_{n+1} - \varphi_{n+1} H_n \right) \tag{10.16}$$

Rearranging and taking expected values, we have

$$E\{H_{n+1}\} = \left[I - 2\mu \, E\{\varphi_{n+1}^H \, \varphi_{n+1}\} \right] E\{H_n\} + 2\mu \, E\{\varphi_{n+1}^H y_{n+1}\} \tag{10.17}$$

If we define $H_{opt} = E\{(\varphi^H \varphi)^{-1} \varphi^H y\}$ and $R^x = E\{\varphi^H \varphi\}$ as seen in

$$E\{H_{n+1}\} = \left[I - 2\mu \, R^x \right] E\{H_n\} + 2\mu \, R^x H_{opt} \tag{10.18}$$

we can write the LMS recursion as a coefficient vector error as given in

$$
\begin{aligned}
E\{H_{n+1}\} - H_{opt} &= \left[I - 2\mu R^x \right] \left[E\{H_n\} - H_{opt} \right] \\
&= \left[I - 2\mu R^x \right]^2 \left[E\{H_{n-1}\} - H_{opt} \right] \\
&\quad \vdots \qquad\quad \vdots \qquad\quad \vdots \\
&= \left[I - 2\mu R^x \right]^{n+1} \left[E\{H_0\} - H_{opt} \right]
\end{aligned}
\tag{10.19}
$$

We note that R^x, the input signal autocorrelation matrix (see Equations 9.13 and 9.14), is positive definite and invertible. In a real system, there will always be some residual white noise due to the successive approximation error in the A/D converters for the least significant bit. The autocorrelation matrix can be diagonalized to give the signal eigenvalues in the form $D = Q^H R^x Q$, where the columns of Q have the eigenvectors which correspond to the eigenvalues λ_k, $k = 0, 1, 2, \ldots, M$ on the main diagonal of D. Since the eigenvectors are orthonormal, $Q^H Q = I$ and Equation 10.19 becomes [4]

$$E\{H_{n+1}\} - H_{opt} = Q \begin{bmatrix} (1-2\mu\lambda_0)^{n+1} & 0 & 0 & \ldots & 0 \\ 0 & (1-2\mu\lambda_1)^{n+1} & 0 & \ldots & 0 \\ \vdots & \vdots & \vdots & \ldots & \vdots \\ 0 & & \ldots & 0 & 0 & (1-2\mu\lambda_M)^{n+1} \end{bmatrix} Q^H \left[E\{H_0\} - H_{opt} \right]$$

$$(10.20)$$

We can now define an upper limit for μ, the LMS step size to ensure a stable adaptive recursion where the coefficient vector error is guaranteed to converge to some small stochastic value. The estimated eigenvalues will range from a minimum value λ_{min} to some maximum value λ_{max}. For each coefficient to converge to the optimum value the expression $|1-2\,\mu\lambda_k|$ must be less than unity. Therefore, $0 < \mu < 1/(2\,|\lambda_{max}|)$. Setting μ to the maximum allowable step size for the fastest possible convergence of the LMS algorithm, one can expect the kth coefficient of H to converge at a rate proportionally to

$$h_{n+1,k} - h_{opt,k} \propto e^{-(\lambda_k / \lambda_{max})n} \tag{10.21}$$

Therefore, the coefficient components associated with the maximum eigenvalue (signal component with the maximum amplitude) will converge the fastest while the components due to the smaller eigenvalues will converge much more slowly. Let us emphasize, very slowly. For input signals consisting of many sinusoids or harmonics with wide separation in power levels, the time-domain LMS algorithm performs very poorly.

Determining λ_{max} in real time in order to set μ for the fastest possible stable convergence performance of the LMS algorithm requires a great deal of computation, defeating the advantages of using the LMS algorithm. If the input signals are completely known and stationary one could solve for μ once (in theory at least), set up the LMS algorithm, and enjoy the fastest possible LMS performance. However, in practice, one usually does not know the input signal statistics *a priori* requiring some efficient method to make μ adaptive. We note that for the white-noise input signal case, the eigenvalues will all be identical to the signal variance, making the sum of the eigenvalues equal to the signal power times the adaptive filter model order. For sinusoidal inputs, the eigenvalues (and eigenvalue spread) depend on the amplitudes and frequencies of the input sinusoids. However, if one kept the amplitudes the same and varied the frequencies, the sum of the eigenvalues stays the same while the eigenvalue spread changes as seen in the following examples. Therefore, we can be guaranteed that λ_{max} is always less than the adaptive filter model order times the input signal variance. Adaptive tracking of the input signal power using an exponential memory data window as described in the simple integrator (beginning of this chapter) provides a real-time input signal power estimate which can be used to very simply compute μ for reasonable LMS performance. The *normalized LMS algorithm* is the name most commonly used to describe the use of an input signal power tracking μ in the adaptive filter.

Consider a system identification problem where the input signal consists of unity variance ZMG white noise plus a 25 Hz sinusoid of amplitude 5 where the sample rate is 1024 samples/s. To further illustrate the parameter tracking abilities of the LMS and RLS adaptive filter algorithms, we twice change the parameters of the so-called unknown system to test the adaptive algorithms' ability to model these changes using only the input and output signals from the system. The "unknown

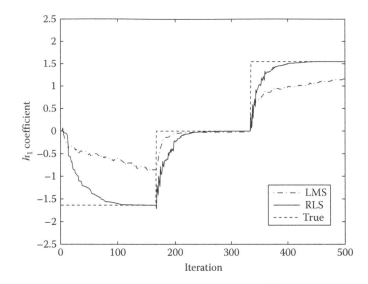

FIGURE 10.2 LMS and RLS h_1 coefficient tracks for an input signal x_n consisting of unity variance white Gaussian noise plus a 25 Hz sinusoid of amplitude 5 with a sample rate f_s of 1024 Hz.

system" simply consists of a pair of complementary zeros on the unit circle which start at ±100 Hz, then at iteration 166 immediately changes to ±256 Hz, then again at iteration 333 immediately changes to ±400 Hz. Since the zeros are on the unit circle, only h_1 changes in the unknown system as seen in Figure 10.2.

Figure 10.2 shows poor convergence performance of the LMS algorithm relative to the more complicated RLS algorithm in the beginning and end of the trial, where the unknown system's zero is away from the center of the Nyquist band. In the center region, the two algorithms appear comparable, even though the input signal autocorrelation matrix has a fairly wide eigenvalue spread of −11.2 and 38.2. Note that the sum of the eigenvalues is 27, or 2(1 + 12.5) which is the model order times the noise variance plus the sinusoid power (25/2). The parameter μ_{rel} in Equation 10.15 was set to 0.05 for the simulation which translates to an equivalent exponential memory window of about 20 samples ($\alpha = 0.95$ in the RLS algorithm). Note that the time constant for the RLS algorithm (and LMS algorithm in the middle range) in response to the step changes is also about 20 samples. The error signal response corresponding to the trial in Figure 10.2 is shown in Figure 10.3. Separating the LMS step size into a factor μ_{max} set adaptively to inversely track the input signal power, and μ_{rel} to control the effective data memory window length (approximately equal to $1/\mu_{rel}$) for the algorithm is indeed very useful. However, the LMS convergence properties will only be comparable to the RLS algorithm for cases where the input signal is white noise or has frequencies in the center of the Nyquist band.

If we move the 25 Hz sinusoid to 256 Hz in the input signal we get the far superior coefficient tracking shown in Figure 10.4 which shows essentially no difference between the RLS and LMS algorithms. The eigenvalues are both 13.5 for the ZMG white noise only input signal autocorrelation matrix. Note that the sum of the eigenvalues is still 27 as was the case for the 25 Hz input. The factors μ_{rel} and α are set to 0.05 and 0.95 as before. The exponential response of the LMS and RLS coefficient convergence is completely evident in Figure 10.4. Yet, the LMS is slightly slower than the RLS algorithm for this particular simulation. Further analysis is left to the student with no social life. Clearly, the simplicity of the LMS algorithm and the nearly equivalent performance make it highly advantageous to use LMS over RLS is practical applications where the input signal is white. The error signal for this case is shown in Figure 10.5.

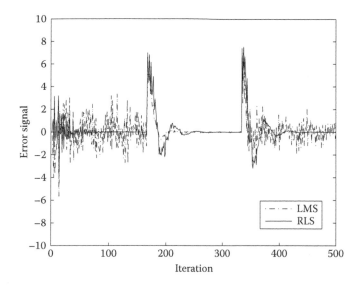

FIGURE 10.3 Error signal responses for the LMS and RLS algorithm and the trial shown in Figure 10.2.

10.2.2 SIGNAL MODELING USING ADAPTIVE SIGNAL-WHITENING FILTERS

Signal modeling using adaptive signal-whitening filters is another basic adaptive filtering operation. The big difference between system modeling and signal modeling is that the input to the unknown system generating the available output signal is not available. One assumes a ZMG unity variance white-noise signal as the *innovation* for the available output signal. The signal can be thought to be generated by a digital filter with the signal innovation as input. The frequencies of the signal are thus the result of poles in the generating digital filter very near the unit circle. The relative phases of the frequencies are determined by the zeros of the digital filter. Therefore, our signal model allows the parameterization of the signal in terms of poles and zeros of a digital filter with the white-noise innovation. By adaptively filtering the signal to remove all the spectral peaks and dips (whitening

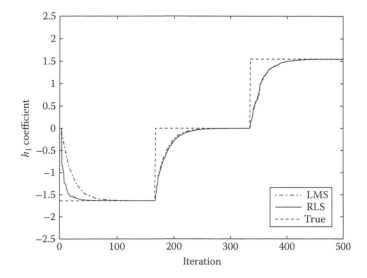

FIGURE 10.4 LMS and RLS h_1 coefficient tracks for an input signal x_n consisting of unity variance white Gaussian noise plus a 256 Hz sinusoid of amplitude 5 sampled at 1024 Hz.

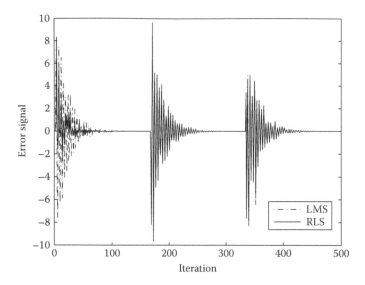

FIGURE 10.5 Error signal responses for the LMS and RLS algorithm and the trial shown in Figure 10.4.

the signal to reproduce the ZMG white-noise innovation as the error signal), one can recover the modeled signal parameters in the form of the poles and/or zeros of the whitening filter. The converged adaptive whitening filter models the inverse of the signal generation filter. In other words, inverting the converged adaptive whitening filter provides the digital generation filter parameters which are seen as the underlying model of the signal of interest. Figure 10.6 depicts the signal flow and processing for signal modeling using adaptive whitening filters.

The block diagram in Figure 10.6 shows a typical whitening filter arrangement where, rather than a full blown ARMA signal model (see Section 3.2), a simpler AR signal model is used. The whitening filter is then a simple MA filter and the resulting error signal will be approximately (statistically) $b_0 w_n$ if the correct model order M is chosen and the whitening filter is converged to the least-squared error solution. Chapter 11 will formulate the ARMA whitening and other filter forms. Note the sign change and the exclusion of the most recent input y_n to the linear prediction signal y'_n. For the AR signal model, the prediction is made solely from past signal outputs. The input signal innovation is not available for a prediction by definition.

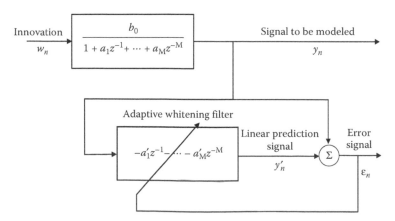

FIGURE 10.6 Block diagram showing an adaptive whitening filter to recover signal model parameters in terms of the linear prediction coefficients of a signal generating filter.

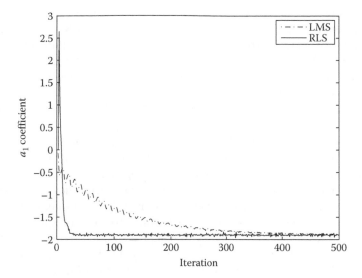

FIGURE 10.7 Whitening filter coefficient a_1 tracks for the RLS and LMS algorithms and a 50 Hz sinusoid in white noise.

Development of an LMS whitening filter also provides an opportunity to illustrate an alternative heuristic approach to the development of the LMS algorithm. We start with the linear prediction model.

$$y'_n = -a_1 y_{n-1} - \cdots - a_M y_{n-M} \tag{10.22}$$

In the RLS whitening filter, one simply makes $\varphi_n = [-y_{n-1} \ -y_{n-2} \ \cdots \ -y_{n-M}]$ and omits the zeroth coefficient (which is not used in whitening filters) from the coefficient vector. But in the LMS algorithm the sign must be changed on the coefficient update to allow $a_{k,n} = a_{k,n-1} - 2\,\mu_{max}\mu_{rel}\,y_{n-k}\,\varepsilon_n$ rather than the plus sign used in Equation 9.16. One can see the need for the sign change by looking at the gradient of the error, $\varepsilon_n = y_n - y'_n$, which is now positive with respect to the whitening filter coefficients. To achieve a gradient descent algorithm, one must *stepwise adapt the LMS coefficient*

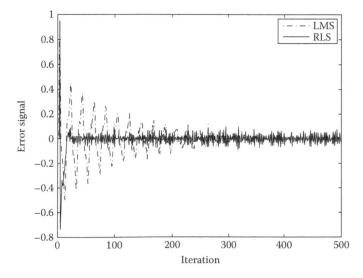

FIGURE 10.8 Error signal for the 50 Hz whitening filter example in Figure 10.7.

weights in the opposite direction of the error gradient. Whitening filters have a positive error gradient with respect to the coefficients, while a system identification LMS application has a negative error gradient. Therefore, an LMS whitening filter has a negative sign to the coefficient update while a system identification LMS coefficient update has a positive sign. *Setting this sign incorrectly leads to very rapid divergence of the LMS algorithm.* Note that the model orders for the system identification and whitening filter examples are both two, even though the system identification filter has three coefficients while the whitening filter only has two. The third whitening filter coefficient is fixed to unity by convention for an AR process.

A numerical example of a whitening filter is shown in Figure 10.7 where a 50 Hz sinusoid (1024 Hz sample rate) of amplitude 1.0 is mixed with ZMG white noise of standard deviation 0.01 (40 dB SNR). To speed up convergence, $\mu_{rel} = 0.2$ and $\alpha = 0.8$, giving an approximate data memory window of only five data samples. The signal autocorrelation matrix eigenvalues are -0.4532 and 1.4534, giving a slow convergence of the whitening filter. The sum of the eigenvalues gives the model order times the signal power as expected. Figure 10.8 shows the error signal responses for the whitening filter for the LMS and RLS algorithms. Simply moving the sinusoid frequency up to 200 Hz (near the midpoint of the Nyquist band) makes the eigenvalue spread of the signal autocorrelation matrix closer at 0.1632 and 0.8370. A much faster convergence is shown in Figure 10.9 for the LMS algorithm. The convergence of the RLS and LMS filters are identical for the sinusoid at 256 Hz. LMS whitening performance is best for signal frequencies in the center of the Nyquist band.

The convergence properties of the LMS and RLS algorithms have been presented for two basic adaptive filtering tasks: system identification and signal modeling. The LMS algorithm is much simpler than the RLS algorithm and has nearly identical convergence performance when the input signal autocorrelation matrix for the adaptive filter have narrow eigenvalue spreads (ZMG white noise or sinusoids in the middle of the Nyquist band). When the signal frequency range of interest resides at low or high frequencies in the Nyquist band, the LMS performance becomes quite slow, limiting applications to very stationary signal processes where slow convergence is not a serious drawback. The reason why the RLS algorithm performs gracefully regardless of input eigenvalue spread is that the Kalman gain vector is optimized for each coefficient. In the more simplified LMS algorithm, the upper limit for a single step size μ is determined by the maximum eigenvalue and this "global" step size limits the convergence performance of the algorithm. Generally, one approximates

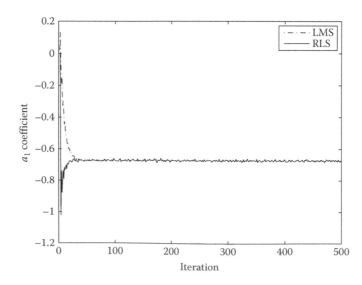

FIGURE 10.9   Whitening filter coefficient a_1 tracks for the RLS and LMS algorithms and a 200 Hz sinusoid near the middle of the Nyquist band showing little convergence difference.

the maximum step size for μ by the inverse of the adaptive filter model order times the variance of the input signal as described for the normalized LMS algorithm.

10.3 LATTICE AND SCHUR TECHNIQUES

Lattice filters are of interest because they offer the fast convergence properties of the RLS algorithm with a significant reduction in computational complexity for large model order adaptive filter applications. The lattice filter structure contains a series of nearly independent stages, or sections, where the filter coefficients are called partial correlation, or PARCOR, coefficients. One stage is needed for each model order increment. Each stage passes two signals: a forward error signal is passed directly; and a backward error signal is passed through a time delay of one sample. The cross-correlation of the forward and delayed backward error signals are calculated in the PARCOR coefficients. The PARCOR coefficients times their respective forward and backward error signals are used to subtract any measurable cross-correlations from the error signals before they exit the stage. The structure of successive lattice stages, each removing the correlation between forward and backward error signals, is unique in that adding an $(M + 1)$th stage has no effect on any of the lower-order stage PARCOR coefficients. Increasing the model order for an RLS or LMS filter requires that all the adaptive filter coefficients change. While the PARCOR coefficients (also called Schur coefficients) and the LMS or RLS filter coefficients are not the same, the equivalent FIR filter coefficients for the lattice are computed by using a Levinson recursion. New concepts such as forward and backward prediction errors, PARCOR coefficients, and the Levinson recursion are introduced below.

The "forward" prediction error is the error signal defined in the previous sections. It represents the error in making a future prediction for our signal y_n using an Mth-order linear combination of past samples. This is called a *forward prediction error*.

$$\varepsilon_n = y_n - y'_n = y_n + a_{M,1} y_{n-1} + a_{M,2} y_{n-2} + \cdots + a_{M,M} y_{n-M} \tag{10.23}$$

The predicted signal in Equation 10.23 is simply

$$y'_n = -a_{M,1} y_{n-1} - a_{M,2} y_{n-2} - \cdots - a_{M,M} y_{n-M} \tag{10.24}$$

The backward prediction error represents a prediction backward in time $M + 1$ samples using only a linear combination of the available M samples from time n to $n - M + 1$.

$$r_{n-1} = a^r_{M,M} y_{n-1} + a^r_{M,M-1} y_{n-2} + \cdots + a^r_{M,1} y_{n-M} + y_{n-M-1} \tag{10.25}$$

Clearly, the backward prediction error given in Equation 10.25 is based on the prediction backward in time of

$$y'_{n-M-1} = -a^r_{M,M} y_{n-1} - a^r_{M,M-1} y_{n-2} - \cdots - a^r_{M,1} y_{n-M} \tag{10.26}$$

where y'_{n-M-1} is a linear prediction of the data $M + 1$ samples in the past. Why bother? We have the sample y_{n-M-1} available.

The reason for backward and forward prediction can be seen in the symmetry for linear prediction for forward and backward time steps. Consider a simple sinusoid as y_n. Given a second-order linear predictor for y_n two past samples, that is, $y'_n = -a_{2,1} y_{n-1} - a_{2,2} y_{n-2}$, one can show that $a_{2,1}$ is approximately $-2\cos(2\pi f_0/f_s)$ and $a_{2,2}$ is approximately unity. The approximation is to ensure that the magnitude of the poles is slightly less than unity to insure a stable AR filter for generating our modeled estimate of y_n. Sinusoids have exactly the same shape for positive and negative time. Therefore, one could predict y'_{n-3} using the linear predictor coefficients in opposite order, that is, $y'_{n-3} = -a_{2,2} y_{n-1} - a_{2,1} y_{n-2}$. Note that the highest indexed coefficient is always the farthest from the predicted sample. An important aspect of the forward–backward prediction symmetry is that for a stationary signal the backward prediction coefficients are always equal to the forward predictor coefficients in reverse order. For complex signals, the backward coefficients are the complex conjugate of the forward coefficients in reverse order. Therefore, one of the reasons why adaptive lattice filters have superior convergence performance over transversal LMS filters is that for nonstationary signals (or during convergence of the algorithm) the forward–backward prediction symmetry is not the same, and the algorithm can use twice as many predictions and errors to "learn" and converge. A sketch of an adaptive lattice filter and the equivalent LMS transversal filter is shown in Figure 10.10.

Figure 10.10 shows that the lattice filter is clearly much more complicated than the LMS filter. The complexity is actually even worse due to the need for several divide operations in each lattice stage. The LMS algorithm can be executed with simple multiplies and adds (and fewer of them) compared to the lattice. However, if the model order for the whitening task is chosen higher than the expected model order needed, the lattice PARCOR coefficients for the extraneous stages will be approximately zero. The lattice filter allows the model order of the whitening filter to be estimated from the lattice parameters when the lattice model order is overdetermined. The fast convergence properties and the ability to add on stages without affecting lower-order stages makes the lattice filter a reasonable choice for high-performance adaptive filtering without the computational requirements and singularity concerns from overdetermined model orders of the RLS algorithm.

To derive the PARCOR coefficients from the signal data, we need to examine the relationship between the signal autocorrelation and the prediction error cross-correlation. We begin with the prediction error in Equation 10.23 and proceed to derive the respective correlations.

$$\varepsilon_{M,n} = y_n + \sum_{i=1}^{M} a_{M,i} y_{n-i} \tag{10.27}$$

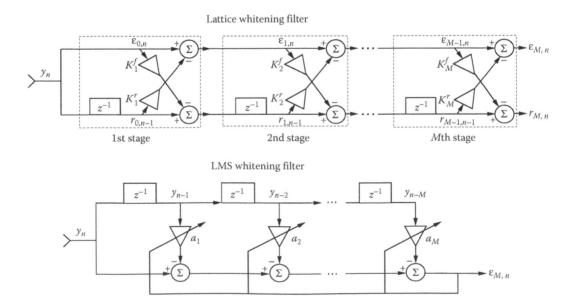

FIGURE 10.10 *M*th-order recursive least-squares lattice and the corresponding FIR transversal LMS whitening filter.

The subscripts M,n mean that the error at time n is from a whitening filter model order M. For the linear prediction coefficients, the subscripts M,i refers to the ith coefficient of an Mth-order whitening filter model. Multiplying both sides of 10.27 by y_{n-j} and taking expected values (averaging) yield

$$E\{\varepsilon_{M,n}\,y_{n-j}\} = E\{y_n\,y_{n-j}\} + \sum_{i=1}^{M} a_{M,i}\,E\{y_{n-i}\,y_{n-j}\}$$

$$= R_j^y + \sum_{i=1}^{M} a_{M,n} R_{j-i}^y \qquad (10.28)$$

$$= \begin{cases} 0, & j > 0 \\ R_M^\varepsilon, & j = 0 \end{cases}$$

where R_M^ε is the Mth-order forward prediction error variance. The reason why the forward prediction error cross-correlation with the signal y_n is zero for $j > 0$ is that $\varepsilon_{M,n}$ is uncorrelated with y_{n-j} as well as $\varepsilon_{M,n-j}$ since the current signal innovation cannot appear in past signal samples. Following a similar approach for the backward prediction error in Equation 10.25,

$$E\{r_{M,n-1}y_{n-j}\} = E\left\{y_{n-m-1}y_{n-j}\right\} + \sum_{i=1}^{M} a_{M,M+1-i}^r\,E\left\{y_{n-i}y_{n-j}\right\}$$

$$= R_{j-M-1}^y + \sum_{i=1}^{M} a_{M,M+1-i}^r\,R_{j-i}^y \qquad (10.29)$$

$$= \begin{cases} 0, & j > 0 \\ R_M^r, & j = M+1 \end{cases}$$

As with the forward prediction error, the current backward prediction error is uncorrelated with future backward prediction errors. We now may write a matrix equation depicting the forward and backward error correlations and their relationship to the signal autocorrelation matrix.

$$\begin{bmatrix} 1\,a_{M,1}\dots a_{M,M-1}a_{M,M} \\ a_{M,M}^r a_{M,M-1}^r \dots a_{M,1}^r\,1 \end{bmatrix} \begin{bmatrix} R_0^y & R_1^y & & \cdots & R_M^y \\ R_{-1}^y & R_0^y & R_1^y & \cdots & R_{M-1}^y \\ \vdots & \vdots & & & \vdots \\ R_{-M}^y & & \cdots & R_{-1}^y & R_0^y \end{bmatrix} = \begin{bmatrix} R_M^\varepsilon & 0 & 0 & \cdots & 0 \\ 0 & \cdots & 0 & 0 & R_M^r \end{bmatrix} \qquad (10.30)$$

The signal autocorrelation matrix in Equation 10.30 has a symmetric Toeplitz structure where all the diagonal elements are equal on a given diagonal. For a complex signal, the positive and negative lags of the autocorrelation are complex conjugates. This can also be seen in the covariance matrix in Equation 9.13. If we neglect the backward error components, Equation 10.30 is also seen as the Yule–Walker equation, named after two British astronomers who successfully predicted periodic sunspot activity using an autoregressive filter model around the beginning of the twentieth century. Note that Equation 10.30 is completely solvable by simply assuming a white prediction error for the given signal autocorrelations. However, our motivation here is to examine the algebraic relationship between successive model orders as this will lead to an algorithm for generating the $(p + 1)$th-order whitening filter from the pth model.

Suppose we make a trial solution for the $(M + 1)$th-order whitening filter coefficients by simply letting $a_{M+1,k} = a_{M,k}$ and let $a_{M+1,M+1} = 0$. As seen in Equation 10.31, this obviously leads to a nonwhite prediction error where the autocorrelation of the error is no longer zero for the nonzero time lags [5].

$$
\begin{bmatrix} 1 & a_{M,1} \cdots a_{M,M-1}\, a_{M,M} & 0 \\ 0 & a^r_{M,M}\, a^r_{M,M-1} \cdots a^r_{M,1} & 1 \end{bmatrix}
\begin{bmatrix} R^y_0 & R^y_1 & \cdots & R^y_{M+1} \\ R^y_{-1} & R^y_0\, R^y_1 & \cdots & R^y_M \\ \vdots & \vdots & \vdots & \vdots \\ R^y_{-M-1} & \cdots & R^y_{-1} & R^y_0 \end{bmatrix}
=
\begin{bmatrix} R^\varepsilon_M & 0 & 0 & \cdots & \Delta^\varepsilon_{M+1} \\ \Delta^r_{M+1} & \cdots & 0 & 0 & R^r_M \end{bmatrix}
\tag{10.31}
$$

However, we can make the $(M+1)$th-order prediction error white through the following multiplication:

$$
\begin{bmatrix} 1 & -\Delta^\varepsilon_{M+1}/R^r_M \\ -\Delta^r_{M+1}/R^\varepsilon_M & 1 \end{bmatrix}
\begin{bmatrix} 1 & a_{M,1} \cdots a_{M,M-1}\, a_{M,M} & 0 \\ 0 & a^r_{M,M}\, a^r_{M,M-1} \cdots a^r_{M,1} & 1 \end{bmatrix}
\begin{bmatrix} R^y_0 & R^y_1 & \cdots & R^y_{M+1} \\ R^y_{-1} & R^y_0\, R^y_1 & \cdots & R^y_M \\ \vdots & \vdots & \vdots \\ R^y_{-M-1} & \cdots & R^y_{-1} & R^y_0 \end{bmatrix}
$$

$$
=
\begin{bmatrix} 1 & -\Delta^\varepsilon_{M+1}/R^r_M \\ -\Delta^r_{M+1}/R^\varepsilon_M & 1 \end{bmatrix}
\begin{bmatrix} R^\varepsilon_M & 0 & 0 & \cdots & \Delta^\varepsilon_{M+1} \\ \Delta^r_{M+1} & \cdots & 0 & 0 & R^r_M \end{bmatrix}
\tag{10.32}
$$

$$
=
\begin{bmatrix} R^\varepsilon_{M+1} & 0 & 0 & \cdots & 0 \\ 0 & \cdots & 0 & 0 & R^r_{M+1} \end{bmatrix}
$$

From Equation 10.32 we have a Levinson recursion for computing the $(M+1)$th model order whitening filter from the Mth model coefficients and error signal correlations.

$$
\begin{bmatrix} 1 & -\Delta^\varepsilon_{M+1}/R^r_M \\ -\Delta^r_{M+1}/R^\varepsilon_M & 1 \end{bmatrix}
\begin{bmatrix} 1 & a_{M,1} \cdots a_{M,M-1} a_{M,M} & 0 \\ 0 & a^r_{M,M} a^r_{M,M-1} \cdots a^r_{M,1} & 1 \end{bmatrix}
=
\begin{bmatrix} 1 & a_{M+1,1} \cdots a_{M+1,M} a_{M+1,M+1} \\ a^r_{M+1,M+1} a^r_{M+1,M} \cdots a^r_{M+1,1} & 1 \end{bmatrix}
\tag{10.33}
$$

The error signal cross-correlations are seen to be

$$
\begin{aligned}
\Delta^\varepsilon_{M+1} &= R^y_{M+1} + \sum_{i=1}^{M} a_{M,i}\, R^y_{M-i-1} \\
&= R^y_{-M-1} + \sum_{i=1}^{M} a^r_{M,i}\, R^y_{-M+i-1} \\
&= \Delta^{r^{\mathrm{H}}}_{M+1}
\end{aligned}
\tag{10.34}
$$

where for the complex data case Δ^ε_{M+1} equals the complex conjugate of Δ^r_{M+1}, depicted by the H symbol for Hermitian transpose in Equation 10.34. Even though we derive a scalar lattice algorithm, carrying the proper matrix notation will prove useful for reference by latter sections of the text. To more clearly see the forward and backward error signal cross-correlation, we simply write the expression in terms of expected values.

$$
\begin{aligned}
\Delta_{M+1} &= \begin{bmatrix} 1 & a_{M,1} \cdots a_{M,M} & 0 \end{bmatrix} E\left\{ \begin{bmatrix} y_n \cdots y_{n-M-1} \end{bmatrix}^H \begin{bmatrix} y_n \cdots y_{n-M-1} \end{bmatrix} \right\} \begin{bmatrix} 0 & a^r_{M,M} \cdots a^r_{M,n} & 1 \end{bmatrix}^H \\
&= E\left\{ \varepsilon_{M,n} r_{M,n-1} \right\}
\end{aligned}
\tag{10.35}
$$

The PARCOR coefficients are defined as

$$K_{M+1}^{\varepsilon} = \frac{\Delta_{M+1}^{H}}{R_M^{\varepsilon}}$$

$$K_{M+1}^{r} = \frac{\Delta_{M+1}}{R_M^{r}} \tag{10.36}$$

We also note that from Equation 10.32 the updates for the $(M + 1)$th forward and backward prediction error variances are, respectively,

$$R_{M+1}^{\varepsilon} = R_M^{\varepsilon} - K_{M+1}^{r} \, \Delta_{M+1}^{H} \tag{10.37}$$

and

$$R_{M+1}^{r} = R_M^{r} - K_{M+1}^{\varepsilon} \, \Delta_{M+1} \tag{10.38}$$

It follows from Equations 10.35 through 10.38 that the forward and backward error signal updates are

$$\varepsilon_{M+1,n} = \varepsilon_{M,n} - K_{M+1}^{r} \, r_{M,n-1} \tag{10.39}$$

and

$$r_{M+1,n} = r_{M,n-1} - K_{M+1}^{\varepsilon} \, \varepsilon_{M,n} \tag{10.40}$$

The structure of a lattice stage is determined by the forward and backward error signal order updates given in Equations 10.39 and 10.40. Starting with the first stage, we let $\varepsilon_{0,n} = r_{0,n} = y_n$ and estimate the cross-correlation Δ_1, between the forward error $\varepsilon_{0,n}$, and the delayed backward error signal $r_{0,n-1}$. The zeroth stage error signal variances are equal to R_0^y, the mean square of the signal y_n. The PARCOR coefficients are then calculated using Equation 10.36 and the error signals and error signal variance are updated using Equations 10.37 through 10.40. Additional stages are processed until the desired model order is achieved, or until the cross-correlation between the forward and backward error signals becomes essentially zero.

The PARCOR coefficients are used to calculate the forward and backward linear predictors when desired. The process starts off with $-K_1^r = a_{1,1}$ and $-K_1^{\varepsilon} = a_{1,1}^r$. Using the Levinson recursion in Equation 10.33 we find $a_{2,2} = -K_2^r$ and $a_{2,2} = a_{1,1} - K_2^r a_{1,1}^r$, and so on. A more physical way to see the Levinson recursion is to consider making a copy of the adaptive lattice at some time instant allowing the PARCOR coefficients to be "frozen" in time with all error signals zero. Imputing a simple unit delta function into the frozen lattice allows the forward linear prediction coefficients to be read in succession from the forward error signal output of the stage corresponding to the desired model order. This view makes sense because the transversal FIR filter generated by the Levinson recursion and the lattice filter must have the same impulse response. One of the novel features of the lattice filter is that using either the Levinson recursion or frozen impulse response technique, the linear prediction coefficients for all model orders can be easily obtained along with the corresponding prediction error variances directly from the lattice parameters. Figure 10.11 shows the detailed algorithm structure of the $(p + 1)$th lattice filter stage. The lattice structure with its forward and backward prediction allows an adaptive Gram–Schmidt orthogonalization [2] of the linear prediction filter. The backward error signals are orthogonal between stages in the lattice allowing each stage to adapt as rapidly as possible independent of the other lattice stages.

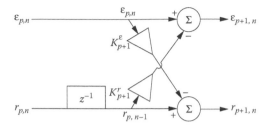

FIGURE 10.11 Detailed structure of an adaptive lattice filter stage.

The orthogonality of the stages in the lattice structure is seen as the key feature which makes adaptive lattice filters so attractive for fast convergence without the computational complexity of the RLS algorithm for transversal FIR-type filters. The error signal variances and cross-correlations can be estimated using simple expected values such as that suggested for the exponentially weighted data memory window. Since the PARCOR coefficients are the error signal cross-correlation divided by the variance, a common bias in the cross-correlation and variance estimates is not a problem. For an exponentially decaying data window of effective length N, the cross-correlation is estimated using

$$\Delta_{p+1,n} = \alpha \Delta_{p+1,n-1} + \varepsilon_{p,n} \, r_{p,n-1} \tag{10.41}$$

where N is effectively $1/(1 - \alpha)$ and $\Delta_{p+1,n} \cong N \, E\{\varepsilon_{p,n} \, r_{p,n-1}\}$. The forward error signal variance is

$$R^{\varepsilon}_{p,n} = \alpha \, R^{\varepsilon}_{p,n-1} + \varepsilon_{p,n} \, \varepsilon_{p,n} \tag{10.42}$$

and the backward error signal variance is

$$R^{r}_{p,n-1} = \alpha \, R^{r}_{p,n-2} + r_{p,n-1} \, r_{p,n-1} \tag{10.43}$$

The PARCOR coefficients are calculated using the expressions in Equation 9.44.

$$K^{\varepsilon}_{p+1,n} = \frac{\Delta^{H}_{p+1,n}}{R^{\varepsilon}_{p,n}}$$
$$K^{r}_{p+1,n} = \frac{\Delta_{p+1,n}}{R^{r}_{p,n-1}} \tag{10.44}$$

We show the lattice cross-correlation, variances, and PARCOR coefficients as time dependent in Equations 10.41 through 10.44 to illustrate the recursive operation of the lattice algorithm. For stationary data and time periods long after the start-up of the lattice Equations 10.41 through 10.44 are optimum. However, for nonstationary signals and/or during the initial start up of the lattice, some additional equations are needed to optimize the lattice algorithm. Consider that the delay operators in the backward error of the lattice do not allow the residual for the data into the pth stage until $p + 1$ time samples have been processed. The transient effects on the PARCOR coefficients will persist, slowing the algorithm convergence. However, if we use the orthogonal decomposition of the Hilbert space spanned by the data outlined in Section 9.2, we can make the lattice recursions least-squared error even during the transients from start-up or nonstationary data. While somewhat complicated, subspace decomposition gives the lattice algorithm an effective convergence time on the order of the number of lattice stages on startup and N for tracking nonstationary data.

10.4 ADAPTIVE LEAST-SQUARES LATTICE ALGORITHM

Projection operator subspace decomposition can also be used to derive the complete least-squares lattice algorithm including the optimal time updates for the error signal variances and cross-correlations. As shown in Figure 10.11, the PARCOR coefficients may be computed any number of ways including even an LMS-type gradient descent algorithm. One of the interesting characteristics about adaptive processing is that even if the algorithm is not optimal, the result of the learning processing still gives a sometimes reasonable result; they're just not the best result possible. We now make use of the projection operator orthogonal decomposition given in Section 9.2 and in particular in Equations 9.20 through 9.22 to optimize the lattice equations for the fastest possible convergence to the least-squared error PARCOR coefficients. Exploring these details may be a bit intimidating, but it does point out some fascinating signal processing details and it only adds one more equation to the lattice algorithm, plus offers some options for replacing other equations. All this comes from the subspace decompositions in time and order and there are no compromises on performance and speed. By the time the first signal input sample into the lattice exits the last stage of a least-squares lattice filter, the algorithm can be converged, albeit with a very short memory, but converged to the least squared error solution nonetheless.

Application of orthogonal subspace decomposition to the development of fast adaptive filtering can be attributed to a PhD dissertation by Martin Morf [6]. The lattice structure was well known in network theory, but its unique processing properties for adaptive digital filtering generally were not well appreciated until the 1970s. By the early 1980s a number of publications on lattice filters (also called ladder or even "wave" filter structures) appeared, one of the more notable by Friedlander [5]. A slightly different notation is used below from what has previously been presented in Section 9.2 to facilitate the decomposition matrix equations. We start by constructing a vector of signal samples defined by

$$y_{n-N:n-1} = \begin{bmatrix} y_{n-N} \cdots y_{n-2} \, y_{n-1} \end{bmatrix} \tag{10.45}$$

For a time block of N samples and a whitening filter model order p, we define a p-row, N-column data matrix.

$$Y_{p,N} = \begin{bmatrix} y_{n-N:n-1} \\ y_{n-N-1:n-2} \\ \vdots \\ y_{n-p-N+1:n-p} \end{bmatrix} = \begin{bmatrix} y_{n-N} & \cdots & y_{n-2}\,y_{n-1} \\ y_{n-N-1} & \cdots & y_{n-3}\,y_{n-2} \\ \vdots & & \vdots \\ y_{n-p-N+1} & \cdots & y_{n-p-1}\,y_{n-p} \end{bmatrix} \tag{10.46}$$

The linear prediction error from time $n - N + 1$ to time n can now be written as

$$\varepsilon_{p,n-N+1:n} = y_{n-N+1:n} + \begin{bmatrix} a_{p,1} \, a_{p,2} \cdots a_{p,p} \end{bmatrix} Y_{p,N} \tag{10.47}$$

where $a_{p,2}$ is the second linear prediction coefficient in a pth-order whitening filter. The least-squared error linear prediction coefficients can be written as

$$\begin{bmatrix} a_{p,1} \, a_{p,2} \cdots a_{p,p} \end{bmatrix} = y_{n-N+1:n} Y_{p,N}^{H} \left(Y_{p,N} Y_{p,N}^{H} \right)^{-1} \tag{10.48}$$

which is essentially the transpose form of Equation 9.11. Equation 10.47 is expressed in the form of an orthogonal projection

$$\varepsilon_{p,n-N+1:n} = y_{n-N+1:n} \left(I - P_{Y_{p,N}} \right) \tag{10.49}$$

where $P_{Yp,N} = Y_{p,N}^H \left(Y_{p,N} Y_{p,N}^H \right)^{-1} Y_{p,N}$ is the projection operator. Equations 10.45 through 10.49 can be compared to Equations 9.15 through 9.19 to see that the effect of the change in variable definitions is really only a simple transpose. We can "pick out" the prediction error at time n by adding a postmultiplication of the form

$$\varepsilon_{p,n} = y_{n-N+1:n} \left(I - P_{Y_{p,N}} \right) \pi^H \tag{10.50}$$

where $\pi = [0\ 0\ \ldots\ 0\ 1]$. Shifting the data vector $y_{n-N+1:n}$ $p + 1$ samples to the right yields a data vector suitable for computing the backward prediction error

$$r_{p,n-1} = y_{n-N+1:n}^{p+1} \left(I - P_{Y_{p,N}} \right) \pi^H \tag{10.51}$$

where

$$y_{n-N+1:n}^{p+1} = y_{n-N-p:n-p-1} \tag{10.52}$$

We have gone to some extent to present the projection operator back in Section 9.2 and here to cast the lattice forward and backward prediction errors into an orthogonal projection framework. After presenting the rest of the lattice variables in the context of the projection operator framework, we will use the orthogonal decomposition in Equation 9.22 to generate the least-squared error updates. The error signal variances for the order p are given by postmultiplying Equations 10.50 through 10.51 by their respective data vectors.

$$R_{p,n}^\varepsilon = y_{n-N+1:n} \left(I - P_{Y_{p,N}} \right) y_{n-N+1:n}^H \tag{10.53}$$

$$R_{p,n-1}^r = y_{n-N+1:n}^{p+1} \left(I - P_{Y_{p,N}} \right) y_{n-N+1:n}^{p+1\,^H} \tag{10.54}$$

The error signal cross-correlation is simply

$$\Delta_{p+1,n} = y_{n-N+1:n} \left(I - P_{Y_{p,N}} \right) y_{n-N+1:n}^{p+1\,^H} \tag{10.55}$$

However, a new very important variable arises out of the orthogonal projection framework called *the likelihood variable*. It is unique to the least-squares lattice algorithm and is responsible for making the convergence as fast as possible. The likelihood variable $\gamma_{p-1,n-1}$ can be seen as a measure of how well recent data statistically matches the older data. For example, as the prediction error tends to zero, the main diagonal of the projection operator tends toward unity, as seen in the example described in Equations 9.23 through 9.28. Premultiplying and postmultiplying the orthogonal projection operator by the π vector gives $1 - \gamma_{p-1,n-1}$.

$$1 - \gamma_{p-1,n-1} = \pi \left(I - P_{Y_{p,N}} \right) \pi^H \tag{10.56}$$

If the projection operator is nearly an identity matrix, the likelihood variable $\gamma_{p-1,n-1}$ approaches zero (the parameter $1 - \gamma_{p-1,n-1}$ will tend toward unity) for each model order, indicating that the data-model fit is nearly perfect. As will be seen for the update equations below, this makes the lattice adapt very quickly for statistically new data. If the data are noisy or from a changing statistical distribution, $1 - \gamma_{p-1,n-1}$ approaches zero ($\gamma_{p-1,n-1}$ approaches unity) and the lattice will weigh the recent data much more heavily, thereby "forgetting" the older data which no longer fits the current distribution. This "intelligent memory control" is independent of the data memory window, which also affects the convergence rate. The ingenious part of the likelihood variable is that it naturally arises from the update equation as an independent adaptive gain control for optimizing the nonstationary data performance in each lattice stage [7].

Consider the update equation for the Hilbert space orthogonal decomposition back in Section 9.2. We now pre- and postmultiply by arbitrary vectors V and W^H, respectively, where the existing subspace spanned by the rows of $Y_{p,N}$ is being updated to include the space spanned by the rows of the vector S. If you are wondering why just hang on, we develop a general decomposition formula that can be used to derive all the lattice equations.

$$V\left(I - P_{\{Y_{p,N}+S\}}\right)W^H = V\left(I - P_{Y_{p,N}}\right)W^H - V\left(I - P_{Y_{p,N}}\right)$$
$$\times S\left[S^H\left(I - P_{Y_{p,N}}\right)S\right]^{-1}S^H\left(I - P_{Y_{p,N}}\right)W^H \tag{10.57}$$

If we add a row vector on the bottom of $Y_{p,N}$ we have an order update to the subspace spanned by the rows of $Y_{p,N}$.

$$Y_{p+1,N} = \begin{bmatrix} Y_{p,N} \\ y_{n-N+1:n}^{p+1} \end{bmatrix} \tag{10.58}$$

Therefore, choosing $S = y_{n-N+1:n}^{p+1}$ allows the following order updates when $V = y_{n-N+1:n}$ and $W = \pi$,

$$\varepsilon_{p+1,n} = \varepsilon_{p,n} - \Delta_{p+1,n}\left[R_{p,n-1}^r\right]^{-1}r_{p,n-1}$$
$$= \varepsilon_{p,n} - K_{p+1,n}^r r_{p,n-1} \tag{10.59}$$

and when $V = y_{n-N+1:n}$ and $W = y_{n-N+1:n}$,

$$R_{p+1,n}^\varepsilon = R_{p,n}^\varepsilon - \Delta_{p+1,n}\left[R_{p,n-1}^r\right]^{-1}\Delta_{p+1,n}^H$$
$$= R_{p,n}^\varepsilon - K_{p+1,n}^r \Delta_{p+1,n}^H \tag{10.60}$$

Setting $V = W = \pi$ yields one of several possible expressions for the likelihood variable.

$$1 - \gamma_{p,n-1} = 1 - \gamma_{p-1,n-1} - r_{p,n-1}^H [R_{p,n-1}^r]^{-1} r_{p,n-1} \tag{10.61}$$

The likelihood variable gets its name from the fact that the PDF for the backward error is

$$p\left(r_{p,n-1}\right) = \frac{1}{\sqrt{2\pi R_{p,n-1}^r}} e^{-\frac{1}{2}r_{p,n-1}^H\left[R_{p,n-1}^r\right]^{-1}r_{p,n-1}} = p\left(y_n\right) \tag{10.62}$$

where the probability density for the backward error signal is the same as the probability density for the data because they both span the same subspace and share the same innovation. When $1 - \gamma_{p,n-1}$

is nearly unity ($\gamma_{p,n-1}$ is small), the exponent in Equation 10.62 is nearly zero, making the PDF a very narrow Gaussian function centered at zero on the ordinate defined by the backward error signal for the pth lattice stage. If the lattice input signal y_n changes level or spectral density distribution, both the forward and backward error signal PDFs become wide, indicating that the error signals have suddenly grown in variance. The likelihood variable detects the error signal change before the variance increase shows up in the error signal variance parameters, instantly driving $1 - \gamma_{p-1,n-1}$ toward zero and resulting in rapid updates of the lattice estimates for error variance and cross-correlation. This will be most evident in the time-update recursions shown below, but it actually occurs whether one uses the time update forms, or the time and order, or order update forms for the forward and backward error signal variances.

Consider a time-and-order subspace update which amounts to adding a row to the top of $Y_{p,N}$ which increases the model order and time window by one.

$$Y_{p+1,N+1} = \begin{bmatrix} y_{n-N:n} \\ 0 \; Y_{p,N} \end{bmatrix} = \begin{bmatrix} y_{n-N} & y_{n-N-1} & \cdots & y_{n-1} & y_n \\ 0 & y_{n-N} & \cdots & y_{n-2} & y_{n-1} \\ 0 & y_{n-N-1} & \cdots & y_{n-3} & y_{n-2} \\ & & & & \vdots \\ 0 & y_{n-p-N+1} & \cdots & y_{n-p-1} & y_{n-p} \end{bmatrix} \tag{10.63}$$

Therefore, for the subspace spanned by the rows of $Y_{p,N}$ we can choose $S = y_{n-N+1:n}$ allowing the following order updates when $V = y_{n-N+1:n}^{p+1}$ and $W = \pi$,

$$\begin{aligned} r_{p+1,n} &= r_{p,n-1} - \Delta_{p+1,n}^{H} \left[R_{p,n}^{\varepsilon} \right]^{-1} \varepsilon_{p,n} \\ &= r_{p,n-1} - K_{p+1,n}^{\varepsilon} \varepsilon_{p,n} \end{aligned} \tag{10.64}$$

and when $V = y_{n-N+1:n}^{p+1}$ and $W = y_{n-N+1:n}^{p+1}$,

$$\begin{aligned} R_{p+1,n}^{r} &= R_{p,n-1}^{r} - \Delta_{p+1,n}^{H} \left[R_{p,n}^{\varepsilon} \right]^{-1} \Delta_{p+1,n} \\ &= R_{p,n-1}^{r} - K_{p+1,n}^{\varepsilon} \Delta_{p+1,n} \end{aligned} \tag{10.65}$$

Setting $V = W = \pi$ yields another one of the several possible expressions for the likelihood variable.

$$1 - \gamma_{p,n} = 1 - \gamma_{p-1,n-1} - \varepsilon_{p,n}^{H} \left[R_{p,n}^{\varepsilon} \right]^{-1} \varepsilon_{p,n} \tag{10.66}$$

Finally, we present the time update form of the equations. A total of nine equations are possible for the lattice, but only six update equations are needed for the algorithm. The error signal cross-correlation can only be calculated using a time update. The importance and function of the likelihood variable on the lattice equation will become most apparent in the time update equations.

$$Y_{p,N+1} = \begin{bmatrix} y_{n-N:n} \\ y_{n-N-1:n-1} \\ \vdots \\ y_{n-p-N+1:n-p+1} \end{bmatrix} = \begin{bmatrix} y_{n-N} & \cdots & y_{n-2} y_{n-1} y_n \\ y_{n-N-1} & \cdots & y_{n-3} y_{n-2} y_{n-1} \\ \vdots & & \\ y_{n-p-N+1} & \cdots & y_{n-p-1} y_{n-p} y_{n-p+1} \end{bmatrix}$$

$$= \begin{bmatrix} Y_{p,N} & \begin{bmatrix} y_n \\ y_{n-1} \\ \vdots \\ y_{n-p+1} \end{bmatrix} \end{bmatrix} \tag{10.67}$$

It not clear from our equations how one augments the signal subspace to represent the time update. However, we can say that the projection operator $P_{Yp,N+1}$ is $N+1$ by $N+1$ rather than N by N in size. So, again for the subspace spanned by the rows of $Y_{p,N}$ a time update to the projection operator can be written as

$$\begin{aligned} P_{\{Y_{p,n}+\pi\}} &= P_{\{Y_{p,N-1}\}} + P_{\pi} \\ &= \begin{bmatrix} P_{\{Y_{p,N-1}\}} & 0 \\ & \vdots \\ 0 & \cdots & 0 \end{bmatrix} + \pi^{H} (\pi^{H}\pi)^{-1} \pi \\ &= \begin{bmatrix} P_{Y_{p,N-1}} & 0 \\ & \vdots \\ 0 & \cdots & I \end{bmatrix} \end{aligned} \tag{10.68}$$

Equation 10.68 can be seen to state that the projection operator for the subspace augmented by π is actually the previous time iteration projection operator! However, recalling from Section 9.3, as the data are fit more perfectly to the model, the projection operator tends toward an identity matrix. By defining the time update to be a perfect projection (although backward in time) we are guaranteed a least-squared error time update. The orthogonal projection operator for the time-updated space is

$$\left(I - P_{\{Y_{p,N}+\pi\}}\right) = \begin{bmatrix} \left(I - P_{Y_{p,N-1}}\right) & 0 \\ & \vdots \\ 0 & \cdots & 0 \end{bmatrix} \tag{10.69}$$

which means $V(I - P_{\{Yp,N+\pi\}})W^{H}$ in the decomposition equation actually corresponds to the old parameter projection, and the term $V(I - P_{Yp,N})W^{H}$ corresponds to the new time updated parameter. This may seem confusing, but the time updates to the least-squares lattice are by design orthogonal to the error and therefore optimum. Choosing $S = \pi$, $V = y_{n}-_{N+1:n}$ and $W = y_{n-N+1:n}^{p+1}$ we have the time update for the cross-correlation coefficient. A forgetting factor α has been added to facilitate an exponentially decaying data memory window. Note the plus sign which results from the rearrangement of the update in Equation 10.57 to place the new parameter on the left-hand side.

$$\Delta_{p+1,\,n} = \alpha \Delta_{p+1,\,n-1} + \varepsilon_{p,n} [1-\gamma_{p-1,\,n-1}]^{-1} r_{p,\,n-1}^{H} \tag{10.70}$$

Choosing $S = \pi$ and $V = W = y_{n-N+1:n}$ gives a time update for the forward error signal variance.

$$R_{p,n}^{\varepsilon} = \alpha\, R_{p,n-1}^{\varepsilon} + \varepsilon_{p,n} [1-\gamma_{p-1,\,n-1}]^{-1} \varepsilon_{p,n}^{H} \tag{10.71}$$

Choosing $S = \pi$ and $V = W = y_{n-N+1:n}^{p+1}$ gives a time update for the backward error signal variance.

$$R_{p,n-1}^r = \alpha R_{p,\,n-2}^r + r_{p,\,n-1}[1 - \gamma_{p-1,\,n-1}]^{-1} r_{p,\,n-1}^H \tag{10.72}$$

Comparing Equations 10.70 through 10.72 we immediately see the role of the likelihood variable. If the recent data does not fit the existing model, the likelihood variable (and corresponding main diagonal element for the projection operator) will be small, causing the lattice updates to quickly "forget" the old outdated data and pay attention to the new data. Equations 10.61 and 10.66 can be combined to give a pure time update for the likelihood variable which is also very illustrative of its operation.

$$\left(1 - \gamma_{p,n}\right) = \alpha\left(1 - \gamma_{p,n-1}\right) + r_{p,n-1}^H \left[R_{p,n-1}^r\right]^{-1} r_{p,n-1} - \varepsilon_{p,n}^H \left[R_{p,n}^\varepsilon\right]^{-1} \varepsilon_{p,n} \tag{10.73}$$

Clearly, if either the forward or backward error signals start to change, $1 - \gamma_{p,n}$ will tend toward zero giving rise to very fast adaptive convergence and tracking of signal parameter changes. Figure 10.12 compares the whitening filter performance of the least-squares lattice to the RLS and LMS algorithms for the 50 Hz signal case seen previously in Figure 10.7.

The amazingly fast convergence of the lattice can be attributed in part to the initialization of the error signal variances to a small value (1×10^{-5}). Many presentations of adaptive filter algorithms in the literature performance comparisons which can be very misleading due to the way the algorithms are initialized. If we initialize the signal covariance matrix P^{-1} of the RLS algorithm to be an identity matrix times the input signal power, and initialize all of the forward and backward error signal variances in the lattice to the value of the input signal power, a more representative comparison is seen in Figure 10.13.

Figure 10.13 shows a much more representative comparison of the RLS and Lattice algorithms. The lattice filter still converges slightly faster due to the likelihood variable and its ability to optimize each stage independently. The effective memory window is only five samples, and one can see from Figures 10.12 and 10.13 that the convergence time is approximately the filter length (three samples) plus the memory window length, but also depends on the initialization of the algorithms. Initialization

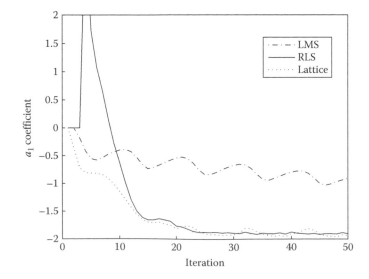

FIGURE 10.12 Comparison of RLS, LMS, and lattice for the case of the 50 Hz whitening filter signals shown in Figure 10.7.

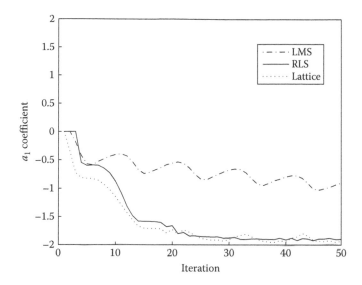

FIGURE 10.13 Comparison of RLS, LMS, and lattice for the case of the 50 Hz whitening filter signals when the initial covariances are set equal to the input signal power.

of the lattice error signal variances to some very small value makes any change in error signals rapidly drive the convergence by the presence of the likelihood variable in the updates, making the lattice converge astonishingly fast in time intervals on the order of the filter length. Initialization of the lattice error variances at higher levels makes the start-up transients in the error signals seem less significant and the parameters converge at a rate determined mainly by the memory window length.

Table 10.2 summarizes the least-squares lattice implementation. There are nine possible update equations where only six are actually needed. Naturally, there is some flexibility available to the programmer as to which equation forms to choose based on storage requirements and convenience. The equation sequence in Table 10.2 is just one possible implementation. Note that the error signal variances cannot be initialized to zero due to the required divisions in the algorithm. Divisions are much more computationally expensive than multiplications in signal processing engines. However, the availability of fast floating-point processors makes the lattice filter an attractive choice for high performance adaptive filtering in real time.

The linear prediction coefficients for all model orders up to the lattice whitening filter model order M can be calculated from an updated set of PARCOR coefficients using the Levinson recursion in Table 10.3. Another way to visualize how the Levinson recursion works is to consider the impulse response of the lattice. As the impulse makes its way down each stage of the lattice, the forward error signal at the Mth stage yields the Mth order linear prediction coefficients. It can be very useful for the student to calculate this sequence manually for a small lattice of say three stages and compare the results to the Levinson recursion. Checking the impulse response is also a useful software debugging tool for lattice Levinson recursion algorithms.

10.4.1 Wiener Lattice

Let us consider the case where both the input and output are known and we use a lattice filter for system identification, rather than just signal modeling. The system identification application of adaptive filters is known as Wiener filtering. It is timely to introduce the lattice Wiener filter here, as the changes necessary from the whitening filter arrangement just presented are very illustrative. Since we have both x_n and y_n and wish to fit an FIR filter model to the unknown linear system relating x_n and y_n, recall that x_n is the input to the LMS Wiener filter in Section 10.2. The same is true for the Wiener

TABLE 10.2
Least-Squares Lattice Algorithm

Description	Equation
Initialization at time $n = 0$ for $p = 0, 1, 2, \ldots, M$	$R_{p,0}^{\varepsilon} = R_{p,0}^{r} = E\left\{y_n\, y_n^H\right\} 1 - \gamma_{-1,-1} = 1$
Input error signals at the first stage	$\varepsilon_{0,n} = r_{0,n} = y_n$
Lattice stage update sequence $\alpha = (N-1)/N$	for $p = 0, 1, 2, \ldots, M$
Time update for error signal cross-correlation	$\Delta_{p+1,n} = \alpha\,\Delta_{p+1,\,n-1} + \varepsilon_{p,n}\, r_{p,\,n-1}^{H} / (1 - \gamma_{p-1,\,n-1})$
Time update of forward prediction error variance	$R_{p,n}^{\varepsilon} = \alpha\,R_{p,\,n-1}^{\varepsilon} + \varepsilon_{p,n}\,\varepsilon_{p,n}^{H} / (1 - \gamma_{p-1,\,n-1})$
Time update of backward prediction error variance	$R_{p,\,n-1}^{r} = \alpha\,R_{p,\,n-2}^{r} + r_{p,\,n-1}\,r_{p,\,n-1}^{H} / (1 - \gamma_{p-1,\,n-1})$
Order update of the likelihood variable	$(1 - \gamma_{p,\,n-1}) = (1 - \gamma_{p-1,\,n-1}) - r_{p,\,n-1}^{H}\,R_{p,\,n-1}^{-r}\,r_{p,\,n-1}$
Forward PARCOR coefficient	$K_{p+1,n}^{\varepsilon} = \Delta_{p+1,n}^{H}\,R_{p,n}^{-\varepsilon}$
Backward PARCOR coefficient	$K_{p+1,n}^{r} = \Delta_{p+1,n}\,R_{p,\,n-1}^{-r}$
Order update of the forward error signal	$\varepsilon_{p+1,n} = \varepsilon_{p,n} - K_{p+1,\,n}^{r}\,r_{p,\,n-1}$
Time and order update of the backward error signal	$r_{p+1,n} = r_{p,\,n-1} - K_{p+1,\,n}^{\varepsilon}\,\varepsilon_{p,n}$

lattice as seen in Figure 10.14. To see what to do with the output y_n we have to revisit the backward error signal definition in Equation 10.26, replacing y_n (used in the whitening filter problem) with x_n.

$$
\begin{bmatrix} r_{0,\,n} \\ r_{1,\,n} \\ \vdots \\ r_{M-1,\,n} \\ r_{M,\,n} \end{bmatrix} = \begin{bmatrix} 1 & 0 & \cdots & & 0 \\ a_{1,1}^{r} & 1 & 0 & \cdots & 0 \\ \vdots & & \vdots & & \vdots \\ a_{M-1,M-1}^{r} & & \cdots & 1 & 0 \\ a_{M,M}^{r} & a_{M,M-1}^{r} & \cdots & a_{M,1}^{r} & 1 \end{bmatrix} \begin{bmatrix} x_n \\ x_{n-1} \\ \vdots \\ x_{n-M+1} \\ x_{n-M} \end{bmatrix} \tag{10.74}
$$

$$
= \bar{r}_{0,\,n:M,n} = L\,\varphi_n^{\mathrm{T}}
$$

The lower triangular matrix L in Equation 10.74 also illustrates some interesting properties of the lattice structure. Postmultiplying both sides by $\bar{r}_{0,n:M,n}^{H}$ and taking expected values leads to $R_M^r = L\,R_M^x\,L^H$ where $R_M^r = \mathrm{diag}\,\{R_0^r\,R_1^r \ldots R_M^r\}$ and $R_M^x = E\,\{\varphi_n^{H}\,\varphi_n\}$, the covariance matrix for

TABLE 10.3
Levinson Recursion

Description	Equation
Lowest forward and backward prediction coefficients	$a_{p,0} = a_{p,0}^{r} = 1$ for $p = 0, 1, 2, \ldots, M$
Highest forward and backward prediction coefficients	$a_{p+1,\,p+1} = -K_{p+1}^{r} \quad a_{p+1,\,p+1}^{r} = -K_{p+1}^{\varepsilon}$
Linear prediction coefficient recursion	$\left. \begin{aligned} a_{p+1,\,j} &= a_{p,j} - K_{p+1}^{r}\,a_{p,\,p-j+1}^{r} \\ a_{p+1,\,p-j+1}^{r} &= a_{p,\,p-j+1}^{r} - K_{p+1}^{\varepsilon}\,a_{p,j} \end{aligned} \right\} \; j = 1, 2, \ldots, p$

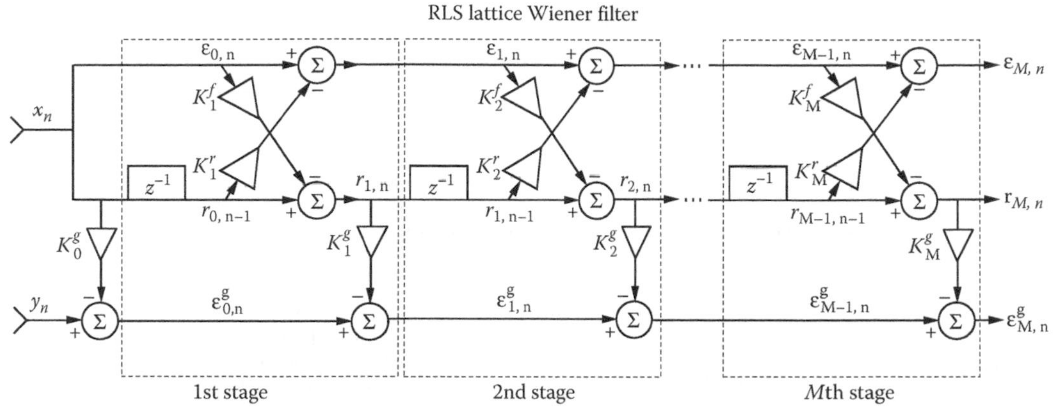

RLS lattice Wiener filter

FIGURE 10.14 Wiener lattice filter structure for fast adaptive system identification given both the input x_n and output y_n of an unknown system.

the input data signal. The importance of this structure is seen when one considers that the covariance matrix inverse is

$$\left[R_M^x\right]^{-1} = L^H \left[R_M^r\right]^{-1} L \tag{10.75}$$

where the inverse of the diagonal matrix R_M^r is trivial compared to the inverse of a fully populated covariance matrix. The lower triangular backward error predictor matrix inverse L^{-1} and inverse Hermitian transpose L^{-H} are the well-known LU Cholesky factors of the covariance matrix. The orthogonality of the backward error signals in the lattice makes the Cholesky factorization possible. The forward error signals are not guaranteed to be orthogonal.

Also recall from Equations 9.4 and 10.74 that the predicted output is $y_n' = \varphi_n H = H^T L^{-1}\, \bar{r}_{0,n:M,n}^T$. We can therefore define a PARCOR coefficient vector $K_{0:M}^g = \left[K_0^g\, K_1^g \cdots K_M^g\right]$ which represents the cross-correlation between the backward error signals and the output signal y_n as seen in Figure 10.14.

The prediction error for the output signal is

$$\varepsilon_{p,n}^g = y_n - K_p^g\, r_{p,n} \quad p = 0, 1, \ldots, M \tag{10.76}$$

Since the backward error signals for every stage are orthogonal to each other, we can write an error recursion similar to the forward and backward error recursions in the lattice filter.

$$\varepsilon_{p,n}^g = \varepsilon_{p-1,n}^g - K_p^g\, r_{p,n} \quad p = 1, \ldots, M; \quad \varepsilon_{-1,n}^g = y_n \tag{10.77}$$

The PARCOR coefficients for the output data are found in an analogous manner to the other lattice PARCOR coefficients. The output error signal cross-correlation is

$$\Delta_{p,n}^g = \alpha\, \Delta_{p,n-1}^g + \varepsilon_{p-1,n}^g\, r_{p,n} / (1 - \gamma_{p-1,n}) \quad p = 0, 1, \ldots, M \quad \varepsilon_{-1,n}^g = y_n \tag{10.78}$$

and $K_{p,n}^g = \Delta_{p,n}^g / R_{p,n}^r = \Delta_{p,n}^g\, R_{p,n}^{-r}$. The updates of the additional equations for the Wiener lattice are very intuitive as can be seen in Figure 10.14. To recover H one simply computes $H = L^H K_{0,M}^g$ in

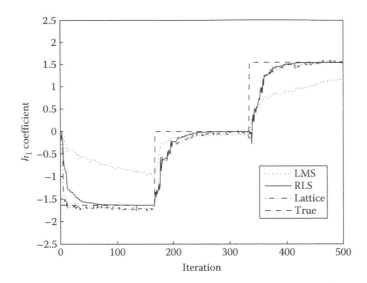

FIGURE 10.15 Comparison of LMS, standard RLS, and Weiner lattice performance on the nonstationary system identification previously shown in Figure 10.2.

addition to the Levinson recursion in Table 10.3 for the whitening filter problem. The transpose of the lower-triangular backward predictors times the output PARCOR vector reduces to

$$h_p = K_p^g + \sum_{i=p+1}^{M} a_{i,\,i-p}^{r\,\,H}\, K_i^g \quad p = 0, 1, 2, \ldots, M \tag{10.79}$$

which is a very straightforward recursion using the backward error prediction coefficients and lattice output signal PARCOR coefficients.

Figure 10.15 shows the Wiener lattice performance compared to the standard RLS and LMS algorithms for the exact same data case as presented in Figure 10.2 where the input signal is unity variance white noise plus a 25 Hz sinusoid of amplitude 5. The Wiener lattice and RLS algorithms have clearly nearly identical parameter tracking performance for this case. However, the lattice results in Figure 10.15 do indicate some additional numerical noise is present, probably due to the sensitivity in the likelihood variable to numerical noise with a short memory window.

10.4.2 DOUBLE/DIRECT WEINER LATTICE

While generally not seen for a whitening filter application, the additional divides in the Wiener lattice can be seen as a source of numerically generated noise due to roundoff error. It is fortunate that this particular example happens to show an excellent example of numerical error processing noise in an adaptive algorithm. We can suppress the noise somewhat by eliminating some of the division operations. Believe it or not, even double precision floating-point calculations are susceptible to roundoff error. Implementing an adaptive filter on a signal processing chip requires careful attention, because double precision floating-point is not usually available without resorting to very slow software-based double precision calculations. Roundoff error is a major concern for any fixed-point signal processing implementation. Figure 10.16 shows the same comparison but with the use of the Double/Direct Weiner lattice algorithm developed by Orfanidis [8], which demonstrates superior numerical performance compared to the conventional lattice algorithm.

The Double/Direct lattice gets its name from the use of double error signal updates (one set updated with the previous PARCOR coefficients and one updated the conventional way) and a

TABLE 11.2
The Kalman Filter

Description	Equation
Basis function (measurement matrix)	$H(N+1)$
State transition matrix	$F(N+1)$
Process noise covariance matrix	$Q(N+1)$
Measurement noise covariance matrix	$R(N+1)$
A priori state prediction	$x'(N+1 \mid N) = F(N) x(N \mid N)$
A priori state error prediction	$P'(N+1 \mid N) = F(N) P(N \mid N) F^H(N) + Q(N)$
A priori measurement prediction	$z'(N+1 \mid N) = H(N+1) x'(N+1 \mid N)$
Innovation covariance update	$S(N+1) = H(N+1) P'(N+1 \mid N) H^H(N+1) + R(N+1)$
Kalman gain update	$K(N+1) = P'(N+1 \mid N) H^H(N+1) [S(N+1)]^{-1}$
Measurement prediction error	$\zeta(N+1) = z(N+1) - H(N+1) x'(N)$
Optimal state vector update using prediction error	$x(N+1 \mid N+1) = x'(N+1 \mid N) + K(N+1) \zeta(N+1)$
Inverse state covariance matrix update	$P(N+1 \mid N+1) = P'(N+1 \mid N) - K(N+1) S(N+1) K^H(N+1)$

An example of the Kalman filter is presented following the simple rocket height tracking example given in Section 5.2 for the α–β–γ tracker. As will be seen below, the α–β–γ tracking filter is a simple Kalman filter where the Kalman gain has been fixed to the optimal value assuming a constant state error covariance. In the rocket, for example, we have height measurements every 100 ms with a standard deviation of 3 m. The maximum acceleration (actually a deceleration) occurs at burnout due to both gravity and drag forces and is about 13 m/s^2. Setting the process noise standard deviation to 13 is seen as near optimal for the α–β–γ tracking filter, allowing nonadaptive tracking of the changes in acceleration, velocity, and position of the rocket's height. The adaptive Kalman gain, on the other hand, depends on the ratio of the state prediction error covariance to the measurement noise covariance. The Kalman gain will be high when the state error is large and the measurements accurate (σ_w and R small), and the gain will be low when the state error is small and/or the measurements inaccurate. The adaptive gain allows the Kalman filter to operate with a lower process noise than the nonadaptive α–β–γ tracking filter, since the gain will automatically increase or decrease with the state prediction error. In other words, if one knows the maneuverability of the target (e.g., the maximum acceleration), and the measurements are generally noisy, the α–β–γ tracking filter is probably the best choice, since the measurement noise will not affect the filter gain too much. But if one does not know the target maneuverability, the Kalman filter offers fast convergence to give the least-squared error state vector covariance. However, when the measurements are noisy, the Kalman filter also suffers from more noise in the state vector estimates when the same process noise is used as in the α–β–γ tracking filter. Figures 11.1 and 11.2 illustrate the noise sensitivity for the rocket example given in Section 5.2.

Comparing the filter gains used in Figures 11.1 and 11.2, we see that the α–β–γ tracking filter used [0.2548 0.3741 0.2746] while the Kalman filter calculated a gain vector of [0.5046 1.7545 3.0499]. The higher Kalman gain fit the state vector more accurately to the measurements, but resulted in more velocity and acceleration noise. In the simulation, if we let the measurement noise approach zero, the responses of the two filters become identical and the gain vector converges to [1 1/T 1/($2T^2$)], or [1 20 200] when T = 100 ms. By comparing the gain vectors for a given measurement noise (and process noise for the α–β–γ tracking filter), we can adjust the Kalman filter process noise so that the filter gains are roughly the same when the measurement noise is still a relatively high 3 m. We find that the Kalman filter process noise can be reduced to about 1.0 to give a Kalman gain vector of [0.2583 0.3851 0.2871], which is very close to the gain vector for the α–β–γ tracking filter. Figure 11.3 shows the Kalman filter tracking results which are nearly identical to the α–β–γ tracking filter with a high process noise of 13.

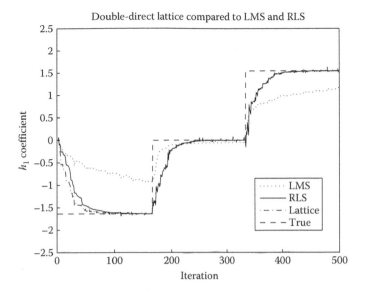

FIGURE 10.16   Comparison of the double/direct Wiener lattice to the RLS and LMS for the signal case shown in Figures 10.2 and 10.15 showing improved numerical properties of the double/direct Weiner lattice algorithm.

direct PARCOR update equation. The Double/Direct lattice eliminates the divisions involving the likelihood variable by making error signal predictions *a priori* to the PARCOR coefficient updates, and combining them with *a posteriori* error signals (as computed in the conventional lattice) in the updates for the PARCOR coefficients. An important formulation is the direct PARCOR update, rather than the computation of a cross-correlation involving a divide, and then the PARCOR coefficient involving a second divide. The *a priori* forward and backward error signals are

$$\varepsilon_{p,n}^{+} = \frac{\varepsilon_{p,n}}{(1-\gamma_{p-1,n-1})} \quad \varepsilon_{p,n}^{+g} = \frac{\varepsilon_{p,n}^{g}}{(1-\gamma_{p,n})} \quad r_{p,n-1}^{+} = \frac{r_{p,n-1}}{(1-\gamma_{p-1,\,n-1})} \tag{10.80}$$

and the updates are done using the previous PARCOR coefficients.

$$\varepsilon_{p+1,n}^{+} = \varepsilon_{p,n}^{+} - K_{p+1,n-1}^{r}\, r_{p,n-1}^{+} \tag{10.81}$$

$$\varepsilon_{p+1,n}^{+g} = \varepsilon_{p,n}^{+g} - K_{p+1,n-1}^{g}\, r_{p+1,n}^{+} \tag{10.82}$$

$$r_{p+1,n}^{+} = r_{p,n-1}^{+} - K_{p+1,n-1}^{\varepsilon}\, \varepsilon_{p,n}^{+} \tag{10.83}$$

The *a posteriori* error signals are updated the conventional way.

$$\varepsilon_{p+1,n} = \varepsilon_{p,n} - K_{p+1,n}^{r}\, r_{p,n-1} \tag{10.84}$$

$$\varepsilon_{p+1,n}^{g} = \varepsilon_{p,n}^{g} - K_{p+1,n}^{g}\, r_{p+1,n} \tag{10.85}$$

$$r_{p+1,n} = r_{p,n-1} - K_{p+1,n}^{\varepsilon}\, \varepsilon_{p,n} \tag{10.86}$$

The *a priori* error signals are mixed into the conventional time, order and time and order updates to essentially eliminate the likelihood variable and its divisions from the algorithm. The likelihood variable can still be retrieved if desired by dividing the conventional *a posteriori* error signals by the *a priori* error signals as can be seen from Equation 9.80. The direct update for the forward PARCOR coefficient is derived in Equation 9.87. Elimination of the divides and the direct form of the PARCOR update are seen as the reasons for the improved numerical properties.

$$
\begin{aligned}
K^{\varepsilon}_{p+1,\,n} &= \Delta_{p+1,n}\ R^{-\varepsilon}_{p,n} = \alpha\ \Delta_{p+1,n-1} + \varepsilon_{p,n}\ r^{+H}_{p,n-1}\ R^{-\varepsilon}_{p,n} \\
&= \left[\alpha\ K^{\varepsilon}_{p+1,n-1}\ R^{\varepsilon}_{p,n-1} + \varepsilon_{p,n}\ r^{+H}_{p,n-1}\right] R^{-\varepsilon}_{p,n} \\
&= \left[\alpha\ K^{\varepsilon}_{p+1,n-1}\ \left\{R^{\varepsilon}_{p,n} - \varepsilon_{p,n}\varepsilon^{+H}_{p,n}\right\}\alpha^{-1} + \varepsilon_{p,n}\ r^{+H}_{p,n-1}\right] R^{-\varepsilon}_{p,n} \\
&= K^{\varepsilon}_{p+1,n-1} + \varepsilon_{p,n}\ \left(r^{+H}_{p,n-1} - \varepsilon^{+H}_{p,n}\ K^{\varepsilon\,H}_{p+1,\,n-1}\right)\ R^{-\varepsilon}_{p,n} \\
&= K^{\varepsilon}_{p+1,n-1} + \varepsilon_{p,n}\ r^{+H}_{p+1,n}\ R^{-\varepsilon}_{p,n}
\end{aligned}
\tag{10.87}
$$

The complete set of updates for the Double/Direct RLS Wiener lattice algorithm is given in Table 10.4. Considering the effect of the likelihood variable, which is not directly computed but rather *embedded* in the algorithm, it can be seen that there will be very little difference between the *a priori* and *a posteriori* error signals when the data is stationary and the lattice PARCOR coefficients have converged. When the data undergoes statistical or underlying model changes the PARCOR's adapt rapidly due to the difference in *a priori* and *a posteriori* error signals.

The Double/Direct algorithm may at first appear to be more complicated than the conventional least-squares lattice algorithm, but we are trading computationally expensive divides for far more efficient (and numerically pure) multiplies and adds. Most modern signal processing and computing chips can execute a combination multiply and accumulate (MAC operation) in a single clock cycle. Since a floating point divide is generally executed using a series approximation (integer divides often use a able look-up technique), a divide operation can take from 16 to 20 clock cycles to execute. The Double/Direct Wiener lattice in Table 10.4 requires 13 MACs and 3 divides per stage while the conventional Wiener lattice requires only 11 MACs, but 8 divides. The Double/Direct lattice algorithm is not only more numerically accurate, but is also more computationally efficient than the conventional lattice algorithm.

10.5   MATLAB® EXAMPLES

A total of 5 m-scripts were used to make the plots in this chapter for the LMS, RLS, and lattice filter structures and are listed in Table 10.5. The scripts share a common input signal design of a sinusoid in white noise sampled at 1024 Hz. With the sinusoid frequency at 25 Hz the eigenvalue spread of the input signal is wide, showing the weakness of the LMS algorithm under those conditions.

We also use a simple 2-zero FIR filter where we can move the conjugate zero pair easily just by changing one of the filter coefficients. By moving the middle coefficient from −1.5 to 0.0, and then to +1.5, we can see the convergence of the adaptive system identification process using the RLS, LMS and Weiner lattice algorithms. Table 10.6 shows just how simple the LMS algorithm is to code given an input "$x(n)$" output "$y(n)$" and a proper setting of the step size "mu" to avoid divergence of the algorithm.

Table 10.7 shows the power of MATLAB's embedded matrix math for the RLS algorithm, where the coding can handle the equations directly without the need for looping and storing data as one would normally for scalar implementation of matrix equations. Note that "alpha" is α and "alphai" is α^{-1} from Table 10.1. Also note the careful use of "./" for element-by-element divides, and ".*" for element-by-element multiplies. The other multiplies depicted by "*" are full matrix multiplies. In particular note the denominator of the Kalman gain in Table 10.5. The term "phi*P*phi'" only

TABLE 10.4
Double/Direct Wiener Lattice Algorithm

Description	Equation
Initialization at time $n = 0$ for $p = 0, 1, 2, \ldots, M$	$R_{p,0}^{\varepsilon} = R_{p,0}^{r} = E\left\{x_n\, x_n^H\right\}$
Input error signals at the first-stage input	$\varepsilon_{0,n} = r_{0,n} = \varepsilon_{0,n}^{+} = r_{0,n}^{+} = x_n$
A priori output error signal at the first-stage input	$\varepsilon_{0,n}^{+g} = y_n - K_{0,n-1}^{g}\, r_{0,n}^{+}$
Output PARCOR update	$K_{0,n}^{g} = K_{0,n-1}^{g} + \varepsilon_{0,n}^{+g}\, r_{0,n}^{H}\, R_{0,n}^{-r}$
A posteriori output error signal update at the first stage	$\varepsilon_{0,n}^{g} = y_n - K_{0,n}^{g}\, r_{0,n}$
Lattice-stage update sequence $\alpha = (N-1)/N$	for $p = 0, 1, 2, \ldots, M$
A priori forward error	$\varepsilon_{p+1,n}^{+} = \varepsilon_{p,n}^{+} - K_{p+1,n-1}^{r}\, r_{p,n-1}^{+}$
A priori backward error	$r_{p+1,n}^{+} = r_{p,n-1}^{+} - K_{p+1,n-1}^{\varepsilon}\, \varepsilon_{p,n}^{+}$
Forward PARCOR update	$K_{p+1,n}^{\varepsilon} = K_{p+1,n-1}^{\varepsilon} + r_{p,n}^{+}\, \varepsilon_{p,n}^{H}\, R_{p,n}^{-\varepsilon}$
Backward PARCOR update	$K_{p+1,n}^{r} = K_{p+1,n-1}^{r} + \varepsilon_{p+1,n}^{+}\, r_{p,n-1}^{H}\, R_{p,n-1}^{-r}$
A posteriori forward error update	$\varepsilon_{p+1,n} = \varepsilon_{p,n} - K_{p+1,n}^{r}\, r_{p,n-1}$
A posteriori backward error update	$r_{p+1,n} = r_{p,n-1} - K_{p+1,n}^{\varepsilon}\, \varepsilon_{p,n}$
Forward error variance time update	$R_{p+1,n}^{\varepsilon} = \alpha\, R_{p+1,n-1}^{\varepsilon} + \varepsilon_{p+1,n}\, \varepsilon_{p+1,n}^{+H}$
Backward error variance time update	$R_{p+1,n}^{r} = \alpha\, R_{p+1,n-1}^{r} + r_{p+1,n}\, r_{p+1,n}^{+H}$
A priori output error update	$\varepsilon_{p+1,n}^{+g} = \varepsilon_{p,n}^{+g} - K_{p+1,n-1}^{g}\, r_{p+1,n}^{+}$
Output PARCOR coefficient update	$K_{p+1,n}^{g} = K_{p+1,n-1}^{g} + \varepsilon_{p+1,n}^{+g}\, r_{p+1,n}^{H}\, R_{p+1,n}^{-r}$
Output error *a posteriori* update	$\varepsilon_{p+1,n}^{g} = \varepsilon_{p,n}^{g} - K_{p+1,n}^{g}\, r_{p+1,n}$
Execute Levinson recursion, then for $p = 0, 1, 2, \ldots, M$	$h_p = K_p^{g} + \displaystyle\sum_{i=p+1}^{M} a_{i,\,i-p}^{r}{}^{H}\, K_i^{g}$

TABLE 10.5
m-Scripts Used in This Chapter

TABLE 10.6
LMS System Identification Algorithm

```
yp = x(n) + a1p*x(n-1) + a2p*x(n-2);     % prediction
ep = y(n) - yp;                          % pred error
a0p = a0p + 2*mu*x(n)*ep;                % update a0
a1p = a1p + 2*mu*x(n-1)*ep;              % update a1
a2p = a2p + 2*mu*x(n-2)*ep;              % update a2
```

TABLE 10.7
RLS Algorithm Using MATLAB's Embedded Matrix Math

```
phi; = [x(n) x(n-1) x(n-2)];            % basis fcn (vector)
K = P*phi'./(alpha + phi*P*phi');       % Kalman gain (vector)
P = alphai.*(Id - K*phi)*P;             % covariance (matrix)
yp = phi*H;                             % prediction (scalar)
ep = y(n) - yp;                         % error (scalar)
H = H + K*ep;                           % filter update (vector)
```

becomes a scalar if you do full matrix multiplies. While you have to be careful, for the MATLAB's embedded matrix math is a huge time saver and allows clean scripts to implement many signal processing algorithms.

The lattice algorithms are a bit too long to conveniently expose the details here, but they essentially follow the equations listed in Tables 10.2 through 10.4 in scalar form. However, the order of these equation updates is important and must be followed. This is particularly true of the Double/Direct lattice, because *a priori* and *a posteriori* error signal estimates are both used and updated in the same stage. One would think that with double-precision arithmetic there would be no advantage to using the Double/Direct lattice, but the nature of the divides in the lattice algorithm is very subtle. The error signals in a converged lattice are very small; meaning that ratio of signal to any mathematical roundoff error noise (SNR) is at its lowest. These low SNT signals can leave the lattice coefficients essentially "stuck" near the optimal value but with some residual noise. The double/direct lattice uses two sets of error signals, one before PARCOR updating and one after. This is what kicks the PARCOR coefficients out of the noise and into the correct parameter value. It is an elegant development to an already amazing adaptive algorithm.

10.6 SUMMARY

One of the truly fun aspects of digital signal processing is that the computer can be used not only to execute a digital filter for the signals, but also to implement a learning algorithm to adaptive the filter coefficients toward the optimization of some criteria. Frequently, this criterion is expressed in terms of a cost function to be minimized. Since the digital filtering operation is linear, an error signal which is linear can be easily expressed in terms of the difference between the filter model's output and some desired output signal. The squared error can be seen as a quadratic surface which has a single minimum at the point which corresponds to the optimum set of filter coefficients. Solving for the least-squared error set of filter coefficients can be done using a "block" of N signal samples, or recursively using a sliding memory window where the optimum solution is valid only for the most recent signal samples. The latter technique is extremely useful for intelligent signal processing algorithms which adapt in real-time to changing environments or even commands (from humans or even other machines) changing the optimum criteria or cost function to be minimized.

The tradeoff between computational complexity and optimality is presented in detail. For the block solution with model order M and N data points, computing the optimal FIR filter coefficients requires $NM(1 + NM)$ multiplies and adds just to compute the covariance matrix and cross-correlation vector. Inverting the covariance matrix requires approximately M^3 plus an expensive divide operation. For a model order of 10- and a 100-sample rectangular data memory window, the block least-squares algorithm requires 1,002,000 multiplies and adds, plus one divide. Repeating this calculation for every time sample of the signals (to create a sliding rectangular memory window) would be extremely computationally expensive. For example, a 50 MFLOPS DSP chip could only run real-time for an unimpressive sample rate of about 49 Hz. Using the RLS algorithm with an exponentially forgetting data memory window, only $M^3 + 2M^2 + 2M$ multiplies plus 2 divides are needed per time update. For the example model order of 10, this is about 1220 plus 2 divides per time update. The 50 MFLOPS DSP can now run real-time at a sample rate of about 40,000 samples per second using the RLS algorithm getting almost exactly the same results.

Switching to the RLS Wiener lattice algorithm, we get exactly the same performance as the RLS algorithm at a cost of 13 multiplies plus 3 divides per stage, plus 4 multiplies and a divide at the input stage. For the model order of 10 example, and that a divide equals about 16 multiplies in terms of operations, the RLS Wiener lattice requires $61M+20$ operations, or 630 operations per time update. The 50 MFLOPS DSP can now run at about 79,000 samples per second giving exactly the same results as the RLS algorithm. The improved performance of the lattice is not only due to its orthogonal order structure, it is due to the fact that the Levinson recursion (which requires M^2 operations) and output linear predictor generation (which requires M^2-M operations) algorithms need not be computed in real time. If the linear predictors are needed in real time, the RLS Wiener lattice and RLS algorithms are nearly equal in computational complexity.

If convergence speed is less of a concern, we can trade adaptive performance for significant reductions in computational complexity. The shining example of this is the very popular LMS algorithm. Assuming a recursive update is used to estimate the signal power, the step size can be calculated with three multiples and a divide, or about 19 operations. The LMS update requires three multiplies and an addition for each coefficient, or about $3M$ operations since multiplies and additions can happen simultaneously. The normalized LMS filter can be seen to require only $3M + 19$ operations per time update. For the example model order of 10, a 50 MFLOPS DSP can run real-time at a sample rate of over 1 million samples per second. If the input signal is white noise, the LMS algorithm will perform exactly the same as the RLS algorithm, making it the clear choice for many system identification applications. For applications where the input signal always has a known power, the step size calculation can be eliminated making the LMS algorithm significantly more efficient.

PROBLEMS

1. Prove the matrix inversion lemma.
2. How many multiplies and adds/subtractions are needed to execute the Levinson recursion in Table 10.3 in terms of model order M?
3. Compare the required operations in the RLS, lattice, and LMS algorithms, in terms of multiplies, additions/subtractions, and divides per time update for an $M = 64$ tap FIR whitening filter (including Levinson recursion for the lattice).
4. Compare the required operations in the RLS, Wiener Lattice, and LMS algorithms for Wiener filtering, in terms of multiplies, additions/subtractions, and divides per time update for an $M = 64$ tap FIR Wiener filter (including Levinson recursion and output predictor generation for the lattice).
5. An LMS Wiener filter is used for system identification of an FIR filter system. Should one use a large number of equally spaced frequency sinusoids or random noise as system input signals?

6. Given the autocorrelation data for a signal, calculate directly the whitening PARCOR coefficients.
7. Given the whitening filter input signal with samples $\{-2 + 1\ 0 - 1\ +2\}$, determine the PARCOR coefficients for the first five iterations of a single-stage least-squares lattice.
8. Compare the result in Problem 7 to a single-coefficient LMS whitening filter with $\mu = 0.1$.
   a. Compare the PARCOR coefficient to the single LMS filter coefficient.
   b. What is the maximum μ allowable (given the limited input data)?
   c. Compare the PARCOR and LMS coefficients for $\mu = 5$.
12. Show that for a well-converged whitening filter lattice, the forward and backward PARCOR coefficients are approximately equal (for real data) and the likelihood variable approaches unity.
13. Can a Cholesky factorization of the autocorrelation matrix be done using the forward prediction error variances in place of the backward error variances?

REFERENCES

1. P. Strobach, Recursive triangular array ladder algorithms, *IEEE Trans Signal Process*, 1991, 39(1), pp. 122–136.
2. R. F. Stengel, *Stochastic Optimal Control* (New York, NY: Wiley, 1986).
3. B. Widrow and S. D. Sterns, *Adaptive Signal Processing* (New York, NY: Prentice-Hall, 1985).
4. B. Widrow, J. M. McCool, M. G. Larimore, and C. R. Johnson, Stationary and nonstationary learning characteristics of the LMS adaptive filter, *Proc IEEE*, 1976, 64, pp. 1151–1162.
5. M. Morf, Fast algorithms for multivariate systems (PhD dissertation, Stanford University, Stanford, CA, 1974).
6. B. Friedlander, Lattice filters for adaptive processing, *Proc IEEE*, 1982, 70(8), pp. 829–867.
7. M. Morf and D. T. L. Lee, Recursive least-squares ladder forms for fast parameter tracking, *Proceedings of the 17th IEEE Conference on Decision Control*, p. 1326 (San Diego, CA, 1979).
8. S. J. Orfanidis, *Optimum Signal Processing*, 2nd ed. (New York, NY: McGraw-Hill, 1988).

11 Recursive Adaptive Filtering

In this chapter, we develop and demonstrate the use of some important applications of adaptive filters as well as extend the algorithms in Chapter 10 for multichannel processing and frequency domain processing. This book has many examples of applications of adaptive and nonadaptive signal processing, not just to present the reader with illustrative examples, but also as a vehicle to demonstrate the theory of adaptive signal processing. Chapter 10 contains many examples of adaptive whitening filters and Wiener filtering for system identification. Part IV will focus entirely on adaptive beamforming and related processing. There are of course, many excellent texts which cover each of these areas in even greater detail. This chapter explores some adaptive filtering topics which are important to applied adaptive signal processing technology.

In Section 11.1, we present what is now well-known as adaptive Kalman filtering. Signal processing has its origins in telephony, radio, and audio engineering. However, some of the most innovative advances in processing came with Bode's operational amplifier and early feedback control systems for Naval gun stabilization. Control systems based on electronics were being developed for everything from electric power generation to industrial processes. These systems were and still are quite complicated and are physically described by a set of signal states such as, temperature, temperature rate, temperature acceleration, and so on. The various derivatives (or integrals) of the measured signal are processed using +6 dB/oct high pass filters (for derivatives) and or −6 dB/oct low pass filters (for integration). The ±6 dB/oct slope of the filter represents a factor of jω (+6 dB/oct for differentiation), or 1/jω (−6 dB/oct for integration), the plant to be controlled can be physically modeled with a set of partial differential equations. Then sensors are attached to the plant and the various sensor signal states are processed using high or low pass filters to create the necessary derivatives or integrals, respectively. Simple amplifiers and phase shifters supplied the required multiplies and divides. Current addition and subtraction completed the necessary signal processing operations. Signals based on control set points and sensor responses could then be processed in *an analog computer* literally built from vacuum tubes, inductors, capacitors, resistors, and transformers to execute closed-loop feedback control of very complex physical systems. This very elegant* solution to complex system control is known as a "white-box" control problem, since the entire plant and input–output signals are completely known.

By the end of the 1950s, detection and tracking of satellites were a national priority in the United States due to the near public panic over the Soviet Union's Sputnik satellite (the world's first man-made satellite). The implications for national defense and security were obvious and scientists and engineers throughout the world began to focus on new adaptive algorithms for reducing noise and estimating kinematic states such as position, velocity, acceleration, and so on. This type of control problem is defined as a "black-box" control problem because nothing is known about the plant or input–output signals (there is also a "gray box" distinction, where one knows a little about the plant, a little about the signals, but enough to be quite dangerous). In 1958 R. E. Kalman [1] published a revolutionary paper in which sampled analog signals were actually processed in a vacuum tube analog computer. The work was funded by E. I. du Pont de Nemours & Co., a world leader in chemical manufacturing. Many of the complex processes in the manufacture of plastics, explosives, nylon, man-made textiles, and so on, would obviously benefit from a "self-optimizing" controller. This is especially true if the modeled dynamics as described by the physical parameters in a bench top process do not scale linearly to the full-scale production facility. Physical parameters such as viscosity, compressibility, temperature, and pressure do not scale at all while force, flow, and mass do. Raw materials also vary as inputs to the

---

* An elegant solution is defined as a solution one wishes one thought of first.

plant. This kind of technical difficulty created a huge demand for what we now refer to as the Kalman filter where one simply commands a desired output (such as a temperature set point), and the controller does the rest, including figuring out what the plant is. Section 11.1 presents a derivation of the Kalman filter unified with the least-squared error approaches of Chapters 9 and 10. The RLS tracking solution is then applied to the derivation of the α–β–γ tracker (presented in Section 5.2) which is shown to be optimal if the system states have stationary covariances.

In Section 11.2 we extend the LMS and lattice filter structures to IIR forms, such as the all-pole and pole–zero filters presented in Sections 3.1 and 3.2. Pole–zero filters, also known as autoregressive moving average ARMA filters, pose special convergence problems which must be carefully handled in the adaptive filters which whiten ARMA signals, or attempt to identify ARMA systems. These constraints are generally not a problem so long as stability constraints are strictly maintained in the adaptive algorithms during convergence and, the signals at hand are stable signals. Transients in the input–output signals can lead to incorrect signal modeling as well as inaccurate ARMA models. ARMA models are particularly sensitive to transients in that their poles can move slightly on or outside the unit circle on the complex z-plane, giving an unstable ARMA model. An embedding technique will be presented for both the LMS and lattice filters which also show how many channels of signals may be processed in the algorithms to minimize a particular error signal.

Finally in this chapter, we present frequency domain adaptive processing with the LMS algorithm. Frequency domain signals have the nice property of orthogonality, which means that an independent LMS filter can be assigned to each FFT frequency bin. Very fast convergence can be had for frequency domain processing except for the fact that the signals represent a time integral of the data. However, this integration is well known to suppress random noise, making frequency domain adaptive processing very attractive to problems where the signals of interest are sinusoids in low SNR. This scenario applies to a wide range of adaptive signal processing problems. However, the frequency domain adaptive filter transfer functions operations are eventually converted back to the time domain to give the filter coefficients. Spectral leakage from sinusoids which may not be perfectly aligned to a frequency bin must be eliminated using the technique presented in Section 6.4 to ensure good results. Presentation of the Kalman filter, IIR, and multichannel forms for adaptive LMS and lattice filters, and frequency domain processing complement the applications of signal processing presented throughout the rest of this book.

11.1 ADAPTIVE KALMAN FILTERING

Consider the problem of tracking the extremely high-velocity flight path of a low-flying satellite from its radio beacon. It is true that for a period of time around the launching of Sputnik, the world's first satellite, many ham radio enthusiasts provided very useful information on the time and position of the satellite from ground detections across the world. This information could be used to refine an orbit state model which in turn is used to predict the satellites position at future times. The reason for near public panic was that a satellite could potentially take pictures of sensitive defense installations or even deliver a nuclear bomb, which would certainly wreck one's day. So with public and government interest in high gear, the space race would soon join the arms race, and networks of tracking radio stations were constructed not only to detect and track enemy satellites, but also to communicate with one's own satellites. These now-familiar satellite parabolic dishes would swivel as needed to track and communicate with a satellite passing overhead at an incredible speed of over 17,500 miles per hour (for an altitude of about 50 miles). At about 25,000 miles altitude, the required orbital velocity drops to about 6500 miles/h and the satellite is *geosynchronous*, meaning that it stays over the same position on the ground.

For clear surveillance pictures, the satellite must have a very low altitude. But, with a low altitude, high barometric pressure areas on the ground along the flight path will correspond to the atmosphere extending higher up into space. The variable atmosphere, as well as gravity variations due to the earth not being perfectly spherical, as well as the effects of the moon and tides, will cause the satellite's orbit to change significantly. A receiver dish can be pointed in the general direction of the satellite's path, but the precise path is not completely known until the tracking algorithm "locks in" on the trajectory. Therefore, today adaptive tracking systems around the world are used to keep up to date the trajectories of where satellites (as well as orbital debris) are at all times.

Tracking dynamic data are one of the most common uses of the Kalman filter. However, the distinction between Kalman filtering and the more general RLS become blurred when the state vector is replaced by other basis functions. In this text, we specifically refer to the Kalman filter as a RLS estimator of a state vector, most commonly used for a series of differentials depicting the state of a system. All other "nonstate vector" basis functions in this text are referred to in the more general terms of a RLS algorithm. The Kalman filter has enabled a vast array of twenty-first century mainstay technologies such as: worldwide real-time communications for telephones, television, and the Internet; the global positioning system (GPS) which is revolutionizing navigation and surveying; satellite wireless telephones and computer networks; industrial and process controls; and environmental forecasting (weather and populations, diseases, etc.); and even monetary values in stock and bond markets around the world. The Kalman filter is probably even more ubiquitous than the FFT in terms of its value and uses to society.

The Kalman filter starts with a simple state equation relating a column vector of n_z measurements, $z(t)$, at time t, to a column vector of n_x states, $x(t)$, as seen in Equation 11.1. The measurement matrix, $H(t)$, has n_z rows and n_x columns and simply relates the system states linearly to the measurements [1]. $H(t)$ will usually not change with time but it is left as a time variable for generality. The column vector $w(t)$ has n_z rows and represents the measurement noise, the expected value of which will play an important part in the sensitivity of the Kalman filter to the measurement data.

$$z(t) = H(t)\, x(t) + w(t) \tag{11.1}$$

It is not possible statistically to reduce the difference between $z(t)$ and $Hx(t)$ below the measurement noise described by the elements of $w(t)$. But, the states in $x(t)$ will generally have significantly less noise than the corresponding measurements due to the "smoothing" capability of the Kalman filter. We can therefore define a quadratic cost function in terms of the state to measurement error to be minimized by optimizing the state vector.

$$J(N) = \frac{1}{2} \left[\begin{bmatrix} z(1) \\ z(2) \\ \vdots \\ z(N) \end{bmatrix} - \begin{bmatrix} H(1) \\ H(2) \\ \vdots \\ H(N) \end{bmatrix} x \right]^H \begin{bmatrix} R(1) & 0 & \cdots & 0 & 0 \\ 0 & R(2) & 0 & \cdots & 0 \\ & & & \vdots & \\ 0 & 0 & \cdots & 0 & R(N) \end{bmatrix}^{-1} \left[\begin{bmatrix} z(1) \\ z(2) \\ \vdots \\ z(N) \end{bmatrix} - \begin{bmatrix} H(1) \\ H(2) \\ \vdots \\ H(N) \end{bmatrix} x \right] \tag{11.2}$$

The cost function is defined over N iterations of the filter and can be more compactly written as

$$J(N) = \frac{1}{2} \left[z^N - H^N x \right]^H \left[R^N \right]^{-1} \left[z^N - H^N x \right] \tag{11.3}$$

where $J(k)$ is a scalar, z^N is kn_z rows by 1 column ($kn_z \times 1$), H^N is ($kn_z \times n_x$), R^N is ($kn_z \times kn_z$), and x is ($n_x \times 1$). Clearly, R^N can be seen as $E\{w(t)w(t)^H\}$ giving a diagonal matrix for uncorrelated Gaussian white noise. Instead of solving for the FIR filter coefficients which best relates an input signal and output signal in the block least-squares algorithm presented in Section 9.1, we are solving for the optimum state vector which fits the measurements $z(t)$. Granted, this may not be very useful for sinusoidal states unless the frequency is very low such that the N measurements cover only a small part of the wavelength. It could also be said that for digital dynamic system modeling, one must significantly oversample the time updates for Kalman filters to obtain good results. This was presented for a mass-spring dynamic system in Section 5.1.

The cost function is a quadratic function of the state vector to be optimized as well as the error. Differentiating with respect to the state, we have

$$
\frac{\partial J(N)}{\partial x} = -\frac{1}{2}\left[H^N\right]^H\left[R^N\right]^{-1}\left[z^N - H^N x\right] - \frac{1}{2}\left[z^N - H^N x\right]^H\left[R^N\right]^{-1}\left[H^N\right]
$$

$$
= \left[H^N\right]^H\left[R^N\right]^{-1}\left[H^N\right]x - \left[H^N\right]^H\left[R^N\right]^{-1}z^N
$$

(11.4)

where a second derivative with respect to x is seen to be positive definite, indicating a "concave up" error surface. The minimum error is found by solving for the least-squared error estimate for the state vector based on N observations, $x'(N)$, which gives a zero first derivative in Equation 11.4.

$$
x'(N) = \left\{\left[H^N\right]^H\left[R^N\right]^{-1}\left[H^N\right]\right\}^{-1}\left[H^N\right]^H\left[R^N\right]^{-1}z^N
$$

(11.5)

The matrix term in the braces which is inverted in Equation 11.5 can be seen as the covariance matrix of the state vector x prediction error. The state prediction error is

$$
e(N) = x - x'(N)
$$

$$
= x - \left\{\left[H^N\right]^H\left[R^N\right]^{-1}\left[H^N\right]\right\}^{-1}\left[H^N\right]^H\left[R^N\right]^{-1}\left[H^N x + w^N\right]
$$

$$
= x - \left[H^N\right]^{-1}\left[R^N\right]\left[H^N\right]^{-H}\left[H^N\right]^H\left[R^N\right]^{-1}\left[H^N x + w^N\right]
$$

$$
= -\left\{\left[H^N\right]^H\left[R^N\right]^{-1}\left[H^N\right]\right\}^{-1}\left[H^N\right]^H\left[R^N\right]^{-1}w^N
$$

(11.6)

making the covariance matrix of the state prediction error

$$
P(N) = E\left\{e(N)e(N)^H\right\}
$$

$$
= \left\{\left[H^N\right]^H\left[R^N\right]^{-1}\left[H^N\right]\right\}^{-1}\left[H^N\right]^H\left[R^N\right]^{-1}R^N\left[R^N\right]^{-1}H^N\left\{\left[H^N\right]^H\left[R^N\right]^{-1}\left[H^N\right]\right\}^{-1}
$$

(11.7)

$$
= \left\{\left[H^N\right]^H\left[R^N\right]^{-1}\left[H^N\right]\right\}^{-1}
$$

For a recursive update, we now augment the vectors and matrices in (11.2 and 11.3) for the $N + 1$st iteration and write the state prediction error as an inverse covariance matrix.

$$
P(N+1)^{-1} = \left[H^{N+1}\right]^H\left[R^{N+1}\right]^{-1}\left[H^{N+1}\right]
$$

$$
= \left[[H^N]^H\ H(N+1)^H\right]\begin{bmatrix}R^N & 0 \\ 0 & R(N+1)\end{bmatrix}^{-1}\begin{bmatrix}H^N \\ H(N+1)\end{bmatrix}
$$

$$
= \left[H^N\right]^H\left[R^N\right]^{-1}\left[H^N\right] + H(N+1)^H R(N+1)^{-1} H(N+1)
$$

$$
= P(N)^{-1} + H(N+1)^H\ R(N+1)^{-1}\ H(N+1)
$$

(11.8)

Equation 11.8 is of the form where the matrix inversion lemma (see Section 10.1) can be applied to produce a recursive update for $P(N + 1)$, rather than its inverse.

$$
P(N+1) = P(N) - P(N)H^H(N+1)\left[R(N+1) + H(N+1)P(N)H^H(N+1)\right]^{-1}H(N+1)P(N)
$$

(11.9)

The quantity in the square brackets which is inverted is called the covariance of the measurement prediction error. The measurement prediction error is found by combining Equations 11.1 and 11.5.

$$
\begin{aligned}
\zeta(N+1) &= z(N+1) - z'(N+1) \\
&= z(N+1) - H(N+1)\,x'(N) \\
&= H(N+1)\,x + w(N+1) - H(N+1)\,x'(N) \\
&= H(N+1)[x - x'(N)] + w(N+1) \\
&= H(N+1)\,e(N) + w(N+1)
\end{aligned}
\tag{11.10}
$$

The covariance of the measurement prediction error, $S(N+1)$ (also known as the covariance of the innovation), is seen to be

$$
\begin{aligned}
S(N+1) &= E\left\{\zeta(N+1)\,\zeta^H(N+1)\right\} \\
&= E\left\{\left[H(N+1)e(N) + w(N+1)\right]\left[e^H(N)H^H(N+1) + w^H(N+1)\right]\right\} \\
&= H(N+1)P(N)H^H(N+1) + R(N+1)
\end{aligned}
\tag{11.11}
$$

We can also define the update gain, or Kalman gain, for the recursion as

$$
K(N+1) = P(N)H^H(N+1)\left[S(N+1)\right]^{-1}
\tag{11.12}
$$

where Equations 11.11 and 11.12 are used to simplify and add more intuitive meaning to the state error covariance inverse update in Equation 11.9.

$$
P(N+1) = P(N) - K(N+1)S(N+1)K^H(N+1)
\tag{11.13}
$$

There are several other forms that update the inverse state error covariance. The state vector update is

$$
\begin{aligned}
x'(N+1) &= P(N+1)\left[H^{N+1}\right]^H\left[R^{N+1}\right]^{-1}z^{N+1} \\
&= P(N+1)\left[H^{N^H}\ H(N+1)^H\right]\begin{bmatrix} R^N & 0 \\ 0 & R(N+1) \end{bmatrix}^{-1}\begin{bmatrix} z^N \\ z(N+1) \end{bmatrix} \\
&= P(N+1)\left[H^N\right]^H\left[R^N\right]^{-1}z^N + P(N+1)H(N+1)^H\,R(N+1)^{-1}\,z(N+1) \\
&= \left[I - K(N+1)H(N+1)\right]P(N)\left[H^N\right]^H\left[R^N\right]^{-1}z^N + P(N+1)H(N+1)^H\,R(N+1)^{-1}z(N+1) \\
&= \left[I - K(N+1)H(N+1)\right]x'(N) + K(N+1)z(N+1)
\end{aligned}
\tag{11.14}
$$

where it is straightforward to show that $P(N+1) = [I - K(N+1)H(N+1)]\,P(N)$ and that the Kalman gain can be written as $K(N+1) = P(N+1)H^H(N+1)R^{-1}(N+1)$. The update recursion for the state vector can now be written in a familiar form of old state vector minus the Kalman gain times an error.

$$
x'(N+1) = x'(N) + K(N+1)\zeta(N+1)
\tag{11.15}
$$

All we have done here is applied the RLS algorithm to a state vector, rather than an FIR coefficient vector, as seen in Section 10.1. The basis function for the filter here is the measurement matrix defined in Equation 11.1, and the filter outputs are the measurement predictions defined in Equation 11.10. The RLS algorithm solves for the state vector which gives the least-squared error for measurement predictions. As seen in Equation 11.5, the least-squares solution normalizes the known measurement error R^N, and as seen in Equation 11.7, as the number of observations N becomes large, the variance of the state error becomes smaller than the variance of the measurement error. This smoothing action of the RLS algorithm on the state vector is a valuable technique to obtain good estimates of a state trajectory, such as velocity or acceleration, given a set of noisy measurements. A summary of the RLS state vector filter is given in Table 11.1.

The Kalman filter is just a recursive kinematic time update form of an RLS filter for dynamic state vectors. Examples of dynamic state vectors include the position, velocity, acceleration, jerk, and so on, for a moving object such as a satellite or an aircraft. However, any dynamic data, such as process control information (boiler temperature, pressure, etc.), or even mechanical failure trajectories can be tracked using a dynamic state Kalman filter. The Kalman filter updates for the dynamic state vector are decidedly different from RLS, which is why we will refer to the dynamic state vector RLS filter specifically as Kalman filtering. It is assumed that an underlying kinematic model exists for the time update of the state vector. Using Newtonian physics, the kinematic model for a position–velocity–acceleration type state vector is well known. The modeled state vector update can be made for time $N+1$ using the information available at time N [2].

$$x(N+1) = F(N+1)x(N) + v(N) + G(N)u(N) \tag{11.16}$$

The term $v(N)$ in Equation 11.16 represents the *process noise* of the underlying kinematic model for the state vector updates, while $G(N) u(N)$ represent a control signal input, which will be left to Part V of this text. The matrix $F(N+1)$ is called *the state transition matrix* and is derived from the kinematic model. As with the measurement matrix $H(N+1)$, usually the state transition matrix is constant with time, but we allow the notation to support a more general context. Consider the case of a state vector comprised of a position, velocity, and acceleration. Proceeding without the control input, we note that our best "*a priori*" (meaning before the

TABLE 11.1
The RLS State Vector Filter

Description	Equation
Basis function (measurement matrix)	$H(N+1)$
Kalman gain update	$K(N+1) = \dfrac{P(N)H^H(N+1)}{R(N+1)+H(N+1)P(N)H^H(N+1)}$
	$= P(N)H^H(N+1)\left[S(N+1)\right]^{-1}$
Inverse autocorrelation matrix update	$P(N+1) = \left(H^{N+1^H}\left[R^N\right]^{-1}H^{N+1}\right)^{-1}$
	$= P(N) - K(N+1)S(N+1)K^H(N+1)$
Measurement prediction error	$\zeta(N+1) = z(N+1) - z'(N+1)$
	$= z(N+1) - H(N+1)x'(N)$
Optimal state vector update using prediction error	$x'(N+1) = x'(N) + K(N+1)\zeta(N+1)$

state is updated with the least-squared error recursion) update for the state at time $N+1$, given information at time N is

$$x'(N+1\mid N) = F(N)x(N\mid N)$$

$$= \begin{bmatrix} 1 & T & \frac{1}{2}T^2 \\ 0 & 1 & T \\ 0 & 0 & 1 \end{bmatrix} \begin{bmatrix} x_{N\mid N} \\ \dot{x}_{N\mid N} \\ \ddot{x}_{N\mid N} \end{bmatrix} \tag{11.17}$$

where T is the time interval of the update in seconds. The process noise is not included in the *a priori* state vector prediction since the process noise is assumed uncorrelated from update to update. However, the state prediction error variance must include the process noise covariance. A typical kinematic model for a position, velocity, and acceleration state vector would assume a random change in acceleration from state update to state update. This corresponds to a constant white jerk variance* that results in an acceleration which is an integral of the zero-mean Gaussian white jerk. Assuming a piecewise constant acceleration with a white jerk allows the state transition in Equation 11.17, where the process noise variance for the acceleration is $E\{v(N)v^H(N)\} = \sigma_v^2$. The acceleration process variance scales to velocity by a factor of T, and to position by a factor of $1/2\,T^2$ leading to the process noise covariance $Q(N)$ in the following equation:

$$Q(N) = \begin{bmatrix} \frac{1}{2}T^2 \\ T \\ 1 \end{bmatrix} \sigma_v^2 \begin{bmatrix} \frac{1}{2}T^2 & T & 1 \end{bmatrix} = \begin{bmatrix} \frac{1}{4}T^4 & \frac{1}{2}T^3 & \frac{1}{2}T^2 \\ \frac{1}{2}T^3 & T^2 & T \\ \frac{1}{2}T^2 & T & 1 \end{bmatrix} \sigma_v^2 \tag{11.18}$$

The process noise Q, state transition F and measurement matrix H are usually constants, but our notation allows time variability. If one were to change the update rate T for the Kalman filter, Q must be rescaled. There are quite a number of process noise assumptions which can be implemented to approximate the real situation for the state kinematics. Bar-Shalom and Li provide a detailed analysis of various process noise assumptions. An *a priori* state error covariance can be written as

$$P'(N+1\mid N) = F(N)P(N\mid N)F^H(N) + Q(N) \tag{11.19}$$

The *a priori* state error estimate is then used to update the innovation covariance

$$S(N+1) = H(N+1)P'(N+1\mid N)H^H(N+1) + R(N+1) \tag{11.20}$$

and finally to update the Kalman gain.

$$K(N+1) = P'(N+1\mid N)H^H(N+1)\left[S(N+1)\right]^{-1} \tag{11.21}$$

Given the latest measurement $z(N+1)$ at time $N+1$, an *a posteriori* update of the least-squared error state vector and state vector error covariance are completed as given in Equations 11.22 and 11.23.

$$x(N+1\mid N+1) = x'(N+1\mid N) + K(N+1)\{z(N+1) - H(N+1)x'(N+1\mid N)\} \tag{11.22}$$

---

* Not to be confused with a relentless obnoxious Caucasian (see Section 5.2).

$$P(N+1 \mid N+1) = P'(N+1 \mid N) - K(N+1)S(N+1)K^H(N+1) \tag{11.23}$$

Equations 11.17 through 11.23 constitute a typical Kalman filter for tracking a parameter along with its velocity and acceleration as a function of time. There are countless applications and formulations of this fundamental processing solution which can be found in many excellent texts in detail well beyond the scope of this book. But it is extremely useful to see a derivation of the Kalman filter in the context of the more general RLS algorithm, as well as its similarities and differences with other commonly used adaptive filtering algorithms.

One of the more useful aspects of Kalman filtering is found not in its ability to smooth the state vector (by reducing the affect of measurement noise), but rather in its predictive capabilities. Suppose we are trying to intercept an incoming missile and are tracking the missile trajectory from a series of position measurements obtained from a radar system. Our antimissile requires about 30 s lead time to fly into the preferred intercept area. The preferred intercept area is determined by the highest probability of hit along a range of possible intercept positions defined by the trajectories of the two missiles. Equations 11.17 through 11.19 are used to determine the incoming missile trajectory and variance of the trajectory error, which in three dimensions is an ellipsoid centered around a given predicted future position for the incoming missile. The further ahead in time one predicts the incoming missile's position, the larger the size of prediction error ellipsoid becomes. The state error prediction can be seen in this case as a three-dimensional Gaussian probability density "cloud," where the ellipsoid is a one-standard deviation contour. Using Baye's rule and other hypothesis scoring methods, one can solve the kinematic equations in a computer for the launch time giving rise to the most likely intercept. This is not always easy or straightforward because the tracking filter is adaptive and the states and measurement noise are dynamically changing with time.

Some of the best evidence of the difficulty of missile defense was seen with the Patriot system during the Gulf War of 1991. To be fair, the Patriot was designed to intercept much slower aircraft, rather than ballistic missiles. But, in spite of the extreme difficulty of the task, the missile defense system brought to the attention of the general public the concept of an intelligent adaptive homing system using Kalman filters. Track-intercept solutions are also part of aircraft collision avoidance systems and may soon be part of intelligent highway systems for motor vehicles. Using various kinematic models, Kalman filters have been used routinely in financial markets, prediction of electrical power demand, and in biological/environmental models for predicting the impact of pollution and species populations. If one can describe a phenomenon using differential equations and measure quantities systematically related to those equations, tracking and prediction filters can be of great utility in the management and control of the phenomena.

Table 11.2 summarizes the Kalman filter. The type of measurements used determine the measurement matrix $H(N+1)$ and measurement noise covariance $R(N+1)$. The state kinematic model determines the process noise covariance $Q(N+1)$ and state transition matrix $F(N+1)$. As noted earlier, H, R, Q, F, σ_v^2, σ_w^2, and T are typically constant in the tracking filter algorithm. As can be seen from Table 11.2, the Kalman filter algorithm is very straightforward. One makes *a priori* state and state covariance predictions, measures the resulting measurement prediction error, and adjusts the state updates according to the error and the define measurement and process noise variances. However, the inverse state error covariance update given in the table can be reformulated to promote lower numerical roundoff error. The expression in Equation 11.24 is algebraically equivalent to that given in Table 11.2, but its structure, known as Joseph form, promotes symmetry and reduces numerical noise.

$$P(N+1 \mid N+1) = \left[I - K(N+1)\,H(N+1) \right] P(N+1 \mid N) \left[I - K(N+1)\,H(N+1) \right]^H$$
$$+ K(N+1)\,R(N+1)\,K(N+1)^H \tag{11.24}$$

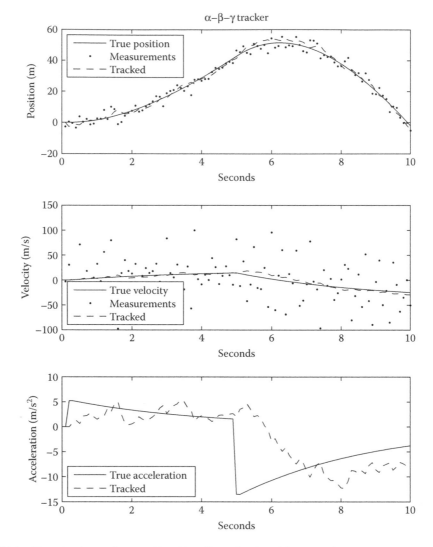

FIGURE 11.1   Nonadaptive α–β–γ tracking filter (from Section 5.2) where the measurement noise is 3 and the process noise is 13.

The significance of determining the Kalman filter process noise which gives the same filter gain as the α–β–γ tracking filter when the measurement noise is high is that this gives an indication of the minimum process noise needed to track the target maneuvers without significant overshoot. Note that for a near zero measurement noise, the gain vectors are only identical for the same process noise. Given a reasonable minimum process noise for the Kalman filter, it gives superior tracking performance to the α–β–γ tracking filter when the measurement noise is low as seen in Figures 11.4 and 11.5.

The examples in Figure 11.1 through 11.5 and in Section 5.2 indicate that the α–β–γ tracking filter is the algorithm of choice when the measurements are noisy. Unfortunately, there is not an algebraic expression which can tell us simply how to set the process noise in a Kalman filter for optimum performance. The reason is that the Kalman gain is not only a function of measurement noise, process noise, and time update (as is the case with the α–β–γ tracking filter), but it also depends on the state prediction error covariance, which is a function of target maneuvers. In most situations, one knows a reasonable expected target track maneuverability and the expected

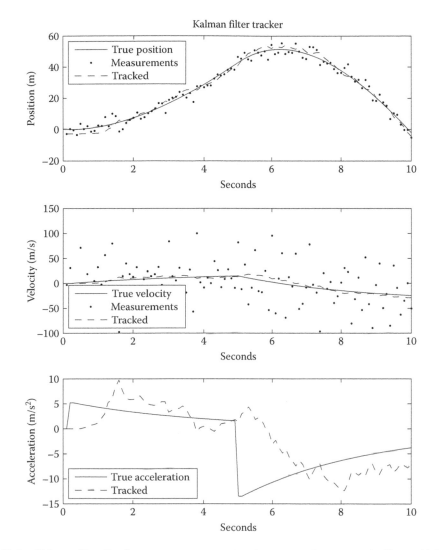

FIGURE 11.2 Kalman filter for the same measurement and process noise as seen in Figure 11.1 showing slightly noisier, but more reactive response.

measurement noise *a priori*. This information allows one to execute simulations for a covariance analysis of the Kalman filter's performance on the expected data. With the measurement noise small (or near zero), one can set the Kalman filter process noise to almost any large value and the filter will perform nearly perfectly. With high measurement noise, it is easier to determine the minimum Kalman process noise empirically for a given target maneuver. Setting the process noise to the maximum expected acceleration is a good starting point (and ending point for the α–β–γ tracking filter), but the Kalman filter generally allows a lower process noise due to the enhanced ability of the adaptive updates to help the state error converge rapidly to target maneuvers.

The α–β–γ tracking filter is a Kalman filter with fixed state error covariance. In Section 5.2 we introduced the α–β–γ tracking filter as an important application of a state-variable filter and its derivation was referenced to this section. Its derivation is rather straightforward, but algebraically tedious. We show the approach of Bar-Shalom [2] to demonstrate the technique. But more importantly, the solution for the α–β–γ tracking filter shows that α, β, and γ cannot be chosen

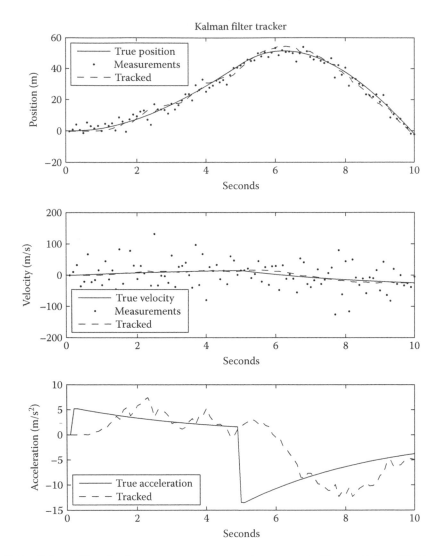

FIGURE 11.3 Kalman filter for the same measurement noise as seen in Figures 11.1 and 11.2 but this time with a process noise of 1 showing smoother, less reactive response.

independently. For a given amount of position noise reduction (as determined by $\alpha < 1$), β and γ are systematically determined from the measurement error σ_w and update time T. Setting α therefore also sets the process noise, and the state prediction error covariances to constant "steady-state" values. However, the filter will only converge to the prescribed steady-state error covariance if all the kinematic assumptions are true. If the tracking target makes an unexpected maneuver, the state error will increase for a while as the state vectors "overshoot" the measurements temporarily. This can be seen in Figure 5.5 where a "sluggish" track is produced from using too small a process noise in the α–β–γ tracking filter.

Combining Equations 11.19, 11.21, and 11.23 we can write an expression for the updated state prediction error covariance P assuming steady-state conditions.

$$P = F^{-1}\left(P' - Q\right)F^{-H} = \left(I - KH\right)P' \tag{11.25}$$

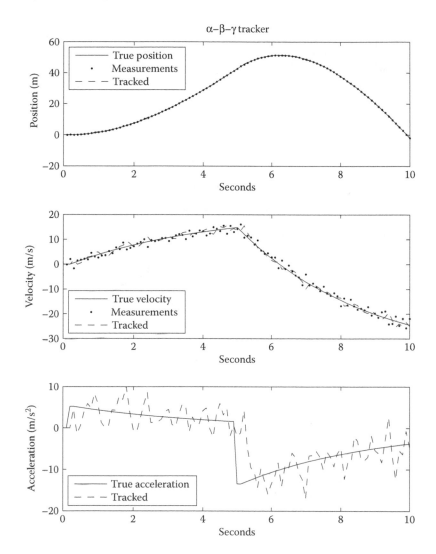

FIGURE 11.4 Reducing the measurement noise to 0.1 improves the α–β–γ tracking filter with the same process noise of 13 as seen in Figure 11.1.

If we simplify the problem by assuming only the α and β tracking gains and a piecewise constant acceleration model, we have the following matrix equation to solve [2]:

$$\begin{bmatrix} 1 & -T \\ 0 & 1 \end{bmatrix} \begin{bmatrix} p_{1'1} - \dfrac{T^4}{4}\sigma_v^2 & p_{1'2} - \dfrac{T^3}{2}\sigma_v^2 \\ p_{2'1} - \dfrac{T^3}{2}\sigma_v^2 & p_{2'2} - T^2\sigma_v^2 \end{bmatrix} \begin{bmatrix} 1 & 0 \\ -T & 1 \end{bmatrix} = \begin{bmatrix} 1 & -k_1 & 0 \\ -k_2 & 1 \end{bmatrix} \begin{bmatrix} p_{1'1} & p_{1'2} \\ p_{2'1} & p_{2'2} \end{bmatrix} \tag{11.26}$$

Equation 11.26 is best solved by simply equating terms. After some simplification, we have

$$k_1 p_{1'1} = 2T\, p_{1'2} - T^2 p_{2'2} + \dfrac{T^4}{4}\sigma_v^2 \tag{11.27}$$

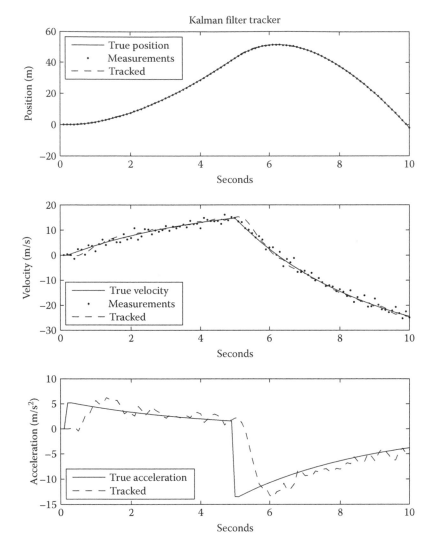

FIGURE 11.5   The Kalman filter with 0.1 measurement noise and a process noise of 1 gives improved performance (lower state error yet fast tracking) over the α–β–γ tracker on the same measurements but with a process noise of 13 as seen in Figure 11.4.

$$k_1 p_{1'2} = T p_{2'2} - \frac{T^3}{2} \sigma_v^2 \qquad (11.28)$$

and

$$k_2 p_{1'2} = T^2 \sigma_v^2 \qquad (11.29)$$

Solving for the predicted state covariance elements we have

$$p_{1'1} = \frac{k_1}{1 - k_1} \sigma_w^2 = \frac{\alpha}{1 - \alpha} \sigma_w^2 \qquad (11.30)$$

and

$$p_{1'2} = p_{2'1} = \frac{k_2}{1 - k_1} \sigma_w^2 \tag{11.31}$$

Finally, solving for the p'_{22} term we have

$$p_{2'2} = \left(\frac{k_1}{T} + \frac{k_2}{2} \right) p_{1'2} \tag{11.32}$$

Combining Equations 11.27 through 11.32 and after some algebraic cancellations, we arrive at the following bi-quadratic equation for the filter gains.

$$k_1^2 - 2Tk_2 + Tk_1k_2 + \frac{T^2}{4}k_2^2 = 0 \tag{11.33}$$

The filter gains in terms of α and β are simply $k_1 = \alpha$ and $k_2 = \beta/T$. Equation 11.33 reduces to

$$\alpha^2 - 2\beta + \alpha\beta + \frac{\beta^2}{4} = 0 \tag{11.34}$$

Solving for α in terms of β using the quadratic formula we have

$$\alpha = \sqrt{2\beta} - \frac{\beta}{2} \tag{11.35}$$

and solving for β in terms of α we have

$$\beta = 4 - 2\alpha - 4\sqrt{1 - \alpha} \tag{11.36}$$

For the α–β–γ filter, it can be shown (with some algebraic complexity) that $\gamma = \beta^2/\alpha$. Another important relation is seen when equating terms for p'_{12}.

$$
\begin{aligned}
p_{1'2} &= \frac{T^2 \sigma_v^2}{k_2} = \frac{k_2}{1 - k_1} \sigma_w^2 \\
\frac{T^2 \sigma_v^2}{\beta/T} &= \frac{\beta/T}{1 - \alpha} \sigma_w^2
\end{aligned}
\tag{11.37}
$$

Equation 11.37 leads to an expression for the track maneuverability index λ_M.

$$\lambda_M = \frac{T^2 \sigma_v}{\sigma_w} \tag{11.38}$$

The results of the derivation above are also presented in Equations 5.28 through 5.32. The maneuverability index determines the "responsiveness" of the tracking filter to the measurements. For the fixed gain α–β–γ filter one estimates the maximum target acceleration to determine the process noise standard deviation σ_v. Given the measurement noise standard deviation σ_w and track update time T, the maneuverability index λ_M is determined as well as α, β, and γ (see Section 5.2). The α–β–γ filter is an excellent algorithm choice for cases where the kinematics are known and the measurements are relatively noisy. Given unknown target maneuverability and/or low noise measurements, the adaptive Kalman filter is a better tracking algorithm choice, because the process noise can be significantly lowered without sacrificing responsiveness to give more noise reduction in the tracking state outputs. While derivation of the α–β–γ filter is quite tedious, knowledge of the Kalman filter equations allows a straightforward solution. Tracking filters are an extremely important and powerful signal processing tool. Whenever one measures a quantity with a known measurement error statistic, and observes the quantity changing deterministically over time, one is always interested in predicting when that quantity reaches a certain value, and with what statistical confidence.

11.2 IIR FORMS FOR LMS AND LATTICE FILTERS

The least-squared error system identification and signal modeling algorithms presented in Chapters 9 and 10 were limited to FIR filter structures. However, modeling systems and signals using infinite impulse response (IIR) digital filter is quite straightforward. Because an IIR filter with adaptive coefficients has the potential to become unstable, one must carefully constrain the adaptive IIR filter algorithm. If we are performing system identification where both the input and output signals are completely known, stability is less of a concern, provided that the unknown system to be identified is stable. When only the output signal is available and we are trying to model the signal as the output of an IIR filter driven by white noise, the stability issues can be considerably more difficult. Recall from Section 3.1 that FIR filters are often referred to as moving average, or MA, and are represented on the complex z-plane as a polynomial where the angles of the zeros determine the frequencies where the FIR filter response is attenuated, or has a spectral dip. In Section 3.2 the IIR filter is presented as a denominator polynomial in the z-domain. The angles of the zeros of the IIR polynomial determine the frequencies where the filter's response is amplified, or has a spectral peak. The IIR zeros are called the filter's poles because they represent the frequencies where the response has a peak. IIR filters with only a denominator polynomial feedback the past values of the output signal in the calculation of the current output, and as a consequence, are often called autoregressive, or AR filters.

The most general form of an IIR filter has both a numerator polynomial and a denominator polynomial, and is usually referred to as a pole–zero or ARMA filter. The zeros of the numerator polynomial are the filter zeros and the zeros of the denominator polynomial are the filter poles. For the IIR filter to be stable, the magnitude of all the poles must be less than unity. This ensures that the output feedback of some particular output signal sample eventually reverberates down to zero as it is fed back in the autoregressive generation of the current IIR output. Therefore, the denominator part of a stable IIR filter must be a minimum phase polynomial with all its zeros (the system poles) inside the unit circle.

Figure 11.6 depicts system identification of some unknown system represented by the ratio of z-domain polynomials $B[z]/A[z]$ where the zeros of $B[z]$ are the system zeros and the zeros of $A[z]$ are the system poles. The adaptive system identification is executed by a pair of adaptive filters which are synchronized by a common error signal $e[n]$, and the input $x[n]$ and output $y[n]$. Recall from Equation 3.19 the difference equation for an ARMA filter assuming a numerator polynomial order of Q (Q zeros) and a denominator polynomial order P (P poles).

$$x[n]b_0 + x[n-1]b_1 + \cdots + x[n-Q]b_Q = y[n] + y[n-1]a_1 + \cdots + y[n-P]a_P \qquad (11.39)$$

We can model the actual ARMA filter in Equation 11.39 to make a linear prediction of the output $y'[n]$ using the available signals except the most recent output sample $y[n]$ (to avoid a trivial solution).

$$y'[n] = x[n]b_0' + x[n-1]b_1' + \cdots + x[n-Q]b_Q' - y[n-1]a_0' - \cdots - y[n-P]a_P' \qquad (11.40)$$

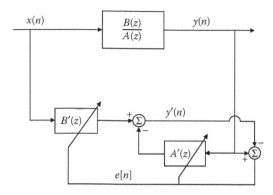

FIGURE 11.6 An ARMA LMS system identification (Wiener filter) operation given both the input and output signals of the unknown system to be modeled.

The prediction error is simply

$$
\begin{aligned}
e[n] &= y[n] - y'[n] \\
&= y[n] - x[n]b'_0 - x[n-1]b'_1 - \cdots - x[n-Q]b'_Q \\
&\quad + y[n-1]a'_1 + \cdots + y[n-P]a'_P
\end{aligned}
\tag{11.41}
$$

The gradient of the error with respect to $B'[z] = b'_0 + b'_1 z^{-1} + \cdots + b'_Q z^{-Q}$ is negative and the gradient with respect to $A'[z] = a'_1 z^{-1} + a'_2 z^{-2} + \cdots + a'_P z^{-P}$ is positive. This sign distinction will be important to the LMS coefficient updates. However, we note that the model orders P and Q are assumed known and the AR model polynomial $A'[z]$ does not have the leading coefficient of unity seen in $A[z]$. The squared error for this case is a quadratic function of both sets of coefficients, and the second derivative is also positive indicating that the least-squared error solution corresponds to the set of ARMA coefficients which gives a zero gradient. When the input and output signals are known, the least-squared error solution applies only to the chosen model orders P and Q. If both P and Q are chosen not to correspond to the actual unknown system's polynomial orders, one only has a least-squared error solution (best fit) for the chosen model orders, not overall. Recall that for the FIR system identification case, one could choose a model order much higher than the actual system model order, and the unneeded coefficients would nicely converge to zero. This unfortunately is not the case for ARMA filter models. Each combination of P and Q will give a least-squared error solution, but only for that particular case. If all possible combinations of model orders are tried, the actual unknown system's polynomial orders will correspond to the overall least-squares solution. Obviously, it would be very valuable to have a straightforward methodology to determine the optimal ARMA model efficiently. It can be done manually using a careful analysis of the ARMA output autocorrelation sequence. Beyond the actual MA model order Q, the autocorrelation time lags will only have the AR part of the system. One can first whiten the AR part of the signal and then work back to estimate the MA part of the system when only the output signal is available.

But since we have both input and output signals with known variances, we can directly apply an LMS algorithm to find the coefficients for the chosen model order. The FIR or MA part of the model is updated at time n using

$$
b_{i',n} = b_{i',n-1} + 2\mu_b \, x[n-i]e[n] \quad \mu_b \leq \frac{1}{Q \, E\{x[n]^2\}}
\tag{11.42}
$$

and the IIR or AR part of the ARMA model is updated using

$$a_{j',n} = a_{j',n-1} - 2\mu_a y[n-j]e[n] \quad \mu_a \leq \frac{1}{P\,E\{y[n]^2\}} \tag{11.43}$$

The sign of the updates in Equations 11.42 and 11.43 are due to the need to adjust the coefficients in the opposite direction of the gradient as a mean to move to the minimum squared error. Stability of the algorithm is simplified by knowing the variances of the input and output, and by the structure in Figure 11.6 where $y'[n]$ is not fed back into the adaptive filter (it is only used in calculating the error signal). Doing so could cause instability if one or more of the zeros of $A'[z]$ move outside the unit circle during adaptation. So long as the input $x[n]$ and output $y[n]$ are stationary allowing an accurate μ_b and μ_a, we are assured stable LMS updates and convergence to something very close to the unknown system, exactly if the model orders are known *a priori*.

Error signal bootstrapping is used for ARMA signal modeling [3, 4]. When only the output signal $y[n]$ is available, we can model the signal as an ARMA process defined as an ARMA filter with unity variance white-noise input, or innovation. This procedure is somewhat tricky, since both an input and output signal are needed for an ARMA model. The input is estimated using a heuristic procedure called *error bootstrapping*. Since the ARMA process innovation is always a unity-variance white-noise signal, we can estimate the input by calculating the prediction error in the absence of the actual input.

$$\begin{aligned} e^b[n] &= y[n] - y^{b'}[n] \\ &= y[n] - e^b[n-1]b_1' - \cdots - e^b[n-Q]b_Q' \\ &\quad + y[n-1]a_1' + \cdots + y[n-P]a_P' \end{aligned} \tag{11.44}$$

The bootstrapped error signal in Equation 11.44 is very close, but not identical to, the linear prediction error in Equation 11.41, except the ARMA input $x[n]$ is replaced by the bootstrapped error $e^b[n]$. Note that the bootstrapped output linear prediction $y^b[n]$ assumes a zero input for $e^b[n]$, allowing a *prediction error prediction* based on a least-squared error assumption. The bootstrapped ARMA input signal can be seen as an *a priori* estimate of the prediction error. The adaptive update for the MA coefficients must use an LMS step size based on the output power since the bootstrapped error will initially be of magnitude of the order of the ARMA output.

$$b_{i,n}' = b_{i,n-1}' + 2\mu_a\, e^b[n-i]\,e[n] \quad \mu_a \leq \frac{1}{Q\,E\{y[n]^2\}} \tag{11.45}$$

The prediction error for the bootstrapped ARMA model $e[n]$, is calculated as

$$\begin{aligned} e[n] &= y[n] - y'[n] \\ &= y[n] - e^b[n]b_0' - e^b[n-1]b_1' - \cdots - e^b[n-Q]b_Q' \\ &\quad + y[n-1]a_1' + \cdots + y[n-P]a_P \end{aligned} \tag{11.46}$$

which is exactly as in the Wiener filtering case in Equation 11.41 except the actual ARMA input $x[n]$ is replaced by the bootstrapped error signal $e^b[n]$. For an AR signal Model, the MA order becomes $Q = 0$, and the bootstrapping procedure reduces to a whitening filter algorithm.

Rescaling Bootstrapped MA Coefficients—To recover the MA coefficients for the ARMA bootstrapped model we must *obtain the proper scaling by multiplying the MA coefficients by the standard deviation of the prediction error.* This compensates for the premise that the ARMA signal is the output of a pole–zero filter driven by unity-variance zero-mean white Gaussian noise. For the LMS and RLS bootstrapped ARMA algorithms, a whitening filter structure is used where the bootstrapped error signal is used to simulate the unavailable input to the unknown ARMA system. Minimizing this error only guarantees a match to the real ARMA system within a linear scale factor. The scale factor is due to the variance of the bootstrapped error not being equal to unity. Multiplying the converged MA coefficients by the square-root of the bootstrapped error variance scales the whitening filter, giving an ARMA signal model assuming a unity-variance white-noise innovation. When the ARMA input signal is available (the Wiener filtering system identification problem), scaling is not necessary for the LMS and RLS algorithms. However, one will see a superior MA coefficient performance for the RLS bootstrapped ARMA algorithm when the bootstrapped error is normalized to unity variance. For the embedded ARMA lattice, a little more complexity is involved in calculating the MA scale.

Figures 11.7 and 11.8 show the results of an ARMA LMS system identification and error bootstrapping simulation. The actual ARMA filter has a pair of complex conjugate poles at ±200 Hz with magnitude 0.98 (just inside the unit circle). A pair of conjugate zeros is at ±400 Hz also of magnitude 0.95. In the simulation, the sample rate is 1024 Hz making the pole angles ±1.227 radians and the zero angle ±2.454 radians on the complex z-plane (see Sections 2.1 through 2.3 and Chapter 3 for more detail on digital filters) [5]. A net input–output gain of 5.0 is applied to the actual ARMA filter giving a numerator polynomial in z of $B[z] = 5.0 + 7.3436z^{-1} + 4.5125z^{-2}$. The actual

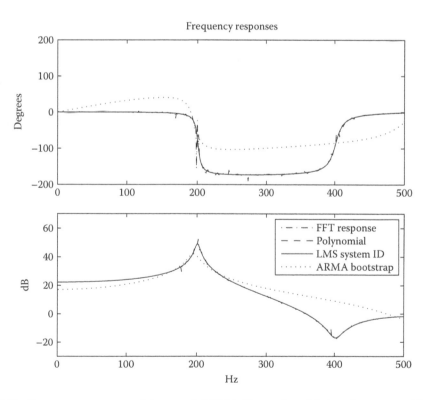

FIGURE 11.7 Frequency responses of the actual ARMA filter and models based on system identification using Weiner filtering (both input and output signals available) and error bootstrapping (only the output signal is available) showing a poor phase response and missed zero for the bootstrapped algorithm.

ARMA denominator polynomial is $A[z] = 1.0 - 0.6603z^{-1} + 0.9604z^{-2}$ and 2000 output samples are generated using a unity variance white-noise input $x[n]$ and the equation below

$$y[n] = 5.0x[n] + 7.3436x[n-1] + 4.5125x[n-2] + 0.6603y[n-1] - 0.9604y[n-2] \qquad (11.47)$$

The frequency responses shown in Figure 11.7 are generated two ways. First, the FFT of the output signal is divided by the FFT of the input signal and the quotient is time averaged (see Section 7.2 for details on transfer functions). Second, the response is calculated directly from the ARMA coefficients by computing the ratio of the zero-padded FFTs of the coefficients directly. This technique is highly efficient and is analogous to computing a z-transform where z is $e^{-j\Omega n}$ (see Section 3.1 for the frequency responses of digital filters). Figure 11.7 clearly shows the input–output signal FFT, the actual ARMA polynomial response, and the LMS Wiener filter response overlaying nearly perfectly. The LMS Wiener filter calculated polynomials are $B'[z] = 5.0000 + 7.3437z^{-1} + 4.5126z^{-2}$ and $A'[z] = 1.0000 - 0.6603z^{-1} + 0.9604z^{-2}$ that are very close to the actual ARMA coefficients.

The bootstrapped ARMA signal model is close near the ARMA peak created by the pole at 200 Hz, but completely misses the zero at 400 Hz. It will be seen below that proper normalization of the bootstrapped error signal to better model the ARMA innovation will result in improved MA coefficient performance for the RLS algorithm. In theory, the bootstrapped ARMA estimate may improve depending on the predictability of the innovation and the effective length of the data memory window. Performance limitations due to slow LMS algorithm convergence are the main reason for high interest in fast adaptive algorithms. The coefficient estimates as a function of algorithm iteration are seen in Figure 11.8.

A fast algorithm allows one to get the most information for a given input–output signal observation period. The LMS bootstrapped ARMA filter had to be slowed considerably by reducing the step

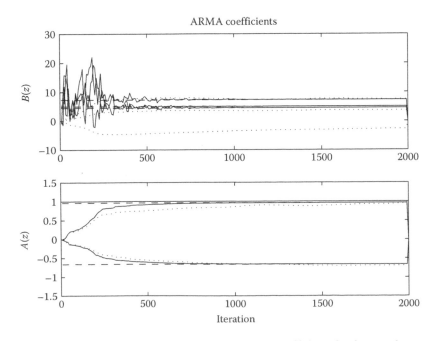

FIGURE 11.8 ARMA numerator ($B[z]$) and denominator ($A[z]$) coefficients for the actual system (--), LMS system identification (–), and ARMA bootstrapped (···) algorithms showing a slight bias to the numerator coefficients of the bootstrapped algorithm.

size to about 1% of its theoretical maximum defined in Equations 11.42, 11.43, and 11.45 to get the performance seen in Figures 11.7 and 11.8. This corresponds to a data memory window for the LMS algorithm of about 100 samples. The bootstrapped ARMA performance seen in Figure 11.8 shows a reasonable AR coefficient response and a very poor MA coefficient response. This is also seen in Figure 11.7 in the reasonable peak alignment but completely missed zero. However, a close examination of the ARMA coefficient response in Figure 11.8 shows a good convergence on the pole magnitude (0.98), but some drift in the pole angle (or frequency). This can be seen as caused by an interaction of the pole with the zero. Indeed, the LMS algorithm for both the AR and MA parts are driven by a single error signal where the input data is not orthogonal in order. The interaction between the AR and MA part for the bootstrapped ARMA signal model is exacerbated by the coupling (due to the least-squared error approximations) in the LMS ARMA algorithm. The poorly-converged bootstrapped ARMA coefficients after 2000 iterations are $B'[z] = 6.7311 + 3.3271z^{-1} - 2.4981z^{-2}$ and $A'[z] = 1.0000 - 0.7046z^{-1} + 0.9537z^{-2}$. As noted earlier, the AR part of the match is reasonable. The zeros of the system cannot be identified until the poles of the system are identified correctly otherwise the poles and zeros interfere with each other during convergence. Given enough iterations, a reasonable ARMA model can be obtained using the LMS, but still we would rather use faster converging algorithms.

The embedding technique allows multiple signal channels to be in the RLS or Lattice algorithms by vectorizing the equations [6,7]. The RLS algorithm of Section 10.1 (see Table 10.1) is the most straightforward to embed an ARMA model. One simply extends the basis vector to include the ARMA filter input and output sequences and correspondingly, extends the coefficient vector to include the AR coefficients.

$$\varepsilon_n = y_n - \varphi_n H$$

$$= y_n - \begin{bmatrix} x_n \ x_{n-1} \ \cdots \ x_{n-Q} \ -y_{n-1} \ -y_{n-2} \ -\cdots \ -y_{n-P} \end{bmatrix} \begin{bmatrix} b_0 \\ b_1 \\ \vdots \\ b_Q \\ a_1 \\ a_2 \\ \vdots \\ a_P \end{bmatrix} \tag{11.48}$$

Using the basis function and coefficient vector in Equation 11.48, one simply executes the RLS algorithm as seen in Table 10.1. For error bootstrapping for ARMA signal models, the bootstrapped error is simply

$$\varepsilon_n^b = y_n - \varphi_n^b H$$

$$= y_n - \begin{bmatrix} 0 \ \varepsilon_{n-1}^b \ \cdots \ \varepsilon_{n-Q}^b \ -y_{n-1} \ -y_{n-2} \ -\cdots \ -y_{n-P} \end{bmatrix} \begin{bmatrix} b_0 \\ b_1 \\ \vdots \\ b_Q \\ a_1 \\ a_2 \\ \vdots \\ a_P \end{bmatrix} \tag{11.49}$$

where ε_n^b is zero on the right-hand side of Equation 11.49. Given the bootstrapped error to model the ARMA process innovation, the linear prediction error is

$$\varepsilon_n = y_n - \varphi_n H$$

$$= y_n - \left[\varepsilon_n^b \ \varepsilon_{n-1}^b \ \cdots \ \varepsilon_{n-Q}^b - y_{n-1} - y_{n-2} - \cdots - y_{n-P} \right] \begin{bmatrix} b_0 \\ b_1 \\ \vdots \\ b_Q \\ a_1 \\ a_2 \\ \vdots \\ a_P \end{bmatrix} \tag{11.50}$$

An example of the enhanced performance of the RLS ARMA algorithm is seen in Figures 11.9 and 11.10, where the RLS ARMA algorithm is applied to the same data as the LMS ARMA example seen in Figures 11.7 and 11.8. The RLS algorithm converges much faster than the LMS algorithm and appears to give generally better ARMA modeling results. This is true even for the bootstrapped error case, which still suffers from MA coefficient modeling difficulty. However, we see that there is much less coupling between the AR and MA parts of the model. Recall that the true ARMA polynomials are $B[z] = 5.0 + 7.3436z^{-1} + 4.5125z^{-2}$ and $A[z] = 1.0 - 0.6603z^{-1} + 0.9604z^{-2}$. The RLS Wiener filter ARMA model is $B[z] = 5.0 + 7.3436z^{-1} + 4.5125z^{-2}$ and $A[z] = 1.0 - 0.6603z^{-1} + 0.9604z^{-2}$. The RLS bootstrapped error ARMA signal model is $B[z] = 8.1381 + 1.6342z^{-1} + 0.4648z^{-2}$ and $A[z] = 1.0 - 0.7002z^{-1} + 0.9786z^{-2}$. The Wiener filter RLS results are practically zero error while the bootstrapped ARMA results still show a weak zero match in the MA part of the ARMA signal model. What is going wrong?

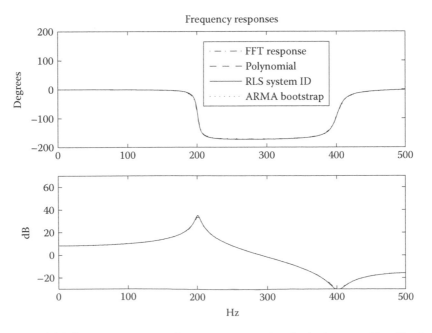

FIGURE 11.9 ARMA frequency responses for the RLS algorithm for both system identification (Weiner filtering) and error bootstrapping showing a phase and zero error for the bootstrapped case.

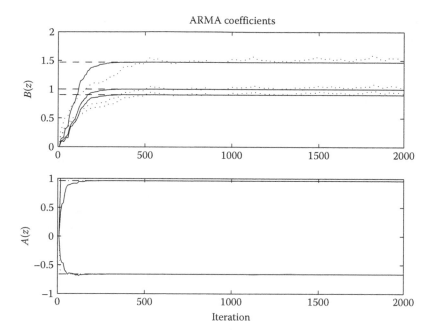

FIGURE 11.10 ARMA numerator ($B[z]$) and denominator ($A[z]$) coefficients for the actual system (- -), RLS system identification (–), and ARMA bootstrapped (· · ·) algorithms showing a significant bias to the numerator coefficients of the bootstrapped algorithm.

Normalizing the bootstrapped error signal to unity variance in the RLS algorithm constrains the bootstrap process to better model the ARMA innovation [8]. This also conditions the prediction error for the algorithm to be better balanced between the MA and AR coefficients, thus allowing vastly improved ARMA signal modeling. The normalized bootstrap error performance can also be seen in the embedded ARMA lattice, presented below. The lattice PARCOR coefficients are naturally normalized by their respective error signal variances. *The ARMA lattice requires simply multiplying the MA coefficients by the square-root of the linear prediction error to produce properly normalized MA coefficients (for $b_0 \neq 1$).* Multiplying the MA coefficients by the standard deviation of the prediction error certainly scales the model so the spectral peaks match well. However, applying this technique directly to the LMS and RLS ARMA bootstrap algorithm does not result in a very good MA coefficient match. *Normalizing the bootstrap error signal to unity variance achieves the same scaling effect in the RLS ARMA model and greatly improves the MA coefficient match.* It can be seen that normalizing the bootstrap error leads to improved conditioning in the adaptive algorithm to balance the adaptive effort between the MA and AR parts of the model. Equation 11.51 shows the basis function φ_n^b used in calculating the bootstrap error $\varepsilon_n^b = \varphi_n^b H$, similar to the unnormalized bootstrap error in Equation 11.49, except σ_b is the square root of the bootstrap error variance.

$$\varphi_n^b = \left[\; 0 \;\; \frac{\varepsilon_{n-1}^b}{\sigma_b} \cdots \frac{\varepsilon_{n-Q}^b}{\sigma_b} \; -y_{n-1} - y_{n-2} - \cdots - y_{n-P} \;\right] \qquad (11.51)$$

The bootstrap error is then used in a recursive estimate for the bootstrap error variance, and subsequently, standard deviation can be calculated. The bootstrap error variance recursion is best initialized to unity as well. The RLS algorithm is very sensitive to the bootstrap error amplitude.

The basis function used in the RLS algorithm is seen in Equation 11.52 as is used to generate the linear prediction error in the RLS as seen in Equation 11.50.

$$\varphi_n = \left[\begin{array}{cccccccc} \dfrac{\varepsilon_n^b}{\sigma_b} & \dfrac{\varepsilon_{n-1}^b}{\sigma_b} & \cdots & \dfrac{\varepsilon_{n-Q}^b}{\sigma_b} & -y_{n-1} & -y_{n-2} & -\cdots & -y_{n-P} \end{array} \right] \tag{11.52}$$

It is both surprising and interesting that simply normalizing the bootstrap error improves the RLS ARMA model result as significantly as shown in Figures 11.11 and 11.12. Why should it matter whether $b_0 = 5$ and $\sigma_b = 1$, or if we have $b_0 = 1$ and $\sigma_b = 5$? Consider that the RLS basis vector and linear prediction error must be separate for the net gain of the MA coefficients to be properly calculated. Since the RLS linear prediction error is not normalized, the MA coefficients are scaled properly in the RLS normalized ARMA bootstrap algorithm without the need to multiply the coefficients by some scale factor afterward. The reason bootstrap error normalization does not help the LMS performance significantly is that much of the optimization in the RLS algorithm is lost in the approximations used to create the simple and robust LMS algorithm.

In this example of normalized bootstrap error RLS ARMA modeling, it is noted that the RLS bootstrap algorithm is very sensitive to initial conditions compared to the highly reproducible results for Wiener filtering, where both input and output are known. Initializing the bootstrap error variance to, say the output signal y_n variance, also gave improved MA coefficient matching results, but not as good as that with an initial unity variance. The ARMA normalized bootstrap results are $B[z] = 5.0114 + 7.5178z^{-1} + 4.3244z^{-2}$ and $A[z] = 1.0000 - 0.6738z^{-1} + 0.9149z^{-2}$, which are nearly as good as some of the Wiener filtering ARMA results.

Embedding an ARMA filter into a lattice structure is very straightforward once one has established a matrix difference equation. Recall that the projection operator framework in Chapter 8 was completely general and derived in complex matrix form (the superscript H denotes transpose plus

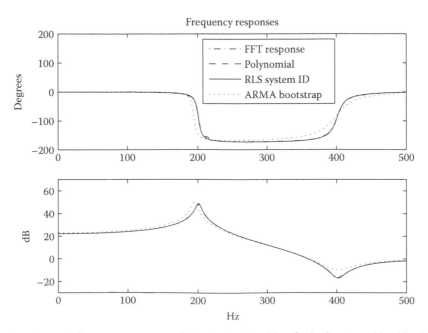

FIGURE 11.11 ARMA frequency responses for the RLS algorithm for both system identification (Weiner filtering) and normalized error bootstrapping showing a near perfect match in magnitude and phase for the normalized error bootstrapping case.

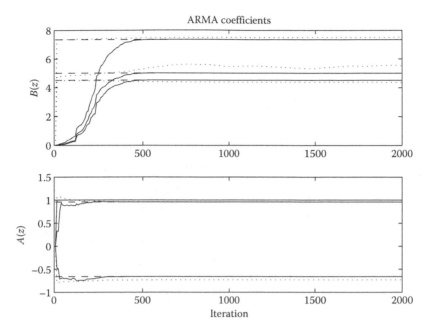

FIGURE 11.12 ARMA numerator ($B[z]$) and denominator ($A[z]$) coefficients for the actual system (--), RLS system identification (–), and ARMA normalized error bootstrapped ($\cdots$) algorithms showing a precise coefficient match even for the normalized error bootstrapped case where only the output signal is available.

complex conjugate). All of the lattice and Levinson recursions in Chapter 10 are also presented in complex matrix form. We start by writing the ARMA difference equations in matrix form.

$$\begin{bmatrix} \varepsilon_n^y \\ \varepsilon_n^x \end{bmatrix} = \begin{bmatrix} y_n \\ x_n \end{bmatrix} + \sum_{i=1}^{M} \begin{bmatrix} a_i & -b_i \\ -c_i & d_i \end{bmatrix} \begin{bmatrix} y_{n-i} \\ x_{n-i} \end{bmatrix} \tag{11.53}$$

It is straightforward to show that $d_i = a_i/b_0$ and $c_i = b_i/b_0$ where $i = 1, 2, \ldots, M$, $d_0 = 1/b_0$, and $a_0 = c_0 = 1$. The forward prediction error ARMA lattice recursion is seen to be

$$\begin{bmatrix} \varepsilon_{p+1,n}^y \\ \varepsilon_{p+1,n}^x \end{bmatrix} = \begin{bmatrix} \varepsilon_{p,n}^y \\ \varepsilon_{p,n}^x \end{bmatrix} - \begin{bmatrix} K_{p+1,n}^{ryy} & K_{p+1,n}^{rxy} \\ K_{p+1,n}^{ryx} & K_{p+1,n}^{rxx} \end{bmatrix} \begin{bmatrix} r_{p,n-1}^y \\ r_{p,n-1}^x \end{bmatrix} \tag{11.54}$$

and the backward error ARMA lattice recursion is

$$\begin{bmatrix} r_{p+1,n}^y \\ r_{p+1,n}^x \end{bmatrix} = \begin{bmatrix} r_{p,n-1}^y \\ r_{p,n-1}^x \end{bmatrix} - \begin{bmatrix} K_{p+1,n}^{\varepsilon yy} & K_{p+1,n}^{\varepsilon xy} \\ K_{p+1,n}^{\varepsilon yx} & K_{p+1,n}^{\varepsilon xx} \end{bmatrix} \begin{bmatrix} \mu_{p,n}^y \\ \mu_{p,n}^x \end{bmatrix} \tag{11.55}$$

Equations 11.54 and 11.55 lead to the ARMA lattice structure shown in detail for the $(p + 1)$th stage in Figure 11.13. The three-dimensional sketch of the ARMA lattice stage shows how the 2-channel matrix equations can be laid out into a virtual circuit. The importance of the symmetry within each lattice stage, and among the various stages of the lattice filter, is that a very complicated set of matrix updates can be executed systemically, dividing the total operational load between multiple processors in a logical manner. This is called systolic processing for highly parallel computing. Unless one is bootstrapping, the lattice stages do not have to be updated sequentially. One

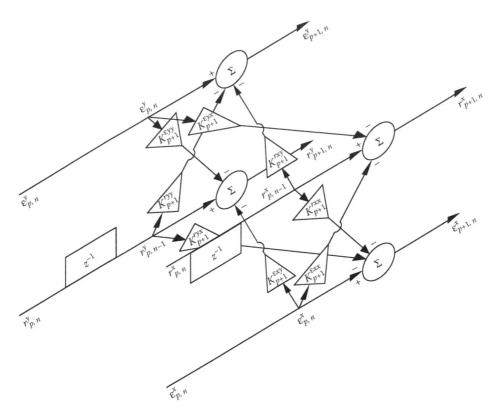

FIGURE 11.13 The ARMA lattice ($p + 1$)th stage section showing 2-channel forward and backward prediction error signal paths for the embedded ARMA model.

could process them in parallel with a delay latch between stages and have a small delay in the PARCOR coefficients responses of the higher stages relative to the lower stages. The stage orthogonality of the lattice allows for some very high performance computing architectures.

Figure 11.14 shows the layout of three lattice stages to form a third-order ARMA filter. The bootstrap error is shown for ARMA signal modeling where the innovation must be modeled. In this case of whitening the signal $y_n = e_{0,n}^y$, the stages cannot be updated in parallel. If the unknown ARMA filter input and output are available, Wiener filtering can be used to identify the system where the ARMA output y_n enters the lattice through $\varepsilon_{0,n}^y$ and the ARMA input x_n enters through $\varepsilon_{0,n}^x$ replacing the bootstrap error ε_n^b. For the Weiner lattice case, the processing can be made systolic.

Figure 11.15 shows the frequency responses for the original ARMA system polynomials, the lattice results using Wiener filtering, and the ARMA lattice bootstrap error technique. Clearly, the ARMA modeling results using the lattice are quite good. Figure 11.16 shows the ARMA coefficient results. Note how even the ARMA lattice bootstrap, with only the signal y_n available, does a pretty good job at identifying all the coefficients, although it takes longer than the Weiner RLS Lattice where both unknown system input x_n and output y_n are available. This is an illustrative example of why fast processing can also lead to high performance.

The ARMA embedded lattice is essentially a 2-channel whitening filter. For the Wiener filtering case, the Levinson recursion supplies the coefficients of $A[z]$ and $B[z]$ assuming the lattice order M is greater than or equal to both P and Q, the ARMA polynomial orders, respectfully. The value of any coefficients from order P or Q up to M should converge to zero for the Wiener filtering case. However, we are missing b_0 (and $d_0 = 1/b_0$).

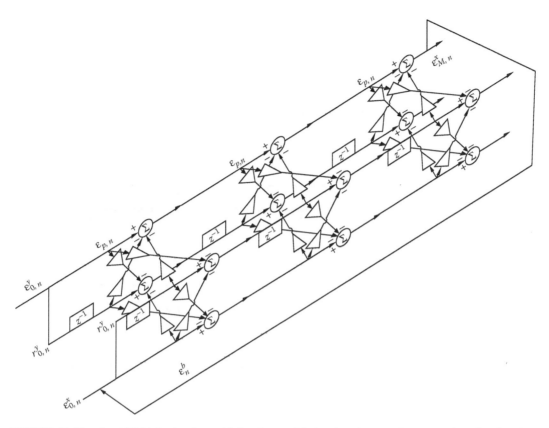

FIGURE 11.14 An ARMA lattice for a third-order model showing the error bootstrapping signal path to model the innovation.

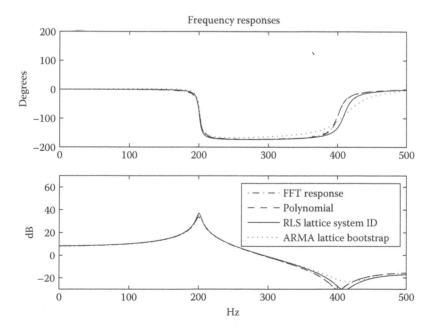

FIGURE 11.15 ARMA frequency responses for the embedded ARMA lattice algorithm for both system identification (Weiner filtering) and normalized error bootstrapping showing a near perfect match in magnitude and phase for the normalized error bootstrapping case.

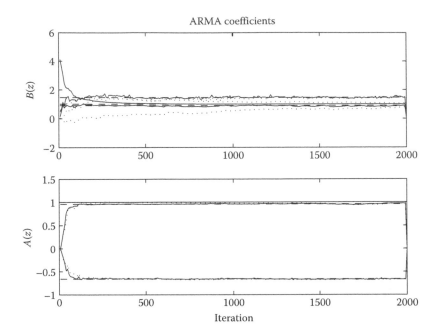

FIGURE 11.16 ARMA numerator ($B[z]$) and denominator ($A[z]$) coefficients for the actual system (--), embedded ARMA lattice system identification (–), and ARMA lattice normalized error bootstrapped (···) algorithms showing a precise coefficient match even for the normalized error bootstrapped case where only the output signal is available.

Given the forward prediction error variances leaving the Mth stage, the scale factor for b_0 is

$$b_0 = \sqrt{\frac{R_M^{\varepsilon yy}}{R_M^{\varepsilon xx}}} = \sqrt{\frac{R_M^{\varepsilon}(1,1)}{R_M^{\varepsilon}(2,2)}} \tag{11.56}$$

which only is required for b_0 in the Wiener filtering case. The rest of the coefficients in $B[z]$ are actually properly scaled by the lattice and Levinson recursions. For the ARMA bootstrap error case, $B'[z]$ (b_0' is assumed unity) is simply multiplied by the standard deviation of the bootstrap error.

$$B[z] = \sqrt{\frac{R_M^{\mu yy}}{\bar{N}}}\, B'[z] \tag{11.57}$$

The prediction error is divided by $\bar{N}$ before the square root because of the recursion

$$R_{M,n}^{\varepsilon} = \alpha R_{M,n-1}^{\varepsilon} + \varepsilon_{M,n}\, \varepsilon_{M,n}^{+\,H} \tag{11.58}$$

where $\alpha = 1 - 1/\bar{N}$. This removes a bias of $\bar{N}$ from the lattice forward prediction error variance. As seen in Figure 11.15, the scaling technique works quite well for the ARMA lattice.

Table 11.3 compares all of the ARMA filter and signal modeling performance results. By inspection of the converged ARMA coefficients, one can see that for Wiener filtering where both input and output signals are known, the ARMA coefficients can be estimated with great precision using either LMS, RLS, or the least-squares lattice. For ARMA bootstrapping, where one models the signal as an ARMA process with unity variance white-noise innovation, the results using LMS and RLS with unnormalized bootstrap error are unacceptable (as indicated by the shaded table cells).

By normalizing the bootstrap error to unity variance, good RLS results are obtained as indicated in the table by the "RLSN" algorithm. The ARMA bootstrapped lattice also gives good ARMA signal modeling performance although Figure 11.16 clearly shows the sensitivity of the lattice to noise. The memory window $\bar{N}$ for the trials was generally kept at 200 samples for all cases except the LMS algorithm, which was too slow to converge in the space of 200 data samples. Reducing the LMS data memory window to 100 samples was necessary for comparable convergence.

Multichannel Embedding: Generalized embedding of l signal channels in the lattice algorithm can be done using the ARMA embedding methodology. This type of multichannel signal processor is useful when ARMA models are estimated simultaneously between many signals, multidimensional data (such as 2D and 3D imagery) is modeled, or when arrays of sensor data are processed. Examination of the signal flow paths naturally leads to a processor architecture which facilitates logical division of operations among multiple processors. To examine the general embedded multichannel lattice algorithm, consider ℓ input signals to be whitened $\bar{y}_n = [y_n^1 \ y_n^2 \ldots y_n^\ell]^T$. The forward error signal vector update is $\bar{\varepsilon}_{p+1,n} = \bar{\varepsilon}_{p,n} - \bar{K}^r_{p+1,n} \ \bar{r}_{p,n-1}$, or in gory detail in

$$
\begin{bmatrix} \varepsilon_{p+1,n}^1 \\ \varepsilon_{p+1,n}^2 \\ \vdots \\ \varepsilon_{p+1,n}^\ell \end{bmatrix} = \begin{bmatrix} \varepsilon_{p,n}^1 \\ \varepsilon_{p,n}^2 \\ \vdots \\ \varepsilon_{p,n}^\ell \end{bmatrix} - \begin{bmatrix} K_{p+1,n}^{r11} & K_{p+1,n}^{r12} & \cdots & K_{p+1,n}^{r1\ell} \\ K_{p+1,n}^{r21} & K_{p+1,n}^{r22} & \cdots & K_{p+1,n}^{r2\ell} \\ \vdots & \vdots & & \vdots \\ K_{p+1,n}^{r\ell1} & K_{p+1,n}^{r\ell2} & \cdots & K_{p+1,n}^{r\ell\ell} \end{bmatrix} \begin{bmatrix} r_{p,n-1}^1 \\ r_{p,n-1}^2 \\ \vdots \\ r_{p,n-1}^\ell \end{bmatrix}
\tag{11.59}
$$

and a similar expression for the backward prediction error vector.

$$
\begin{bmatrix} r_{p+1,n}^1 \\ r_{p+1,n}^2 \\ \vdots \\ r_{p+1,n}^\ell \end{bmatrix} = \begin{bmatrix} r_{p,n-1}^1 \\ r_{p,n-1}^2 \\ \vdots \\ r_{p,n-1}^\ell \end{bmatrix} - \begin{bmatrix} K_{p+1,n}^{\varepsilon11} & K_{p+1,n}^{\varepsilon12} & \cdots & K_{p+1,n}^{\varepsilon1\ell} \\ K_{p+1,n}^{\varepsilon21} & K_{p+1,n}^{\varepsilon22} & \cdots & K_{p+1,n}^{\varepsilon2\ell} \\ \vdots & \vdots & & \vdots \\ K_{p+1,n}^{\varepsilon\ell1} & K_{p+1,n}^{\varepsilon\ell2} & \cdots & K_{p+1,n}^{\varepsilon\ell\ell} \end{bmatrix} \begin{bmatrix} \varepsilon_{p,n}^1 \\ \varepsilon_{p,n}^2 \\ \vdots \\ \varepsilon_{p,n}^\ell \end{bmatrix}
\tag{11.60}
$$

While embedding the RLS basis vector with additional channel results in significant increase in the dimension (effective RLS model order) of the matrices and vectors, embedding additional signal channels into the lattice simply expands the size of each stage, leaving the effective lattice model order the same. Therefore, we expect significant computational efficiencies using the lattice structure over an RLS structure. The signal flow for an 8-channel lattice stage is seen in Figure 11.17.

TABLE 11.3
ARMA Filter Modeling Comparison

	Algorithm	b_0	b_1	b_2	a_1	a_2	N
	Actual	5.0000	7.3436	4.5125	−0.6603	0.9604	x
Wiener filter	LMS	5.0000	7.3437	4.5126	−0.6603	0.9604	100
	RLS	5.0000	7.3436	4.5125	−0.6603	0.9604	200
	Lattice	5.0047	7.5809	4.8011	−0.6565	0.9678	200
Bootstrapped	LMS	6.7311	3.3271	−2.4981	−0.7046	0.9537	100
ARMA error	RLS	8.1381	1.6342	0.4648	−0.7002	0.9786	200
	RLSN	5.0114	7.5178	4.3244	−0.6738	0.9149	200
	Lattice	5.3203	7.2554	4.4180	−0.6791	0.9881	200

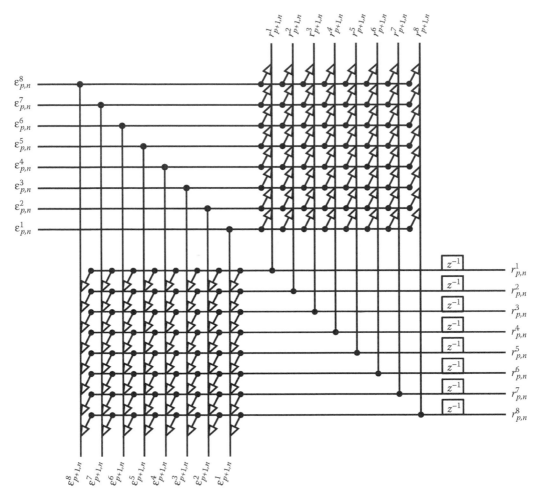

FIGURE 11.17 Signal flow block diagram for an 8-channel embedded lattice showing the interconnections of the PARCOR matrix elements and error signals showing the forward error signals entering from the left and leaving on the bottom left, backward error signals entering through the delay latches in the right and leaving at the top right.

The multichannel lattice stage in Figure 11.17 is a completely general adaptive processing structure allowing any number of signal channels to be embedded. The solid "dots" represent connections between signal wires, or soldier joints, while the open triangles are the elements of the PARCOR coefficient matrix. The forward error signals enter at the top left side and are multiplied by the elements of $\overline{K}^{\varepsilon}_{p+1,n}$ in the upper right corner and summed with the backward error which exits at the top right. At the bottom left, the updated forward error signal vector is produced from the forward error input (top left side) summed with the product of $\overline{K}^{r}_{p+1,n}$ and the backward error which enters at the bottom right side. This curiously symmetric processing structure also has the nice property that successive stages can be easily "stacked" (allowing straight buss connections between them) by a simple 90° counterclockwise rotation as seen in Figure 11.18.

Clearly, future developments in adaptive signal processing will be using structures like the multichannel lattice in Figure 11.18 for the most demanding processing requirements. The orthogonal decomposition of the signal subspace directly leads to the highly parallel processing architecture. It can be seen that correlations in both time and space (assuming the input signal vector is from an array of sensors) are processed allowing the PARCOR coefficient matrices to harvest the available signal information. Besides, the graphic makes a nice book cover.

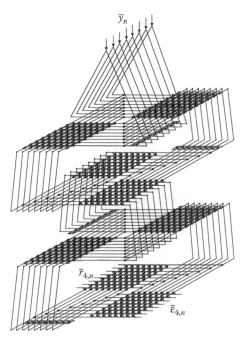

FIGURE 11.18 Graphical depiction of a 4-stage 8-channel embedded lattice showing the input signals entering from the top and the stages rotating 90° counterclockwise as the signals flow through from stage to stage.

11.3 FREQUENCY DOMAIN ADAPTIVE FILTERS

Adaptive LMS processing in the frequency domain offers some advantages over the time domain LMS algorithm in performance [9], yet it is not as computationally complex as the RLS or lattice algorithms. The main advantage for the frequency domain LMS (FDLMS) comes from the orthogonality of spectral data, allowing the ability to implement a single complex coefficient LMS filter for each FFT bin independently. This results in the same fast convergence rate for all the FFT bins as compared to the time domain algorithm which only converges fast for the dominant eigenvalue of the input data. One can maintain a time domain filter for execution in parallel with the FDLMS, such that the input–output data and prediction error are transformed into the frequency domain, the adaptive FDLMS algorithm updates the optimal filter's frequency response, and the time domain filter coefficients are calculated using an inverse FFT and replace the existing filter coefficients. This sounds like, and is, a lot of computation. But, it does not need to be executed in real time to provide a filter coefficient update with every new input and output signal sample.

Because the FFT of a signal represents an integral of the signal over a fixed time interval, one should only consider "overlapping" the FFT buffers in time by no more than 50%. The reason for limited overlapping is that the error spectrum which causes the filter frequency response adaptation will not respond immediately to the new coefficients due to the time integral. If spectral updates to the filter coefficients continue before the new error response is seen in the error spectrum, the FDLMS algorithm will "over adapt" which generally will lead to oscillations (even instability) in the adaptive algorithm. For stationary input data, it can be seen that the point of diminishing return is about a 50% overlap (50% of the data buffer is old data). However, the error signal is not very stationary due to the changing filter coefficients. Therefore, a 50% data buffer overlap for the FFTs is seen as the maximum advisable overlap. For our work, we use an even more conservative 0% overlap so that the residual error from the previous adaptation is completely flushed from the error spectrum time buffer for a particular FDLMS filter frequency response update operation [10]. While the FFT and update operations do not happen very often, the updates are quite spectacular in terms of the filter coefficient convergence. The FDLMS does not converge as fast as the RLS or lattice algorithms, but it is much less computationally complex and is seen as a reasonable choice for simplified LMS processing when the eigenvalue spread (frequencies and power levels) of the input data is wide.

The FDLMS algorithm is also of great interest for applications where the error and/or other signals are best expressed in the frequency domain. A good example of an appropriate FDLMS application for a spectral error is for an application of intensity error minimization. Sections 7.3 and 7.4 show some representations of wave intensity as a spectral measurement. Other applications could be in the area of transfer function error spectra or even physical (electrical, mechanical, acoustic) impedance error spectra. These are largely adaptive control issues and will be discussed further in Section 15.3. However, some image processing applications could use frequency domain adaptive control on wavenumber data. Finally, important modeling information such as spectral error, SNR, coherence, and confidence can be easily extracted from frequency domain data allowing the adaptive process to apply spectral weighting and other additional controls not readily available in the time domain.

Consider the standard filtered-x type LMS adaptive filter update on the FIR filter coefficients $h_{k,n}$, $k = 0, 1, 2, \ldots, M$ at time n.

$$h_{k,n} = h_{k,n-1} - 2\mu x_{n-k}\varepsilon_n \tag{11.61}$$

The parameter μ is μ_{max} times μ_{rel}, where $\mu_{max} = 1/\{\sigma_x^2\}$ and σ_x^2 is the variance of the input data x_n. As noted in Chapter 10, μ_{rel} is the step size component which creates an effective "data memory window" with an exponential forgetting property of approximate length $1/\mu_{rel}$. It can be seen that the shortest possible memory window length for a stable LMS algorithm is the filter length M + 1 and occurs if $\mu_{rel} = 1/(M + 1)$. The frequency domain version of Equation 11.61 will provide the complex frequency response of the FIR filter, to which we subsequently apply the inverse Fourier transform to get the time-domain FIR filter coefficients. To facilitate Fourier processing, we expand Equation 11.61 to include a block of N + 1 input and error data samples and the entire FIR filter vector of M + 1 coefficients to be updated as a block every N_0 samples. N_0 is the sample offset between data blocks and was noted earlier to be no smaller than half the FFT buffer size (N_0 equal to the FFT buffer size is recommended). This prevents the FDLMS from over adapting to the error spectrum.

$$
\begin{bmatrix} h_{0,n} \\ h_{1,n} \\ h_{2,n} \\ \vdots \\ h_{M,n} \end{bmatrix} =
\begin{bmatrix} h_{0,n-N_0} \\ h_{1,n-N_0} \\ h_{2,n-N_0} \\ \vdots \\ h_{M,n-N_0} \end{bmatrix} - 2\mu_{max}\mu_{rel}
\begin{bmatrix} x_{n-N} & \cdots & x_{n-2} & x_{n-1} & x_n \\ x_{n-N-1} & \cdots & x_{n-3} & x_{n-2} & x_{n-1} \\ x_{n-N-2} & \cdots & x_{n-4} & x_{n-3} & x_{n-2} \\ \vdots & & \vdots & \vdots & \vdots \\ x_{n-N-M} & \cdots & x_{n-2-M} & x_{n-1-M} & x_{n-M} \end{bmatrix}
\begin{bmatrix} \varepsilon_{n-N} \\ \vdots \\ \varepsilon_{n-2} \\ \varepsilon_{n-1} \\ \varepsilon_n \end{bmatrix} \tag{11.62}
$$

It can be seen that the product of the M + 1 by N + 1 input data matrix and the error can be conveniently written as a cross correlation.

$$
\begin{bmatrix} h_{0,n} \\ h_{1,n} \\ h_{2,n} \\ \vdots \\ h_{M,n} \end{bmatrix} =
\begin{bmatrix} h_{0,n-1} \\ h_{1,n-1} \\ h_{2,n-1} \\ \vdots \\ h_{M,n-1} \end{bmatrix} - 2\mu_{max}\mu_{rel}
\begin{bmatrix} R_0^{x\varepsilon} \\ R_{-1}^{x\varepsilon} \\ R_{-2}^{x\varepsilon} \\ \vdots \\ R_{-M}^{x\varepsilon} \end{bmatrix}, \quad
R_{-k}^{x\varepsilon} = \frac{1}{M}\sum_{j=0}^{M-1} x_{n-j-k}\varepsilon_{n-j} \tag{11.63}
$$

Equation 11.63 is expressed in the frequency domain by applying a Fourier transform. Equation 11.64 shows the Fourier transformed equivalent of Equation 11.63.

$$H_n(\omega) = H_{n-N_0}(\omega) - 2\mu_{max}\mu_{rel}X_n^*(\omega)E_n(\omega) \tag{11.64}$$

The cross spectrum in Equation 11.64 should raise some concern about the possibility of circular correlation errors when the input signal has sinusoids not bin-aligned with the Fourier transform. See Section 6.4 for details about circular correlation effects. The potential problem occurs because the finite length signal buffers in the FFT provide a spectrum which assumes periodicity outside the limits of the signal buffer. This is not a problem for random noise signals or sinusoidal signals where the frequencies lay exactly on one of the FFT frequency bins. Practical considerations make it prudent to develop an algorithm which is immune to circular correlation errors. As seen in Section 6.4, the precise correction for circular correlation error is to double the FFT input buffer sizes, while zero padding one buffer to shift the circular correlation errors to only half of the resulting inverse transformed FIR coefficient vector. Since we want the first $M + 1$ coefficients of the FIR impulse response to be free of circular correlation errors, we double the FFT buffer sizes to $2M + 2$ samples, and replace the oldest $M + 1$ error signal samples with zeros [11].

$$X_n^c(\omega) = \Im\{ x_{n-2M-1}\ldots x_{n-M-1} x_{n-M} \cdots x_{n-1}\ x_n\} \tag{11.65}$$

$$E_n^c(\omega) = \Im\{0\ 0\ldots0\ \varepsilon_{n-M}\ldots\varepsilon_{n-1}\varepsilon_n\} \tag{11.66}$$

Using the double-sized buffers in Equations 11.65 and 11.66 in the FFTs, we can update a more robust filter frequency response (in terms of its inverse FFT) and also include a frequency dependent step size $\mu_{max}(\omega)$, which is the inverse of the signal power in the corresponding FFT frequency bin. It will be shown below that some spectral averaging of the power for the bin step size parameter will improve robustness. The parameter μ_{rel} can be set to unity since the memory window need not be longer than the integration already inherent in the FFTs.

$$H_n^c(\omega) = H_{n-N_0}^c(\omega) - 2\mu_{max}(\omega)\mu_{rel}\{X_n^c(\omega)\}^* E_n^c(\omega) \tag{11.67}$$

To recover the time domain FIR filter coefficients, an inverse FFT is executed which, due to the circular correlation correction described in Section 6.4 and Equations 11.65 through 11.67, provides robust FIR filter coefficients in the leftmost $M + 1$ elements of the FFT output buffer. The symbol $\mathbb{N}$ in Equation 11.68 depicts that one discards the corresponding coefficient which by design may have significant circular correlation error.

$$H_n^c = \Im^{-1}\{H_n^c(\omega)\} = [h_{0,n}h_{1,n}\ldots h_{M,n}\ \mathbb{N}\ \mathbb{N}\ldots\mathbb{N}] \tag{11.68}$$

An example is presented below comparing the RLS, LMS, and FDLMS algorithms in a Weiner filtering application where the input, or reference data, has sinusoids plus white noise. With a white noise only input, the three algorithms perform the same (there is only one eigenvalue in that case). The Weiner filter to be identified is FIR and has nine coefficients resulting from zeros at (magnitude and phase notation) $0.99\angle \pm 0.2\pi$, $1.40\angle \pm 0.35\pi$, $0.90\angle \pm 0.6\pi$, and $0.95\angle \pm 0.75\pi$ where the sample rate is 1024 Hz. The frequencies of the zeros are 109.4 Hz, 179.2 Hz, 307.2 Hz, and 384 Hz, respectively. The reference input signal has ZMG white noise with standard deviation 0.001, and three sinusoids of amplitude 10 at 80 Hz, amplitude 30 at 250 Hz, and amplitude 20 at 390 Hz. To facilitate the narrowband input, the FIR filter model has 18 coefficients, and the effective data memory window length is set at approximately 36 samples. This way, the RLS, LMS, and FDLMS algorithms all share the same effective memory window (the FDLMS doubles the FFT buffers to 36

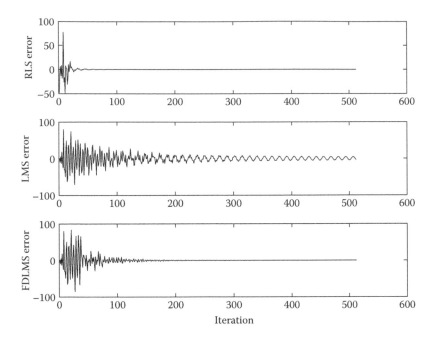

FIGURE 11.19   Error responses of the RLS, LMS, and FDLMS algorithms for the three sinusoid signal input to the nine coefficient Weiner filter system identification problem.

samples). The FDLMS is updated every 36 samples so that no input or error data is shared between successive updates (0% buffer overlap). Because the input consists mainly of three sinusoids in a very small amount of broadband noise, we expect the model frequency response to be accurate only at the three sinusoid frequencies. As seen in Figure 11.19 for the RLS, LMS, and FDLMS algorithms, all three converge but the RLS shows the best results. Figures 11.20 through 11.22 show the short-time Fourier spectra of the error signals processed as 64-point blocks for each iteration. In Figure 11.20 the RLS algorithm's superior performance is clear, but the FDLMS's performance

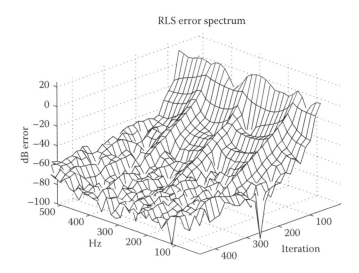

FIGURE 11.20   The RLS system error spectra from 0 to 512 Hz for the first 450 iterations of the algorithm showing a consistent fast convergence for all frequencies.

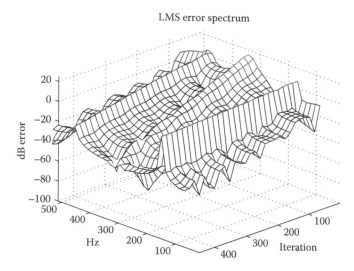

FIGURE 11.21   The LMS system error spectra from 0 to 512 Hz for the first 450 iterations showing convergence rates depending of sinusoid power and frequency (eigenvalue spread).

in Figure 11.22 is also quite good. The LMS performance in Figure 11.21 again shows the convergence faster for the signals near the center of the Nyquist band, which is analogous to the more dominant eigenvalues of the input signal as discussed in Section10.2. It was mentioned earlier that while the FDLMS allows each individual bin to operate as an independent adaptive filter with its own optimized step size, that this is generally not very robust for narrowband inputs. Recall that for a broadband ZMG input, the RLS, LMS, and FDLMS all perform the same. For the ZMG case, the FDLMS would have identical step sizes for each FFT bin equal to the inverse of the total power of the input signal at that bin frequency. When the input consists of narrowband sinusoids, the power for some bins will be near zero. The small bin power gives a large gain for the adaptive update to that particular bin. Since the power estimate should be averaged over several adaptive updates (to minimize random fluctuations in the model estimate), occasional spectral leakage from a nearby bin

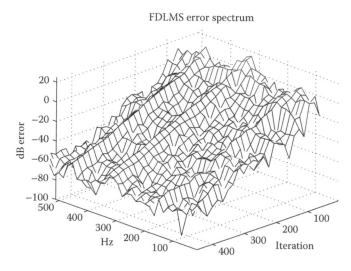

FIGURE 11.22   The FDLMS algorithm system error spectra from 0 to 512 Hz for the first 450 iterations showing a consistent convergence rate for all frequencies yet less computation than the RLS algorithm.

with a high-amplitude sinusoid will drive the bin with a large step size to instability. This problem is solved by averaging the bin power with the power in adjacent bins. The number of adjacent bins to include in the average depends on the frequency spacing of the narrowband sinusoids in the input. If one cannot *a priori* determine the approximate spectral content of the input, a single step size can be used for the entire FDLMS based on the total power of the input signal. For a single step size for the entire spectrum, the FDLMS will perform with convergence rates similar to the LMS algorithm, where the strongest peaks and those in the middle of the Nyquist band converge the fastest, while the weaker peaks and those at high and low frequencies converge much more slowly. Figure 11.23 shows the input spectral power, the bin-average power, and the step sizes used based on the bin-averaged power.

The frequency domain also offers us some very insightful ways to characterize the "goodness of fit" for the model given the available error and input signals. The FDLMS is not required to make this measurement, but the FDLMS allows a convenient way to implement a weighted least-squares algorithm using the frequency domain error. Consider the equation for the FDLMS error.

$$E_n(\omega) = X_n(\omega)\left\{H_{\mathrm{opt}} - H'(\omega)\right\} \tag{11.69}$$

Multiplying both sides of Equation 11.69 by $X^*(\omega)$ and rearranging, we can estimate the optimal filter from our current estimate and an error estimate based on the prediction error spectra.

$$|H_{\mathrm{opt}}(\omega)| = |H'(\omega)| \pm \left|\frac{X^*(\omega)\,E(\omega)}{X^*(\omega)\,X(\omega)}\right| \tag{11.70}$$

For the modeling error depicted in Equation 11.70 to be meaningful, $X(\omega)$ should be broadband, rather than narrowband. A broadband input ensures that the complete frequency response of $H_{\mathrm{opt}}(\omega)$ is excited and observed in the error spectrum $E(\omega)$. For a narrowband input, we only expect the response of the system to be excited and observed at the frequencies on the input. For the example

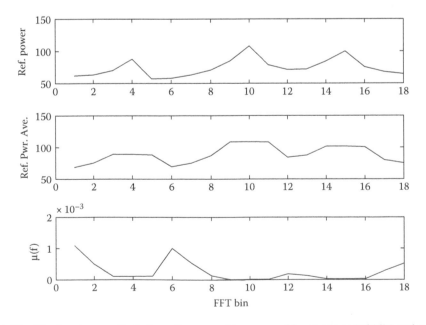

FIGURE 11.23 The input power (top), three frequency bin-averaged input power (middle) and step size per bin for the FDLMS algorithm.

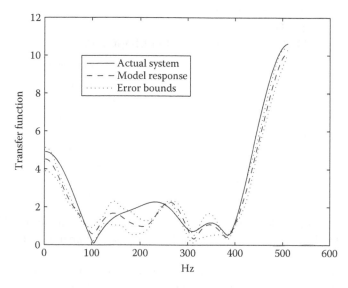

FIGURE 11.24 Comparison of the actual system, FDLMS-modeled response, and estimated spectral prediction error responses showing the actual system lying within the predicted error bounds.

seen in Figures 11.19 through 11.23 we expect precise modeling at 80, 250, and 390 Hz and a poor match between the frequency response of the actual system and the model elsewhere. However, increasing the broadband noise from amplitude (standard deviation) 0.001–5.0 yields the modeling result in Figure 11.24, which also includes and error measure as given in Equation 11.70. Increasing the broadband noise further helps the spectral match and reduces the uncertainty, especially in the middle of the Nyquist band around 250 Hz. The absolute values of the error are used as a "worst case" estimate error, but the error is not strictly abound in the sense of a Cramer–Rao lower bound. The model is seen to be very precise at 80, 250, and 390 Hz and the error model correctly tracks the model errors in other frequency regions. However, if the input is not broadband, then the error measure will not be meaningful.

The spectral modeling error in Figure 11.24 is very useful for determining how close the model is to the optimum filter response. Both the input and error signals (the error signal in particular) can have interfering noise which will widen the error bounds as seen in Figure 11.24. Applying the probability density distributions presented in Sections 7.1 and 7.2, we can estimate a probability for the model to be some percentage above the actual, and so on. The statistic representation can be used to associate the probability of algorithm divergence, convergence, or confidence for the model estimate. While these measures can be applied to any adaptive algorithm, the FDLMS very conveniently allows adaptive weighing of the error spectrum to optimize performance in specific frequency ranges.

11.4 MATLAB® EXAMPLES

This chapter used six separate m-scripts to make the figure plots listed in Table 11.4. The 4 "SYSIDx.M" scripts, where "x" is "LMS, RLS, RLN, or LAT" are very similar except for the particular adaptive algorithm used, and were not combined just to keep the scripts small and manageable, but much of the code to generate the system models and plots is identical. While in Chapter 10 we examine convergence rates and tracking of parameter changes, in the examples for Chapter 11 we look at the different algorithms and focus on the adaptive modeling performance particularly for ARMA modeling and the bias effects of ARMA error bootstrapping when only the output signal from the unknown system is available. We also revisit the tracking filter and derive the Kalman filter

TABLE 11.4
MATLAB m-Scripts Used in Chapter 11

Figure	m-Script Name
11.1–11.5	ABKTRACK.M
11.7–11.8	SYSIDLMS.M
11.9–11.10	SYSIDRLS.M
11.11–11.12	SYSIDRLN.M
11.15–11.16	SYSIDLAT.M
11.19–11.24	WFDLMS.M

based on a state kinematic model and the matrix inversion lemma in the script "ABKTRACK.M," which uses the rocket flight model from Section 5.2 again and compares an α–β–γ tracker to the Kalman filter.

The Kalman filter equations are all vectors and matrices, which make some very clean MATLAB codes as seen in Table 11.5. Again, there is nothing like clean readable algorithms in your software models.

The experiments in "ABKTRACK.M" allow one to see what Kalman filtering offers over a fixed gain, albeit optimized, tracking filter. In general the Kalman filter is a bit more reactive with the same process noise. To see this, turn on the noise burst feature (set burston = 1 on line 40) and run a case with a measurement noise of 3 and process noise of 1.

The "SYSIDx.M" scripts generate a simple ARMA pole–zero filter using the MATLAB "poly()" function with pairs of conjugate poles and zeros. The ARMA input is ZMG noise and the output is generated in for-loops the old school way so one can see every detail. One could use the MATLAB "filter()" function as well and get the same signals. The frequency response is generated via a transfer function using the MATLAB function "tfestimate()" which saves a lot of work and gives us a well-understood baseline for comparisons. We also compute the transfer function from the ratio of the FFTs of the numerator over denominator polynomials. These techniques are lots of fun to compare and are quite easily calculated and plotted in the examples. For each algorithm, we save the estimated coefficients and error at each iteration, plotting them for comparison as well. These plots of the coefficients are very illustrative of some very nonobvious points about how to properly error bootstrap the respective algorithms in ARMA modeling.

The frequency domain LMS algorithm is compared to RLS and time-domain LMS in the m-script "WDLMS.M" for an input of ZMG noise plus several sinusoids across the Nyquist band.

TABLE 11.5
A Kalman Filter in MATLAB

```
xp = F*xu;
Pp = F*Pu*F' + Q;
zp = H*xp;
ep = xmeas(k) − zp;
S = H*Pp*H' + R;
K = Pp*H'*S^(−1);
xu = xp + K*ep;
Pu = Pp − K*S*K';
```

The idea is to see in the frequency domain the convergence properties by watching these sinusoids decay in the error spectrum. The RLS and FDLMS decay all the frequencies at the same rate while the LMS is sensitive to the eigenvalue spread of the input signal. By increasing the white noise relative to the sinusoid amplitudes, the spectral error can provide a useful bound for the system model accuracy, as seen in Figure 11.24.

11.5 SUMMARY

There are several major algorithm classes in adaptive filtering which can be seen to be RLS, Kalman filtering, least-squares lattice algorithms, and the simple and robust LMS algorithm. Each of the classes can be further broken down into algorithms optimized for specific applications including parameter tracking and prediction, system identification using Weiner filtering, and signal modeling and parameterization using ARMA filters. For each particular application, the literature has a substantial volume of articles with very detailed derivations and results limited to the specific application. The student can be overwhelmed by the number of different least-squared error algorithms published if attempting to study them individually. The approach of this book (in particular Chapters 10 and 11) is to assimilate least-squared error modeling of linear systems into three basic forms: recursive algorithms derived for a particular model order using the matrix inversion lemma, orthogonal projection based algorithms, and least-squared error modeling involving a state transition as part of the prediction which we refer to as Kalman filtering. The RLS and LMS algorithms are clearly based on application of the matrix inversion lemma to implement a RLS error solution for a specific model order. Using a projection operator framework, orthogonal updates in both time and/or model order produce the least-squares lattice algorithm. The projection operator update equation is surprisingly similar to the matrix inversion lemma equation (as seen in Chapter 10) but allows model order expansion in a least-squared sense as well as time updates. The Kalman filter is distinguished from RLS simply by the inclusion of the state transition before the prediction error is estimated and the state parameters adjusted to maintain a least-squared error state modeling performance.

Kalman filtering is one of the most important and prevalent adaptive filtering algorithms of the twentieth century and there are literally hundreds of texts and thousands of articles available in the literature. The scope of this text cannot possibly address the many subtle technical points of Kalman filtering and state variable theory. However, it is extremely useful to contrast a basic Kalman filter algorithm to RLS and other adaptive filtering derivations. The intended result is to present all least-squared algorithms in as unified and concise an approach as possible. In Section 11.1 we see that the algebraic Riccati equation combination of Equations 11.19 and 11.23 for the update of the state error covariance, is really just an application of the matrix inversion lemma to the state error covariance including the state transition. The steady-state error covariance represents Kalman filter tracking performance when the underlying assumptions for the state transition (and number and type of states) and measurement noise are correct for the observed data.

In a steady-state situation, the required Kalman filter gain for the desired state process noise is a constant vector. We can calculate this gain vector in the form of, say, an α–β–γ filter (for three states). When the measurement noise is near zero, the α–β–γ and Kalman filters, each give identical performance. However, the Kalman filter gain is computed not only from the measurement and process noises, but also by adaptive minimization of the state error covariance. Therefore, for target maneuvers outside the range of the kinematic model used to determine the number and type of states, the Kalman filter gives superior performance compared to the α–β–γ tracker. This allows the Kalman filter to rapidly track target maneuvers with a smaller process noise than the α–β–γ tracker, giving more accurate state estimates overall. For high measurement noise levels, the Kalman filter will tend to overcompensate to minimize the state error, making the state vector rather noisy. Because of the added power of the Kalman filter, a covariance performance analysis should be executed on simulated target maneuvers to determine the best designed process noise for the adaptive tracking system.

Section 11.2 introduced the embedding technique for the RLS and lattice algorithms allow the underlying FIR, or MA filter, to be enhanced to a pole–zero, or ARMA filter. In the RLS algorithm, the embedding of an Mth-order ARMA filter (M poles and M zeros) effectively doubles the length of the RLS algorithm coefficient vector, whose computations increase with the cube of the coefficient vector length. In the lattice algorithm, the PARCOR coefficients become 2×2 matrices resulting in an increase in complexity of approximately 4, where the lattice algorithm overall complexity increases linearly with model order. The lattice's huge advantage in complexity reduction is due in part to the fact that the Levinson recursions required to extract the linear prediction coefficients from the PARCOR coefficients are not required for each filter update. If the linear prediction coefficients are computed with each lattice filter update, the computations required is closer to the required RLS computations, but still less. Proper bootstrapping and prediction error normalization allows good performance for both Wiener filtering for system identification given input and output signals, and ARMA signal modeling. An interesting result of the examples given in Section 11.2 is that the LMS filter performs extremely well for the Weiner filtering problem and extremely poor for the bootstrapped ARMA signal model. It is also seen that the Double/Direct lattice form demonstrates superior numerical properties while also reducing computational load through the elimination of some divisions in the lattice updates.

The lattice structure allows simple embedding of multichannel data from sensor arrays or even image data. Figure 11.13 illustrates topology of time and order expansion in a lattice structure in three dimensions graphically. This elaborate adaptive filter structure can clearly be seen as a parallel process, where individual processors are assigned to each "layer" or lattice stage. However, for true parallel-execution, a process time delay must be added between each stage making all the PARCOR coefficient matrices M time samples old for an Mth-order parallel process lattice. A pure spatial adaptive filter could be computed using only a single multichannel lattice stage, as depicted in Figures 11.17 and 11.18. The multichannel lattice architecture represents the future for practical large-scale adaptive processing. While hardly the epitome of simplicity, the multichannel lattice structure completes our unified presentation of basic adaptive filtering structures.

Section 11.3 introduces the FDLMS algorithm which offers several interesting features not available in the time-domain. First, we see that near RLS convergence performance is available for signals with wide eigenvalue spreads which would converge slowly in a time-domain LMS algorithm. The LMS step size can be set independently for each FDLMS filter frequency bin. However, the number of independent frequency bands where an independent step size can be used is best determined by the number of dominant eigenvalues in the input signal. This prevents signal bands with very low power from having overly sensitive adaptation from too large a step size. The bandwidth for each power estimate and corresponding step size can be simply determined from the spectral peak locations of the input signal, where one simply applies a step size for a particular peak over the entire band up to a band for an adjacent spectral peak. For a single sinusoid input, a fixed step size is used for the entire frequency range. For multiple sinusoids, individual bands and step sizes are determined for each peak band. Because of the integration effects of the Fourier transform, the FDLMS algorithm need not be updated with each input time sample, but rather with a limited (say 50%) overlap of the input data buffers to prevent overadaptation and amplitude modulation in the error-adaptation loop. The output of the FDLMS algorithm is a frequency response of the FIR optimal filter. To prevent circular correlation errors from spectral leakage, doubling the buffer sizes and zero-padding the latter half of the error buffer shifts any circular correlation errors into the latter half of the resulting FIR impulse response calculated from the inverse Fourier transform of the converged frequency response for the filter. Careful application of the FDLMS algorithm with circular correlation corrections, proper step size bandwidth, and update rates which do not allow over adaptation of the error spectrum, yields a remarkable and unique adaptive filter algorithm. Since the error is in the form of a spectrum, a wide range of unique applications can be done directly using physical error signals such as impedance, intensity, or even wavenumber spectra.

PROBLEMS

1. An airplane moves with constant velocity at a speed of 200 m/s with a heading of 110° degrees (North is 0°, East is 90°, South 180°, and West 270°). The radar's measurement error standard deviation is 10 m in any direction. Calculate the α–β gains for a tracking filter designed with a 1 s update rate and a process noise standard deviation of 1 m.

2. Determine the east–west and north–south (i.e., x and y components) of velocity state error for the airplane tracking data in problem 1.

3. Derive the Joseph form in Equation 11.24.

4. Show that for $\mu_{rel} = 0.01$ the effective exponential data memory window is about 100 samples and for the lattice this corresponds to a forgetting factor α of 0.99.

5. For the joint process ARMA system model in Equation 11.53 show that $d_i = a_i/b_0$ and $c_i = b_i/b_0$ where $i = 1, 2, \ldots, M$, $d_0 = 1/b_0$, and $a_0 = c_0 = 1$.

6. Prove Equation 11.56 deriving the b_0 from the ratio of the forward prediction errors at the Mth ARMA Wiener lattice stage and from the normalized standard deviation of the forward prediction error for the bootstrapped error case.

7. Derive the likelihood parameter for a multichannel least-squares lattice algorithm.

8. The time buffers in the FDLMS algorithm are 2M samples long where the oldest M error samples are replaced with zeros to correct for circular correlation errors due to possible spectral leakage. What other time buffer forms produce the same effect in the FDLMS algorithm?

9. Explain why updating an FDLMS adaptive filter update every time sample could lead to slow convergence and that updating the FDLMS adaptive filter less often speeds convergence.

REFERENCES

1. A. Gelb, *Applied Optimal Estimation*, Cambridge, MA: MIT Press, 1974.

2. Y. Bar-Shalom and X.-R. Li, *Estimation and Tracking: Principles, Techniques, and Software*, Norwood, MA: Artech House, 1993.

3. D. T. L. Lee, M. Morf, and B. Friedlander, Recursive least-squares ladder estimation algorithms, *IEEE Trans Acoust Speech Signal Process*, 1981, 29(3), Part III, pp. 627–641.

4. D. T. L. Lee, M. Morf, and B. Friedlander, Recursive ladder algorithms for ARMA modeling, *IEEE Proceedings 19th Conference on Decision and Control*, 1980 (Albuquerque, MN), pp. 1225–1231, December 10–12.

5. W. S. Hodgkiss, Jr. and J.A. Presley, Jr., Adaptive tracking of multiple sinusoids whose power levels are widely separated, *IEEE Trans Acoust Speech Signal Process*, 1981, 29(3), Part III, pp. 710–721.

6. B. Friedlander, System identification techniques for adaptive noise cancelling, *IEEE Trans. Acoust Speech Signal Process*, 1981, 30(5), Part III, pp. 627–641.

7. B. Friedlander, Lattice filters for adaptive processing, *Proc IEEE*, 1982, 70, pp. 829–867.

8. S. J. Orfanidis, *Optimum Signal Processing*, New York, NY: McGraw-Hill: 1988.

9. F. A. Reed and P. L. Feintuch, A comparison of LMS cancellers implemented in the frequency domain and the time domain, *IEEE Trans Acoust Speech Signal Process*, 1981, 29(3), Part III, pp. 770–775.

10. K. M. Reichard and D.C. Swanson, Frequency-domain implementation of the filtered-x algorithm with on-line system identification," in *Proceedings of the Second Conference on Recent Advances in Active Control of Sound and Vibration*, 1993, pp. 562–573.

11. J. J Shynk, "Frequency-domain multirate adaptive filtering," *IEEE Signal Proces Mag*, 1992, 9(1), pp. 14–37.

Part IV

Wavenumber Sensor Systems

By the title "wavenumber sensor systems," we mean a sensor system with an array of sensors that measure both the amplitude and the relative phase of sinusoid waves in space. This applies to radar, sonar, and seismic sensing in general. A wavenumber is a spatial representation of a propagating sinusoidal wave. It is generally referred to in the electromagnetic, vibration, and acoustics community with the symbol $k = \omega/c$, where ω is the radian frequency and c is the wave propagation speed in m/s, or $k = 2\pi/\lambda$, where λ is the wavelength in meters. Wavenumber sensor systems typically consist of arrays of sensors in a system designed to filter and detect waves from a particular direction (bearing estimation) or of a particular type (modal filtering). The most familiar device to the general public which employs wavenumber filtering is probably the medical ultrasound scanner. Medical ultrasound has become one of the most popular and inexpensive medical diagnosis tools due to the recent advancements in low-cost signal processing and the medical evidence suggesting that low-level ultrasonic acoustic waves are completely safe to the body (as compared to x-rays). The ability to see anatomical structures as well as measure blood velocity and cardiac output without insult to the body have saved millions of lives. Another example widely seen by the public at most airports is radar, which operates with nearly the same system principles as ultrasound, but typically scans the skies mechanically using a rotating parabolic reflector. Earlier, medical ultrasound scanners also operated with a mechanical scan, but this proved too slow for real-time imaging in the body with its many movements. Electronic scanning of the beam requires no mechanical moving parts and thus is much faster and more reliable. The most-advanced radar systems in use today also employ electronic scanning techniques.

The ultrasonic scanner is quite straightforward in operation. The transmitted frequency is quite high (typically in the 1–6 MHz range) so that small structures on the order of 1 mm or larger will scatter the wave. A piezoelectric crystal on the order of 1 cm in size (15–100 wavelengths across the radiating surface) tends to transmit and receive ultrasound from a direction normal to its face. In other directions, the waves on the crystal surface tend to sum incoherently, greatly suppressing the amplitudes transmitted and received. An acoustic lens (typically a concave spherical surface) helps shape the sound beam into a hyperbola which narrows the beam further in the region from about 1–6 cm from the face. The lens has the effect of sharpening the resulting ultrasound image in the region below the skin/fat layer. If one mechanically scans a single ultrasonic transceiver, the familiar gray-scale image could be constructed from the distance normalized backscattered ultrasound received. However, using a line array of transceivers, the beam direction is controlled by delaying the transmitted signal appropriately for each array element, as well as delaying the received element signals before summing the beamformed output (see Section 8.1). The beam steering can be electronically done quite fast enabling a clear image to be reconstructed in real-time for the medical practitioner. There exist a wide range of industrial applications of ultrasound as well ranging from welding of plastics, cleaning, nondestructive testing and evaluation, and inspection.

Sonar is another ultrasonic technique, but is widely seen by the public mainly through movies and literature. The "ping" depicted in movies would never actually be heard by anyone except the sonar operator because in order to steer the sound beam to a specific direction, the frequency needs to be at a frequency higher than would be easily heard by a human (if at all). A frequency demodulator in the sonar operator's equipment shifts the frequency down into the audible range. Again, the ratio of transmitting array size (the apertures) to the sound wavelength determines the beam width, and therefore, the angular resolution of the scanning operation. Small unmanned vehicles such as torpedoes/missiles use much higher sonar/radar transmitting frequencies to maintain a reasonable resolution with the smaller aperture transceiver array.

For very-long-range scanning, the time between transmitted pulses must be quite long to allow for wave propagation to the far-off object called the scatterer, and back to the receiver. The long distance to the scatterer also means that the echo signal received will generally have a low SNR. By increasing the length of the transmitted sonar pulse, more energy is transmitted increasing the received SNR, but the range resolution is decreased. The transmitted amplitude for active sonar is limited by transducer distortion, hydrostatic pressure, and the formation of bubbles (cavitation) when large acoustic pressures are generated near the surface where the hydrostatic pressure is low. Radar transmitting power is only limited by the quality of the electrical insulators and available power. Low-SNR radar and sonar processing generally involves some clever engineering of the transmitted signal and cross-correlating the transmitted and received signals to find the time delay, and thus range, of the scatterer in a particular direction. Cross-correlating the transmitted and received signals is called a *matched filter*, which is the optimum way to maximize the performance of a detection system when the background interference is ZMG noise (spectrally white noise in the receiving band). For nonwhite noise in the receiver frequency band, an Eckhart filter (Section 6.4) can be used to optimize detection. This becomes particularly challenging in propagation environments which have multipath. Our presentation here focuses on detection, bearing estimation, and field reconstruction and propagation techniques.

Wavenumber sensor systems detect and estimate a particular signal's spatial wave shape. Foremost, this means detecting a particular signal in noise and depicting the associated probability confidences that you have, in fact, detected the signal of interest and not a noise waveform of the same amplitude. Chapter 12 describes techniques for CFAR detection for both narrowband and broadband stationary signals using matched filter processing. For typical radar and sonar applications, wavenumber estimation provides an estimate of the DOA of a plane wave front from a distance source or scatterer. However, as depicted in Chapter 13, wavenumber estimation can also be done close to the source for spherical waves, as a means to reconstruct the sound field in areas other than where the sensor array is located. Wave field reconstruction is correctly described as holography, and is extremely useful as a tool to measure how a particular source radiates waves. Some of these waves do not propagate to the farfield as a complex exponential, but decay rapidly with distance as a real exponential. These "nearfield" or evanescent waves are to be studied because they can affect the efficiency of a source or array of sources, they contain a great deal of information about the surface response of the radiator or scatterer, and if ignored, the presence of evanescent waves can grossly distort any measurements close to the source. Using wave-number-domain Green's functions and a surface array of field measurements, the field can be propagated toward the source or away from the source, which provides a very useful analysis tool.

The presence of multiple wavenumbers for a particular temporal frequency is actually quite common and problematic in wavenumber detection systems. In the author's opinion, there are two "white lies" in most adaptive beamforming mathematical presentations. The first white lie is that the background noise at each sensor position is incoherent with the background noise at the other sensor positions in the array. In turbulent fluids for acoustic waves, noise independence is not guaranteed and neither is signal coherence. The second white lie is that one can have multiple sources radiating the same frequency and yet be considered "incoherent" from each other. Sources at the same frequency are by definition coherent, but in practice, if the true source generation mechanisms are independent, such as separate motors, the frequencies do vary just enough that over time they can be processed as incoherent sources. Incoherent background noise and sources make the adaptive beamforming algorithm mathematical presentation clean and straightforward, but this is simply not real in a physical sense. One must actually proceed further to achieve the mathematical appearance of incoherence by spectral averaging of time or space array data "snapshots" to enforce noise independence and spatial coherence in the signal. The practical assumption is that over multiple data snapshots, each of the sources drift enough in phase with respect to each other to allow eigenvector separation of the wavenumbers (bearing AOAs). Chapter 14 presents modern adaptive beamforming with an emphasis on the physical application of the algorithms in real environments with real sources and coherent multipaths. The problems due to coherent noise and source multipaths are very significant issues in adaptive wavenumber processing and the techniques presented are very effective. Chapter 14 also discusses active localization techniques using correlation and frequency modulation (FM) techniques.

There are many other types of wavenumbers other than nearfield, spherical, and plane waves that require wavenumber filters to measure. These include resonant modes of cavities, waveguides, and enclosures, propagation modes in nonuniform flow or inhomogeneous media, and structural vibration responses. Although, for a single frequency excitation, the modes with resonant frequency closest to the excitation frequency will be excited the most, and the other modes are also excited due to damping in the system. Since the excitation is in many cases controllable, one can design a scanning system to detect and monitor the system modes using wavenumber sensing, or modal filtering. This new wavenumber filtering technology will likely be very important in industrial process controls and system condition-based maintenance (CBM) technology. Wavenumber filtering is also useful in long-range wave propagation models. In a real environment far from the source, the wave front may not be planar due to wave speed variations (e.g., wind and temperature for atmospheric acoustics), refraction, and diffraction around objects in the ray path. The nonplanar wave front can be modeled as a Fourier sum of "modes" or wavenumbers which each reflect, refract, or diffract independently allowing a stepwise field reconstruction and propagation algorithm for very complex environments. It is unconventional to put beamforming, holography, and wave propagation together, but they really do share the same mathematics of wavenumber fields.

12 Signal Detection Techniques

The probability of detection (P_d) and the probability of a false alarm (P_{fa} or false detection) define the essential receiver operating characteristic (ROC) for any wavenumber or other signal detection system. The ROC for a particular detection system is based on practical models for signals and noise using assumptions of underlying Gaussian probability density functions (PDFs). When many random variables are included in a model, the central limit theorem leads us to assume an underlying Gaussian model because a large number of separate random events with arbitrary density functions will tend to be Gaussian when taken together (see Section 7.1). Therefore, it is reasonable to assume that the background noise is Gaussian at a particular sensor (but not necessarily independent of the background noise at the other sensors in the array). We will begin by assuming that our signal has a constant amplitude and is combined with the background noise in the receiver data buffer. The job of the signal detector is to decide whether or not a signal is present in the receiver buffer. The decision algorithm will be based on the assumption that the receiver buffer data have been processed to help the signal stand out from the noise such as filtering or averaging, and so on. A simple threshold above the estimated noise power defines a simple robust boundary between strong signals and noise. But where to best set this detection threshold? The underlying Gaussian noise process model will allow us to also label the signal-or-noise decision with an associated probability. The larger the signal in the receiver buffer is in relation to the decision threshold, the greater the likelihood will be that it is in fact signal and not a false detection. However, the closer the decision threshold is to the noise level, the greater the likelihood that we may detect as signal something that is actually a noise signal, giving a false alarm output from the detector. Using straightforward statistical models for signals in noise we can define the threshold in the objective terms of SNR, P_d and P_{fa}. The detection algorithm is then designed to operate with a prescribed P_{fa} associated with the threshold above the background noise that is used to make the detection decision. The P_d then depends on SNR and this P_{fa}-determined threshold. Therefore, one can trace a curve of P_d (y-axis) versus P_{fa} (x-axis) for a given SNR called the ROC curve. Any operating point on this curve defines a P_{fa}, detection threshold, and P_d for the SNR corresponding to the ROC curve.

An ROC curve is essentially defined by the decision threshold level relative to the noise power, which along with the noise mean and variance defines a probability of false alarm, P_{fa}. The probability of signal detection, P_d, depends on the receiver data buffer processing and the strength of the signal in relation to the decision threshold. Processing of the receiver data such as averaging, cross-correlations, Fourier transforms, and so on are used to improve the P_d (the same as improving SNR) without raising the P_{fa}. Section 12.2 describes techniques for constant false alarm rate (CFAR) detection which is highly desirable in environments with nonstationary background noise characteristics. CFAR signal detection is adaptive to the changing environment and permits a very robust ROC for system design.

However, in a multipath environment, the signal power may be either enhanced or reduced in level depending on the propagation path length differences. For active systems, the time delay and corresponding range estimate can be made ambiguous by a multipath environment. When the transceiver and/or a scatterer move with respect to the multipath, or if the multipath changes as is the case with turbulence, the multipath can be described statistically as outlined in Section 12.3. With a statistical description of the multipath, we can enhance the value of a signal detection decision with statistical confidences or probabilities based on the ROC and on the underlying noise and multipath statistics. While the multipath is a problem to be overcome in radar, ultrasonic imaging, and sonar applications, it can be a source of information for other applications such as meteorology, chemical process control, composite material evaluation, and combustion control.

12.1 RICIAN PDF

The Rician PDF describes the magnitude envelope (or RMS) representation of a signal mixed with noise [1]. Using statistical models to represent the signal and noise, one can design a signal detection algorithm with an associated P_d, for a given signal level, and a particular P_{fa}, for the particular background noise level and detection threshold. Before we present the Rician density function we will examine the details of straightforward pulse detection in a real waveform. Consider the decision algorithm for detecting a 1 μs burst of a 3.5 MHz sinusoid of amplitude 2 in unity variance ZMG noise propagating in a lossless waveguide. This is a relatively low SNR situation but the received signal is aided by the fact that wave spreading and other wave losses do not occur in our theoretical waveguide. Figure 12.1 shows the transmitted waveform, the received waveform including the background noise, and the normalized cross-correlation between the transmitted and received waveforms. Figure 12.2 shows the magnitude of the waveforms in Figure 12.1 that help to distinguish the echo at 10 μs delay. Note that for a water-filled waveguide (assume the speed of sound is 1500 m/s), the 10 μs delay corresponds to the scatterer being about 7.5 mm away (5 μs) from the source. Clearly from Figure 12.2, the cross-correlation of the transmitted and received waveforms significantly improves the likelihood of detecting the echo at 10 μs. Why is this true? Let the transmitted source waveform be $s(t)$ and the received waveform be $r(t)$. The normalized cross-correlation of the transmitted and receiver waveforms is defined as

$$R(-\tau) = \frac{1}{\sigma_s^2} E\{ s(t) r(t+\tau)\} \tag{12.1}$$

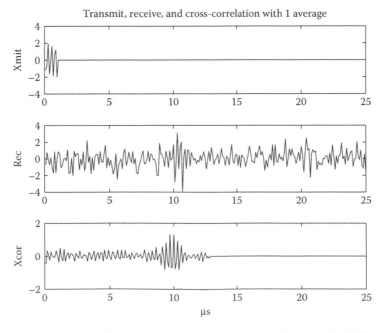

FIGURE 12.1 Transmitted, received, and cross-correlated signals for a simultated 3.6 MHz ultrasound pulse traveling 7.5 mm in a 1500 m/s water column and reflecting back to the receiver.

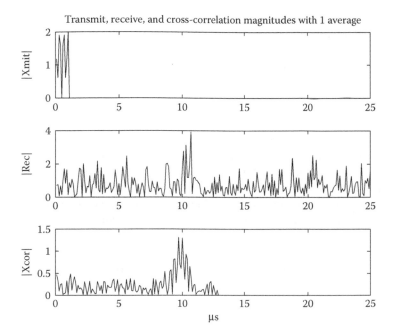

FIGURE 12.2   Magnitude envelope for the transmitted and received signals using only a single data buffer.

where σ_s^2 is the estimated variance of the transmitted waveform $s(t)$ over the recording time interval. If the background noise is ZMG, the correlation operation in Equation 12.1 is often called a *matched filter*, since the frequency response of this "filter" (the cross-correlation corresponds to a cross-spectrum in the frequency domain) matches the frequency response of the transmitted signal. The data in Figures 12.1 and 12.2 are simulated using 256 samples and a sampling rate of 10 MHz. Each correlation lag τ is computed with $n/2$, or 128, sample averages. This averaging effect gives the result of reducing the randomness of the correlation of the background noise with the signal while maintaining the coherence of the correlation between the received and transmitted signals.

12.1.1   TIME-SYNCHRONOUS AVERAGING

Time-synchronous averaging is quite effective at improving the SNR in periodic signals such as repetitive pulses in the sonar or radar. The background noise in our received waveform is ZMG with unity variance and is independent pulse to pulse. We need to develop a few useful relationships for averaged random variables first to see statistically what averaging does to the underlying PDF and the associated probabilities of detection and false alarm.

The general expression for PDF of a Gaussian random variable x with mean M_x and variance σ_x^2 is

$$p(x) = \frac{1}{\sigma_x \sqrt{2\pi}} \, e^{-(x-M_x)^2/2\sigma_x^2} \tag{12.2}$$

Suppose we scale our random variable as $y = ax$. We can easily determine the probability density for y using the following chain rule relationship [2,3]:

$$p(y) = \sum_{i=1}^{N} p(x_i) \left| \frac{dx_i}{dy} \right| \tag{12.3}$$

where $x_1, x_2, \ldots, x_N$ are all solutions of $y = f(x)$, which in our case is only one solution $y = ax$. Since $y/a = x$, $dx/dy = 1/a$, and the probability density $p(y)$ is

$$p(y) = \frac{1}{a\sigma_x \sqrt{2\pi}} e^{-(y/a - M_x)^2/2\sigma_x^2} = \frac{1}{a\sigma_x \sqrt{2\pi}} e^{-(y - aM_x)^2/2a^2\sigma_x^2} \tag{12.4}$$

where clearly, $\sigma_y^2 = a^2\sigma_x^2$ and $M_y = aM_y$ in Equation 12.4. Note that we are using a slightly different approach here to get the same statistical results given in Section 7.1 for averaged power spectra.

Consider the sum of a number of random variables $z = x_1 + x_2 + \cdots + x_N$. The PDF for z is the convolution of the N density functions for $x_1, x_2, \ldots, x_N$. Noting that multiplications in the frequency domain are equivalent to convolutions in the time domain, the PDF for z may be found easily using the density function's characteristic function. A Fourier integral kernel defines the relation between probability density $p(x)$ and its corresponding characteristic function $Q_x(\omega)$ [4].

$$Q_x(\omega) = E\{e^{j\omega x}\} = \int_{-\infty}^{+\infty} p(x) e^{+j\omega x} \, dx$$

$$p(x) = \frac{1}{2\pi} \int_{-\infty}^{+\infty} Q_x(\omega) e^{-j\omega x} \, d\omega \tag{12.5}$$

Note that the sign of the exponential is positive for the forward transform of the density function. One of the many useful attributes of the characteristic function is that the moments of the random process (mean, variance, skewness, kurtosis, etc.) are easily obtainable from derivatives of the characteristic function. The nth moment for x is simply

$$M\{x^n\} = \frac{1}{j^n} \frac{d^n Q_x(\omega)}{d\omega^n} \bigg|_{\omega=1} \tag{12.6}$$

Returning to our new random variable $z = x_1 + x_2 + \cdots + x_N$, the density function $p(z)$ has the characteristic transform

$$Q_z(\omega) = \int_{-\infty}^{-\infty} p(z) e^{+j\omega z} \, dz$$

$$= \int_{-\infty}^{+\infty} \left[p(x_1) \otimes p(x_2) \otimes \cdots \otimes p(x_N) \right] e^{+j\omega z} \, dz$$

$$= Q_{x_1}(\omega) Q_{x_2}(\omega) \cdots Q_{x_N}(\omega) \tag{12.7}$$

which is the product of each of the characteristic functions for each independent random variable x_i; $i = 1, 2, \ldots, N$. To derive the density function for z, we first derive the characteristic for the Gaussian density in Equation 12.2.

$$Q_x(x_i) = \int_{-\infty}^{+\infty} \frac{1}{\sigma_{x_i} \sqrt{2\pi}} e^{-(x_i - M_{x_i})^2/2\sigma_{x_i}^2} e^{+j\omega x_i} \, dx_i$$

$$= e^{\left(jM_{x_i}\omega - \left(\sigma_{x_i}^2 \omega^2/2\right)\right)} \tag{12.8}$$

As can be seen when combining Equations 12.7 and 12.8, the means and variances add for z, giving another Gaussian density function

$$p(z) = \frac{1}{\sigma_z \sqrt{2\pi}} \, e^{-(z-M_z)^2/2\sigma_z^2} \tag{12.9}$$

where

$$\sigma_z^2 = \sum_{i=1}^{N} \sigma_{x_i}^2$$

and

$$M_z = \sum_{i=1}^{N} M_{x_i}.$$

We can combine the summing and scaling to derive the density function for a random variable which is the average of a set of independent Gaussian variables with identical means and variances, such as is the case in the cross-correlation plot in Figure 12.1 in the regions away from the echo peak at 10 μs (microseconds). Let $z = 1/N \sum_{i=1}^{N} x_i$. The mean for the resulting density stays the same as one of the original variables in the average, but the variance is reduced by a factor of N [5]. Therefore, for the 128 samples included in the normalized cross-correlation in Figures 12.1 through 12.4, it can be seen that the variance of the background noise squared is reduced by about 1/128, or the standard deviation by about 1/11.3 or $128^{-1/2}$.

Time-synchronous averaging of periodic data, such as that from a repetitive pulsing sonar or radar scanning system, is an important technique for reducing noise and improving detection statistics. Synchronous averaging can also be used to isolate vibration frequencies associated with a particular shaft in a transmission or turbine. With the transmitted and received waveforms recorded at the same time and synchronous to the transmit time, simply averaging the buffers coherently in the time domain before processing offers a significant improvement in SNR. Figure 12.3 shows the same time and correlation traces as in Figure 12.1, but with 10 synchronous averages. Figure 12.4 shows the result for the magnitudes of the transmitted, received, and cross-correlated waveforms for 10 averages. With the signal coherently averaging and the noise averaging toward its mean, which is zero for ZMG noise, it is clear that this technique is extremely important to basic signal detection in the time domain. Comparing Figures 12.1 and 12.3, the 10 synchronous averages reduce the noise by about a factor of $10^{-1/2}$, or a little over 3. The same synchronous averaging noise reduction can be seen when comparing the magnitude plots in Figures 12.2 and 12.4. Figure 12.5 shows a block diagram of a pulse detection system using time-synchronous averaging and the cross-correlation envelope.

12.1.2 Envelope Detection of a Signal in Gaussian Noise

Detection of amplitude changes for dynamic signals requires a very straightforward analysis of the underlying signal statistics. We develop the statistical models using a general complex waveform and its magnitude. Consider a zero-mean signal (the real part or imaginary part of a complex waveform) with ZMG noise, which after time-synchronous and RMS averaging has the statistics of an RMS signal level of S_0 and a noise standard deviation of σ_x. For the noise-only case we apply the Gaussian density in Equations 7.21 and 12.2 and Figure 7.1, with zero mean, to the magnitude-squared

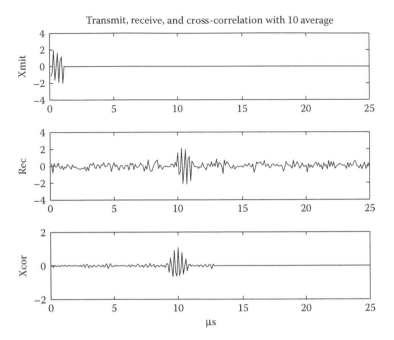

FIGURE 12.3 Time-synchronous-averaged transmit, receive, and cross-correlation signals using 10 synchronous averages to reduce the incoherent noise before calculating the cross-correlation showing a significant enhancement of SNR.

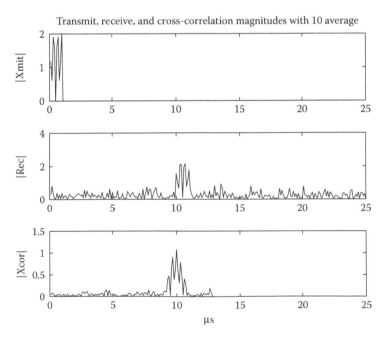

FIGURE 12.4 Applying a magnitude envelope operation to the signals in Figure 12.3 allow for a straightforward detection model based on probability.

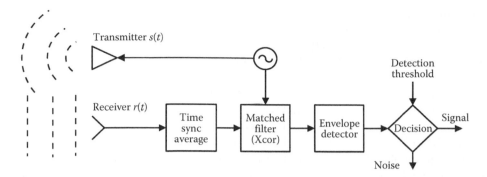

FIGURE 12.5 Block diagram of a matched-filter envelope detector showing both time-synchronous averaging and the matched-filter cross-correlation followed by the probabilistic detection decision.

random variable $y = x^2$ using Equation 12.3 to give the *Chi-Square 1 degree-of-freedom density function*

$$p(y) = \frac{p(x)}{2x}\bigg|_{x=+\sqrt{y}} + \frac{p(x)}{2x}\bigg|_{x=-\sqrt{y}}$$

$$= \frac{1}{\sigma_x\sqrt{2\pi y}}e^{-y/2\sigma_x^2} \tag{12.10}$$

where the mean is $M_y = E\{y\} = E\{x^2\} = \sigma_x^2$. This is the same density function as in Equation 7.25 where the variance, $E\{y^2\}$ was found to be $\sigma_y^2 = 2\sigma_x^4$ using the even central moments of a Gaussian density function in Equations 7.26 and 7.27. This density function is known as a *Chi-Square probability density function* with one degree of freedom, as shown in Figure 7.2 with several other Chi-Square densities with more degrees of freedom.

For summing the real part squared and imaginary part squared of a complex number, $y = x_R^2 + x_I^2$, to get the magnitude squared of a complex random signal, we can either convolve two Chi-Squares one degree-of-freedom densities or multiply the two corresponding characteristic functions. To keep matters simple, let us keep σ_y^2 the same, so that $\sigma_{xR}^2 + \sigma_{xI}^2 = \sigma_y^2 = 2\sigma_x^2$ and so the distribution is for the sum of two squared ZMG variables. This results in a Chi-Square density with two degrees of freedom, better known as an *exponential probability density function*.

$$p(y) = \frac{1}{2\sigma_x^2}e^{-y/2\sigma_x^2} = \frac{1}{\sigma_y^2}e^{-y/\sigma_y^2} \tag{12.11}$$

To find the density function for the complex noise magnitude, we need to substitute a square root for y in Equation 12.11 as $z = y^{1/2} = (x_R^2 + x_I^2)^{1/2}$. Using Equation 12.3, where $dy/dz = 2\sqrt{y}$, we have

$$p(z) = 2\sqrt{y}\,\frac{1}{2\sigma_x^2}e^{-y/2\sigma_x^2}\bigg|_{y=z^2}$$

$$= \frac{z}{\sigma_x^2}e^{-z^2/2\sigma_x^2} \tag{12.12}$$

which is known as a *Rayleigh probability density function*. This is the density function that describes the magnitude waveforms in Figures 12.2 and 12.4. Even though the figures display real rather than complex signals, the Rayleigh density function applies because the underlying mathematical time-harmonic signal structure is $e^{j\omega t}$. The difference for a real signal converted to complex (say in an FFT bin) is that the input noise variance is split between the real and imaginary parts of the complex bin, dropping the variance of the Rayleigh distribution by half.

Why did we sum two squared random variables to get the Rayleigh density? Consider the RMS for a moment. It is computed by squaring and averaging a signal, then computing the square root to provide an "RMS-level" representation of the zero-mean random signal. If we averaged the two squared random variables leading to Equation 12.11, σ_y^2 would be reduced by 1/2 which is a simple scaling factor. It is helpful to keep this in mind through the derivations presented here. You could go through the chain rule in Equation 12.3 for the case $y = x^2$ and again for $z = y^{1/2}$ to simply get a Gaussian PDF times 2 for positive z and zero for negative z. The mean and standard deviation would both work out to be σ_x. If one averages N squared-ZMG random variables and then takes the square root for an RMS level, one would have another Gaussian PDF but this time with mean σ_x and standard deviation $\sigma_x/\sqrt{N}$. We will return to a discussion on RMS later but continue now with discussion on the complex FFT bin case.

The mean, mean square, and variance for the Rayleigh density are found by evaluating the first and second moments of the PDF. Unless you are really smart, a table of definite integrals will provide a relation such as the following [6].

$$\int_0^\infty x^n e^{-ax^p}\, dx = \frac{\Gamma(n+1/p)}{p\, a^{(n+1)/p}}; \quad \Gamma(n+1) = n!; \quad \Gamma\left(m+\frac{1}{2}\right) = \frac{1 \cdot 3 \cdot 5 \ldots (2m-1)}{2\,m}\sqrt{\pi} \tag{12.13}$$

The mean of the Rayleigh density works out to be

$$M_z = \bar{z} = \int_0^\infty z\, \frac{z}{\sigma_x^2}\, e^{-z^2/2\sigma_x^2}\, dx = \sqrt{\frac{\pi}{2}}\sigma_x \approx 1.2533\,\sigma_x \tag{12.14}$$

and the mean square value is

$$\overline{z^2} = \int_0^\infty z^2\, \frac{z}{\sigma_x^2}\, e^{-z^2/2\sigma_x^2}\, dx = 2\sigma_x^2, \tag{12.15}$$

so the variance of z is simply

$$\sigma_z^2 = \overline{z^2} - M_z^2 = \left(2 - \frac{\pi}{2}\right)\sigma_x^2 \approx 0.4292\,\sigma_x^2 \tag{12.16}$$

which is not quite half the variance of the ZMG random variable x. Figure 12.6 shows the unity-variance zero-mean Gaussian density, and the corresponding one degree-of-freedom Chi-Square, exponential, and Rayleigh density functions. Note that the peak of the Rayleigh density is at the noise standard deviation σ_x^2, but the mean of the Rayleigh distribution is slightly greater (to the right of the peak in Figure 12.6).

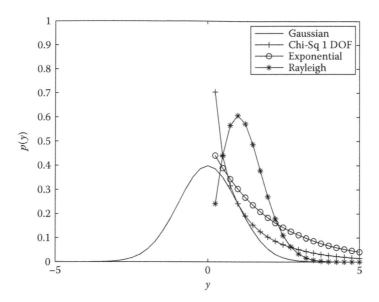

FIGURE 12.6 Gaussian, one degree-of-freedom Chi-Square, exponential, and Rayleigh PDFs.

The Rician* PDF is based on the envelope of a complex signal with RMS amplitude S_0 and real noise x_R and imaginary noise x_I, where the envelope can be defined as

$$
\begin{aligned}
r &= \sqrt{(S_0 + x_R)^2 + x_I^2} \\
x_1 &= r\cos\theta = S_0 + x_R \\
x_2 &= r\sin\theta = x_I \\
\theta &= \tan^{-1}\left(\frac{x_2}{x_1}\right)
\end{aligned}
\tag{12.17}
$$

where the angle θ is really of no consequence at this point. The expected value of x_1 is S_0, while the expected value of x_2 is zero in Equation 12.17. Assuming that x_R and x_I are ZMG with identical variances σ_x^2, the variances of x_1 and x_2 are both σ_x^2. With x_1 and x_2 independent, the joint probability density for x_1 and x_2 can be written as

$$
p(x_1, x_2) = \frac{1}{2\pi\sigma_x^2} e^{-\left[(x_1 - S_0)^2 + x_2^2\right]/2\sigma_x^2}
\tag{12.18}
$$

To get the envelope density function, we convert the joint density in Equation 12.18 to r and θ using the following Jacobian relationship (no sum is required because there is only one solution for r and θ in terms of x_1 and x_2).

$$
p(r, \theta) = p(x_1, x_2)
\begin{Vmatrix}
\dfrac{\partial x_1}{\partial r} & \dfrac{\partial x_1}{\partial \theta} \\
\dfrac{\partial x_2}{\partial r} & \dfrac{\partial x_2}{\partial \theta}
\end{Vmatrix}_{\substack{x_1 = r\cos\theta \\ x_2 = r\sin\theta}}
$$

* The Rice-Nakagami PDF is often referred to as "Rician."

$$
= p(x_1, x_2) \begin{vmatrix} \cos\theta & -r\sin\theta \\ \sin\theta & r\cos\theta \end{vmatrix}_{\substack{x_1 = r\cos\theta \\ x_2 = r\sin\theta}}
$$

$$
= r p(x_1, x_2) \Big|_{\substack{x_1 = r\cos\theta \\ x_2 = r\sin\theta}} \tag{12.19}
$$

Evaluating the joint density in Equation 12.19 gives the Rician probability density [3]

$$
p(r, \theta) = \frac{r}{2\pi\sigma_x^2} e^{-(r^2 + S_0^2 - 2rS_0\cos\theta)/2\sigma_x^2} \tag{12.20}
$$

where θ may be integrated out to give the envelope PDF since we are not interested in phase detection. This is why we were not concerned with the particular value of θ when the envelope detection model was set up.

$$
p(r) = \frac{r}{\sigma_x^2} e^{-(r^2 + S_0^2)/2\sigma_x^2} \left[\frac{1}{2\pi} \int_0^{2\pi} e^{-S_0 r \cos\theta/\sigma_x^2} \, d\theta \right]. \tag{12.21}
$$

The square-bracketed term in Equation 12.21 is recognized as a modified Bessel function of the first kind.

$$
I_0(z) = \frac{1}{\pi} \int_0^{\pi} e^{\pm z\cos\theta} \, d\theta = \frac{1}{2\pi} \int_0^{2\pi} e^{\pm z\cos\theta} \, d\theta = J_0(jz); \quad j = \sqrt{-1}. \tag{12.22}
$$

The Rician envelope PDF is therefore

$$
p(r) = \frac{r}{\sigma_x^2} e^{-(r^2 + S_0^2)/2\sigma_x^2} I_0\left(\frac{S_0 r}{\sigma_x^2}\right). \tag{12.23}
$$

Revisiting RMS, we note that the Rician PDF in Equation 12.23 is appropriate for signals with magnitude S_0 in a complex FFT bin and the real and imaginary bins each have an underlying ZMG random signal with standard deviation σ_x. As the SNR in the bin changes, the shape of the Rician PDF changes significantly as shown in Figure 12.7, which is another reason why the Rician PDF is important. If the underlying signal started as a real signal in the time domain and the Rician PDF is for an N-point FFT bin, one should calibrate the noise appropriately noting that the noise is divided evenly over the N FFT bins and that half the bin noise is in the real and imaginary components of the bin. This reduces the signal and noise in Equation 12.23 by half for a real input signal, but the SNR is the same.

For RMS estimation with many samples averaged, we can use the central limit theorem to get the appropriate PDF, mean, and variance estimates which will tend to be Gaussian for large number of averaged samples. The Rician PDF is most appropriate for statistical modeling of the complex data in an FFT frequency bin with no spectral averaging. It can be seen that for $S_0 \ll \sigma_x$, the Rician density becomes a Rayleigh density (the modified Bessel function approaches unity). For high SNR, $S_0 \gg \sigma_x$, the modified Bessel function of the first kind may be approximated by

$$
I_0\left(\frac{S_0 r}{\sigma_x^2}\right) \approx \frac{1}{\sqrt{2\pi(S_0 r/\sigma_x^2)}} e^{+S_0 r/\sigma_x^2}; \quad S_0 \gg \sigma_x \tag{12.24}
$$

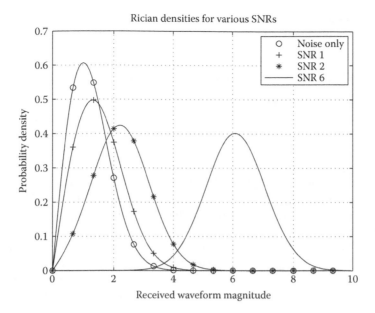

FIGURE 12.7 Rician PDF for SNRs of 0 (noise only), 1, 2, and 6 showing how the PDF changes shape with SNR and no averaging.

Inserting the high-SNR approximation for the modified Bessel function in the Rician density gives a very interesting result.

$$p(r) \approx \sqrt{\frac{r}{S_0}} \frac{1}{\sigma_x \sqrt{2\pi}} e^{-(r-S_0)^2/2\sigma_x^2}; \quad S_0 \gg \sigma_x \qquad (12.25)$$

The high-SNR (say SNR > 4) approximation to the Rician density shown in Equation 12.25 indicates that the Rician has the shape of a Gaussian density function with mean at $r = S_0$ and variance σ_x^2, approximately. Figure 12.7 shows the Rician density for various SNRs.

The Rician density function applies to complex signals where the envelope magnitude is found by summing the real and imaginary components of the signal. If a real signal S_0 in ZMG noise with variance σ_x^2 is converted to complex, such as in an N-point FFT bin, the noise amplitude of the complex bin is $\sigma_x/\sqrt{N}$ and the signal amplitude is still S_0 assuming that the sinusoid matches the FFT bin. Remember we integrated over phase between Equations 12.20 and 12.21? The phase integration sums up the real and imaginary parts of the FFT bin. The N-point FFT enhances the amplitude SNR by a factor of about $1/\sqrt{N}$ (window dependent). So, when using a Rician density to describe a signal in an FFT, do not forget to include the SNR enhancement of the FFT in the noise level. For real signals, one can just use Equation 12.23 as it is, because both the signal and noise magnitudes for a real signal are knocked down by 1/2 compared to a complex signal. This is why the Rician density function is so important for signal detection and the associated statistical metrics in spectral processing. But most importantly, when the SNR is low (below around 4), the peak of the Rician PDF is not at the signal level, but somewhere between the signal level and signal plus noise-only mean, as can be seen in Figure 12.7. The peak of the Rician can be seen to approximately follow $\sqrt{\sigma_x^2 + S_x^2}$ which converges to something slightly higher than the signal level for high SNR and to something slightly higher than the noise standard deviation at very low SNR. This can be seen in Figure 12.7 upon careful examination of the peaks of the PDFs. As soon as one computes M averages of a random variable with low SNR and a Rician PDF, the PDF is essentially Gaussian with a mean at the RMS noise standard deviation level σ_x, and a standard deviation of $\sigma_x M^{-1/2}$ which is significantly narrower after only a few averages.

12.2   RMS, CFAR DETECTION, AND ROC CURVES

In this section, we develop a straightforward decision algorithm to detect RMS signals in noise with a CFAR. CFAR detection means that the magnitude threshold for a detection decision is not set at an absolute level, but rather is set relative to the estimated noise level. This allows a receiver operating characteristic (ROC) curve to be developed for a wide range of false alarm (erroneous detection of a noise transient as "signal") rates and an expected SNR. ROC curves for numerous SNRs can be overlaid allowing one to view, in a single graph, the complete performance of the detection system available. By selecting a false alarm rate, a corresponding threshold-to-noise ratio (TNR) is defined, and the statistical detection performance can be read of the ROC curve for a given SNR received. This is a completely general detection algorithm.

RMS signals generally involves a significant number of sample averages with an effective time constant of the averaging long enough to capture the signal features or periods. This is still done using the LPF of a resistor and a capacitor in series with a full-wave-rectified signal as input (kind of like the square of the input and square-rooted, or signal instantaneous amplitude) and output read across the capacitor to ground. The capacitor charges and discharges with an exponential time constant of $\tau = RC = 1/2\pi f_c$ where f_c is the –3 dB cutoff frequency for the LPF. The time constant is the time it takes for the capacitor to discharge to $1/e$ ($e = 2.718$), or about 37%, of its initial value once the input voltage is switched from this initial value to zero. If you were charging the capacitor from an initial voltage of 0 V, the time constant is the time it takes to get to 63%, or $1 - e^{-1}$ of the final value. Our simple integrator seen in the beginning of Chapter 10 has exactly the same type of time constant, but this time defined in terms of the number of samples used in the effective sliding window with exponential decay. Recall that for some desired exponential window length N, we used an averaging recursion of

$$\langle y^2 \rangle_n = \alpha \langle y^2 \rangle_{n-1} + \beta x_n^2; \quad \alpha = \frac{N-1}{N}; \quad \beta = \frac{1}{N}, \tag{12.26}$$

such that the mean-square average is recursively updated from sample to sample, and the averaging window extends backward in time (in a $1/e$ sense) approximately N samples. Samples of x_n^2 older than N samples quickly fade away from the recursive estimate, allowing the mean-square estimate to adapt to signal changes in amplitude.

Figure 12.8 shows this recursive average with exponential decay at work. The top curve shows a ZMG signal with unity variance. The sample rate is 4096 Hz and at about 0.25 s into the plot, a signal burst occurs as a constant offset of amplitude 10 for about 200 ms until about 0.45 s, when the signal again becomes ZMG, still with unity variance. The middle curve has a fast RMS time average with a time constant of 10 ms, or about 41 samples. The bottom curve is a longer time-constant average of 100 ms, or about 410 samples. Figure 12.9 shows the same RMS averaging but on a signal burst of amplitude 1.0 rather than 10.0. Note how the middle plot for the fast RMS average converges approximately as $\sqrt{\sigma_x^2 + S_x^2}$ for both high and low SNR (it is about $\sqrt{2}$ in Figure 12.9). In both Figures 12.8 and 12.9 the ZMG noise is squared and then averaged using the recursive update in Equation 12.26, then the square root of the mean-square signal is plotted in the lower two plots of the figures. For the high-SNR case in Figure 12.8, one can clearly see the exponential window artifacts in the bottom plot for the 100 ms time constant. It generally takes over three time constants for the RMS estimate to converge near the expected RMS value, which is evident in the middle plot of Figure 12.8.

Clearly, the more the averaging one does, the smaller the standard deviation of the RMS signal will be. This can be most easily seen for the low-SNR case since the plots are autoscaled. Recall that

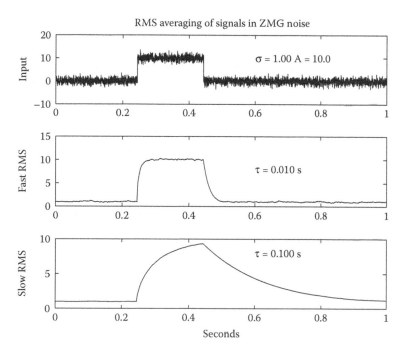

FIGURE 12.8 RMS averaging of a unity variance ZMG noise signal with a signal pulse of amplitude 10 from about 0.25 to 0.45 s (top plot) showing RMS averaging with a short and long time constant (middle and lower plots).

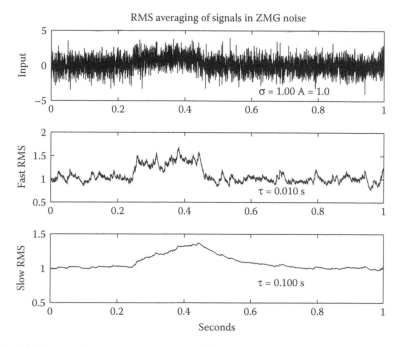

FIGURE 12.9 RMS averaging of a unity variance ZMG noise signal with a signal pulse of amplitude 1 from about 0.25–0.45 s (top plot) showing RMS averaging with a short- and long-time constant (middle and lower plots).

the peak of the Rician PDF is at $\sqrt{\sigma_x^2 + S_x^2}$ and the shape of the PDF is quite different from the Gaussian, but only two random samples were summed in its derivation. If we sum random variables the means and variances add in the result. If we scale a random variable (multiply by a constant as part of an RMS average) the resulting mean and standard deviation—not the variance) scales with the constant. Also, the shape of the PDF will tend toward Gaussian as N gets large in the average. By large we mean $N > 5$ for the Rayleigh PDF since its shape is fairly close to Gaussian to begin with. Figure 7.2 gives a good representation of the central limit theorem for a number of Chi-Square distributions of order higher than 2 (two random variables squared and summed).

For the RMS signal in the middle plot of Figure 12.9 for the low-SNR case, the mean of the RMS signal initially matches the expected RMS level of 1.0 for unity variance ZMG noise, but the standard deviation of the RMS signal, thanks to the approximately 41 samples averaged, is reduced to about 0.156 which is $41^{-1/2}$. This corresponds to the wiggles seen throughout the middle plot in Figure 12.9. For the bottom plot, again we have the expected mean, but the variance is much smaller, thanks to the approximately 410 samples in the RMS estimate, and is about 0.05. The long average of the bottom plot nearly removes the noise randomness entirely, but takes a very long time to detect the signal burst as well as recover after the burst has gone. This is the general trade-off for most estimation and detection problems. How does one best ignore the noise in the signal?

Figure 12.10 performs a very simple analysis on the amplitude of the high-SNR signal in the top plot of Figure 12.8, found by squaring the signal and then taking the square-root of each sample. We start with a very low detection threshold Λ and count the number of samples above this threshold in a 1-s signal recording 4096 samples long. Dividing this count by the sample total gives a reasonable P_d for the chosen threshold Λ. With Λ set below zero, 100% of the samples are detected. As Λ is elevated, we start missing samples, seen as the declining curves in Figure 12.10 as one progresses to the left. With Λ set well above the signals, none of the samples are detected, seen as the curves approach zero on the left side of Figure 12.10. These scans of the detection threshold represent an integration of the PDF from the detection threshold out to a high threshold value (approaching infinity theoretically). One can clearly see two "shelves" in the curves of Figure 12.10, indicating that there are two distinct statistical states in the duration of the signal. One could examine the noise-only part of the recording and the signal burst part separately to more clearly define the mean and

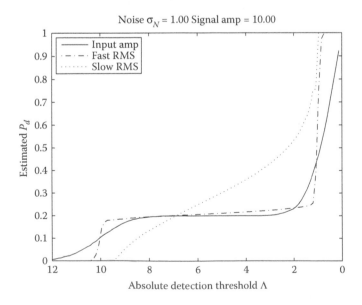

FIGURE 12.10 Brute-force-integrated detections as a function of detection threshold Λ for the high-SNR signals in Figure 12.8.

variance of these distinct sections of the recording. Or one could use the section around the transition to the signal burst to examine where one might set a detection threshold for good performance. The "brute force" integration of the signal will work for any PDF, even where the background noise contains sinusoids and non-Gaussian noise events. But since we are doing RMS averaging, there is a pretty good reason to expect the RMS noise signal to have a nonzero mean and a Gaussian variance reduced from the original signal by the amount of averaging done. Figure 12.11 shows the RMS averaging for a low-SNR signal where the signal burst and noise mean are both unity. Note that the RMS level seen in the middle plot of Figure 12.11 converges to around $2^{1/2}$ during the signal burst as predicted by $\sqrt{\sigma_x^2 + S_x^2}$.

There is an analytical solution for the integration of a Gaussian PDF from $-\infty$ to some point x called the *error function* or "erf(x)" [7]. The integral from x to $+\infty$ is called the *complimentary error function* or *erfc(x)*. The *erfc(x)* is the most useful to us because we are interested in the probability of the noise signal exceeding our chosen threshold and being falsely called "detection." This is called a *false alarm*, or false detection, when you are looking at noise only. We would like to set the detection threshold high enough so that the probability of a false alarm P_{fa} is managed at a very low rate, but not at such a low rate that we do not detect low-SNR signals because the detection threshold Λ is set too high [8]. To optimize the detection threshold, we cannot use the brute force integration because we need to examine very low probabilities which would take a very long time to integrate. We begin by examining the integration of a squared exponent.

$$\int_{-\infty}^{+\infty} e^{-x^2}\, dx = 2\int_{0}^{+\infty} e^{-x^2}\, dx$$

$$= 2\int_{0}^{+\infty} \frac{1}{2} e^{-t} t^{-(1/2)}\, dt$$

$$= \Gamma\left(\frac{1}{2}\right) = \sqrt{\pi} \tag{12.27}$$

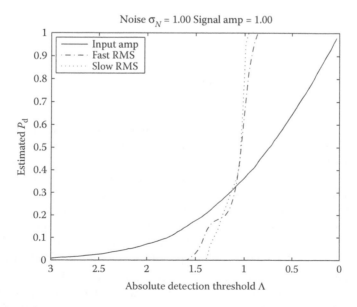

FIGURE 12.11 Brute-force-integrated detections as a function of detection threshold Λ for the low-SNR signals in Figure 12.9.

Following the same change of variable procedure for the Gamma function in Equation 12.27 we find

$$\int_{-\infty}^{+\infty} e^{-x^2/2\sigma_x^2}\, dx = \sigma_x \sqrt{2\pi} \tag{12.28}$$

and since we require that the integral of the Gaussian PDF be unity, we now have a derivation for the Gaussian PDF.

$$p(x) = \frac{1}{\sigma_x \sqrt{2\pi}} e^{-x^2/2\sigma_x^2} \tag{12.29}$$

The error function is defined as

$$erf(x) = \frac{2}{\sqrt{\pi}} \int_0^x e^{-t^2}\, dt, \tag{12.30}$$

and the complimentary error function is defined as

$$erfc(x) = 1 - erf(x)$$
$$= \frac{2}{\sqrt{\pi}} \int_x^{+\infty} e^{-t^2}\, dt. \tag{12.31}$$

The P_d of a sample above the threshold is therefore

$$P_d(\Lambda) = \int_\Lambda^{+\infty} \frac{1}{\sigma_x \sqrt{2\pi}} e^{-(x-M_x)^2/2\sigma_x^2}\, dx$$
$$= \frac{1}{2} erfc\left(\frac{\Lambda - M_x}{\sigma_x \sqrt{2}}\right). \tag{12.32}$$

Equation 12.32 provides a model to examine very small P_d without having to brute-force integrate enormous amounts of single samples. Figure 12.12 shows the P_d for a wide range of SNRs and detection thresholds for unity variance noise and four RMS averages. The SNR = 0 case is the noise-only case. We call detection for the noise-only case a false alarm, and so the corresponding P_d for the noise-only case is the P_{fa}. Setting the absolute detection threshold to about 3.5 would produce a P_{fa} of about 10^{-6}. But when you consider the recording sample rate of 4096 Hz, only one false alarm is produced every 244 s on average, or a false alarm about every 4 min. One may wish to set the threshold even higher for one false alarm per day, per week, or even per month. This is why we need the error function model to make performance estimates for the detector. To model higher SNRs we simply shift the mean of the PDF according to $\sqrt{\sigma_x^2 + S_x^2}$.

A much more general plot called the receiver operating characteristic (ROC) is shown in Figure 12.13 for the four-RMS average case. This plot compares the nonzero SNR P_d with the P_{fa} for the SNR = 0 case. ROC curves can be generated for any SNR and for any number of RMS averages.

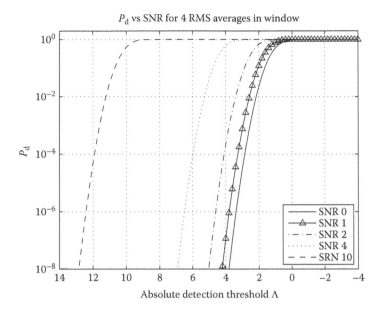

FIGURE 12.12 Calculated (P_d) versus detection threshold Λ for various SNRs and four RMS averages.

One can select the desired P_{fa} and evaluate the expected P_d performance for any SNR. Once an acceptable P_{fa} is selected, the threshold Λ is determined.

$$\Lambda = M_x + \sigma_x \sqrt{2} erfc^{-1}\left(2P_{fa}\right) \tag{12.33}$$

Figure 12.14 shows the absolute threshold level and P_{fa} for the four-RMS average case. Since we had unity variance noise, the mean M_x in Equation 12.33 is unity and the threshold in Figure 12.14

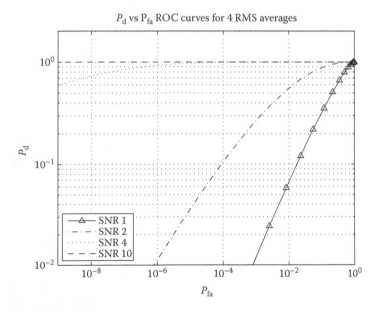

FIGURE 12.13 ROC curve comparing noise-only false detections (P_{fa}) to signal detections for various SNRs and four RMS averages (some averaging required for the PDF to be Gaussian in shape).

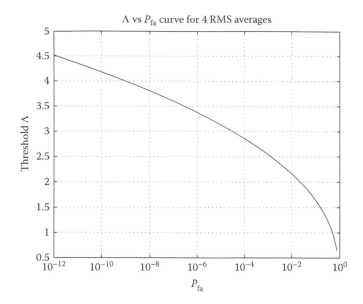

FIGURE 12.14   Curve showing the fixed relationship between the detection threshold Λ and the false alarm rate for the signals with four RMS averages.

can be interpreted as a "threshold- to-noise ratio" or TNR. A fixed TNR has a CFAR independent of the actual noise amplitude.

$$\text{TNR} = \frac{\Lambda}{M_x} = 1 + \frac{\sigma_x \sqrt{2} erfc^{-1}(2P_{fa})}{M_x} \tag{12.34}$$

The TNR is only a function of the desired P_{fa} and the amount of RMS averaging done to the signal. Figures 12.15 through 12.17 show the P_d versus Λ, P_d versus P_{fa}, and TNR versus P_{fa} for the 41-RMS average case, respectively. The effect of the 41 averages is to make it extremely easy to detect any signal burst above an SNR of 2 with virtually no false alarms.

CFAR detection is very desirable for situations where the background noise varies over time and the detection threshold must be maintained so that the false alarm performance remains at its designed operating point. Otherwise, an increase in background noise could lead to an explosion of false alarms. A decrease in background noise without a threshold adjustment means that the system fails to exploit the environment improvement. So how does one keep the TNR from jumping around with a fast RMS estimate? This is where the longer estimate for background noise is used (the bottom plots in Figures 12.8 and 12.9). Using 410 samples in the RMS average produces a very smooth RMS signal in Figures 12.8 through 12.9 that is very slow to react to the signal burst. Assuming that the detector is supposed to alert us when the onset of a signal burst, an ultraslow RMS average is desired to keep the detection threshold stable, but also able to react to slow changes in background noise. This too comes at the expense of increased "numbness" after repeated signal bursts which will elevate the estimated background. If the slow return of the long-term average is a concern, one can simply reset the long-term average to the short-term average when the short-term average is below the long term, or use a tracking filter such as an α–β–γ tracker or Kalman filter. The tracking filter, with its states and covariances available, is an ideal algorithm for linking to CFAR detection. One can set process noise levels to achieve short- and long-term RMS average goals, yet have the states rapidly respond to signal transients one is trying to detect. Linking these powerful algorithms is strongly encouraged.

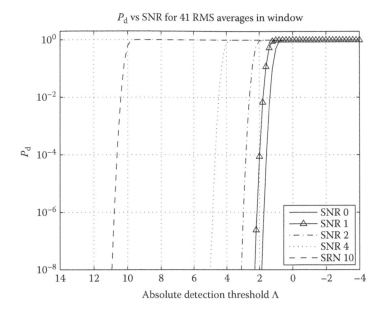

FIGURE 12.15   Calculated probability of detection (P_d) versus detection threshold Λ for various SNRs and 41 RMS averages corresponding to the fast RMS averaging in Figures 12.8 and 12.9.

12.3   STATISTICAL MODELING OF MULTIPATH

For any wavenumber sensor system, whether it be a radar, sonar, ultrasonic images, or structural vibration modal filter, underlying assumptions of the wave shape, wavelength (or wave speed), and temporal frequency are required to reconstruct the wave field. A wave radiated from a distant point source (a source much smaller than the wavelength) in a homogeneous infinite-sized medium is essentially and plane-shaped wave front beyond a hundred wavelengths as described in Section 7.4.

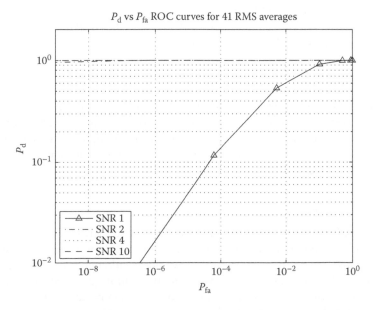

FIGURE 12.16   ROC curve comparing noise-only false detections (P_{fa}) to signal detections for various SNRs and 41 RMS averages corresponding to the fast RMS averaging in Figures 12.8 and 12.9.

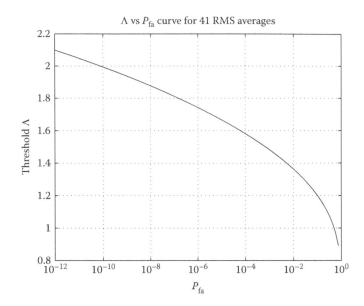

FIGURE 12.17 Curve showing the fixed relationship between the detection threshold Λ and the false alarm rate for the signals with 41 RMS averages.

Given a single plane wave propagating across a 3D array of sensors, any AOA can be easily determined. For a line array of sensors, the wave frequency and speed are needed to determine the AOA relative to the line array axis. For a planar array, the AOA can be determined in a half-space, and if the source is in the same plane as the sensors, the AOA and wave speed can be determined explicitly. For a 3D sensor array in an infinite homogeneous medium, any AOA and wave speed can be measured explicitly for a simple plane wave. However, when more than one plane wave (from correspondingly different directions), but of the same frequency, arrive at the sensor array, the direction finding problem becomes far more complicated. We term this situation as multipath propagation, which occurs in real-world applications of sensor technology whenever the propagation medium is either inhomogeneous or has reflecting surfaces.

There are two general classes of multipath which can exist at any given instant of time. The first we call "coherent multipath" and results from a single-source radiator and multiple ray paths to the sensors from either an inhomogeneous medium or reflectors or scatterers in the propagation medium. The second class of multipath is from multiple sources radiating the same frequency (or wavelength—the sources could be moving and radiating slightly different frequencies before Doppler) at the sensor array. This "multisource" multipath carries an important physical distinction from a coherent multipath, in that over time, the phases of the sources will become incoherent, allowing the arrival angles to be measured. The same is true for coherent multipath when the source is moving relative to the sensor array, or if the multipath is changing due to a nonstationary inhomogeneous medium, reflecting surface, or scatterers.

12.3.1 MULTISOURCE MULTIPATH

Multisource multipath results in statistically independent phases across the array: Chapter 14 presents adaptive beamforming algorithms to deal specifically with multisource multipaths. It is very important to fully understand the physics of the wave propagation before designing the sensor and signal processing machine to extract the propagation information. Both the coherent multipath and the multisource multipath have a significant impact on detection because the multiple waves at the sensor site interfere with each other to produce signal enhancement at some sensor site and signal cancellation at other sensor sites in the array. The wave interference throughout the sensor array allows one to detect the presence of multipath. However, even in a homogeneous medium, one cannot resolve the angles of arrival unless one knows the amplitude, phase, and distance of each source (if multisource), or the amplitude and phase differences between ray paths (if coherent multipath). Section 14.4 examines the case of known sources, where one is not so interested in localizing sources, but rather localizing reflecting objects. From a signal detection point of view, the possibility of multipath interference impacts the detection algorithm with additional signal fluctuations at the receiver independent of the background noise [9].

12.3.2 COHERENT MULTIPATH

Coherent multipath is characterized by the phases of the arrivals being dependent solely on the propagation channel, which in many cases is relatively stationary over the detection integration timescale. Consider the simple two-ray multipath situation depicted in Figure 12.18 where a point source and a receiver are a distance h from a plane boundary and separated by a distance R in a homogeneous medium with constant wave speed c in all directions (if the field has a nonzero divergence, or net flow, the wave speed is directional). The field at the receiver can be exactly expressed using the sum of a direct and reflected ray path. For simplicity, we will assume an acoustic wave in air where the boundary is perfectly reflecting (infinite acoustic impedance). This allows the incident and reflected waves to have the same amplitude on the boundary and our analysis to be considerably more straightforward. Figure 12.18 shows the reflected ray path using an equivalent "image source," which along with the real source, would produce the same field at the receiver with the boundary absent. If the impedance of the boundary were finite and complex, the image source would have an amplitude and phase shift applied to match the boundary conditions on the reflecting surface. The equation for the field at the receiver is

$$p(R) = Ae^{j\omega t}\left\{ \frac{1}{R}e^{-jkR} + \frac{1}{R_f}e^{-jkR_f} \right\} R_f = \sqrt{R^2 + 4h^2}, \tag{12.35}$$

where A is the source amplitude, R is the direct path, R_f is the reflected path length, k is the wavenumber ($k = \omega/c = 2\pi/\lambda$), and ω is the radian frequency. Figure 12.19 shows the multipath frequency response at the receiver compared to the response without the reflected path assuming $R = 100$ m, $h = 30$, and an acoustic wave speed of 345 m/s.

Equation 12.35 is exact for a perfectly reflecting planar surface. The model includes a $1/R$ attenuation for a spherical wave, but it does not include the nearfield phase differences between the pressure and velocity of a true acoustic spherical wave, which would allow for proper diffraction modeling. The reader should be advised that the model here has limitations when the number of wavelengths between the source and the receiver is small. Before we consider a random surface (or random distribution of wave scatterers), we examine the case where we have a line array of sensors rather than a single receiver. Figure 12.20 shows how two waves with the same temporal frequency,

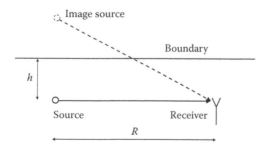

FIGURE 12.18 Direct and reflected path (dotted) ray paths depicted using an image source analogy.

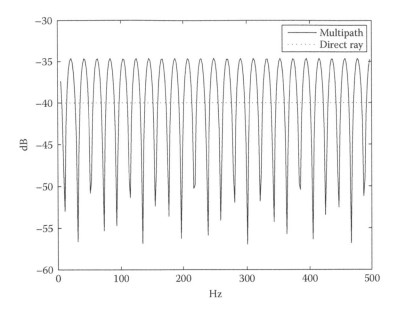

FIGURE 12.19 Direct and multipath (direct plus reflected) responses for a source and a receiver separated by 100 and 30 m from an acoustically reflecting boundary in air ($c = 344$ m/s).

arriving from two angles simultaneously, will cause an interference pattern spatially across a sensor array. In the time domain, each sensor detects the same frequency. But spatially, the amplitude and the phase vary across the sensor locations due to the sum of the waves.

For example, if a line array observes a plane wave, the spatial amplitude will be constant and the wavelength "trace" will be representative of the frequency, wave speed, and AOA. A plane wave passing the line array from a broadside direction will have a wavelength trace which looks like an infinite wavelength, while the same wave from the axial direction will have a wavelength trace equal to the free wave wavelength ($c/f = \lambda$). Now if two plane waves of the same temporal frequency arrive at the line array from different directions, the wavelength traces sum, giving an interference pattern where the "envelope wavelength" trace is half the difference of the respective traces for the two waves, and the "carrier wavelength" trace is half the sum (or the average) of the two respective wavelength traces. This is exactly the same mathematically as an AM signal. If a direction-finding algorithm uses the spatial phase to detect the AOA, it will calculate an angle exactly in between the two actual arrival angles (this would be weighted toward the stronger wave for two unequal amplitude plane waves). If a particular sensor happens to be near a cancellation node in the trace envelope, it will have very poor detection performance. This is often experienced with cordless or

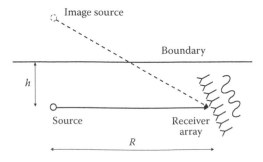

FIGURE 12.20 Coherent multipath showing wave interference at a receiving array of sensors showing a spatial envelope of amplitude variations across the array.

cellular telephones as well as automobile radios in a multipath environment. If the line array length is long enough to observe a significant portion of the envelope peaks and dips, the individual arrival angles can generally be determined using beamforming as described in Section 8.1.

12.3.3 STATISTICAL REPRESENTATION OF MULTIPATH

Statistical representation of multipath is useful for situations where coherent reflections are coming from random surfaces or refraction in a random inhomogeneous wave propagation medium. Good examples of this are when the scanning system (radar or sonar) is traveling over or near a rough surface, or if the medium has turbulence or scatterers. Physically, we need to describe the variance of the multipath phase for a given frequency, and the timescale and/or spatial scale for which ensemble observations will yield the modeled statistics for the multipath. When the medium is nonstationary, one also has to consider an outer timescale, beyond which one should not integrate to maintain a statistical representation of the multipath medium.

Consider a multipath due to a large number of small scatterers with perfect reflecting properties. Using Babinet's principle in optics, our point scatterer will reradiate the energy incident on its surface equally in all directions. If the scatterer is large or on the order of the wavelength, or if the scatterer is not perfectly reflecting, the subsequent reradiation is quite complex and beyond the scope of this book. We can exploit Babinet's principle by approximating the reflection of a wave from a complicated boundary by replacing the boundary with a large number of point scatterers. The magnitude and phase of the waves reradiated by the point scatterers is equal to the magnitude and phase of the incident field at the corresponding position along the boundary. This technique is well known in acoustics as Huygens' principle which states that any wave can be approximated by an appropriate infinite distribution of point sources. Summing the responses of the Huygens' point sources over a surface, albeit a reflecting surface or a radiating surface, is called a Helmholtz integral. Therefore, we will refer to approximating the reflection from a geometrically complicated boundary as a Helmholtz–Huygen technique.

Figure 12.21 compares the exact image source solution to the Helmholtz–Huygens' technique using 2000 point sources on the boundary $dl = 0.3$ m apart, where the wave speed is 345 m/s, the

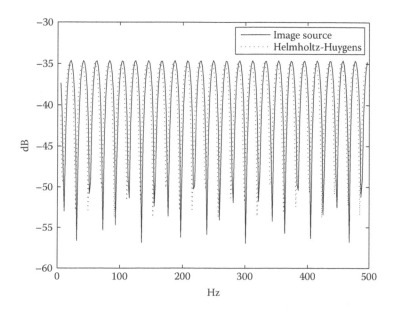

FIGURE 12.21 Comparison of an image source and Helmholtz integral approximation method for modeling the multipath at the receiver showing good agreement for the approximation.

boundary is 30 m from both the source and the receiver which are separated by $R = 100$ m as shown in Figure 12.18. Because 2000 point sources is hardly infinite and we are interested in only a frequency range from 0 to 500 Hz, we also apply an empirical loudness adjustment over frequency so that the source energy per wavelength is constant over frequency at any one position on the boundary. This normalizes the channel frequency response in Figures 12.21 and 12.22. Using a Cartesian coordinate system with the direct path along the x-axis and the normal to the boundary along the y-axis, the magnitude and phase of each point source modeled on the boundary is

$$A_n = \frac{\lambda}{2\,dlR_{1,n}} e^{-jkR_{1,n}} \quad R_{1,n} = \sqrt{x^2 + h^2} \quad x = ndl, \tag{12.36}$$

where $R_{1,n}$ is the ray from the main source to the nth point source model on the boundary. The Helmholtz–Huygens' pressure approximation at the receiver is

$$p(f) = \frac{A}{R} e^{-jkR} + \sum_{n=1}^{N} \frac{A_n}{R_{2,n}} e^{-jkR_{2,n}} \quad R_{2,n} = \sqrt{(R-x)^2 + h^2}, \tag{12.37}$$

where $R_{2,n}$ is the ray from the nth boundary source to the receiver and R is the direct path length. $N = 2000$ sources, and they were spaced evenly $dl = h/R$ m apart symmetrically about the midpoint of the physical source and the receiver to give a reasonable approximation for $h = 30$ m. Note that both R and h can be seen as large compared to wavelength for frequencies above around 150 Hz.

Figure 12.22 shows the utility of the Helmholtz–Huygens' technique for modeling rough boundaries. In this figure, a ZMG distribution with a standard deviation of 0.25 m is used to randomly distribute the point sources on the y-axis centered around $y = h$, the boundary. The boundary sources therefore have a y-coordinate of $y + \xi$. Figure 12.22 clearly shows the transition from deterministic to stochastic wave interference somewhere in the 200 to 300 Hz range. This simple example of

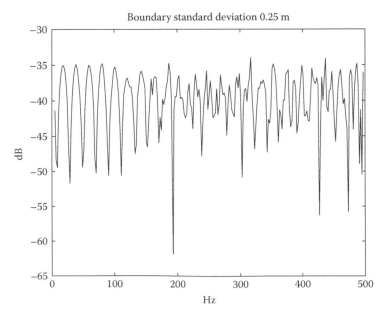

FIGURE 12.22 Direct and scatter ray paths approximated using the Helmholtz integral showing a randomized surface with 0.25 m standard deviation begins to have an effect on the multipath for frequencies above about 100 Hz where the wavelength λ is 3.4 m (three standard deviations is about $\lambda/4$).

rough boundary scattering exposes interesting wave phenomena where the "roughness" of the reflecting boundary depends on wavelength along with the variance of the roughness. At low frequencies, the wavelength is quite large compared to ξ and the scattering is not detectable. But when the boundary roughness is approaching a quarter wavelength or more, the effect of the rough boundary is readily seen. When the boundary roughness is not ZMG, one would calculate a spatial Fourier transform of the rough boundary and use a power spectrum estimate to get the variance at a wavenumber appropriate for the boundary reflection to determine the scattering effect.

We can develop a model that relates the ZMG distribution in the y-direction along the reflecting surface to the reflected path length.

$$R_\xi = \sqrt{R^2 + 4(h+\xi)}$$
$$= R_f \sqrt{1 + \frac{8h\xi + 4\xi^2}{R^2 + 4h^2}}. \tag{12.38}$$

Equation 11.64 is easily approximated noting that the second term in the square root is quite small.

$$\sqrt{1+\varepsilon} \approx 1 + \frac{\varepsilon}{2}; \quad \varepsilon \gg 1. \tag{12.39}$$

The reflected path length including the random variation R_ξ is given in Equation 11.66 in terms of the vertical distance to the boundary h, the reflected path length with no roughness R_f, and the roughness random variable ξ.

$$R_\xi \approx R_f + \frac{4h\xi}{R_f} \tag{12.40}$$

Clearly, the standard deviation of the reflected path length σ_ξ can be equated to the standard deviation of the boundary along the y-axis using

$$\sigma_R \approx \frac{\sigma_\xi 4h}{R_f} \tag{12.41}$$

we can model the statistics of the acoustic pressure at the receiver by adding our random variable to Equation 11.61.

$$p(R) + p' = \frac{A}{R} e^{j\omega t - kR} \left\{ 1 + \frac{R}{R_f} e^{-jk(\Delta R - \xi)} \right\} \Delta R = R_f - R \tag{12.42}$$

The term p' in Equation 12.42 represents the random part of the pressure at the receiver. We note that for the pressure to change from the midpoint loudness at −40 dB to the peak loudness at −34 dB in Figure 12.22, a phase change of $\pi/2$ is needed in the reflected path. If we consider a noticeable pressure fluctuation to be corresponding to a phase fluctuation of $\pi/4$, and note that $4h/R_f$ is about unity for $R = 100$ and $h = 30$, we need k on the order of π to have significant scattering effects. This corresponds to a frequency of about 170 Hz. Figure 12.22 clearly shows the stochastic effect of the random reflection boundary in this frequency range. If the source and the receiver are moved closer to the boundary, the path randomness decreases. If h is increased, the path length variances increases to a limit of $2\sigma_\xi$ when $h \gg R$. However, for large h, the reflected path is much larger than the direct path that the pressure fluctuations again decrease. The maximum pressure fluctuations are for a ratio of $h/R = 0.5$. For low frequencies, k (the wavenumber) is small (wavelength large) and the fluctuations are also correspondingly small.

One might be misled into thinking that wave scattering at boundaries is generally a bad thing. Consider that a coherent reflection can lead to a severe cancellation of the wave at the receiver. When the wave is scattered at the boundary these paths of total cancellation do not exist at any frequency or position, which makes it easier to separate from the direct path by coherent averaging and the fact that the randomly scattered wave is weaker at the receiver than the direct wave. This is why speech communication is much easier in an old stone Cathedral than in a racquetball court. The surfaces in the old Cathedral are not random, but contain enough detail from carvings and statues to randomize the reflections and eliminate coherent echoes. This is not the case for the flat walls of a racquetball or squash court where the coherent echoes make simple speech communication extremely difficult.

In architectural acoustics there is a device specifically designed to scatter waves in all directions called a "quadratic residue diffuser" or QRD panel [10]. It is based on a short pseudorandom number sequence which is precisely white, but is also periodic. The autocorrelations of the pseudorandom sequence produce an exact delta-like impulse at the 0th lag with a slight amplitude offset for the whole sequence. The Fourier transform of the QRD surface produces a random beam pattern independent of the incident wave angle of frequency within the design bandwidth. QRD panels are very effective at eliminating problem reflections in buildings and can also be used for problematic reflections near transmitting antennas and receivers. Designing a QRD is simple. One begins by selecting N as an odd prime number where $N = 1 + f_{high}/f_{low}$ where f_{high} and f_{low} define the design bandwidth for the panel. A maximal length sequence (MLS) is generated as $s_n = (n^2 \bmod N)$, and so for $N = 11$, the sequence would be [0 1 4 9 5 3 3 5 9 4 1 0] and could be repeated multiple, but complete, times for a larger panel if needed. The QRD panel has wells (holes) of depth $d_n = (\lambda/2N)$ s_n where s_n is from the MLS sequence. The low-end design wavelength $\lambda = c/f_{low}$ and the well width is $w = c/2f_{high}$. A 2D QRD panel is designed by summing the QRD sequences for each direction and reducing the design depth accordingly.

12.3.4 Random Variations in Refractive Index

Random variations in refractive index (changes in wave speed relative to the mean) are also a very important concern in multipath propagation. We are not only interested in how a refractive multipath affects detection, but also how received signal fluctuations can be used to measure the propagation medium. This physical effect occurs in electromagnetic propagation due to variations in propagation speed due to humidity, fluctuations in the earth's magnetic field, solar activity, and turbulence in the ionosphere. At night, when solar interference is low, long-range AM and short-wave broadcasts have characteristic waxing and fading, and at times, unbelievably clear reception even from distances of thousands of kilometers. The complexity of the propagation is due to multipath interference and the fluctuations are due to time-varying factors such as winds, turbulence, and fluctuations in the earth's magnetic field (due to magma movement), rotation, and solar electromagnetic waves interacting with the ionosphere. One can think of the earth's atmosphere, with its charged particles in the ionosphere and fluctuating ground plane due to rain and weather at the surface, as a huge waveguide defined by two concentric spheres. Changes in the boundary conditions affect which modes propagate in the waveguide and at what effective speed (a ray which reflects off the ionosphere and ground travels slower than a ray traveling more parallel to the boundaries). Therefore, the receiver experiences multipath due to refraction of the waves as well as reflection. These magnetic fields are what keep the sun's solar wind from ionizing and stripping away much of the earth's atmosphere.

In sonar, the refractive index of a sound wave in seawater is affected by the warm temperature near the surface, and changes in temperature, salinity, and pressure with depth and with ocean currents. Sound channels or waveguides can form around layers of slow propagation media. The channel happens physically by considering that a plane wave leaving the layer will have the part of the wave in the faster media (outside the layer) outrun the part of the wave inside the layer, thus refracting the wave back into the layer. Whales are often observed using these undersea channels for long-range communication. Clearly, it is a great place to be for a quiet surveillance submarine but not a good place to be if you are in a noisy submarine and want to avoid being detected acoustically. The seasonal changes in undersea propagation conditions are much more gradual than that in the atmosphere, where dramatic changes certainly occur diurnally, and can even occur in a matter of minutes due to a shift in wind direction. In general, sound propagation from sources on the ground to receivers on the ground is much better at night than during the day, and is always better in the downwind propagation direction. This is because at night the colder heavier air from the upper atmosphere which settles near the ground is unheated by the sun. The slower air near the ground traps sound waves since any wave propagating upward has the upper part of the wave outrunning the lower part, thus refracting the wave back down to the ground. This effect is easily heard on a clear night after a warm sunny day. For example, most readers can try, on a cool night after a sunny day, to listen to a distant sound source such as truck traffic on an interstate highway a few kilometers away. This would not be impossible during a hot afternoon due to the upward sound refraction. However, even the truck tire noise (a much higher frequency) can be heard to fluctuate over periods of 10–15 s at night. This is likely due to nocturnal turbulence from heavy parcels of air displacing warmer air near the ground. The same effect can be seen in the twinkle of lights from a distance at night. The same refractive effect happens when sound propagates downwind, since the wind speed increases with height and adds to the sound speed. However, like in the long-range short-wave radio broadcasts, the atmospheric turbulence due to wind and buoyancy will cause acoustic propagation fluctuations due to stochastic variations in index of refraction.

Consider the case of acoustic propagation at night in the atmosphere. We will greatly simplify the propagation problem greatly by eliminating wind and splitting the atmosphere into two layers: the lower layer near the ground with a constant temperature, and an upper layer with a positive temperature gradient to represent the lighter warmer air which supports faster, downward refracting, sound propagation. Figure 12.21 shows graphically the direct and refracted sound rays where the direct ray is propagating in an air layer with constant sound speed, while the refracted ray propagates in an upper air layer where the air temperature (and sound speed) increases with increasing height. For our simplified model of a constant sound speed gradient, we can express the sound velocity profile as

$$c(z) = c_0 + z \frac{dc(z)}{dz} \tag{12.43}$$

where c_0 is the sound speed in the constant lower layer near the ground and z is the height in the upper layer. It is straightforward to show that for a linear gradient shape, the ray travels along the arc of a circle. The radius of this circle, R in Figure 12.23, is found by solving for $c(z) = 0$.

$$R = \frac{c_0}{dc(z)/dz} \tag{12.44}$$

For any length of direct ray x in Figure 12.23, there is a corresponding refracted ray s (it could be more than one ray for more complicated propagation) which intersects both the source and the receiver at a "launch angle" of $\theta/2$.

$$\frac{\theta}{2} = \sin^{-1}\left(\frac{x}{2R}\right) \tag{12.45}$$

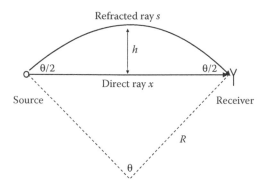

FIGURE 12.23 Sketch of simple downward refraction for outdoor sound propagation showing the direct ray path (horizontal ray x) and the curved refracted ray for a simple linear sound speed gradient vertically in the air, typical of early-evening temperature inversions.

The refracted ray length is simply

$$s = \theta R \tag{12.46}$$

and the ray's maximum height is found by solving

$$h = R\left[1 \pm \sqrt{1 - \frac{1}{4}\left(\frac{x}{R}\right)}\right] \tag{12.47}$$

and can be useful in practical applications to see whether noise barriers of buildings effectively block the sound path. This is one reason why noise barriers are really only helpful in blocking noise in the immediate vicinity of the barrier and not at long distances. As a practical example, if $c_0 = 345$ m/s, $dc/dz = +0.1$ m/s/m, and $x = 1$ km, we find $R = 3450$ m, $\theta/2 = 8.333°$, $h = 36.4$ m, and $s = 1003.5$ m, giving a path length difference for the two rays of only 3.5 m. These are very small differences and launch angles. However, when we consider a fluctuation in the sound speed gradient due to turbulence, a significant change in path length is possible.

Using a Taylor-series approximation for the arc sine function, we can express a function for the refracted ray length as

$$s = \frac{2c_0}{(dc/dz)}\left\{\frac{x}{2R}\left(1 + \frac{1}{6}\left(\frac{x}{2R}\right)^2\right)\right\} \tag{12.48}$$

which can be reduced to show a path length difference of

$$s - x = \frac{x^3}{24c_0^2}\left(\frac{dc}{dz} + \zeta\right)^2 \tag{12.49}$$

where ζ is introduced as a random variable to account for fluctuations in the sound speed gradient. For the above practical example, Equation 12.49 provides an estimated path difference of $s - x = 3.5 + 70\zeta + 350\zeta^2$ m. Clearly, a very small fluctuation in sound speed gradient leads to a very significant refracted ray path length.

As with the coherent reflection from the boundary, the sound pressure fluctuation scales with frequency since the phase difference between the direct and refracted path is the wavenumber times the path difference in meters. This means that low frequencies are much less impacted than high frequencies for a given overall distance. If we express the effect in terms of wavelengths, we can say that the pressure fluctuations will become severe when the path lengths exceed about a quarter wavelength. For small sound speed gradients this might require over a 1000 wavelengths propagation distance. But for larger sound speed gradients due to flow near a boundary, the effect may be seen over only a few dozen wavelengths. This technique of stochastically characterizing multipaths could prove useful in medical imaging of arteries, chemical process sensors and control, and advanced sensor tasks such as pipe corrosion and lagging inspection by sensing the structure of the turbulent boundary layer. It may also be valuable as a signal processing tool for environmental sensing in the atmosphere and undersea, or for electromagnetic propagation channel studies.

12.4 MATLAB® EXAMPLES

We used six m-scripts in this chapter for the figures, summarized in Table 12.1. The m-scripts are available from the book website or from the author and are very straightforward demonstrations of the presented concepts. "DETSIG.m" is a simple routine to generate a ZMG noise signal with unity variance and place a basic dc-offset type of signal pulse in a short recording so that we can evaluate the statistical properties. For a high SNR like 10, the results are quite obvious and plots show the dramatic effect RMS averaging has on one's ability to detect where the signal is on and off. At a lower SNR like 1.0, the detection is slightly less obvious even with RMS averaging. The m-script puts the RMS averaging of Equation 12.26 into a time-constant framework and so it is directly comparable to the RC time constant of a resistor–capacitor circuit for the low-pass filter-rectified waveforms.

The m-scripts "PDFS.m" and "RICIAN.m" are straightforward plots of the PDF s for comparison, given an initial real noise signal with unity variance. The Rician PDF clearly shows the shape dependence on SNR which is especially important if doing detection on an FFT frequency bin with no spectral RMS averaging. If any averaging is done, the PDF for the noise will have a Gaussian shape with mean equal to the standard deviation for the noise in the bin, and standard deviation for the PDF declining by the square of the number of effective samples in the RMS average. Also, for averaging of signals in noise, the Rician PDF takes on a Gaussian shape where the mean is the square root of the sum of the squares of the signal level and the noise standard deviation. The standard deviation of the averaged signal PDF also declines with the square root of the number of averages. As such, just a small amount of RMS averaging separates the signal and noise-only PDFs into two Gaussian-shaped PDFs separated by the SNR. This makes it fairly obvious where to place a decision threshold for whether there is signal present or not.

The m-script "PdvsLambda.m" contains a very interesting technique for examining signals and noise where you have no knowledge of the underlying statistics or when you are fairly sure that the

TABLE 12.1
MATLAB m-Scripts Used in This Chapter

m-Script	Figures
DETSIG.m	12.1–12.4
PDFS.m	12.6
RICIAN.m	12.7
PdvsLambda.m	12.8–12.11
CFAROC.m	12.12–12.17
HELMGEOM.m	12.19, 12.21, 12.22

background noise is not Gaussian. Examples of this might be where the background contains spurious noise, sinusoid harmonics, or other types of interferants. Recall that the Gaussian assumption emerges from the central limit theorem where a large number of independent events lead to a Gaussian PDF; this is not always the case. So in "PdvsLambda.m" we do a dumb brute force integration of the P_d and P_{fa} by literally setting a threshold and going back and counting the number of samples that exceed the threshold in the record. Comparing the "noise-only" and "signal" cases can lead to an obvious detection threshold choice and provide some metrics of what the false alarm rate might be. Even for the low SNR, the fast RMS averaging case in Figure 12.11, two statistical distributions can be seen by the bump in the fast RMS curve. This indicates where the noise and signal separate. One could draw a comparison of two sections of the recording, one where the signal is present and another where the signal is absent, to get the probability distributions to make an ROC curve. The P_d when only noise is present is the probability of false alarm. If each probability (signal and noise-only) is evaluated at the same set of thresholds, the two probability curves can be plotted against each other to give the estimated ROC curve—with no models for the underlying statistics! This is a very useful experimental technique and also a great way to independently evaluate a detection system.

Once some RMS averaging is employed to improve the SNR, the PDF for both signal and noise tend to have a Gaussian shape. As such, we can use well-known analytical functions such as the complementary error function $erfc(x)$ to evaluate the integral of a Gaussian PDF from x to infinity to estimate the P_d if the signal is present or P_{fa} if only noise is present. Based on the $erfc(x)$, these probabilities can be evaluated at extremely low false alarm rate to determine the optimum thresholds for signal detection. Figures 12.13 and 12.16 plot these ROC curves for 4 and 41 RMS averages, respectively. Then we can estimate the P_d for various SNRs to report the system's overall detection performance. Given system specifications for false alarm rate, the threshold is set and can be maintained as a ratio of the estimated background noise, and so the threshold "floats" as a fixed ratio to the background, thus giving a CFAR, or CFAR detection performance. We also show a long-term RMS average for this purpose, but an even better way of making a CFAR detector is to use a Kalman filter with really low process noise to estimate the noise background. CFAR detection in this way works fine for impulsive or transitory signals in slow-changing background noises, such as for a burglar alarm. In the frequency domain, one might use a nearby FFT bin known to only have background noise for the purpose of estimating the background noise for the CFAR threshold.

The m-script "HELMGEOM.m" illustrates a simple example for how a signal can become statistical in amplitude, not from additive background noise, but from changing propagation paths. The reader should be advised that this example is in fact oversimplified for the acoustic case, but is a good approximation for high frequencies. Signal scattering can occur in urban environments, outdoors, and indoors, whenever the signal propagation path contains objects $\lambda/4$ in size or larger that are reflective and moving relative to the source and receiver or have nonstationary frequencies. The signal amplitude can also be fluctuating from refractive turbulence which will be demonstrated in Section 13.4. The signal fluctuation statistics and the background noise statistics are additive at the receiver, and so both must be taken into account when designing a detector to meet a given performance requirement.

12.5 SUMMARY

Chapter 12 covers the important technology of a signal detection system, its design, and most importantly, the confidence of the detected signal based on signal and noise statistical models and direct measurements. The notion of *statistical confidence* is critical to an intelligent sensor system because it puts the detected information into proper context. When combined with other pieces of detected information, the confidence associated with each detection is extremely valuable to the process of *data fusion*, or the blending of pieces of information together to produce a *situational awareness*. Since we can robustly define the probability of signal detection P_d, and probability of false alarm (false detection) P_{fa}, for a particular signal in noise, it is prudent to compute these quantities even if they are not readily required for the detection process. It is even most beneficial to link detection statistic to tracking state vectors and their respective state prediction errors. This creates an algorithm suite where signal dynamics, set points, thresholds, and so on can all be part of a predictable decision performance for the sensor system.

A *sentient* adaptive signal processing system is defined by having the power of perception by the senses, which includes a capability to combine and place weight on particular sensor information as well as assess the validity of a particular sensor's output in the context of the overall perceived situation. Our bodies provide a straightforward analogy; after all, we are sentient beings. If one is sunburned on the face and arms, one resolves the local skin temperature differences from an awareness that (1) there has been recent exposure to sunlight or heat increasing the likelihood of sensor damage; (2) the temperature differences are independent of body position or place; and (3) actual skin temperature in the exposed areas is elevated and appears to be more red (an inflammation symptom of burn damage). One realizes and deals with the situation of being sunburned by reducing confidence in temperatures from the burned areas, supporting healing of the burned skin, and avoiding any more exposure or damage. Our bodies are temperature regulated and thus, feeling too hot or too cold can sometimes be a very good indicator of pathology. Does one feel hot yet objects are cool to the touch? Does one feel cold yet objects do not feel cold? When feeling cold does the body perspire? One resolves these sensory contradictions by association with a known state of health. When the sensors are in agreement and match a pattern with other sensor types, one associates a known environmental state from experience and also that one's own sensors are operating correctly. In this self-perceived "healthy" state, one associates a very high confidence in the sensed information. This sentient process can be constructed, albeit crudely, in machines through the use of statistical measures of confidence and data and information fusion using techniques such as fuzzy logic discussed in Chapter 17. But one cannot even attempt this level of situational awareness unless the basic parameters of signal and noise can be put into the context detection and false alarm probabilities.

Section 12.1 presented the Rician probability distribution for a sinusoid in white noise along with the virtues of time-synchronous averaging. By averaging multiple buffers of the waveform which includes a known signal and random zero-mean noise over a period which is an integer multiple of the signal period, the noise in the averaged buffer tends toward its mean of zero while the signal is unaffected. *Time-synchronous averaging is one of the most simple and effective methods for signal-to-noise improvement in signal processing and should be exploited wherever possible.* When the signal is a sinusoid, the matched detection filter is also a sinusoid of the same frequency. When the signal has some bandwidth, the matched filter for detection also has the same spectrum as the signal when the background noise is white (see the example in Section 6.4 for nonwhite noise). Given an estimate of the background noise and the matched filter signal detection output, one can set a detection threshold above the mean background noise for a desirable low false alarm rate. The probability of detection, and thus signal detection confidence, can be estimated directly from the SNR, as shown in Section 12.2. When the signal is really strong compared to the noise, the signal plus noise PDF is essentially a Gaussian distribution with mean shifted up to the signal RMS level. When the SNR is closer to unity, the density function is Rician requiring greater care in estimating the probability of detection. However, when a lot of RMS averaging is done on the random signal, the underlying probability density tends to be Gaussian. We can use the complimentary error function to aid us in modeling where detection thresholds should be to get a very low false alarm rate.

Section 12.3 examines the case where the signal propagation path can be modeled as a stochastic process. Such is the case with ground clutter in radar systems, scattering from tissue in medical ultrasound, and the twinkling of stars in astronomical observations due to turbulence multipath. In the situation of a statistical multipath, the signal level is described by a nonzero mean PDF. This impacts directly the calculation of the P_d because of the broadening of the density function (due to the convolution of the multipath density function with the signal-plus-noise density function). Given the physical distribution of scatterers, we have shown how to calculate the signal PDF. This is of course a very complicated task beyond the framework presented here for most real-world propagation situations.

PROBLEMS

1. A 1-s real-analog recording of a 1 V RMS (vrms) 50 Hz sinusoid signal in 1 vrms Gaussian white noise is available for digital processing using a spectrum analyzer.
   a. If we low-pass filter at 400 Hz, sample the recording at 1024 samples/s, and calculate a single 1024-point FFT, what is the spectral SNR?
   b. If we low-pass filter at 3.2 kHz, sample the signal at 8192 samples/s, and calculate a single 8192-point FFT, what is the spectral SNR?
2. A seismic array geophones (measure surface velocity) is used for detecting rock slides and needs to have no more than one false alarm per month on detection trial every 50 ms. Assume a signal magnitude detector in ZMG background noise with standard deviation of 1×10^{-7} m/s. What is the false alarm rate in % and how would you determine the detection threshold?
3. A rock slide occurs and a peak signal at the geophone is 10×10^{-7} m/s where the background noise is 1×10^{-7} m/s. If the detection threshold were set at 10×10^{-7} m/s as well, what is the P_d?
4. Derive the Rician PDF. Show that for very high SNR, the Rician PDF can be approximated by a Gaussian density function with mean equal to the signal level.
5. Show that for M averages of a Gaussian random variable, the mean stays the same while the standard deviation decreases by a factor of $M^{1/2}$.
6. Suppose one has a large number of complex FFT spectral buffers (of the same size and resolution) of a sinusoid in white noise with unknown gaps of time between each FFT buffer.
   a. If one simply added all the FFT buffers together into a single buffer, would one expect an increase in SNR?
   b. If one first multiplied each FFT buffer by a complex number to make the phases at the frequency bin of the sinusoid identical and then added the buffers together, would the SNR increase?
   c. If one first multiplied each FFT buffer by a complex number corresponding to the linear phase shift due to the time the FFT input buffer was recorded and then added the buffers together, would the SNR increase?
7. If one has a 1% false alarm rate on a magnitude detector of a signal peak in Gaussian noise with no averaging, how much averaging will reduce the false alarm rate to below 0.1% keeping the absolute detection threshold the same?
8. Describe qualitatively how one could model the detection statistics of a sinusoid in white Gaussian zero-mean noise where a reflected propagation path fluctuates with probability density $p(r)$ while the direct path fluctuates with density function $p(x)$. Neither density function is Gaussian, but histograms are available to numerically describe the path statistics.
9. A system requires a P_{fa} of 1×10^{-6} and a $P_d > 99\%$ for an SNR of 1.0. How many RMS averages are needed?

REFERENCES

1. K. Sam Shanmugam, *Digital and Analog Communication Systems*, Chapter 3, New York, NY: Wiley, 1979.
2. W. H. Press, B. P. Flannery, S. A. Teukolsky, and W. T. Vetterling, *Numerical Recipes: The Art of Scientific Computing*, Chapter 7, Cambridge: Cambridge University Press, 1986.
3. W. S. Burdic, *Underwater Acoustic System Analysis*, Chapters 5, 9, 13, 15, 2nd Ed, Englewood Cliffs, NJ: Prentice-Hall, 1991.
4. H. L. Van Trees, *Detection, Estimation, and Modulation Theory,* New York, NY: Wiley, 1968.
5. A. Papoulus, *Signal Analysis*, Chapters 9–11, New York, NY: McGraw-Hill, 1977.

6. D. Zwillinger, ed., *Standard Mathematical Tables and Formulae*, Chapter 7, New York, NY: CRC Press, 1996.

7. M. Abramowitz and I. A. Stegun, *Handbook of Mathematical Functions*, Chapter 26, New York, NY: Dover, 1972.

8. G. R. Cooper and C. D. McGillem, *Probabilistic Methods of Signal and System Analysis,* New York, NY: HRW, 1971.

9. K. Attenborough, K. Li, and K. Horoshenkov, *Predicting Outdoor Sound,* New York, NY: Taylor & Francis, 2007.

10. M. Long, *Architectural Acoustics,* New York, NY: Academic Press, 2006, pp. 282–283.

13 Wavenumber and Bearing Estimation

In this chapter, we examine the fundamental techniques of measuring the spatial aspects of waves which can be propagated in a reflection free space from a distant source, reverberating in a confined space, or represent the complicated radiation in the nearfield of one or more sources. The types of wave of interest could be either mechanical (seismic, structural vibration, etc.), acoustic (waves in fluids), or electromagnetic. When the source of the waves is distant from the receiver array we can say that the wavefront is planar and the receiving array of sensors can estimate the direction of arrival, or bearing. This technique is fundamental to all passive and active sonar and radar systems for measuring the direction to a distant target either from its radiated waves or its reflections of the actively transmitted sonar or radar wave. However, when more than one target is radiating the same frequency, the arriving waves at the receiver array can come from multiple directions at a given frequency.

To resolve multiple DOA at the same frequency, the receiving array can process the data using a technique commonly known as beamforming. Beamforming is really an application of spatial wavenumber filtering (see Section 8.1). The waves from different directions represent different sampled wavelengths at the array sensor locations. An array beam pattern steered in a particular "look direction" (LD), corresponds to a wavenumber filter which will pass the corresponding wavenumber to the LD while attenuating all other wavenumbers. The "beam" notion follows from the analogy to a search light beam formed by a parabolic reflector or lens apparatus. The array beam can be "steered" electronically, and with parallel array processors, multiple beams can be formed and steered simultaneously, all without any mechanical systems to physically turn the array in the LD. Electronic beam steering is obviously very useful, fast, and the lack of mechanical complexity is very robust. Also, electronic beamforming and steering allows multiple beams each in different LDs to exist simultaneously on the same sensor array.

The array can also be "focused" to a point in its immediate vicinity rather than a distant source. This application is fairly novel and useful, yet the technology simply involves the derivation of different wavenumber filters from the classic beamforming problem. We could simply refer to array nearfield focusing as "spherical beamforming" since we are filtering spherical, rather than planar wavenumbers. But, a more descriptive term would be holographic beamforming because the array sensor spatial sampling of the field for a source in the vicinity of the array can allow reconstruction of the wave field from measurements of both the propagating and nonpropagating (evanescent) wavenumbers. Holographic beamforming implies measurement and reconstruction of the full 3D field from scanning a surface around the source with the array. Analysis of the observed wavenumbers using wavenumber filtering is of great interest in the investigation of how a source of interest is radiating wave energy. For example, changes in the observed electromagnetic fields of a motor, generator, or electronic component could be used to pinpoint a pending problem from corrosion, circuit breakdown, or component wear out.

Wavenumber processing for fields in confined spaces is generally known in structural acoustics as modal filtering. The vibration field of the bound space can be solved analytically in terms of a weighted sum of vibration modes, or structural resonances. Each structural resonance has a frequency and associated mode shape. When the structure is excited at a point with vibration (even a single frequency) all of the structural modes are excited to some extent. Therefore, in theory, a complete analysis of the structural response should allow one to both locate the source and filter out all the structural "reverberation," or standing wave fields. If there are changes in the structural integrity (say from corrosion or fatigue), changes in structural stiffness should be observable as changes in the mode shapes and frequencies. This should also be true for using microwaves to investigate corrosion or fatigue in metal structures. In theory, acoustical responses of rooms could be used by robotic vehicles to navigate interior spaces excited by known sources and waveforms. While these application ideas are futuristic, the reader should consider that they are all simply applications wavenumber filtering for various geometries and wave types.

Section 13.1 presents the Cramer–Rao lower bound (CRLB) for parameter estimation. This general result applies not only to beamforming estimates, but actually any parameter estimate where one can describe the observable in terms of a PDF. This important technique spans the probabilistic models of Chapter 12 and the adaptive filtering models in Chapters 9 and can be applied to any parameter estimate. For our immediate purposes, we present the CRLB for bearing estimates. This has been well developed in the literature and is very useful as a performance measure for beamforming algorithms. In Section 13.2, we examine precision bearing estimation by array phase directly or as a "split-beam." In the split-beam algorithm, the array produces two beams steered close together, but not at exactly the same LD. By applying a phase difference between the two beams, the pair can be "steered" precisely to put the target exactly in between the beams, thus allowing a precision bearing estimate. Section 13.3 presents the holographic beamforming technique and shows application in the analysis of acoustic fields, although this could be applied to any wave field of interest. Reconstruction of fields away from the sensing surface is often called tomographic inversion, implying a geo-spatial reconstruction of the field from Green's theorem in field theory. Finally, we look at one-way field propagation modeling using wavenumber spectra. This may at first seem out-of-place because propagation modeling is traditionally separate from beamforming, but they both involve wavenumber filtering, and so the flow is straightforward mathematically. By modeling propagation in the wavenumber domain, one can easily incorporate source and receiver directivities, scattering from turbulence, surface impedances, and nonhomogeneous wave speeds.

13.1 CRAMER–RAO LOWER BOUND

The CRLB [1,2] is a statistically based parameter estimate measure which provides a basis for stating the best possible accuracy for a given parameter estimate based on the statistics of the observables and the number of observables used in the estimate. As discussed below, the CRLB is very closely related to the least-squared error of a parameter estimate. The main difference between the CRLB and the least-squared error of a linear parameter estimate is that the CRLB represents the predictability of an estimate of a function's statistical value based on N observations. For example, one starts with a probability density model for the function of interest; say the bearing angle measured by a linear array of sensors. The array processing algorithm produces a time difference of arrival (or phase difference for a narrowband frequency) between various sensors that have a mean and a variance. If there are $N + 1$ sensors, we have N observations of this time or phase difference. Because the SNR is not infinite, the time delay or phase estimates come from a well-defined PDF (see Section 12.1). The derivation of an AOA, or bearing, requires translation of the PDF from the raw sensor measurements, but this is also straightforward, albeit a bit tedious. With N statistical observations of the bearing for a given time interval from the array, we seek the mean bearing as the array output, and use the CRLB to estimate the minimum expected standard deviation of our mean bearing estimate. The CRLB provides an important measure of the *expected accuracy* of a parameter estimate. We will then carry the CRLB through the calculations for AOA to provide an expectation of accuracy for it as well. The derivation of the CRLB is quite interesting and also contains some rather innovative thinking on how to apply statistics to signal processing.

Consider a vector of N scalar observations, where each observation is from a normal probability distribution with mean m and variance σ^2.

$$Y = [y_1 \; y_2 \; \dots \; y_N]; \quad p(y_i) = \frac{1}{\sigma\sqrt{2\pi}} e^{-(y_i - m)^2/2\sigma^2} \quad i = 1, 2, \dots, N \tag{13.1}$$

We designate the parameter vector of interest to be $\lambda = \{m\sigma^2\}$ and the joint PDF of the N observations to be

$$p(Y, \lambda) = \frac{1}{\sqrt{(2\pi\sigma^2)^N}} e^{-\sum_{i=1}^{N}(y_i - m)^2/2\sigma^2} \tag{13.2}$$

Suppose we have some arbitrary function $F(Y, \lambda)$, for example, the bearing, for which we are interested in estimating the mean. Recall that the first moment is calculated as

$$m_F = E[F(Y, \lambda)] = \int_{-\infty}^{+\infty} p_F(Y, \lambda) F(Y, \lambda) \, dY \tag{13.3}$$

For the statistical models of the observables and our arbitrary function described in Equations 13.1 through 13.3, we will be interested in the gradient of m_F with respect to the parameter vector λ as well as the second derivative. This is because we are constructing a linear estimator that should have a linear parameter error. Following Chapter 9, we note that the error squared will be quadratic where the least-squared error will be the parameter values where the gradient is zero. The gradient of the expected value of F is

$$\frac{\partial}{\partial\lambda} E[F] = E\left[\frac{\partial F}{\partial\lambda}\right] + E[F\psi] \tag{13.4}$$

where

$$\psi(Y, \lambda) - \frac{\partial \ln p_F}{\partial\lambda} = \frac{1}{p_F}\frac{\partial p_F}{\partial\lambda} \tag{13.5}$$

is the gradient of the log-likelihood function for the arbitrary function $F(Y, \lambda)$. The second derivative of Equation 13.5 has an interesting relationship with the first derivative.

$$\frac{\partial\psi}{\partial\lambda} = \frac{\partial^2 \ln p_F}{\partial\lambda^2} = \frac{\partial}{\partial\lambda}\left(\frac{1}{P_F}\frac{\partial p_f}{\partial\lambda}\right)$$

$$= \frac{-1}{p_F^2}\frac{\partial^2 p_F}{\partial\lambda^2} = -\psi^2 = \psi\psi^T \tag{13.6}$$

Proof of Equation 13.4 follows from a simple application of the chain rule.

$$\frac{\partial}{\partial\lambda} E[F] = \frac{\partial}{\partial\lambda}\left[\int p_F F \, dY\right] = \int\left\{p_F \frac{\partial F}{\partial\lambda} + F\frac{\partial p_F}{\partial\lambda}\right\} dY$$

$$= \int p_F \frac{\partial F}{\partial\lambda} dY + \int p_F\left(F\frac{\partial \ln p_F}{\partial\lambda}\right) dY$$

$$= E\left[\frac{\partial F}{\partial\lambda}\right] + E[F\psi] \tag{13.7}$$

Equation 13.4 and its proof in Equation 13.7 show the intuitive nature of using the gradient of the log-likelihood function for our arbitrary function F. Since F is functionally a constant with respect to λ, it can be seen that for $F = 1$, $E\{\psi\} = 0$. For $F = \psi$, we obtain another important relation.

$$\frac{\partial}{\partial \lambda} E[\psi] = \frac{\partial}{\partial \lambda}\left[\int p_F \psi \, dY\right]$$

$$= \int \left\{p_F \frac{\partial \psi}{\partial \lambda} + \psi \frac{\partial p_F}{\partial \lambda}\right\} dY$$

$$= \int p_F \frac{\partial \psi}{\partial \lambda} \, dY + \int p_F \left(\psi \frac{\partial \ln p_F}{\partial \lambda}\right) dY$$

$$= E\left[\frac{\partial \psi}{\partial \lambda}\right] + E[\psi \psi^T] = 0 \tag{13.8}$$

Therefore,

$$-E\left[\frac{\partial \psi}{\partial \lambda}\right] = E[\psi \psi^T] = J \tag{13.9}$$

where J in Equation 13.9 is referred to as the *Fisher Information Matrix.*

We note that if the slope of the PDF is very high in magnitude near the mean, there is not much "randomness" to the distribution. This corresponds to the elements of J being large in magnitude, and the norm of the Fisher information matrix to be large. Hence, the observations contain significant information. Conversely, a broad PDF corresponds to relatively low information in the observations. The elements of J are defined as

$$J_{i,j} = E\left[\frac{-\partial^2 \ln p_F}{\partial \lambda_i \, \partial \lambda_j}\right] \tag{13.10}$$

The reason as to why we took the effort to derive Equations 13.4 and 13.9 is we need to evaluate the statistics of a parameter estimate $\hat{\lambda}(Y)$ for the parameter $\lambda(Y)$. Since $\lambda(Y)$ and $\psi(Y)$ are correlated, we can write a linear parameter estimation error as

$$e(Y) = \lambda(Y) - \beta \psi(Y) \tag{13.11}$$

Recall from Section 9.1 that minimization of the squared error is found by setting the gradient of the squared error with respect to β to zero and solving for β.

$$E[\beta] = E[\lambda \psi^T] E[\psi \psi^T]^{-1} \tag{13.12}$$

making the parameter estimation error

$$e = \lambda - E[\lambda \psi] E[\psi \psi^T]^{-1} \psi \tag{13.13}$$

We note that the error in Equation 13.13 is uncorrelated with ψ, and since $E\{\psi\} = 0$, $E\{e\} = E\{\lambda\}$. Since we are interested in the variation between the actual parameter value and its expected value, we define the following two variational parameters:

$$\Delta \lambda = \lambda - E[\lambda] \tag{13.14}$$

$$\Delta e = e - E[e] \tag{13.15}$$

Note that subtracting a constant (the expected values λ) does not affect the cross-correlation

$$M = E[\lambda \psi^T] = E[\Delta \lambda \psi^T] \tag{13.16}$$

where M in Equation 13.16 is known as the *bias of the parameter estimation* and equals unity for an unbiased estimator. This will be described in more detail below. The variational error is therefore

$$\Delta e = \Delta \lambda - E[\lambda \psi^T] E[\psi \psi^T]^{-1} \psi$$
$$= \Delta \lambda - M J^{-1} \psi \tag{13.17}$$

The expected value of the variance of the parameter estimation variational error is

$$E\left[\Delta e \Delta e^T\right] = E\left[(\Delta \lambda - M J^{-1} \psi)(\Delta \lambda^T - \psi^T J^{-1} M^T)\right]$$
$$= E\left[\Delta \lambda \Delta \lambda^T\right] - M J^{-1} E\left[\psi \Delta \lambda^T\right] - E\left[\Delta \lambda \psi^T\right] J^{-1} M^T + M J^{-1} \psi \psi^T J^{-1} M^T$$
$$= E\left[\Delta \lambda \Delta \lambda^T\right] - M J^{-1} M^T - M J^{-1} M^T + M J^{-1} M^T$$
$$= E\left[\Delta \lambda \Delta \lambda^T\right] - M J^{-1} M^T \tag{13.18}$$

Since $E\{\Delta e \Delta e^T\} \geq 0$ we can write a lower bound on the variance of our parameter estimate.

$$\sigma^2(\hat{\lambda}) \doteq E\left[\Delta \lambda \Delta \lambda^T\right] \geq M J^{-1} M^T \tag{13.19}$$

Equation 13.19 is the result we have been looking for—a measure of the variance of our parameter estimate based on the statistics of the observables, with the exception of the bias term.

$$M = E\left[\lambda' \psi^T\right] = \frac{\partial}{\partial \lambda} E[\lambda'] - E\left[\frac{\partial \lambda'}{\partial \lambda}\right]$$
$$= I \quad \text{for the unbiased case} \tag{13.20}$$

The bias in Equation 13.20 is unity for the case where λ' has no explicit dependence on λ, making the partial derivative in the rightmost term zero. The unbiased parameter estimate will converge to the true parameter, given an infinite number of observables. A biased estimate will not only converge to a value offset from the true parameter value, but the bias will also affect the variance of the parameter estimate depicted in Equation 13.19. For the unbiased case

$$\sigma^2(\hat{\lambda}) \doteq E\left[\Delta \lambda \Delta \lambda^T\right] \geq J^{-1} \tag{13.21}$$

and in terms of N observations of the scalar PDF in Equation 13.1,

$$\sigma^2(\hat{\lambda}) = \frac{-1}{NE\left[(\partial^2 / \partial \lambda^2) \ln p(\lambda)\right]} = \frac{1}{NE\left[((\partial / \partial \lambda) \ln p(\lambda))^2\right]} \tag{13.22}$$

Equation 13.22 provides a simple way to estimate the CRLB for may parameter estimates. When the bias is unity, the estimator is called *efficient* because it meets the CRLB.

Consider an example of N observations of a Gaussian process described as a joint N-dimensional Gaussian probability density.

$$p(Y,\lambda) = \frac{1}{\sqrt{(2\pi\sigma^2)^N}} e^{-\sum_{i=1}^{N}(y_i-m)^2/2\sigma^2} \tag{13.23}$$

Our parameter vector is $\lambda = \{m \; \sigma\}^T$. To find the CRLB we first take the log of $p(Y,\lambda)$

$$\ln p(Y,\lambda) = -\frac{N}{2}\ln(2\pi) - N\ln\sigma - \frac{1}{2\sigma^2}\sum_{i=1}^{N}(y_i-m)^2 \tag{13.24}$$

and then differentiate

$$\psi(Y,\lambda) = \frac{\partial}{\partial\lambda}\ln p(Y,\lambda) = \begin{bmatrix} \dfrac{\partial\ln p}{\partial m} \\[2mm] \dfrac{\partial\ln p}{\partial(\sigma^2)} \end{bmatrix} = \begin{bmatrix} \dfrac{1}{\sigma^2}\sum_{i=1}^{N}(y_i-m) \\[4mm] -\dfrac{N}{2\sigma^2} + \dfrac{1}{2\sigma^4}\sum_{i=1}^{N}(y_i-m)^2 \end{bmatrix} \tag{13.25}$$

Differentiating again yields the elements of the matrix Ψ

$$\Psi(Y,\lambda) = \frac{\partial}{\partial\lambda}\psi(Y,\lambda) = \begin{bmatrix} \dfrac{\partial^2\ln p}{\partial m \partial m} & \dfrac{\partial^2\ln p}{\partial m\,\partial(\sigma^2)} \\[3mm] \dfrac{\partial^2\ln p}{\partial m\,\partial(\sigma^2)} & \dfrac{\partial^2\ln p}{\partial(\sigma^2)\,\partial(\sigma^2)} \end{bmatrix}$$

$$= \begin{bmatrix} \dfrac{N}{\sigma^2} & \dfrac{1}{\sigma^4}\sum_{i=1}^{N}(y_i-m) \\[4mm] \dfrac{1}{\sigma^4}\sum_{i=1}^{N}(y_i-m) & -\dfrac{N}{2\sigma^4} + \dfrac{1}{\sigma^6}\sum_{i=1}^{N}(y_i-m)^2 \end{bmatrix} \tag{13.26}$$

and taking expected values gives the Fisher information matrix.

$$J = E\{\Psi(Y,\lambda)\} = \begin{bmatrix} \dfrac{N}{\sigma^2} & 0 \\[3mm] 0 & \dfrac{N}{2\sigma^4} \end{bmatrix} \tag{13.27}$$

The CRLB for an unbiased estimate of the mean and variance is therefore

$$\sigma^2(\hat{\lambda}) = \begin{bmatrix} E\{\Delta m\Delta m\} & E\{\Delta m\Delta\sigma^2\} \\[2mm] E\{\Delta m\Delta\sigma^2\} & E\{\Delta\sigma^2\Delta\sigma^2\} \end{bmatrix} \geq \begin{bmatrix} \dfrac{\sigma^2}{N} & 0 \\[3mm] 0 & \dfrac{2\sigma^4}{N} \end{bmatrix} \tag{13.28}$$

An even more practical description can be seen if we consider a Gaussian distribution with say mean 25 and variance 9. How many observations are needed for an unbiased estimator to provide estimates within 1% of the actual values 63% of the time (i.e., one standard deviation of the estimate is 1% of its value)?

Solution: We note that the variance is 9 and that the variance of the mean estimate is $9/N$. The standard deviation of the mean estimate is $3/\sqrt{N}$. Therefore, to get a mean estimate where the standard deviation is 0.25 or less, N must be greater than 144 observations. To get a variance estimate with standard deviation 0.09, N must be over 1.6 million observations. To get a variance estimate where the variance of the variance estimate is 0.09, N must be greater than about 145,800 observations, and so on. The CRLB provides an objective way to estimate how accurate a statistical estimate can be, at best. So given an estimated mean and variance based on N observations, we can say that the true mean and variance are at least ± the CRLB of the estimated values, but could be worse than that.

13.2   BEARING ESTIMATION AND BEAM STEERING

In this section, we apply the technique of establishing the CRLB for a statistical representation of wave front bearing, or the observed wavenumber by an array of sensors. This is presented both in terms of a direct bearing measurement for single arrival angles and later in this section by way of array beamforming and beam steering to determine source bearing. Direct bearing estimation using wave front phase differences across an array is a mainstay process of passive narrowband sonar engineering, but also finds application in other areas of acoustics, phased array radar processing, as well as seismology and radio astronomy. By combining measured phase and/or time delay information from a sensor array with the array geometry and wave speed one can determine the direction of arrival of the wave from a single distant source. If more than one target is radiating the same frequency, or if propagation multipath exists, a beamforming and beam steering approach must be used to estimate the target bearings. This is the general passive sonar problem of determining the bearing of a plane wave passing the array to provide a direction to a distant target [3].

If the array is large compared to the distance to the source, it can be "focused" on the origin of a spherical wave allowing the source location to also be observed with some precision. When a complete surface enclosing the source(s) of interest is scanned by an array coherently, Gauss's theorem provides that the field can then be reconstructed on any other surface enclosing the same sources. The spherical and 3D field measurement representations will be left to the next section. This section will deal specifically with plane wave fields, which is always the case when the source is so far from the receiving array that the spherical wave front observed by the array is essentially planar.

13.2.1   BEARINGS FROM PHASE ARRAY DIFFERENCES

Let us begin by considering a simple three-element array and a 2D bearing estimation problem for a single distant source (plane wave) of a single sinusoid in white noise. To simplify our analysis even further, we place sensor 1 at the origin, sensor 2 is placed d units from the origin on the positive x-axis, and sensor 3 is placed d units from the origin on the positive y-axis. Figure 13.1 depicts the array configuration and the bearing of the plane wave of interest. For the plane wave arriving at the Cartesian-shaped array from the angle θ, one can write very simple expressions for the phase differences across the array of sensors

$$\Delta\phi_{21} = \phi_2 - \phi_1 = kd\cos\theta \tag{13.29}$$

$$\Delta\phi_{31} = \phi_3 - \phi_1 = kd\sin\theta \tag{13.30}$$

where ϕ_1, ϕ_2, and ϕ_3 are the phases of the particular FFT bin corresponding to the radian frequency ω and the wavenumber $k = \omega/c$, c being the wave propagation speed. The convenience of using a

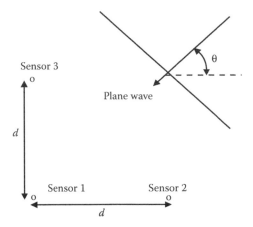

FIGURE 13.1 Array configuration for 2D phase-difference estimation of the bearing for a plane wave.

Cartesian-shaped array and the expressions for sine and cosine of the arrival angle are evident in the solution for θ in Equation 13.31.

$$\theta = \tan^{-1}\left\{\frac{\Delta\varphi_{31}}{\Delta\varphi_{21}}\right\} \tag{13.31}$$

Note that for the Cartesian equal-spaced array, the bearing angle is calculated independent of wave speed, frequency, and wavenumber. This can be particularly useful for dispersive waves, such as shear waves where high frequencies travel faster than low frequencies making bearing estimates from time delay estimation problematic. Clearly, one can also provide an estimate of bearing uncertainty given the probability density of the signal and noise in the FFT bin of interest. We will first generalize the bearing estimate to an arbitrary shaped array and then examine statistics of the bearing estimate.

Consider an arbitrary-shaped planar array where each sensor position is defined by a distance and angle relative to the origin. Recall from Problem 8.1 that

$$\Delta\phi_{jk} = \phi_j - \phi_k = \frac{\omega}{c}\left\{\cos\theta\,(r_j\cos\theta_j - r_k\cos\theta_k) + \sin\theta\,(r_j\,\sin\theta_j - r_k\sin\theta_k)\right\} \tag{13.32}$$

where r_j is the distance from the origin of sensor j and θ_j is the angle of sensor j position relative to the x-axis. Thus

$$\begin{bmatrix}\cos\theta \\ \sin\theta\end{bmatrix} = \begin{bmatrix}(r_3\cos\theta_3 - r_1\cos\theta_1) & (r_3\sin\theta_3 - r_1\sin\theta_1) \\ (r_2\cos\theta_2 - r_1\cos\theta_1) & (r_2\sin\theta_2 - r_1\sin\theta_1)\end{bmatrix}^{-1}\begin{bmatrix}\dfrac{\Delta\varphi_{31}c}{\omega} \\ \dfrac{\Delta\varphi_{21}c}{\omega}\end{bmatrix} \tag{13.33}$$

The inverted matrix in Equation 13.33 contains terms associated with the position of the three sensors. The sine and cosine are evaluated as an inverse tangent to produce a bearing angle relative to the positive x-axis. To covert this angle to the common "compass angle" where 0 degrees is due North (positive y-axis) and compass bearing is measured clockwise from due North, use an inverse cotangent instead of the inverse tangent.

It can be shown that the inverse exists if the three sensors define a plane (they are not in a line nor all at one point). An arithmetic mean can be calculated for the sine and cosine of the arrival angle using a number of sensor pairings.

$$\begin{bmatrix} \cos\theta \\ \sin\theta \end{bmatrix} = \frac{1}{N^3 - 2N^2 + N} \sum_{i=1}^{N} \sum_{i=1}^{N} \sum_{k=1}^{N} \begin{bmatrix} (r_i\cos\theta_i - r_j\cos\theta_j) & (r_i\sin\theta_i - r_j\sin\theta_j) \\ (r_k\cos\theta_k - r_j\cos\theta_j) & (r_k\cos\theta_k - r_j\cos\theta_j) \end{bmatrix}^{-1} \begin{bmatrix} \dfrac{\Delta\varphi_{ij}c}{\omega} \\ \dfrac{\Delta\varphi_{kj}c}{\omega} \end{bmatrix}$$

(13.34)

Equation 13.34 uses all possible pairings of sensors assuming that all sensor pairs are separated by a distance less than one-half wavelength. For large arrays and relatively high wavenumbers (frequencies), this is not possible in general. However, averaging N-pairings which do meet the requirement of less than a half-wavelength spacing will greatly reduce the variance of the bearing error.

To determine the CRLB for bearing error, we first make a systematic model for the statistics and then consider combining N observations for a bearing estimate. Consider the FFT bin complex number for each of the three sensors in Figure 13.1 for a particular frequency of interest. Figure 13.2 graphically shows how the standard deviation of the phase σ_φ, can be expressed as

$$\sigma_\varphi = \tan^{-1}\left\{\frac{\sigma_N}{S}\right\} = \tan^{-1}\left\{\frac{1}{\text{SNR}}\right\} \tag{13.35}$$

where S is the signal amplitude and σ_N is the noise standard deviation in the FFT bin. While the envelope of the complex FFT bin probability density has been shown to be a Rician PDF, we can approximate the probability density for the complex bin as a 2D Gaussian density where the mean is simply the complex number representing the true magnitude and phase of the signal. This approximation is generally true if the SNR is greater than about 4.

The phase-difference probability density results from the convolution of the two Gaussian densities for the real and imaginary part of the FFT bin. Therefore, the FFT phase variances for each sensor add to give the variance of the phase difference. Assuming that the noise densities for the sensors are identical in a given FFT bin, the standard deviation for the phase difference is seen as

$$\sigma_{\Delta\varphi} = \sqrt{2}\,\sigma_\varphi \tag{13.36}$$

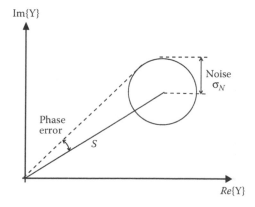

FIGURE 13.2 Graphical representation of the phase error in an FFT bin due to background noise showing that the phase error can be seen as the arctangent of the inverse of the SNR.

Normalizing the phase differences along the x and y axes by kd for our simple three-sensor Cartesian-shaped array, we can consider the PDFs for our estimates of the sine and cosine of the arrival angle θ for the plane wave. Figure 13.3 graphically shows the standard deviation for the bearing error.

Normalization by kd naturally scales the phase difference variance relative to the observable phase difference for the sensor separation. For example, suppose the standard deviation of the phase difference is 0.01 rad. If d/λ is smaller than but close to 0.5, the random bearing error is quite small. But for a closer spacing, or lower frequency, d/λ might be 0.05, making the random bearing error considerably larger for the same phase-difference variance. The standard deviation for the bearing error is therefore

$$\sigma_\theta = \tan^{-1}\left\{\frac{\sqrt{2}\,\sigma_\varphi}{kd}\right\} = \tan^{-1}\left\{\frac{\sqrt{2}}{kd}\tan^{-1}\left\{\frac{1}{\mathrm{SNR}}\right\}\right\} \tag{13.37}$$

If the SNR is large (say > 10), and $\sigma_\varphi \ll kd$, the bearing error standard deviation is approximately

$$\sigma_\theta \approx \frac{\sqrt{2}}{\mathrm{SNR}\,kd} = \frac{1}{\sqrt{2}}\frac{1}{\pi}\frac{1}{\mathrm{SNR}}\frac{\lambda}{d} \tag{13.38}$$

which is in agreement with the literature [3]. Note that the unit of Equation 13.38 is radians. If we combined M sensor pairings with the same separation d as in Figure 13.1, the bearing CRLB is

$$\sigma_\theta \approx \frac{1}{\sqrt{M}}\frac{1}{\sqrt{2}\,\pi}\frac{1}{\mathrm{SNR}}\frac{\lambda}{d} \tag{13.39}$$

One can also average the expressions for sine and cosine in Equations 13.34 using different frequencies and sensor spacings, but a corresponding SNR and wavelength/aperture weighting must be applied for each unique sensor pair and frequency bin. The expression for the CRLB tells us that to improve bearing accuracy one should increase SNR (integrate longer in time and/or space), choose sensor spacings near, but necessarily less than, $\lambda/2$ for the frequency, and combine as many sensor pairs and frequencies as is practical. When a wide range of frequencies are used in a

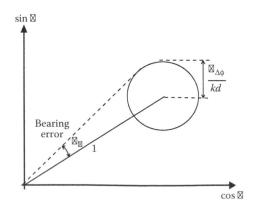

FIGURE 13.3 Graphical representation of the bearing random error given phase differences normalized by kd to provide for the sine and cosine of the bearing angle.

nonwhite background noise, the Ekhart filter (see Section 6.5) can be applied to equalize SNR over a range of interest. This is equivalent to weighting various frequencies according to their SNR in the bearing estimate.

It can be shown that the CRLB for time delay estimation where the average of M observations used is

$$\sigma_\tau \approx \frac{1}{\sqrt{M}\, \text{SNR}\, \beta^2} \qquad (13.40)$$

where β is the bandwidth in Hertz of the signal used in the time delay estimate. For estimation of Doppler frequency shift from a moving source based on the average of M observations, the CRLB can be shown to be

$$\sigma_{\Delta f} \approx \frac{1}{\sqrt{M}\, \text{SNR}\, T^2} \qquad (13.41)$$

where T is the period of one observation in seconds. Time delay and frequency shift estimation are generally associated with target range and velocity along the bearing direction. However, it can be seen that for a given signal's time-bandwidth product and SNR, there are definable parameter estimation errors which cannot be less than the CRLB. Note that the SNR enhancement of an FFT defines a time-bandwidth product.

13.2.2 MULTIPLE ANGLES OF ARRIVAL

Suppose we have two or more sources at different bearing angles radiating the same frequency? What bearing would a direct phase-based bearing calculation predict? Clearly, the array would be exposed to a propagating wave field and an interference field from the multiple sources. If one were to decompose the waves from the sources into x- and y-axis components, summing the field of the sources results in the sum of various wavelengths on the x and y axes. Therefore, one can argue that along the x and y axes, one has a linear combination of wave components from each source. In theory, an array with M sensors can resolve $M - 1$ sources at the same frequency, but this will be left to Chapter 14. For two sources of equal amplitude, phase, and distance from the array and bearings θ_a and θ_b, the phase difference between two sensors separated by a distance d is

$$\frac{P_2(\omega)}{P_1(\omega)} = e^{j\Delta\varphi_{21}} = e^{jkd\cos\theta_a} + e^{jkd\cos\theta_b} \qquad (13.42)$$

Recalling that adding two sinusoids of different wavelengths gives an AM signal where the "carrier wave" is the average of the two wave frequencies (or wavelengths) and the envelope is half the difference of the two wave frequencies.

$$\frac{P_2(\omega)}{P_1(\omega)} = e^{j(kd/2)(\cos\theta_a + \cos\theta_b)} 2\cos\left(\frac{kd}{2}\left[\cos\theta_a - \cos\theta_b\right]\right)$$

$$= e^{j(kd/2)\cos(\theta_a + \theta_b/2)\cos(\theta_a - \theta_b/2)} 2\cos\left(\frac{kd}{2}\left[\cos\theta_a - \cos\theta_b\right]\right) \qquad (13.43)$$

Equation 13.43 shows that for two arrival angles close together ($\theta_a \approx \theta_b$), the estimated bearing will be the average of the two arrival angles since the cosine of a small number is nearly unity.

However, as the bearing difference increases a very complicated angle of arrival (AOA) results. When the two waves are of different amplitudes, the average and differences are weighted proportionately. Therefore, we can say with confidence that a direct bearing estimate from an array of sensors using only spatial phase will give a bearing estimate somewhere between the sources and biased toward the source wave of higher amplitude. It would appear that measuring both the spatial phase and amplitude (interference envelope) should provide sufficient information to resolve the bearing angles. However, in practice, the sources are both moving and not phase synchronized making the envelope field highly nonstationary and not linked physically to the direction of arrivals for the sources. The only practical way to resolve multiple source directions at the same frequency is to apply array beamforming and to steer the beam around in a search pattern to detect sources.

Using an array of sensors together to produce a beam-shaped directivity pattern, or a beam pattern, requires that all or a number of sensor output signals be linearly filtered with a magnitude and phase at the frequency of interest such that the sensor response to a plane wave from a particular bearing angle produces a filter output for each sensor channel that has the same phase. This phase coherent output from the sensor array has the nice property of very high signal output for the wave from the designed look-direction angle, and relatively low incoherent output from other directions. Thus, the sensor array behaves somewhat like a parabolic dish reflector. However, the array beamforming can "steer" the beam with no moving parts simply by changing the magnitude and phase for each frequency of interest to produce a summed output which is completely coherent in the new LD. Even more useful is constructing beams which have zero output in the direction(s) of unwanted sources. Array "null-forming" is done adaptively, and several methods for adaptive beamforming will be discussed in Chapter 14. This is also called spectral estimation, but usually refers to wavenumber spectra from sensor arrays.

Consider the beam pattern response for a simple theoretical line sensor as shown in Figure 13.4. The distant source essentially radiates a plane wave across the line sensor where, relative to the origin, we have early arrivals on the right and late arrivals on the left. If we sum all the differential

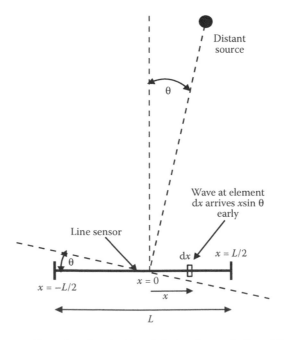

FIGURE 13.4 Line sensor configuration showing length L, bearing angle θ, and the response at differential element dx where the wave arrives early compared to the line center at $x = 0$.

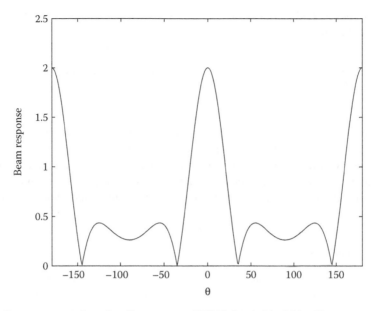

FIGURE 13.5 Beam response for a 2 m line sensor at 300 Hz in air ($c = 344$ m/s).

elements along the line sensor, we obtain the spatial response of the line sensor, relative to a single point source.

$$D(\theta) = \int_{x=-L/2}^{x=L/2} e^{jkx\sin\theta}\,dx = \frac{e^{jkx\sin\theta}}{jk\sin\theta}\Bigg|_{x=-L/2}^{x=+L/2} = L\,\frac{\sin\big((kL/2)\sin\theta\big)}{(kL/2)\sin\theta} \tag{13.44}$$

For an example using acoustic in air, $c = 344$ m/s, $L = 2$ m, $f = 300$ Hz, and $k = 5.4795$ m^{-1}. The beam response is shown in Figure 13.5, and in Figure 13.6 in polar form. Electronically steering the

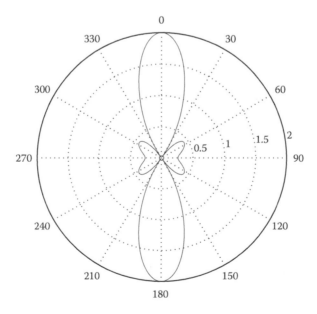

FIGURE 13.6 Polar plot of the 2 m line sensor showing the North–South beams expected to be symmetric about the horizontal axis of the line sensor.

beam to a LD θ' requires adjusting the phase at 300 Hz for each element so that the line sensor response at θ' is coherent.

$$D(\theta) = \int_{x=-L/2}^{x=L/2} e^{jkx\sin\theta} \, e^{-jkx\sin\theta'} \, \mathrm{d}x = L \frac{\sin\left((kL/2)[\sin\theta - \sin\theta']\right)}{(kL/2)[\sin\theta - \sin\theta']} \qquad (13.45)$$

Figure 13.7 shows the 30° steered beam response in polar form. Note how the southern lobe also moves toward east by 30°. This is because of the symmetry of the line sensor which is oriented along an east-west line. The reason the "south" beam also moves around to the East is that a line array cannot determine which side of the line the sources is on. In 3-dimensions, the beam response of the line array is a hollow cone shape which becomes a disk shape when no steering is applied.

13.2.3 WAVENUMBER FILTERS

We can explain beam forming in a much more interesting way using spatial Fourier transforms such as what was done in Section 8.2 for images to show high- and low-pass filtering. Consider a 256-point × 256-point spatial grid representing 32 m × 32 m of physical space. Placing our 2 m line sensor in the center of this space at row 128, we have 16 "pixels," each numerically unity, extending from $\langle x, y \rangle$ coordinates $\langle 56, 128 \rangle$ to $\langle 72, 128 \rangle$ representing the line sensor. Figure 13.8 shows a sketch of our input Fourier space. Figure 13.9 shows the magnitude of a 2D FFT of the spatial array data with a dashed circle centered on the wavenumber space origin representing $k = 5.4795$ m⁻¹. If we sample the wavenumber response along the $k = 5.4795$ circle, we get the response shown in Figure 13.5! For frequencies below 300 Hz, the wavenumber response circle is smaller and the beam response is wider. At high frequencies well above 300 Hz, the circle defined by k is much larger making the beam width much narrower. Is not this fun? It will work for any array shape, set of amplitudes, and phases. However, when you apply phases and amplitudes to the array elements,

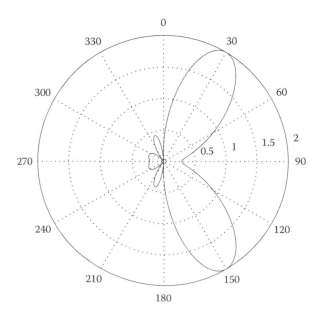

FIGURE 13.7 Beam response at 300 Hz for the 2 m line sensor steered to 30° East, again showing the symmetry about the horizontal x-axis.

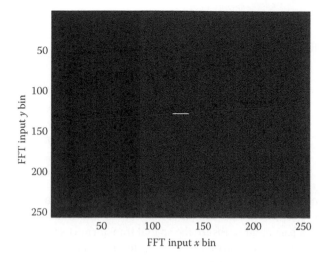

FIGURE 13.8   Input grid for the 2D FFT to represent the wavenumber filter response of the line array.

there is an implied temporal frequency. The evaluation circle of radius k gives the array response to a plane wave with a frequency and directions corresponding to the points on the circle. Figures 13.8 and 13.9 provide a nice way to see the relationship between array size and shape and the corresponding wavenumber response.

Table 13.1 compares the DFT on temporal and spatial data. Determining the wavenumber range from the number of samples and the spatial range of the input can sometimes be confusing. Note that for, say, 1 s of a real-time signal sampled 1024 times, the sample rate is 1024 Hz and a 1024-point FFT will yield discrete frequencies 1 Hz apart from −512 to +512 Hz. For our spatial FFT on the 2 m line sensor, we have a space 32 m × 32 m sample 256 times in each direction. This gives a spatial sample rate k_s of 8 samples/m. Since the wavenumber $k = 2\pi/\lambda$ and the FFT output has a digital frequency $\Omega = k/k_s$ span of $-\pi \leq \Omega \leq +\pi$, the physical wavenumber span is $-\pi k_s \leq k \leq +\pi k_s$, or

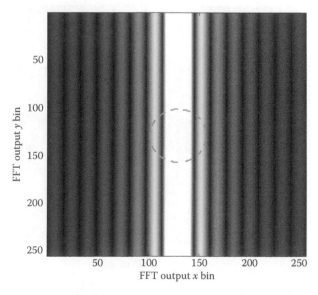

FIGURE 13.9   2D 256-point FFT of the 2 m line sensor showing wavenumbers from −8π to +8π where the spatial sample rate is 256/32 m or 8 samples/m.

TABLE 13.1
Comparison of Spatial and Temporal FFT Parameters

	Spatial FFT	Temporal FFT
Input buffer	Xmax meters N samples	T seconds N samples
Sample rate	$k_s = N/$Xmax samples/m	$f_s = N/T$ samples/s
Digital frequency range	$-\pi \le \Omega \le +\pi$, $\Omega = k/k_s$ $k = 2\pi/\lambda = \omega/c$, $\omega = 2\pi f$	$-\pi \le \Omega \le +\pi$, $\Omega = 2\pi f/f_s$ $\Omega = 2\pi\omega/\omega_s$, $\omega = 2\pi f$
Physical frequency range	$-\pi k_s \le k \le +\pi k_s$	$f_s/2 \le f \le +f_s/2$, $-\omega_s/2 \le \omega \le \omega_s/2$

-8π to $+8\pi$. This is perfectly analogous to the physical frequency span for temporal FFTs of real data being $-f_s/2 \le f \le +f_s/2$. Table 13.1 compares the spatial and more familiar temporal Fourier transform parameters in particular to insure the spatial transform is scaled properly.

Our 2D wavenumber domain approach to beamforming is interesting when one considers the wavenumber response for a steered beam, such as the 30° beam steer in Figure 13.7. Figure 13.10 shows the effect of steering the line sensor's LD beam 30° to the East. Note how the wavenumber response circle appears shifted to the left relative to the main lobe of the array. It can be shown that for the "compass" (rather than trigonometric) bearing representation, the x and y wavenumber components are

$$k_x = k \sin\theta - k \sin\theta'$$
$$k_y = k \cos\theta \tag{13.46}$$

where k is the wavenumber and θ' is the steered direction. This approach to beam forming is quite intuitive because one can define a wavenumber response for the array shape and then separately evaluate the beam pattern for a specific wavenumber (temporal frequency and propagation speed) and steering direction.

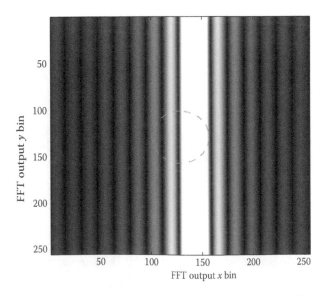

FIGURE 13.10   2D wavenumber response for the 2 m line sensor steered to 30° East showing the wavenumber circle for 300 Hz shifted leftward relative to the main beam which results in the beam response in Figure 13.7.

We can also use the wavenumber approach to examine the effects of grating lobes which arise from the separation between array elements. With our line sensor 2 m long and FFT input space 32 m with 256 samples, our line array is actually 16 adjacent elements in the digital domain. This makes the response closely approximate a continuous line sensor. Figure 13.11 shows the wave-number response for a four-element line array still covering 2 m total aperture. The separation between sensors gives rise to "grating lobes" in the wavenumber transform. For low frequencies, the wavenumber is small and the circle representing the beam response is not significantly affected by the sensor element spacing. Figure 13.12 shows the polar response for 300 Hz, 30° steering, for the four-element array. This response shows some significant leakage around 270° (due West). Note that the sound speed is 344 m/s, the wavelength at 300 Hz is 1.467 m while the element spac-ing is 2 m divided by four elements, or 0.5 m. In other words, the sensor spacing is slightly less than half a wavelength.

At 600 Hz, the array response steered to 30° is shown in Figure 13.13 where the larger circle represents the bigger wavenumber of $k = 10.959$ m^{-1}. At 600 Hz, the sensor spacing of 0.5 m is greater than a half wavelength (0.2867 m). The circle clearly traverses the grating lobes meaning that the array response now has multiple beams. Figure 13.14 shows the grating lobes at 600 Hz for the 2 m four-element line array with the steering angle set to 30°. Clearly, a beam pattern with grat-ing lobes will not allow one to associate a target bearing with a large beam output when steered in a specific LD.

Perhaps the most interesting application of our wavenumber approach to beamforming is seen when we consider 2D planar arrays. Consider an 8 element × 8 element square grid, 2 m on each side. Taking FFTs we have the wavenumber response shown in Figure 13.15 where the circle repre-sents 300 Hz and a steering angle of 30°. Figure 13.16 shows the corresponding polar response in the circle in Figure 13.15. Note that the effect of steering is to shift the wavenumber circle up and to the left of the main array lobe in Figure 13.15. This is because the FFT data are displayed such that 0° is down, 90° is to the right, 180° is up, and 270° is to the left. In the polar plot, we display the beam using a "compass" bearing arrangement, which corresponds to how the bearing data are used in real-world systems. Therefore, we can calculate a generic 2D spatial FFT of the array, and place

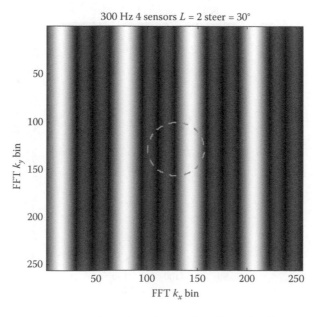

FIGURE 13.11 Wavenumber response for a four-element 2 m line array (not a continuous line sensor), steered to 30° at 300 Hz showing multiple periodic beams due to the discrete element spacing.

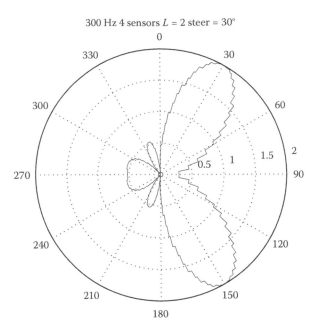

FIGURE 13.12　Polar response for the four-element 2 m line array steered to 30° at 300 Hz showing an increase in the small lobes on the left side of the beam pattern due to grating lobes.

a circle on the array wavenumber response representing a wavenumber and steering direction of interest to the observed beam pattern. The wavenumber shifts follow the case for the line array

$$k_x = k \sin \theta - k \sin \theta'$$
$$k_y = k \cos \theta - k \cos \theta'$$

(13.47)

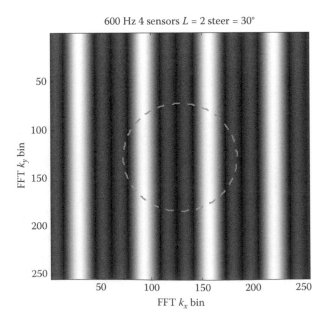

FIGURE 13.13　Four-element 2 m line array response circle for 600 Hz showing strong grating lobe interference due to spatial aliasing from undersampling spatially.

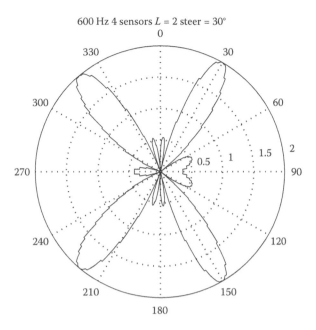

FIGURE 13.14 Polar beam pattern response corresponding to evaluation of the 600 Hz wavenumber circle from Figure 13.13 showing strong grating lobes on the left side of the beam pattern.

where k is the wavenumber, θ is the bearing, and θ' is the steered direction. Note that if one prefers a trigonometric circular coordinate system rather than compass bearings, all one needs to do is switch the sines to cosines in Equations 13.44 and 13.45.

Another physical effect which can cause the wavenumber circle to shift is a flow field which causes the wave propagation speed to be directional. This physical effect of flow is that the wave speed is now a function of direction. It also affects incoming waves differently than outgoing waves. For example, if winds are out of the East (90°), the "listening" response of the array will be skewed

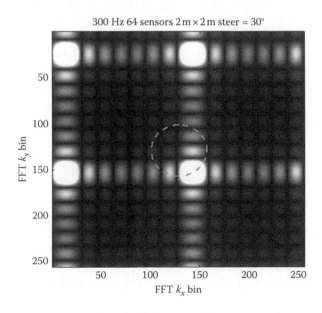

FIGURE 13.15 Wavenumber response for an 8×8 element grid sensor array 2 m on a side and steered to 30° at 300 Hz showing the beam response circle shifted up and to the left.

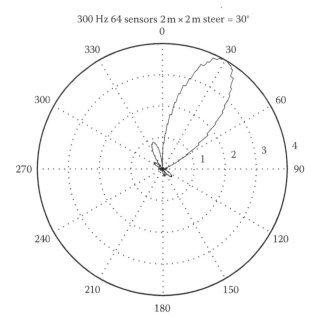

FIGURE 13.16 Polar response for the 8 element × 8 element 2 m × 2 m grid array at 300 Hz and a steering angle of 30° showing a single beam with no backside leakage due to the 2D array grid.

slightly in the direction of the wind. Waves traveling toward the array will arrive faster from the East, just as if the beam were steered in that direction. For outgoing waves, an array of sources would beam slightly more toward the downwind direction, if effect, blowing the transmitted beam downstream somewhat. Outdoors, this effect is very slight because the speed of sound is very high relative to a typical wind speed. But, in a jet airplane, this is the reason why so much of the engine's noise radiates behind the plane. For a moving car more tire noise radiates in the forward direction for a different reason. the car is moving slow compared to the speed of sound, but the air pushed out of the tire in the front has this velocity added to the sound speed to create more efficient radiation than in the direction behind the tire [4].

Now that beamforming physics and processing have been established, we need to revisit the CRLB to include the SNR gains available from beamforming. For a beamforming-based bearing estimate, the beam is swept around while the output is monitored for directions with high SNR. The CRLB for bearing is therefore tied to the array SNR gain in the LD and the beam width. The SNR gain can be numerically computed by calculating a parameter called the directivity index [5].

$$DI = 10 \log_{10} \left\{ \frac{|D(\theta')|^2}{(1/4\pi) \int_{4\pi} |D(\theta)|^2 \, d\Omega} \right\} \tag{13.48}$$

Equation 13.45 must be integrated in three dimensions to properly model the isotropic noise rejection of the beam pattern. It represents the total response of the beam normalized by the gain in the LD θ'. An omnidirectional point sensor has a DI of 0 dB. The higher the DI, the more directional the beam and the greater the background noise rejection is. The DI can be numerically calculated, but in general, an analytic expression is foreboding without at least some approximations. We can use the following approximation based on a heuristic physical argument.

$$d_I = 10^{(DI/10)} \approx \frac{L}{\lambda/2} \tag{13.49}$$

A *DI* of 0 dB ($d_I = 1$) is nearly the case when the aperture of the line source $L = \lambda/2$ or less. One can argue that a sensor smaller in size than a half wavelength is essentially a point sensor with omnidirectional response. Since our line array has a maximum gain L in the LD, we can assume $d_I = 2L/\lambda$, but it will be somewhat less than that for LDs steered off broadside to the array. This is because the effective aperture of the line array gets small for LDs away from broadside (0°). However, the directivity index actually improves for a line array when the beam is steered along the axis (90° from broadside). This is because the backward beam collapses leaving only one lobe pointing out from one end of the line array. The directivity has a similar effect on SNR as does an FFT for a sinusoid in white noise. This is because summing the array elements increases the amplitude of the spatially coherent wave from the LD while not amplifying waves from other directions and the spatially incoherent background noise. This is also where the term "wavenumber filtering" originates.

Finally, the CRLB is affected by beam width. The broader the beam width, the higher the CRLB because it will be more difficult to pin-point the precise bearing angle with a beam pattern which does not vary much with angle. Consider the angle off broadside where the directivity power gain is down by 1/2. We can estimate this by noting that (sin x)/x = 0.707 for x = 1.4 approximately. Therefore,

$$\frac{\pi L}{\lambda} \sin \theta \approx 1.4$$

$$\theta \approx \sin^{-1}\left\{\frac{1.4\,\lambda}{\pi L}\right\} \qquad (13.50)$$

$$\theta \approx 0.4456 \frac{\lambda}{L} \approx \frac{\lambda}{2L}$$

and we interestingly pick up another factor of $2L/\lambda$ from the beam width. Therefore, our approximate estimate for the CRLB for bearing error for a line array of length L and wavelength λ (near broadside arrival angle) is

$$\sigma_\theta \approx \frac{1}{\text{SNR} \cdot (2L/\lambda)^2} \qquad (13.51)$$

Comparing the CRLB for phase difference in Equation 13.11 to the CRLB for a line array, one might think that for $L > \lambda$, the bearing estimates for the line array beam pattern are better than a direct measurement of phase to get bearing. But, the CRLB for the beamforming estimator is actually not as good as a direct measurement. Recall that there are many elements available for phase-difference pairings and the CRLB in Equation 13.11 is for M observations for only one sensor element pair. When one has more than one arrival angle for a particular frequency, only beamforming techniques can provide correct bearing answers.

There are other physical problems that make large arrays with closely spaced sensors underperform the theoretical CRLB for bearing error. First, when the sensors are closely spaced, the noise is no longer incoherent from sensor to sensor, especially where the sound speed is relatively slow compared to fluid flow speeds, such as in air. Thus, the beamforming algorithm does not reduce the background noise as much as planned. Second, for large arrays outdoors in air in inhomogeneous media (say acoustic arrays with flow and turbulence), the signal coherence from one end of the array to the other is not guaranteed to be unity. Therefore, it is unlikely that the array gain will be as high as theory predicts possible throughout the CRLB. For underwater acoustic arrays, the sound speed is nearly five times faster than in air and the fluid flow is generally very much less, and so coherence is much less of a concern. For electromagnetic arrays, the signals are completely coherent and the noise incoherent across the array, unless a hostile alien sprays the array with plasma. The CRLB is *a lower bound* by definition, and is used to form a confidence estimate on the bearing error to go along with the bearing estimate itself. As many physicists and engineers know all too well, data with some confidence measure (such as error bars, variance, probability density, etc.) is far more useful than raw data alone. For intelligent sensor and control systems, confidence measures are even more important to ensure correctly weighted data fusion to produce artificially measured information and knowledge. Information is data with confidence metrics while knowledge is an identifiable pattern of information which can be associated with a particular state of interest for the environment. The CRLB is essential to produce bearing information, rather than bearing data.

13.3 FIELD RECONSTRUCTION TECHNIQUES

Sensor arrays can be used for much more than determining the AOAs of plane wave radiated from distance sources. In this section we examine the use of array processing to measure very complicated fields in the vicinity of a source. Some useful applications are investigation of machine vibrations from radiated acoustic noise, condition monitoring electrical power generators or components, or even optical scattering from materials as a means of production quality control sensing. In all cases, a sensor array scans a surface to observe the field and to translate the measurements from the array surface to what is happening where the sources are. For example, an acoustic intensity scan over a closed surface enclosing a sound source of interest (such as a diesel engine) can provide the net Watts of radiated power (Gauss's theorem). But the acoustic pressure and velocity on the scanning surface could also be used to reconstruct the acoustic field much closer to the source surface, allowing surface vibrations to be mapped without contact. It can be seen that this technique might be useful in the investigation of things like engine or tire noise in cars. For electromagnetic equipment, changes in the field could provide valuable precursors to component failure, allow one locate areas of leakage/corrosion/damage, or measure the dynamic forces governing the operation of a motor or generator, for example, when internal windings fail.

Field reconstruction is possible because of Green's integral formula [6,7]. Green's integral formula can be seen as an extension to 3D fields of the well-known formula for integration by parts.

$$\int u \, dv = uv - \int v \, du \tag{13.52}$$

When we develop a measurement technique for the radiated waves from a source or group of sources, it is useful to write the field equations as a balance between the radiated power from the sources and the field flux through a surface enclosing the source(s) and the field space of interest. This follows from Gauss's law, which simply stated says that the net electric flux through a surface enclosing a source(s) of charge is equal to the total charge enclosed by that surface. Gauss's law for electric fields is

$$\oint_S D_s \, dS = \int_{vol} \rho_e \, dv \tag{13.53}$$

where D_s is the electric field flux, dS is the surface area element, ρ_e is the charge density, and dv is the volume element. For the case of the wave field inside a closed surface due to a number of sources also inside the surface we have the following 3D equation for acoustic waves called the Helmholtz–Huygens integral.

$$\oint_S \left(p(X) \frac{\partial g(X \mid X')}{\partial n} - g(X \mid X') \frac{\partial p(X)}{\partial n} \right) dS = \int_{vol} g(X \mid X') F(X') dv \tag{13.54}$$

where $X = \langle x, y, z \rangle$ and $X' = \langle x', y', z' \rangle$ are the surface field and source points, respectively, (∂/∂_n) is the gradient normal to the surface, $p(X)$ is the acoustic pressure on the surface field point of interest X, and $g(X \mid X')$ is the free space Green's function for a source at X' and a receiver at X in three dimensions given in

$$g(X \mid X') = \frac{e^{jk|X-X'|}}{4\pi \mid X - X' \mid} \tag{13.55}$$

The term "free-space" means that there are no reflections from distant boundaries, that is, a reflection-free or anechoic space. However, if there were a reflection boundary of interest, the left-hand side of Equation 13.54 would be used to define the pressure and velocity on the boundary surface allowing the field to be reconstructed on one side or the other. One could substitute any field quality such as velocity, electric potential, and so on, for $p(X)$ with the appropriate change of units in $F(X')$.

Huygen's principle states that a wave can be seen to be composed of an infinite number of point sources. The Helmholtz–Huygens integral establishes a field mathematical representation by an infinite number of sources (monopole velocity sources and dipole force sources) on a closed surface to allow the reconstruction of the field on one side of the boundary surface or the other. If we know the point source locations, strengths, and relative phases, one would simply use the right-hand side of Equation 13.54 and sum all the source contributions from the locations X' for the field point of interest X. From an engineering perspective, we would like to measure the source locations, strengths, and relative phases from a sensor array which defines the field on a surface. However, this surface must separate the field point and the sources of interest to be of value mathematically, but this is easily achieved mathematically by separating a source or field point from the surface with a narrow tube and infinitely small sphere surrounding the field point. As will be seen later, the definition of the integration surface in Equation 13.54 mainly has an effect on the sign of the derivatives with respect to the normal vector to the surface. In the midst of these powerful field equations, the reader should keep in mind foremost that the Green's function can be used with the field measured by an array of sensors to reconstruct the field on another surface of interest.

The physical significance of the left side of Equation 13.31 is that the surface has both pressure $p(X)$ and velocity $(\partial_p(X)/\partial_n)$ (actually a quantity proportional to velocity) which can describe the field inside the surface (between the sources and the bounding surface) due to the sources depicted by the right side of the equation. *For a specific source distribution, there are an infinite number of combinations of pressure and velocity on the enclosing surface which give the same field inside.* However, assuming that one knows the approximate location of the sources (or a smaller volume within the integration volume where all sources are all contained), and that one has measurements of the field on the outer surface, then an "image field" can be reconstructed on any surface of interest not containing a source. This 3D field reconstruction from a set of surface measurements is known in acoustics as acoustical holography [4]. The physics are in fact quite similar when one considers the wave interference on the measurement surface and the spatially coherent processing to reconstruct the field. Field reconstruction using sensor arrays is an extremely powerful technique to analyze radiated waves.

Acoustic fields provide us with a nice example of how holographic reconstruction can be useful in measuring sound source distributions. For example, when an automobile engine has a loose or defective valve, a tapping sound is easily heard with every rotation of the cam shaft. It is nearly impossible to determine which valve using one's ears as detectors, because the sound actually radiates to some extent from all surfaces of the engine. Mechanics sometimes use a modified stethoscope to probe into the engine to find the problem based on loudness. Given the valve tap radiates sound most efficiently in a given frequency range, one could use acoustic holography and a large planar array of microphones to find the bad valve. By measuring the field at a number of sensor positions in a plane, the field could be reconstructed in the plane just above the engine surface, revealing a "hot spot" of acoustic energy over the defective valve position. To do a similar mapping using intensity, one would have to measure directly over the radiating surface in a very fine spacing. Using holographic field reconstruction, all field components can be calculated, which for acoustics means velocity, intensity, and impedance field can all be calculated from the pressure measurements.

Unlike the time-average field intensity, one needs the precise magnitude and phase spatially for each frequency of interest in the measurement plane to reconstruct the field accurately in another plane. In acoustics, it is very useful in noise control engineering where one must locate noise "hot-spots" on equipment. The technique will likely also find uses in electromagnetics and structural vibrations, although full vector field reconstruction is significantly more complicated than the scalar–vector fields for acoustic waves in fluids. In machinery failure prognostics, field holography can be used as a measurement tool to detect precursors to failure and damage evolution from subtle changes in the spatial response. Changes in spatial response could provide precursors to equipment failure well before detectable changes in wave amplitude are seen at a single sensor.

We begin our development of the holographic field reconstruction technique by simply examining the free-space Green's function and its Fourier transform on a series of x–y planes at different distances z from the point source location. In Cartesian coordinates, the 3D free-space Green's function for a receiver at $X = \langle x, y, z \rangle$ and point source at $X' = \langle x', y', z' \rangle$ is

$$g(x, y, z \mid x', y', z') = \frac{e^{j\sqrt{(k_x^2 + k_y^2 + k_z^2)}\sqrt{(x-x')^2 + (y-y')^2 + (z-z')^2}}}{4\pi\sqrt{(x-x')^2 + (y-y')^2 + (z-z')^2}} \tag{13.56}$$

Figure 13.17 shows the spatial responses for the acoustic case of a 700 Hz ($c = 350$ m/s, $k = 12.65$ rad/m) point source at $X' = \langle 0, 0, 0 \rangle$ where the field surfaces are planes located at $z = 0.001$ m, $z = 1.00$ m, and $z = 50.00$ m. The measurement planes are 10 m × 10 m and sampled on a 64 × 64 element grid. A practical implementation of this measurement would be to physically scan the plane with a smaller array of sensors, keeping on additional sensor fixed in position to provide a reference phase. One would calculate temporal FFTs and process the 700 Hz bin spatially as described here. On the right side of Figure 13.17 one sees the spatial Fourier transform of the Green's function, $G(k_x, k_y, z)$, on the corresponding z-plane for the left-hand column of surface plots. The spatial real-pressure responses are plotted showing maximum positive pressure as white and maximum negative pressure as black to show the wave front structure. The wavenumber plots on the right-hand column are shown for the magnitude where white corresponds to maximum amplitude and black minimum amplitude. Each of the six plots are independently scaled.

The spatial and corresponding wavenumber plots in Figure 13.17 are extremely interesting and intuitive. In the spatial plane just in front of the source seen in the upper left plot, the singularity of the point source dominates the response at $x = y = 0$. There is, however, a wave structure in this plane where $k^2 \approx k_x^2 + k_y^2$ and $k_z \approx 0$. This is clearly seen in the wavenumber transform in the upper right plot. The bright ring corresponds to the wavenumber $k = 12.7856$ if the speed of sound is taken as 344 m/s. Since we are practically in the plane with the point source, there is very little wave energy at wavelengths longer than 0.491 m (the wavelength of $k = 12.7856$ and 700 Hz). There is wave energy at wavelengths shorter than 0.491 m, mainly due to the "point-like" spatial structure at $z = 0.001$ m. This wave energy is called evanescent because it will not propagate very far (the waves self-cancel). For propagation of these waves in the z-direction we have

$$\frac{e^{jk_z z}}{4\pi X} = \frac{e^{-\sqrt{k_x^2 + k_y^2 - k^2}\, z}}{4\pi X} \qquad \sqrt{k_x^2 + k_y^2} > k \tag{13.57}$$

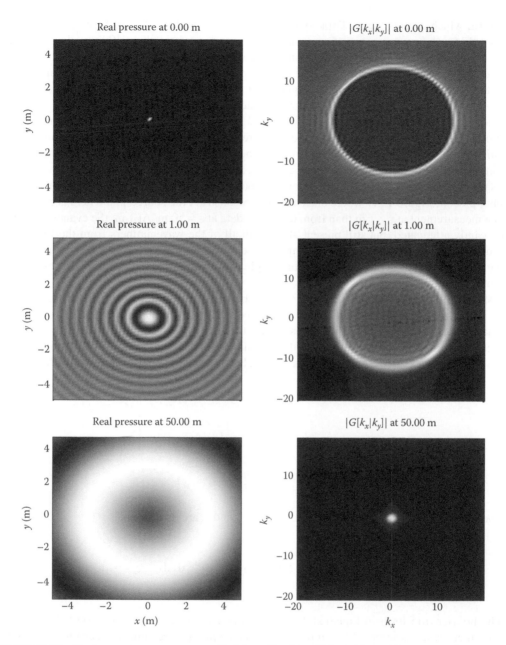

FIGURE 13.17 Real pressure responses sampled in a 64 × 64 grid in a 10 m × 10 m space of an acoustic free space Green's function in three dimensions showing the wavenumber transforms for three different z-axis planes at 700 Hz.

causing the waves with wavelengths shorter than the free propagating wavelength to exponentially decay in the positive z-direction. The middle row of plots clearly shows the rapid decay of the evanescent field and the "leakage" of longer wavelengths (smaller wavenumbers) into the center region of the wavenumber spectrum. One can see in the spatial plot at $z = 1.00$ m that the wavelengths get longer slightly as one approaches the center of the measurement plane (middle left plot). This is caused by the angle between the normal to the surface of the spherical wave and the measurement plane. The diameter of the bright ring in the wavenumber plot also is smaller. As z becomes quite large ($z = 50$ m is 100 wavelengths), one sees a near uniform pressure spatially and a wavenumber transform which reveals a near Dirac delta function (bottom plots). Recall Section 7.3 where we showed that a plane wave assumption could only be used at ranges about 100 wavelengths from a point source using both geometry and intensity theory.

The ring-shaped peak energy "ridge" in the wavenumber plots in Figure 13.17 collapses into a delta function as z approaches infinity (right plots). The amount of the ring diameter collapse is a function of the measurement plane aperture and the distance from the point source. If we call the equivalent wavenumber for this ridge diameter k_d, it can be expressed as $k_d = k \sin \theta$, where $\theta = \tan^{-1}(1/2L/z)$ and L is the width of the aperture. The aperture angle is an important physical quantity to the signal processing. For example, if one wants to keep the ridge diameter wavenumber within about 10% of the source plane value, the measurement aperture width needs to be over four times the measurement plane distance z from the source plane. This can be clearly seen in the wavenumber plots of Figure 13.17 where one can better reconstruct the source-plane field at $z = 0.001$ m from a measurement at $z = 1$ m than from measured data at $z = 50$ m. At 1 m, the evanescent field is significantly attenuated, but still present along with all of the wavenumbers from the source. The aperture angle for a 10 m × 10 m measurement plane can be seen as 90° at $z = 0.001$ m, about 79° at 1.00 m, and about 5.7° at 50°m. This aperture angle will be important for defining the resolution of the holography measurement.

It should be possible to define a transfer function between the measurement plane and the "image" plane so long as the SNR and dynamic range of the measurement plane wavenumber spectrum is adequate. Suppose our measurement plane is parallel to the $x - y$ plane at $z = z_m$. We wish to reconstruct the field in an image plane also parallel to the measurement plane at $z = z_i$. Given the measured pressure wavenumber spectrum $P(k_x, k_y, z_m)$, the image plane wavenumber spectrum is found to be

$$P(k_x, k_y, z_i) = H(k_x, k_y, z_i, z_m) \, P(k_x, k_y, z_m) \tag{13.58}$$

where

$$H(k_x, k_y, z_i, z_m) = \frac{G(k_x, k_y, z_i)}{G(k_x, k_y, z_m)} \tag{13.59}$$

The Green's function wavenumber transform at z_m and z_i are defined as

$$G(k_x, k_y, z_k) = \int_{-L_x/2}^{+L_x/2} \int_{-L_y/2}^{+L_y/2} g(x, y, z_k) e^{-jk_x \hat{x}} e^{-jk_y \hat{y}} \, d\hat{x} \, d\hat{y}, \quad z_k = z_i, z_m \tag{13.60}$$

The Fourier transform in Equation 13.60 is efficiently carried out using a 2D FFT where the measurement plane is sampled k_s samples per meter giving a wavenumber spectrum from $-k_s\pi$ to $+k_s\pi$. The larger the L_x and L_y are, the finer the wavenumber resolution will be. However, one should have at least the equivalent of 2 samples/wavelength in the source plane to avoid aliasing. Even though the 2D wavenumber FFT has complex data as input from the temporal FFT bin corresponding to the frequency of interest, the original time-domain signals from the array sensors are generally sampled as real digital numbers.

Figures 13.18 through 13.21 show results of holographic imaging of a quadrapole in a plane adjacent to the source plane from measurement planes a further distance away. As will be seen, the spatial resolution of the holographically reconstructed image depends on wavelength, distance between the measurement and imaging planes, and the aperture of the measurement plane. We will use these four figures to help visualize the issue of holographic resolution. In Figure 13.18, the field from a symmetric 700 Hz quadrapole is measured at 1 m distance. Green's function wavenumber transforms for a point source at the origin are calculated numerically as seen in Equation 13.60 for

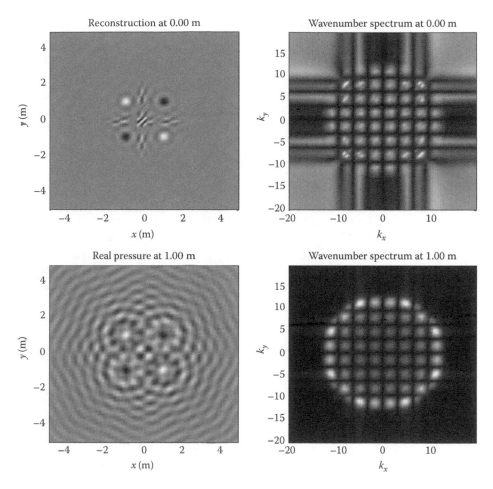

FIGURE 13.18 Equal-spaced quadrupole reconstruction from a measurement plane 1 m away in z for a 64×64 measurement over a 10 m × 10 m field area at 700 Hz.

the measurement plane at $z = 1$ m and the image plane taken as $z = 0.001$ m. The transfer function $H(k_x, k_y, z_i, z_m)$ is computed as seen in Equation 13.59 and the image wavenumber spectrum is computed using Equation 13.57. The reconstructed field is seen in the upper left plot in Figure 13.18. The measurement field at 1 m distance is seen in the lower left plot. The corresponding wavenumber spectra are seen in the right-hand column. Note the substantial amount of evanescent field in the image plane. The equal spacing (the point sources are at $\langle 1, 1 \rangle$, $\langle -1, 1 \rangle$, $\langle -1, -1 \rangle$, and $\langle 1, -1 \rangle$ meters) and strengths of the quadrupole sources create symmetric interference patterns which are barely visible in the pressure field plots, but clearly seen in the wavenumber spectra. In Figure 13.19 we move the sources around (just to show arrogance) to the positions $\langle 1.2, 1.0 \rangle$, $\langle -1.0, 1.3 \rangle$, $\langle -1.8, -1.5 \rangle$, and $\langle 1.5, -0.5 \rangle$ meters. Actually, what is seen in the wavenumber spectra is an asymmetry to the wavenumber peaks and great complexity to the evanescent field. Also shown in Figures 13.18 and 13.19 are the fact that at 1 m it is still possible to determine that four sources are present in the measurement plane (lower left plot). In Figure 13.20 the measurement field is moved back to 5 m distance from the sources. In the reconstruction, the sources are still detectable, but there is considerable loss of resolution. The "ridge diameter" for the wavenumber spectrum at 5 m is also smaller, indicating that the aperture angle may be too small. Figure 13.21 also has the measurement plane at 5 m distance, but these data are for a frequency of 1050 Hz, which translates into 30 wavelengths for the 10 m aperture of the measurement array. With the resolution nicely restored, one can see that there

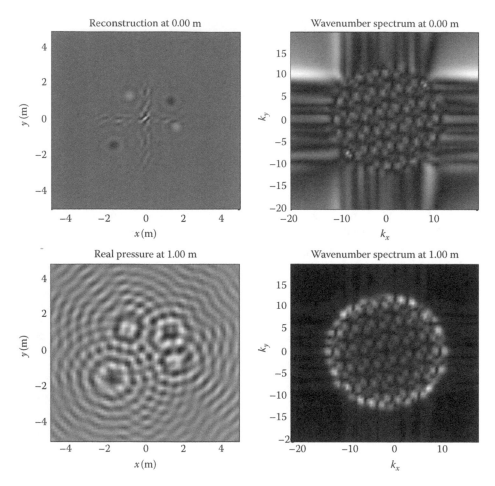

FIGURE 13.19 Quadrapole reconstruction at 700 Hz when the sources are not symmetrically spaced showing the changes in wavenumber field complexity in 1 m-distant measurement plane.

is an interesting, albeit nonobvious, relationship between aperture, measurement distance, frequency, and reconstruction resolution.

A model can be developed to estimate the available resolution for a particular array aperture, source wavelength, and measurement plane distance. We start by noting that the apparent wavelength in the measurement plane gets long as this plane is moved farther from the source plane. The wavelength in the measurement plane is $\lambda' = \lambda/\sin\theta$, where $\theta = \tan^{-1}(1/2L/z)$ and L is the width of the array aperture. As z gets large, θ tends to zero and λ' tends to infinity. From a beamforming point of view (see Section 13.2), the long wavelength trace in the measurement plane will translate into a limited ability to spectrally resolve the wavenumbers in the 2D wavenumber transform. We can estimate this resolution "beam width" as approximately $\beta = 2\sin^{-1}[\lambda/(L\sin\theta)]$. Given this beam width, the spatial resolution in the measurement plane is approximately

$$\Delta = z\sin\left(\frac{\beta}{2}\right) \approx z\,\frac{\lambda}{L\sin\theta} = \frac{2z^2\lambda}{L^2} \tag{13.61}$$

where Δ is the resolution in meters of the image plane available for the reconstruction geometry. Equation 13.61 clearly shows reconstruction resolution improving (a smaller number) for higher frequencies, larger apertures, and measurement planes closer to the image plane. However, it is not

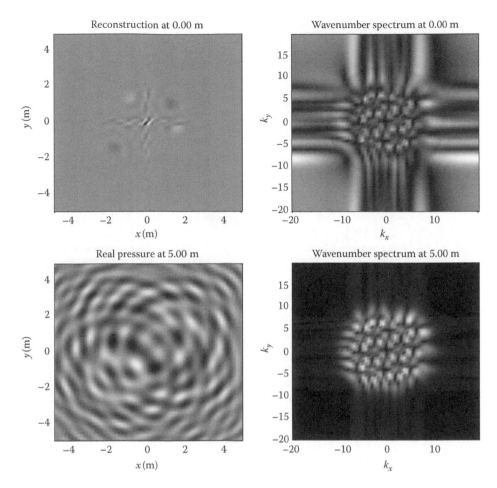

FIGURE 13.20 Moving the measurement plane out to 5 m for the 700 Hz quadrupole is close to the limits for resolution of the sources.

that simple because the measurement field is sampled spatially with a resolution of L/N, N being the number of samples across the $N \times N$ spatial FFTs, which in our case is 64 samples. One cannot get a higher resolution in the image reconstruction then available in the measurements. It takes a larger aperture and more samples to get more resolution at the same frequency. However, for a given frequency and aperture one can find the distance z where the resolution begins to degrade significantly.

The resolution question then focuses (literally) on determining the maximum measurement plane distance where the resolution significantly starts to decrease. This is analogous to the depth of field (the depth where the view stays in focus) of a camera lens system. The f-stop of a camera lens is the ratio of the lens focal length to the aperture of the lens opening. The larger the lens aperture, the smaller the depth of field (and smaller the f-stop) will be for a given focal length. To find the distance z_0 where resolution in the image plane is equal to the measurement plane resolution, we set Equation 13.61 equal to L/N and solve for z_0.

$$z_0 = L\sqrt{\frac{L}{2N\lambda}} \tag{13.62}$$

Note that as N gets large z_0 approaches zero, which is counter intuitive. Think of z_0 as the depth of field for the holography system, outside of which the reconstructed image will be out of

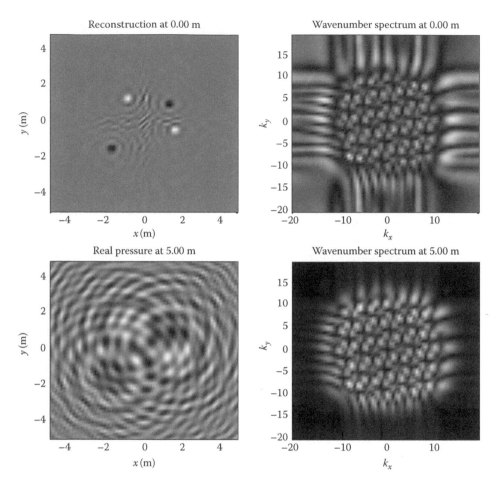

FIGURE 13.21 Increasing the frequency to 1050 Hz actually restores much of the reconstructed details from a 5 m measurement due to the array's narrower beam response at high frequency.

focus spatially. Usually more FFT samples improve resolution, but in this case only if the array aperture also gets larger. Using the camera analogy, increasing the lens aperture reduces the depth of field, allowing the lens to filter out image details outside of the depth of field. This is essentially what we are doing in holography—getting a highly detailed image inside the depth of field. So a small depth of field translates into a precise holographic reconstruction with high resolution (small Δ).

This is good news for wave field holography where the measurement plane is relatively far from the imaging plane, because fewer sensors in the array and smaller FFT sizes can actually improve resolution when the measurement plane is relatively far from the image plane, but only to a point. Again it comes down to the physics of aperture. Once the measurement plane is far enough away from the sources so that the sources fall within the beam width for the array, there is little hope for resolving the sources. Another way to see this is that the measured field will be a mix of very low wavenumber waves yielding little spatial information.

Figure 13.22 shows the 700 Hz quadrapole reconstructed from a 5 m measurement plane using a 32×32 measurement grid as compared to the 64×64 grid used in Figure 13.20. The reconstruction resolution actually improves with the smaller FFT as predicted by Equation 13.61 and 13.62. For 700 Hz and 64×64 point FFTs, z_0 is about 3.98 m. For 1050 Hz z_0 increases to 4.88 m. Using 32×32 points at 700 Hz, z_0 is about 4.14 m which means that the resolution is degraded more for the reconstructed field using the larger spatial FFTs. However, using too small an FFT will again limit resolution at a particular frequency. Note how with 32×32 points we have 3.2 samples/m, reducing the wavenumber range to $\pm 3.2\pi$, or ± 10 m^{-1} from $\pm 6.4\pi$ with 64 points, or ± 21 m^{-1}. If our frequency were higher, there would be serious spatial aliasing. In fact, some aliasing is going on since our spatial samples are 0.3125 m apart and the wavelength is 0.491 m at 700 Hz. The aliasing is seen as the ripples near the origin in the reconstruction in the upper left of Figure 13.22 and also in the nearly out-of-band evanescent energy in the right-hand column of plots. It does not completely fail because the measurement plane happens to be far enough away from the sources, but not so far that the sources are within the beam width of the array. Also, the sources in our example are well separated spatially to begin with, but what happens outside the depth of field of the holography array is that the source energy gets spread across adjacent bins making it harder to see in the reconstructed image.

An analytical expression for the Green's function-based wavenumber transfer function can be approximated and is useful when one does not know where the source plane is. In this regard, the field in the measurement plane can be translated a distance $d = |z_m - z_i|$ along the z-axis. This is not exactly the same as the transfer function method described above, but it is reasonable for many applications. The analytical solution is found by applying the Helmholtz–Huygens integral where

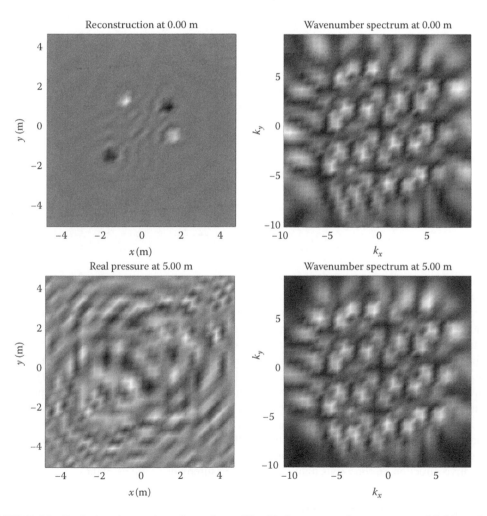

FIGURE 13.22 Reducing the number of samples to 32×32 also restores the reconstructed field resolution for nonobvious signal processing reason of the source energy being spread over fewer bins in the reconstructed field.

the boundary between the source and the field point is an infinite plane. We then assign a Green's function (a particular solution to the wave equation) of the form

$$g(x, y, z) = -\frac{1}{2\pi} \frac{\partial}{\partial \alpha} \left\{ \frac{e^{jk\sqrt{x^2 + y^2 + \alpha^2}}}{\sqrt{x^2 + y^2 + \alpha^2}} \right\}_{\alpha=z} \tag{13.63}$$

The 2D Fourier transform of Equation 13.63 can be found analytically (skudzryk) as

$$G(k_x, k_y, d) = e^{jd\sqrt{k^2 - k_x^2 - k_y^2}} \tag{13.64}$$

and we note that for the evanescent part of the wavenumber field, Equation 13.64 is equivalent to Equation 13.57. The distance d in Equation 13.64 represents the distance from the measurement plane in the direction away from the source one is calculating the wavenumber field in the image plane. This is considered wave propagation modeling and the pressure field in the image field plane which is farther away from the sources than the measurement plane is

$$p(x, y, z_i) = \Im\{G(k_x, k_y, d) \, P(k_x, k_y, z_m)\}^{-1} \quad d = z_i - z_m \tag{13.65}$$

where $\Im\{\}^{-1}$ denotes an inverse 2D F $p(x, y, z_i) = \Im\{G(k_x, k_y, d) \, P(k_x, k_y, z_m)\}^{-1}, \quad d = z_i - z_m$ FT to recover the spatial pressure in the image field from the wavenumber spectrum.

For holographic source imaging, we are generally interested in reconstructing the field very close to the source plane. Equation 13.66 shows the inverse Green's function wavenumber spectrum used to reconstruct the field.

$$p(x, y, z_i) = \Im\left\{G(k_x, k_y, d)^{-1} \, P(k_x, k_y, z_m)\right\}^{-1}, \quad d = z_i - z_m \tag{13.66}$$

where

$$G(k_x, k_y, d)^{-1} = e^{jd\sqrt{-k^2 + k_x^2 + k_y^2}} \tag{13.67}$$

Note the sign changes in the square-root exponent in Equation 13.67. This very subtle change is the result of using the inverse wavenumber. Therefore, the region inside the free wavenumber circle is actually propagated back toward the source in a nonphysical manner by having an evanescent-like exponential increase. This explains the apparent high frequency losses seen in the reconstructions in Figures 13.23 and 13.24. It also can explain why we had to use 750 Hz for the example rather than 700 Hz (readers should try this in the m-script example). It appears that as long as the array aperture and N is large, this can work, but it is more sensitive to errors for long wavelengths. The analytic Green's function technique works fairly well, even though some approximations are applied to its derivation. It can be seen as a "wavenumber filter" rather than a physical model for wave propagation, although its basis for development lies in the Helmholtz–Huygens integral equation. To calculate the wavenumber field a distance toward the source from the measurement plane, the sign of the exponent is changed simply by letting $d = z_i - z_m$, assuming that the source plane is further in the negative direction on the z-axis than either the measurement or image planes.

13.4 WAVE PROPAGATION MODELING

Wave propagation modeling [8] is an important area for acoustics, electromagnetics, and material characterization. While this topic is generally considered a physics topic, the Green's function

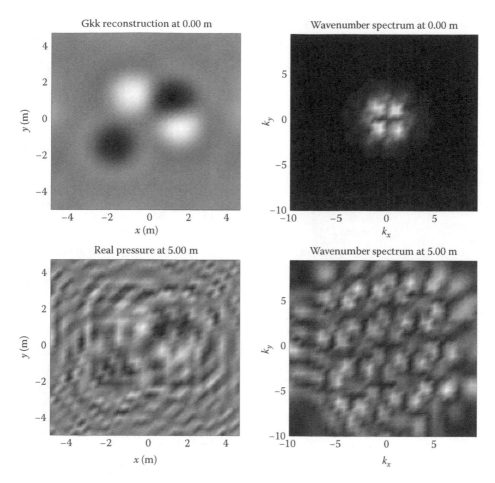

FIGURE 13.23 750 Hz quadrupole reconstruction for 32×32 samples at 5 m using the analytic Green's function wavenumber filter showing strong reconstructed signals but with less spatial resolution and much higher sensitivity to aliasing errors (why frequency increased to 750 Hz).

approach of the previous section brings us to the doorstep of presenting this important technique. Of particular interest is wave propagation in inhomogeneous media, that is, media where the impedance or wave speed is not constant. From the sensor system perspective, wave propagation modeling along with spatial sensor measurement of the waves offer the interesting opportunity of *tomographic measurement of the media in homogeneity*. For example, surface layer turbulence could be mapped in real-time for airport runway approaches using an array of acoustic sources and receivers. Another obvious example would be to use adaptive radio transmission frequencies to circumvent destructive interference predicted by a communications channel propagation model. Perhaps the most interesting application of wave propagation modeling may be in the area of materials characterization and even chemical process sensors and control. Chemical reactions and phase changes of materials results in wave speed and impedance changes whether the wave is vibration, acoustic, electromagnetic, or thermal/optical. Inexpensive and robust sensor systems that can tomographically map the state of the wave propagation media can be of enormous economical importance. But the main application here is predicting outdoor sound.

Our discussion begins by considering 2D propagation in cylindrical coordinates where our propagation direction is generally in the r-direction and our wave fronts are generally aligned in the

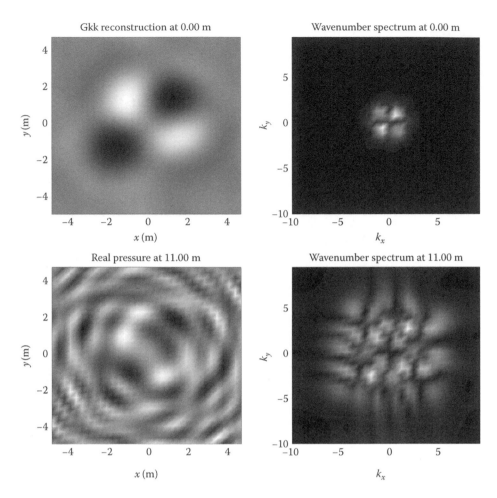

FIGURE 13.24 Even at 11 m and 750 Hz (well beyond the resolution range of the transfer function method) the analytic Green's function allows one to determine that a quadrupole is present in the source plane.

z-direction [9]. The homogeneous wave equation assuming field symmetry about the origin along the θ circle is

$$\frac{\partial \psi}{\partial z^2} + \frac{\partial \psi}{\partial r^2} + k^2(r, z), \quad \psi = 0 \tag{13.68}$$

where $k(r, z)$ is the wavenumber $\omega/c(r, z)$ which varies with both r and z for our inhomogeneous wave propagation media. The wavenumber can be decomposed into a wave component in the r-direction which varies with z and a component in the z-direction which varies with r.

$$k^2(r, z) = k_r^2(z) + k_z^2(r) \tag{13.69}$$

If the wave speed is inhomogeneous, the wave front from a point source will distort with refraction as some portions propagate faster than others. We note that if media impedance also changes, one would also have to include wave scattering in both forward and backward (toward the source) directions. Our discussion will be limited to the refraction case which is certainly important but will also allow a brief presentation. We also note that for 3D wave propagation in which the $r - z$ plane represents symmetry for the source radiation, one can divide by the square root of r as $\psi_{2D} = \sqrt{r}\, \psi_{3D}$.

The wave equation in Equation 13.68 can be written in terms of an operator Q as

$$\left(\frac{\partial}{\partial r}+j\sqrt{Q}\right)\left(\frac{\partial}{\partial r}-j\sqrt{Q}\right), \quad \psi = 0 \tag{13.70}$$

where $Q = (\partial^2/\partial_z^2) + k^2(z)$ and we will assume that $k(r,z)$ has only z-dependence for the moment. Given the homogeneous solution to the wave equation, one can find the complete particular solution by applying boundary, source, and initial conditions. However, our interests here assume that one has the field measured over the z-direction at range r and is interested in the field over the z-direction at a farther distance $r + \Delta r$ away from the source. We will show that the solution to Equation 13.68 will be in the same form as the Green's function used for holographic reconstruction in Section 13.3. Equation 13.70 is generally known as a parabolic wave equation because of the operator product. Clearly, one can see that

$$\frac{\partial \psi}{\partial r} = \pm j\sqrt{Q}\,\psi \tag{13.71}$$

where waves represented temporally as $e^{j\omega t}$ traveling away from the source have the minus sign. Therefore, to propagate the wave a distance Δr, one must add the phase

$$\psi(r+\Delta r, z) = e^{j\Delta r\sqrt{Q}}\,\psi(r, z) \tag{13.72}$$

where the operator Q is constant over the range step Δr. But in the wavenumber domain, the spectral representation of the operator Q allows the wavenumber spectra in the z-planes at r and $r + \Delta r$ to be written as

$$\Psi(r+\Delta r, k_z) = e^{j k_r \Delta r}\,\Psi(r, k_z) \tag{13.73}$$

where

$$\Psi(r, k_z) = \int_{-\infty}^{+\infty} \psi(r, z)e^{-jk_z z}\,dz \tag{13.74}$$

Substituting Equation 13.69 into Equation 13.73 we have the same Green's function propagation expression as seen in Section 13.3 in Equations 13.64 and 13.65.

$$\Psi(r+\Delta r, k_z) = e^{j\Delta r\sqrt{k^2-k_z^2}}\,\Psi(r,k_z) \tag{13.75}$$

Equation 13.75 is not all that impressive in terms of inhomogeneous media propagation because we assumed that Q is constant over the range step Δr. The wavenumber k_z variation in z due to an inhomogeneous medium is not explicitly seen in Equation 13.75. One could inverse Fourier transform the wavenumber spectrum at $r + \Delta r$ and then multiply the field at each z by the appropriate phase to accommodate the wave front refraction if a simple expression for this could be found. To accomplish this, we apply what is known as a "split-step approximation" based on the assumption that we have propagation mainly in the r-direction and that the variations in wavenumber along the z-direction are small. This is like assuming that the domain of interest is a narrow wedge with the waves propagating outward. We note that a wavenumber variation in the r-direction will speed up or slow down the wave front, but will not refract the wave's direction.

Let the variation along the z-direction of the horizontal wavenumber $k_r(z)$ be described as

$$k_r^2(z) = k_r^2(0) + \delta k^2(z) \tag{13.76}$$

where $k_r^2(0)$ is simply a reference wavenumber taken at $z = 0$. Assuming that we have mainly propagation in the r-direction, $k^2(r, z) \approx k_r^2(z)$, we can make the following approximation:

$$k_r(z) \approx \sqrt{k_r^2(0) + \delta k^2(z) - k_z(r)}$$
$$\approx \sqrt{k_r^2(0) + k_z(r)} + \frac{\delta k^2(z)}{k_r(0)} \tag{13.77}$$

Applying the split-step approximation to Equation 13.72 we have

$$\psi(r + \Delta r, z) = e^{j \Delta r \sqrt{Q_r}} \psi(r, z) \tag{13.78}$$

where $\sqrt{Q_r} = \sqrt{(\partial^2/\partial z^2) + k_r^2(0) - k_z^2(r)} + (\delta k^2(r)/2k_r(0))$. In the wavenumber domain, the expression is a bit more straightforward, but requires an inverse Fourier transform.

$$\psi(r + \Delta r, z) = e^{j \Delta r \frac{\delta k^2(z)}{2 k_r(0)}} \left\{ \frac{1}{2\pi} \int_{-\infty}^{+\infty} \left[\Psi(r, k_z) e^{j \Delta r \sqrt{k_r^2(0) - k_z^2(r)}} \right] e^{+jk_z z} \, dk_z \right\} \tag{13.79}$$

Equation 13.79 defines an algorithm for modeling the propagation of waves through an inhomogeneous medium. The wavenumber spectrum is computed at range r and each wavenumber is multiplied by the appropriate phase in the square brackets to propagate the spectrum a distance Δr. Then the inverse Fourier transform is applied to give the field at range $r + \Delta r$ including any phase variations due to changes in horizontal wavenumber k_r. Finally, the phase is adjusted according to the wavenumber variations in z depicted in the term $(\delta k^2(z)/2k_r(0))$. To begin another cycle, the left-hand side of Equation 13.79 would be Fourier transformed into the wavenumber spectrum, and so on.

This "spectral marching" solution to the problem of wave propagation has three significant numerical advantages over a more direct approach of applying a finite-element method on the time–space wave equation. The first advantage is that the range steps Δr can be made larger than a wavelength and in line with the variations in the media, rather than taking many small steps per wavelength in a finite-element computing method. The second advantage is that with so much less computation in a given propagation distance, the solution is much less susceptible to numerical errors growing within the calculated field solution. Finally, the wavenumber space approach also has the advantage of a convenient way to include an impedance boundary along the r-axis at $z = 0$. This is due to the ability to represent the field as the sum of a direct wave, plus a reflected wave, where the reflection factor can also be written as a Laplace transform.

Considering the case of outdoor sound propagation, we have a stratified but turbulent atmosphere and a complex ground impedance $Z_g(r)$. The recursion for calculating the field is

$$\psi(r + \Delta r, z) = e^{j \Delta r \frac{\delta k^2(z)}{2 k_r(0)}} \left\{ \frac{1}{2\pi} \int_{-\infty}^{+\infty} \left[\Psi(r, k_z) e^{j \Delta r \sqrt{k_r^2(0) - k_z^2(r)}} \right] e^{+jk_z z} \, dk_z \right.$$
$$+ \frac{1}{2\pi} \int_{-\infty}^{+\infty} \left[\Psi(r, -k_z) R(k_z) e^{j \Delta r \sqrt{k_r^2(0) - k_z^2(r)}} \right] e^{+jk_z z} \tag{13.80}$$
$$+ \left. 2 j\beta(r) e^{-j\beta(r)z} \, e^{j \Delta r \sqrt{k_r^2(0) - \beta^2(r)}} \, \Psi(r, \beta) \right\}$$

where $\beta(r) = k_r(0)/Z_g(r)$ and $R(k_z) = [k_z(r)Z_g(r) - k_r(0)]/[k_z(r)Z_g(r) + k_r(0)]$. The ground impedance is normalized to $\rho c = 415$ Rayls, the specific acoustic impedance for air. As $Z_g \to 1$, $R(k_z) \to 0$ and no reflection occurs. This condition also drives $\beta(r) \to k_r(0)$, the reference horizontal wavenumber at ground level. This essentially places the horizontal wavenumber in the vertical direction and no phase change occurs as the result of the range step Δr. This high-frequency spatial wave is then completely canceled in the next range step. Thus, as $Z_g(r) \rightarrow 1$ we have the no-boundary propagation case of Equation 13.79.

The first term on the right-hand side of Equation 13.80 represents the direct wave from the source, the second term is the reflected wave from the boundary (or the wave from the image source), and the third term is a surface wave which results from the ground impedance being complex, and a spherical wave interacts with a planar boundary surface representing the ground. The physics behind the third term are intriguing. One can see that the term $\Psi(r, \beta)$ is a single complex number found from the Laplace transform of $\psi(r, z)$ for the complex wavenumber β (assuming Z_g is also complex). This complex number is then phase shifted and attenuated for the range step (note that with β complex one has real and imaginary exponent elements). The complex result is then multiplied by $2j\beta$ and finally the function $e^{-j\beta z}$, which adds to the field at every elevation z. Generally, the amplitude of the surface wave decreases exponentially as one moves away from the surface. At the next range step iteration, the entire field is again used to derive the surface wave, which is then in turn added back into the field. It has been observed both numerically and experimentally that for some combinations of point source height, frequency, and ground impedance, the surface wave can propagate for substantial distances. We also note that because of the complex nature of β, the surface wave does not propagate as fast as the direct wave. All of this complexity makes measurement of inhomogeneous fields and impedance boundaries using arrays of sensors and propagation modeling an area for future scientific and economic exploitation.

Figure 13.25 shows the result of a Green's function-based parabolic wave equation in a "downward refracting" atmosphere. That is, one where sound travels faster in the propagation direction at higher altitudes than near the ground. This situation arises typically at night when cold air settles near the ground (the wave speed is slower in colder air) or when sound propagates in a downwind direction. The wind speed adds to the sound speed and winds are generally stronger at higher elevations. In the lower right corner, one can see the interaction of the downward refracting sound rays with the surface wave, causing a standing wave pattern. In the nonvirtual world,

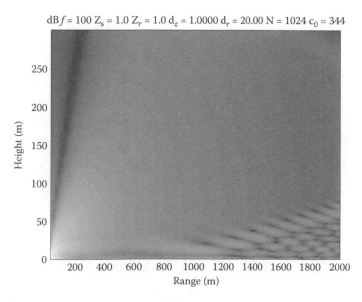

FIGURE 13.25 Sound propagation model of a 100 Hz point source at 1 m elevation over a ground impedance of $Z_g = 12.81 + j11.62$ in a downward refracting atmosphere where $c(z) = 340 + 0.1z$ m/s (typical of early evening after a sunny day) showing strong long-range reception.

turbulent fluctuations cause this standing wave field to move around randomly. To the listener on the ground, the source is heard to fade in and out of detection. The fluctuating multipath effect can also be experienced with short-wave or AM radio where the ionosphere and ground form a spherical annular duct which results in nonstationary multipath from the transmitter to the receiver.

Figure 13.26 shows a 100 Hz source in an upward refracting atmosphere. Such is the case during a sunny afternoon or when sound is propagating in the upwind direction. For upwind propagation, the wind speed is subtracted from the sound speed and the wind speed generally increases with elevation. During a sunny afternoon, the air near the ground is heated significantly. Since the part of the sound wave close to the ground is opposing a slower wind and has a faster sound speed due to the high temperature, the near-ground part of the wave outruns the upper level part of the wave. Thus the wave refracts away from the ground, hence upward refracting propagation. Clearly, one can see a rather stark falling off of the wave loudness on the ground as compared to the downward refracting case. The dark, "quiet" area in the lower right side of Figure 13.26 is known as a shadow zone where in theory, little or no sound from the source can penetrate. This is of extreme importance to submarine operations because shadow zones provide natural sonic hiding places. In the atmosphere, common sense tells us that it should be easier to hear a source when it is upwind of the listener, and that long range detection of sound in the atmosphere should be easier at night if for no other reason the background noise is low.

Figure 13.27 shows the effect of turbulence on propagation in an upward refracting environment. Turbulence not only affects detection performance (improving it in the upward refracting case), but also has an impact of bearing estimation [10]. This is typical during a hot afternoon because the hot air near the ground becomes buoyantly unstable, and plumes upward drawing in cooler air from above to replace it. Thermal pluming generally leads to a buildup of surface winds as various heat areas at differing rates. These surface winds are very turbulent due to the drag caused by the ground and its objects (trees, buildings, etc.). The parabolic wave equation propagation modeling technique allows the inclusion of turbulence as a simple "phase screen" to be added in at each range step in the algorithm. This alone illustrates the power of wavenumber filtering to achieve what otherwise would likely be an extraordinarily difficult and problematic modeling effort.

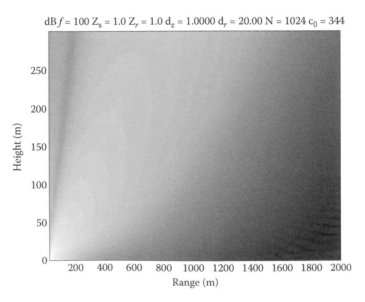

dB $f = 100$ $Z_s = 1.0$ $Z_r = 1.0$ $d_z = 1.0000$ $d_r = 20.00$ $N = 1024$ $c_0 = 344$

FIGURE 13.26 Propagation results for 100 Hz in an upward refracting atmosphere where $c(z) = 340 - 0.1z$ m/s (typical of a hot sunny afternoon) showing poor long-range reception.

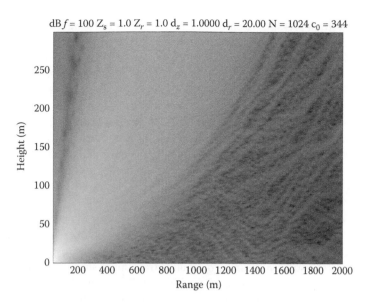

dB $f = 100$ $Z_s = 1.0$ $Z_r = 1.0$ $d_z = 1.0000$ $d_r = 20.00$ $N = 1024$ $c_0 = 344$

FIGURE 13.27 100 Hz source in an upward refracting atmosphere including turbulence effects using a Gaussian phase distribution with standard deviation of only 1% of the sound speed.

Physical considerations for wave propagation modeling can be categorized into timescales, environmental models for wave speed profiles, knowledge of wave scattering attributes, and knowledge of impedances of the dominant boundaries. The model is only as good as the physical assumptions incorporated. For example, underwater sound propagation will have timescales that are mostly seasonal while atmospheric sound propagation will certainly have diurnal (daily) cycles, but also significant changes with local weather conditions. Radio communications channels will have a diurnal cycle, but also one synchronized with solar flares and the earth's magnetic field activity. For any given environmental state, one must have detailed models that allow an accurate wave speed profile. This requires expertise in surface-layer meteorology for atmospheric sound propagation, oceanography for underwater sound propagation, and oceanography, meteorology, and astrophysics and radio wave propagation modeling. Refraction is defined as an effect which changes the direction of a ray in a spatially coherent manner, as shown in Figures 13.25 and 13.26 and in other devices such as a lens for an optical wave. Scattering is defined as an effect where the ray tends to be redirected in multiple directions in a spatially incoherent manner. In the nonvirtual world, the physics which cause diffraction and scattering are often the same and it is very difficult to prove the precise nature of wave propagation in inhomogeneous media. However, from a wave processing perspective, we can say that modeling these effects with a wavenumber filter which "leaks" incoherent wave energy is very reasonable to include the effects of random diffraction and scattering for inhomogeneous wave propagation modeling.

Numerical considerations of the Green's function parabolic equation are extremely important because the wave "step marching" approach carries a significant danger of numerical error accumulation. One of our initial assumptions was that the wave propagation is largely in the horizontal r-direction. With the point source near the left boundary in Figures 13.25 through 13.27, this means that the results in the upper left corner of the figures are likely in error due to the steep angle of the waves relative to the horizontal step direction. The assumption that the variations in wave speed are small compared to the mean wave speed is generally true, but one should certainly check to see if the square-root approximation in Equation 13.77 is reasonable.

Another numerical consideration is the wave amplitude along the horizontal boundary, if present in the problem. For undersea acoustic propagation modeling, the boundary condition at the sea surface is zero pressure (a pressure release boundary condition). However, for atmospheric sound propagation, the acoustic pressure along the boundary is not zero and it is advisable to use a trapezoidal or higher-order integration to prevent propagation of significant numerical error. At the upper elevation of the virtual wave propagation space, an artificial attenuation layer is needed to suppress wave "reflections" from the top of the space. Again, in underwater propagation one usually has a natural attenuation layer at the bottom sea floor which is usually a gradual mixture of water and sediment which does not reflect sound waves well. There are likely analogous electromagnetic boundary conditions which also require special handling of numerical error. These extremely important aspects of wave propagation are best handled by controlled experiments for validation.

13.5 MATLAB® EXAMPLES

The seven m-scripts used to make the figures in this chapter cover a diverse range of useful algorithms for both demonstrating concepts as well as actual use for solving real signal processing challenges. Starting with "LINERESP.m," we have a classic beam pattern response of a line source or sensor. By line source (sensor) we mean a continuous transducer shaped as a line, not an array of transducers that form a line. The beam response given in Equation 13.35 is programmed literally into the script to produce Figures 13.5 and 13.6. With no precedent other than curiosity, the 2D Fourier transform of the line source/sensor for the spatial data in Figure 13.8 produces the very interesting plot of Figure 13.9. This required the use of the MATLAB® function "fftshift()" to arrange the 2D FFT output so that the origin is in the center of the image, although the image is actually displayed upside down because of the way the function "image()" displays data in a matrix. There are probably better ways to display the wavenumber data. It is important to keep in mind that we have eight samples of the field per meter, and so the wavenumber range in Figure 13.9 is $\pm 8\pi$ over 256 FFT bins, or 0.19635 rad/bin. This lets us scale the wavenumber circle centered in the spectrum for a 300 Hz signal. We then read off the nearest pixels around this circle to get the same beam pattern as shown in Figure 13.6, albeit a little more jaggy due to the fewer pixels than points used in the analytic solution in Figure 13.6. Comparing Figures 13.7 and 13.8 shows the wavenumber shift from the 30° steering. This is a different way of thinking about beam patterns and steering in wavenumber space, but is completely valid. In the m-script "LIN8RESP.m," we simply replace the line source/sensor with four discrete point sensors in a line. The software allows any number of elements and four showed the most interesting effect of grating lobes at 300 and 600 Hz in Figures 13.11 through 13.14. Here one can see the jaggy polar plots read off the wavenumber image. The same idea follows in two dimensions using "GRIDRESP.m" seen in Figures 13.15 and 13.16. The m-scripts also print out the directivity index on the command line using the approximation given in Equation 13.49 (see Table 13.2).

In the m-script "HOLOG.m," we generate three parallel planes of x–y coordinates for our field grid where planes are at $z = 0.0001$ m, $z = 1.0$ m, and $z = 50$ m relative to the source plane at $z = 0.0$.

TABLE 13.2
MATLAB m-Scripts Used in This Chapter

m-Script	Figure(s)
LINERESP.m	13.5–13.10
LIN8RESP.m	13.11–13.14
GRIDRESP.m	13.15–13.16
HOLOG.m	13.17
RECONST.m	13.18–13.22
RECONSTH.m	13.23–13.24
GFPEDEMO.m	13.25–13.27

In the source plane, we place a simple point source at the origin and compute the complex field at each of the grid points for each of the three planes, which the real part is displayed on the left side of Figure 13.17. The corresponding wavenumber spectra for each field plane are shown on the right side of Figure 13.17. These wavenumber plots are very illustrative of how a slice through a spherical wave field changes with distance from the source, and how these changes appear in the wavenumber domain. The "ring" in the wavenumber plots corresponds to the dominant wavelength seen in the corresponding slice and collapses as one moves far away from the source and the waves in the slide look more like plane waves.

Using the m-script "RECONST.m" we attempt to reconstruct the sound field near the source plane from a measurement plane further away using a wavenumber transfer function filter. This transfer function is a point-by-point spectral division and is really no different than dividing an output spectrum by an input spectrum to get the transfer function of an unknown system. We examine the resolution using Figures 13.18 through 13.22 to show how the processing is like the depth of field in a lens system. The m-script "RECONSTH.m" uses the more popular acoustic holography method based on the Green's function inversion in Equation 13.67. This method also works well by some slightly different artifacts from the evanescent field (wavenumbers larger or wavelengths shorter than the free wave for a given frequency). These holography scripts can be very useful for many 3D field reconstruction challenges.

Finally, the m-script "GFPEDEMO.m" is a very valuable tool for predicting sound propagation outdoors with temperature and wind gradients and even turbulence. Implementation of the core algorithm is given in Table 13.3 and contains several very subtle but important implementation features. First, note the comment "trapezoidal integration" and the multiplication of the first point (the one on the ground) by 1/2. This prevents a numerical error from accumulating. The term "B" is for the ground wave and the terms "e_vari()" and "e_step()" are wavenumber filters that contain an attenuation window near the top of the numerical space (to prevent reflections from up above) and the split-step wavenumber approximation used in Equation 13.78 to simplify the propagation as

TABLE 13.3
The Greens Function Parabolic Equation GFPE Algorithm

```
for nr = 1:Nr,
    psi(1) = .5*psi(1);                          % trapezoidal integration
    psif = fft(psi);                             % FFT
    T = 0;
    for i = 1:npts/4,                            % k-space integration
        T = psif(i)/kb(i) + psif(npts-i+1)/kb(npts-i+1) +T;
    end;
    B = (-dk*2*beta*T*e_step_b/(2*pi)).*e_jbz;   % ifft
    u = zeros(size(psif));
    for n = 2:npts/2,                            % Xio Shuffle
        t1 = psif(n);
        t2 = psif(npts+2-n);
        u(n) = (t1 + t2*Rkp(n)).*e_step(n);
        u(npts+2-n) = (t2 + t1*Rkpm(n)).*e_step(n);
    end;
    u(1) = psif(1)*(1+Rkp(1)).*e_step(1);
    u(npts/2+1) = psif(npts/2+1)*(1+Rkp(npts/2+1)).*e_step(npts/2+1);
    A = ifft(u);                                 % inverse FFT
    cturb = exp(j.*pi.*cturblev.*randn(size(e_vari))); psi_temp = cturb.*
        e_vari.*(A+B);                           % direct & reflected summed
P_grid(1:nattn,nr) = psi_temp(1:nattn).'./sqrt(r(nr));   % save
    psi = psi_temp;                             % get fft's ready for next
                                                 loop
end;
```

mainly horizontal. That simplification is sensible, given that we are most interested in the model predictions away from the source and near the ground. The section with the comment "Xio Shuffle" is attributed to Dr. Xio Di who coded the original GFPE in Fortran and used these steps to access the negative wavenumber spectrum from the positive wavenumber spectrum to save computation. The wavenumber components are all summed up and then an inverse FFT is done to get the pressure fields at the next step. Turbulence is just a little randomization added to the phase.

Finally, note in Table 13.3 that the saved field is divided by $\sqrt{r}$. This produces a field with the spherical spreading component returned and corrected for a 2D circular spreading model, which is what is actually being updated. In addition, the way one sets the step size in range r and height z can greatly affect the accuracy for a given wavelength and ground impedance [11].

13.6 SUMMARY

This chapter brings together the theoretical and practical aspects of processing wavenumber signals as measured coherently by spatial arrays of sensors. Section 13.1 presented the derivation of the CRLB for error on statistical parameter estimates. We then used these results to analyze the precision of wave bearing angle estimates both as a direct phase measurement and as a beamformed result in Section 13.2. A new physics-based approach to describing array beamforming is seen using 2D wavenumber transforms of the array shape and then applying a wavenumber contour directly to the wavenumber spectrum to determine the array response. The interesting aspect of this presentation is that one can directly observe the beam response over a wide range of frequencies and steering angles in a single graph. Section 13.3 carries beamforming further to the problem of field reconstruction using wave field holography, a scanning array of sensors, and a wavenumber domain Green's function. By proper inversion of this Green's function, one can reconstruct the field in a plane in the region between the source(s) and the sensor array. This is useful for identifying wave sources and direct observations of the wave-radiating mechanisms. Section 13.4 applies the wavenumber domain Green's function to model the propagation of waves in the direction away from the sources and the sensor array surface. Wave propagation modeling has obvious value for many currently practical applications.

Fundamentally, measurements of spatial waves are all based on one's ability to observe spatially the time-of-arrival or phase at a given frequency. These types of measurements all have some SNR limitation which can be objectively determined. As one then computes a physical quantity such as bearing, or beam output at a given LD, the confidence of the time-of-arrival or phase is scaled according to the SNR enhancement of the array, array shape, and LD. When the array beam is focused in a near range, rather than infinity (for plane waves from a distance source) one can apply the same principles to determine source localization error, or resolution of the holographic process. For the analytical Green's function, this resolution is mainly a function of array aperture and distance between the measurement plane and the reconstruction plane. However, simply increasing the number of scanning elements does not improve long distance resolution but rather improves resolution in the region close to the measurement array. Increasing sensor array aperture generally improves source localization accuracy at long distances for the same reason that it also narrows beam width for a plane wave beamforming array. A consistent framework for wavenumber-based measurements and processing in sensor systems is presented in this chapter. Since these measurements are ultimately used in an "intelligent" decision process, the associated statistical confidences are of as much importance as the measurement data itself to ensure proper weighting of the various pieces of information used. What makes a sensor system output "intelligent" can be defined as the capability for high degrees of flexibility and adaptively in the automated decision algorithms.

But why the CRLB, 2D wavenumber filters for beamforming, acoustic holography, and the Greens function parabolic equation approximation for sound propagation in a single chapter? These topics may not even have appeared in the same book, let alone the same chapter. They are exquisitely linked through signal processing, not application. All signal estimates have some measurable level of random noise and the CRLB provides a way to quantify the lower bound for estimation precision given the number of observations used and the estimated statistics for the signal. The CRLB can then be scaled through the physical models to also provide error estimates in the AOA for a beamforming wavenumber filter. We can also describe the wavenumber response of a simple array of sensors by taking the 2D Fourier transform of the sensor positions, magnitude, and phase and examine the array response at a frequency of interest by evaluating the 2D Fourier transform along a circle defined by the wavenumber for the frequency. One could even extend this technique to 3D arrays for analysis of array shape and potential beamforming response. When measuring a 2D wavenumber field, a wavenumber transfer function can be used to translate the measurements to another surface in much the same way a camera lens system can focus a narrow depth of field onto an image plane. The Green's function for these field translations can be used as well to model wave propagation paths, such as one-way outdoor sound propagation based on a parabolic wave equation. Each step of the field equations can be seen as a wavenumber filter which can introduce refraction, turbulence, and surfaces wave along the ground impedance boundary. These applications are shown for acoustics, but the concepts can also be applied to structural vibrations, electromagnetic, and seismic waves.

PROBLEMS

1. A noisy dc signal is measured to obtain the mean dc value and the rms value of the noise. The variance is assumed to be 100 mv². How many independent samples of the voltage would be needed to be sure that the estimated dc voltage mean has a variance less than 1 mv²?

2. A 5 m line array of 24 sensors is used to estimate the arrival angle of a 500 Hz sinusoid in air which has an SNR of 20 dB. Assuming the speed of sound is 345 m/s and all sensors are used to estimate the bearing (assumed to be near broadside), what is the rms bearing error in degrees for one data snapshot?

3. A circular array 2 m in diameter has 16 equally spaced sensors in water ($c = 1500$ m/s).
   a. What is the highest frequency one can estimate bearings for?
   b. Define an orientation and calculate the relative phases at 1 kHz for a bearing angle of 90°

4. Using the array in problem 3(b), how many sources can be resolved at the same frequency?

5. What is the resolution beam width (assume −3 dB responses of the beam determine beam width) in degrees at 2 kHz, 1 kHz, and 200 Hz for the arrays in problem 2 and 3?

6. An air conditioner is scanned by a line array of seven microphones where one microphone is held at a constant position as a reference field sensor to obtain an 8×8 grid of magnitude and phases at 120 Hz covering a 2 m × 2 m area. What is the spatial resolution of the holography system?

7. A survivor in a lifeboat has an air horn which is 100 dB at 1 m and 2 kHz. There are foggy (downward refracting conditions) such that there is spherical wave spreading for the first 1000 m, and then circular spreading of the wave in the duct just above the water. If the background noise ashore is 40 dB, how far out to sea can the horn be heard by a rescuer?

REFERENCES

1. H. Cramer, *Mathematical Methods of Statistics*, Section 32.3, Princeton, NJ: Princeton University Press, 1951.
2. C. R. Rao, *Linear Statistical Inference and Its Applications*, 2nd ed., New York, NY: Wiley, 1973.
3. V. H. MacDonald and P. M. Schultheiss, Optimum passive bearing estimation in a spatially incoherent noise environment, *J Acoust Soc Am*, 1969, 46(1), pp. 37–43.
4. R. J. Ruhala and D. C. Swanson, Planar nearfield holography in a moving medium, *J Acoust Soc Am*, 2002, 112(2), pp. 420–429.
5. W. S. Burdic, *Underwater Acoustic System Analysis*, Englewood Cliffs, NJ: Prentice-Hall, 1991.

6. E. G. Williams, J. D. Maynard, and E. Skudrzyk, Sound source reconstructions using a microphone array, *J Acoust Soc Am*, 1980, 68(1), pp. 340–344.

7. E. Skudrzyk, *The Foundations of Acoustics*, New York, NY: Springer-Verlag, 1971.

8. K. Attenborough, K. Li, and K. Horoshenkov, *Prediction Outdoor Sound*, New York, NY: Taylor & Francis, 2007.

9. K. E. Gilbert and X. Di, A fast Green's function method for one-way sound propagation in the atmosphere, *J Acoust Soc Am*, 1993, 94(4), pp. 2343–2352.

10. D. K. Wilson, Performance bounds for direction-of-arrival arrays operating in the turbulent atmosphere, *J Acoust Soc Am*, 1998, 103(3), pp. 1306–1319.

11. J. Cooper and D. C. Swanson, Parameter selection in the Green's function parabolic equation, *Appl Acoust*, 2007, 68(4), pp. 390–402.

14 Adaptive Beamforming and Localization

Adaptive beamforming [1] is used to optimize the signal-to-noise ratio (SNR) in the look-direction (LD) by carefully steering nulls in the beam pattern toward the direction(s) of interference sources. From a physical point of view, one should be clear in understanding that at any given frequency the beamwidth and SNR for spatially incoherent noise (random noise waves from all directions) is dictated by physics, not signal-processing algorithm. The larger the array, the greater the SNR enhancement will be in the LD, provided all array elements are spaced less than half a wavelength. The more array elements one has, the wider the temporal frequency bandwidth will be where wavenumbers can be uniquely specified without spatial aliasing. However, when multiple sources are radiating the same temporal frequency from multiple directions, it is highly desirable to resolve the sources and directions as well as provide a means to recover the individual source wavenumber amplitudes. Examples where multiple wavenumbers need to be resolved can be seen in sonar when a large boundary is causing interfering reflections, in radio communications where an antenna array might be adapted to control multipath signal cancellation, and in structural acoustics where modes of vibration could be isolated by appropriate processing of accelerometer data.

Beamforming can be done on broadband as well as narrowband signals. The spatial cross-correlation between array sensors is the source of directional (wavenumber) information, so it makes no physical difference whether the time-domain waveform is broadband (having many frequencies) or narrowband (such as a dominant sinusoid). However, from a signal-processing point of view, narrowband signals are problematic because the spatial cross-correlation functions are also sinusoidal. For spectrally white broadband signals, the spatial cross-correlation functions will yield a Dirac delta function representing the time difference of arrivals across the sensor array. A narrowband spatial cross-correlation is a phase-shifted wavenumber, where multiple arrival angles correspond to a sum of wavenumbers, or modes. This is why an eigenvalue solution is so straightforward. However, the sources radiating the narrowband frequency which arrive at the sensor array from multiple directions must be independent in phase for the spatial cross-correlation to be unique. From a physical point of view, source independence is plausible for, say, acoustics where the two sources are vehicles that coincidentally happen to have the same temporal frequency for a period of time. This might also be physically true for multipath propagation from a single source where the path lengths are randomly fluctuating. But, for vibration source multipath in a mechanical structure, or radio transmission multipath from reflections off buildings, the multipath is *coherent*, meaning that the phases of the wave arrivals from different directions are not statistically independent. Thus, the coherent multipath situation gives sensor spatial cross-correlations which depend on the phases of the wave arrivals and, thus, does not allow adaptive beamforming in the traditional sense.

We present adaptive beamforming using two distinct techniques that parallel the two adaptive filtering techniques (block least squares and projection-based least-squared error). The block technique is used in Section 14.1 to create a spatial whitening filter where an finite impulse response (FIR) filter is used to predict the signal output of one of the array sensors from the weighted sum of the other array sensors. The wavenumber response of the resulting spatial FIR whitening filter for the array will have nulls corresponding to the angle-of-arrival (AOA) of any spatially coherent source. By relating the wavenumber of a null to a AOA through the geometry of the array and the wave speed, a useful multipath measurement tool is made. If the sources and background noise are all

broadband, such as Gaussian noise, the null-forming and AOA problem is actually significantly easier to calculate numerically. This is because the covariance and cross-correlation matrices are better
conditioned. When the sources are narrowband, calculation of the covariance and cross-correlation
matrices can be problematic due to the unknown phases and amplitudes of the individual sources.
This difficulty is overcome by compiling a random set of data "snapshots" in the estimation of the
covariance and cross-correlation matrices. This reduces the likelihood that the phases of the individual sources will present a bias to the covariance and cross-correlation matrices, which should be
measuring the amplitude phase due to the spatial interference, not the relative phases and amplitudes
of the sources.

To be able to construct a beam for a desired LD while also nulling any sources in other directions,
we need to process the eigenvectors of the covariance matrix as seen in Section 14.2. This parallels
the projection operator approach in Section 9.2 for calculating the least-squared error linear predictor.
The covariance matrix has a Toeplitz structure, meaning that all the diagonals have the same number
in each element. This numerical structure leads to a unique physical interpretation of the array signal
covariance eigenvectors and eigenvalues. Assuming that the signals from the sources in question
have a reasonably high SNR (say 10 or better) the solved eigenvalues can be separated into a "signal
subspace" and a "noise subspace," where the largest eigenvalues represent the signals. Each eigenvector represents an array FIR filter with a wavenumber response with nulls for all source locations
except the one corresponding to itself. The "noise eigenvectors" will all have nulls to the signal
source wavenumber s and have additional spurious nulls in their wavenumber responses if there are
fewer than $M - 1$ sources for a M-element array. To obtain the whitening filter result one can simply
add all the noise eigenvectors, each of which has the nulls to the sources. To obtain a beam in the
desired LD with nulls to all the other "interfering sources" one simply postmultiplies the inverse of
the covariance matrix by the desired steering vector! This very elegant result is known as a minimum
variance beamformer because it minimizes the background noise for the desired LD.

We present another technique in Section 14.3 which is highly useful when one can control the
transmitted signal such as in active radar/sonar and in communications. By encoding the phase of the
signal with a random periodic signal, or maximal length sequence (MLS), one can uniquely define
all the wavenumbers at the receiving array whether the effect is multisource multipath or propagation
multipath. This is one reason why spread-spectrum signals are so useful in communication systems.
By modulating the narrowband signal phase according to a prescribed unique repeating sequence, we
are effectively spreading the signal energy over a wider bandwidth. A Fourier transform actually
shows a reduced SNR, but this is recoverable at the receiver if one knows the phase modulation
sequence. By cross-correlating the known signal's unique spread-spectrum sequence with the
received signal in a matched detection filter, the SNR is recovered and the resulting covariance is well
conditioned to extract the multipath information. This technique not only allows us to recover low
SNR signals but also allows us to make multipath propagation measurements with an element of
precision not available with passive narrowband techniques. This is inherently useful for separating
multiple source of reflection from a transmitted MLS-encoded signal, because each channel of the
array can have the multipath defined independently. In addition, one can exploit the reproducibility
of the MLS encoding algorithm for bistatic systems where the transmitter and the receiver are separated, but can still sync to the same sequence since it can be precisely generated in both places even
synchronized to a common clock such as that broadcast by GPS satellites.

In Section 14.4, we present another technique for active localization which uses a frequency-
modulated continuous wave (FMCW) signal that tends to maximize the SNR in a monostatic (transmitter and receiver colocated) sensing system. The frequency modulation is generally a linear
frequency shift over time, or "chirp." The received signal is also a chirp, but delayed depending on
the distance to the reflector. By multiplying the received and transmitted signals and low-pass filtering (this is called mixing) one gets frequencies proportional to the distance of the reflectors. By
doing a Fourier transform on the mixing output, one gets peaks in frequency bins where the bin
number is proportional to range. Since the operation is synchronous, one can greatly enhance the

SNR through averaging if desired. The FMCW technique provides a way to have excellent range resolution and very high SNR. If one uses a reasonable line array to focus the transmitted and received beams, the beams can be coherently scanned over a large surface, say from an airplane or ship traveling in a line over a surface such as the ground or seafloor, respectively. Since the transmitted and received signals are coherent, the responses of adjacent cycles can be combined as if the array were longer. This has the effect of making the beam narrower along the axis of platform motion. Extending the beam summing to 1000s of array lengths creates a "synthetic aperture" for a very narrow beam at the midpoint of the summing window. This enables the cross-range resolution on the surface (depends on beamwidth and range) to closely match the down-range resolution which is already high from the FMCW process. This is the technique used in synthetic aperture radar mapping and "side-scan" sonar mapping.

14.1 ARRAY "NULL-FORMING"

Fundamental to the traditional approach to adaptive beamforming is the idea of exploiting some of the received signal information to improve the SNR of the array output in the desired LD. Generally, this is done by simply steering an array "null," or direction of zero-output response, to a useful direction other than the LD where another "interfering" source can be suppressed from the array output to improve SNR (actually signal-to-interference ratio). The array SNR gain and LD beamwidth are still defined based on the wavelength-to-aperture ratio, but an additional interference rejection can be obtained by adaptively "nulling" the array output in the direction of an interfering source [2]. When sweeping the LD beam around to detect possible target signals, the application of adaptive null-forming allows one to separate the target signals quite effectively. However, the spatial cross-correlation of the array signals must have source independence and no source dependence on background noise for the process to work.

A straightforward example of the issue of source phase independence is given to make the physical details as clear as possible. This generally applies to narrowband temporal signals (sinusoids) although it is possible that one could have coherent broadband sources and apply the narrowband analysis frequency by frequency [3]. Generally speaking, incoherent broadband sources such as multiple fluid-jet acoustic sources or electric arc RF electromagnetic sources will have both temporal and spatial correlation function approximating a Dirac delta function. For the narrowband case, consider two distant sources widely separated which radiate the same frequency to a sensor at the origin from two directions θ_1 and θ_2 (measured counterclockwise from the horizontal positive x-axis) with received amplitudes A_1 and A_2 and phases φ_1 and φ_2:

$$X_0 = A_1 e^{j\varphi_1} + A_2 e^{j\varphi_2} \tag{14.1}$$

If we have a line array of sensors along the positive x-axis each at location $x = d_\ell$, we can write the received signals from the two sources relative to the X_0 in terms of a simple phase-shift:

$$X_{d_\ell} = A_1 e^{j(\varphi_1 + kd_\ell \cos\theta_1)} + A_2 e^{j(\varphi_2 + kd_\ell \cos\theta_2)} \tag{14.2}$$

Clearly, an arrival angle of 90° (relative to the positive x-axis) has the wavefront arriving simultaneously at all sensors from the broadside direction. The spatial cross-correlation between X_0 and an arbitrary sensor out of the line array X_{dl} is defined as

$$
\begin{aligned}
R_{-d_\ell} &= E\{X_0 X_{d_\ell}^*\} = E\{(A_1 e^{j\varphi_1} + A_2 e^{j\varphi_2})(A_1 e^{-j(\varphi_1 + kd_\ell \cos\theta_1)} + A_2 e^{-j(\varphi_2 + kd_\ell \cos\theta_2)}\} \\
&= E\{[A_1^2 + A_2 A_1 e^{-j(\varphi_1 - \varphi_2)}] e^{-jkd_\ell \cos\theta_1} + [A_2^2 + A_1 A_2 e^{+j(\varphi_1 - \varphi_2)}] e^{-jkd_\ell \cos\theta_2}\}
\end{aligned} \tag{14.3}
$$

Note that the spatial cross-correlation is a function of the amplitudes and phases of the two sources! This means that the spatial information for the two arrival angles and amplitudes is not recoverable unless one already knows the amplitude and phases of the two narrowband sources. However, if the sources are statistically independent, the phase difference between them estimated over time will be a random angle between $\pm\pi$. Therefore, given a sampling of "snapshots" of spatial data from the array to estimate the expected value in Equation 14.3 one obtains

$$R_{-d_t} = A_1^2 e^{-jkd_t\cos\theta_1} + A_2^2 e^{-jkd_t\cos\theta_2} \tag{14.4}$$

since the expected value of a uniformly distributed complex number of unity magnitude is zero. The wavenumber response of the spatial cross-correlation given in Equation 14.4 has the Dirac delta function structure (delta functions at $d_1 \cos\theta_1$ and $d_1 \cos\theta_2$) in the time–space domain as expected of two incoherent broadband sources. One can see that the linear combination of two or more sources for each spatial cross-correlation can be resolved by either linear prediction or an eigenvalue approach without much difficulty so long as there is at least one more sensor than sources. For an array with M sensors, there are $M - 1$ spatial correlations available, therefore $M - 1$ degrees of freedom corresponding to modes from spatially separated sources with independent phases [4].

In practice, one has to put in some effort to achieve this phase independence between sources. The term "spatially incoherent radiating the same frequency" is actually a contradiction. Since the time derivative of phase is frequency, one would do better to describe the desired situation as having "sources of nearly identical frequency processed in a wider frequency band with multiple observations to achieve spatial phase independence in the average." For example, if one had two sources within about 1 Hz of each other to be resolved spatially using signals from a large sensor array with a total record length of a second, one would segregate the record into, say, 10 records of 100 ms length each, and then average the 10 available snapshots to obtain a low-bias spatial cross-correlation for a 10 Hz band containing both sources. This is better than trying to resolve the sources in a 1 Hz band from a single snapshot because the phases of the sources will present a significant bias to the spatial cross-correlation for the array. Another alternative (if the array is large enough) would be to divide the array snapshot into a number of "subarrays" and calculate an ensemble average spatially of each subarray's cross-correlation to get an overall cross-correlation with any dependence of the source phases suppressed. This is known as spatial smoothing to achieve phase independence of the sources. It is very valuable for the system design to understand the physical reason the correlation bias arises for coherent sources in the first place and the techniques required to suppress this coherence. For the rest of this section and Section 14.2, we will assume spatially incoherent sources. A better way to depict this assumption is to assume that the sensor system has provided the additional algorithms necessary to ensure incoherent sources in the covariance matrix data.

Consider the spatial "whitening" filter in Figure 14.1 which uses a weighted linear combination of the signals from sensors 1 through M to "predict" the signal from sensor 0. The filter coefficients (or weights in the linear combiner) are determined by minimizing the prediction error. Over N time snapshots, the prediction error is

$$\begin{bmatrix} \varepsilon_t \\ \varepsilon_{t-1} \\ \varepsilon_{t-2} \\ \vdots \\ \varepsilon_{t-N+1} \end{bmatrix} = \begin{bmatrix} x_{0,t} \\ x_{0,t-1} \\ x_{0,t-2} \\ \vdots \\ x_{0,t-N+1} \end{bmatrix} - \begin{bmatrix} x_{1,t} & x_{2,t} & x_{3,t} & \cdots & x_{M,t} \\ x_{1,t-1} & x_{2,t-1} & x_{3,t-1} & \cdots & x_{M,t-1} \\ x_{1,t-2} & x_{2,t-2} & x_{3,t-2} & \cdots & x_{M,t-2} \\ \vdots & \vdots & \vdots & \ddots & \vdots \\ x_{1,t-N+1} & x_{2,t-N+1} & x_{3,t-N+1} & \cdots & x_{M,t-N+1} \end{bmatrix} \begin{bmatrix} h_1 \\ h_2 \\ h_3 \\ \vdots \\ h_M \end{bmatrix} \tag{14.5}$$

where our array signals could be either broadband real-time data or complex narrowband FFT bin values for a particular frequency of interest. For the narrowband frequency-domain case, the coefficients are all complex. The relationship between these coefficients and the wavenumbers and

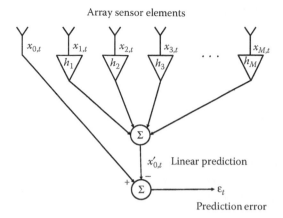

FIGURE 14.1 A spatial linear prediction filter can be used to determine the wavenumber corresponding to the AOA of spatially incoherent source signals.

physical arrivals angles of interest will be discussed momentarily. Equation 14.5 can be written compactly in matrix form as

$$\overline{\varepsilon} = \overline{x_0} - \overline{X}\,H \tag{14.6}$$

Recalling from Section 8.1, the block least-squared error solution for the coefficients is

$$H = \left(\overline{X}^{\mathrm{H}}\overline{X}\right)^{-1}(\overline{X})^{\mathrm{H}}\overline{x_0}^{\mathrm{H}} \tag{14.7}$$

Applying the well-known least-squared error result in Equation 14.7 to the spatial whitening problem can be seen by writing the array output as the spatial prediction error in one row of Equation 14.6

$$\varepsilon_t = x_{0,t} - \sum_{m=1}^{M} h_m x_{m,t} \tag{14.8}$$

Note that for the narrowband case, the t subscripts in Equation 14.8 can refer to the complex value of an FFT bin at time t without any loss of generality. The array output can be written as the output of a spatial FIR filter:

$$\varepsilon_t = \sum_{m=0}^{M} A_m x_{m,t}, \quad A_0 = 1, \ A_m = -h_m \tag{14.9}$$

The spatial whitening FIR filter $A(z)$ can now be evaluated as a function of a complex variable $z = e^{-jkd_{\ell m}\cos\theta}$ where $d_{\ell m}$ is the x-coordinate of the mth sensor in the line array, k is the wavenumber, and θ is the direction of interest. This has an analogy in the frequency response being evaluated on the unit circle on the z-plane defined by $z = e^{j\omega T}$. However, it was convenient to write $A(z)$ as a polynomial with integer powers of z when the time-domain signal samples are evenly spaced.

But, for our implementation of a spatial FIR filter, the sensors are not necessarily evenly spaced. The wavenumber response of our spatial whitening FIR filter is found by a straightforward Fourier sum using the appropriate wavenumber for the array geometry. For the line array, the wavenumber simply scales with the cosine of the arrival angle, which is limited to the unique range of 0–180° by symmetry:

$$D(\theta) = \sum_{m=0}^{M} A_m e^{-jk d_{\ell_m} \cos\theta} \tag{14.10}$$

If the array is an evenly spaced line array along the x-axis, the most efficient way to get the wave-number response is to compute a zero-padded FFT where the array coefficients are treated as an FIR (see Section 3.1). The digital-domain wavenumber response from $-\pi$ to $+\pi$ is first divided by kd (wavenumber times element spacing) and then an inverse cosine is computed to relate the k to angle θ. If the array sensors are part of a 2D array and each located at $x = d_{x_m}$ and $y = d_{y_m}$, the directional response of the spatial whitening filter is

$$D(\theta) = \sum_{m=0}^{M} A_m e^{-jk d_{x_m} \cos\theta} e^{-jk d_{y_m} \sin\theta} \tag{14.11}$$

where k represents the wavenumber in the x–y plane. If the plane wave can arrive from any angle in three dimensions and we have a 2D planar sensor array, the wavenumber in the x–y plane scales by $k' = k \sin \gamma$, where γ is the angle from the positive z-axis to the ray normal to the plane wavefront and k is the free propagation wavenumber for the wave ($k = \omega/c = 2\pi/\lambda$). If the wave source is on the x–y plane ($\gamma = 90°$), $k' = k$. For a 3D array and plane wave with arrival angles θ (in the x–y plane) and γ (angle from the positive z-axis), the directional response of the spatial whitening filter is given in Equation 14.12 where d_{z_m} is the z-coordinate of the mth sensor in the 3D array and γ is the arrival angle component from the positive z-axis.

$$D(\theta,\gamma) = \sum_{m=0}^{M} A_m e^{-jk d_{x_m} \cos\theta \sin\gamma} e^{-jk d_{y_m} \sin\theta \sin\gamma} e^{-jk d_{z_m} \cos\gamma} \tag{14.12}$$

The coefficients of $A(z)$ are found by the least-squares whitening filter independent of the array geometry or wavenumber analysis. The spatial whitening filter is simply the optimum set of coefficients which predict the signal from one sensor given observations from the rest of the array. For the linear, evenly spaced array, one can compute the roots of the polynomial where the zeros closest to the unit circle produce the sharpest nulls, indicating that a distinct wavenumber (wave from a particular direction) has passed the array allowing cancellation in the whitening filter output. The angle of the polynomial zero on the z-plane is related to the physical arrival angle for the line array by the exponent in Equation 14.10. In two and three dimensions where the array sensors are not evenly spaced, root finding is not so straightforward and a simple scanning approach is recommended to find the corresponding angles of the nulls. For example, the array response is spatially whitened at a particular free wavenumber k corresponding to radian frequency ω. The FFT bins corresponding to ω are collected for N snapshots and the spatial whitening filter coefficients are computed using Equations 14.5 through 14.9. To determine the source arrival angles, one evaluates Equation 14.12 for $0° < \theta < 360°$ and $0° < \gamma < 180°$, keeping track of the dominant "nulls" in the array output. A straightforward way to find the nulls is to evaluate the inverse of the FIR filter for strong peaks. This has led to the very misleading term "super resolution array processing" or "spectrum estimation" since the nulls are very sharp and the inverse produces a graph with a sharp peak. But this is only graphics! The array beamwidth is still determined by wavelength and aperture and angle measurement is still a function of SNR and the CRLB.

To illustrate an example of adaptive "null-forming," Figure 14.2 shows the result of three 100 Hz sources at 20°, 80°, and 140° bearing with a 10 m eight-element line array along the x-axis. The

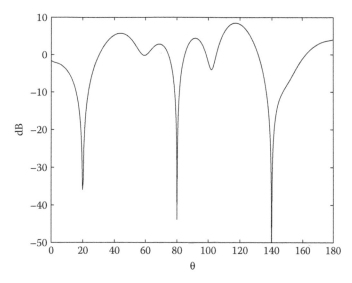

FIGURE 14.2 Spatial whitening filter directional response for three 10 dB SNR 100 Hz sources at bearing angles of 20°, 80°, and 140° on an 8-element 10 m line array with 10 snapshots.

SNR in the 100 Hz FFT bin is 10 dB and the whitening coefficients are estimated using 10 snapshots. The bearings of the three 100 Hz sources are clearly seen and the relative phases for the three sources are assumed to be uniformly distributed between 0 and 2π for each of the 10 snapshots. With the source phases distributed over a narrower range for each snapshot, the results are not nearly as consistent. One might also think that the higher the SNR, the better the result. But this too is not entirely true. Since for an $(M + 1)$-element line array one can resolve a maximum of M sources, some noise is required so that the $M \times M$ matrix inverse in Equation 14.7 is well conditioned. But in low SNR (0 dB or less), the magnitude and phase of the signal peak are not strong enough relative to the random background noise to cancel. Using more snapshots improves the whitening filter performance, since the background noise is incoherent among snapshots. Finally, the depth of the null (or height of the peak if an inverse FIR response is evaluated) is not reliable because once the source wavenumber is cancelled, the residual signal at that wavenumber is random background noise causing the null depth to fluctuate.

The spatial resolution for the nulls is amazingly accurate. Figure 14.3 shows the detection of two sources only 2° apart. However, when one considers the CRLB (Section 13.1) for a 10 m array with eight elements, 10 dB SNR, and 10 snapshots, the bearing accuracy should be quite good. In practical implementations, the bearing accuracy is limited by more complicated physics, such as wave scattering from the array structure, coherent phase between sources, and even coherent background noise from wave turbulence or environmental noise.

14.2 EIGENVECTOR METHODS OF MUSIC AND MVDR

One can consider the wavenumber traces from multiple sources arriving at an array of sensors from different angles at the same temporal frequency as a sum of modes, or eigenvectors. In this section, we explore the general eigenvector approach to resolving multiple (independent phase) sources [5]. Two very popular algorithms for adaptive beamforming are MUSIC (MUltiple SIgnal Classification) and MVDR (Minimum Variance Distortionless Response). A simple example of the physical problem addressed by adaptive beamforming is a line array of evenly distributed sensors where the arriving plane waves are measured by the magnitude and phase of each array sensor output for the temporal FFT bin for the frequency of interest. If a plane wave from one distant source in a direction normal to the array (broadside) line is detected, the magnitude and phase observed from the array sensors

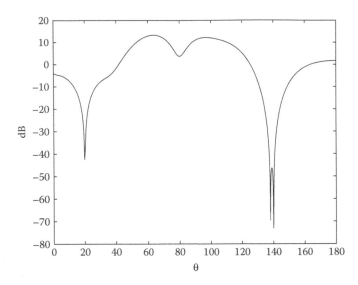

FIGURE 14.3 Spatial whitening filter directional response for three 10 dB SNR 100 Hz sources at bearing angles of 20°, 138°, and 140° on an 8-element 10 m line array with 10 snapshots.

should be identical. From a practical but very important standpoint, this assumes that the sensors are calibrated in magnitude and phase and that no additional phase difference between sensor channels exists from a shared ADC. The spatially constant magnitude and phase can be seen as a wavenumber at or near zero (wavelength infinite). But, if the array is rotated so that the propagation direction is along the line-array axis (endfire), the wavenumber trace observed by the array matches the free wavenumber for the wave as it propagates from the source. Therefore, we can say in general for a line array of sensors that the observed wavenumber trace $k' = k \cos \theta$, where k is the free wavenumber and θ is the AOA for the wave relative to the line-array axis. Figure 14.4 shows the spatial phase response for a 150 Hz plane wave arriving from 90° (broadside), 60°, and 0° (endfire).

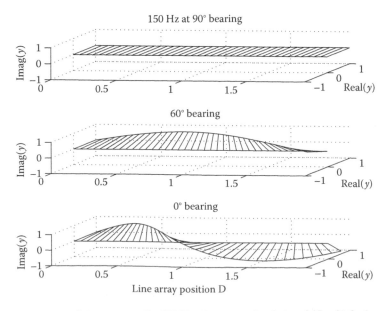

FIGURE 14.4 Complex spatial response of a 150 Hz plane wave in air ($c = 345$ m/s) for broadside (top), 60° (middle) and endfire (bottom) directions.

Even though multiple sources may be radiating the same temporal frequency, the wavenumber traces observed by the sensor array will be different depending on the AOA for each source. From a spatial signal point of view, the multiple sources are actually producing different wavenumber traces across the array, which are observed as different wavelength traces. Therefore, the spatial response of the array can be seen as a sum of eigenvectors representing the mode "shapes" and corresponding eigenvalues representing the spatial frequency of the corresponding wavenumber trace observed by the array. Figure 14.5 shows the spatial complex response for the case where all three plane waves in Figure 14.4 are active simultaneously. Both the magnitude and the phase are distorted by the combination of waves. This distortion changes over time as a result of the phase differences between the multiple sources as described in Equation 14.3 for the spatial cross-correlation. Clearly, the complexity of the mixed waves is a challenge to decompose. However, if we assume either time or spatial averaging (moving the array in space while taking snapshots) where the multiple source bearings remain nearly constant, the covariance matrix of the received complex array sensor signals for a given FFT bin will represent the spatial signal correlations and not the individual source phase interference. Equation 14.13 depicts the covariance of the array signals:

$$
\overline{R} = \begin{bmatrix} R_0^x & R_1^x & R_2^x & \cdots & R_M^x \\ R_{-1}^x & R_0^x & R_1^x & \cdots & R_{M-1}^x \\ \vdots & \vdots & & \ddots & \vdots \\ R_{-M}^x & R_{-M+1}^x & \cdots & R_{-1}^x & R_0^x \end{bmatrix}, R_{i-j}^x = \sum_{k=0}^{N} x_{i,t-k}^* x_{j,t-k} \tag{14.13}
$$

Note that $x_{i,t}$ could be either a broadband time-domain signal where the N time snapshots happen at the sampling frequency, or a narrowband FFT bin where the N snapshots correspond to N overlapped FFT blocks of time data. As suggested earlier, the N snapshots could also be accomplished by either moving the array around or sampling "subarrays" from a much larger array. The snapshot-averaging process is essential for the covariance matrix to represent the spatial array response properly. Note also that the covariance matrix in Equation 14.13 is $M+1$ rows $\times$ $M+1$ columns for our $M+1$ sensor array. For the null-forming case using a spatial whitening filter in Section 14.1, the

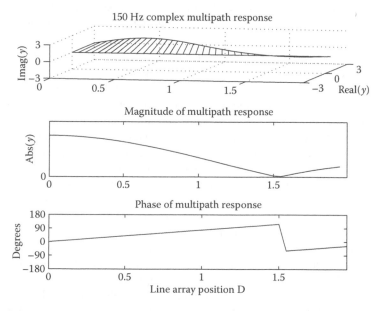

FIGURE 14.5 Complex spatial response of the multipath interference caused by three (equal amplitude and phase at sensor 1) 150 Hz plane waves arriving broadside, 60°, and endfire (all three waves added together in plots).

inverted covariance matrix in Equation 14.7 was M rows × M columns. In both cases, the covariance matrix has a Toeplitz structure, meaning that all elements on a given diagonal (from upper left to lower right) are the same. This structure is not necessary for matrix inversion, but does indicate that the matrix is well conditioned for inversion so long as R_0^x on the main diagonal is nonzero. If there is any residual noise in the signals the main diagonal is always nonzero. Given that the inverse exists (the matrix is full rank), one can solve for the eigenvalues (which correspond approximately to the signal "power levels" for each mode) and eigenvectors (which approximately correspond to the mode shape for each mode). One can think of each eigenvector as an array beam steering weight vector which, when multiplied by the array outputs and summed as a vector dot product, gives an array spatial response (beam pattern) with the LD aligned to the corresponding source direction. The corresponding eigenvalue represents the approximate signal power times the number of array elements and snapshots.

However, both of these physical claims are in practice affected negatively by a low SNR or phase coherence of the sources. "Incoherent sources radiating the same narrowband frequency" is physically impossible unless the "narrowband" frequency is really drifting around in the band randomly. Narrowband frequency modulation is also known as phase modulation (frequency is the time derivative of phase) because very small changes in frequency are seen as phase shifts in the time waveform. Therefore, if the sources or receiving array are moving, or if the sources are physically independent (such as separate vehicles), it is not a bad assumption to assume source independence after N snapshots so long as the narrowband bandwidth is not too narrow. In fact, the broader the frequency bandwidth for the covariance matrix, the shorter the required time will be to observe a snapshot, and the faster one will be able to compile a number of snapshots to ensure source independence.

Perhaps the single most useful feature of using an eigenvalue approach for adaptive beamforming is that the eigenvectors are *orthonormal*. The dot product of an eigenvector with itself gives unity and that the dot product of any two different eigenvectors is zero. Physically, this means that the spatial response of each eigenvector will have a main lobe in the corresponding source LD and zeros in the directions of other sources (in theory) or, in practice, a response that sums to near zero for other source directions. This is exactly the desired result for adaptive beamforming where one optimizes array performance by getting array SNR gain in the LD while suppressing sources from other directions [6]. The classic phased-array beam steering of Section 14.1 also provides SNR gain in the LD of the main lobe, but beam response in the direction of other sources may not be zero, leading to interference in the beamformer output.

Assuming that the number of sensors is greater than the number of phase-independent sources (this intuitively implies a solution since there are more equations than unknowns), and each eigenvector and eigenvalue correspond to a source, what do the "extra" eigenvectors and eigenvalues correspond to? The extra eigenvector and eigenvalues will correspond to "spurious" sources that appear in the background noise. There will be some amount of spatially independent noise on each sensor output from electronic noise, turbulence (acoustic arrays), and even the LSB of the A/D conversion. This spatially independent noise can be seen as created by an infinite number of sources, so there is no way for the eigenvector approach to resolve them. The "noise eigenvectors" each have an LD which is random, but will have a zero beam response in the directions of the real signal sources. This is a very nice property because if one sums all the noise eigenvectors, one is left with a beam response where the only nulls are in the real source directions, just like the null-forming spatial whitening filter of Section 14.1. This technique is called MUSIC. One can then, as in the spatial whitening filter null-forming case, evaluate the sharp nulls in the beam response for the sum of the noise eigenvectors to determine the DOA of the sources. Subsequently, the optimal beam to look in a given direction while nulling all other sources is computed from the covariance matrix in the MVDR algorithm. To have a beam look in the direction of one physical signal source while nulling the other sources, one would use the corresponding eigenvector for the source of interest, or one could synthesize a steering vector based on knowledge of the source directions.

The general eigenvalue problem is defined by the solution of the following equation:

$$\overline{R}v_i = v_i \lambda_i, \quad i = 0, 1, \dots, M \tag{14.14}$$

where v_i is a column eigenvector and λ_i is the corresponding scalar eigenvalue. The problem is taught universally in all engineering and science undergraduate curricula because it provides an elegant numerical technique for inverting a matrix, an essential task for matrix linear algebra. From an algorithm point of view, Equation 14.14 has three parameters and only one is known. However, the structure of the eigenvectors is constrained to be orthonormal as depicted in Equation 14.15 and each of the eigenvalues is constrained to be a scalar:

$$\begin{aligned} v_{iv}^{H} v_j &= 1, \quad i = j \\ &= 0, \quad i \neq j \end{aligned} \tag{14.15}$$

we can write Equation 14.14 for all the eigenvectors and eigenvalues

$$\overline{R}A = A\Lambda \quad \text{where } A = \begin{bmatrix} v_0 & v_1 & v_2 \cdots v_M \end{bmatrix} \tag{14.16}$$

and the diagonal matrix of eigenvalues Λ is defined as

$$\Lambda = \begin{bmatrix} \lambda_0 & 0 & 0 & 0 & \dots & 0 \\ 0 & \lambda_1 & 0 & 0 & \dots & 0 \\ \vdots & & & \vdots & & \vdots \\ 0 & \dots & 0 & 0 & 0 & \lambda_M \end{bmatrix} \tag{14.17}$$

Note that $A^H A = I$, which allows $A^{-1} = A^H$. The superscript "H" means Hermitian transpose where the elements are conjugated (imaginary part has sign changed) during the matrix transpose. We also refer to a Hermitian matrix as one which is conjugate symmetric.

We are left with the task of finding the eigenvectors in the columns of the matrix A defined in Equation 14.16. This operation is done by "diagonalizing" the covariance matrix:

$$A^H \overline{R}A = A^H A\Lambda = \Lambda \quad \text{since } A^H A = I \tag{14.18}$$

The matrix A which diagonalizes the covariance matrix can be found using a number of popular numerical methods. If the covariance is real and symmetric, one simple sure-fire technique is the Jacobi transformation method. This is also known in some texts as a Givens rotation. The rotation matrix $A_{kl'}$ is an identity matrix where the zero in the kth row and lth column is replaced with sin θ, lth row and kth column with $-\sin \theta$, and the ones on the main diagonal in the kth and lth rows are replaced with cos θ. Pre- and postmultiplication of a (real symmetric) covariance matrix by the rotation matrix has the effect of zeroing the off-diagonal elements in the (k,l) and (l,k) positions when θ is chosen to satisfy

$$\tan 2\theta_{kl} = \frac{2R_{kl}}{R_{ll} - R_{kk}} \tag{14.19}$$

If the covariance matrix is two-row, two-column, the angle θ can be seen as a counterclockwise rotation of the coordinate axis to be aligned with the major and minor axis of an ellipse (a circle if $R_{11} = R_{22}$ requires no rotation). The rotation also has the effect of increasing the corresponding main diagonal elements such that the norm (or trace) of the matrix stays the same. The residual covariance with the zeroed off-diagonal elements from the first Jacobi rotation is then pre- and postmultiplied by another rotator matrix to zero another pair of off-diagonal elements, and so on, until all the off-diagonal elements are numerically near zero. The eigenvector matrix A is simply the product of all the Jacobi rotation matrices and the eigenvalue matrix in Equation 14.17 is the diagonalized covariance matrix.

How can Jacobi transformations be applied to complex matrices? Suppose our covariance matrix is not composed of the broadband spatial cross-correlation among element signals in the time domain, but rather the narrowband spatial cross-correlation among the elements using a narrowband FFT bin complex signal (averaged over N snapshots of course). The covariance matrix is Hermitian (conjugate symmetric) and can be written in the form

$$\bar{R} = \bar{R}^R + j\bar{R}^I \tag{14.20}$$

where the real part is symmetric and the imaginary part is skew-symmetric (corresponding off-diagonals have same magnitude but opposite sign). The complex eigenvalue problem is constructed as a larger-dimension real eigenvalue problem:

$$\begin{bmatrix} \bar{R}^R & -\bar{R}^I \\ \bar{R}^I & \bar{R}^R \end{bmatrix} \begin{bmatrix} v_i^R \\ v_i^I \end{bmatrix} = \lambda_i \begin{bmatrix} v_i^R \\ v_i^I \end{bmatrix} \tag{14.21}$$

The eigenvalues for the matrix in Equation 14.21 are found in $M + 1$ pairs of real and imaginary components. For large matrices, there are a number of generalized eigenvalue algorithms with much greater efficiency and robustness. This is particularly of interest when the matrix is ill-conditioned. A good measure of effectiveness to see if the eigenvector/eigenvalue solution is robust is to simply compute the covariance matrix inverse and multiply it by the original covariance matrix to see how close the result comes to the identity matrix. The covariance matrix inverse is found trivially from the eigenvalue solution.

$$\bar{R}^{-1} = A\Lambda^{-1}A^H \quad \text{where } \Lambda^{-1} = \text{diag}\left\{\lambda_0^{-1}\ \lambda_1^{-1} \cdots \lambda_M^{-1}\right\} \tag{14.22}$$

There are also a number of highly efficient numerical algorithms for matrix inversion based on other approaches than the generalized eigenvalue problem. One of the more sophisticated techniques has already been presented in Section 10.4.1 of this book. Recall Equation 10.75 relating the backward prediction coefficients and backward error covariance to the inverse of the signal covariance matrix R_M^x.

$$\left[R_M^x\right]^{-1} = L^H \left[R_M^r\right]^{-1} L \tag{14.23}$$

where L is a lower triangular matrix of backward predictors defined by Equations 10.74 and 14.24 below and R_M^r is a diagonal matrix of the orthogonal backward prediction error variances.

$$\begin{bmatrix} r_{0,n} \\ r_{1,n} \\ \vdots \\ r_{M-1,n} \\ r_{M,n} \end{bmatrix} = \begin{bmatrix} 1 & 0 & & \cdots & 0 \\ a_1^r,1 & 1 & 0 & \cdots & 0 \\ & & \ddots & \vdots & \vdots \\ a_{M-1,M-1}^r & & \cdots & 1 & 0 \\ a_{M,M}^r & a_{M,M-1}^r \cdots a_{M,1}^r & & & 1 \end{bmatrix} \begin{bmatrix} x_n \\ x_{n-1} \\ \vdots \\ x_{n-M+1} \\ x_{n-M} \end{bmatrix} = \bar{r}_{0,n:M,n} = L\varphi_n^T \tag{14.24}$$

Note that L^H is upper triangular and L is lower triangular. If the elements of L and L^H are appropriately normalized by the square-root of the corresponding backward prediction error variance, the expression in Equation 14.23 can be written as an *LU Cholesky factorization*, often used to define the square-root of a matrix. However, the L matrix in Equation 14.24 differs significantly from the eigenvector matrix. While the backward predictors are orthogonal, they are not orthonormal. Also, the backward prediction error variance is not the eigenvectors of the signal. The eigenvector matrix and eigenvalue diagonal matrix can however be recovered from the Cholesky factorization, but the required substitutions are too tedious for presentation here.

Getting back to the physical problem at hand of optimizing the beam response for an array of sensors to separate the signals from several sources, we can again consider a line array of eight equally spaced sensors and plane waves from several distant sources. Our initial example has a single 100 Hz plane wave arriving at 90° bearing. For simplicity, we will consider a high-SNR case, acoustic propagation in air ($c = 343$ m/s), and an array spacing of 1.25 m (spatial aliasing occurs for frequencies above 137 Hz). A particular eigenvector is analyzed by a zero-padded FFT where the resulting digital spatial (wavenumber) frequency covers the range $-\pi \leq \varphi \leq +\pi$. To relate this digital spatial angle to a physical bearing angle it can be seen that $\varphi = kd \cos \theta$, where θ is the physical bearing angle. For unevenly spaced arrays, there is not such a direct mapping between the physical bearing angle and the digital spatial frequency. For random arrays, one uses the UFT (an FFT is not possible for uneven spaced samples) of Section 6.6 to evaluate the beam response wavenumber spectrum, but then uses an average spacing d' to relate the digital and physical angles. This is analogous to relating the digital frequency in the UFT to a physical frequency in Hertz by using the average sample rate.

Figure 14.6 shows the eigenvector beam response calculated by simply zero-padding the eigenvector complex values to 1024 points and executing an FFT. The physical angle on the x-axis is recovered by dividing the digital frequency by kd and taking its inverse cosine. Obviously, $\theta = \cos^{-1}(\varphi/kd)$ is not a linear function, so relating the digital and physical bearing angles requires careful consideration even for evenly spaced arrays. A total of 500 independent snapshots are used where the signal level is 3 and the noise standard deviation is 1×10^{-4}. The calculated eigenvalue for the signal is divided by the number of snapshots times the number of array elements (500 times 8 or

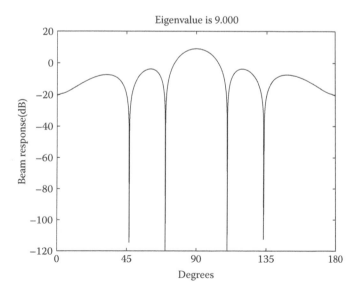

FIGURE 14.6 The signal eigenvector beam response corresponding to a 100 Hz sinusoidal plane wave of amplitude 3 (eigenvalue is 9.0) at 90° (broadside) in a low Gaussian background noise of 0.0001 standard deviation.

4000) to give a normalized value of 9.000006, or the signal amplitude-squared. For fewer snapshots and lower SNR there is less precision in this calculation. The normalized noise eigenvalues are all on the order of 1×10^{-8} or smaller. Clearly, one can simply examine the eigenvalues and separate them into "*signal eigenvalues*" and "*noise eigenvalues*," each with respective eigenvectors. There are some algorithms such as the Akaike information criterion (AIC) and minimum description length (MDL) which can help identify the division into the subspaces when SNR is low or an eigenvalue threshold is difficult to automatically set or adapt in an algorithm [6]:

$$[v_0 \ v_1 \ \cdots \ v_M] = A_S + A_N = \left[v_{S_1} \ v_{S_2} \cdots v_{S_{N_s}} | v_{N_1} \ v_{N_2} \cdots v_{N_{N_n}} \right] \tag{14.25}$$

The eigenvectors v_{Si}, $i = 1, 2, \ldots, Ns$, define what is called the *signal subspace* A_S while $v_{Nj}, j = 1, 2, \ldots, Nn$, define the *noise subspace* A_N spanned by the covariance matrix where $Nn + Ns = M + 1$, the number of elements in the array. Since the eigenvectors are orthonormal, the eigenvector beam response will have its main lobe aligned with the physical AOA of the corresponding signal source (whether it be a real physical source or a spurious noise source) and nulls in the directions of all other corresponding eigenvector sources. If an eigenvector is part of the noise subspace, the associated source is not physical but rather "spurious" as the result of the algorithm trying to fit the eigenvector to independent noise across the array. Figure 14.7 shows the seven noise eigenvector beam responses for our single physical source case. *The noise eigenvectors all have the same null(s) associated with the signal eigenvector(s).* Therefore, if one simply averages the noise eigenvectors, the spurious nulls will be filled in while the signal null(s) will remain. This is the idea behind the MUSIC algorithm. Figure 14.8 shows the beam response of the MUSIC vector and Equation 14.26 shows the MUSIC vector calculation:

$$v_{\text{MUSIC}} = \frac{1}{Nn} \sum_{j=1}^{Nn} v_{N_j} \tag{14.26}$$

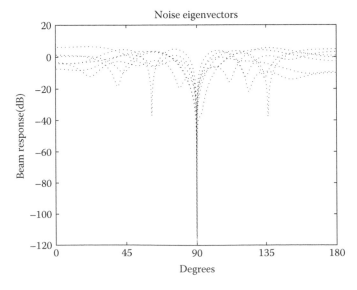

FIGURE 14.7 Beam responses of each of the "noise" eigenvectors which correspond to eigenvalues on the order of 1×10^{-8} (the square of the RMS noise) showing a common null (orthogonality) to the signal eigenvector for the source at 90°.

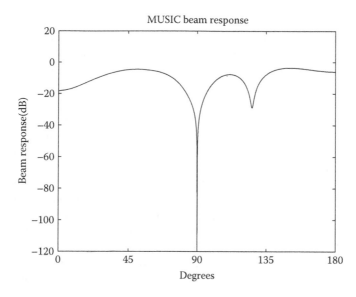

FIGURE 14.8 MUSIC beam response from the sum of all the noise eigenvectors leaving strong nulls in the signal direction, see at 90°.

Consider the case of three physical sources radiating plane waves of amplitudes 2, 3, and 1, at the array from the angles 45°, 90°, and 135°, respectively, in the same low SNR of 1×10^{-4} standard deviation background noise. Figure 14.9 shows the eigenvector beam responses for the three signals. The normalized eigenvalues are 9.2026, 3.8384, and 0.9737 corresponding to the arrival angles 90°, 45°, and 135° as seen in Figure 14.9. Most eigenvalue algorithms sort the eigenvalues and corresponding eigenvectors based on amplitude. The reason the eigenvalues do not work out to be exactly the square of the corresponding signal amplitudes is that there is still some residual spatial correlation, even though 500 independent snapshots are in the covariance simulation. It can be seen that the

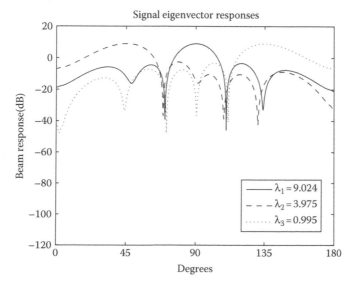

FIGURE 14.9 The three signal eigenvector beam responses for the three 100 Hz sources of amplitude 2, 3, and 1 and angles 45°, 90°, and 135° showing the estimated eigenvalues close to the square of the signal amplitudes for each source.

sum of the signal eigenvalues (normalized by the number of array elements times the number of snapshots, or $8 \times 500 = 4000$) is 14.0147 which is quite close to the expected 14. One should also note that the signal eigenvector responses do not show sharp nulls in the directions of the other sources due to the limited observed independence of the sources. It can be seen that for practical implementations, the signal eigenvectors are only approximately orthonormal. It is very important for the user of MUSIC and all eigenvector methods to understand the issue of source independence. The whole approach simply fails when the plane waves all have the same phase with respect to each other. Such is the case when multiple arrivals are from coherent reflections of wave from a single source mixing with the direct wave. Figure 14.10 shows the MUSIC vector beam response which clearly shows the presence of three sources at 45°, 90°, 135°.

If one wanted to observe source 1 while suppressing sources 2 and 3 with nulls, simply using the source 1 eigenvector, v_{S1}, would provide the optimal adaptive beam pattern if the source signals were completely independent. To steer an optimal adaptive beam to some other direction θ' we can use the inverse of the covariance matrix to place a normalized (to 0 dB) gain in the direction θ' while also keeping the sharp nulls of the MUSIC response beam. This is called an MVDR-adaptive beam pattern. The term "minimum variance" describes the maintaining of the sharp MUSIC nulls while "distortionless response" describes the normalization of the beam to unity gain in the LD. The normalization does not affect SNR; it just lowers the beam gain in all directions to allow LD unity gain. If the delay-sum, or Bartlett beam, steering vector is defined as

$$S(\theta') = \left[1 e^{-jkd_1 \cos\theta'} \, e^{-jkd_2 \cos\theta'} \cdots e^{-jkd_M \cos\theta'} \right]^H \tag{14.27}$$

where d_i, $i = 1, 2, \ldots, M$, is the coordinate along the line array and "H" denotes Hermitian transpose (the exponents are positive in the column vector), the MVDR steering vector is defined as

$$S_{MVDR}(\theta') = \frac{\overline{R}^{-1} S(\theta')}{S(\theta')^H \overline{R}^{-1} S(\theta')} \tag{14.28}$$

Note that the denominator of Equation 14.28 is a scalar which normalizes the beam response to unity (0 dB gain) in the LD. Even more fascinating is a comparison of the MVDR steering vector to

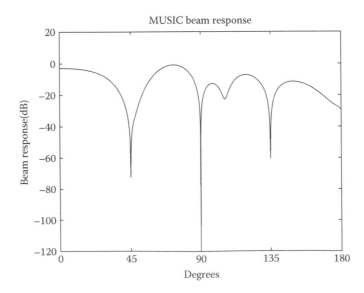

FIGURE 14.10 MUSIC beam response for the three 100 Hz sources at 45°, 90°, and 135°.

the Kalman gain in the RLS algorithm summarized in Table 10.1. In this comparison, our "Bartlett" steering vector $S(\theta')$ corresponds to the RLS basis function and the MVDR steering vector represents the "optimal adaptive gain update" to the model, which is our array output. There is also a comparison of the projection operator update in Equation 9.21 which is similar in mathematical structure. Figure 14.11 compares the Bartlett and MVDR responses for a steering angle of 112.5° and our three sources at 45°, 90°, and 135°. The MVDR beam effectively "blinds" the array to these signals while looking in the direction of 112.5°. Note that the Bartlett beam response has a gain of 8 (+18 dBV) in the LD from the eight sensor elements of the linear array.

MVDR beamforming is great for suppressing interference from strong sources when beam steering to directions other than these sources. If one attempts an MVDR beam in the same direction as one of the sharp nulls in the MUSIC beam response, the algorithm breaks down due to an indeterminate condition in Equation 14.28. Using the eigenvector corresponding to the signal LD is theoretically the optimal choice, but in practice some intrasource coherence is observed (even if it is only from a limited number of snapshots) making the nulls in the directions of the other sources weak. However, from the MUSIC response, we know the angles of the sources of interest. Why not manipulate the MVDR vector directly to synthesize the desired beam response?

Like any FIR filter, the coefficients can be handled as coefficients of a polynomial in z where $z = e^{jkd\cos\theta}$. For an evenly spaced linear array, the delay-sum or Bartlett steering vector seen in Equation 14.17 can be written as $S(z) = [1 \; z^{-1} \; z^{-2} \; z^{-3} \ldots z^{-M}]^H$ where $z = e^{jkd\cos\theta}$. For a nonevenly spaced array we can still treat the steering vector as an integer-order polynomial, but the interpretation of the physical angles from the digital angles is more difficult. This is not a problem because our plan is to identify the zeros of the MVDR steering vector associated with the signals, and then suppress the zero causing the null in the LD. The resulting beam response will have a good (but not optimal) output of the source of interest and very high suppression of the other sources. The procedure will be referred to here as the "pseudo reduced-order technique" and is as follows:

1. Solve for the MVDR steering vector.
2. Solve for the zeros of the MVDR steering vector as if it were a polynomial in z.

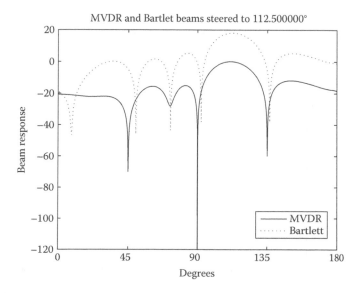

FIGURE 14.11   Comparison of the MVDR beam response steered to 112.5° to a simple delay-sum (Bartlett) beam response steered to 112.5° showing the MVDR maintaining the nulls at the identified sources at 45°, 90°, and 135°.

3. Detect the zeros on or very close to the unit circle on the z-plane. These are the ones producing the sharp nulls for the signal directions. The number of these zeros equals the number of signal sources detected.
4. Noting that the polynomial zero angles will range from $-\pi$ to $+\pi$ while the physical wavenumber phase will range from $-kd$ to $+kd$, select the signal zero corresponding to the desired LD and suppress it by moving it to the origin. Move the other spurious zeros further away from the unit circle (toward the origin or to some large magnitude).
5. Using the modified zeros calculate the corresponding steering vector by generating the polynomial from the new zeros. Repeat for all signal directions of interest.

The above technique can be thought of as similar to a reduced-order method, but less mathematical in development. The threshold where one designates a zero as corresponding to a signal depends on how sharp a null is desired and the corresponding distance from the unit circle. Figure 14.12 shows the results of the pseudo reduced-order technique for the three beams steered to their respective sources. While the beams are rather broad, they maintain sharp nulls in the directions of the other sources that allow separation of the source signals. The corresponding MUSIC beam response using the steering vector modified by the pseudo reduced-order technique presented here is free of the spurious nulls because those zeros have been suppressed.

We can go even farther toward improving the signal beams by making use of the suppressed zeros to sharpen the beamwidth in the LD. We will refer to this technique as the "null synthesis technique" since we are constructing the steering vector from the MVDR signal zeros and placing the remaining available zeros around the unit circle judiciously to improve our beam response. The usefulness of this is that the available (nonsignal related) zeros can be used to improve the beam response rather than simply be suppressed in the polynomial. The null synthesis technique is summarized as follows:

1. Solve for the MVDR steering vector.
2. Solve for the zeros of the MVDR steering vector as if it were a polynomial in z.

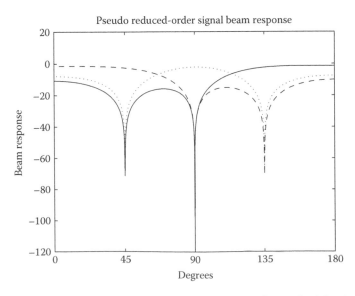

FIGURE 14.12 The pseudo reduced-order technique presented here for synthesizing beams that suppress the noise sources while maintaining the desired LD (45°, 90°, and 135° shown) without spurious nulls.

3. Detect the zeros on or very close to the unit circle on the z-plane. These are the ones producing the sharp nulls for the signal directions. The number of these zeros equals the number of signal sources detected.

4. Distribute the number of remaining available (nonsignal related) zeros plus one around the unit circle from $-\pi$ to $+\pi$ avoiding a double zero at π. The extra zero nearest the main lobe will be eliminated to enhance the main lobe response.

5. For the desired signal LD and corresponding unit circle angle, find the closest zero you placed on the unit circle and eliminate it. Then add the zeros for the other source directions.

6. Using the synthesized zeros calculate the corresponding steering vector by generating the polynomial from the new zeros. Repeat for all signal directions of interest.

Figure 14.13 shows the results using null synthesis. Excellent results are obtained even though the algorithm is highly heuristic and not optimal. The beam steer to 45° and 135° could further benefit from the zeros at around 30° and 150° being suppressed. The effect of the zeros placed on the unit circle is not optimal, but still quite useful. However, the nonlinearity of $\theta = \cos^{-1}(\varphi/kd)$ means that the physical angles (bearings) of evenly spaced zeros on the unit circle (evenly spaced digital angles) do not correspond to evenly spaced bearing angles. To make the placed zeros appear closer to evenly spaced in the physical beam response, we can "premap" them according to the known relationship $\theta = \cos^{-1}(\varphi/kd)$ between the physical angle θ and the digital wavenumber angle φ. If we let our evenly spaced zeros around the unit circle having angles φ', we can premap the z-plane zero angles by $\varphi = \cos^{-1}(\varphi'/kd)$. This was done for the null synthesis example in Figure 14.13 and it improved the beam symmetry and main lobe significantly. The nulls in Figure 14.13 also appear fairly evenly spaced except around the main lobes of the signal steering vectors. While the signal eigenvectors are the theoretical optimum beams for separating the source signals, the affect of source phase dependence limits the degree to which the signals can be separated. Heuristic techniques such as the pseudo reduced-order and null synthesis techniques illustrate what can be done to force the signal separation given the information provided by MUSIC. Both these techniques can also be applied to a spatial whitening filter for signal null-forming as described in Section 14.1.

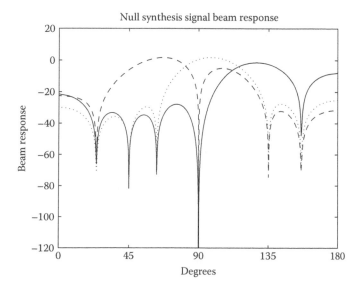

FIGURE 14.13 The null synthesis technique presented here sharpens the beam in the LD (45°, 90°, and 135° shown) by distributing the spurious zeros evenly and maintaining nulls for interfering sources.

For N_S signal source eigenvectors in the matrix A_S, and $N_N = M + 1 - N_S$ noise eigenvectors in the matrix A_N, we note from Equation 14.25 that since $A^H = A^{-1}$

$$
\begin{aligned}
I &= (A_S + A_N)^H (A_S + A_N) \\
&= (A_S + A_N)(A_S + A_N)^H \\
&= A_S A_S^H + A_S A_N^H + A_N A_S^H + A_N A_N^H \\
&= A_S A_S^H + A_N A_N^H
\end{aligned}
\tag{14.29}
$$

represents the sum of the projection matrices (Section 9.2) onto the noise and signal subspaces. This can be seen by considering a signal matrix Y which has $M + 1$ rows and N_S columns representing some linear combination of the signal eigenvectors $Y = A_S Q_S$, where Q_S is any invertible square matrix with N_S rows. Therefore, $Y^H Y = Q_S^H A_S^H A_S Q_S = Q_S^H Q_S$. The projector matrix onto the signal subspace can be seen to be

$$
\begin{aligned}
P_Y &= A_S A_S^H = \left(Y Q_S^{-1} \right)\left(Q_S^{-H} Y^H \right) \\
&= Y (Q_S^H Q_S)^{-1} Y^H \\
&= Y (Y^H Y)^{-1} Y^H
\end{aligned}
\tag{14.30}
$$

which is a fancy way of saying that the signal subspace is expressible as the linear combination of the signal eigenvectors. So, if the signal levels of our plane wave sources change, the subspace spanned by the rows of Y does not change, but the eigenvalues do change according to the power relationship. If the bearing(s) of the sources change then the subspace also changes. The same is true for the noise subspace. If we define our noise as $W = A_N Q_N$,

$$
\begin{aligned}
P_W &= A_N A_N^H = \left(W Q_N^{-1} \right)\left(Q_N^{-H} W^H \right) \\
&= W \left(Q_N^H Q_N \right)^{-1} W^H \\
&= W \left(W^H W \right)^{-1} W^H
\end{aligned}
\tag{14.31}
$$

and the combination of noise and signal projectors is

$$
P_Y + P_W = Y(Y^H Y)^{-1} Y^H + W(W^H W)^{-1} W^H = I
\tag{14.32}
$$

Equation 14.32 can be used to show that the noise is orthogonal to the signal subspace by $P_Y = I - P_W$. This orthogonal projection operator was used in Sections 9.2, 10.3, and 10.4 to define a least-squared error update algorithm for the adaptive lattice filter, which results from the orthogonal decomposition of the subspace in time and model order. Because of the orthonormal nature of the eigenvectors, we could have divided the space spanned by the rows of the covariance matrix by any of the eigenvectors. It makes physical sense to break up the subspace into signal and noise to allow detection of the signal wavenumber s and the design of beam steering vectors to allow signal separation and SNR enhancement.

14.3 COHERENT MULTIPATH RESOLUTION TECHNIQUES

In the previous two sections, we have seen that near-optimal beam patterns can be designed either using adaptive spatial filter or by eigenvector processing. For narrowband phase-independent sources one can use the received signals from a sensor array of known geometry in a medium of known free wave propagation speed to design beams that identify the arrival angles of, and even separate, the source signals. However, the notion of "phase independent narrowband sources" is problematic in most practical applications. Many "snapshots" of the spectral data are averaged in the hope that the random phase differences of the "independent" sources will average to zero in the covariance matrix elements, as seen in Equation 14.3. There are coherent multipath situations where the array records several wavenumber traces at a given temporal frequency where the snapshot-averaging technique simply will not work unless the individual phases and amplitudes of the sources are known. Furthermore, if it requires $1/\Delta f$ seconds of time data to produce the complex FFT bin Δf Hertz wide, and then N FFT snapshots to be averaged to ensure that the spatial covariance matrix for that bin is independent of the source phases, a long net time interval is then required to do the problem. If the sources are moving significantly during that time the arrival angle information may not be meaningful. However, as one widens the FFT bin bandwidth by shorting the time integration, a large number of snapshots can be calculated in a very short time span. If the beam bandwidth approaches the Nyquist band (maximum available bandwidth), the snapshots become a single time sample and the "broadband" covariance matrix is calculated in the shortest possible time without the use of an FFT. Why not do broadband beamforming where the sources are more readily made independent? The answer is seen in the fact that a broadband beam may not be optimal for resolving specific narrowband frequencies, since it is optimized for the entire broad bandwidth. Therefore, a fundamental trade-off exists between narrowband resolution in both frequency and wavenumber (arrival angle) and the amount of integration in time and space to achieve the desired resolution. As with all information processing, higher fidelity generally comes at an increased cost, unless of course human intelligence is getting in the way.

Suppose we know the amplitude and phase of each source and we have an extremely HiFi propagation model which allows us to predict the precise amplitude and phase from each source across the array. The solution is now trivial. Consider an eight-element ($M = 7$) line array and three sources

$$
\begin{bmatrix} \overline{s_1} \\ \overline{s_2} \\ \overline{s_3} \end{bmatrix} = \begin{bmatrix} s_1 & 0 & 0 \\ 0 & s_2 & 0 \\ 0 & 0 & s_3 \end{bmatrix} \begin{bmatrix} 1 & e^{jk_1 d_1} & e^{jk_1 d_2} \cdots e^{jk_1 d_M} \\ 1 & e^{jk_2 d_1} & e^{jk_2 d_2} \cdots e^{jk_2 d_M} \\ 1 & e^{jk_3 d_1} & e^{jk_3 d_2} \cdots e^{jk_3 d_M} \end{bmatrix} \tag{14.33}
$$

where d_i, $i = 1, 2, \ldots, M$, is the position of the ith element along the line array relative to the first element. The wavenumber traces, k_i $i = 1, 2, 3$, are simply the plane wave projections on the line array. The right-most matrix in Equation 14.33 is called a Vandermode matrix since the columns are all exponential multiples of each other (for an evenly spaced line array). Clearly, the signal eigenvectors are found in the rows of the Vandermode matrix and the eigenvalues are the magnitude-squared of the corresponding signal amplitudes and phases s_i, $i = 1, 2, 3$. This is like saying if we know A (the signal amplitudes and phases) and B (the arrival angles or wavenumbers), we can get $C = AB$. But if we are only given C, the received waveform signals from the array, there are an infinite combination of signal amplitudes and phases (AB) which give the same signals for the array elements. Clearly, the problem of determining the AOA from multiple narrowband sources and separating the signals using adaptive beamforming from the array signal data alone requires one to eliminate any coherence between the sources as well as the background noise, which should be independent at each array element.

Suppose one could control the source signal, which has analogy in active sonar and radar where one transmits a known signal and detects a reflection from an object, whose location, motion, and perhaps identity are of interest. Given broadband transducers and good propagation at all frequencies, a broadband (ZMG) signal would be a good choice because the sources would be completely independent at all frequencies. An even better choice would be a broadband periodic waveform which repeats every N sample and has near zero cross-correlation with the broadband periodic waveforms from the other sources. Time-synchronous averaging (Section 12.1.1) of the received waveforms is done by simply summing the received signals in an *N*-point buffer. The coherent additions will cause the background noise signals to average to their mean of zero while coherently building up the desired signals, with period *N* significantly improving SNR over the broad bandwidth. By computing transfer functions (Section 7.3) between the transmitted and received signals, the frequency response of the propagation channel is obtained along with the propagation delay. For multipath propagation, this transfer function will show frequency ranges of cancellation and frequency ranges of reinforcement due to the multipath and the corresponding impulse response will have several arrival times (seen as a series of impulses). Each element in the array will have a different frequency response and corresponding impulse response due to the array geometry and the different arrival angles and times. Therefore, the "coherent" multipath problem is solvable if a broadband signal is used. Solving the multipath problem not only allows us to measure the propagation media inhomogeneities, but also allows us to remove multipath interference in communication channels.

14.3.1 MAXIMAL LENGTH SEQUENCES

MLS are sequences of random bits generated by an algorithm which repeat every $2^N - 1$ bits, where *N* is the order of the MLS generator. The interesting signal properties of an MLS sequence are that its autocorrelation resembles a digital Dirac delta function (the 0th lag has amplitude $2^N - 1$ while the other $2^N - 2$ lags are equal to −1) and the cross-correlation with other MLS sequences is nearly zero. The algorithm for generating the MLS sequence is based on primitive polynomials modulo 2 of order *N*. We owe the understanding of primitive polynomials to a young French mathematical genius named Evariste Galois (1812–1832) who died as a result of a duel* and the application of primitive polynomials to MLS sequence generation has played an enabling role in computer random number generation, digital spread-spectrum communications, satellite GPS, and data encryption standards. Clearly, this is an astonishing contribution to the world by a 20-year-old genius.

To illustrate the power of MLS, consider the following digital communications example. The MLS generating algorithm are known at a transmitting and receiver site, but the transmitter starts and stops transmission at various times. When the receiver detects the transmitter, it does so by detecting a Dirac-like peak in a correlation process. The time location of these peaks allows the receiver to "synchronize" its MLS correlation to the transmitter. The transmitter can send digital data by a simple modulo-2 addition of the data bit-stream to the transmitted MLS sequence. This appears as noise in the receiver's correlation detector. But once the transmitter and receiver are synchronized, a modulo-2 addition (an exclusive or operation or XOR) of the MLS sequence to the received bit-stream gives the transmitted data! Bit errors in the residual stream can be further reduced by repeating the stream and averaging the residual bits and/or by employing error correction algorithms. Since multiple MLS sequences of different generating algorithms are uncorrelated, many communication channels can coexist in the same frequency band, each appearing to the others as uncorrelated background noise. There are limits to how many independent MLS sequences can be generated for a particular length. It is also possible to embed both short and long MLS sequences to allow fast synchronization and low bit errors in the data channel from repeated data sequences. This is often done for very-long-range low-SNR transmissions such as those from deep space probes.

A talented scientist named Gold [7] showed that one could add modulo-2 two generators of length $2^N - 1$ and different initial conditions to obtain $2^N - 1$ new sequences (which is not an MLS but close), plus the two original base sequences. The cross-correlation of these Gold sequences are shown to be bounded, but not zero, allowing many multiplexed MLS communication channels to coexist with the same length in the same frequency channel. The "Gold codes," as they have become

---

* In the Romantic age, many lives were cut short by death from consumption or dueling. Fortunately for us, the night before Galois fate he sent a letter to a friend named Chevalier outlining his theories. Now known as Galios theory, it is considered one of the highly original contributions to algebra in the nineteenth century.

known, are the cornerstone of most wireless digital communications. For GPS systems, a single receiver can synchronize with multiple satellites, each of which also sends their time and position (ephemeris equations), enabling the receiver to compute its position. GPS is another late twentieth-century technology that is finding its way into many commercial and military applications.

A typical linear MLS generator [8] is seen in Figure 14.14. The notation describing the generator is $[N,i,j,k,\ldots]$ where the sequence is $2^N - 1$ "chips" long, and i, j, k, and so on are taps from a 1-bit delay line where the bits are XOR'd to produce the MLS output. For example, the generator depicted as [7,3,2,1] has a seven-stage delay line, or register, where the bits in positions 7, 3, 2, and 1 are modulo-2 added (bit exclusive OR denoted as XOR'd) to produce the bit that goes into stage 1 at the next clock pulse. The next clock pulse shifts the bits in stages 1–6 to stages 2–7, where the bit in stage 7 is the MLS output, copies the previous modulo-2 addition result into stage 1, and calculates the next register input for stage 1 by the modulo-2 addition of the bits in elements 7, 3, 2, and 1. Thanks to Galios theory, there are many, many irreducible polynomials from which MLS sequences can be generated. The Gold codes allow a number of nearly maximal sequences to be generated from multiple base MLS sequences of the same length, but different initial conditions and generator stage combinations. The Gold codes simplify the electronics needed to synchronize transmitters and receivers and extract the communication data. A number of useful MLS generators are listed in Table 14.1 which many more generators can be found in an excellent book on the subject by Dixon [9].

Spread-spectrum sinusoids   Our main interest in MLS originates in the desire to resolve coherent multipath for narrowband signals. This is also of interest in communication systems and radar/sonar where multipath interference can seriously degrade performance. By exploiting time-synchronous averaging of received data buffers of length $2^N - 1$ the MLS sequence SNR is significantly improved and by exploiting the orthogonality of the MLS we can separate in time the multipath arrivals. For communications, one would simply synchronize with one of the arrivals while for a sonar or radar application, one would use the separated arrival times measured across the receiver array to determine the arrival angles and propagation times (if synchronized to a transmitter). Applying MLS encoding to narrowband signals is typically done by simply multiplying a sinusoid by a MLS sequence normalized so that a binary 1 is +1 and a binary 0 is −1. This type of MLS modulation is called *bi-phase modulation* for obvious reasons and is popular because it is very simple to implement electronically.

A bi-phase modulation example is seen in Figure 14.15 for a 64 Hz sinusoid sampled at 1024 Hz. The MLS sequence is generated with a [7,1] generator algorithm and the chip rate is set to 33 chips per second. It can be seen that it is not a good idea for the chip rate and the carrier frequency to be harmonically related (such as chip rates of 16, 32, 64, etc., for a 64 Hz carrier) because the phase transitions will always occur at one of a few points on the carrier. Figure 14.16 shows the power spectrum of the transmitted spread-spectrum waveform. The "spreading" of the narrowband 64 Hz signal is seen as about 33 Hz, the chip rate. At the carrier frequency plus the chip rate (97 Hz) and minus the chip rate (31 Hz) we see the nulls in the "sin x/x" or sinc(x) function. This sinc(x) function is defined by the size of the chips. If the chip rate equals the time sample rate (1024 Hz in this example), the corresponding spectrum is white noise. Figure 14.16 shows a 200 Hz sinusoid with 100 chips per

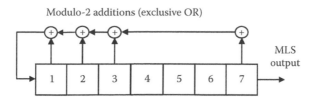

FIGURE 14.14   Block diagram of a [7,3,2,1] MLS generator showing a 7-element 1-bit tapped delay line for modulo-2 additions (XOR) of the taps 7, 3, 2, and 1 to generate a $2^7 - 1 = 127$ bit random sequence of "chips."

TABLE 14.1
Simple MLS Generators

Number of Stages	Code Length	Maximal Tap Formula
4	15	[4,1]
5	31	[5,2]
6	63	[6,1]
7	127	[7,1]
8	255	[8,4,3,2]
9	511	[9,4]
10	1023	[10,3]
11	2047	[11,1]
12	4095	[12,6,4,1]
13	8191	[13,4,3,1]
14	16,383	[14,12,2,1]
15	32,767	[15,1]
16	65,535	[16,12,3,1]
18	262,143	[18,17]
24	16,777,215	[24,7,2,1]
32	4,294,967,295	[32,22,2,1]

second. However, for any chip rate there is actually some energy of the sinusiodal carrier spread to every part of the spectrum from the sharp phase transitions in the time-domain spread-spectrum signal. Note that because we are encrypting the sinusoid's phase with a known orthogonal pseudo-random MLS, we can recover the net propagation time delay as well as the magnitude and phase of the transfer function between a transmitter and receiver. This phase encoding changes significantly what can be done in the adaptive beamforming problem of multipath coherence.

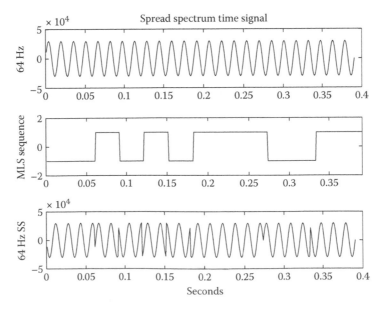

FIGURE 14.15 A 64 Hz sinusoidal carrier multiplied by a [7,1] MLS sequence with 33 chips/s (middle) to produce a bi-phase modulated sinusoid (bottom).

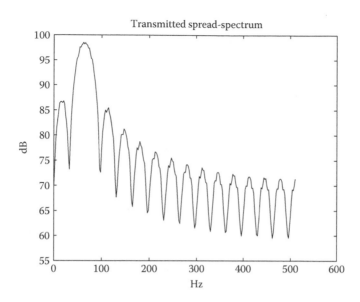

FIGURE 14.16 The transmitted spread spectrum of the bi-phase modulated sinusoid in Figure 14.15.

Resolving Coherent Multipath Clearly, one can associate the propagating wave speed and chip time length (the inverse of the chip rate) to associate a propagation distance for one chip. For example, the acoustic wave speed in air is about 350 m/s (depends on wind and temperature). If one transmitted a 500 Hz spread-spectrum sinusoid with 35 chips per second, the main sinc lobe would go from 465 through 535 Hz in the frequency domain and the length of a chip in meters would be $c/f_c = 350/35$ or 10 m/chip where f_c is the chip rate and c is the wave speed. By cross-correlating the transmitted and received waveforms one can resolve the propagation distance easily to within one chip, or 10 m if the propagation speed is known. If the distance is known then the cross-correlation yields the propagation speed $c(1 \pm 1/[f_c T_p])$ where f_c is the chip rate and T_p is the net propagation time. Resolution in either propagation distance or in wave speed increases with increasing chip rate, or in other words, propagation time resolution increases with increasing spread-spectrum bandwidth. This is in strong agreement with the CRLB estimate in Equation 13.40 for time resolution in terms of signal bandwidth.

An efficient method for calculating the cross-correlation between two signals is to compute the cross-spectrum (Section 7.2) and then computing an inverse FFT. Since the spectral leakage contains information, one is better off not using a data window (Section 6.3) or applying zero-padding to one buffer to correct for circular correlation (Section 6.4). Consider a coherent multipath example where the direct path is 33 m and the reflected path is 38 m. With a carrier sinusoid of 205 Hz and the 100 Hz chip rate (generated using a [11,1] MLS algorithm), the transmitted spectrum is seen similar to the one in Figure 14.17. The cross-correlation of the received multipath is seen in Figure 14.18 where the time lag is presented in m based on the known sound speed of 350 m/s. For this case, the length of a chip in m is $c/f_c = 350/100$ or 3.5 m. The 5 m separating the direct and reflected paths are resolved. Using a faster chip rate will allow even finer resolution in the cross-correlation. However, using a slightly different processing approach will yield an improved resolution by using the entire bandwidth regardless of chip rate. This approach exploits the coherence of the multipath signals rather than try to integrate it out.

The Channel Impulse Response The channel impulse response is found from the inverse FFT of the channel transfer function (Section 7.2) which is defined as the expected value of the cross-spectrum divided by the expected value of the autospectrum of the signal transmitted into the channel. This simple modification gives the plot in Figure 14.19 which shows the broadband spectral effects of coherent multipath. Figure 14.20 shows the inverse FFT of the transfer function which

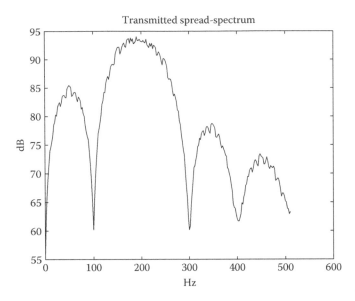

FIGURE 14.17 The transmitted spread-spectrum signal from a 200 Hz carrier with 100 chips/s from a [11,1] MLS sequence where $f_s = 1024$ Hz and 16 s of data are in the estimated power spectrum.

easily resolves the 33 and 38 m paths. The theoretical resolution for the channel impulse response is $c/f_s = 350/1024$ or about 0.342 m, where f_s is the sample rate rather than the chip rate. The transfer function measures the phase across the entire spectrum and its inverse FFT provides a much cleaner impulse response to characterize the multipath. To test this theory, let the reflected path be 34 m rather than 38 m giving a path difference of 1 m where our resolution is about 1/3 of a meter. The result of the resolution measurement for the 205 Hz sinusoid with the 100 Hz chip rate (generated by a [11,1] MLS) is seen in Figure 14.21. This performance is possible for very low chip rates as well. Figure 14.22 shows the 205 Hz sinusoid with a chip rate of 16 Hz while Figure 14.23 shows the corresponding 33 and 34 m paths resolved.

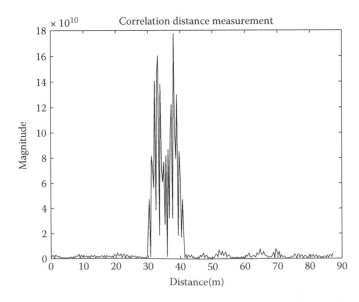

FIGURE 14.18 The cross-correlation of transmitted and received signals for the 200 Hz carrier 100 Hz chip rate showing resolution of a 33 and 38 m ray path.

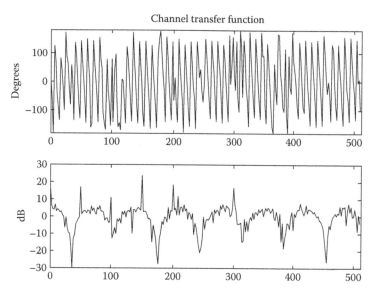

FIGURE 14.19 Channel transfer function for the 200 Hz carrier and 100 Hz chip rate clearly shows the broadband effects of coherent multipath.

In the previous examples, 512-point FFTs were used which have a corresponding buffer length of 175 m due to the 1024 Hz sample rate and the speed of sound being 350 m/s. The 100 Hz chip rate had a chip length of 3.5 m and the 16 Hz chip rate had a chip length of 21.875 m. What if the chip length were increased to about half the FFT buffer length? This way each FFT buffer would have no more than one phase transition from a chip change. As such, there would be no false correlations between phase changes in the transfer function. Consider a chip rate of 4 Hz where each chip is 87.5 m long or exactly half the FFT buffer length of 175 m. Some FFT buffers may not have a phase change if several chips in a row have the same sign. But, if there is a phase change, there is only one phase change in the FFT buffer. For the multipath in the received FFT buffer, each

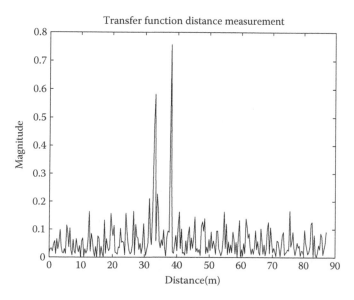

FIGURE 14.20 The channel impulse response found from the inverse FFT of the channel transfer function resolves the 33 and 38 m paths much better than the cross-correlation technique.

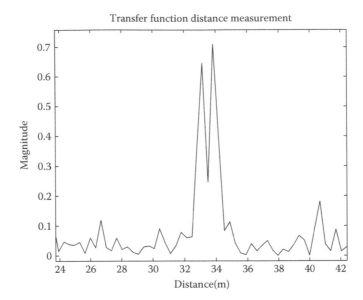

FIGURE 14.21 Resolution of a 33 and 34 m path using the 200 Hz carrier, 100 Hz chip rate spread-spectrum signal sampled at 1024 Hz.

coherent path produces a phase change at slightly different times, each of which is highly coherent with the transmitted signal. Figure 14.24 shows the multipath resolution using a chip rate of 4 Hz.

The multipath resolution demonstration can be carried to an extreme by lowering the carrier sinusoid to a mere 11 Hz (wavelength is 31.82 m) keeping the chip rate 4 Hz. Figure 14.25 shows the spread spectrum of the transmitted wave. Figure 14.26 shows the uncertainty of trying to resolve the multipath using only cross-correlation. The chips have an equivalent length of 87.5 m with a 350 m/s propagation speed. It is no wonder the multipath cannot be resolved using correlation alone. Figure 14.27

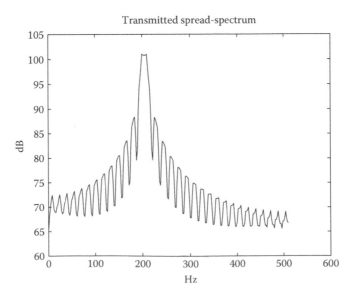

FIGURE 14.22 A 205 Hz carrier with 16 Hz chip rate transmitted spread-spectrum signal for separating paths of 33 and 34 m.

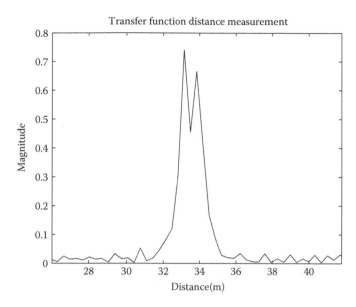

FIGURE 14.23 Resolution of the 33 and 34 m paths using the 205 Hz carrier, 16 Hz chip rate sampled at 1024 Hz.

shows the transfer function for the channel with the transmitted 11 Hz carrier and 4 Hz chip rate. The inverse FFT of the transfer function yields the channel impulse response of Figure 14.28, which still nicely resolves the multipath, even though the wavelength is over 31 times this resolution and the chip length is over 87 times the demonstrated resolution! As amazing as this is, we should be quick to point out that low SNR will seriously degrade resolution performance. However, this can be at least partially remedied through the use of time-synchronous averaging in the FFT buffers before the FFTs and inverse FFTs are calculated. For synchronous averaging in the time domain to work, the transmitted signal must be exactly periodic in a precisely known buffer length. Summing the time-domain signals

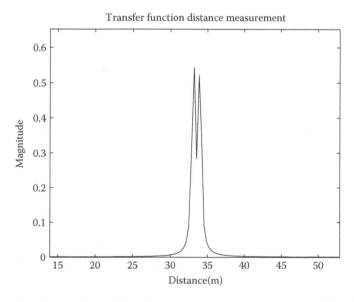

FIGURE 14.24 Multipath resolution of 33 and 34 m using a chip rate of only 4 which places only 1 phase change per FFT buffer still provide useful resolution.

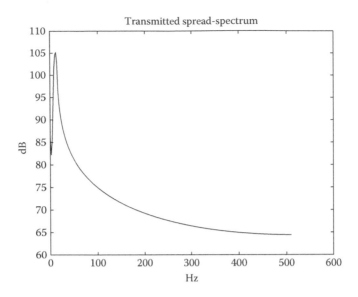

FIGURE 14.25   Transmitted spread spectrum of an 11 Hz carrier with a 4 Hz chip rate generated by an [11,1] MLS sequence.

in this buffer repeatedly will cause the background noise to average to its mean of zero while the transmitted spread-spectrum signal will coherently add.

One final note about MLS and spread-spectrum signals. For very long MLS sequences, one can divide the sequences into multiple "subsequence" blocks of consecutive bits. While not maximal, these sub-MLS blocks are nearly uncorrelated with each other and also have an autocorrelation approximating a digital Dirac delta function. The GPS system actually does this with all 24 satellite transmitters broadcasting the same MLS code which has a period on 266 days. However, each satellite transmits the code offset by about one-week relative to the others. At the receiver, multiple correlation operations run in parallel on the composite MLS sequence received which allows rapid

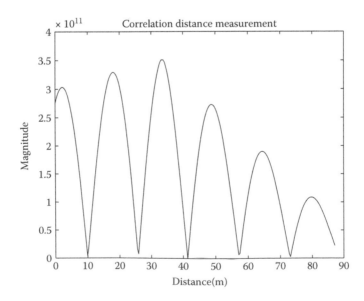

FIGURE 14.26   Cross-correlation of the 11 Hz carrier, 4 Hz chip rate transmitted and received signals incapable of resolving the different path lengths.

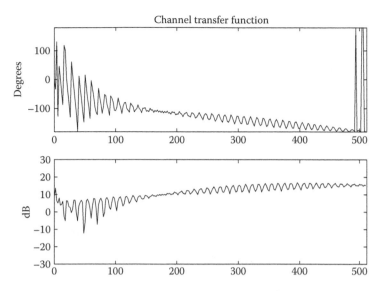

FIGURE 14.27 The transfer function of the channel measured using the 11 Hz carrier, 4 Hz chip rate.

synchronization and simplified receiver circuitry. Once one of the receiver correlation is synchronized with a GPS satellite, it can start receiving the data sent via modulo-2 addition to the MLS sequence. These data identify the satellite, its position (ephemeris), and its precise clock. When three or more satellites are in received by the GPS receiver a position on the ground and precise time is available. Combinations of MLS sequences (modulo-2 added together) are not maximal, but still can be used to rapidly synchronize communication while allowing long enough sequences for measuring large distances. Such is the case with the JPL (NASA's Jet Propulsion Laboratory) ranging codes. Other applications of spread-spectrum technology include a slightly different technique called frequency hopping where the transmitter and receiver follow a sequence of frequencies like a musical arpeggio. Frequency hopping also uses pseudo-random numbers but is somewhat different

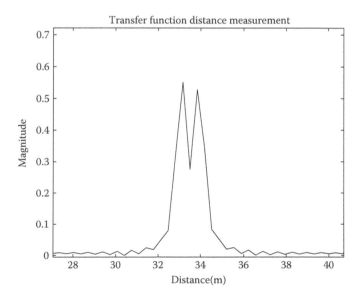

FIGURE 14.28 The 33 and 34 m paths are still resolvable using the channel impulse response for the 11 Hz carrier, 4 Hz chip rate, even though the carrier wavelength is 31.8 m and the chip length is 87.5 m.

from the direct sequence spread-spectrum technique presented here as a means to resolve coherent multipath.

Unfortunately, correlation of long MLS sequences is very intolerant of fluctuations in the propagation time which would cause the sequence to be compressed or stretched in time. This limits the length of a sequence that can be used outdoors acoustically due to wind turbulence. Also, Doppler compression/expansion must be taken into account with the correlation to maintain sequence alignment. However, this difficulty given fast processors could become a feature allowing acoustical spread-spectrum signals to measure turbulence by finding the sequence adjustments necessary to maintain synchronization.

14.4 FMCW AND SYNTHETIC APERTURE PROCESSING

FMCW signals have the advantage of concentrating the transmitted signal in a narrowband to give high SNR and sweeping the signal in frequency precisely to measure propagation times to multiple reflectors in the transmitted beam [10]. The frequency sweep, or chirp, does not have to be linear in time, but it does have to be precisely known and monic within the sweep period. Monic means each frequency in the sweep occurs at only one time during the sweep period. The received signal contains the transmitted signal delayed by the round trip propagation time to the reflector and back to the receiver. This delay can be converted to a frequency by multiplying the transmitted and received signals together and low-pass filtering in a process well-known in communications electronics called "mixing." The most simple way to interpret the low frequencies in the mixed signal is to use a linear chirp. A linear chirp makes the mixed signal low frequency(s) proportional to the range(s) of the reflector(s). We begin by defining a linear frequency chirp

$$f_c(t) = f_1 + \beta t; \quad \beta = \frac{f_2 - f_1}{T} \tag{14.34}$$

where the frequency sweeps at a constant rate from f_1 Hz to f_2 Hz in a time T seconds. To make this happen in a sinusoidal signal, the chirp has to be scaled so that the time derivative of the phase equals Equation 14.34

$$x(t) = A\cos\left(2\pi\left[f_1 + \frac{1}{2}\beta t\right]t\right)\Big|_{t=0}^{T} \tag{14.35}$$

The received signal is delayed and attenuated by the time τ it takes to propagate out a distance R to a given reflector and back to the receiver, assumed to be colocated with the transmitter.

$$y(t) = \frac{A}{2R}\cos\left(2\pi\left[f_1 + \frac{1}{2}\beta(t-\tau)\right](t-\tau)\right)\Big|_{t=0}^{T}$$

$$= \frac{A}{2R}\cos\left(2\pi f_1(t-\tau) + \pi\beta(t-\tau)^2\right)\Big|_{t=0}^{T} \tag{14.36}$$

we assume spherical spreading of the wave to the reflector and back which may not be true depending on the transmitted beamwidth, propagation losses, and the directivity of the reflector.

The mixed signal before low-pass filtering is the product of Equations 14.35 and 14.36

$$z(t) = x(t)y(t)$$

$$= \frac{A^2}{2R}\cos\left(2\pi f_1 t + \pi\beta t^2\right)\cos\left(2\pi f_1\left[t-\tau\right] + \pi\beta\left[t-\tau\right]^2\right) \tag{14.37}$$

we can simplify the reduction of this product using a well-known trigonometry product formula.

$$\cos u \cos v = \frac{1}{2}\left[\cos(u+v) + \cos(u-v)\right] \qquad (14.38)$$

Defining u and v as

$$\begin{aligned} u &= 2\pi f_1 t + \pi \beta t^2 \\ v &= 2\pi f_1 (t - \tau) + \pi \beta (t - \tau)^2 \end{aligned} \qquad (14.39)$$

The high-frequency component of the mixed signal has cosine argument

$$u + v = 2\pi\left[2f_1\right]\left(t - \frac{\tau}{2}\right) + 2\beta\pi\left(t^2 + t\tau + \frac{\tau^2}{2}\right) \qquad (14.40)$$

while the low-frequency component has argument

$$\begin{aligned} u - v &= 2\pi f_1 \tau + 2\pi \beta t \tau - \pi \beta \tau^2 \\ &= 2\pi\left[\beta\tau\right]t + \left(2\pi f_1 \tau - \pi\beta\tau^2\right) \\ &= 2\pi f_R t + \phi_R \end{aligned} \qquad (14.41)$$

It is clear to see that the high-frequency component is at least twice the starting frequency and chirps upward at twice the transmitted rate. The low-frequency component does not chirp and has a frequency proportional to twice the range to the reflector.

$$R = \frac{f_R c}{2\beta} \qquad (14.42)$$

Reflectors closer to the transmitter/receiver will have the lowest frequency and reflectors farther away will have a higher frequency. Removing the frequencies above $2f_1$ with low-pass filtering of $z(t)$ we can do an FFT to resolve the frequencies that correspond to reflectors at different ranges down the beam. Range resolution is proportional to the frequency resolution for the size FFT used.

$$dR = df\,\frac{c}{2\beta} = \frac{f_s}{N}\frac{c}{2\beta} = \frac{c}{T2\beta} \qquad (14.43)$$

Equation 14.43 shows that the range resolution is inversely proportional to the time–bandwidth product of the chirp in the FFT. Increasing the chirp rate β, integration time T, or both improves the range resolution.

Figure 14.29 shows a simple FMCW range estimation example simplified to clearly explain the algorithm. The chirp goes from 1.2 through 2.4 kHz in 2 s ($\beta = 600$ Hz/s) with a wave speed of 1500 m/s (underwater) and the signals are sampled at a rate of 10 kHz. Reflectors are positioned at 30, 150, 400, and 750 m. The top spectrogram in Figure 14.29 shows the transmitted signal spectrum versus time and the bottom plot shows the received signal spectrum versus time. The 2s buffer was sliced into 256-point buffers overlapped 50% with a Hanning window to show the changing frequencies. In the lower plot one can see the delayed and attenuated reflections from the objects in the beam. Figure 14.30 shows the mixed signal without any low-pass filtering. The high-frequency components are easily seen on the right side of the spectrogram and the low-frequency components are the vertical lines on the left side of the spectrogram.

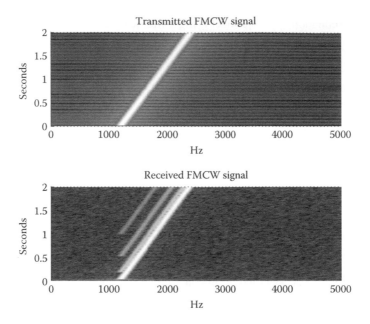

FIGURE 14.29 Transmitted (top) and received (bottom) FMCW signals with four reflectors at ranges of 30, 150, 400 and 750 m displayed a spectrograms with 256-point FFTs, 50% overlapped and a sample rate of 10 kHz and a linear chirp from 1.2 through 2.4 kHz.

Figure 14.31 shows the low-frequency part of a 16384-point FFT of the mixed signal with a frequency scale on the top of the plot and a range scale on the bottom. With 2s of data sampled at 10 kHz, we have 20,000 samples to work with, and a 16384-point FFT provides at great deal of SNR gain (42 dB) and frequency resolution (1.22 Hz). The Hanning window reduces the resolution by a factor of about 1.5 (see Figure 6.10) so $df = 1.8$ Hz. Since $c/\beta = 2.5$, we expect a range resolution for this FMCW setup of about 2.3 m. Figure 14.32 shows the result when we add a 5th reflector at 402.5 m, or 2.5 m farther than the 400 m object. If the maximum chirp time–bandwidth product is $2T\beta = 2f_1$, the

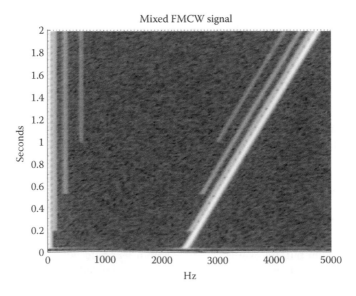

FIGURE 14.30 Mixed signal without low-pass filtering showing the stationary frequencies produced from the reflections and the fast chirping high frequencies as a result of the mixing.

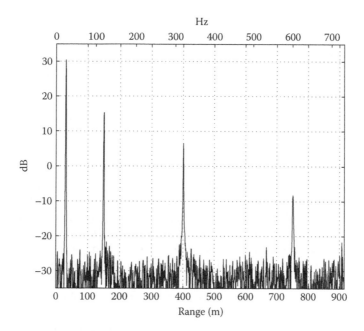

FIGURE 14.31 The low frequencies of the mixed signal evaluated with a 16384-point FFT (Hz on top scale) and the estimated range on the bottom scale.

maximum round trip range is theoretically 3 km (2 s times the wave speed) so the farthest reflector one could detect is 1.5 km and would give a frequency of $f_R = 2f_1$ in Equation 14.42.

How can this extraordinary range resolution be improved? Our example is overly simple and designed to show what the signals look like in time and frequency. Clearly we were only using a fraction of the available FFT bandwidth. The way one would build an FMCW system would be to

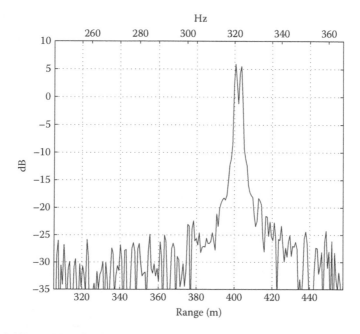

FIGURE 14.32 Adding a 5th reflector at 402.5 m which can be resolved from the 400 m reflectors allows one to see the kind of range resolution available and that can be improved with large time–bandwidth product of the FMCW chirp.

do the mixing and low-pass filtering in the analog domain using analog multipliers and filters designed for the carrier frequencies used. The resulting "baseband" signal is then Fourier transformed at audio signal bandwidths and the frequencies interpreted accordingly. For example, if we chirped from 40 through 50 kHz in 2 s, $\beta = 5000$ and the highest baseband frequency we would see from a reflector 1.5 km away would be 10 kHz, but we would have to continue the sweep beyond 2s to even see this signal. So let us say we integrate for 4 s while the transmitted linear chirp goes from 40 through 60 kHz. The mixed signal is low-pass filtered at 10 kHz and sampled at 32 kHz so there is no significant aliasing. During a 4s chirp, we capture the mixed signal but process the last 2s using a 65536-point FFT, so we have baseband frequencies from every reflector out to 1.5 km. The range resolution is an astonishing 7.5 cm (11.25 cm if a Hanning window is used)! 11 cm resolution over 1.5 km provides over 13 thousand "pixels" in the line of the transmit and receive beam.

The available range resolution far exceeds the cross-range resolution of the beam. For the example of our 40–60 kHz sonar chirp, if we used a 1 m long line-array source/sensor, the beamwidth would be $\lambda/(2L) = 1.1°$ at 40 kHz. But at 1.5 km the cross-range width is about 56 m. It would be nice if we could have a "square" pixel and assemble an image from the sonar. Recall that the transmit and receive are completely coherent and independent from position to position. We can sum two adjacent FMCW range scans from a moving transceiver platform and the beam response would be as if the array length were 2 m. The array would have to move at a slow rate of 25 cm/s, but one could combine N such scans to synthesize an array aperture $N = 200$ m long with the synthetic beam positioned at $N/2 = 100$ m behind the most recent scan and with a beamwidth of $1.1/N = 0.0055°$ giving a cross-range of 14.4 cm at 1.5 km. Now we have a pixel size of 11 cm × 14 cm which is very usable to form high-resolution images of the sea floor. The rate of motion is faster for larger line arrays or shorter maximum ranges. For radar, the scales become even more interesting since the wave speed is about 200,000 times faster. Integration times for radar FMCW scans out to 200 km are only 1.33 ms. A jet aircraft traveling 300 m/s moves only about 0.4 m during the integration time. If we use a wavelength for radar equal to the 3.75 cm wavelength for our 40 kHz sonar, the radar would be in the 8 GHz range. Suppose we chirp from 8 to 8.001 GHz in 2 ms for $\beta = 5 \times 10^8$. At 200 km the return is delayed 1.33 ms for a frequency shift of 665 kHz, a bit more than audio bandwidth but easily digitized at 2 MHz. If we capture 4 ms of data (8000 samples), zero pad to 8192 and do a full FFT we will have a frequency resolution of about 250 Hz, which translates to a range resolution of 75 m, enough for mountains and valleys, but not buildings. Bump up the chirp to go from 8 GHz to 8.1 GHz in 2 ms for $\beta = 5 \times 10^{10}$, the 200 km signal produces 66.5 MHz which we would need to digitize at 200 MHz. The resulting range resolution now is a useable 75 cm for synthetic aperture radar mapping of details on the ground.

14.5 MATLAB® EXAMPLES

5 m-scripts were used to make the 32 figures in this chapter and are listed in Table 14.2. These m-scripts are available from the book website and can be downloaded edited and run in MATLAB to allow the reader to experiment and fully understand the concepts presented. Some m-scripts were used with varying parameter settings which are commented in the code for easy trials.

The script "NULLFORM.m" demonstrate two very important concepts to adaptive beamforming. The first concept is that for multiple sources at the same frequency, the signals from the sources are combined into "snapshots" in time where each snap shot has a random phase from the sources. This is important for the covariance matrix to estimate spatial correlations without bias from coherent source interactions. The script simply computes a linear prediction for one element of the array using the other elements and block least squares. A heavily zero-padded FFT of the linear prediction coefficients yields the beam patterns displayed. The m-script "Eigcase.m" uses some very interesting features of MATLAB graphics to show what interference happens spatially when the sources are not incoherent. One triple plot shows the spatial complex components of each source across the array using primitive line plotting to simulate a 3D real–imaginary–time plot. The second triple plot

TABLE 14.2
MATLAB m-Scripts Used in Chapter 14

m-Script	Figure
NULLFORM.m	14.1–14.3
Eigcase.m	14.4–14.5
NULLEIG.m	14.6–14.13
MLSP2.m	14.15–14.28
fmcw.m	14.29–14.32

shows the magnitude and phase across the array for the three sources summed together. This is designed to illustrate why the snapshots and source need to have a randomized phase.

The script "NULLEIG.m" generates the randomized snapshots for any number of sources where each source can have a unique amplitude. A covariance matrix is calculated and the eigenvalues and eigenvectors are calculated using the function "$[V, D] = eig(R)$" a built-in MATLAB function, likely one of the oldest. We point out here that somewhere between versions 4 and 7 of MATLAB, the order of the eigenvalues on the diagonal of D were switched from largest first to largest last. Regardless of order, each eigenvalue has a corresponding eigenvector in the columns of V. This is incredibly convenient and what MATLAB was originally designed for. The source eigenvectors for the beams steered to the source directions, the square-root of the corresponding eigenvalues give the source amplitudes, and the remaining "noise" eigenvectors are all orthogonal to the sources as well as each other. So a MUSIC spectrum is just the sum of the noise eigenvectors (how easy!). The MVDR beam steered to some desired angle maintains adaptive nulls at the directions of the sources identified in MUSIC, and normalizes the main lobe of the beam to unity in the LD. The main-lobe-to-side-lobe beam gain may not appear to great, but the array nulls have nearly cancelled the identified interfering noise sources. The script goes on to synthesis beams in a sensible way by solving for the zeros of the MUSIC linear prediction polynomial and judiciously placing those zeros for nice LD beams for each source without spurious nulls by suppressing those zeros and reducing the linear prediction order. It also shows a what we call "null synthesis" by using those spurious zeros to sharpen up the main beam in the LD and evenly suppress the side lobes while still maintaining the desired nulls in the interfering source directions. This is a very useful m-script for demonstrating how one can take control of the information provided by MUSIC and the eigenvalue analysis.

Spread-spectrum signals and their use for distance measurement and multipath resolution is demonstrated in the script "MLSP2.m" for acoustic cases. Note that if one spreads the spectrum completely with a chip rate equal to the Nyquist rate, one just has white noise, likely with near Gaussian statistics. However, many situations do not give access to a wide bandwidth. Encoding the phase of a sinusoid with a precise MLS sequence allows the receiver to do the same and thus recover the propagation time as well as multipath channels. This is the principle behind the global positioning system (GPS) so common in cell phones and automobiles today. There are several demonstrations of spread-spectrum signal used for resolving multipath, even on showing an 11 Hz carrier with a 4 Hz chip rate still able to resolve two paths only 1 m apart. This works is because we are able to recover the low-level bandwidth through coherent integration of the spread-spectrum signal at the receiver.

To close the Chapter, we present another time–bandwidth signal ranging concept called frequency-modulated continuous wave, or FMCW, which is demonstrated in the imaginatively named m-script "fmcw.m." Here we use the built-in function "spectrogram()" in MATLAB to save some coding, but one could also do this by filling a matrix of small overlapped FFTs and displaying them so that the frequency changes over time are seen. We use "surf()" to display the transmitted, received, and mixed spectrograms because it provides a slightly cleaner plot than the process built

TABLE 14.3
Making a Double *X*-Axis Plot in MATLAB

```
figure(4);
pltpts = 1200;
plot(Rng(1:pltpts),20.*log10(abs(Rec(1:pltpts))),'k');
ylabel('dB');
xlabel('Range (m)');
axis([0 Rng(pltpts) -35 35]);
grid on;
%title('FMCW Range Estimation');
% let's add a second x-axis on top
ax1 = gca;
ax2 = axes('Position',get(ax1,'Position'),...
    'XAxisLocation','top',...
    'YAxisLocation','left',...
    'Color','none','XColor','k','YColor','k');
h12 = line(F(1:pltpts),20.*log10(abs(Rec(1:pltpts))),...
    'Color','k','Parent',ax2);
xlabel('Hz');
axis([0 F(pltpts) -35 35]);
```

into "spectrogram()." The is one rare graphics technique notable enough to show here, and that is the double *x*-axis in the range plot, seen in Table 14.3.

A double *x*-axis plot is very useful when there are completely different scales to show as the abscissa (x-coordinate) which is the case for mapping frequency to range. The line "ax1 = gca" get the current axis (the first plot) and saves the axis handle into "ax1." Then we make a second axis in the same position as the first with the same size, but placing the *x*-axis on the top. Keeping with primitive graphics calls, we use the "line()" function and add the parameter "'Parent',ax2" to tie the line to the 2nd axis. The axis is the frame around the plotted line with the scales, tick marks, and labels. Note how "xlabel()" at the end places "Hz" on the top axis. That is because all these annotations to an axis always default to the last axis created. The last line fixes the *y*-scale and *x*-scale so the plot exactly overlays the first plot.

14.6   SUMMARY

Chapter 14 presents a baseline set of approaches to adaptive beamforming. There are many additional algorithms in the literature, but they can be seen to fall within one of the three categories presented in Sections 14.1 through 14.3. Adaptive null-forming (Section 14.1) is simply a spatial correlation canceller where the resulting spatial FIR response for a specific narrowband frequency represents a beam pattern which places nulls in the dominant source direction(s). The technique can also work for broadband sources, but the nulls will not be as sharp and precise in angle as in the narrowband case. It is also shown that for the spatial null-forming to be precise, the phases of the sources must be independent so that the covariance matrix for the array has magnitudes and phases only associated with the spatial response and not the phases of the individual sources. The assumption of Phase independent sources is problematic for narrowband signals. To overcome this mathematical limitation, one averages many signal "snapshots" or time-domain signal buffers converted to the specific narrowband FFT bin for the covariance matrix signal input. Some presentations in the literature evaluate the inverse of the spatial FIR filter response, as if it were an IIR filter. The inverse response shows sharp peaks, rather than nulls in the beam response. This approach, sometimes called "super-resolution beamforming" or more simply "spectrum estimation" is specifically avoided here because it is physically misleading. An array's main beam lobe width and SNR gain is defined by the array aperture, number of elements, and the wavelength, which have nothing to do with the choice of mathematical display of the physical beam response with its nulls displayed as peaks.

Section 14.2 analyzes the array covariance data as an eigenvalue problem. This follows quite logically from the Vandermode matrix seen in Equation 14.33 where each source has an associated eigenvector representing the spatial phase due to the AOA and array geometry, and a source eigenvalue associated with the magnitude-squared of the source signal at the array. The phase associated with a particular narrowband source is lost in this formulation, hence, the sources must be phase-independent and the background noise spatially independent for the eigenvalue approach to work. However, given phase independent sources and spatially incoherent background noise, solving for an eigenvalue representation of the spatial covariance matrix for the array provides a set of orthonormal beam steering vectors (the eigenvectors) and associated signal powers (the eigenvalues). If one assumes the source signals of interest are stronger than the spatially incoherent background noise, the eigenvalues can be separated into a signal subspace and a noise subspace. The eigenvectors associated with the signal subspace can be used as optimal beam steering vectors to detect the signal from the corresponding source while nulling the other sources since the eigenvectors are by definition orthonormal. However, in practice this theory is encumbered by the partial coherence between narrowband sources at the same frequency, even with snapshot-averaging attempts to make the sources appear incoherent with each other.

One can safely assert that the noise eigenvectors each have nulls in their beam responses in the direction(s) of the signal source(s), plus "spurious" nulls in the directions of the estimated incoherent noise sources. Therefore, by summing the noise eigenvectors we are left with a beam steering vector which has nulls in the direction(s) of all the source(s) only. This beam response is called the minimum variance, or minimum norm beam response. Section 14.2 presents a technique for steering the array to a LD other than a source direction and maintaining the sharp nulls to the dominant source directions and unity gain in the LD. The beam is called the minimum variance distortionless response, or MVDR beam. It is very useful for suppressing interference noise in specific directions while also suppressing incoherent background noise in directions other than the LD. But, the MVDR beam becomes indeterminate when the LD is in the direction of one of the signal sources. As mentioned earlier, the associated eigenvector makes a pretty good beam steering vector for a source's LD. We also show in Section 14.2 several techniques for synthesizing beams with forced sharp nulls in the other dominant source directions while maintaining unity gain in the desired LD.

Section 14.3 presents a technique for resolving multipath which exploits multipath coherence to resolve signal arrival times at the array using broadband techniques where the transmitted signal is known. This actually covers a wide range of active sensor applications such as sonar, radar, and lidar. The transmitted spectrum is broadened by modulating a sinusoid with an MLS which is generated via a specific pseudo-random bit generator algorithm based on primitive polynomials. The MLS bit sequence has some very interesting properties in that the autocorrelation of the sequence gives a perfect delta function and the cross-correlation between different sequences is zero. These properties allow signals modulated with different MLS codes to be easily separated when they occupy the same frequency space in a signal propagation channel. Technically, this has allowed large number of communication channels to coexist in the same bandwidth. Some of the best examples are seen in wireless Internet, digital cellular phones, and in the Global Positioning System (GPS). For our purposes in wavenumber processing, the spread-spectrum technique allows coherent multipath to be resolved at each sensor in an array. Using straightforward cross-correlations of transmitted and received signals, the propagation time for each path can be resolved to within on bit, or chip, of the MLS sequence. This is useful but requires a very high bandwidth defined by a high chip rate for precise resolution. Section 14.3 also shows a technique where one measures the transfer function from transmit to receive which produces a channel impulse response with resolution equal to the sample rate. This, like the high chip rate, uses the full available bandwidth of the signals. By slowing the chip rate to the point where no more than one phase transition occurs within a transfer function FFT buffer, we can maximize the coherence between the transmitted signal and the received multipath. This results in a very "clean" channel impulse response. The spatial covariance matrix of these measured impulse responses can be either analyzed in a broadband or narrowband sense to construct beams which allow separation of the signal paths.

Section 14.4 discussed the FMCW technique and its application to synthetic aperture processing used is synthetic aperture radar (SAR) and synthetic aperture (side-scan) sonar (SAS). The technique is very well suited for mapping surfaces from an altitude above in a homogeneous medium like air or water. This way the scan out to one side of the platform in motion can be used to highlight reflectivity at the surface with very high precision in range from the side of the moving platform. To enhance resolution along the direction of travel, many scans are summed to extend the aperture such that the resolution across the beam is close to the resolution in range down the beam where this synthetic aperture beam is virtually at the midpoint of the set of successive scans. Repeating this sliding average of the scans literally allows one to trace out a 2D image with very high spatial resolution. The resolution in range down the scanning beam is inversely proportional to the time–bandwidth product of the transmitted chirp signal. As the integration time and bandwidth increase, the range resolution improves as well as SNR from the FFT integration. The limits of this process are defined by the time–bandwidth product of the system. In recent years the cost of wide band digital signal processing has become low enough for SAR and SAS systems to become more affordable for applications in surveillance, mapping, and perimeter security.

PROBLEMS

1. For a linear array of three sensors spaced 1 m apart in a medium where the plane wave speed c is 1500 m/s, derive the spatial filter polynomial analytically for a plane wave of arbitrary frequency f arriving from 60° bearing where 90° is broadside and 0° is endfire.

2. Derive the steering vector to steer a circular (16 elements with diameter 2 m) array to 30°. Evaluate in-plane the beam response for $c = 344$ m/s and 100 Hz.

3. Implement an adaptive null-forming spatial FIR filter for the linear array in problem 1 but using an LMS adaptive filter. How is the step size set to guarantee convergence?

4. Using the Vandenburg matrix to construct an "ideal" covariance matrix, show that the signal eigenvectors do not necessarily have nulls in the directions of the other sources.

5. Show that under ideal conditions (spatially independent background noise and phase independent sources) the MVDR optimal beam steering vector for looking at one of the signal sources while nulling the other sources is the eigenvector for the source in the LD.

6. Given a linear but unequally spaced array of sensors with standard deviation $\sigma_d = 0.01$ m and a mean spacing of 0.125 m, determine the bearing accuracy at 500 Hz if the plane wave speed is 1500 m/s, the wave arrives from a direction near broadside with SNR 10, and if there are 16 sensors covering an aperture of 2 m.

7. Show that the autocorrelation of the ±MLS sequence generated by [3,1] is exactly 7 for the zeroth lag and −1 for all other lags.

8. Show analytically why a chip rate of say 20 chips per second for bi-phase modulation of a sinusoid gives a sinc-function like spectral shape where the main lobe extends ±20 Hz from the sinusoid center frequency.

9. Show that for bi-phase modulation of a sinusoidal carrier, a chip rate equal to the sample rate will produce a nearly white signal regardless of the carrier frequency.

10. What is the relationship between chip rate and resolution, and MLS sequence length and maximum range estimation?

11. An FMCW scanner has a linear frequency sweep 1 s long of 10 kHz in a medium with 344 m/s wave speed. What is the best possible resolution?

REFERENCES

1. R. J. Vaccaro, Ed., The past, present, and future of underwater acoustic signal processing, UASP Technical Committee, *IEEE Signal Process Mag*, July 1998, 15(4), pp. 54–72.
2. B. D. Van Veen and K. M. Buckley, Beamforming: A versatile approach to spatial filtering, *IEEE ASSP Mag*, 1988, 5(2), pp. 4–24.
3. S. J. Orfanidis, *Optimum Signal Processing*, Chapter 6, New York, NY: McGraw-Hill, 1988.
4. W. S. Burdic, *Underwater Acoustic System Analysis*, Chapter 14, Englewood Cliffs, NJ: Prentice-Hall, 1991.
5. T. J. Shan and T. Kailath, Adaptive beamforming for coherent signals and interference, *IEEE Trans Acoust, Speech, Signal Process*, 1985, ASSP-33(3), pp. 527–536.
6. M. Wax and T. Kailath, Detection of signals by information theoretic Criteria, *IEEE Trans Acoust, Speech, Signal Process*, 1985, ASSP-33(2), pp. 387–392.
7. R. Gold, Optimal binary sequences for spread spectrum multiplexing, *IEEE Trans Inform Theory*, October 1967, 13, pp. 619–621.
8. W. H. Press, B. P. Flannery, S. A. Teukolsky, and W. T. Vetterling, *Numerical Recipes: The Art of Scientific Computing*, Sections 7.4 and 11.1, New York, NY: Cambridge, 1986.
9. R. C. Dixon, *Spread Spectrum Systems with Commercial Applications*, 3rd ed., Table 3.7, New York, NY: Wiley, 1994.
10. S. W. McCandless and C. R. Jackson, Principles of synthetic aperture radar, *SAR Marine Users Manual*, Chapter 1, NOAA, 2004.

Part V

Signal Processing Applications

One of the characteristics of an *intelligent sensor system* is its ability to be self-aware of its performance based on physical models, analysis of raw sensor data, and data fusion with other sensor information. While this capability may sound very futuristic, all of the required technologies are currently available. However, very few, if any, sensor systems at the end of the twentieth century exploit a minimal level of *sentient processing*. By the end of the twenty-first century, it is very likely that all sensors will have the capability of *self-awareness*, or at least *situational awareness*. According to most dictionaries, the definition of "sentient" is "having the five senses of sight, hearing, smell, touch, and taste." We avoid the philosophical controversy of the term sentient implying that the machine has a soul, or being aware of one's soul, or at least the reset or off button. Perhaps one could add that to the operating system but it still does not even remotely compare to a human. In general, the term "sentient" applies only to animal life (humans being the most sentient) and not machines as of the end of the twentieth century. However, there is no reason why machines with commensurate sensors of video, acoustics, vibration, and chemical analysis cannot aspire to the same definition of sentient, as long as we are not going as far as comparing the "intelligent sensor system" to a being with a conscience, feelings, and emotions. Some would argue that this too is within reach of humanity within the next few centuries, but we prefer to focus on a more straightforward physical scope of sensor intelligence. In other words, we discuss how to build the sensor intelligence of an "insect" into a real-time computing machine while extracting the sensor information into an objective form which includes statistical metrics, signal features and patterns, as well as pattern dynamics (how the patterns are changing). The information is not very useful unless we have some computer–human interface, and so it can be seen that we build the sensor part of an insect with an Ethernet port for information access. This twentieth century dream is becoming a reality in the twenty-first century.

Where appropriate we will begin our discussions on a given sensor class or processing task by describing how biological systems have solved the technical problem, which is often extremely insightful into the sensing and signal processing physics. This applies mostly to Chapter 16 in our discussions on a broad range of important sensors. In Chapter 17 we look at pattern recognition, networking, signal features, and feature dynamics and prediction. In an animal, these are mostly inherited and associative processes and it is clearly beyond our scope to try and explain how and where these biological processes occur. But biology is proceeding in this direction and already can explain much of the cellular metabolism process in very complex system diagrams. As chemistry and biology become more applied to biomimicry (biomimetic systems), intelligent signal processing will be woven throughout these system models and their robotic counterparts. For networking of sensor information systems, we have already seen the rapid proliferation of the Internet and movement of all sorts of sensor systems to this backbone of communications. Commercial hard disks cost less than $0.10/GB (1 TB disks for less than US $100 are widely available in the year 2010), and the amount of data moved and stored over the Internet is astonishing and growing at an accelerating rate. Search engines help us organize and find our documents, but something far more sophisticated is needed to manage the data mountains from intelligent sensor networks. We complete this edition of the book with a detailed discussion on how sensor networks should organize their data to facilitate efficient data mining and intelligent software agents for automated recognition of situations of interest.

15 Noise Reduction Techniques

In order to make an algorithm which can extract signal features from raw sensor data, match those features with known patterns, and then provide predictive capability of the pattern motion, we must have a solid foundation of sensor physics and electronic noise. Given this foundation, it is possible for the sensor algorithms to identify signal patterns that represent normal operation as well as patterns that indicate sensor failure. The ability to provide a signal quality, or sensor reliability, statistical metric along with the raw sensor data is extremely valuable to all subsequent information processing. The sensor system is broken down into the electronic interface, the physical sensor, and the environment.

Each of these can be modeled physically and can provide a baseline for the expected signal background noise, SNR, and ultimately the signal feature confidence. Our discussion on electronic noise will obviously apply to all types of sensor systems.

We present some very useful techniques for the cancellation of noise coherent with an available reference signal or when the noise can be modeled in a predictive way. This can also be seen as adaptive signal separation, where an undesirable signal can be removed using adaptive filtering. Adaptive noise cancellation is presented in two forms: electronic noise cancellation and active noise cancellation. The distinction is very important. For active noise cancellation, one physically cancels the unwanted signal waves in the physical medium using an actuator, rather than electrically "on the wire" after the signal has been acquired. But where does the energy go? This is a great physical question. When active cancellation occurs, the energy conversion efficiency is reduced, usually by the active system having an effect on the coupling or radiation impedance for the noise. *One can't cancel energy but one can affect the efficiency of energy conversion or propagation.* Active noise cancellation has wider uses beyond sensor systems and can provide a true physical improvement to sensor dynamic range and SNR if the unwanted noise waveform is much stronger than the desired signal waveform. The use of adaptive and active noise cancellation techniques gives an intelligent sensor system the capability to counter poor signal measuring environments. However, one must also model how the SNR and signal quality may be affected by noise cancellation techniques for these countermeasures to be truly useful to an intelligent sensor system. Active noise control is mostly an academic curiosity at the time of the writing of this second edition, but its commercial limitations are primarily a cost and maintenance issue, not a feasibility concern. There are a few new techniques presented along with some very practical guidelines for active noise control.

15.1 ELECTRONIC NOISE

There are four main types of electronic noise to be considered along with signal shielding and grounding issues when dealing with the analog electronics of sensor systems. A well-designed sensor system will minimize all forms of noise including ground loops and cross talk from poor signal shielding. But, an intelligent sensor system will also be able to identify these defects and provide useful information to the user to optimize installation and maintenance. As will be seen below, the noise patterns are straightforwardly identified from a basic understanding of the underlying physics. The four basic types of noise are *thermal* noise, *shot* noise, *contact* noise, and *popcorn* noise. They exist at some level in all electronic circuits and sensors. Another form of noise is interference from other signals due to inadequate shielding and/or ground loops. Ground loops can occur when a signal ground is present in more than one location giving rise to a ground potential difference. One common example of this is when the data acquisition system is grounded and at the sensor casing is also grounded at a different location at the end of a long cable. The ground potential can be significantly different at these two locations causing small currents to flow between the ground points. These currents can cause interference with the signal from the sensor. Usually, ground loop interference is seen at the ac power line frequencies because these voltages are quite high and return to ground. The ac mains are also grounded at some point near the distribution location, and since the earth does not have zero impedance, small ground currents flow creating what we know as signal "hum" or "buzz" or waviness in scanned images. Ground loops can be identified in a number of ways and eliminated by proper shielding and floating of sensor signal grounds.

Recall that all solids can be described as a lattice of molecules or atoms, somewhat like a volume of various-sized spheres representing the molecules. The atoms are electrically charged where the nucleus is positive and the electron shells are negative. These "shells" have fascinating fuzzy orbitals with shapes like 3D beam patterns, depending on how many electrons fill the shell. Metals have nearly empty orbitals while nonmetals are just a few electrons shy of a full orbital, placing them on the far right of the periodic table. "Metaloids" such as boron, silicon, germanium, arsenic, and so on lie in between the metals and nonmetals, making them suitable for semiconductors. In a semiconductor, a group 4 metalloid (four electrons in the outer shell) such as silicon or germanium in its purest form, is doped with a small amount of a group 5 element such as arsenic or phosphorus, which acts as an electron donor in the lattice leaving an extra electron in the silicon electron shell. This makes an n-type semiconductor. Doping the silicon with a group 3 element like boron or aluminum pulls an electron away from the silicon (an accepter rather than donor) leaving a "hole" in the outer shell of the silicon which makes a p-type semiconductor. A semiconductor can act as a metal (conductor) or insulator (nonmetal) depending on the external electrical field applied. Both p- and n-type semiconductors are fairly conductive since they have electrons or holes to act as charge carriers. But if you make a crystal lattice that changes from p-type to n-type, or vice versa, the region between the two types of semiconductors naturally becomes depleted of charge carriers by charge diffusion between the two regions. Apply a positive voltage to the p-type side of the junction and a negative voltage to the n-type side and the depletion zone collapses and the material conducts as a *forward-biased diode*. Apply a positive voltage to the n-type end and a negative voltage to the p-type end and the depletion region actually widens, reducing conductivity in what we call a *reverse-biased diode*. For a type of diode called a zener diode, there is a reversed-bias voltage called the breakdown voltage where there is sufficient electric field for electron–hole pairs to penetrate the depletion zone and cause the diode to conduct but with a fixed voltage. There are also tunneling diodes and avalanche diodes which operate similar to zener diodes but are based on different material physics and provide faster or more linear responses. The basic operation of a diode is a 2-terminal device where current can flow only one way (except for the zener, tunneling, and avalanche diodes).

Put two diodes together by doping a crystal in a p–n–p or n–p–n arrangement and you have a transistor, which comes from "transfer resistance" which is how they are used in circuits. Apply a positive voltage to the p-terminal (the base) of a n-p-n transistor and the deletion zone collapses and the two n-terminals (collector and emitter) conduct. The same is true for a negative voltage applied to the n-terminal of a p–n–p transistor. The middle terminal of these "bipolar" transistors is called the base and it acts like the control of a valve. Only a very small bias current into the base (n-p-n) is needed to drop the resistance across the depletion zones allowing a relatively huge current to flow through the other two terminals, called the collector and emitter. There are also field-effect transistors (FETs) with very high base input impedances, metal oxide semiconductor FET (MOSFET), complementary metal oxide semiconductors (CMOS), and other transistors. We are avoiding a huge amount of material science details, but this is the very basic operation of transistors and diodes. What is important to know from a noise point of view is that semiconductor materials respond to external fields including random electrons freed by thermal energy, electromagnetic radiation, even cosmic radiation. These external sources of "noise" can momentarily switch transistors and diodes, generate random voltages, and even damage semiconductor-based electronics.

Thermal Noise results from the agitation of electrons due to temperature in conductors and resistors. The physical aspects of thermal noise were first reported by Johnson [1] and are often referred to as "Johnson noise." But it was Nyquist [2] who presented a mathematical model for the RMS noise level. A little background in electronic material science is useful in describing how resistors and conductors can generate noise waveforms. Atoms bonded into molecules also have spatial charges. The negative-charged electrons of one atom or molecule are attracted to the positive-charged nucleus of neighboring atoms or molecules, yet repelled by their electrons. One can visualize the "spheres" (atoms or molecules) connected together by virtual springs such that the motion of one atom or molecule affects the others. Because the spheres have thermal energy, they all vibrate randomly in time and with respect to each other. A solid crystalline material will generally have its atoms and/or molecules arranged in a nearly regular pattern, or lattice, the exceptions being either impurities or crystal defects called dislocations. These imperfections in the lattice cause "grain boundaries" in the material and also cause blockages to the natural lattice-formed pathways in the lattice for electrons to move. As two spheres move closer together, the repulsion increases the potential energy of the electrons. If the material is a conductor with unfilled electron shells (such as the metals of the periodic chart, i.e., aluminum Al^{3+}), which tend to lose electrons readily in reactions. Metals tend to be good heat conductors, have a surface luster, and are ductile (bend easily without fracture). The unfilled electron shells allow a higher energy electron enter and be easily transported across the lattice, but with an occasional collision with an impurity, dislocation, or vibrating atom. These collisions cause electrical resistance which varies with material type and geometry, temperature, and imperfections from dislocations and impurities. At absolute zero, a pure metal theoretically forms a perfect lattice with no dislocations or vibration, and therefore should have zero electrical resistance as well as zero thermal noise. Insulators, on the other hand, tend to have filled electron shells, are brittle and often amorphous molecular structures, and are often poor heat conductors. This binds the electrons to their respective atoms or molecules and eliminates the lattice pathways for electron travel. This is why oxides on conductor surfaces increase electrical resistance as well as cause other signal problems.

Nyquist's formula for thermal noise in RMS volts is given in the following equation:

$$V_t = \sqrt{4kTBR} \tag{15.1}$$

The following parameters in Equation 15.1 need to be defined:

k is Boltzmann's constant (1.38×10^{-23} J/K), T the absolute temperature (degrees Kelvin, or K), B the bandwidth of interest (Hz), and R the resistance in ohms (Ω).

The RMS spectral density of the thermal noise in volts per $\sqrt{Hz}$ is

$$\sigma_t = \sqrt{4kTR} \tag{15.2}$$

which can be used to model the spectrum of thermal noise in electronic circuits.

As an example of the importance of thermal noise considerations, at room temperature (290 K) an accelerometer device 1 MΩ resistance will produce about 10 µV RMS thermal noise over a bandwidth of 10 kHz. If the accelerometer's sensitivity is a typical 10 mV/g (1 g is 9.81 m/s^2), the broadband acceleration noise "floor" is only 0.001 g's. This is not acceptable if one needs to measure broadband vibrations in the 100 µg range. One way to reduce thermal noise is to operate the circuitry at very low temperatures. Another more practical approach is to minimize the resistance in the sensor circuit. The structure of thermal noise is a Gaussian random process making the noise spectrally white. Thus, the noise is described using a spectral density (Section 7.1), which is independent of spectral resolution. 10 µV RMS over 10 kHz bandwidth (0.01 pV^2/Hz) power density corresponds to 0.1 µV/$\sqrt{Hz}$ RMS noise density, which is physically reasonably small 100 ng/$\sqrt{Hz}$.

For detection of a narrowband acceleration in a background noise of 1000 μg RMS using a 1024-point FFT (spectral resolution of 10 Hz per bin), one can simply consider that the power spectrum SNR gain is about 30 dB, thus making a narrowband acceleration greater than about 1 μg RMS detectable just above the noise floor. Note that thermal noise can also be generated mechanically from atomic and molecular motion but, instead of RMS volts from electrical resistance to the Boltzmann energy, one gets RMS forces using mechanical resistance.

These background noise considerations are important because they define the best possible SNR for the sensor system. Of course, the environmental noise can be a greater source of noise. If this is the case, one can save money on electronics to achieve the same overall performance. To easily model a thermal noise spectrum, simply use the RMS spectral density in Equation 15.2 times the width of an FFT bin in Hz as the standard deviation of a zero-mean Gaussian (ZMG) random process for the bin (the real and imaginary parts of the bin each have 1/2 the variance of the magnitude-squared of the bin).

Shot Noise is also a spectrally white Gaussian noise process, but it is the result of dc currents across potential barriers [3], such as in transistors (and more so in vacuum tubes due to the heat). The broadband noise results from the random emission of electrons from the base of a transistor (or from the cathode of a vacuum tube). More specifically in a transistor, it comes from the random diffusion of carriers through the base and the random recombination of electron–hole pairs. A model for shot noise developed by Schottky is well-known to be

$$I_{sh} = \sqrt{2qI_{dc}B} \tag{15.3}$$

where I_{dc} is the average direct current across the potential barrier, B is the bandwidth in Hz, and q is the charge of an electron (1.6×10^{-19} C). The shot noise current can be minimized for a given bandwidth by either minimizing the dc current or the number of potential barriers (transistors or vacuum tubes) in a circuit. To model the power spectrum of the shot noise, multiply Equation 15.3 times the circuit resistance and divide by the square root of total bandwidth in Hz to get the standard deviation of the noise voltage. For a power spectrum, each bin of an N-point FFT is modeled as an independent complex ZMG random process with $1/N$ of the noise variance. For a real noise time-domain waveform, the real and imaginary parts of the bin are also independent, each having $1/2N$ of the total noise variance. Figure 15.1 shows a ZMG noise waveform as a function of time and frequency.

Contact Noise is widely referred to as "one-over-f noise" because of the typical inverse frequency shape to the spectral density. Contact noise is directly proportional to the dc current across any contact in the circuit, such as solder joints, switches, wire connectors, and internal component materials. This is the result of fluctuating conductivity due to imperfect contact between any two materials. While connections like solder joints are considered quite robust, after years of vibration, heat cycle stress, and even corrosion, these robust connections can break down leading to contact noise. This has become particularly worrisome of new "environmental safe" solder which replaced the tin–lead alloy traditionally used. Like shot noise, it varies with the amount of dc current in the circuit. But unlike shot noise, its spectral density is not constant over frequency and the level varies directly with dc current rather than the square root of dc current as seen in the model in Equation 15.3

$$I_c = KI_{dc}\sqrt{\frac{B}{f}} \tag{15.4}$$

The constant K in Equation 15.3 varies with the type of material and the geometry of the material. Clearly, this type of noise is very important to minimize in low frequency or dc sensor circuits. If we write the contact noise current as an equivalent voltage by multiplying by the circuit resistance

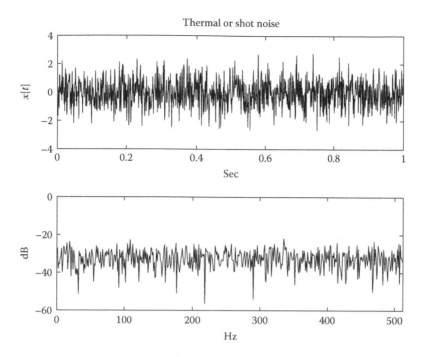

FIGURE 15.1 Both thermal and shot noise are zero-mean Gaussian and spectrally white.

R, and divide by the square-root of bandwidth, we get the RMS spectral density in units of RMS volts per $\sqrt{\text{Hz}}$

$$\sigma_c(f) = RK\,I_{dc}\sqrt{\frac{1}{f}} \tag{15.5}$$

The RMS spectral density in Equation 15.5 is clearly a function of frequency and the variance represents the classic "1/f" noise often seen in real sensor systems. Contact noise is best modeled only in the frequency domain using a power spectrum. The time domain waveform can then be obtained via inverse Fourier transform. To get the power spectrum using an N-point FFT, first square Equation 15.5 and then multiply by f_s/N, the bin width in Hertz to get the total variance of the complex FFT bin magnitude squared. The imaginary and real parts of this bin each are modeled with a ZMG process with 1/2 of the total bin variance for a real contact noise time-domain waveform. To ensure that the time-domain waveform is real, apply Hilbert transform symmetry to bins 0 through $N/2-1$ for the real and imaginary parts of each bin. In other words, the random number in the real part of bin k equals the real part of the random number in bin $N-k$, while the imaginary part of bin k equals the minus the imaginary part of bin $N-k$ for $0 \le k < N/2$. The inverse FFT of this spectrum will yield a real-time noise waveform dominated by the low-frequency 1/f contact noise as seen in Figure 15.2.

Popcorn Noise is also known as burst noise, and gets its name from the sound heard when it is amplified and fed into a loudspeaker. Popcorn noise arises from manufacturing defects in semiconductors where impurities interfere with the normal operation. It is not clear whether popcorn noise is a constant characteristic of a semiconductor or if it can appear over time, perhaps from internal corrosion, damage, or wear-out of a semiconductor. Popcorn noise in the time domain looks like a temporary offset shift in the noise waveform. The amount of shift is fixed since it is due to the band gap potential energy of a semiconductor junction. The length of time the offset is shifted is random as is the frequency of the offset "bursts." The burst itself is a current offset, so the effect of popcorn

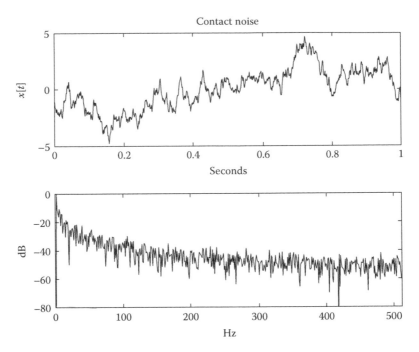

FIGURE 15.2 Contact noise is dominated by low frequencies and has a characteristic "$1/f$" shape in the frequency domain (note the log base-10 scale for dB on the spectral plot).

noise can be minimized by keeping circuit impedances low. In the frequency domain, popcorn noise can have a spectral shape which averages to $1/f^n$, where n is usually 2. Figure 15.3 depicts popcorn noise and the corresponding power spectrum. Popcorn noise is most readily identified in the time domain.

Electronic Design Considerations to reduce intrinsic noise can be summarized into the following practical guidelines for any sensor system:

1. Physically locate the sensor and choose the sensor type to maximize SNR.
2. Minimize circuit resistance to reduce thermal noise.
3. Minimize dc currents and circuit resistance to reduce shot and contact noise.
4. Choose high-quality electronic components and test to verify low noise.
5. Protect components from heat and apply cooling if practical.

Following the above general guidelines, it can be seen that low-noise sensor circuits will generally have low impedance, strong ac signal currents, and minimal dc currents. Low dc currents are obviously not possible when the sensor is very low bandwidth, such as temperature, barometric pressure, and so on. However, for dc-like sensor signals, the noise can be greatly reduced with signal processing such as by simple averaging (integration) of the data or filtering out-of-band noise. Depending on the desired sensor information, one would optimize the sensor circuit design as well as the data acquisition to enhance the SNR. For example, in meteorology, a precise pressure gauge (say $\pm 2''$ H_2O full scale) could measure barometric pressure relative to a sealed chamber as well as pressure fluctuations due to turbulence using two sensor circuits and data acquisition channels. The SNR of the dynamic signals is enhanced by filtering out the static atmospheric pressure (pressure gauge). If one wants the absolute barometric pressure, the turbulence is not important and thus can be averaged out of the measurement. The SNR of the absolute pressure measurement is enhanced by low-pass filtering a dc-coupled circuit while the turbulence SNR is enhanced by low-pass filtering a separately sampled channel.

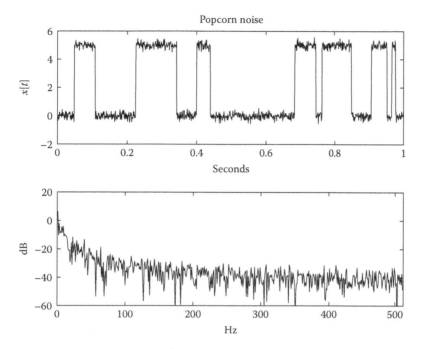

FIGURE 15.3 Popcorn noise (also known as burst noise) is due to a semiconductor defect where the plateau offsets (top plot) are always the same but the duration and spacing are random.

Controlling calibration gain in dc circuits is particularly a challenge because small voltage or current offsets can drift over time, with supply voltage (say from batteries), or with environmental temperature. One power-hungry technique for controlling temperature drift is to locate the entire sensor circuit in a thermal mass which is heated to a constantly controlled temperature above the environmental temperature. Heating is more cost effective than cooling, but either or both can be done so long as the temperature can be controlled. Another approach is to purposely include a temperature or supply voltage sensitive part of the circuit and then compensate the sensor output bias due to temperature in the signal processing software. Such temperature compensation, which is often nonlinear over a wide range of environments, is very easily done in the digital domain so long as the compensation is repeatable. Finally, one should consider the slew rate and bandwidth of the electronic circuitry amplifiers. For example, if the gain-bandwidth product of a particular operational amplifier is 1 million and the required bandwidth is 20 kHz, the maximum theoretical stable gain for that bandwidth is only 50 from the limitation of the device gain-bandwidth product. If a gain of 160 (+44 dBV) is required, two amplifier stages each with a gain of 12.65 (+22 dBV) should be used. The system electronics bandwidth is now about 80 kHz and exceeds the required 20 kHz. This means that the amplifier circuits will be nearly transparent to the sensor signals over the frequency range of interest. In high gain ac circuits, dc offsets may have to be managed at each stage to prevent saturation.

Signal Shielding can improve SNR by reducing cross talk between signal channels as well as suppressing environmental electric fields. A basic understanding of field theory suggests that wrapping the signal wire in a coaxial conductor will completely suppress all environmental electric fields [4]. However, it is actually more complicated due to the wave impedance of the interfering signal and the shield impedance, which has a small amount of resistance. Recall that the characteristic impedance of a plane electromagnetic wave in air is

$$Z_W = Z_0 = \sqrt{\frac{\mu_0}{\varepsilon_0}} = 377\Omega \qquad (15.6)$$

where $\mu_0 = 4\pi \times 10^{-7}$ H/m (H is Henrys) and $\varepsilon_0 = 8.85 \times 10^{-12}$ F/m (F is Farads) are the characteristic permeability and dielectric constants for free space (a vacuum or dry air). This makes air a pretty good insulator given that copper is about 17 nΩ. The impedance of a conductive material ($\sigma \gg j\omega\varepsilon$, where σ is the conductivity in mhos) is

$$Z_s = \sqrt{\frac{2\pi f \mu}{2\sigma}}\,(1+j) \tag{15.7}$$

The loss due to plane wave reflection at the conducting shield surface is

$$R = 20\log\frac{|Z_W|}{4|Z_s|}$$

$$= 168 - 10\log\left(\frac{\mu_r f}{\sigma_r}\right) dB \tag{15.8}$$

where $\mu_r = \mu/\mu_0$ and $\sigma_r = \sigma/\sigma_0$ are the relative permeability and conductivity, respectively, and $\sigma_0 = 5.82 \times 10^7$ mhos is the conductivity for copper. Both μ_r and σ_r are unity for copper in Equation 15.8. At 100 Hz, there is about 148 dB reflection loss to a plane wave impinging on a copper shield. But, at 1 MHz, this drops to 108 dB which is still pretty good shielding. But, a plane wave in free space occurs approximately only after several wavelengths from a point source, a hundred or more wavelengths to be precisely a plane wave (recall Section 7.4.1). Since the wave speed is 300×10^6 m/s, a hundred wavelength at 60 Hz is 5 million meters. Since the radius of the earth is 6.3 million meters, we can safely say that power line fields (50–60 Hz) from the other side of the planet are plane waves, but any power line electromagnetic fields within about 10 km of our circuit are likely not plane waves. However, radio and television signals around 100 MHz can be considered plane waves if the source is 100 m away or more.

Nonplane wave shielding is much more difficult mainly due to currents induced in the signal wire from magnetic field and the asymmetry of the field around the signal wire. The electrical nearfield ($r < \lambda/2\pi$) wave impedance can be approximated by

$$Z_W|_E \approx \frac{1}{2\pi f \varepsilon_0 r} \tag{15.9}$$

where r is the distance in meters from our signal wire to the field source. Applying Equation 15.8 to our nearfield wave impedance, one obtains the approximate electrical refection loss for the shield.

$$R_E = 321 + 10\log\left(\frac{\sigma_r}{\mu_r}\right) - 10\log\left(f^3 r^2\right) dB \tag{15.10}$$

Equation 15.10 tells us that electrical shielding is fairly strong even at close distances. But, one must also consider that the field strength is also increasing as the signal wire is moved closer to the source. Magnetic shielding is not nearly as strong in the nearfield. The magnetic nearfield wave impedance ($r < \lambda/2\pi$) is approximately

$$|Z_W|_H \approx 2\pi f \mu_0 r \tag{15.11}$$

in free space. Again, applying Equation 15.8 gives the approximate nearfield magnetic reflection loss

$$R_H = 14.6 + 10\log\left(\frac{\sigma_r}{\mu_r}\right) + 10\log\left(f r^2\right) dB \tag{15.12}$$

However, for thin shields multiple reflections of the magnetic wave within the shield must be accounted by adding a correction factor which should only be used for thin shields in the nearfield.

$$R_{\mathrm{H}} = 14.6 + 10\log\left(\frac{\sigma_r}{\mu_r}\right) + 10\log\left(f\,r^2\right) + 20\log\left(1 - e^{-2t/\delta}\right)\mathrm{dB} \tag{15.13}$$

Note that for really small thickness t, the correction may cause R_{H} to be negative in Equation 15.13 which is erroneous. For the cases where R_{H} is either negative or near zero in both Equations 15.12 and 15.13, it is reasonable to use $R_{\mathrm{H}} = 0$ dB (no magnetic field shielding) as a further approximation. The skin depth, δ, in meters is defined as

$$\delta = \sqrt{\frac{2}{2\pi f\,\mu\sigma}} \tag{15.14}$$

and describes the exponential decay in amplitude as the field penetrates the conductor. The skin-depth effect also results in an absorption loss to the wave given in

$$A = 20\log(e^{-t/\delta}) = 8.7\left(\frac{t}{\delta}\right)\,\mathrm{dB} \tag{15.15}$$

Equation 15.15 expresses that as a rule of thumb* one gets about 9 dB absorption loss per skin-depth thickness shield. For copper at about 100 Hz, the skin depth is a little over 1 cm, but even so, in the nearfield this is the dominant effect of shielding at low frequencies when close to the source. This is also why for extremely high currents, like those at the output of a power plant, the conductors look like a series of copper plates about 1 cm thick separated some to allow cooling, instead of a huge solid copper wire. The electricity would not flow in the center of the conductor.

Some additional shielding is possible by surrounding the signal wire with a shroud of low-reluctance magnetic material. Magnetic materials, sometimes known as "mu-metals, ferrite beads, donuts, life-savers, and so on," are sometimes seen as a loop around a cable leaving or entering a box. The mu-metal tends to attract magnetic fields conducted along the wire, and then dissipate the energy through the resistance of the currents induced into the loop. They are most effective however, for high frequencies (i.e., MHz) only. Magnetic shielding at low frequencies is extremely difficult. For all shielding of components in boxes where one must have access holes, it is better to have a lot of small holes than one big hole of the same total area. This is because the leakage of the shield is determined by the maximum dimension of the biggest opening rather than the total area of all the openings. In general, for low-frequency signal wires one must always assume that some small currents will be induced by nearby sources even when shielding is used. To minimize this problem, one should make the sensor signal cable circuit as low impedance as practical and make the signal current as strong as possible. This is commonly done in audio engineering where any cable runs longer than a few meters are converted from "unbalanced" 2-wire high-impedance (50 kΩ) to 3-wire "balanced" low-impedance (150 Ω) signals where there is a positive and negative wire for the signal and a separate ground shield. Both the positive and negative signal wires get magnetic-induced interference, but are canceled out when the signal is converted back to a high-impedance 2-wire (signal plus ground) signal using a differential amplifier.

Ground Loops occur when more than one common signal ground is present in the sensor system. This is the most common sensor signal interference problem. Grounding problems typically are

---

* The "rule of thumb" dates back to medieval times and specifies the maximum thickness stick one may use to beat one's spouse.

seen as the power line frequency (60 Hz in North America, 50 Hz in Europe, etc.) and its harmonics. Note that if your sensor is detecting acoustic or vibration waves and you see multiples of *twice* the power line frequency in the power spectrum (i.e., 120, 240, 360 Hz, etc., in North America), you may actually be sensing the magnetic thermal expansion/hysteresis noise from a transformer or motor, which is a real vibration/acoustic wave and not a ground loop. Disconnecting the mechanical mount of the vibration sensor or isolating the sensor will verify this. However, ground loops can be very tricky to solve because the electromagnetic fields can have directivities and change with the power line currents.

Let us take a brief detour and discuss the electrical power distribution typically found at the consumer end of the electric power grid. The typical service available at the distribution point has three phases of ac power and a local ground or neutral wire. The phases are all at the same frequency but shifted so that each line is 120° apart. This gives a constant power flow from the generating plant to the industrial consumer who needs three-phase motors to supply the mechanical power to run the factory. The RMS voltage from one of the three phases to the neutral is typically the countries standard voltage for small appliances and lights, 120 V RMS in North America and 230 V RMS in Europe and most of Asia, and so on. This standard voltage can be delivered as part of a *three-phase* system or as part of a *split phase* system. This distinction is important for ground loop solutions. The split phase supply takes one of the three phases from the power utility distribution lines and drops the voltage down with a single split-tap transformer giving two "legs" and a neutral from the center tap of the output side of the transformer. The customer can have a high-voltage supply that is twice the standard line voltage (460 V RMS in Europe and 240 V RMS in North America), or each leg can supply separate banks of the standard voltage. The two legs in a split-phase system are actually 180° out-of-phase to each other with respect to the neutral. That is a good thing if both legs draw the same current which cancels in the system neutral line voltage. A local earth ground is separate from the neutral within the user's facility but the neutral is usually grounded at the transformer, often less than a few hundred feet away. The ground wires are for safety. If the loads of the two legs are not in balance, there is actually a current that flows into the ground at the point where the neutral is grounded. If there are multiple split-phase circuits in the area, there can be ground currents to flowing between them through the earth. So if your sensor shield is in contact with the earth ground, or anything that can complete a circuit with the earth ground, there can be a small current that is picked up and returned to earth ground via the shield and ground connection on the data acquisition electronics. It may only be present under some unbalanced load conditions or when the earth changes conductivity from moisture. This is why a ground loop can be so difficult to isolate.

Why are the ground and neutral wires separate within the user's facility, but tied together at the utility transformer? This is for safety and to allow smart circuit breakers to monitor the current in the leg and neutral of a particular power circuit. A "ground fault" occurs when the leg and the neutral are not balanced in current, meaning all those electrons are making their way back to earth ground through some other path—*like your body*, instead of the neutral wire. If a short circuit occurs, the local ground wire provide a safe path to earth ground and a ground fault interrupter circuit breaker (GFCI) opens the circuit quickly to prevent a fire or electrocution. Most power line plugs have a 3-wire connection with one power leg, a neutral, and a ground wire pin. If it is a high-voltage circuit (240 V RMS or 460 V RMS), there are typically two legs, a neutral, and a ground wire. Split phase systems are generally desirable since there a little more power available at high voltage unless the user requires very high-power service.

The three-phase power system is a little more complex and common to commercial businesses that use a lot of power, especially high horsepower motors. The power utilities make three-phase power because it is the lowest number of phases where they can have a continuous 100% power transfer from the generator to the customer. This also makes it the most suitable for driving large high horsepower motors. Many consumers without these needs get a two-phase power feed which is just two of the three available phases. We call the lines "phases" and not "legs" as with the split-phase system. The voltage from a single-phase line to neutral is still the standard line voltage, but the high-voltage option is slightly less in voltage because the phases are 120° rather than 180° apart (208 V RMS or 398 V RMS). This is only about 75% of the power of the split-phase system, but most if not all customers will not notice that their oven takes a few minutes longer to heat their dinner. The power company will alternate each customer's pair of phases from the three-phase system to try and balance the load. If the total load happens to be balanced, the net current in the neutral is zero and the distribution system is at peak efficiency. When the load is unbalanced, currents flow in the neutral and back to earth ground from the unbalanced loads. Back at the generating plant, currents can flow back out of the ground to make up the imbalance. Since copper is a better conductor than the earth, it would appear to be more efficient to balance the loads all the time in the distribution system. This could be done in the future using high-speed inverters and switching ac power supplies, but at a greater equipment cost than the present-day simple transformer.

The big problems occur if there are any connection faults with the neutral wire when there is a load imbalance. Without a solid reliable neutral connection, the voltages of the three phases will drift around according to the loads, floating the neutral voltage between them. This can cause very big differences in the line voltages of the phases and even damage equipment in the user's facility from high or low voltages. When high currents flow in the neutral lines there are usually significant ground currents in the earth, which change as the loads change. It is almost impossible to troubleshoot which circuit is causing the ground loop under these conditions. Sometimes the signal interference is reduced by disconnecting the ground pin on a power plug (lifting the ground), but this is usually a bad idea that can lead to mild electric shocks, or worse if there is a real problem in the neutral connection somewhere.

So what is a ground loop from a sensor signal prospective? Recall that each power outlet has three wires (for the standard voltage) of the standard line voltage (three-phase or split phase), a neutral, and a safety ground. All those safety grounds make their way back to the power distribution panel and utility power meter where they connect with the neutral to a copper ground rod hammered into the earth. This is the common ground point. However, a sensor may have a metal case also with an electrical pathway to the earth and back through the data acquisition system to earth. Back at the utility's transformer there is a safety ground rod tied to the neutral as well. The earth is not 0 Ω, so small ac currents always flow between these ground points. *The circuit between two points on an insulated wire having contact with the earth at different locations is called a ground loop.* The difference in earth potential could be due to an imbalance in the three-phase system, changes in conductivity in the soil with currents flowing through it, or a ground fault somewhere else insufficiently strong enough to open a circuit breaker, but significant enough to show up in your sensor signal spectrum on an otherwise nice day.

Direct current-powered circuits (such as those that use batteries) can also suffer from power main ground loops. This is because our power distribution system generates electricity relative to an earth ground which is local to each generating station. The distribution system is grounded at many places, including the main circuit breaker box in any given dwelling. If one simply attaches the positive and negative leads of a spectrum analyzer to two points on a nonconducting floor, one would very likely see these stray power line voltages. These voltages, which may be less than 1 mV, could be due to a current from the local ground rod, or may be a magnetically induced current from a nearby power line. When the chassis of a sensor data acquisition system comes in contact with the ground, or a part of a structure with magnetically induced currents, there will be some small voltages actually on its electrical ground reference. It is for this reason that most electrical measurement and high-fidelity multimedia gear make a separate signal ground connection available. The problem is made even more vexing when multiple sensors cabled to a distance from the data acquisition system come in contact with slightly different ground potentials at their locations. In this situation ground loops can occur between sensors as well as each sensor and the data acquisition system. As noted in the shielding description above, the cable itself becomes a source for magnetically induced power line voltages especially if it is a high-impedance circuit (the signal currents are as small as the magnetically induced ones). Most data acquisition systems have very high input impedance of over 500 MΩ so they can measure very weak sensor signals, such as brain waves measured by an electro-encephalogram (EEG). This force the sensor signal current to be very small, on the order of magnetically induced interference. The quick fix is to place a 10 kΩ across the data acquisition terminals so that the sensor current will be 50 times greater than the noise. This will boost the SNR by over 30 dB. If the impedance of the data acquisition input was 500 MΩ and you dropped it to 150 Ω for a balanced differential input, the SNR improvement can be over 130 dB.

The solution for almost all ground loop and power line isolation problems is seen graphically in Figure 15.4 and starts by selecting a common single point for all sensor signal grounds [5]. This is most conveniently done at the data acquisition system. The grounds to all sensors must therefore be "floated." This means that for a 2-wire sensor, a 2-wire plus shield cable is needed to extend the ground shield all the way out to the sensor, *but this cable shield must not be connected to any conductor at the sensor.* Furthermore, the sensor "ground lead" must not be grounded, but rather brought back along with the signal lead inside the shielded cable to the data acquisition system. The sensor case if conductive must also be insulated from the ground and the cable shield. With both the sensor signal (+) and sensor ground (−) floated inside the shielded cable, any magnetically induced currents are common to both wires, and will be small if the circuit impedance is low. At the front end of the data acquisition system, the common mode rejection of an instrumentation operational amplifier will significantly reduce or eliminate any induced currents, even if the sensor ground (−) is connected to the cable shield and data acquisition common ground at the acquisition input terminals. By floating all the sensor grounds inside a common shield and grounding all signals at the same point ground loops are eliminated and maximum shield effectiveness is achieved.

For a 3-wire sensor such as an electret microphone with FET preamplifier, one needs a 3-wired plus shield cable to float all three sensor wires (power, signal, and sensor "ground"). The FET should be wired as a current source to drive a low-impedance cable circuit. At the data acquisition end, a small resistor is placed between the sensor signal (+) and sensor "ground" (−) wires to create a sensor current loop and to convert the current into a sensible voltage. Without this resistor, the impedance of the sensor signal-to-ground loop inside the cable shield would remain very high with a very weak sensor signal current. This weak sensor current could be corrupted by magnetic field-induced currents, and thus, *it is always advisable to make the sensor signal current as strong as possible.* Shot and contact noise-causing dc currents can be eliminated from the signal path by a simple blocking capacitor at one end of the sensor signal (+) wire. Using an FET at the sensor to transfer the high resistance of a transducer to the low resistance of a signal current loop is the

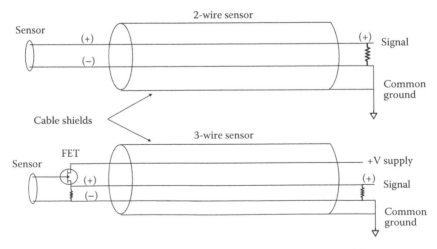

FIGURE 15.4 Proper ground isolation and shielding at the sensor for 2-wire and 3-wire sensors minimizes signal interference and eliminates ground loops.

preferred method for using high-impedance sensors. The resistors seen in Figure 15.4 should be chosen to match the impedance of the sensor (2-wire case) or the FET source resistor (3-wire case with a common drain n-channel FET shown). This maximizes the current in the twisted pair of conductors inside the cable shield, thus minimizing the relative strength of any interference. Note that if the shield ground isolation is lost at the sensor, interference voltages could appear from the potential difference between the ground at the data acquisition end and the sensor end of the cable.

Avoid Unbalanced Sensor Connections. A balanced sensor connection is one where the sensor (+) and (−) signals are separate from the shield and ground. This detailed discussion of ground loops, power, and shielding, is warranted because it is so commonly misunderstood by electronics engineers and ground loop noise interference is sensor data is a black eye on the electrical engineering profession. This issue is well understood by most audio/telephone engineers and also those who manufacture equipment with high voltage, or the potential to produce high voltage, with electrical sensors. The later group equates sensor ground loops with electrocution when an unwieldy arc decides to use the sensor cable shield and data acquisition system as the path to earth ground. Since we generally need to keep the electronics grounded, *thou shall not ground the sensor cable at the sensor location, period.* To convert an unbalanced sensor connection (sensor (+) signal wire plus a ground wire) to a balanced connection, one uses either a transformer such as those used for XLR type audio connections, or an instrumentation line driver. Both these devices create an out-of-phase copy of the sensor (+) signal as the sensor (−) signal and stop the cable shield from connecting to the sensor and earth ground. The transformer passively drops the voltage and increases the current to make the signal current strong relative to interfering magnetically induced power line noise. The balanced 2-wire plus shield line then connects to another transformer or instrumentation amplifier at the data acquisition end of the sensor wire. Electromagnetic interference is common to both the (+) and (−) wires, and so it is not passed through the differential instrumentation amplifier or the transformer if that is used. For audio systems the transformers work adequately for most applications, but the best solution is always to use a powered "direct box" which is essentially an instrumentation amplifier to make the currents strong even from high output impedance sensors.

15.2 NOISE CANCELLATION TECHNIQUES

Noise reduction is both straightforward and useful to a wide range of signal processing applications. We define *noise* simply as any unwanted signal. The noise may not be random, but rather some sort of interference which adversely affects the desired information processing of the signal. One can even describe an FFT as a noise reduction technique because is separates the signal into frequency bins and allows the magnitude and phases of many signals to be processed with little interference from each other as well as random background noise. But, if our *signal* is itself a random waveform and we wish to separate it from other signals, one has an interesting and challenging task.

Signal and noise separation can be achieved in one of two general approaches: by using a statistical whitening filter (see Sections 3.3 and 11.2) or by optimal Wiener filtering (Section 11.2). The optimal "Wiener" separation filter is defined from the knowledge of the signal that one wishes to reject. "Knowledge" of the undesirable noise is defined here as either access to the waveforms or the ability to accurately synthesize either the signal or "noise" waveform coherently. On the other hand, a whitening filter is based on the approach of our "signal" waveform being periodic, and thus, we can synthesize it coherently for a short time into the future (say one sample ahead), and use this linear prediction to suppress the "noise" waveform and therefore enhance our signal. This is done by first suppressing all the periodic components of our waveform, or "whitening" the waveform. The whitened waveform "residual" can then be subtracted from the original waveform to enhance the signal, which is assumed periodic. This is called *Adaptive Signal Enhancement* and is generally used to remove random noise components (or unpredictable noise transients) from a waveform with periodic (predictable) signal components. A good example of this is signal processing algorithm which removes "hiss," "clicks," and "pops" from old audio recordings in music and film archives. Using considerably more numerical processing, the same technique can be applied to images in film or video archives to remove "snow" or "specs" from damaged film.

Optimal Wiener filtering for noise suppression (and even complete cancellation) is more straight-forward than whitening filtering. Suppose our desired signal is 62 Hz and our "noise" waveform is a 60 Hz sinusoid (typical of a ground-loop problem). The "optimal" separation filter has a narrow unity-gain passband at 62 Hz with a zero response at 60 Hz. Obviously, filtering our signal in this way rejects the 60 Hz interference. For separation of steady-state sinusoids of different frequencies, the FFT can be seen as having each bin as an optimal filter to allow separation and measurement of the amplitude and phase of the signal closest to the bin center frequency. Indeed, the convolution of a signal with a sinusoid representing a particular bin frequency can be seen as an FIR filter where the integral of the FIR filter output gives the FFT amplitude and phase for the bin. In general, opti-mal Weiner filtering is best suited for situations where a "reference signal" of interfering noise is available. This noise waveform is also a component of our original waveform of interest, but the net time delay and frequency response of the channel which introduces the noise into our original wave-form is unknown. In Section 11.2, the optimal Wiener filter problem was discussed from the point of view of identification of this transfer function. In the system identification configuration, the "reference noise" is the system input, and our original waveform is the system output corrupted by another interfering waveform (our desired signal). For noise cancellation, one uses a Wiener filter to remove the "reference noise" which is coherent with the original waveform. The residual "error signal," which optimally has all components coherent with the reference noise removed, provides us our "desired signal" with this reference noise interference suppressed. This is sometimes called a *correlation canceller* in the literature because the optimal filter removes any components of our waveform which are correlated with the reference noise. In this section we will describe in detail the adaptive signal processing applications of adaptive signal whitening, adaptive signal enhancement, and adaptive noise cancellation.

Adaptive Signal Whitening describes the technique of removing all predictable signal compo-nents, thus driving the processed signal toward spectrally white Gaussian noise. One begins by assuming a model for our waveform y_t as the output of an infinite impulse response (IIR) filter with white noise input, or *innovation*. Using z-transforms, the waveform is

$$Y[z] = \frac{E[z]}{\left[1 + a_1 z^{-1} + a_2 z^{-2} + \cdots + a_M z^{-M}\right]} \tag{15.16}$$

where $E[z]$ is the z-transform of the white-noise innovation sequence for the IIR process y_t. Rearranging Equation 15.16 yields

$$Y[z]\left[1 + a_1 z^{-1} + a_2 z^{-2} + \cdots + a_M z^{-M}\right] = E[z] \tag{15.17}$$

and taking inverse z-transforms one has

$$y_t + \sum_{i=1}^{M} a_i y_{t-i} = \varepsilon_t \tag{15.18}$$

where ε_t is called the *innovation*. To remove the periodicity in y_t, or, in other words, remove the predictability or correlation of y_t with itself, one can apply an FIR filter $A[z] = 1 + a_1 z^{-1} + a_2 z^{-2} + \cdots + a_M z^{-M}$ to our waveform y_t and the output approximated the innovation, or white noise. The filter $A[z]$ is said to be a "whitening filter" for y_t such that the residual output contains only the "unpredictable components" of our waveform. We can call our approximation to the innovation the "prediction error" and define the linear one-step-ahead prediction for y_t

$$\hat{y}_{t|t-1} = -\sum_{i=1}^{M} a_i \, y_{t-i} \tag{15.19}$$

where the notation "$t|t-1$" describes a prediction for the waveform for time t using past waveform samples up to and including time $t-1$. This "linear predicted" signal by design does not contain very much of the unpredictable innovation. Therefore, the predicted signal is seen as the enhanced version of the waveform where the noise is suppressed.

To estimate the linear prediction coefficients of $A[z]$, a variety of adaptive algorithms can be used as seen in Section 11.2. The most straightforward adaptive algorithm is the least-mean square, or LMS algorithm. If we consider the innovation in Equation 15.18 as an error signal, the gradient of the error is a positive function of the coefficients a_i; $i = 1, 2, \ldots, M$. Therefore, a gradient decent for the LMS algorithm will step the coefficient updates in the opposite direction of the gradient as seen in Equation 15.20. Figure 15.5 graphically depicts a whitening filter.

$$a_{i,t} = a_{i,t-1} - 2\mu\varepsilon_t \, y_{t-i}, \quad \mu \le \frac{\mu_{\text{rel}}}{\text{ME}\{y_t^2\}}, \quad \mu_{\text{rel}} \approx \frac{1}{N} \tag{15.20}$$

The step size μ in Equation 15.20 is defined in terms of its theoretical maximum (the inverse of the model order times the waveform variance) and a relative parameter μ_{rel} which effectively describes the memory window in terms of N waveform samples. As described in Section 11.2, the smaller the μ_{rel} is, the slower the LMS convergence will be, effectively increasing the number of waveform samples influencing the converged signal whitening coefficients. This is a very important consideration for whitening filters because the filter represents a *signal model* rather than a *system model*. By slowing the adaptive filter convergence we can make the whitening filter ignore nonstationary sinusoids. This can be very useful for canceling very stationary sinusoids such as a 60 Hz power line frequency from a ground fault.

Adaptive Signal Enhancement is the technique of enhancing desired predictable signal components by removal of the unpredictable noise components. Signal enhancement usually requires a whitening filter to produce the unpredictable components to be removed. To further illustrate whitening and enhancement filters, Figure 15.6 shows the results of the whitening filter output, and the enhanced (linear predicted) signal output for a case of a stationary 120 Hz sine wave in white noise along with a nonstationary frequency in the 350 Hz range with a standard deviation of 10 Hz. The instantaneous frequency is found for the nonstationary sinusoid by integrating a Gaussian random variable over several thousand samples to obtain a relatively low-frequency random walk for the

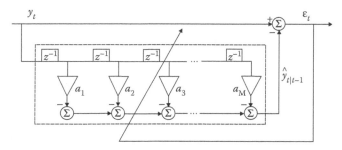

FIGURE 15.5 Signal flow diagram for an adaptive FIR whitening filter for the signal y_t where ε_t is the whitened error signal used to drive the adaptive filter and $\hat{y}_{t|t-1}$ the linear predicted (enhanced) signal at time t given a filter model last updated at time $t-1$.

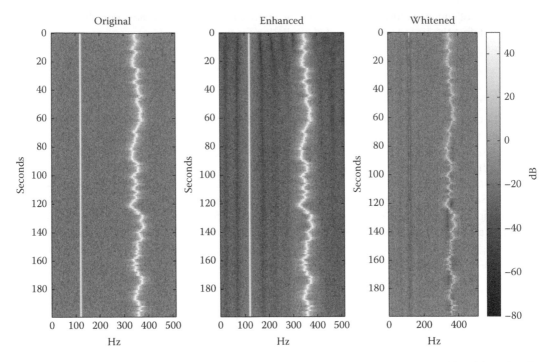

FIGURE 15.6 Adaptive whitening and signal enhancement results for a 120 Hz sinusoid in white noise plus a random frequency-modulated sinusoid to show the effects of signal stationarity.

sinusoid frequency. Since instantaneous frequency is the time derivative of the instantaneous phase, one can synthesize the nonstationary spectra seen in Figure 15.6 on the left spectrogram. The center spectrogram shows the enhanced signal (from the linear prediction) and the right spectrogram shows the whitened error signal output. An LMS adaptive algorithm is used in the time domain with a memory window of about 200 samples. Figure 15.7 shows the time-averaged power spectra of the

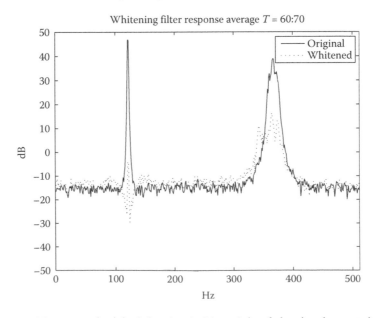

FIGURE 15.7 Ensemble-averaged original signal and whitened signal showing the most significant whitening for the stationary 120 Hz signal.

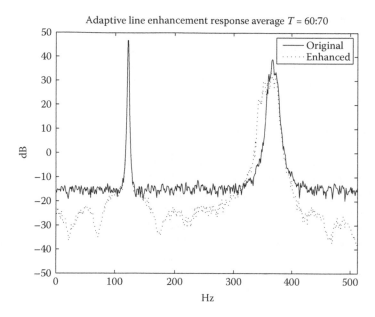

FIGURE 15.8 Ensemble-averaged and original signals for the signal enhancement spectra showing the suppression of the unpredictable white noise, leaving the signals largely intact.

original and whitened signals for the time interval from 60 to 70 s. The whitened signal clear shows over 50 dB of cancellation for the 120 Hz stationary sinusoid and about 20 dB cancellation for the nonstationary sinusoid. Figure 15.8 shows the time average spectra for the enhanced signal which clearly shows in detail a 10–20 dB broadband noise reduction while leaving the sinusoids essentially intact.

Adaptive Noise Cancellation significantly suppresses, if not completely removes, the coherent components of an available noise reference signal from the signal of interest through the use of optimal Wiener filtering. Figure 15.9 shows a block diagram for a straightforward Wiener filter used to remove the interference noise correlated with an available "reference noise," thus leaving the residual error containing mostly the desired signal [6–8]. If coherence is high between the reference noise and the waveform with our desired signal plus the noise interference, the cancellation can be nearly total.

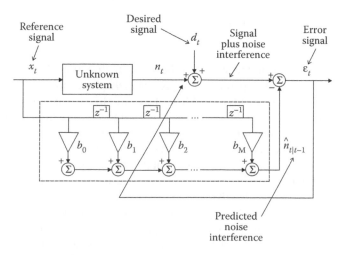

FIGURE 15.9 An adaptive Wiener filter for removing any signal correlated with the reference signal input so that the error signal output is mostly the external desired signal.

This of course requires that we have a very accurate model of the transfer function between the reference noise and our waveform. The Wiener filtering in Section 11.2 is mainly concerned with the identification of this transfer function system, or "system identification." A valuable by-product of precise system identification is the capability of separating the reference noise components from the output waveform to reveal the "desired signal" with the noise interference removed.

For example, suppose the President, not wishing to answer reporters' questions while walking from his helicopter to the White House, instructs the pilot to race the turbine engines creating enough noise to mask any speech signals in the area. However, a gadget-wise reporter places one microphone near the helicopter and uses another to ask a question of the President. A real-time Wiener filter is used to remove the turbine noise from the reporter's microphone allowing the question to be clearly heard on the videotape report. This technique has also been used to reduce background noise for helicopter pilot radio communications, as well as many other situations where a coherent noise reference signal is available. But perhaps the easiest use of this technique is to remove power line frequencies from data recordings produced by hapless engineer who did not have the benefit of Section 15.1. Power line frequencies are very stable and predictable unless there is some sort of power emergency like a brown out of power surge. The frequency is held very stable because all the generating sources must be synchronized in order to deliver power to the grid. This offers an opportunity for us to generate a reference signal with the power line frequencies of concern, and then let an adaptive filter find the phase and amplitude to cancel the unwanted interference.

We begin our description of the adaptive noise cancellation seen in Figure 15.9 by assuming an FIR system model (the model could also be IIR if desired), with the following linear prediction of the noise interference n_t, given a reference signal x_t, and system model last updated at time $t - 1$.

$$\hat{n}_{t|t-1} = \sum_{k=0}^{M} b_k x_{t-k} \tag{15.21}$$

The linear prediction error is then

$$\varepsilon_t = n_t + d_t - \hat{n}_{t|t-1} \tag{15.22}$$

where d_t is our desired signal. The adaptive LMS update for the FIR coefficients is

$$b_{k,t} = b_{k,t-1} + 2\mu\varepsilon_t x_{t-k}, \quad \mu \leq \frac{\mu_{rel}}{ME\{x_t^2\}}, \quad \mu_{rel} \approx \frac{1}{N} \tag{15.23}$$

where a plus sign is used to insure the step is in the opposite direction of the gradient of the error with respect to the coefficients. The parameter μ_{rel} is inversely proportional to the memory window of the LMS algorithm. For a very steady signal line, the power line frequencies, one could make the memory window N very long and let the LMS slowly converge to a very high performance solution. If the reference signal is very dynamic a much shorter memory window would be used. Note that the predicted signal $\hat{n}_{t|t-1}$ which is the filtered reference signal x_t in Figure 15.9 is subtracted from signal plus the interfering noise to give the error signal. This means the error signal gradient with respect to the adaptive filter coefficients will be negative causing the gradient decent to converge at the least-squared error solution for the filter.

Use of the LMS algorithm and an FIR system model makes the processing remarkably simple. If the transmission channel between the reference signal and the noise interference is essentially propagation delays and multipath, the FIR model is very appropriate. If the transmission channel contains strong resonances (the system transfer function has strong peaks in the frequency domain), an IIR filter structure is to be used because it requires fewer coefficients to accurately model the system.

If the reference noise contains signal components which are uncorrelated with our waveform, these components will unfortunately add to the noise interference. This is because the converged system filter responds only to the coherent signals between the reference noise and our waveform. The uncorrelated noise in the reference signal will be "filtered" by the system model, but not removed from the output waveform. This is a particular concern when the reference signal contains random noise from spatially uncorrelated sources, such as turbulence on a microphone. Such situations are problematic when acoustic wind noise interferes with our desired signal. Because the turbules shed and mix with the flow, the pressure fluctuations are spatially uncorrelated after a given distance depending on the turbulence spectrum. The signal plus turbulence at one location (or generated by one microphone in air flow) is in general not correlated with the wind noise at another location. If one attempts to cancel the flow noise using a reference sensor a large distance away one would simply add more random noise to the output signal. However, wind noise on a microphone has been canceled using a colocated hot-wire turbulence sensor signal as a reference noise [9]. When tonal acoustic noise is to be removed, one can design or synthesize a "noiseless" reference signal by using a nonacoustic sensor such as an optical or magnetic tachometer pulse train designed to produce all the coherent frequencies of the noise generating device. This frequency-rich reference signal is then adaptively filtered so that the appropriate gain and phase are applied to cancel the coherent tones without adding more random noise. This is the best strategy for canceling fan noise in HVAC systems (heating, ventilation, and air conditioning), propeller noise in commercial passenger aircraft, and unwanted machinery vibrations.

As an example demonstrating the ease, power, and simplicity of adaptive noise cancellation, a pair of nonstationary sinusoids is seen in the left spectrogram of Figure 15.10. Nonstationary noise (around 100 Hz) and signal (around 300 Hz) sinusoids are used to emphasize that once the system transfer function is identified, cancellation occurs independently of the actual signals. For this case our reference noise is the lower sinusoid near 100 Hz plus white Gaussian noise. The SNR of the signals is 20 dB, the SNR gain of the 1024-point FFT is 30.1 dB, yet the observed SNR of about 40 dB is less than 50.1 dB because of the nonstationarity of the frequencies. The reference noise

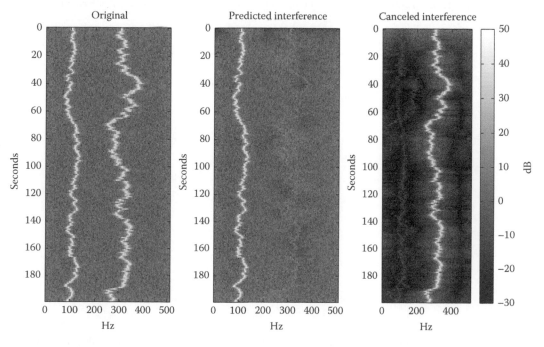

FIGURE 15.10 Signal separation using Wiener filtering on two nonstationary sinusoids where approximately 100 Hz sinusoid in white noise is the reference signal for the cancellation.

mean frequency is 105 Hz and its standard deviation is 10 Hz while our desired signal has a mean of 300 Hz and a standard deviation of 20 Hz. The Gaussian frequency estimates were integrated over 10,000 samples (sample rate for the simulated waveforms is 1024 Hz) to give a low-frequency randomness. The center spectrogram represents the predicted noise interference signal consisting of the system-filtered 105 Hz nonstationary sinusoid plus the random noise. The right spectrogram illustrates the nearly complete removal of the noise interference, even though the signals are nonstationary. Figure 15.11 shows the time-average spectra of the original waveform and the cancellation filter error output averaged over 40–45 s. One can clearly see the cancellation of the random noise as well as the interfering 105 Hz nonstationary sinusoid. Adaptive noise cancellation is a straightforward and impressively simple technique for separating signal and noise interference.

15.3　ACTIVE NOISE ATTENUATION

This last section discusses the physical cancellation of unwanted noise outside of the electronics system known as active noise attenuation [6–8]. We carefully use the term attenuation, rather than cancellation, because in the physical world wave energies are not canceled by these systems, just attenuated. The key question is if you use energy to make a canceling wave with a transducer, the canceling and original waves mix, and the resulting wave is attenuated, where did the energy go? Clearly energy cannot be created nor destroyed but only converted from one form or place to another. Consider a loudspeaker with 1 W of a 100 Hz sinusoid radiating 88 dB (relative to 20 µPa RMS) at 1 m distance in air. Move the loudspeaker into a vacuum and apply the same 1 W power and get no sound. Where did the energy go? The loudspeaker cone is now moving more since there is no air loading it and no cabinet compliance (inverse of stiffness) since there is no air in the cabinet, therefore there is more mechanical dissipation and more electrical heating of the voice coil wire due to resistance. So the 1 W of energy went other places than acoustic radiation. That is what active noise attenuation can do—change where the wave energy goes.

In active noise control, one must take into account the possibility of feedback from the physical actuator back into the reference signal. Active source feedback is not a deal breaker, but it does

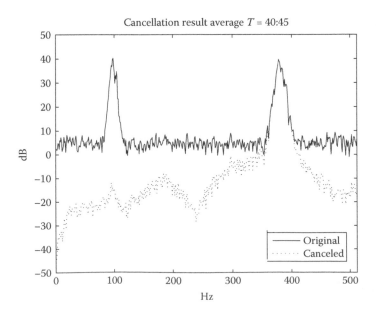

FIGURE 15.11　Average spectra from 40 to 45 s comparing the original and canceled signals for the Wiener filtering signal separation data in Figure 15.10 showing the almost total cancellation of the reference sinusoid and much of the broadband noise.

require compensation in the controller and can limit the effectiveness of the control. In addition, active noise control requires delays in the error plant to be accounted for in the adaptive controller updates. Preferably, the active control source and the point in space where the waves mix is such that the controller can put a filter reference signal wave at that point ahead of the real noise signal, so that an appropriate delay in the control filter would allow any wave to be canceled, even coherent random noise. We call this a *causal active controller*. Note that there is always at least one sample period delay because the data acquisition takes place, then the processing followed by the digital-to-analog converter (DAC), which is usually clocked at the same time as the next analog-to-digital converter (ADC). But even for cases where the minimum time delay through the active controller is longer than the time it takes for the reference signal driving the real noise to get to the wave mixing point, one can still cancel sinusoids, but not random signals.

Finally, we will present a brief note on feedback control approaches for active noise control, including digital ARMAX (autoregressive moving-average with auxiliary input) control. Feedback control loops have been a fundamental part of electronic control systems since World War II. It works extremely well at low frequencies when there is very little delay in the feedback loop. When there is significant delay, care must be taken to ensure that high frequencies and in particular, transients, are attenuated in the feedback loop. Consider that a delay is a linear phase shift in the frequency domain. So, if the feedback loop has high gain and negative (stable) feedback at low frequencies, the delay means that the gain will have positive (unstable) feedback at some high frequencies. If any high frequencies or even a single transient gets into the feedback loop with high gain, the control loop will whistle its way to oblivion at those frequencies where the phase shift results in unstable positive, not stable negative, feedback. So all controllers that use feedback have what is called a compensation filter to limit the bandwidth of the feedback to a range where stable negative feedback control is guaranteed. This applies to all thermostats as well as those "noise canceling headphones" widely available today. Generally, the active noise canceling headphones are effective from about 100 Hz to 2 kHz, and only by around 20–25 dB in the middle of that range, and only for sound pressure levels below about 120 dB, limited by the power capability of small transducers.

Active noise control suppresses the unwanted noise in the physical medium (rather than on the signal wire or in a digital recording of the waveform). This physical cancellation requires that feedback from the actuator to the reference signal be accounted for, and that delays in the actuator to error sensor transfer function be accommodated. There are significant complications due to the additional delays and control loop transfer functions, and we emphasize here that *Active Noise Control* and *Adaptive Noise Cancellation* refer to significantly different signal processing tasks. Adaptive noise cancellation is much easier than active noise control.

Figure 15.12 provides a basic comparison between active noise cancellation and adaptive noise control (ANC). While both algorithms are adaptive controllers, the word *active* signifies that a physical actuator produces a real wave which mixes the interfering noise to produce a physical cancellation. Where does the wave energy go? The answer depends on how the ANC system is configured. If one is canceling an acoustic wave outdoors, the cancellation simply occurs at a point in space from out-of-phase superposition. If the acoustic cancellation occurs in a duct where the wavelength is larger than twice the longest dimension in the duct cross section, the ANC actively causes zero impedance at the active control source. This actually causes a reflection of the acoustic wave back toward the "primary" noise source, and zero-power radiation downstream in the duct [9]. The concept of active control of impedance is extremely powerful because it allows one to control the flow of power and the efficiency of energy conversion. To reduce radiated energy, one drives radiation impedance to zero. To create a "zone-of-silence" at the location of a sensor, one is not concerned so much about impedance and uses ANC to allow sensor signal gains to be increased for a physical increase in SNR.

As seen in Figure 15.12, the ANC control system is considerably more complicated than a Wiener filter for adaptive noise cancellation "on the wire." The block "$S[z]$" symbolizes an active

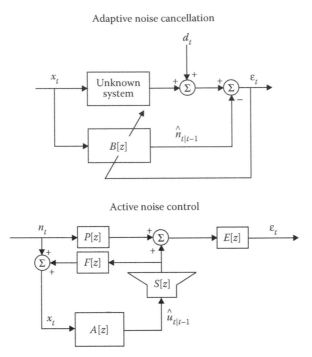

FIGURE 15.12 Comparison of adaptive noise cancellation to active noise control (ANC) showing the added complexity of the active transducer $S[z]$ and feedback $F[z]$ in ANC.

source which could be electrical, mechanical, or acoustical. However, once the ANC "secondary" source produces its canceling wave, it can have an effect on the reference sensor which is symbolized by the feedback block "$F[z]$." The forward propagation noise plant is "$P[z]$." For adaptive noise cancellation seen in the top of Figure 15.12, the effect of the control signal output is seen immediately in the error signal after subtracting the predicted noise interference from the waveform. However, in the ANC system, this control output must first pass through the actuator transfer function "$S[z]$," and then through the error propagation channel and transducer symbolized by "$E[z]$." The error loop transfer function can be seen as "$S[z]E[z]$," or "SE" for short, and can have significant time delays due to the physical separation and the electro-mechanical transfer function of the transducers. Since in the least-squared error adaptive filter, the error and reference signals are cross-correlated to provide the filter coefficient update (no matter which adaptive algorithm is used), any phase shift (or time delay caused by the response of SE will be significantly detrimental to the success of the cancellation). Thus, by filtering the reference signal with SE we make a new signal r_t, which we call the "filtered-x" signal which is properly aligned in time and phase with the error signal. Note that a small amplitude error in the error signal due to SE is not a significant problem since the adaptive algorithm will continue to update the control filter coefficients until the error is zero. But with a phase or time delay error, the coefficients of the controller will be updated with the wrong cross correlation. The "Filtered-X" adaptive ANC algorithm is seen in Figure 15.13.

To see the optimal converged control filter solution for ANC, we break down the system transfer functions into frequency-domain representative blocks which we can manipulate algebraically. These transfer functions could be of FIR or IIR type in terms of model paradigm. For now we will simply leave them as a letter block to simplify our analysis. The error signal can be seen as

$$\varepsilon(f) = n(f)P(f)E(f) + x(f)A(f)S(f)E(f) \tag{15.24}$$

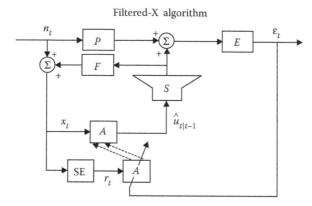

FIGURE 15.13 The signal r_t is the "Filtered-X" signal required for the adaptive control filter "A" to properly compensate for the time delay in the error plant $S[z]E[z]$ and showing how the adaptive filter coefficients are simply copied to "A" each iteration.

where $A(f)$ or $A[z]$ in the digital z-domain can be solved for the control filter "A" which gives zero-error signal explicitly (dropping the f or z for brevity). Note that

$$x = n + x \, \text{ASF} \tag{15.25}$$

so that

$$\frac{x}{n} = \frac{1}{1 - \text{ASF}} \tag{15.26}$$

Now we can rewrite Equation 15.16 in terms of the ANC error relative to the original noise as a transfer function.

$$\begin{aligned} \frac{e}{n} &= \text{PE} + \frac{x}{n} \text{ASE} \\ &= \text{PE} + \frac{\text{ASE}}{1 - \text{ASF}} \end{aligned} \tag{15.27}$$

We now solve for "A" which makes the noise zero.

$$\begin{aligned} -\text{PE} &= \frac{\text{ASE}}{1 - \text{ASF}} \\ -\text{PE}(1 - \text{ASF}) &= \text{ASE} \\ -\text{PE} + A \cdot \text{SE} \cdot \text{FP} &= A \cdot \text{SE} \\ -\text{PE} &= A(\text{SE} - \text{SE} \cdot \text{FP}) \\ A &= \frac{-\text{PE}}{\text{SE}(1 - \text{FP})} \end{aligned} \tag{15.28}$$

Assuming the error sensor transfer function "E" is nonzero for all frequencies, we simplify further as

$$A = \frac{-P}{S(1 - \text{FP})} \tag{15.29}$$

If the error loop response "S" is zero at some frequency(s), the ANC control system will not respond at that frequency(s), but can be held stable by limiting the amplitude of the control filter "A."

One can clearly see from the optimal ANC control filter in Equation 15.25 that to be a completely stable filter, "S" must be nonzero at all frequencies and the forward path "P" times the feedback path "F" must be less than unity when the phase is near zero. For "A" to be a stable causal filter, its poles must lie within the unit circle, and for practical reasons, the overall gain of "A" should be limited to avoid overdriving "S" into its nonlinear range. While this makes "A" a stable filter, the feedback loop from the actuator "S" back into the reference sensor through "F" must also be stable. This is guaranteed if the loop gain through "ASF" is less than unity, or $|FP| < 1/2$. We call the requirement for $|FP| < 1/2$ the "*passivity condition*" for ANC systems which must be met to ensure system stability. In general, the feedback from the active control source back into the reference sensor is undesirable and usually robs the ANC system of performance and dynamic range [10]. It is best eliminated by using a reference sensor which does not respond to the acoustic or mechanical waves from the actuator, such as an optical tachometer for canceling tonal noise interference from rotating equipment. Directional sensor arrays can also be used to ensure passivity. While eliminating the feedback is the best strategy, if present and meeting the passivity constraint, feedback will simply cause the ANC system to cancel much of its own reference signal, leading to poor overall performance. Clearly, applying a basic physical understanding of the system can significantly improve the performance of the applied adaptive signal processing.

We now turn our attention to measurement of the error loop transfer function "SE." Clearly, one could measure this while the ANC system is offline and apply the "SE" Filtered-X operation while hoping that this transfer function does not change during the ANC system operational time. But, the Filtered-X operation is important because it is generally done concurrently with the ANC adaptive filtering in commercially available systems. Figure 15.14 shows an "active" online system identification which injects a white noise signal into the error loop to simultaneously identify the "SE" transfer function while the ANC system is operating [11]. Clearly, this noise has to be strong enough so that the "SE" transfer function can be measured in a reasonable amount of time. However, if the system identification noise is too loud at the mixing position, then the ANC is defeated. Figure 15.15 shows a very clever [12] approach which allows both the "PE" and "SE" transfer functions to be estimated using only the available control signals. This is desirable because it adds no noise to the ANC system or the ANC error output. It works by balancing the models for "PE" and "SE" to predict the error signal, which could be near zero during successful cancellation. The "error of the error," $\varepsilon_t^\varepsilon$, is then used to update the estimates for the "PE" and "SE" transfer functions. Only the

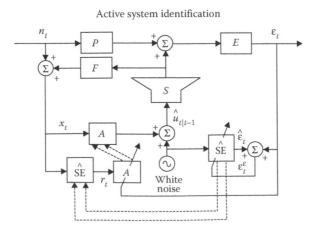

FIGURE 15.14 A Filtered-X ANC system with active system identification of the $S[z]E[z]$ plant using white noise injected into the control output.

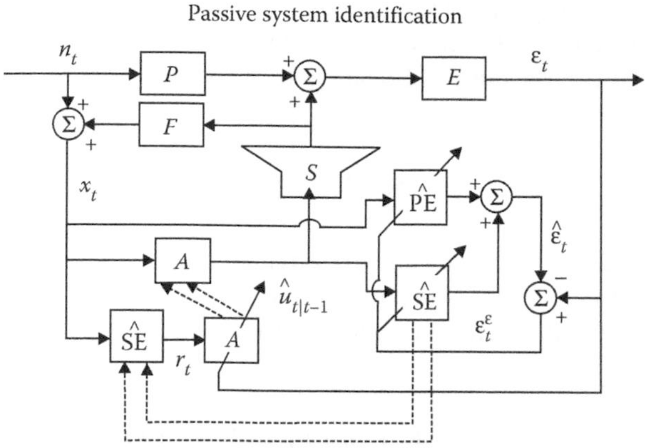

Passive system identification

FIGURE 15.15 A Filtered-X ANC system using passive system identification to model the error signal as the sum of the $P[z]E[z]$ forward plant driven by the reference signal x_t and the error plant $S[z]E[z]$ driven by the control signal $u_{t|t-1}$.

"SE" transfer function is needed for the Filtered-X operation, and only its phase at the frequency components in the noise, reference, and error signals is needed to ensure correct convergence of the ANC system. However, if the reference noise is narrowband, the estimated "SE" transfer function will only be valid at the narrowband frequency.

The passive system identification scheme for the Filtered-X algorithm involves two LMS filters to do the system identification, and a third LMS to find the active control filter. The signals going into the third LMS filter will affect how good the ANC is, coupling the three adaptive filters together. This in some cases can cause convergence sluggishness or the "PE" side of the system identification compensating for the "SE" side, since there is only one zero signal from the sum of two signal paths. Algebraic prudence tells us that it would be nice to have two equations to solve for the two unknown transfer functions "PE" and "SE." Since we will do this using algebra, we will work in the frequency domain where each FFT bin is orthogonal and independent of adjacent bins. The easiest way to get a second equation for the error is to adjust the gain of the control signal slightly [13].

$$\begin{bmatrix} e_1 \\ e_2 \end{bmatrix} = \begin{bmatrix} x_1 & u_1 \\ x_2 & u_2(1+\zeta) \end{bmatrix} \begin{bmatrix} PE \\ SE(1-FP) \end{bmatrix} \tag{15.30}$$

Solving for "PE" and "SE(1 − FP)" in the frequency domain we get

$$\begin{aligned} \begin{bmatrix} PE \\ SE(1-FP) \end{bmatrix} &= \begin{bmatrix} x_1 & u_1 \\ x_2 & u_2(1+\zeta) \end{bmatrix}^{-1} \begin{bmatrix} e_1 \\ e_2 \end{bmatrix} \\ &= \frac{\begin{bmatrix} u_2(1+\zeta) & -u_1 \\ -x_2 & x_1 \end{bmatrix} \begin{bmatrix} e_1 \\ e_2 \end{bmatrix}}{x_1 u_2(1+\zeta) - u_1 x_2} \end{aligned} \tag{15.31}$$

We can now write unique expressions for the forward and error plants, with feedback if it exists, but why bother? We can also solve explicitly the active control filter as seen in

$$A(f) = \frac{u_1(f)e_2(f) - u_2(f)(1+\zeta)e_1(f)}{x_1(f)e_2(f) - x_2(f)e_1(f)} \tag{15.32}$$

Since the desired form of Equation 15.32 in a digital filter in the time domain, one must use the circular convolution corrections given in Section 6.4 by zero padding the oldest error signals in the FFT buffer. If the active control filter in Equation 15.32 is near perfect, then the lower limit for the error spectrum is

$$e_2 \geq u_2 \mathrm{SE}(1 - \mathrm{FP})\zeta \tag{15.33}$$

so ζ can be set very small if desired, or can be set dynamically as the system converges. One can also judge by looking at Equation 15.32 that the estimated controller will be very noisy for random signals. To smooth the estimated active controller in the frequency domain, all one needs to do is some averaging

$$A(f)_t = \alpha A(f)_{t-1} + \beta \frac{u_{t-1}(f)e_t(f) - u_t(f)(1+\zeta)e_{t-1}(f)}{x_{t-1}(f)e_t(f) - x_t(f)e_{t-1}(f)}; \quad \alpha = \frac{N-1}{N}; \quad \beta = \frac{1}{N} \tag{15.34}$$

where N is the exponential memory window length. This is a relatively new development in ANC, although some of the earliest noise cancellers did processing in the frequency domain.

Looking back to Equation 15.28 it is clear that the only way the ANC system can cause a problem with the control filter "A" is if "SE" is zero or if "|FP|" is < 1/2 (passivity). Note that "|FP|" less than unity will make "A" a stable control filter but the control loop "ASF" must be less than unity for system stability, and so "|FP|" is further restricted to < 1/2. When "S" is near or equal to zero, we do not have what is known as *control authority* meaning that no matter what signal level we send to the active transducer, it will not be observable in the error signal. When "E" the transfer function from the active transducer to the error sensor is at or near zero, we do not have what is known as *system observability*, meaning we cannot measure the effect of the control signal. Either of these is a cause for instability where the adaptive controller might require really high gain, driving the active transducer into nonlinear response further causing performance problems in the ANC system. We must have observability to measure what is happening and controllability to have an effect on it. We must have passivity of the feedback loop for the ANC system to remain stable assuming that the ANC control filter is stable.

What could go wrong for a frequency bin in Equation 15.34? We have two sets of spectra, one at time t and one at time $t-1$, which, by definition, the FFT buffer do not overlap. In fact, it is probably a good idea to let the signals settle for a bit of time after the active control gain is set to $(1 + \zeta)$ so that the system identification is all based on stationary signals. As such, if the reference signal $x(f)$ is random, the cross spectra in the denominator of Equation 15.34 will be random or zero. The result might improve if all the cross-spectra in Equation 15.34 are time-averaged with an α–β integrator before the calculation in Equation 15.32. But in all cases, it appears that the reference signal needs to be a periodic signal for the system identification scheme in Equations 15.32 and 15.34 to work properly. This must be evaluated frequency bin by bin, and can be supported by estimating the ordinary coherence function from Section 7.3 for each of the cross spectra in Equation 15.32. If the coherence is too low, the corresponding bin in the ANC control filter can be set to zero.

Multichannel ANC The Filtered-X algorithm can also be applied for multichannel ANC where one has p reference signals, q actuators, and k error sensors [14]. One would choose to use multiple reference signals only when several independently observable coherence reference noise signals are available, and as such, one can optimize the adaptive controllers independently for each type of reference signal summing the control outputs at the actuators. The ANC control filter A is represented in the frequency domain as a *pxq* matrix for each frequency. This way each reference signal has a path to each actuator. In general, when a number of actuators and error sensors are used, the ANC system is designed to produce spatial or modal control. One should note that simultaneously minimizing the squared error at every error sensor location is one spatial solution. One can apply a number of useful modal (wavenumber) filters to the error sensor array for each frequency of interest in much the same way one could design a beam pattern with nulls and directional lobes as desired. The issue of modal control is central to ANC problems of sound radiation from large structures and high-frequency sound cancellation in enclosures such as airplanes, automobiles, and HVAC ducts. Using modal ANC control, one can design the ANC-converged response to minimize vibration energy which radiates from a structure, or excites enclosure modes.

In general, each actuator "channel" should respond adaptively to every error signal, either independently or as part of a modal filter output. When the number of error sensors is greater than the number of actuators ($k > q$), this combining of error sensors allows each actuator signal to be optimized. When the number of actuators is larger than the number of error sensors ($q > k$), the error signals will be repeated for multiple actuators redundantly. This will in general still work, but there is a potential problem with linear dependence of actuator channels. Note that in Equation 15.25 the actuator transfer function S is inverted. For multichannel ANC, S is a $q \times q$ matrix and the error propagation and transducer paths are modeled as a $q \times k$ matrix for each frequency. This allows each actuator a path to each error sensor. However, the actuator matrix S must be invertible, which means that the columns of S must be linearly independent and each element on the main diagonal of S must be nonzero for all frequencies [15]. An example of linear dependence between ANC channels can be seen for a two-channel system where both actuators and both error sensors are closely located. As such, the columns of the S matrix are linearly dependent. Channel 1 can interfere with channel 2 and the error sensors cannot observe the effect. Thus, the control effort can go unstable without correction instigated by the error signals. One can design around the actuator linear dependence problem by careful choice of actuator and error sensor location. Another method is to apply a singular value decomposition and "turn off" the offending actuators at selected frequencies. This is a significant source of complexity in robust design of a multichannel ANC system.

With p reference sensors, q actuators, and k error sensors the multichannel Filtered-X algorithm poses a problem: how does one filter p reference signals with the $q \times k$ SE transfer function estimate to produce p "filtered-x" (r_t in Figures 15.13 through 15.15) signals for use by the adaptive algorithm? We adopt a "copy/combine" strategy to support the matrix interfaces and the ANC design goals. If $p > q$ (more reference channels than actuators) the control filters for the extra reference channels can be calculated in parallel where the outputs are summed at the actuators. For $p < q$ the reference signals are simply copied into the extra channels. Now we have k filtered-x signals (one for each error channel) and we need to adapt the pxq ANC control filter frequency-domain coefficients. For $k > q$ (more error channels than actuators) the extra filtered-x signals are combined for each actuator channel. In other words, more than q filtered-x signals and error signals contribute to the adaptive update for a particular control filter element. When $k < q$ (more actuators than error sensors) the extra actuator channels are updated with the same filtered-x signals and error signals as other channels at the risk of causing linear dependence. This is less risky than it sounds because the actual transfer functions between individual actuators and a particular error sensor maybe different enough to support linear independence of the channels.

Active Noise Attenuation Using Feedback Control is best applied in situations where the closed-loop response has a time constant significantly shorter than the period of the highest frequency to be controlled. This is because the feedback loop time delay determines the available bandwidth for stable active attenuation. Generally speaking, active feedback control is well suited for "smart structures" where the vibration sensor and actuator are closely located, the wave speed is very high, and therefore the control loop delay is very short compared to the period of the frequencies to be actively attenuated. This arrangement can be seen as a basic regulator circuit for very low frequencies. As one attempts to employ active feedback control at higher frequencies, one must limit the control bandwidth in accordance with the loop delay, creating a "tuned regulator." Because of the devastating quickness of divergence in a high-gain tuned regulator, adaptive techniques are best executed offline where one can apply stability gain margin analysis before updating the feedback controller.

ARMAX feedback control

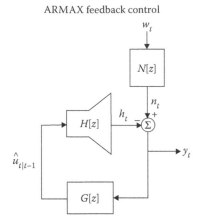

FIGURE 15.16 ARMAX Control can used to optimize the design of a digital feedback control system for active attenuation of noise using no reference signal, but it is well advised to constrain the adaptation and feedback gain to maintain some margin of stability.

Figure 15.16 depicts a basic feedback control system which can be optimized using what is known as ARMAX control [16–18]. ARMAX stands for Auto Regressive Moving Average with auxiliary input and describes a modern digital adaptive control technique for feedback systems (the Filtered-X algorithm is a feedforward system). Before we get into the details of ARMAX control, let us examine the basic feedback control system where the plant ($H[z]$ in Figure 15.16) has some gain H_0 and delay τ. This delay is inherent to all transducers but is particularly large in electromagnetic loudspeakers and air acoustic systems, where the sound propagation speed is relatively slow. The delay response for most sensors in their passband is usually quite small if the frequency response is flat. For piezoelectric transducers embedded into structures, the delay between sensor and actuator is especially small. This delay determines the available bandwidth and amount of stable active attenuation. The delay in the plant will cause a linear phase shift where the phase becomes more negative as frequency increases. At some high frequency, what was designed as low-frequency stable negative feedback becomes unstable positive feedback. Therefore, one typically designs a compensator filter ($G[z]$ in Figure 15.16), which attenuates the feedback out-of-band to maintain stability.

To examine the feedback control system stability problem in its most basic incarnation (a whole textbook has been devoted to this subject countless times), we will consider a "perfect transducer" model where the net response from the electrical input to the actuator to the electrical output of the error sensor is simply $H(s) = H_0 e^{-j\omega\tau}$. Our compensator $G(s)$ is a simple first-order low-pass filter defined as

$$G(s) = \frac{1}{1 + j\omega/\omega_c}, \quad \omega_c = \frac{1}{RC} \tag{15.35}$$

which can be implemented in a variety of ways. The compensator has a flat frequency response up to ω_c and then decreases at a rate of −6 dB per octave as frequency increases. Placing two such filters in series gives a second-order (−12 dB/oct) filter, four filters in series yields a fourth-order (24 dB/oct), and eight filters in series yields a sharp cutoff of (48 dB/octave). Figure 15.17 compares the magnitude and phase of these filters designed with a 300 Hz cutoff frequency. While the higher order compensators provide a sharp cutoff and a lot of attenuation for high frequencies, there is also a considerable phase change in the cutoff region. As will be seen, this phase change is the source of considerable difficulty (and art) in designing stable feedback control systems.

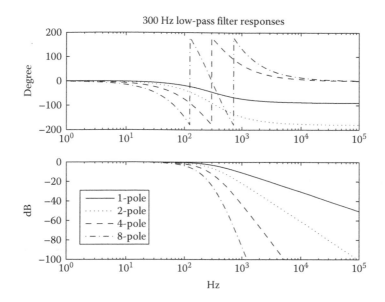

FIGURE 15.17 Comparison of the magnitude and phase of a first-, second-, fourth-, and eighth-order low-pass filter to attenuate feedback signal above 300 Hz in a compensation filter.

The open-loop response is simple the $H(s)G(s)$ product, or

$$H(s)G(s) = \frac{H_0 e^{-j\omega \text{Å}}}{\left(1 + j(\omega/\omega_c)\right)^n} \tag{15.36}$$

where "n" is the order of the filter (second order, etc.) also denoted as the number of poles in the open-loop system. For the open-loop response to be stable, all its poles must lie in the left-half s-plane. However, if the open-loop system is unstable, the closed-loop response can still be stable, which is a useful and interesting aspect of feedback control. *Open-loop stability implies that there are no poles of $H(s)G(s)$ in the right-half s-plane while closed-loop stability implies that there are no zeros of $1 + H(s)G(s)$ in the right-half s-plane.* The closed-loop response of the feedback control system is seen to be

$$\frac{y(s)}{n(s)} = \frac{1}{1 + H(s)G(s)} \tag{15.37}$$

where the denominator of Equation 15.37 is known as the *characteristic equation* for the closed-loop response. The closed-loop responses for our system with either first-, second-, fourth-, or eighth-order low-pass filters in seen in Figure 15.18. The eighth-order (8-pole) response is actually unstable, which will be explained shortly. The zeros of the characteristic equation are the closed-loop system poles. Therefore, the characteristic equation can have no zeros in the right-half s-plane if the closed-loop response is stable.

Solving for the characteristic equation zeros and testing to determine their location is quite tedious, but can be done in a straightforward manner using numerical techniques. The Routh–Hurwitz criterion is a method for determining the location of the zeros (in either the left- or right-half planes) for a real-coefficient polynomial without actually having to solve for the zeros, explicitly. However, the Nyquist criterion is a more appealing graphical solution to determining closed-loop system stability. For our simple system with no open-loop poles in the right-half plane, one plots the

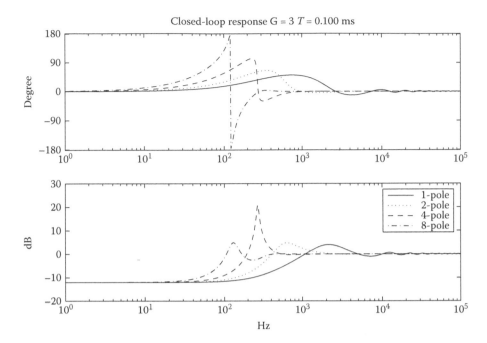

FIGURE 15.18 Closed-loop responses with a feedback loop gain of 3.00 and 100 μs loop delay showing the 4-pole compensation filter marginally stable and the 8-pole compensator unstable due to the 180° phase shift and gain greater than unity near 100 Hz.

real versus imaginary parts of the open-loop response $H(s)G(s)$ from $s = +j\infty$ to $s = j0$ taking care to integrate around any poles of the characteristic equation on the $j\omega$ axis.

The "Nyquist plots" for our system can be seen in Figure 15.19. Since we have a low-pass compensator $G(s)$, all the open-loop response plots start at the origin corresponding to $\omega = +j\infty$. As the contour integral (counter clockwise around the right-half s-plane) progresses down the $s = j\omega$ axis, the $H(s)G(s)$ plot spirals out in a general counterclockwise direction. Obviously, if it intersects the critical point $[-1, j0]$, the characteristic equation is exactly zero and the closed-loop response diverges. What is less obvious is that if the Nyquist plot *encircles* (curves around to the left of) the critical point, the closed-loop system is unstable because there will be at least one zero of the characteristic equation in the right-half s-plane. Figure 15.19 clearly shows the fourth-order system is stable, but nearly intersecting the critical point while the eighth-order system encircles the critical point indicating instability.

But, the closed-loop response seen in Figure 15.18 shows no sign of instability in the magnitude of the eighth-order compensator closed-loop response. Why? The answer can be seen in the fact that frequency response magnitudes provide no information about time causality. The phase response clearly shows a sizable phase shift of nearly 2π through the cutoff region. Recall from Section 3.3 that positive phase shifts imply a time advance (due to perhaps a time delay in the transfer function denominator). The eighth-order response in Figures 15.18 and 15.19 is unstable because it is non-causal. The effect of increasing the gain is to expand the Nyquist plot of Figure 15.19 in all directions. The effect of increasing the time delay is to rotate the Nyquist plots clockwise, significantly increasing the chances of encirclement of the critical point. Lowering the compensator cutoff frequency will move the Nyquist plots toward the origin at high frequencies which buys some gain and phase margin. *Phase margin* is the angle between the point at high frequencies where the open-loop gain is unity and the real axis on the Nyquist plot. *Gain margin* is 1 over the magnitude of the open-loop gain where it crosses the real axis, hopefully to the right of the point $[-1, j0]$. If the compensator does not attenuate sharply the high frequencies, the Nyquist plot widely encircles the origin

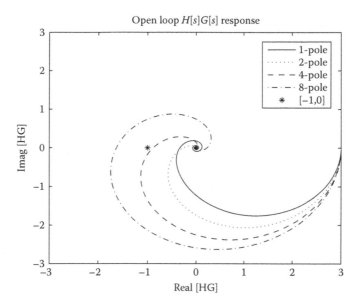

FIGURE 15.19 Nyquist plot of the open-loop $H[s]G[s]$ responses showing the eighth-order compensator encircling the [−1.0 0.0] coordinate indicating instability.

many times as frequency increases due to the plant delay τ. These wide encirclements cause ripples and overshoot to be seen in the closed-loop responses, all of which with a gain of 3.0 show a little over 12 dB active attenuation at low frequencies. Clearly, the design goal for robust active attenuation using feedback control is to have as short a time delay as possible in the control loop and to use a low-order compensator to avoid too much phase shift in the cutoff region. This is why embedded piezoelectrics with extremely short response delays work so well in smart structures designed to attenuate vibrations using active electronics. Even active hearing protectors, with nearly colocated loudspeakers and error microphones get only about 15 dB of active attenuation over a bandwidth of up to 1–2 kHz. An additional 20–25 dB of high-frequency attenuation can occur passively if the hearing protectors are of the large-sealed type.

Consider the mass–spring–damper (MSD) back in Section 2.2

$$M\,\ddot{y}_t + R\,\dot{y}_t + K\,y_t = f_t \tag{15.38}$$

where $\ddot{y}_t$ is the acceleration, $\dot{y}_t$ is the velocity, y_t is the displacement of the mass, and f_t is an applied external force on the mass. In Section 2.2, we looked at the response of the system to a force impact delta function, using the Laplace transform and then designing digital filters to respond the same way. Recall that the frequency of resonance is $\omega_0 = \sqrt{K/M}$ and the natural damped frequency is $\omega_d = \sqrt{\omega_0^2 - \zeta^2}$; $\zeta = R/(2M)$. Now we will look at putting a sensor on the mass and feeding back a signal as the external force. We will have to compensate this feedback so that only low frequencies are feedback and in a way which will be stable. But this "active feedback control" of the MSD will have a profound effect on the dynamics.

Suppose the MSD is for vibration isolation of the mass from the floor. This works well above the resonance of the MSD which is $f_0 = \sqrt{K/M}$ Hz. How far above resonance depends on the damping R. Near resonance, the vibration of the mass is actually greater than the vibration of the floor on the other side of the spring, a lot more if the damping is low. Below resonance the MSD is not effective at vibration isolation at all. So we have a couple of practical problems to solve. The springs need to be stiff enough to hold the weight of the mass due to gravity, and the mass has to be large enough so that the resonance is as low as possible. But the larger you make the mass, the stiffer the springs have to be. What should a signal-processing geek do in this situation? Suppose we sense the acceleration and apply a force $f_t = A \ddot{y}_t$ to the mass. We now can rewrite Equation 15.38 as a passive MSD.

$$(M - A)\ddot{y}_t + R\dot{y}_t + K y_t = 0 \tag{15.39}$$

If we make the acceleration feedback negative, the MSD active system will behave as if its mass were much heavier without the added static weight. Furthermore, the MSD frequency of resonance is now tunable by adjusting the feedback gain and sign. Positive acceleration feedback makes the mass appear smaller and raises the frequency of resonance. However, one cannot have the acceleration coefficient in Equation 15.39 become negative, or the system response will become unstable. Recall from Section 2.2 that the damping factor is $-R/(2[M - A])$ which if becomes negative will cause the impulse response of the system to be unstable. This can be deceiving because $[M - A]$ becomes small under high-positive feedback gain, the system damping becomes large along with the frequency of resonance. The active MSD begins to "lock up" and is a poor vibration isolator. With high negative acceleration feedback $[M - A]$ becomes large and the damping factor very small. This gives good vibration isolation but low-frequency oscillations will take a long time to die down.

A very common option for active feedback control is to use a velocity sensor to make the applied force where $f_t = B\dot{y}_t$. This makes the active MSD behave as

$$M \ddot{y}_t + (R - B)\dot{y}_t + K y_t = 0 \tag{15.40}$$

where now the amount of velocity feedback does not shift the frequency of resonance, but only affects the system damping. Damping is sometimes difficult to do with passive materials at low frequencies, so the active damping seen in Equation 15.40 can be attractive where low weight is required, such as earthquake stabilization in buildings and vibration control on spacecraft. It is also done for elite sports equipment such as high-performance skis and tennis rackets where the active damping reduces vibrations in stiff low-mass structures. The important stability concern is to ensure that the feedback loop bandwidth is small enough for the applied gain to be stable.

Applying digital ARMAX control offers no relief from the stability physics of feedback control systems, only an approach that allows one to exploit (with some risk) the known closed-loop response to gain greater attenuations over a wider bandwidth for periodic noise disturbances. If one knows the loop delay precisely, and can be sure that the loop delay will be unchanged, a periodic noise waveform can be predicted "d" samples into the future. This follows a whitening filter where one predicts the waveform sample into the future to derive the whitened error signal. The compensator $G(s)$ is designed to invert the minimum phase response of $H(s)$ and provide a "d-step ahead" linear prediction of the control signal $u_{t+d|t-1}$. Clearly, this is a risky business because if the control loop delay or noise disturbance suddenly changes, the transient signals in the control loop can quickly lead to divergence and/or nonlinear response of the actuators. ARMAX control has been successfully applied to very low-bandwidth problems in industrial process controls [astrom], among other applications. One of the nice features of ARMAX control is that one can, with least-squares error, drive the closed-loop response to produce any desired waveform or have any stable frequency response, assuming that a high-fidelity plant actuator is used and the system loop delays are small or known and constant.

ARMAX control starts by assuming an ARMA process for the noise disturbance n_t. The ARMA process has innovation w_t, which is ZMG white noise.

$$N[z] = W[z] \frac{C[z]}{D[z]} \tag{15.41}$$

Taking inverse z-transforms one obtains

$$n_t = \sum_{k=0}^{K} c_k w_{t-k} - \sum_{m=1}^{M} d_m n_{t-m} \tag{15.42}$$

The plant transfer function is also assumed to be a pole – zero system, but we will separate the delay out so that we can say the pole – zero system is in the minimum phase. The delay τ corresponds to d samples in the digital system.

$$H[z] = U[z] \frac{B[z]}{A[z]} z^{-d} \tag{15.43}$$

Taking inverse z-transforms one gets

$$h_t = \sum_{p=0}^{P} b_p u_{t-p-d} - \sum_{q=1}^{Q} a_q h_{t-q} \tag{15.44}$$

The ARMAX output y_t is the noise of Equation 15.42 minus the plant output of Equation 15.44, we can write a closed-loop system transfer function relating the ARMAX output to the noise innovation.

$$\frac{Y[z]}{W[z]} = \frac{C[z]A[z]}{A[z]D[z] + B[z]D[z]z^{-d}G[z]} \tag{15.45}$$

Solving for the compensator $G[z]$, which drives $Y[z]$ to $W[z]$ (in other words, completely whitens the ARMAX output), one obtains

$$G[z] = \frac{A[z]}{B[z]} z^{+d} \left(\frac{C[z]}{D[z]} - 1 \right) \tag{15.46}$$

But, for active noise attenuation purposes, one can do slightly better by not completely whitening the noise, but rather just whitening the noise poles and leaving any zeros in the noise alone (Swanson's thesis). The result is that ARMAX output will be driven to the MA process defined by $W[z]C[z]A[z]$. The controller for driving the noise to an all-zero process is as follows [16]:

$$\begin{aligned}
G_{CA}[z] &= \frac{A[z]}{B[z]} z^{+d} \left(\frac{1}{A[z]D[z]} - 1 \right) \\
&= z^{+d} \left(\frac{1 - B[z]D[z]}{B[z]D[z]} \right)
\end{aligned} \tag{15.47}$$

The result in Equation 15.47 becomes even more interesting when one considers that the polynomial product $B[z]D[z]$ is just the AR part of the control signal u_t.

$$U[z] = \frac{W[z]C[z]A[z] - Y[z]A[z]D[z]}{B[z]D[z]} \tag{15.48}$$

Equation 15.48 tells us that one can simply whiten the control signal u_t to obtain the necessary system identification to synthesize $G_{CA}\{z\}$. However, because $B[z]$ contains the net gain of the minimum phase plant $H[z]$, and a whitening filter operation to recover $B[z]D[z]$ will have a leading coefficient of zero, one must have an estimate of the plant gain and delay in order to properly scale $B[z]D[z]$ and employ the feedback control for active attenuation of periodic noise.

The ARMAX characteristic equation is

$$1 + H[z]G_{\mathrm{CA}}[z] = \frac{1}{A[z]D[z]} \tag{15.49}$$

which by design has no zeros to cause closed-loop instability. The polynomials in the denominator are also in the minimum phase, which helps stability. But the closed-loop gain is very high if the plant has sharp resonances ($A[z]$ has zeros near the unit circle), or if the noise disturbance is nearly sinusoidal ($D[z]$ has zeros near the unit circle). With this high-loop gain, any miscalculation of the plant delay or change in the plant response can lead to near instant divergence of the signals in the feedback loop. If we use the compensator given in Equation 15.46, we have to be concerned that the noise is *persistently exciting* such that $C[z]$ has no zeros on the unit circle and that the poles of the plant $H[z]$ are inside the unit circle. This is another reason why the compensator given in Equation 15.43 is desirable. Figure 15.20 shows an example ARMAX active attenuation result for a 50 Hz sinusoid noise disturbance and a slow-adapting LMS whitening filter. The control effort is deliberately slowed initially with a small open-loop gain to prevent the start-up transient from driving the feedback loop into instability. ARMAX active attenuation systems are specifically tuned to the plant transfer function and noise disturbance characteristics and are not well suited for nonstationary plants or disturbance signals.

One easy way to deal with the stability issues of a feedback loop is to examine the gain–bandwidth product of the control loop. As a simple guideline, one cannot exceed the gain–bandwidth product of a feedback loop without creating instability. Consider the Nyquist plot in Figure 15.19. If we know the net loop delay and consider the widest possible bandwidth transfer functions for the plant $H(s)$ and control compensator $G(s)$, we can determine the highest frequency where unity transfer functions pass through the $[-1, 0]$ coordinate on the $H(s)G(s)$ plot. This is the point where stable negative feedback for lower frequencies becomes unstable positive feedback, thus defining the maximum bandwidth for unity gain stability. If the feedback gain must be increased beyond unity,

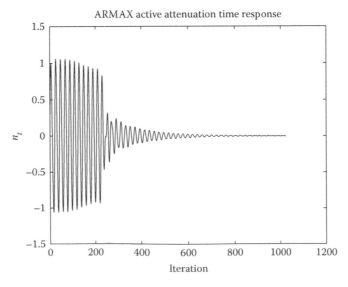

FIGURE 15.20 ARMAX attenuation of a 50 Hz sinusoid sampled at 1024 Hz using a 10-coefficient LMS whitening filter and slow responding adaptive feedback control algorithm.

the bandwidth must be reduced. But here is the challenge. When the control compensator is modified to reduce the bandwidth and increase the gain, if it adds phase to the control loop for frequencies where the gain is attenuated. This reduces the available bandwidth further so that optimizing the feedback loop requires testing the gain–bandwidth with a Nyquist plot until an acceptable response is found. But if the control plant (the actuator in an active attenuation system) has small delay, it is straightforward to obtain high negative feedback gain at low frequencies. Physically, there is an analogous gain–bandwidth product for every operational amplifier determined by the electronic phase shifts and delays in the transistor circuits of the operational amplifier. As such, using too much feedback gain over a wide bandwidth leads to instability and oscillations. When ARMAX control is used, the bandwidth of the controller is minimized so that the gain can be maximized, but instability will still occur for any frequency where the phase shift causes positive feedback with a gain greater than or equal to unity. If the feedback gain is greater than unity, the system will likely diverge beyond the dynamic range of the transducers so fast that the adaptive processing will not be able to recover stable control. It is for this reason that one should maintain some gain and phase margin at all times and avoid adaptive feedback control for plants and bandwidths that have significant phase shift or can change significantly over the time of the feedback delay.

15.4 MATLAB® EXAMPLES

We used 7m-scripts to produce the figures in this chapter, summarized in Table 15.1. The modeled electronic noise was generated in "contnz.m" where the nonwhite noise was generated by loading up a complex spectrum with random Gaussian numbers in the real and imaginary bins, and then scaling the magnitudes of each bin according to the modeled spectral shape. For the popcorn noise, the dc-offset "pops" were just manually applied to a noise waveform, and then the spectral average was estimated. It takes some special equipment and a very low-noise environment to precisely measure these types of noises. These signals are by definition at the noise floor for instrumentation; so your noise floor measuring instruments should better have lower noise than the device noise you are measuring. Typically one does this using cooled electronics in an isolation chamber free of magnetic, electrical, and even acoustical interference. So for our figures, we just synthesized the noise waveforms.

The whitening filter demonstration uses the m-script "fwiggle.m" to generate the two sinusoids in noise, so you will need to run this m-script before running "whiten.m" do execute the simulation. A stationary 120.5 Hz signal plus a randomly varying frequency sinusoid are generated. The random frequency generation is done by averaging a random frequency waveform considerable using an α–β exponential integrator with a time constant of 10,000 samples at 1024 samples/s, or about 10 s. This makes the nonstationary sinusoid meander in frequency slow so that there is some chance for the LMS filter processing it or not depending on the step size. The 120.5 Hz sinusoid is a great example of how to remove ground loop noise, but you could easily synthesize a reference for this

TABLE 15.1
MATLAB m-Scripts Used in Chapter 15

m-Script	Figures
contnz.m	15.1–15.3
whiten.m	15.6–15.8
fwiggle.m	15.6–15.8
weiner2f.m	15.10–15.11
fwiggle2f.m	15.10–15.11
feedback.m	15.17–15.19
ARMAXdemo.m	15.20

since the power lines are so stable and predictable in frequency. The m-script "weiner2f.m" uses a random frequency sinusoid in white noise as the reference signal, and also has a signal with both the reference signal and yet another random frequency. These two random frequency sinusoids are generated in the m-script "fwiggle2f.m" using independent random numbers filtered down to a time constant of the order of 10 s. Here one can see in Figure 15.10, the complete control the LMS Weiner filter can have on the two signals, separating them with relative ease using a simple LMS algorithm.

The m-script "feedback.m" evaluates some simple compensation filters and then asks the user to enter a loop gain and delay in seconds. It then produces the closed-loop frequency response showing only a modest amount of low-frequency attenuation and "ringing" for all the circuits and instability for the eighth-order compensator. The reason, the eighth-order compensator is unstable whereas it suppresses more high frequencies in the transition band, is because of the added phase introduced by the sharper filter cutoff. It is a bit of an art to design the active feedback circuit in today's active noise canceling headsets that will not become an oscillator. These devices have been possible at modest cost in recent years because of very high-performance analog electronics, and very careful measurements and design. Finally, the m-script "ARMAXdemo.m" uses an LMS filter to optimize the feedback compensation. It is getting about 46 dB attenuation at 50 Hz, with a loop delay of about 1 ms. This is considerably better than the analog example, but the bandwidth is actually quite narrow around the frequency canceled and the performance can very easily be made unstable by adjusting the parameters slightly. Adaptive feedback control is used most often for industrial plant controls where the actuator responses are fast compared to the sample rate, which may be many seconds or even minutes per sample on a large boiler controller. Increasing the bandwidth to audio rate is not a signal processing problem, as long as the actuator delays scale to very short times.

15.5 SUMMARY

This chapter presents some very important models for electrical noise, transducer performance, and noise cancellation techniques. These concepts are very much characteristic of intelligent sensor systems, which recognize signal patterns, have self-awareness of system health, and have the ability to adapt to environmental changes such as changes in noise interference. Building a physical foundation for this requires an examination of how noise is generated, how transducers work, and how to minimize or even remove the effects of noise. In Section 15.1, we present well-accepted models for several types of electrical noise. This is quite interesting, both practically and physically. An understanding of the signal features of the various types of noise (thermal, shot, contact, and popcorn) allows an intelligent sensor system to determine the cause of a given noise spectrum on any one of its sensors.

We also discuss in detail the topic of ground loops and the nature of ac power mains. By following the ac voltages and currents through the distribution system and into the sensor grounds, this often overlooked source of design error can be explained. But this does not make troubleshooting a ground loop any easier since the loop could be any number of a combination of things. The types of ac power distributed in the world are three-phase, split phase, and two-phase, and understanding how these systems are grounded and how significant currents can occur through the earth ground are import system features to understand the nature of ground loops. Fortunately, only a few general guidelines are usually sufficient to prevent ground loops and other sources of electrical interference. First is to use transducers where the case is electrically isolated allowing control of the ground and cable shield at the sensor–cable interface. Second, a balance differential signal from the sensor to the data acquisition system should be used with low electrical impedance. Third, even if the data acquisition input has a differential input (instrumentation amplifier), a low-impedance shunt resistor should be used at this input to make the sensor signal current as strong as possible. This way the electromagnetic-induced currents are insignificant compared to the signal currents. Sensor cabling should be a twisted–shielded pair so that the sensor signal is carried over separate (+) and (−) wires surrounded by the shield, grounded at the data acquisition ground point and not grounded at the sensor. Ideally, a balanced line driver amplifier is physically part of the sensor so that a low-impedance connection to the data acquisition system can be used. Following these three straightforward principles should prevent ground loops as well as all other electromagnetic interference.

Section 15.2 presents some applications of adaptive signal processing in canceling or attenuating unwanted noise signals. We must understand the nature of the signal and what we call noise in all cases where we wish to separate the two. For example the noise may be spectrally white while the signal is narrowband sinusoids. The signal may be the result of a long time average where the noise is the real-time fluctuations. Wind speed and direction is a good example of this. Or, the background noise could be found by a long time average and the signal is the transient with respect to the noise. Turbulence measurements are a good example of this. By knowing the nature of the signal and noise, we can employ filters, either fixed or adaptive, to separate the signal, defined as to what we want from the sensor and from the noise, defined as everything we do not care about the sensor. A key concept here is that the sensor "noise" may extend beyond electrical noise and the signal has expected characteristics to be exploited in the signal processing to separate it from the noise.

The signal can be canceled adaptively "on-the-wire" in what we call adaptive noise cancellation, or physically in the medium which is being sensed, in the more difficult active noise cancellation presented in Section 15.3. With feedforward adaptive controllers, the reference signal that is coherent with the error signal can be completely canceled. Using feedback control systems, a reference signal is not required but the unwanted noise can only be attenuated by the feedback control, not completely canceled. For feedforward cancellers, one often seeks a reference signal which is well time-advanced relative to the error signal. For feedback control, the error sensor and actuator response must have as little delay as possible to allow significant cancellation over a wide bandwidth as possible.

Active noise control is a fine example of an intelligent sensor system physically interacting with its environment by introducing cancellation waves to improve its performance. Such integrated adaptive control also provides the added benefit of transducer transfer functions, which can provide valuable information about system health. By identifying patterns in the electrical noise and system transfer functions, the intelligent sensor algorithm gains a level of "self-awareness" that can aid in separating signal and noise better, or reporting a sensor system failure condition. It can also determine the limits of controllability; the active control possible based on the active transducer and error transfer function, and the limits of observability based on the estimated forward error plant transfer function. For feedback active attenuation, the open-loop response of the compensator and active transducer can be evaluated as a Nyquist plot to determine the gain margin and phase margin available between the current system and instability. The feedback loop delay is the primary cause for limited stable feedback control bandwidth in these systems. ARMAX control can maximize the stable gain by narrowing the bandwidth for a particular disturbance signal, but this also minimizes the stable gain and phase margin making this approach suitable for very stable narrowband noise disturbances in nearly fixed plant transfer functions with small loop delays. If either the transducer plant response of the noise disturbance suddenly changes, the ARMAX control loop goes unstable with very rapid certainty.

PROBLEMS

1. Sketch the spectral shapes and typical time-domain plots of thermal, contact, and popcorn noise.
2. What is the amount of dB attenuation provided by a 1 mm-thick cylindrical copper shield on a coaxial cable for a 120 vrms 60 Hz cable located 0.1 m away?
3. Explain how two single-conductor coaxial cables can be used to provide a balanced differential sensor connection where the shield is floated out to a "two-wire" type sensor.
4. The coherence between a reference signal and error signal sensor is only 0.90 in a potential feedforward ANC system. Assuming no feedback to the reference signal and that the

error plant is known or can be measured with great precision, what is the maximum dB cancellation possible?

5. For the "active SE plant identification" what would the "SE" estimate converge to if the error signal were completely driven to zero? Assuming the SE identification injected noise is uncorrelated with the reference signal, if the reference signal noise is completely canceled, what is the residual error signal level?

6. For a multichannel feedforward ANC system, explain why the transfer functions between actuators and error sensor of different channels should be as independent as possible.

7. A velocity sensor is used to provide positive force feedback to a MSD system with a resonant natural frequency of 10 Hz and a damped frequency of 8 Hz and a mass of 1000 kg. How much feedback gain (assume zero delay) is required to prevent any oscillations?

8. A hydraulic actuator is used to stabilize a heavy gun barrel on a ship. The transfer function of the actuator to a vibration sensor on the gun barrel reveals a delay of 200 ms over a bandwidth from 0.1 to 3 Hz. Using feedback control, how much stable attenuation of the ships motion can be achieved over the 3 Hz bandwidth? How about 1 Hz bandwidth?

REFERENCES

1. J. B. Johnson, Thermal agitation of electricity in conductors, *Phys Rev*, 1928, 32, pp. 97–99.
2. H. Nyquist, Thermal agitation of electrical charge in conductors, *Phys Rev*, 1928, 32, pp. 110–113.
3. H. W. Ott, *Noise Reduction Techniques in Electronic Systems*, New York, NY: Wiley, 1976.
4. W. H. Hayt, Jr., *Engineering Electromagnetics*, 3rd ed. New York, NY: McGraw-Hill, 1974.
5. D. F. Stout and M. Kaufmann, *Handbook of Operational Amplifier Design*, New York, NY: McGraw-Hill, 1976.
6. P. A. Nelson and S. J. Elliott, *Active Control of Sound*, New York, NY: Academic Press, 1992.
7. C. R. Fuller, S. J. Elliott, and P. A. Nelson, *Active Control of Vibration*, San Diego, CA: Academic Press, 1996.
8. C. H. Hansen and S. D. Snyder, *Active Control of Noise and Vibration*, New York, NY: E & FN Spon, 1997.
9. R. S. McGuinn, G. C. Lauchle, and D. C. Swanson, Low flow-noise microphone for active control applications, *AIAA J*, 35(1), pp. 29–34.
10. D. C. Swanson, S. M. Hirsch, K. M. Reichard, and J. Tichy, Development of a frequency-domain Filtered-X intensity ANC algorithm, *Applied Acoustics*, 1999, 57(1), pp. 39–50.
11. L. J. Erikkson, Active noise control, in *Noise and Vibration Control Engineering*, eds., L. L. Beranek and I. L. Vér, Chapter 15, New York, NY: Wiley, 1992.
12. S. D. Sommerfeldt and L. Bischoff, Multi-channel adaptive control of structural vibration, *Noise Control Eng J*, 1991, 37(2), pp. 77–89.
13. B. J. Kim and D. C. Swanson, Linear independence method for system identification/secondary path modeling for active control, *JASA*, 2005, 118(3), pp. 1452–1468.
14. D. C. Swanson and K. M. Reichard, The generalized multichannel Filtered-X algorithm, in *Second Conference Recent Advances in Active Control of Sound and Vibration*, ed. R. A. Burdisso, Lancaster: Technomic, 1993, pp. 550–561.
15. F. Asano, Y. Suzuki, and D. C. Swanson, Optimization of control source configuration in active control systems using Gram–Schimdt orthogonalization, *IEEE Trans Speech Audio Proc*, 1999, 7(2), pp. 213–220.
16. D. Swanson, Active attenuation of acoustic noise using adaptive ARMAX control, PhD dissertation, The Pennsylvania State University, Dec 1986.
17. G. C. Goodwin and K. S. Sin, *Adaptive Filtering, Prediction, and Control*, Englewood Cliffs, NJ: Prentice-Hall, 1984.
18. K. J. Åstom and B. Wittenmark, *Adaptive Control*, New York, NY: Addison-Wesley, 1989.

16 Sensors and Transducers

The difference between sensors and transducers is simply that a transducer has the capability to produce the wave it senses. In this chapter, we examine the link between the signals one measures and the physical wave or energy that is being measured by a sensor, or generated by an actuator. We must also consider the signal conditioning and transmission from the sensor in order to properly characterize the signal information and background noise. Sensor signals are processed in all sorts of different ways before the voltages ever get to the data acquisition electronics and it can be very important to understand the physics of the signal chain to obtain the correct signal and noise estimates.

We group sensors and transducers into three main categories: simple sensors; electromechanical transducers; and spectrometers. We do not have space to explain every type of sensor or transducer, but we can explain the most common types of signals from these sensors and how to interpret them. We also can organize the discussion toward the kinds of sensors and transducers most often used or of high importance in the early twenty-first century. Simple sensors are "dc-like" in that we do not examine their information as a function of frequency, but rather as a slowly varying parameter, such as temperature. For sensors and actuators where we are concerned with their frequency response, we analyze linear transducers using electrical circuit equivalents. This technique is very useful for acoustic and vibration transducers, but also applies directly to electromagnetic sensors. At very high frequencies of electromagnetic radiation, we can use a spectrometer to separate the light wavelengths geometrically so that we can sample them using an array of simple optical sensors. Spectrometers are very common for chemical and explosive detection as the light wavelengths of radiation and/or absorption are characteristic of the atomic bonds in the molecules. As we examine large molecules that make up amino acids, proteins, and nucleic acids, the optical spectrometer with ultraviolet light illuminations can be used to detect fluorescence as the huge molecules transfer energy from higher wavelengths to lower. Biological sensing is largely based on this principle and that one can make a marker that will stick to one type of biological molecule and be detected optically. Spectrometer is a name given to any device that can decompose something into its core components, otherwise unseen. The term originates from "specter" which refers to a ghost or spirit (that which is unseen). Spectrometers also apply to devices that categorize the received energy, such as a gamma ray spectrometer (GRS). Nuclear spectrometers really are not physical devices, but rather a signal processing device where the voltage level is binned to be associated with an energy level of a nuclear particle or gamma ray captured by a sensing material.

All of these sensor and transducers taken together for a basis for detecting all known waves in our world such as sound, vibration, heat, pressure, smell, light, and high-energy radiation of nuclear sources, for which the human body possesses no sensing ability (except observing the damage done to tissue by radiation). We will discuss biomimetic sensing wherever appropriate that is how our man-made sensors work relative to those in our five senses in the human body. Comparing biomimetic sensing to the body is fascinating and very illustrative for future sensors, transducers, and the associated information systems to be build around them. Section 16.1 focuses on simple sensors and common signal formats. Section 16.2 explains transducers, mainly used in sound and vibration sensing and actuation. Section 16.3 discusses sensors for chemical detection and identification of larger biological materials. Section 16.4 focuses on detection of nuclear radiation. Together these sensors form the basis for chemical, biological, radiological (harmful electromagnetic radiation), nuclear, and explosive (CBRNE) sensors essential for the protection of modern society against terrorism and war. This is the underlying theme of this textbook: how will we build the sensors and central nervous system for the planet?

16.1   SIMPLE TRANSDUCER SIGNALS

It is very straightforward to view a simple sensor as a dc-like device that produces a slowly varying voltage proportional to the particular energy level of interest, such as pressure, temperature, or light intensity. There are literally hundreds of types of sensors like this [1], all with various interesting physics. The salient point is the output of such sensors, at their most basic level, is a dc-like voltage signal linearly proportional (or nonlinearly with a known response) to the thing being sensed and some background noise, which could either be electronic or physical, or both. In the simplest scenario, we view the sensor signal as a Gaussian distribution where the mean is the sensor signal level and the variance represents a useful noise metric. We can tell when the signal is above the background noise by assuming a Rician distribution for signal and noise and comparing the signal and noise levels (see Section 12.1). Since we have processed the sensor signal with RMS averaging to get the mean and the variance, the Rician PDF is essentially a Gaussian for the signal plus noise, regardless of SNR. This enables one to define thresholds, probabilities of false alarm and detection (see Section 12.2), and ROC curves.

In the field, additional things are often done to these simple sensor signals to enhance the SNR, allow multiple sensors to be multiplexed on the same signal wire, or when transmitting the signal over a telemetry link or recording the signal to magnetic storage tape as an analog signal, although rare today with the advent of digital recording and communications. These sensor signal processing schemes still exist inside the sensor device as well as in older deployed sensor networks. How does one determine the SNR for these processing schemes? What about the noise and bandwidth of the signals? We define the "channel" as a transfer function with additive noise that can represent the communications channel or the recording storage response. For digital transmissions and recordings, the channel response and noise can affect the bit error rate, but with bit error correction algorithms, one either gets perfect data streams or a complete dropout of the channel. However, if the digital data stream is compressed, in a nonlossless format, we can evaluate the signal SNR by examining the compressed data packets (see Section 4.4). In this section we present straightforward models for the signals, noise, and bandwidth for a generic simple sensor, and expand the concept for sensors with bandwidth and modulation transmission.

Recall that for the most simple dc-like sensor signal S_0 in white noise (from the sensor and environment) with standard deviation σ_S when we include N samples in an RMS average the resulting mean is S_0 and the standard deviation is $\sigma_{RMS} = \sigma_s/\sqrt{N}$ where the sensor noise is independent from sample to sample. The sensor signal PDF gives us an estimate for the signal (the mean) and the background noise σ_S from knowing N and σ_{RMS} even though the data samples contain signal and noise mixed together. We describe the sensor signal as "dc-like" meaning that it slowly varies with the energy wave being sensed, but is essentially constant during the RMS averaging time interval defined by the sample rate and effective number of samples used N. This sensor signal is a very low bandwidth signal. As the signal goes from the sensing transducer to the data acquisition and detection system, it travels through a "channel" which could be a circuit, a cable, a waveguide, a radio transmission path, or a recording medium like the magnetic tape. The channel has a frequency response $H(f)$ and can add noise to the sensor signal. This is depicted in Figure 16.1a (top sketch) where sensor and channel noises simply add when the channel response is flat and the channel noise is white. For very low bandwidth signals only the response of the channel near 0 Hz matters. For the baseline sensor SNR improvement at the detector is

$$\text{SNR}_0 = \frac{1}{N}\frac{S_0^2}{\sigma_S^2+\sigma_c^2} \tag{16.1}$$

Suppose the channel noise is not white, but has contact and/or popcorn noise raising the noise levels near 0 Hz? The $1/f^n$ shape of the channel noise suggests one could greatly improve the SNR by moving the low bandwidth sensor signal up to a frequency range in the channel where the channel SNR is much higher. This is illustrated in the signal flow path of Figure 16.1b (bottom sketch). The amplitude

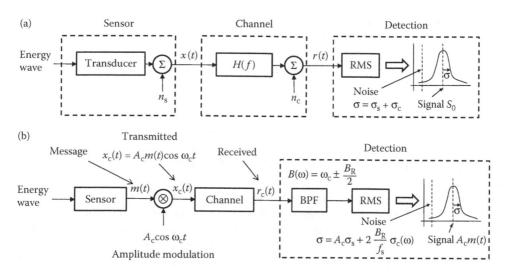

FIGURE 16.1　Simple transmission of a sensor signal plus environmental noise (a) through a channel (wire or transmission) which adds noise, and into the data acquisition and detection system, and (b) using transmission AM and detection to reduce channel noise.

of the cosine wave is scaled by the sensor "message" waveform $m(t)$, the "dc-like" signal from the sensor with its noise. Now a sinusoid is transmitted over the channel and the received signal at the data acquisition and detection system is band-pass filtered (BPF) first and then an RMS average produces an improved SNR. The modulated sensor signal SNR improves in three ways. First, the channel noise bandwidth B_R is much less from the BPF so that the noise spectral density is ηB_R. Second, the SNR of the channel is increased as sinusoidal carrier amplitude is increased above the channel noise. And third, the SNR of the channel is further increased because we moved away from the high channel noise near 0 Hz. The low bandwidth sensor signal SNR in the modulated channel is

$$\text{SNR}_{AM} = \frac{1}{N} \frac{A_c^2 S_0^2}{A_c^2 \sigma_S^2 + 2\eta B_R} = \frac{1}{N} \frac{S_0^2}{\sigma_S^2 + 2\left(\dfrac{\eta B_R}{A_c^2}\right)} \tag{16.2}$$

Clearly from Equation 16.2 one can make the channel noise insignificant by using a strong carrier amplitude, using narrow BPF bandwidth B_R, and using a carrier frequency in a low-noise frequency range of the channel. When the sensor signal is not "dc-like" and has alternating voltages over a wider bandwidth B_S all one needs to do is offset the amplitude modulation (AM) so that the carrier amplitude does not go negative, which would cause severe distortion.

$$x_c(t) = A_c \left[1 + k_{AM} m(t)\right] \cos \omega_c t \tag{16.3}$$

where scale factor $k_{AM} = 1/A_m$ and A_m is the maximum amplitude of the sensor message signal. Also, the BPF in Figure 16.1b must be at least as wide as the signal bandwidth B_S and the RMS integrator must have a time constant short enough to respond to the highest frequency of interest from the sensor. The wider band AM SNR is

$$\text{SNR}_{AM} = \frac{1}{2N} \frac{A_c^2 S_0^2}{A_c^2 \sigma_S^2 + 2\eta B_S} = \frac{1}{2N} \frac{S_0^2}{\sigma_S^2 + 2\left(\dfrac{\eta B_S}{A_c^2}\right)} \tag{16.4}$$

The factor of 1/2 comes from the need to offset and scale the alternating voltage message signal to avoid over modulation. For a "dc-like" sensor signal the factor of 1/2 in Equation 16.4 would not be needed.

A better way to process the amplitude modulated signals is seen in Figure 16.2. Here the received waveform is demodulated to move the BPF signal back down to base band centered at 0 Hz with bandwidth B_S. The low-pass filter (LPF) is used to remove the signal components at twice the modulation frequency. This is the standard modulation-demodulation of the AM done for AM radio broadcasts. However, a big source of performance problem is when the demodulation frequency ω_d does not exactly match the modulation carrier frequency ω_c. There are a few schemes for transmitting and recovering the modulation carrier frequency [2], but the most common approach seen in Figure 16.2 is to use a *phase-locked loop* (PLL). The PLL is very important circuit that allows the demodulator to be synchronized to the modulator precisely. It works by generating a sinusoid using another circuit called a voltage controlled oscillator (VCO) which basically adjusts its frequency based on a dc-like input voltage control signal. A phase detection circuit is used to produce the control voltage input to the VCO. The phase detector can be made a number of ways. A reasonable amount of precision is obtained by first limiting the two sinusoid inputs (the received modulation and the VCO output) to square waves, inverting the VCO square wave, and then comparing the two square waves [3]. When they are matched, the output of the comparator is zero exactly. When the received modulation frequency is slightly higher than the VCO, the comparator outputs positive pulses, which are integrated to a positive dc voltage, which then drive the VCO to a higher frequency. If the VCO frequency is too high, the output of the comparator is mostly negative pulses, which when integrated provide a small negative dc control voltage to the VCO to lower the VCO frequency until they match. No synchronization is perfect, but when the modulation and demodulation sinusoids are closely matched in phase, we say the VCO is locked-in. This then allows the AM signal to be precisely recovered, and if the AM carrier amplitude is well above the channel noise, the added noise of the recovered sensor signal from the channel is negligible. PLLs are the critical circuit in a *lock-in amplifier*, which is the demodulator/detector side of Figure 16.2.

Assuming a strong AM carrier and nearly constant channel frequency response and noise characteristics, one of the nice things about AM modulation is the well-defined bandwidth for a given

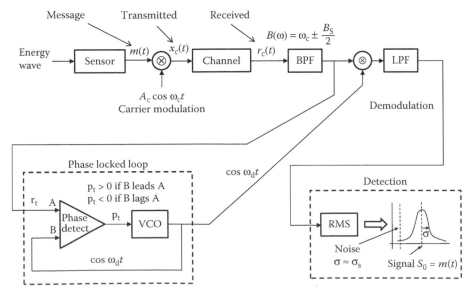

FIGURE 16.2 AM of a band-limited sensor signal where the receiver uses a phase-locked loop (PLL) to synchronize the demodulation.

carrier. If the sensor bandwidth is moderate-to-small compared to the available channel bandwidth, the channel can be sliced into multiple channels so that many sensors can share the same wire or radio link, just like broadcasting stations, and not interfere with each other. This is called *frequency division multiplexing* (FDM) and is still used widely on land-line telephone wires which carry hundreds of 3 kHz voice channel bands on the same twisted pair wires. If the transfer function and background noise in a given FDM channel is not flat, a *pre-emphasis filter* such as an Eckhart filter (see Section 6.5) can be used to boost the transmitted signal and best optimize the FDM channel SNR.

Figure 16.3 shows the concept of angle modulation, which unlike AM, is completely insensitive to the amplitude of the received carrier signal. As the name implies, angle modulation wiggles the carrier phase angle to carry the sensor message signal. This is limited to a modulation of 180° in either direction to avoid severe distortion from a phase ambiguity and is called phase modulation (PM). A PLL circuit can be used to lock-in and measure the carrier PM relative to a stable oscillator frequency standard. The SNR for a PM signal is different from an AM signal by a ratio of the carrier to the peak message amplitude, and because the transmission is not affected by changes in the channel transfer-function magnitude. The SNR improves quite a bit based on how the message signal is scaled to use the full range of PM (it gets worse than AM if you do not use the full range). In general, optimized PM is better than AM in terms of SNR because only the channel noise in-phase with the PM carrier is received coherently. PM is also known as narrowband frequency modulation (NBFM). The PM carrier signal is

$$x_c(t) = A_c \cos[\omega_c t + \phi_m(t)]$$

$$= A_c \cos[\omega_c t + k_p m(t)]; \quad k_p = \frac{\pi}{A_m} \tag{16.5}$$

The SNR of a PM signal is [2]

$$\text{SNR}_{\text{PM}} = k_p^2 S_m \frac{A_c^2}{2\eta B_P} \tag{16.6}$$

where S_m is the message signal power and the bandwidth for the phase modulation B_P is about twice that of the original message signal. *Note that we are no longer considering the original sensor SNR, or the averaging at the receiver, part of the calculation, only the SNR of the channel.* This is because the spectrum of the phase modulation extends outside the AM-required

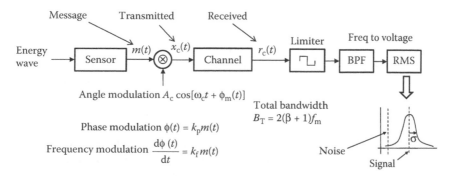

FIGURE 16.3 An FM showing the phase angle modulation of the carrier and the simple limiter and filter of the demodulation process to recover the sensor message signal.

bandwidth. We can improve the SNR even further by widening the channel bandwidth using *frequency modulation* (FM).

$$x_c(t) = A_c \cos[\omega_c t + \phi_m(t)]$$

$$= A_c \cos\left[\omega_c t + k_f \int_{-\infty}^{t} m(\tau)d\tau\right] \quad (16.7)$$

where k_f is the frequency modulation scale factor that maps the sensor message voltage to a frequency. Recall that frequency is the time derivative of phase, hence the integral in angle of Equation 16.7. For a tonal sensor message signal with an amplitude A_m, we can write the frequency deviation relative to the sensor signal bandwidth.

$$D = \frac{k_f A_m}{B_S} \quad (16.8)$$

Using Carson's rule [2], the bandwidth required of the FM channel is approximately

$$B_F = 2(D+1)B_S = 2(f_\Delta + B_S) \quad (16.9)$$

and the maximum frequency deviation is $f_\Delta = DB_S$. Again, this is approximate due to the Bessel function series involved in the spectrum of an FM signal. For $D \ll 1$ the transmitted signal is NBFM or PM with a transmitted bandwidth of about $2B_S$ and when $D \gg 1$ we call the signal wideband FM, or WBFM, and its transmitted bandwidth is approximately $2f_\Delta$.

For a wider bandwidth signal in a channel, one would expect the noise to increase, and since the signal is spread over a wider bandwidth, one would also expect the SNR in a narrow sub-band of the channel to decrease, and it does. However, what matters is the SNR after demodulation. FM demodulation is fairly straightforward. The received signal in a given channel band is limited to make a square wave (this removes amplitude dependence). Then the limited signal is filtered on the falling edge of a sharp band pass filter. This eliminates the harmonics of the square wave and creates a signal with amplitude depending on the carrier frequency. A simple envelope (AM type) detector filters out the carrier entirely restoring the original sensor message signal, which is then scaled by $k_d = k_f^{-1}$ or appropriately to reset the signal amplitude as desired. The SNR of the FM signal is a bit complicated to sort out and is

$$SNR_{FM} = 3D^2 S_m \frac{A_c^2}{2\eta B_F} \quad (16.10)$$

The key feature of FM signals is that the SNR of the channel can be improved by using more bandwidth as seen in Equation 16.10. Equation 16.10 is left similar to the SNR for PM in Equation 16.6 and AM in Equation 16.4 for comparison, but note how the FM signal also requires more channel bandwidth. However, the SNR still improves as bandwidth is increased. This feature allows very low power WBFM to be used where power is limited, such as deep space satellites. It also allows for improved SNR of magnetic recording where the magnetization saturates at a certain amplitude. The details of the PM and FM SNR derivation are beyond this book, but unless we briefly discuss the types of signals and their respective SNRs one may not know how to interpret the sensor signal statistics. For example, FM noise increases with frequency deviation, which corresponds to the original sensor message signal maxima and minima. This is why it works so well. The received noise increases only when the original signal gets large. While the mathematics of FM are fairly complex, the electronics required to make it work are not. A VCO can be used to create the FM transmission and the receiver is not much more complicated than an AM receiver. This was important for the early days of radio electronics, as was keeping the cost of the receivers to a minimum. The markets for AM and FM broadcasts emerged during the great depression in the United States during the 1930s. In the 1950s, a mix of AM and FM multicasts were used for television broadcast, which by the 1960s multiplexed in color video signals and later stereo audio signals. The brilliance of this engineering is seen in the way they maintained backward compatibility to black and white mono television.

Current digital television broadcasts employ quadrature amplitude modulation (QAM) where a sine and a cosine at the same frequency are amplitude modulated simultaneously. The in-phase part is the cosine modulation and the quadrature phase part is the sine modulation. Technically, the digital signals transmitted using AM are called "Amplitude Shift Keyed" (ASK). The counterpart for FM is "Frequency Shift Keyed" or FSK. ASK in QAM allows a "constellation" of states. Typical protocols are QAM-16, QAM-64, and QAM-256. QAM-16 would have four states in each quadrant for the in-phase (real or x-axis) and quadrature (imaginary or y-axis). A 4-bit "nibble" (a nibble is half of an 8-bit byte) can then be transmitted all at once by ASK to the corresponding coordinate in the 16 possible coordinates for QAM-16. A whole byte can be sent at once using QAM-256. When you combine these efficient digital transmission schemes with MPEG compression (see Section 8.5) the bandwidth required for digital television broadcasts is actually less than that required for the old analog transmission. The SNR required at the receiver is also less, and since the transmission is digital reception and is "all or nothing" meaning either a perfect picture and sound, or a blank screen. When the SNR is on the edge of detection the transmission is intermittent showing the partial reconstruction of the MPEG image and sound.

Another type of simple sensor signal is called the *pulse width modulated* (PWM) signal. PWM signals are read using a very high-speed counter that allows the time between and duration of square pulses to represent the sensor signal message. The PWM "duty cycle" is another measure of the pulse width. A 99% duty cycle would have a nearly "dc-like" digital signal with short off spikes while a 1% duty cycle would be off at zero volts and have short on spikes at 5 V. Simple circuits can measure these pulse times relative to a very high-speed internal clock thousands of times the nominal speed of the pulses and can report data such as the pulse leading edge rise times, pulse duration, as well as calculate averages, and so on. The noise in a PWM signal is the minimum clock period, which can be as low as ns (10^{-9} s per clock counter period). So for relatively low-bandwidth sensors which have a dedicated wire to the data acquisition system, PWM can provide a great deal of bits of precision. Since it is a digital signal, the SNR of the channel only needs to be a few dB. PWM sensor signals are very common in high noise environments or when a 1-wire solution is needed with minimal hardware electronics. Most microcontrollers have PWM inputs and outputs for sensing and control.

A variant of PWM is *time division multiplexing* (TDM) where the pulse duration does not change, but the time between pulses is used to convey the information, similar to Morse code. This is common for optical transmission where one can transmit intense pulses with extremely high speed, and you can measure precisely the time between the pulses with high precision. TDM transmissions are of very wide bandwidth. This means that they are less susceptible to interference from poor SNR in one part of the channel or another. But if a pulse is lost the TDM stream is broken and must be resynchronized. To circumvent transmission dropouts the data contain repeated patterns for frame marking, error correction encoding, and repeated data packets, in much the same way straight serial bit streams correct errors. For digital sensor bit streams, an intelligent sensor system must be tolerant of intermittent signals but detecting the break in the data stream and managing the data statistics either side of the break. This avoids the break in communications biasing the signal features such as the signal and noise levels. A noted intermittent sensor with corrected information is much more desirable than a sensor providing erroneous data due to an intermittent connection.

16.2 ACOUSTIC AND VIBRATION SENSORS

Acoustics and vibration sensors are really only a small subset of sensors in use today, but they are by far the most interesting insofar as magnitude and phase calibration of wideband signals, transducer physics, and environmental effects. Our attention focuses on acoustics and vibration because antenna elements for electromagnetic wave sensing are basically a simple unshielded wire or coil, video sensors are just large arrays of photodiodes which integrate an RMS measure of the photons, and low-bandwidth sensors generally do not require a lot of sensor-signal processing. The color correction/calibration problem for video signals is certainly interesting, but will be left for a future revision of this book. Correction of sensor nonlinearities (low bandwidth or video) is also an interesting and worthy problem, but it is really beyond the scope of this book, which is based almost totally on linear system signal processing. Acoustic and vibration sensors are used in a wide range of sensor signal processing applications and also require calibration periodically to insure sensor-data integrity.

We must make a very useful distinction between sensors and transducers. A sensor detects a physical energy wave and produces a voltage proportional to the energy. But, sensors are not reversible, meaning that an electrical energy input also produces a proportional energy wave. A transducer passively converts physical wave energy from one form to another. Furthermore, a transducer is reversible, meaning for example, that if a mechanical wave input produces a voltage output, a voltage input produces a mechanical wave output. All linear passive transducers are reversible in this way. Transducers which are linear (meaning very low distortion but not necessarily constant frequency response) obey the laws of *reciprocity* for their respective transmitting and receiving sensitivities. In a conservative field if one transmits a wave at point A of amplitude X and receives a wave at point B of amplitude Y, the propagation loss being Y/X, one would have the same propagation loss Y/X by transmitting X from point B and receiving Y at point A. For reciprocal transducers, one could input a current I at point A and measure a voltage v at point B, and by subsequently inputting a current i in the same transducer at point B one would measure the same voltage v from the transducer still at point A.

Our discussion focuses initially on two main types of transducers. The first is the *electromagnetic mechanical* transducer and the second is the *electrostatic mechanical* transducer. These two types of transducer mechanisms cover the vast majority of microphones, loudspeakers, vibration sensors, and sonar systems in use today. Later, a brief discussion of *micro-electromechanical system (MEMS)* sensors will be given, which will also help clearly distinguish the physics between transducers and sensors.

16.2.1 ELECTROMAGNETIC MECHANICAL TRANSDUCER

The electromagnetic mechanical transducer is most commonly seen as a moving coil loudspeaker [4]. There are literally billions of moving coil loudspeakers in the world today and has probably had one of the most significant impacts on society of any electronic device until the digital computer. The transduction mechanism in a loudspeaker is essentially the same as that in a "dynamic microphone," a geophone (used widely in seismic sensing and low-frequency vibration), as well as electric motors and generators. Figure 16.4 shows the electromotive force (voltage) generated when a conductor has velocity u in a magnetic flux B in part (a), and the mechanical force f generated when a current i flows through the conductor in the magnetic field B in part (b). The orthogonal arrangement of the "voltage equals velocity times magnetic field" or "force equals current times magnetic field," follow the "right-hand-rule" orientation. This is where with the right-hand index finger, middle finger, and thumb pointing in orthogonal directions, the index finger represents velocity or current, the middle finger represents magnetic field, and the thumb points in the direction of the resulting voltage or force, respectively. The physical equations describing the relationship between electrical and mechanical parameters are

$$f = Bli, \quad e = Blu \tag{16.11}$$

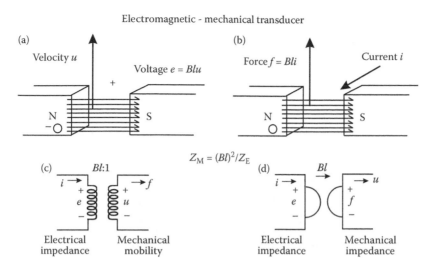

FIGURE 16.4 Voltage generated by a moving conductor in a magnetic flux (a), force generated by a current in a magnetic flux (b), transformer representation (c), and gyrator representation (d).

where i is the current in amperes, B is the magnetic field in Tesla (1 Tesla equals 1 Weber per square meter, 1 Weber equals 1 Newton meter per ampere), l is the effective length of the conductor in the magnetic flux, e is the velocity-generated voltage, and u is the conductor velocity. One can see a very peculiar and important relationship by examining the impedances defined by

$$Z_M = \frac{f}{u} = (Bl)^2 \frac{1}{Z_E} = (Bl)^2 \frac{i}{e} \qquad (16.12)$$

The mechanical impedance Z_M is seen to be proportional to the inverse of electrical impedance Z_E. The inverse of impedance is called *mobility*. This is not required for a transducer to be reversible; it is just the result of the relationship between force and current, and voltage and velocity. It does create some challenging circuit analysis of the transducer as seen in Figure 16.5 for a complete loudspeaker system in a closed cabinet. Figure 16.6 shows a cutaway view of a typical loudspeaker.

Analysis of the loudspeaker is quite interesting. To maximize the efficiency of the force generation, the conductor is wound into a solenoid, or linear coil. This coil has electrical resistance R_e and inductance L_e as seen on the left in Figure 16.5a. The movable coil surrounds either a permanent magnet or a ferrous pole piece, depending on the design. When a ceramic permanent magnet is used, it is usually a flat "washer-shaped" annular ring around the outside of the coil where a ferrous pole piece is inside the coil as seen in Figure 16.6. When a rare earth magnet such as Alnico (aluminum nickel and cobalt alloy) is used, its favored geometry usually allows it to be placed inside the coil and a ferrous return magnetic circuit is used on the outside. In some antique radios, an electromagnet is used in what was called an "electrodynamic loudspeaker." Clever circuit designs in the mid-1930s would use the high inductance of the electromagnet coil as part of the high-voltage dc power supply, usually canceling ac power line hum in the process. A gyrator (there is no physical equivalent device of a "gyrator") is used in part (a) and (b) of Figure 16.5 to leave the mechanical and acoustical parts of the circuit in an impedance analogy [5]. The coil is mechanically attached to a diaphragm, usually cone-shaped for high stiffness in the direction of motion. The diaphragm and coil assembly have mass M_d. The diaphragm structure is attached to the frame of the loudspeaker, called the basket, at the coil through an accordion-like structure called the spider, and at the outer rim of the diaphragm through a structure called the surround, usually made of foam rubber. The

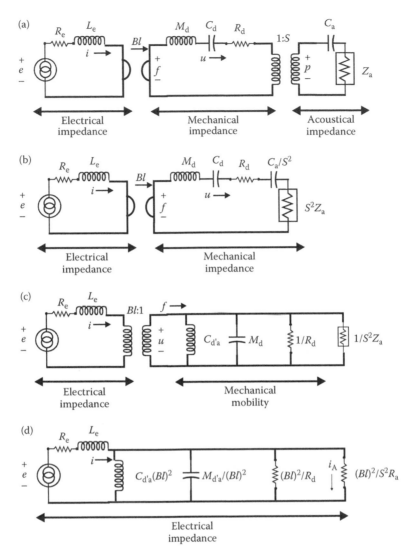

FIGURE 16.5 Equivalent electrical circuit representations of a moving coil loudspeaker in a closed cabinet. (a) Impedance models; (b) combining acoustic into mechanical impedance; (c) converting mechanical impedance to mechanical mobility; and (d) expressing all components in an electrical impedance analogy.

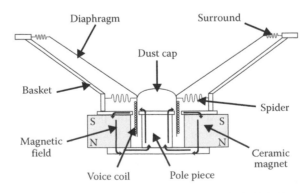

FIGURE 16.6 A typical loudspeaker cutaway showing the magnetic field, voice coil, and mechanical diaphragm suspension where the driving force is created in the small gap around the top of the pole piece.

spider and surround together has a compliance (inverse of mechanical stiffness) C_d, and mechanical damping R_d. The velocity of the diaphragm u produces an acoustic volume velocity $U = Su$ cubic meters per second, where S is the surface area of the diaphragm. The acoustic pressure p is the mechanical force f divided by S. The air in the closed cabinet of the loudspeaker has acoustic compliance $C_a = V_b/\rho c^2$, where V_b is the volume in cubic meters, the density $\rho = 1.21$ kg/m³ for air, and c is the speed of sound in air (typically 344 m/s). Note that for an infinite-sized cabinet the capacitor equivalent is short circuited while if the cabinet were filled with cement (or other stiff material) the capacitor equivalent becomes an open circuit signifying no diaphragm motion. Finally, we have the acoustic radiation impedance Z_a defined as

$$Z_a = \frac{\rho c}{S}\left(\frac{k^2 a^2}{4} + j 0.6\, ka\right) = R_a + j\omega M_a \tag{16.13}$$

where $M_a = 0.6\, \rho a/S$ and $S = \pi a^2$ (diaphragm is assumed circular with radius a). Multiplying acoustical impedance by S^2 gives mechanical impedance, which can be seen on the right in part (b) of Figure 16.5. The mechanical compliance (the inverse of stiffness) of the cabinet air volume C_a/S^2 is combined with the mechanical compliance of the diaphragm C_d to give the total mechanical compliance $C_{d'a}$.

$$C_{d'a} = \frac{C_d\, C_a}{S^2\, C_d + C_a} \tag{16.14}$$

The combination of diaphragm compliance with the compliance of the air in the closed cabinet adds some stiffness to the loudspeaker system, raising the resonance frequency and giving rise to the popular term "acoustic suspension loudspeaker," owing to the fact that the air stiffness is the dominant part of the compliance.

To continue in our transducer analysis, we must convert the mechanical impedance to a mechanical mobility (inverse of impedance) as seen in part (c) of Figure 16.5. In a mobility analogy, the impedance of each element is inverted and the series circuit is replaced by a parallel circuit. Using a mechanical mobility analogy, a transformer with turns ratio $Bl:1$ represents a physical circuit device where the force in the mechanical system is a current proportional to the electrical input current i by the relation $f = Bli$ and the diaphragm velocity u is a voltage in the equivalent circuit directly to the physical applied voltage e by the relation $e = Blu$. Finally, part (d) of Figure 16.5 shows the equivalent electrical analogy circuit with the transformer proportionality reflected in the mechanical mobility elements. One literally could build this electrical circuit, using the physics-based element values for the mechanical parts, and measure exactly the same electrical input impedance as the actual loudspeaker. Figure 16.7 shows the modeled electrical input impedance for a typical 30 cm loudspeaker where $R_e = 6\ \Omega$, $L_e = 3$ mH, $Bl = 13.25$, $M_d = 60$ g, $R_d = 1.5$ mechanical Ω, the box volume V_b is 1 m³, and the driver resonance is 55 Hz. Figure 16.7 shows the electrical input impedance to the equivalent circuit for the loudspeaker in its closed cabinet.

The loudspeaker input impedance is a very interesting and useful quantity because it provides features indicative of many important physical parameters and it can be monitored while the loudspeaker is in use [6]. One can simply monitor the voltage across a 1 Ω resistor in series with the loudspeaker to get the current, and the voltage across the voice coil terminals and calculate a transfer function of voltage divided by current, or electrical impedance. In electrical impedance units, the input impedance is simply $R_e + j\omega L_e + (Bl)^2$ divided by the mechanical impedance of the loudspeaker. As easily seen in Figure 16.7 the impedance at 0 Hz is simply the voice coil resistance R_e. At high frequencies, the slope of the impedance curve is L_e. The total system mechanical resonance is also easily seen as the peak at 57.7 Hz. The system resonance is slightly higher than the

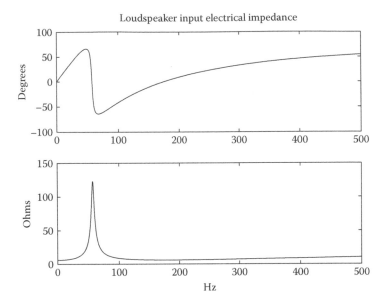

FIGURE 16.7 Electrical input impedance of a typical loudspeaker in a closed cabinet showing the system resonance at 58 Hz, a dc resistance of 6 Ω, and a voice coil inductance of 3 mH.

loudspeaker resonance of 55 Hz because of the added stiffness of the air inside the 1 m³ cabinet. The total system resonance quality factor, Q_T, can be measured by experimentally determining the frequencies above and below resonance, where the impedance drops from Z_{res} (the value at resonance) to $Z(f_2^g, f_1^g) = \sqrt{Z_{res}R_e}$. This "geometric mean" is used because the total system Q_T is the parallel combination of the mechanical Q_{ms} and electrical Q_{es}. The system Q_T is correspondingly

$$Q_T = \frac{f_s}{f_2^g - f_1^g} \sqrt{\frac{R_e}{Z_{res}}} \tag{16.15}$$

where f_2^g and f_1^g are the frequencies above and below resonance, respectively, where the impedance is $\sqrt{Z_{res}R_e}$.

One can measure the other loudspeaker parameters by simple experiment [7]. To determine the diaphragm mass and compliance, one can simply add a known mass M' of clay to the diaphragm. If the resonance of the loudspeaker is f_0 and with the added mass is f_0', the mass and compliance of the diaphragm are

$$M_d = \frac{M' f_0^2}{f_0^2 - f_0'^2}, \quad C_d = \frac{1}{M_d 2\pi^2 f_0^2} \tag{16.16}$$

A similar approach would be to place the loudspeaker in a closed cabinet of precisely known air volume and note the resonance change. To measure the magnetic motor force, one can apply a known dc current to the coil and carefully measure the deflection of the driver. The force is known given the compliance of the loudspeaker and measured displacement. Thus the "Bl factor" is obtained by direct experiment. However, it can also be indirectly measured by the following equation:

$$Bl = \sqrt{\frac{\omega_s M_d R_e}{Q_T} - R_e R_d} \tag{16.17}$$

where R_d can be found from

$$R_d = \frac{\omega_s M_d R_e}{Q_T Z_{res}}$$ (16.18)

Given all the physical parameters of a loudspeaker, one can design an efficient HiFi loudspeaker system. We start by determining an analytical expression for the total system Q_T.

$$Q_T = \frac{2\pi f_s M_d R_e}{B^2 l^2 + R_e R_d}$$ (16.19)

The Q_T factor is an important design parameter because it determines system response. If $Q_T = 1/2$ the system is critically damped. If lower frequency response is desired, one may select a loudspeaker with $Q_T = 0.707$ which will align the pressure response to a maximally flat "2nd order Butterworth" LPF polynomial, as seen in Figure 16.8. The pressure response is analytically given as

$$\frac{p}{e} = \frac{Bl\, S\, R_a}{B^2 l^2 + (R_e + j\omega L_e)\left(j\omega M_d + \left(\dfrac{1}{j\omega C_{d'a}}\right) + R_d + S^2 Z_a\right)}$$ (16.20)

For a given loudspeaker, Q_T can be raised by raising the system resonance frequency f_s, which is easily done by using a smaller cabinet volume. However, as Q_T is raised above 0.707, the alignment is technically a Chebyshev, or "equal ripple" high-pass filter response. The response will peak at the system resonance as Q_T is increased above 1.0, but the effective low-frequency bandwidth does not increase significantly. This suggests that for a given driver, there is an optimal cabinet size. There are similar fourth-order polynomial alignments for vented loudspeaker systems where an acoustical port is designed to extend the low-frequency response [8].

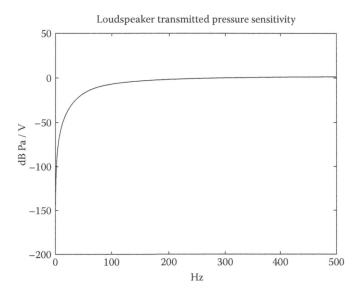

FIGURE 16.8 Transmitted pressure response for the loudspeaker example the $Q_T = 0.707$ for a maximally flat 2nd order Butterworth alignment.

While loudspeakers are definitely fun and certainly extremely common, one can also use this true transducer as a sensor. Since the diaphragm is relatively heavy compared to air, the motion of the diaphragm is caused mainly by pressure differences across the diaphragm. If the diaphragm is large, these forces can be quite significant. The force on the diaphragm due to an acoustic pressure wave is $F = Sp$, where S is the diaphragm area. The diaphragm velocity is this force divided by the mechanical impedance. Since $e = Blu$, we can simply write the voltage output of the loudspeaker due to an acoustic pressure on the diaphragm.

$$\frac{e}{p} = \frac{Bl\,S}{\dfrac{B^2 l^2}{R_e + j\omega L_e} + R_d + j\omega M_d + \dfrac{1}{j\omega C_{d'a}}} \tag{16.21}$$

Figure 16.9 shows the voltage response relative to acoustic pressure for our loudspeaker example. Even though the loudspeaker Q_T is 0.707 the receiving response is anything but uniform. This is why loudspeakers are rarely used as microphones. However, Sir Paul McCartney of the Beatles did exactly that [9] to boost his bass guitar in the recordings of "Rain" and "Paperback Writer" where the microphone for the bass was a 10 inch loudspeaker in a sealed cabinet. From Figure 16.9 one can see how this would boost the low frequencies of the bass guitar.

However, moving coil "dynamic" microphones are quite popular in the music and communication industries mainly because of their inherent nonflat frequency response. For example, in aviation a microphone "tuned" to the speech range of 500 Hz to 3 kHz naturally suppresses background noise and enhances speech. The simple coil and magnetic design also provides an extremely robust microphone for the harsh environment (condensation, ice, high vibration, etc.) of an aircraft. The "bullet shape" of aviator microphones is still heard today and the unique frequency response has become the signature of many blues harmonica players.

A variant of the dynamic microphone even more important to the broadcast and music industries is the ribbon microphone. These large microphones originally introduced in the 1930s are still in high demand today for very good reasons well beyond historical stylistic trends. The ribbon microphone is seen in a sketch in Figure 16.10 and works by direct sensing of acoustic velocity

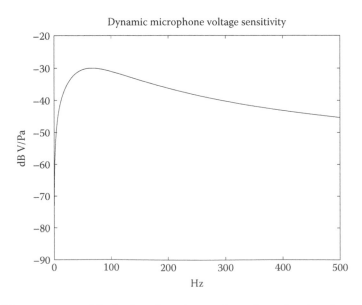

FIGURE 16.9   Voltage response of the loudspeaker when used as a dynamic microphone.

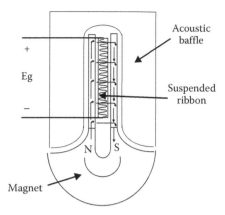

FIGURE 16.10 A ribbon microphone is used to sense the acoustic velocity directly and also has a unique frequency response, proximity effect, and dipole directivity pattern.

with a light-weight metal ribbon in a magnetic field. As the ribbon moves with the acoustic waves, an electrical current is created along the length of the ribbon from the magnetic field traversing across the face of the ribbon. A typical ribbon size is about 3 cm long and only about 1–2 mm wide. It is used above its resonance and essentially has a perfectly flat frequency response. The deep rich sound of a voice from a ribbon microphone can show any radio announcer sound like a giant. Also contributing to the low frequencies, are nearfield acoustic velocity waves (refer to Section 7.3) which add to the volume of wave energy in the proximity of the person talking into the microphone. This is well known to audio technicians as the "proximity effect" and is present for microphones with a nonuniform beam pattern. Finally, since the ribbon has a dipole response pattern aided by the acoustic baffle around the edges of the ribbon, it can be oriented to suppress some unwanted sounds in the studio. However, because of its strong low-frequency velocity response, the ribbon microphone is an extremely poor choice for outdoor use due to high wind noise susceptibility.

The ribbon microphone senses velocity directly, so that the force on the ribbon is proportional to the ribbon mass times the acceleration due to the velocity $j\omega U/S$. The resulting mechanical velocity of the ribbon is this force divided by the mechanical impedance. Therefore, the velocity response is

$$\frac{e}{U} = \frac{Bl\,j\omega M_d}{\left(\dfrac{B^2l^2}{R_e + j\omega L_e}\right) + R_d + j\omega M_d + \left(\dfrac{1}{j\omega C_{d'a}}\right)} \tag{16.22}$$

which can be seen in Figure 16.11 for a transducer widely known as a geophone. We also call this a "geophone" response because it represents the transducer's response to velocity. Geophones are the standard transducer used for seismic surveys and for oil exploration. Obviously, one would isolate a ribbon microphone well from vibration and design a geophone for minimal acoustic sensitivity by eliminating the diaphragm and completely enclosing the coil.

16.2.2 ELECTROSTATIC TRANSDUCER

The electrostatic transducer is commonly used for wide frequency response condenser microphones, accelerometers for measuring vibrations over a wide frequency range, force gauges, electrostatic loudspeakers, and in sonar and ultrasonic imaging [5]. The Greek word *piezen* means "to press." The

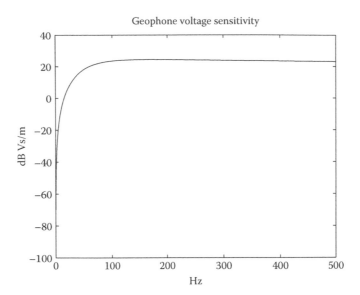

FIGURE 16.11 Voltage response of a loudspeaker coil as a pure mechanical velocity sensor, often called a geophone for its common application in seismic sensing.

piezoelectric effect is found in crystal and polymer transducers, on which when stress is applied, a voltage is produced. Microphones and electrostatic loudspeaker transducers can be made based on the electrostatic charge on a lightweight movable plate capacitor (also known as a condenser). With one plate constructed as a lightweight diaphragm, movement of the diaphragm results in a corresponding change in capacitance, thus producing a subsequent change in voltage for a fixed amount of charge in the capacitor. Both piezoelectric and condenser-based transducers can be physically modeled using equivalent circuits. From Coulomb's law, we know that the force f between two charges q_1 and q_2 is directly proportional to their products, and inversely proportional to the square of the distance between them $f = kq_1q_2/r^2$. This force can also be written as a charge times the electric field $f = qE$, where E has units N/C or V/m. For constant charge, there is a hyperbolic relation between the electric field and distance separating the charges. Therefore, one can linearize the relation between electric field and displacement for small changes, and to a greater extent, if a bias voltage or static force has been applied to the transducer. Since pressing into the piezoelectric crystal or on the capacitor plates causes a voltage increase, the linearized constitutive equations for small displacements and forces are

$$e = -\psi\xi, \quad f = \psi q \tag{16.23}$$

where ψ has physical units of Volts per m (V/m) or Newton per Coulomb (N/C). It can be seen that if one supplies charge to the device (input an electrical current $i = dq/dt$) it will produce an outward positive force. But, if the voltage potential is increased, there will be an inward displacement ξ. Assuming a sinusoidal excitation, the constitutive equations become

$$e = -\psi\frac{u}{j\omega}, \quad f = \psi\frac{i}{j\omega} \tag{16.24}$$

which can be seen as similar to the relations between mechanical and electrical impedances for the electromagnetic transducer ($f = Bli$ and $e = Blu$) with the exception of the sign.

Piezoceramic materials are materials on which when a static electric field is applied, produce a dynamic voltage in response to stress waves. *Piezoelectric* materials are polarized either naturally or by manufacturing process so that only a small dc voltage bias is needed, mainly to insure no net charge leakage from the transducer. Piezoelectric materials require a little more detailed description of physics. Natural piezoelectrics like quartz are crystals of molecules of silicon bonded to a pair of oxygen atoms denoted chemically as SiO_2. In the crystal lattice of atoms these molecules form helical rings. When the ring is compressed along the plane of the silicon atoms, an electric field is produced along the orthogonal direction from the extension of silicon's positive charge on one side (above the stress plane) and O_2's negative charge on the other side (below the stress plane). Clearly, there is a complicated relation between the various stress planes available in the crystal as well as shear planes. For our simplistic explanation, it is sufficient to say that the crystal can be cut appropriately into a parallelepiped such that the voltage measured across two opposing faces is proportional to the stress along an orthogonal pair of faces. Piezoelectric effects are also naturally found in Rochelle salt, but its susceptibility to moisture is undesirable. Man-made ceramics dominate the marketplace for piezoelectric transducers because they can be molded into desirable shapes and have their electrodes precisely bonded as part of the ceramic firing process. Polarization of the ceramic is done by heating the material above a temperature known as the Curie temperature where the molecular orientations are somewhat mobile, and then applying a strong electrostatic field while cooling the transducer. Polarization can thus be applied in the desired direction allowing the transducer to respond electrically to orthogonal stress (called the 1–3 mode) or even to longitudinal stress (in-line or 1–1 mode). However, when the ceramic is stressed in one direction, the other two orthogonal directions also change, affecting the coupling efficiency. Some of the formulas for man-made piezoceramics are PZT, $BaTiO_3$ (barium-titanate), and a very unique plastic film called PVDF (polyvinylidene fluoride), which also has a pyroelectric (heat sensing) effect used commonly in infrared motion sensors.

With the exception of PVDF, piezoelectrics are generally very mechanically stiff and have very high electrical resistance due to the strong covalent bonds of the molecules. This means that a generated force has to overcome the high stiffness of the ceramic to produce a displacement. Thus piezoelectrics can produce a high force but only if very low displacements occur. Conversely, a force applied to a piezoelectric will produce a high voltage but very little current. If the electricity produced is harvested as a power source or dissipated as heat from a resistor, the work done is seen mechanically as damping. An example of electronic material damping can be seen in some high-performance downhill skis, which have embedded piezoelectrics and light-emitting diodes to dissipate the vibrational energy. Another common application of piezoelectrics is a mechanical spark generator on for gas-fired barbeque grills. But perhaps the most technically interesting application of piezoelectric ceramics is as a force, strain, acceleration, or acoustic sensor, or as an acoustic pressure or mechanical force generator.

To analyze the transducer response, we introduce the *Mason equivalent circuits* for plane waves in a block of material with cross-sectional area S, thickness l, density ρ, and wave speed c in Figure 16.12 where the mechanical wavenumber k_m is ω/c. If material damping is significant,

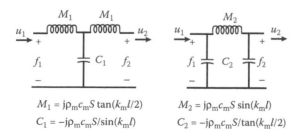

FIGURE 16.12 Mason equivalent circuits for a block of material of density ρ_m, plane wave speed c_m, area S, and length l showing a "T" and a "π" type of circuit section and the corresponding equations for the inductors and capacitors (damping assumed small).

the mechanical wavenumber is complex. For relatively low frequencies where the wavelength λ is much larger than any dimension of the material block, the inductors and capacitor equivalents are shown in Figure 16.12. However, at high frequencies where the block becomes resonant, these parameters will change sign accordingly representing the material as a waveguide rather than as a lumped element. The equation for the "mass" component M_1 seen on the left in Figure 16.12 is

$$Z_1 = j\rho_m c_m S \tan\left(\frac{k_m l}{2}\right) \approx j\omega \frac{\rho_m S l}{2}, \quad \lambda \gg l \tag{16.25}$$

such that M_1 in Figure 16.12 represents actually half the mass of the block of material. The complementary impedance for the left circuit is

$$Z_1^c = \frac{-j\rho_m c_m S}{\sin\left(\dfrac{k_m l}{2}\right)} \approx \frac{1}{j\omega\left(\dfrac{l}{\rho_m c_m^2 S}\right)}, \quad \lambda \gg l \tag{16.26}$$

such that the mechanical compliance is seen to be $l/\{\rho_m c_m^2 S\}$, or the stiffness is $\rho_m c_m^2 S/l$ which has the proper units of kg/s. Either circuit in Figure 16.12 can be used, the choice being made based on which form is more algebraically convenient. The low-frequency approximations given in Equations 16.25 and 16.26 are quite convenient for frequencies where geometry of the block is much smaller than a wavelength. For high frequencies, the block behaves as a waveguide where for any given frequency it can be either inductive or capacitive depending on the frequency's relation to the nearest resonance.

Applying the Mason equivalent circuit to an accelerometer is straightforward. As seen in Figure 16.13, it is quite typical to mount a proof mass (sometimes called a seismic mass) on the transducer material (referred to here as PZT) using a compression bolt or "stud." The compression stud makes

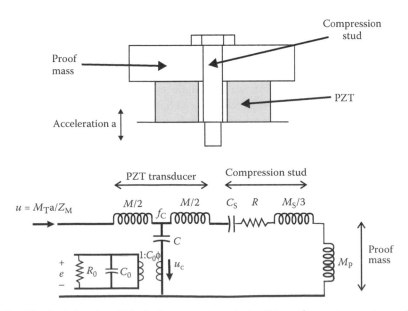

FIGURE 16.13 Physical sketch and block diagram of a typical PZT accelerometer constructed with a compression stud and showing the electrical circuit interface through a transformer with turns ratio $1:C_0\phi$.

a robust sensor structure and is considerably less stiff than the PZT. Its presence does contribute to a high-frequency resonance, but with clever design, this resonance can be quite small and well away from the frequency range of use.

   The electrical part of the accelerometer is seen in Figure 16.13 as a capacitor C_0 connected through a transformer with turns ratio 1:$C_0\phi$. For our example here, we will assume the resistance of the piezoceramic crystal R_0 is similar to quartz at 75×10^{16} Ω or greater (750 peta-Ω, therefore we can ignore the resistance) and the electrical capacitance is 1600 pF. If the resistance R_0 is less than about 1 GΩ, it will significantly cut the ultra low-frequency response which can be seen as well for condenser microphones later in this section. One can appreciate the subtleties of piezoelectric transducers by considering that the crystal contributes both an electrical capacitance and a mechanical compliance which are coupled together. The unusual turns ratio in the transformer in Figure 16.13 provides a tool to physically model the piezoelectric circuit. For example, if one applies a current i to the electrical terminals, a voltage appears across the capacitor C_0 of $i/j\omega C_0$, which subsequently produces an open-circuit force of $f = (\varphi C_0)\ i/j\omega C_0 = \varphi\ i/j\omega$, as expected assuming the mechanical velocity is blocked ($u = 0$). Going the other way, a mechanical velocity u_c on the right side of the transformer produces an open circuit current through C_0 of $-u C_0\phi$. This generates an open circuit voltage of $e = -u\ C_0\phi/j\omega C_0 = \varphi\ i/j\omega$, as expected. The effect of the electrical capacitance on the mechanical side of the transformer is a series compliance of value $1/(C_0\phi^2)$. In other words, the mechanical impedance effect of the electrical capacitor is $C_0\phi^2/j\omega$. For our example we assume a value of φ of 1×10^9 N/C or V/m. The coupling coefficient $d_{31} = 1/\phi$ is therefore 1000 pC/N. The coupling coefficient d_{31} is a common piezoelectric transducer parameter signifying the orthogonal relation between the voltage and stress wave directions. The crystal actually is sensitive to and produces forces in the same direction as the voltage giving a d_{11} coupling coefficient as well. Manufacturing process and mechanical design generally make the transducer most sensitive in one mode or the other.

   The parameters M and C in Figure 16.13 represent the mass and compliance of the PZT material block. We assume a density of 1750 kg/m³, a plane longitudinal wave speed of 1×10^7 m/s, cylindrical diameter of 1/4 in and length of 1/4 in.. This gives a mechanical compliance of 1.1458×10^{-15} and a mass of 3.5192×10^{-4} kg. The compression stud is assumed to be stainless steel with a density of 7750 kg/m³ and wave speed of 7000 m/s. The stud is cylindrical, 1/8 in. in diameter, and 1/4 in. long, giving a mass of 3.8963×10^{-4} kg and a compliance of 2.112×10^{-9} s/kg. It can be shown using kinetic energy that exactly 1/3 of the spring mass contributes to the mass terms in the mechanical impedance. A small mechanical damping of 400 Ω is used and the proof mass is 2.5 g. As the base of the accelerometer accelerates dynamically, the force applied to the PZT is simply the total mass times the acceleration. The velocity into the base is this applied force divided by the total mechanical impedance Z_M.

$$u = \frac{F}{Z_M} = \frac{a\left(M + M_p + \dfrac{M_s}{3}\right)}{Z_M} \tag{16.27}$$

The mechanical impedance looking to the right at the node defined by f_c in Figure 16.13 is the parallel combination of the branch with C and the electrical mobilities and the right-hand branch.

$$Z_c = \frac{\left(\dfrac{1}{j\omega C} + \dfrac{C_0\,\phi^2}{j\omega}\right)\left(j\omega\left[\dfrac{M}{2} + \dfrac{M_s}{3} + M_p\right] + R + \dfrac{1}{j\omega C_s}\right)}{\dfrac{1}{j\omega C} + \dfrac{\phi^2 C_0}{j\omega} + \dfrac{1}{j\omega C_s} + R + j\omega\left[\dfrac{M}{2} + \dfrac{M_s}{3} + M_p\right]} \tag{16.28}$$

The total input mechanical impedance from the base of the accelerometer is $Z_M = j \omega M/2 + Z_c$. It can be seen that the velocity u_c in the PZT is

$$u_c = \frac{a\left[M + M_p + \frac{M_s}{3}\right] Z_c}{Z_M \left(\frac{1}{j\omega C} + \frac{C_0 \phi^2}{j\omega}\right)} \tag{16.29}$$

where a is the base acceleration. Using Equation 16.24 we can solve for the voltage sensitivity in Volts per m/s² (multiplying by 9.810 gives the sensitivity in mV/g).

$$\frac{e}{a} = \frac{-\phi\left[M + M_p + \frac{M_s}{3}\right] Z_c}{j\omega Z_M \left(\frac{1}{j\omega C} + \frac{C_0 \phi^2}{j\omega}\right)} \tag{16.30}$$

Figure 16.14 shows the calculated voltage sensitivity for our example in mV per g.

Piezoelectric materials are also widely used as transmitters and receivers for underwater sound and as force sensors and actuators. To convert from force to acoustic pressure, one simply divides the force by the diaphragm area radiating the sound. For the underwater acoustic radiation impedance, apply Equation 16.13 with $\rho = 1025$ kg/m³ and $c = 1500$ m/s. It can be seen that a force gauge and hydrophone receiver are essentially the same transducer hydrophone transmitter and a force actuator will have a response depending on the load impedance. To convert from an accelerometer to a hydrophone or force transducer, replace the compression stud and proof mass with a general load impedance Z_L. This has the effect of shifting the transducer resonance much higher in frequency. If the transducer is blocked on the left, the PZT mass inductor on the left can be ignored. Otherwise, if

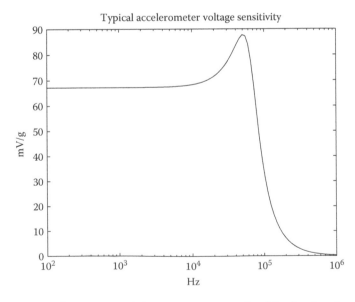

FIGURE 16.14 Typical accelerometer receiving voltage response in mV/g showing the resonance from the stud and proof mass and an extremely flat response at low frequencies.

the left side is free floating, the PZT mass inductor is attached to ground on the left. To simplify the algebra, we'll assume we have a blocked transducer. Also to simplify our algebra, let

$$Z_M = \frac{1}{j\omega C} + j\omega \frac{M}{2} + R \tag{16.31}$$

where C is the PZT mechanical compliance, M is the PZT mass, and R is the transducer mechanical damping. Applying a current i, the force f_c in Figure 16.13 is seen to be

$$f_c = \frac{i \frac{\phi}{j\omega}(Z_M + Z_L)}{\frac{C_0 \phi^2}{j\omega} + Z_M + Z_L} \tag{16.32}$$

The resulting mechanical velocity through the load is

$$u_M = \frac{i \frac{\phi}{j\omega}}{\frac{C_0 \phi^2}{j\omega} + Z_M + Z_L} \tag{16.33}$$

The resulting force across the load impedance relative to the current is

$$\frac{f_L}{i} = \frac{Z_L}{C_0 \phi + \frac{j\omega}{\phi}(Z_M + Z_L)} \tag{16.34}$$

Figure 16.15 shows the force actuation response when the load is a 1 kg mass. As seen in the response plot, at resonance a substantial amount of force can be generated. For a hydrophone

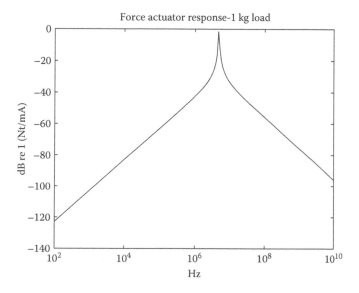

FIGURE 16.15 Transmitting force response relative to applied current for the PZT driving a 1 kg mass.

transmitter, we replace Z_L with Z_a/S^2 for water. However, it is more interesting to examine the pressure response relative to applied voltage. To convert current to voltage we need the total electrical input impedance.

$$Z_{ET} = \frac{1}{j\omega C_0} + \frac{\dfrac{\phi^2}{\omega^2}}{Z_M + Z_L + \dfrac{C_0 \phi^2}{j\omega}} \tag{16.35}$$

The third term in the denominator of Equation 16.35 is relatively small and can be ignored. Actually, the input electrical impedance is dominated by the electrical capacitance C_0. The acoustic pressure radiated to the far field assuming a narrow beam width is

$$\frac{p}{e} = \frac{R_a}{S Z_{ET} \left[C_0 \phi + \left(\dfrac{j\omega}{\phi} \right)(Z_M + Z_a) \right]} \tag{16.36}$$

where R_a is the real part of the acoustic radiation impedance defined in Equation 16.13. Figure 16.16 presents a plot of the transmitted pressure response in water.

For a PZT force gauge or hydrophone receiver, one calculates the applied force on the PZT (across the load impedance) and solves for the velocity through the crystal dividing by the mechanical impedance plus the mechanical compliance of the electrical capacitor. Given the velocity, it is straightforward to derive the voltage on the transducer's terminals due to the applied force. Further dividing by the diaphragm area S gives the voltage response to acoustic pressure, rather than mechanical force.

$$\frac{e}{p} = \frac{-\phi}{j\omega S \left[\dfrac{C_0 \phi^2}{j\omega} + Z_M \right]} \tag{16.37}$$

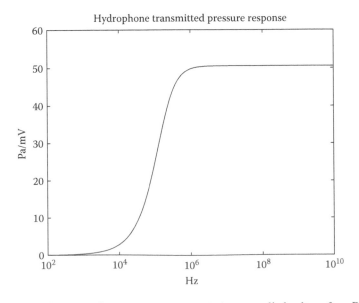

FIGURE 16.16 Transmitting acoustic pressure response relative to applied voltage for a PZT hydrophone.

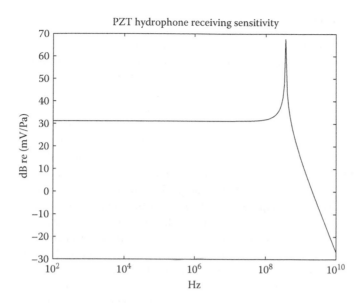

FIGURE 16.17   Receiving pressure sensitivity for the PZT hydrophone (no damping applied).

Figure 16.17 shows the voltage response for our example PZT in a receiving hydrophone configuration. Clearly, one would want to add significant damping for the hydrophone receiver and force actuator if a broadband response is desired. The hydrophone transmitter does not need any additional mechanical damping because of the high resistive component of the radiation impedance into the water. Since the radiation resistance increases as ω^2, the transmitting pressure response of the hydrophone remains constant above the resonance, just as is the case with a common loudspeaker.

It is useful to discuss why the resonance of the accelerometer is so different from the receiving hydrophone and why adding a compression stud to a hydrophone or force transducer has little affect on the frequency response. Without the compression stud, the accelerometer's response would resemble its electrical input impedance, which is essentially a capacitor. The stud-less response would thus show the maximum response at 0 Hz and then fall as $1/\omega$ as frequency increases. This makes our accelerometer a very sensitive gravity sensor provided a dc-coupled amplifier is used which does not drain any net charge from the accelerometer. The compression stud, if infinitely stiff, would block any velocity through the crystal and if infinitely compliant would vanish from the equivalent circuit. Thus, a series capacitor between the PZT and the proof mass is the appropriate circuit model. The effect of the compression stud is to shift the accelerometer resonance up from 0 Hz while also bringing down the low-frequency voltage sensitivity to a value that is constant for acceleration frequencies below the resonance. Note that for vibrations with a given velocity, at low frequencies the displacement will be large and at high frequencies the acceleration will be large. Generally, this makes force, displacement, and strain gauges the natural choice for dc and ultra low-frequency measurements, geophones the good choice for measurements from a few Hz to a few hundred Hz, and accelerometers the best choice for almost higher frequency measurements. However, accelerometer technology is a very rapidly developing field and new technologies are extending the bandwidth, environmental robustness, and dynamic range. For example, some "seismic" accelerometers are recently on the market with voltage sensitivities as high as 10 V/g with excellent response down to less than 1 Hz. Generally these low-frequency accelerometers configure the PZT into a layered cantilever sandwich with the proof mass on the free end of the cantilever. Stacking the PZT layers and constructively adding the voltages from both compression and tension makes a very sensitive transducer. PZT accelerometers are very reliable and inexpensive, but because acceleration favors high-frequency vibration, the SNR is generally best near the upper end of the measurement spectrum.

16.2.3 Condenser Microphone

The condenser microphone is physically a close cousin to the piezoelectric transducer [10]. It has a nearly identical equivalent electromechanical circuit as the piezoelectric transducer and in actuator form is known as an electrostatic loudspeaker. The "condenser" transducer is optimized for air acoustics by simply replacing the crystal with a charged electrical capacitor with one plate configured as a light-weight movable diaphragm. Configuring the movable diaphragm as one surface of a sealed air-filled chamber, the diaphragm will move from the applied force due to acoustic waves. As the diaphragm moves, the capacitance changes, so for a fixed amount of charge stored in the condenser one obtains a dynamic voltage which is linear and constant with the acoustic pressure for frequencies up to the resonance of the diaphragm. This is analogous to the response curve in Figure 16.17 for the receiving hydrophone or force gauge. The stiffness of the condenser microphone is found from the tension of the diaphragm and the acoustic compliance of the chamber. However, since 1 atm is 101,300 Pa, and 1 Pa RMS is equivalent to about 94 dB relative to 20 μPa, a small "capillary vent" is added to the microphone chamber to equalize barometric pressure changes and to allow for high acoustic sensitivity. We also include a shunt resistor for heating in the circuit for a very good reason. Since the capillary vent is open to the atmosphere, the condenser microphone cannot be hermetically sealed from the environment. Water condensation or direct moisture can get into the microphone cavity and significantly change the electrical resistance. It will be seen that moisture, condensation, or damage to the vent or cavity can cause a very significant change to the condenser microphone's low-frequency response, and in particular, its low-frequency phase response. We note that temperature is also a significant environmental factor affecting diaphragm tension, air density, and resistance. Some scientific-grade condenser microphones are heated slightly above room temperature as a means to hold the sensitivity response constant during calibrations and measurements.

Figure 16.18 depicts a typical condenser microphone design and its equivalent circuit showing a thin tensioned membrane as the diaphragm and the back plate with an applied dc voltage bias. This back plate usually has small holes or annular slots to let the air easily move. Some felt or other damping material is used to suppress these and other high-frequency acoustic resonances of the cavity structure, and we limit the fidelity of our model to the basic physics. The small capillary vent in the cavity allows for barometric pressure equalization and also causes a low-frequency sensitivity reduction. The force acting on the movable part of the cavity (the diaphragm and vent hole) is approximately $S_d p$, where p is the acoustic pressure and $S_d = 4 \times 10^{-6}$ m^2 is the diaphragm area for our example condenser microphone. This force results in a diaphragm velocity u_d, and vent velocity u_v, while a much smaller force $S_v p$, where $S_v = \pi a_v$ is the vent area that causes a vent velocity u_v to

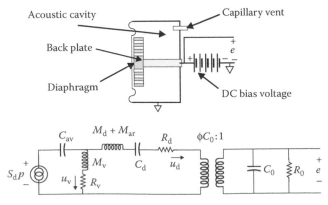

FIGURE 16.18 Condenser microphone and equivalent circuit showing the acoustic cavity, capillary vent, and mechanical velocities in the equivalent circuit.

be compensated slightly. The vent velocity takes away from the diaphragm velocity at low frequencies, which directly weakens the voltage sensitivity e at the microphone terminals at low frequencies. The mechanical compliance of the cavity behind the diaphragm is $C_{av} = V/(\rho c^2 S_d^2)$ where V is 5×10^{-7} m³ for our example. The mechanical response of the diaphragm contains the damping $R_d = 100\ \Omega$, diaphragm compliance $C_d = 6.3326 \times 10^{-8}$ (free space diaphragm resonance is 20 kHz), diaphragm mass $M_d = 1$ g, and the mass of the air which moves with the diaphragm $M_{ar} = 0.6\rho a_d S_d$, where $S_d = \pi a_d^2$ as seen from the radiation reactance in Equation 16.13. The air-mass is included because its mass is significant relative to the lightweight diaphragm.

The capillary vent requires that we present some physics of viscous damping. For a cylindrical vent tube of length L_v and area $S_v = \pi a_v$, some of the fluid sticks to the walls of the tube affecting both the resistance and mass flow. When the tube area is small the frictional losses due to viscosity and the moving mass are seen in acoustic impedance units to be

$$Z_{va} = \left[\frac{8\eta L_v}{\pi a_v^4}\right] + j\omega \left[\frac{4\rho L_v}{3\pi a_v^2}\right] \tag{16.38}$$

where $\eta = 181$ μPoise (1 P = 1 g cm^{-1} s^{-1}) for air at room temperature. There are forces acting on both ends of the capillary vent, $S_v\,p$ on the outside and the much stronger force $S_d\,p$ through the cavity. We can simplify our analysis by ignoring the small counteracting force $S_v\,p$. The mechanical impedance of the vent is therefore

$$Z_v = R_v + j\omega M_v = \left[\frac{8\eta L_v S_d^2}{\pi a_v^4}\right] + j\omega \left[\frac{4\rho L_v S_d^2}{3\pi a_v^2}\right] \tag{16.39}$$

because the volume velocity through the vent is determined by the diaphragm, not the capillary vent area. The length of our example capillary tube is one mm and several vent areas will be analyzed. In our example, the electromechanical coupling factor is 360,000 V/m, which may sound like a lot, but it is generated by a dc bias voltage of 36 V over a gap of 0.1 mm between the diaphragm and the back plate. This gives an electrical capacitor value of 0.354 pF since our diaphragm area is 4×10^{-6} m².

Figure 16.19 shows our example of a "typical" condenser microphone frequency response from 1 Hz to 100 kHz in units of mV/Pa and degrees phase shift for a range of shunt resistances. The 10 GΩ shunt is typical of a high-quality professional audio condenser microphone where the resistance is provided by an audio line transformer. The 100 GΩ shunt is typical of a scientific-grade condenser microphone and is usually the gate impedance of a FET (field-effect transistor) preamplifier attached directly to the microphone terminals. The infrasonic condenser microphone requires an extraordinarily high shunt resistance to maintain its low-frequency response. Infrasonic microphones can be used to measure earthquakes, nuclear explosions, rocket launches, and even tornadoes from thousands of kilometers distance. This is because there is very little absorption in the atmosphere below 10 Hz. However, it is so difficult to maintain the required high shunt resistance and the SNR is naturally low at infrasonic frequencies. Alternative parametric techniques such as carrier wave modulation are used, as described in Section 16.1. To observe the diaphragm motions in the infrasonic range, one can measure the phase shift of a radio-frequency electromagnetic signal passed through the capacitor. This straightforward modulation technique is widely used in MEMS as an indirect means to observe the motions of silicon structures.

Figure 16.20 shows our example condenser microphone frequency responses as a function of capillary vent size. The capillary vents are all 1 mm long and the cross-sectional area is varied to explicitly show the sensitivity of the magnitude and phase responses. The shunt resistance R_0 is taken as $1 \times 10^{15}\ \Omega$ to insure that it is not a factor in the simulation. The most important observation

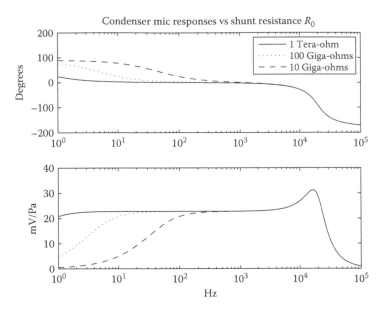

FIGURE 16.19 Magnitude and phase response of our example condenser microphone showing the effect of input shunt resistance changes which can occur as a result of moisture or dirt.

seen in both Figures 16.19 and 16.20 is the very significant phase change seen all the way up to about 1 kHz. If the condenser microphone is part of an acoustic intensity sensor, or part of a low-frequency beamforming array, this phase response variation can be very detrimental to system performance. The effect of the phase error is compounded by the fact that the observed acoustic phase differences at low frequencies across an array might well already be quite small. A small phase difference signal with a large phase bias will likely lead to unusable measurement results. Again, the most common threat to condenser microphones is moisture. But, the vent size sensitivity also indicates that

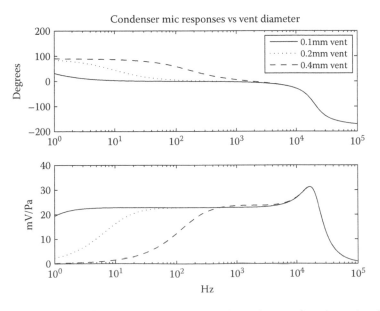

FIGURE 16.20 Magnitude and phase response of our example condenser microphone showing the effect of various capillary vent diameters on the low-frequency response.

damage to the microphone housing or vent can have a profound effect on the microphone response. Note that the shunt resistance and all sources of mechanical and acoustic damping contribute to the thermal "Johnson" noise of the transducer.

One of the major engineering breakthroughs in audio which led to widespread inexpensive condenser-type microphones was the development of the electret microphone [10]. There are a number of possible configurations and respective patents, but the "electret" material is essentially a metalized foil over an insulating material which is permanently polarized. The metallic side of the electret sheet serves as an electrode. The electret sheet can be directly stretched over a back plate (see Figure 16.18) of a condenser microphone and the charge of the electret reduces the required bias voltage. The electret material can also be place on the back plate with the charged insulated side of the sheet facing the metal diaphragm of the condenser microphone in the so-called "back-electret" configuration. The most inexpensive configuration to manufacture is the electret material which is placed directly on an insulating sheet attached to the back plate. As with the condenser microphone, holes in the back plate allow sound to enter a cavity behind the back plate, thus allowing the diaphragm to move in places. No tension is required in the diaphragm and this extremely simple arrangement can produce a reasonable sensitivity, SNR, and frequency response at very low cost. Electret style condenser microphones can be found is almost every small microphone, particularly in telephones.

16.2.4 MICRO-ELECTROMECHANICAL SYSTEMS

Micro-electromechanical systems (MEMS) are a fascinating product of advancements in silicon manufacturing at the nanometer scale during the last 20 years of the twentieth century. Recall that integrated circuits are manufactured as layers of p-doped silicon, n-doped silicon, insulating glass (silicon dioxide), and wires (typically aluminum). The actual circuit weaves its way through the layers on the chip as part of the artful design which has brought us inexpensive microprocessors with literally hundreds of millions of transistors. Micromachining is a variant of large-scale integrated circuits where 3D structures are created out of silicon. Micromachining is not done using a microlathe any more than the first table-top radios had microsized musicians inside. Micromachining is done by etching away layers of material to create 3D structures. Analogy would be to create a 3D casting of various metals by making a sand mold, digging out the desired shape, casting the metal, then stacking another sand mold on top, digging out the desired shape for that layer, casting the metal, and so-on. When the casting is complete, the sand is washed away to reveal the complex 3D structure. For a silicon structure, the layering is done by diffusion doping (kiln process), vacuum vapor deposition, electron beam epitaxial deposition, to name just a few methods. In micromachining, some of these layers are made from materials which can be dissolved using various acids. One can subsequently remove layers from the structure, analogous to removal of the sand from the metal casting, to produce remarkable structures on a nano-meter scale. Scientists have built small gears, pumps, electric motors, gyros, and even a turbine! One of the most widely used MEMS sensors today is the micromachined accelerometer which is used as a crash sensor to deploy a driver air bag in an automobile.

The MEMS accelerometer essentially has a movable silicon structure where the motion can be sensed electronically. The electronic interface circuitry is integrated directly into the surrounding silicon around the MEMS sensor. The sensing of motion is typically done by either measuring changes in capacitance between the movable and nonmovable parts of the silicon, or by sensing changes in resistance of the silicon members undergoing strain. Figure 16.21a shows a MEMS accelerometer which measures the changes in resistance in several thin members attached to a movable trapezoidal cross-section of silicon. By combining the resistance of the silicon members as part of a Wheatstone bridge, temperature compensation is also achieved. This MEMS sensor is a true dc device and therefore can measure gravity and orientation of an object relative to gravity. However, because it has relatively high resistance and requires a net dc current to operate, the intrinsic electronic noise (see Section 15.1) of the device can be a design concern. The more common micromachined

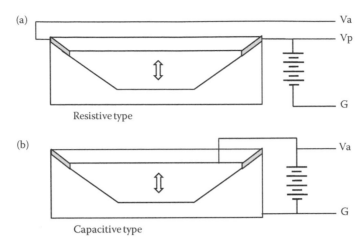

FIGURE 16.21　Sketch of a resistive type MEMS accelerometer (a) and (b), a capacitive type MEMS accelerometer showing the electrical connections to the moving mass.

accelerometer is the capacitance sensing type, seen in Figure 16.21b, where the capacitance between the movable and nonmovable silicon is measured in much the same way as the operation of a condenser microphone. An FET or high input impedance operational amplifier is used in current amplifier mode to produce a strong sensor voltage. Again, because of the electronic elements and high resistance, electronic background noise is a design concern and is often addressed for low-frequency operation using the modulation techniques discussed in Section 16.1.

MEMS sensors are going to be important for the foreseeable future even if they are not as good as a traditional transducer for reasons of cost, shock robustness, and obviously, their small size. Because the layering and etching processes can be highly automated, large numbers of MEMS sensors can be made for very low cost and with very high mechanical precision. Shock robustness is a slightly more elusive feature of MEMS, because it requires thoughtful mechanical design. However, if a material can withstand a certain stress before breaking, the amount of acceleration tolerable can be significantly increased if the mass of the structure can be reduced. Therefore, very small stiff structures can be inherently very difficult to break. This is one reason why the micromachined accelerometer is used as a crash impact detector in automobiles. The initial shock of the crash can conceivably crack the piezoelectric crystal before deceleration signal of the collapsing car body is sensed to cause the proper deployment of the air bag. MEMS sensors will likely have a role in many unusual and extreme applications ranging from sports equipment to heavy manufacturing and construction process controls as well as small low-cost sensors for game controllers and cellular phones. Finally, the small size of MEMS sensors has several intriguing aspects worth noting. Small size allows for use in less invasive medical operations such as catheterization sensors, implantable sensors, and cellular biology technology. Another artifact of small size is the ability to generate very high-voltage fields as well as high-frequency mechanical resonances. Very recently, material scientists have been able to grow piezoelectric materials on silicon MEMS structures which opens a whole new avenue for development.

16.2.5　Charge Amplifier

Electronic sensor interface circuits are worthy of discussion especially for the capacitive-type sensors of piezoelectric transducers, condenser and electret microphones, and MEMS sensors. Because these sensors all have very high output impedance, they can deliver only a very small current. Under these conditions, a cable connecting the transducer to the data acquisition system is subject to electromagnetic interference. As noted in Section 15.1, magnetic fields will induce small currents into these high-impedance cables, which are comparatively large to the small sensor currents. There are two practical choices to correct the weak signal current: (1) use a transformer to increase the current (the sensor voltage will be proportionately stepped down), or; (2) convert the line impedance at the sensor using a junction field effect transistor (JFET). The JFET has extremely high input resistance and thus will draw little current from the sensor. An n-channel JFET works by controlling the conductivity of a bar of n-type silicon by inducing a charge depletion region with an applied "gate" (p-type gate) voltage across the path of current flow along the bar. The current through the source and drain (at their respective ends of the bar) can be varied by the gate voltage between the gate and source. As the gate voltage is driven negative, the depleted charge carriers in the bar cause the n-type silicon to change from a conductor to an insulator, or more appropriately, behave as a semiconductor. By supplying a dc current and appropriately biasing the gate voltage of the JFET, small fluctuations of the gate voltage directly vary the drain current, which is of course much stronger than the current from the high-impedance sensor. By locating the JFET as close as possible to the high-impedance sensor, the size of the "antenna" which can pick up electromagnetic fields is greatly reduced, making for inefficient reception of unwanted electromagnetic interference. Proper voltage biasing of a JFET is an art, made more difficult by the variations in transconductance and the sensitivity of the JFET to temperature. To avoid saturating or even damaging the JFET, the supply current must be limited to a fairly narrow range. This is usually done using a device known as a current-limiting diode.

When the sensor has a high output impedance which is predominantly capacitive, special care should be taken to provide an interface circuit which will not oscillate at high frequencies. Figure 16.22 depicts what is known as a *charge amplifier*. The charge amplifier is really just a modified current amplifier with the addition of a feedback capacitor. The negative feedback resistor and capacitor have the effect of defining a low-frequency limit of

$$f_{\mathrm{L}} = \frac{1}{2\pi R_{\mathrm{f}} C_{\mathrm{f}}} \tag{16.40}$$

where one makes R_{f} and C_{f} as large as practical for extended low-frequency response. Above this lower corner frequency, the "charge gain" is $V_0/\Delta q = -1/C_{\mathrm{f}}$. But since the dynamic voltage e of the transducer output is $e = \Delta q/C_0$, one can also write that the net voltage gain of the charge amplifier is $V_0/e = -C_0/C_{\mathrm{f}}$. The resistor R_1 also serves the important purpose of creating a high-frequency cutoff at

$$f_{\mathrm{H}} = \frac{1}{2\pi R_1 C_0} \tag{16.41}$$

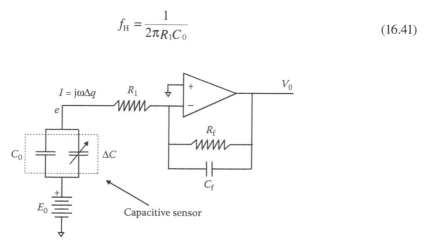

FIGURE 16.22 A charge amplifier circuit showing the capacitive sensor, sensor bias voltage E_0, and a current amplifier with feedback capacitor C_{f} and $V_0 = -E_0\Delta C/C_{\mathrm{f}}$.

which helps keep the charge amplifier stable at very high frequencies. With R_1 nonexistent, the operational amplifier loses its ability to suppress oscillations at high frequencies. Note that for a given amount of amplifier response delay (slew rate), what is stable negative feedback at low frequencies can become unstable positive feedback at high frequencies. This is because the phase shift due to delay τ increases with frequency as $\theta = \omega\tau$. For the given sensor capacitance C_0, one simply increases R_1 until an acceptable amount of stability margin is implemented. The output voltage of the charge amplifier is $V_0 = -E_0 \Delta C / C_f$, which is conveniently not dependent on C_0, which includes the cable capacitance to the sensor. This allows repeatable and controllable sensor sensitivity.

16.2.6 RECIPROCITY CALIBRATION TECHNIQUE

The reciprocity calibration technique is used to obtain an absolute calibration of a sensor or transducer by using a reversible transducer and the associated reciprocity physics to obtain the receiving voltage sensitivity [11]. It should be noted that there are really two types of calibration, *relative calibration* and *absolute calibration*. With relative calibration one transducer or sensor is compared to another previously calibrated transducer or sensor, which has a traceable calibration to national metrology standards. The second type of calibration is called absolute calibration because the calibration can be derived from traceable nontransducer quantities, such as mass, length, density, volume, and so on. For the absolute calibration to be traceable, these other quantities must also be traceable to the national or international standard.

Figure 16.23 depicts reciprocity calibration in a free field (top) and in a cavity (bottom). Both configurations require two setups to allow a measurement with transducer B used as a sensor (setup A) and as an actuator (setup B). For setup A, we required that both sensor A and transducer B receive the same pressure field from C. Co-locating transducer B with sensor A some distance from C or locating all three transducers in a small cavity (small compared to wavelength) is adequate insurance of a uniform received pressure for transducers A and B. In setup B, transducer B is used

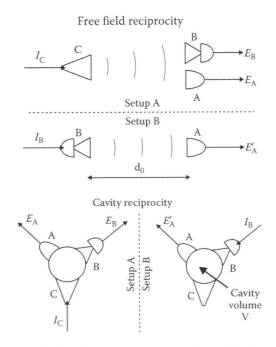

FIGURE 16.23 Reciprocity calibration in a free field (top) or in a cavity (bottom) requires two setups to exploit the reversible transduction of transducer B to obtain and absolute voltage calibrations of sensor A.

as a transmitter (one notes the distance d_0 for the free-field reciprocity case). For our example, we consider acoustic pressure and that transducer A is only used as a sensor for which we seek the receiving voltage sensitivity. But, the reciprocity technique is powerful and completely general for any type of linear transducer. It requires that transducer B be reciprocal, meaning that it can either transmit or receive.

We define the receiving voltage sensitivity for a microphone as M (V/Pa), the transmitting voltage sensitivity as S (Pa/V), and the reciprocity response function as $J = M/S$, which has the units of mobility (inverse impedance). This too applies for whatever domain the transducers are used (mechanical, acoustical, or electrical). For the case of electromagnetic transducers, recall that the transduction equations are $f = Bli$ and $e = Blu$ while for electrostatic or piezoelectric transducers the transduction equations are $f = \varphi i/j\omega$ and $e = -\varphi u/j\omega$, thus the "Bl" can be replaced by "$\varphi/j\omega$" provided the minus sign is handled correctly for the piezoelectric/electrostatic transducer case. When a microphone diaphragm is subjected to acoustic pressure p, the applied force is pS_d, where S_d is the diaphragm area. The velocity of the diaphragm u is this applied force divided by the mechanical impedance of the diaphragm Z_M. The voltage generated by a velocity u is $e = Blu$. For an electromagnetic transducer, the receiving voltage is the velocity times Bl.

$$e = \frac{Bl\, pS_d}{Z_M} \tag{16.42}$$

Therefore, the receiving voltage sensitivity is

$$M = \frac{e}{p} = \frac{Bl\, S_d}{Z_M} \tag{16.43}$$

The transmitting sensitivity is found for a free space by noting the pressure generated by a point source with volume velocity Q a distance d_0 meters away

$$p = \frac{jk\rho cQ}{4\pi d_0}\, e^{-jkd_0} = \frac{\rho f Q}{2d_0}\, e^{-j(kd_0 - \pi/2)} \tag{16.44}$$

where k is the wavenumber (ω/c) and the time harmonic dependence has been suppressed. To get the mechanical velocity of the diaphragm, we note that the applied force due to the current i is $f = Bli$, the velocity is the applied force divided by the mechanical impedance, or $u = Bli/Z_M$. Multiplying the mechanical velocity u times the diaphragm area S_d gives the acoustic volume velocity Q in Equation 16.42. The free-space transmitting voltage sensitivity is therefore

$$S = \frac{p}{i} = \frac{\rho f Bl S_d}{2d_0 Z_M}\, e^{-j(kd_0 - \pi/2)} \tag{16.45}$$

and the "free-space" reciprocity response function is

$$J = \frac{M}{S} = \frac{2d_0}{\rho f}\, e^{-j(kd_0 - \pi/2)} \tag{16.46}$$

When all three transducers are in a small closed cavity of volume V as seen in the bottom of Figure 16.23, we note that the acoustical compliance of the cavity C_A is $V/\rho c^2$, and that the mechanical equivalent $C_M = C_A/S_d^2$. Since the cavity is small, this compliance is low (the stiffness is high) and the transducer diaphragm motion is affected by the stiff cavity impedance. The receiving voltage sensitivity for the cavity is therefore

$$M = \frac{e}{p} = \frac{Bl\,S_d}{\left(Z_M + \dfrac{1}{j\omega C_M}\right)} \tag{16.47}$$

For an applied force due to a current i, the pressure in the cavity is the diaphragm velocity (applied force over total mechanical impedance) times the acoustic impedance of the cavity. The transmitting voltage sensitivity is

$$S = \frac{p}{i} = \frac{Bl\,S_d}{\left(Z_M + \dfrac{1}{j\omega C_M}\right)} \frac{1}{j\omega C_A} \tag{16.48}$$

where C_A is the acoustic compliance of the cavity. The reciprocity response function for the cavity is

$$J = \frac{M}{S} = j\omega C_A \tag{16.49}$$

where the result can be seen as the acoustic mobility of the cavity. The reciprocity response in Equations 16.46 and 16.49 are called absolute because they do not have any terms associated with the transducers, only dimensions (V and d_0), parameters of the medium (ρ and c), and frequency, which can be measured with great precision, traceable to standards.

How does one make use of reciprocity to calibrate a sensor? Consider setup A in Figure 16.23 where both sensor A and transducer B receive the same pressure from transducer C. We measure a spectral transfer function of E_A over E_B for the frequency range of interest and can write the following equation for the receiving voltage sensitivity of sensor A:

$$M_A = \frac{E_A}{E_B} M_B \tag{16.50}$$

For setup B the voltage output of sensor A due to current I_B driving transducer B is

$$E'_A = pM_A = I_B S_B M_A \tag{16.51}$$

where S_B is the transmitting voltage sensitivity of transducer B. Therefore, we can write another equation for M_A, the receiving voltage sensitivity for sensor A, in terms of the transfer function E'_A over I_B for the frequency range of interest.

$$M_A = \frac{E'_A}{I_B} \frac{1}{S_B} \tag{16.52}$$

Independently, Equations 16.50 and 16.52 may not appear useful, but if we multiply them together and take the square root (the positive root) we get an absolute expression for sensor A's receiving voltage sensitivity.

$$M_A = \sqrt{J\left(\frac{E'_A}{I_B}\right)\left(\frac{E_A}{E_B}\right)} \tag{16.53}$$

One simply uses the appropriate reciprocity response function J (free space or cavity) depending on the situation in which the reversible transducer B is employed, measures the required transfer functions, and calculates the receiving voltage sensitivity as given in Equation 16.53. Given good transfer function estimates (i.e., linearity enforced, low-background noise, and large numbers of spectral averages used) and an accurate reciprocity response function model, the calibration technique is both broadband and accurate. The calibration seen in Equation 16.50 is referred to as a "relative" calibration where the sensitivity of one transducer is gauged against another. Usually, a calibration transducer is traceable to national standards (i.e., NIST in the United States) is used to provide a relative calibration metric for a number of other transducers in a system. However, the reciprocity technique provides an absolute calibration, provided one can measure the needed dimensions, medium parameters, and frequency in an accurate and traceable way.

Reciprocity is the preferred technique for intelligent sensor systems to self-calibrate their transducers. It requires one or more reversible transducers in addition to the typical sensor. It requires some environmental sensing (such as temperature and static pressure for acoustical measurements), and some control over the environment (fixed dimensions or masses in a mechanical system). There are two or more transducers available to check measurements as well as check relative calibration. But if the reciprocity technique is employed, all the transducers can be calibrated absolutely and automatically by the intelligent sensor system. Sensor damage, poor electrical connections, or breakdowns in sensor mounting can be identified using calibration techniques. This "self-awareness" capability of an intelligent sensor system is perhaps the most intriguing aspect of what we define as a truly "smart" sensor system. The obvious benefit is that sensor data from a questionable or out-of-calibration sensor can be discounted automatically from the rest of the intelligent sensor system's pattern recognition and control algorithms allowing automatic re-optimization of the system's mission given a new situation.

16.3 CHEMICAL AND BIOLOGICAL SENSORS

As with the previous section, one could write a detail textbook on the subject of chemical and biological sensors alone, but here we provide a very brief synopsis highlighting only the major types of sensors in common use in the early twenty-first century. We also discuss this at the system level, so the reader can get an idea of the physics and signal processing involved along with the general capabilities of these sensors. After the terrorist event of September 11, 2001, there is a broad interest in chemical and biological (CB) sensors and we will avoid comment regarding performance or vulnerability against this important threat to society. Our discussion will focus on the science and signal processing associated with CB sensors, such as the size of the molecules detected and the different techniques used for the larger molecules of biological targets.

Thanks to the solar radiation and the presence of vast quantities of water on our plant, we tend to have small diatomic (two atom) and triatomic molecules such as N_2, O_2, CO_2, H_2O, and so on dominating the free atmosphere. Of these, H_2O is the most peculiar in that it is one of the only triatomic molecules with such high attraction to itself that it is in liquid form at room temperature. This is due to the strong polarity and bond angle of the hydrogen atoms to the oxygen. But it also makes water reactive to just about everything. When larger compounds come along, water and the ozone produced by solar radiation in the upper atmosphere tend to break them down into acids and bases which then dissolve in the Earth's oceans or react with minerals on the Earth's surface. So the air we all breathe is naturally pretty clean thanks to the ecosystem. Another remarkable fact is that the 21% of the atmosphere that is oxygen was all created from photosynthesis by plants. Early in the Earth's life, the atmosphere had a lot more CO_2 which plants consumed producing oxygen and the hydrocarbons we use today as fuels. There is much debate today about whether the CO_2 released from burning hydrocarbons is the cause of global warming. There are signs of enormous global cooling and warming of the Earth from the dozen or so "ice ages" in the distant past. It is also true that air pollution in recent years has become more invisible, allowing more sunlight to heat the earth, while at the same time, more tree covered areas have been cleared for farming reducing the amount of CO_2 reabsorbed into the biosphere, and that the developing world will greatly expand the use are carbon-based fuels in the immediate future. To make things even more complex, as the globe warms, more CO_2 emerges from the oceans and in the frozen tundra of the polar regions of the Earth, there is methane ice which once sublimated is an even stronger greenhouse gas than CO_2. The political pressure on this science problem is enormous because of the implications internationally, the tax revenue available to governments, restrictions of carbon fuels, and the more sinister idea that if mankind can control the climate, would this lead to yet another cause for wars. We are not going to solve this dilemma in this book, but it is worth noting when we engage in the technical task of monitoring the atmosphere with networks of intelligent sensors. This we can and ought to do. It is a surprisingly difficult engineering problem as well as a challenging signal processing problem, since the fluctuations of the Earth's temperatures are quite large compared to the trends in global warming measurable in a human lifetime.

16.3.1 Detection of Small Chemical Molecules

Air pollution is generally a high localized concentration of an unwanted chemical, or mix of chemicals. It is expected in areas of urban industrialization but it also occurs indoors, on farms, and for chemicals we cannot smell or taste. We are born with a capability to associate the smell and taste of foods good for us from poisons that would harm us. How does this work? How did nature solve the problem? Can we mimic this with artificial sensors for detecting chemicals? Animals recognize the materials around them using sight, smell, feel, and eventually taste. Using the color one sees, it is actually an optical chemical detector that discriminates between the ripe red tomato and the green one, or the yellow banana from the unripe green one or the brown rotten one. The strong odor of ammonia is easily smelled to prevent one from getting sickened by rotting fish. In meat the odor of hydrogen sulfide is repulsive, and so on. These odors keep us from eating bad food, but they also signal other animals to come quickly for a meal. There is an entire chemical ecology going on involving microbes, volatile gases produced as a byproduct of microbial respiration, and those gasses triggering other animals to get involved, from other microbes, to insects, fish, birds, and mammals. The ability to taste and smell is the biological process of olfaction, and a very brief discussion of it offers some incredible insights into how chemical sensing can be done artificially.

To understand the details of olfaction in general, insects have provided the greatest research details to date because of the long history of pesticide research, the robustness of insects to neurological probing while alive, and the lack of ethical concerns by society for such research, since it directly effects food production. Removing the head of an insect, the nerves in the antennae and brain continue to respond to chemicals for many hours. On a typical antenna, there are thousands of small hairs and some 100,000 or more olfaction sensors, *or sensilla*, and maybe 50 or so classes of sensilla [12]. The sensilla classes are defined mostly by how they look under high magnification as the pore structures and chemical coatings emerging from the sensilla pores vary by class. The pores are rather small, on the order of 5–20 nm across and the cells around the pore manufacture chemicals which ooze out the pore and help attract the class of environmental chemicals associated with the pore function, and not other classes of chemicals. These structures evolve with each generation of insect and develop into collections of sensillae tuned to the odors of foods, mate pheromones, predators, and environmental odors produced from the chemical ecology. Each class of sensilla are wired to specific outgrowths of the brain called mushroom bodies and are the same for every insect in the species and shared across some species. The sensilla work by differential diffusion, allowing soluble molecules to pass through and be detected and blocking interfering molecules. Detection happens very fast, on the order of milliseconds or less, when the molecule being detected is transported to a nerve dendrite, triggering an electrochemical voltage pulse transmitted to the brain. But then a slower process of self-cleaning commences, taking on the order of seconds to break down the molecule and disperse it using enzymes. This is why we are most sensitive at the onset of an odor but become numb to it after a period of time. Also, the animal can control the uptake of molecules in the pores through the production of the surface chemicals, in much the same way one's iris narrows in bright light and dilates in the dark. *A sensilla is a chemically-functionalized, self-cleaning, single-molecule detector.* Collections of diverse sensilla, when associated with important molecules or mixtures of molecules, provide the olfaction patterns that alert the animal into behaviors that help it survive and reproduce [12].

Figure 16.24 shows a high-resolution atomic force microscopy (AFM) scan of the surface of a hair from a moth antenna, showing the various pores and structures. The AFM has a stylus with a sharp tip that vibrates near its resonance as it is scanned over the surface. The tip material is attracted or repulsed by the chemical makeup of the surface, causing small changes in the stylus resonance. This is detected as a phase shift in the sinusoidal response of the stylus. Recall from Section 16.1 that amplitude or frequency modulation is an effective strategy for improving the SNR of very low-frequency signals in noise. The AFM is a great example of this strategy and allows the imaging of the space surrounding a single molecule to be formed. The pore structure in Figure 16.24 are different due to both physical shape and surface chemical makeup, and both of these characteristics likely play a role in making the pore chemically functionalized to only respond to certain classes of chemicals. AFM scans using different styli tips can be used to investigate the chemical properties differentiating the pores.

The response time is also an important characteristic which is a function of the solubility and diffusivity of the sensed molecule with respect to the surface chemicals of the pore. A complex mixture of hydrocarbons, such as the aromatic odor from a flower, will stimulate various classes of sensilla with different uptake rates, giving rise to an "initial taste, a middle taste, and an after taste" not unlike what we experience when we discern between different beers or wines. The different avenues of uptake for different components of the chemical mixture to be captured, sensed, and cleared creates an "olfaction language" that allows the animal to associate patterns of smell and taste on a useful time scale, and without having to completely separate the constitutive chemicals first. Most man-made chemical sensors offer a detected signature of the chemical in its "neat" state, meaning pure chemical with no other interferants. This is reasonable scientifically, especially for validating the quality and purity of manufactured chemical products, but problematic for environmental chemical surveillance, where mixtures are always present. Herein lays the signal processing challenge for chemical sensing (note that we did not use the word sensors). We need a diversity of sensors and data fusion to have a chance of detecting complex chemical mixtures of interest in real-world natural environments.

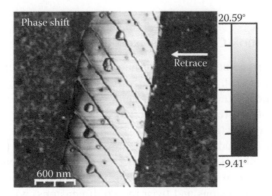

FIGURE 16.24 An atomic force microscope scan of a hair from a moth antenna showing the chemically functionalized pores that only pass certain chemicals for olfaction detection.

A man-made analogy to the sensilla pore is a polymer substrate that has been chemically functionalized with a coating that promotes absorption of some classes of chemicals while repelling other chemicals. The substrate can be part of a capacitor or resistor, or even be the gate of a transistor. As the substrate absorbs the chemical of interest, it material properties change from the added adsorbed mass, or from swelling from the absorbed chemical which could change the electrical resistance. Integrating such materials into an amplifier circuit create a chemically-responsive "switch" which can detect the presence of the chemical analytes of interest. The first step is to make the electronics sensitive to the chemical of interest. The next step is the challenge of getting the switch not to respond to everything else. These devices are sometimes called "chem-FETs" or electronic noses, but they are really quite simple in concept. An array of chem-FETs with diverse material coatings can collect a more sophisticated pattern of chemical analyte absorption to better distinguish the analyte of interest from false alarms from other analytes which "look" like the target on one or more chem-FETs individually. The concentration which flips the switch may be different from analyte to analyte, so the pattern of "voting" by each of the chem-FETs in an array is helpful in rejecting many false alarms by simple data fusion. A more robust scheme for data fusion is to make the switch more of a potentiometer for analyte concentration and to look for consistency across the chem-FET array for a given analyte at a given concentration.

Another electronic device capable of this more detailed concentration response is the surface acoustic wave (SAW) detector. The SAW transmits a high-frequency surface wave across the substrate absorbing the analyte where the wave speed is affected by the added mass of the absorbed analyte. The easiest way to detect small changes in wave speed is to use the SAW delay in a feedback loop to create an oscillator. As the delay increases the oscillation frequency decreases allowing a very precise measurement of the delay, which is proportional to the mass of the analyte absorbed. SAW arrays offer an even more detailed opportunity for data fusion to reject false alarms and to report the concentration for a wide range of analytes.

The down side of the SAW and the chem-FETs is that the polymers have limited shelf life and limited use life. To reset the polymers the devices usually heat them for a few seconds to drive off the absorbed analyte and start a new absorption cycle. Polymers are long chains of molecules where the bonds are somewhat weak and may over time break as a result of these frequency absorption–desorption cycles, may react with the analytes being absorbed, or may breakdown due to the presence of water. These gradual changes mean the calibration of these devices becomes an issue as well as contamination. They are useful where frequent changing of the sensing polymers is manageable such as indoor manufacturing facilities or in the controlled air environments of hospital facilities where deadly gasses are used to sterilize surgical tools and must be carefully monitored to insure the germs on the tools, and not the people, are killed. There are similar applications in food packaging and automatic inspection, monitoring of gasses generated in grain silos and mines, and inspection of aircraft fuel tanks.

16.3.2 Optical Absorption Chemical Spectroscopy

Molecules form when atoms come together and share electrons in a lower energy state from completing their electron shells [13]. Depending on which outer shells are filled or emptied in the bond between a pair of atoms, the electron bond may be strong, weak, a double bond, or even a triple bond. The atomic structure of a molecule has electronic polarity and the forces between atoms are balanced between electronic attraction to each other's protons in the nucleus and repulsion to each other's electron cloud. This makes the molecule, in a simplified linear sense, a structure not unlike masses (nuclei of the atoms) connected to each other with springs (the electron clouds) where the molecules can be seen to have normal modes of vibration [14]. As the atoms vibrate (oscillate in three dimensions and twist in two dimensions) with respect to one another, the outer shell electrons get pushed around, jumping between energy levels and emitting photons when they jump to a lower energy level. The wavelength λ of these photons is proportional to the energy by

$$E = h v = h \frac{c}{\lambda} \qquad (16.54)$$

where h is Planck's constant (6.626×10^{-34} J s), c is the speed of light (299.8×10^{6} m/s), and v is the light frequency in Hertz. Light photons are packets of quantized energy where each quantity is of size hv. This is because the electron shells of atoms represent discrete steps in energy. So if the atom is energized by heating, specific frequencies of light will be emitted as the energized electrons jump back down to their respective shells. The combinations of shells electrons can jump between and provide a unique atomic spectra for a given element. One can also efficiently energize an atom by exciting it with a light frequency that matches one of its spectral lines.

For a molecule it naturally becomes more complex, depending on the atoms involved and which atoms share electrons in a chemical bond. But the same principle applies [15]. A molecule will emit and absorb characteristic discrete light frequencies based on its constitutive atoms and the chemical bonds. However, we do not want to energize the molecule to such a degree that the bonds are broken, breaking the molecule or causing it to react chemically with other molecules. We just want to measure where these characteristic light frequencies are. We also want to keep matters simple by avoiding nonlinear effects such as fluorescence where emitted photon frequencies are well below the excitation frequency. Unfortunately, we cannot easily make any light frequency we wish because the materials we can excite to radiate light are all made from molecules and atoms with these quantum light properties. One way to scan the light absorption of the molecule is to use an incandescent light source and a diffraction grating to filter the light as a means to use one frequency at a time and scan over a range of wavelengths to obtain the *molecular absorption spectra* (MAS). The scanning light is really not monochromatic like a laser, but rather a very narrow range of frequencies defined by the resolution of the filter or diffraction grating. The MAS of a neat chemical is a complete fingerprint that can be associated with only that chemical. To minimize the possibility of fluorescence and to use a range of light wavelengths that are closely spaced in frequency, the MAS is usually done in the infra-red (IR) range of light.

Chemical spectroscopy has an unusual way of noting wavenumber when compared to the rest of the signal processing community. They have had spectrometers to work with long before signal processors got their FFT. Table 16.1 lists a range of useful wave types, wavelengths, wavenumbers, and frequencies. The visible light wavelength range goes from around 400 nm (deep blue) to about 900 nm (deep red), the UV range is from around 150 to 400 nm, and the IR range is around 900 nm

TABLE 16.1
Wavelengths and Frequencies for Spectroscopy

Wave Type	Wavelength	Wavenumber	Frequency	Frequency Pronunciation
γ-rays	10 pm	10^9 cm^{-1}	30 EHz (10^{18} Hz)	Exa-Hertz
x-rays	10 nm	10^6 cm^{-1}	30 PHz (10^{15} Hz)	Penta-Hertz
Ultraviolet (UV)	200 nm	50×10^3 cm^{-1}	1.9 PHz (10^{15} Hz)	Penta-Hertz
Visible light	780 nm	12.8×10^3 cm^{-1}	384 THz (10^{12} Hz)	Tera-Hertz
Infrared (IR)	50 μm	200 cm^{-1}	6 THz (10^{12} Hz)	Tera-Hertz
Microwaves	100 mm	0.1 cm^{-1}	3 GHz (10^9 Hz)	Giga-Hertz

to 1 mm. These wavelength ranges are approximate. 780 nm is typical for most red laser pointers and 532 nm is typical for green laser pointers.

The peculiar wavenumber notation is the cm^{-1} rather than m^{-1} which is a simple matter of dividing the m^{-1} wavenumber by 100. The reason is simply convention and convenience in that the wavenumber range from IR to UV goes from 200 cm^{-1} to 50,000 cm^{-1} making it easier to mark the x-axis of the spectra. The absorption lines for a given analyte can be modeled fairly accurately, the details are well beyond this book, but techniques can be found in recent physical chemistry textbooks (*Phys Chem*. 4th ed., *Normal Mode Analysis, Computational Physics*). Generally, the MAS is experimentally measured for neat chemicals and cataloged in tables. The depth of the absorption lines depends on analyte concentration and on the particular molecular bond and its mode of vibration. Some types of vibration are more observable as photon absorption than others. Again, the details are fascinating but well beyond what we can address in this book. From a signal processing point of view, the MAS is an inverse spectrum with dips representing the molecular absorption frequencies and the depth of the dip proportional to concentration.

To more efficiently measure the IR MAS using a broadband IR source one can create an optical standing wave field using a beam splitter and mirrors and modulate the standing wave pattern by moving one of the mirrors while acquiring brightness data. The standing wave pattern creates a coherent spatial standing wave from the incandescent light. By moving one of the mirrors, each frequency of light is modulated according to its wavelength. Applying a Fourier transform to the record of brightness fluctuations at the detector while the mirror is moving results in a spectrum that is slightly Doppler shifted from the mirror motion. By adding a known laser standard frequency this small Doppler can be corrected and the MAS is recovered. The neat chemical is placed in the IR source beam and the light emerging from the sample is split and remixed to form the standing wave beam at the detector. This technique is known as Fourier transform infrared spectroscopy, or FTIR. The resolution and wavenumber bandwidth of an FTIR depends on the distance the mirror is moved and the length of time and sample rate of the data in the Fourier transform. FTIR is well established and in wide use for chemical identification.

Another MAS technique uses acoustic waves excited by the selective heating of the sample at a characteristic frequency and is called photo-acoustic spectroscopy (PAS). PAS depends on the availability of a narrow band optical source, preferably a laser that has a light emission that closely matches an absorption emission of the chemical sample. By pulsing the laser or light source, the heating of the corresponding chemical will cause thermodynamic expansion with each pulse, thus creating an acoustic pressure wave which can be detected with a microphone. Given one knows what chemical to look for and has a combination of optical sources that match the absorption wavelengths, a very inexpensive detection system can be made. PAS is gaining commercial interest also because variable frequency solid-state lasers are now becoming available with a significant sweepable range. Multiple lasers can pulse simultaneously, each using a specific frequency to multiplex detection of several key absorption wavelengths for the analyte of interest. An FFT of a microphone signal would demodulate the simultaneous interrogation. This could also be done using MLS pulse sequences and correlation to the unique codes to extract the absorption responses at the laser wavelengths. As new tunable laser technologies become available, PAS may become a very powerful and cost effective chemical detector.

16.3.3 Raman Spectroscopy

Raman spectroscopy also provides a unique signature of the molecular normal modes, but not from absorption of photons [13]. Raman spectroscopy is based on the analysis of the scattered light from the molecule. The incoming electromagnetic wave is typically from a laser and imposes a polarization on the molecule, due to the molecule's spatial charge distribution. But this depends on the orientation of the particular molecular bond with the polarization of the incoming laser waveform. It also depends on the molecular vibration of the bond and that is where the useful information emerges. When a conducting object is placed in an electric field E, a dipole moment μ is induced according to the simple relation

$$\mu = \alpha E \tag{16.55}$$

The polarizability α depends on the molecule, orientation, and the mode of vibration of the molecule.

$$\alpha = \alpha_0 + \Delta_k \cos(2\pi \upsilon_k t) \tag{16.56}$$

where υ_k is the frequency in Hz of the kth mode of the molecule and Δ_k is the sensitivity of the mode to the electric field. If the laser transmits an electric field of the form

$$E(t) = E_0 \cos(2\pi \upsilon_0 t) \tag{16.57}$$

the dipole moment of the molecule is time varying as

$$\mu(t) = E_0 \cos(2\pi \upsilon_0 t) \left[\alpha_0 + \sum_{k=1}^{\infty} \Delta_k \cos(2\pi \upsilon_k t) \right]$$

$$= \alpha_0 E_0 \cos(2\pi \upsilon_0 t) + E_0 \sum_{k=}^{\infty} \Delta_k \cos(2\pi \upsilon_k t) \cos(2\pi \upsilon_0 t)$$

$$= \alpha_0 E_0 \cos(2\pi \upsilon_0 t) + \frac{E_0}{2} \sum_{k=}^{\infty} \Delta_k \left\{ \cos\left[2\pi(\upsilon_0 + \upsilon_k)t\right] + \cos\left[2\pi(\upsilon_0 - \upsilon_k)t\right] \right\} \tag{16.58}$$

The dipole moment generates the scattered electric field, which is very weak compared to the incident laser energy, in part because the molecule is generally very small compared to the laser wavelength. The first term in Equation 16.58 is the Rayleigh scattered wave while the second term is called the anti-Stokes backscatter, and the third term is the Stokes backscatter. Unlike in our simplified model, the anti-Stokes spectral lines are actually weaker. This is because of quantum physics and that in order to backscatter a higher energy wave (higher frequency) the electrons must be excited from a higher energy state then they end up at. The Stokes backscatter at $\upsilon_0 - \upsilon_k$ Hz is a lower energy photon where the electron ends up at a higher energy state then where it begun. Given a finite energy from the laser photons, there is stronger Stokes backscatter below the laser frequency than above it. The frequencies of the molecular vibration modes υ_k are independent of the laser frequency, so Raman spectra are typically denoted as a wavenumber shift, subtracting out the excitation frequency. This is why Raman spectra is comparable to the MAS even if the excitation light frequencies are different.

The technical challenge with Raman spectroscopy is to filter out the laser frequency and capture the very weak Stokes backscatter in a high SNR spectrometer. The noise in a spectrometer is called "dark current" and is found by reading the detector output with no light entering the spectrometer. The spectrometer signal can be summed over time to reduce the electronic background noise as the square-root of the number of spectra summed. The amplitude of the Raman spectra is typically in arbitrary units (a.u.) because of the difficulty in obtaining physical units, so the spectra are usually compared relatively, sometimes even using spectral subtraction to emphasize some chemical reaction. Some of the molecular modes are even weaker than others due to the nature of the molecular motion and some modes are not Raman active at all. It depends on the symmetry of the molecule among other things such that some modes are best seen using MAS or FTIR and others are best seen using Raman, while some other modes are extremely difficult to see in either spectrographic method. Using a laser frequency close to a molecular absorption line enhances the molecular vibration and the Raman backscatter and is called "resonance Raman spectroscopy." From the signal processor's point of view, the Raman spectrum is just a spectrum, where the peaks are to be detected and associated with particular target analytes.

The Raman spectrum can be made more sensitive using a specially prepared surface for the analyte, usually a rough surface of gold or silver. The surface must be rough on a near atomic scale, but it can provide a Raman backscatter amplification well over 10^{14} or even higher. The technique is called surface-enhanced Raman spectroscopy (SERS). Clearly, SERS does wonderful things but the surface must be cleverly prepared [16]. How does such a huge amplification occur? At the time of this writing, there is still much debate as neither chemical theory not electromagnetic theory can fully explain this SERS enhancement. When the long wavelength from the laser electrifies the molecule, electrons are pushed back and forth in what is called a *Plasmon Resonance*. This is very weak on the tiny molecule amidst the huge laser wavelength, and it is very sensitive to the orientation of the molecule relative to the laser polarity. But on a rough conducting surface, the large laser wavelength excites many Plasmon resonances on the surface to following the electric field while maintaining a conducting surface. As such there are always components of the field normal to the surface which will radiate back to the spectrometer. The image sources along the rough boundary create Plasmon resonances of moving spatial charge. Enter our molecule adsorbed (adhered to) on the surface. Several beneficial things will now occur to the weak Stokes and weaker anti-Stokes spectral lines. First, the molecule is electrically stimulated more at the boundary and stimulated with spatial variations from the rough surface image sources to a much greater degree than when the molecule is moving in the air. Second, the molecule is fixed in space, at least at one end and as such remains in the beam while the Raman spectrum is summed, rather than moving at hundreds of m/s and stimulated for a short time as it passes through the beam in air. Third, the Stokes and anti-Stokes electromagnetic waves radiated from the molecule reflect from the rough surface back to the spectrometer, adding strength to the Raman signature photons. The surface adsorption will however affect the strength of some of the molecular vibration modes, enhancing some while attenuating others, so the relative intensities of the Raman spectral lines, in particular with SERS, may not be consistent.

16.3.4 Ion Mobility Spectroscopy

Ions are formed when a molecule either accepts or gives up one or more electrons as a result of being in solution, being subjected to high-energy photons (ionizing electromagnetic radiation), or passing through a strong electric field, *corona discharge*. A corona happens where high-energy electrons interact with gas molecules to form charged particles. Ionizing radiation requires a source such as a nuclear isotope. By generating ions in air, we can subject them to electric fields and air flows with reagent gases to separate the ions of interest and detect them on a Faraday plate where the electron-hole pairs are recombined producing a small current that can be measured. The most common ion mobility spectroscopy (IMS) sensor is a smoke detector. The typical ionizing smoke detector works by using α-radiation, which is easily absorbed by smoke particles and just about anything else, to create ions in a passively ventilated air chamber, which in turn creates a sensible current through the ionized air. An isotope of Americium (^{241}Am) is a good α emitter and as mentioned, even the plastic case of the detector completely blocks the radiation from escaping, but it wont be a good idea to take a smoke detector apart because α radiation can give you cancer. When smoke particles enter the ionization chamber, they completely absorb the radiation, shutting down the air ionization current, which is detected and sets off the alarm until the air again clears, or you remove the battery. Another more sophisticated but common IMS detector is found at airports for explosives detection. These devices are incredibly sensitive and are designed to look for specific compounds found in explosives, as well as the nitro-compounds themselves which have extremely low vapor pressure.

There are many variants of IMS but for our focus, we are mainly interested in IMS of air samples at atmospheric pressure chemical ionization (APCI) and nominal relative humidity [17]. Because air is most diatomic and tri-atomic molecules of known proportion, a larger molecule from a harmful chemical is fairly easy to detect. The ions generated are dominated by the effects of water and creates an "ion swarm" of various ion products in equilibrium. Together these ions form a reaction ion pool for separate modes of positive ion detection and negative ion detection. In positive ion mode, the pool of ions is released by dropping a gate voltage and measuring the flight time in a negative electric field to the detector plate, where the amount of ions collected produces a small current. The electric field in the drift tube drives the ions to the detection plate and there is an opposing flow of gas to slow the ions down to a velocity that is a function of the ion size and charge. The ions undergo an enormous number of statistically independent collisions (N_2 for APCI has about 4×10^9 collisions/s) and the resulting time of arrival and distribution of the current from the detection plate (called the Faraday plate) is remarkably Gaussian for a simple type of ion. Figure 16.25 depicts the arrangement as simply as possible. Typical drift times are less than 20 ms so the instrument can enhance SNR through averaging multiple ion "races."

The output of a drift tube IMS is called a *plasmagram* and literally looks like a spectrum where the peaks are Gaussian shaped and represent individual ion swarms. When the air is clean, a single ion peak (the reaction ion pool) is seen in both positive and negative mode, although slightly different in drift times. When an analyte is present and is ionized in the chamber on the left of Figure 16.25, it too forms an ion swarm in equilibrium which then races down the drift tube once the gate is opened, but arrives at a later time than the reaction ion pool because the analyte ion is much larger and has more frequent collisions as it makes its way to the Faraday detection plate. The analyte ion is actually stuck to some of the highly polar reaction ions and pulled along by them, but is slowed by its bulky size and relatively high mass. This single ion from the analyte is called a *monomer*. As the analyte concentration increase, monomers collide together frequently forming *dimmers*, and at even higher concentrations form *trimers* and even *tetramers*. The peak amplitude pattern is very nonlinear with concentration but the drift times for these peaks can be used to identify the analyte. However, there are many analytes that can have a similar plasmagram, or at least a few peaks at the same drift time. The drift time is a function of pressure, temperature, electric field, and the analyte size and charge, so to allow peaks to be compared for different instruments, the drift time is mapped to a reduced mobility index which is inversely proportional to drift time. Figure 16.26 shows an example of the plasmagrams produced by a field-grade IMS showing the analyte monomer and dimer peaks in the positive ion mode plot on the right. It is not unusual for an analyte to have ions in only one or the other. Some analytes may fragment into both positive and negative ions. We can also control which analytes get to form ions by adding a reagent gas to the opposing flow in the drift tube which will capture and neutralize ions with lower electron or proton affinity. This acts like a deodorant to scrub out the ions we do not want to see. By optimizing the ion chemistry, an IMS detector can be tuned to favor one type of analyte, such as explosives, or another, such as toxic industrial chemicals. IMS is already an important field instrument and continues to find new applications even for biological materials such as proteins which can be sprayed into the drift tube and dried into charged particles by the drift gas ionizing them simply from the removal of water. For the

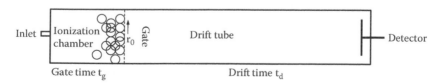

FIGURE 16.25 A simple sketch of an IMS drift tube showing ion chamber, gate, and detection Faraday plate.

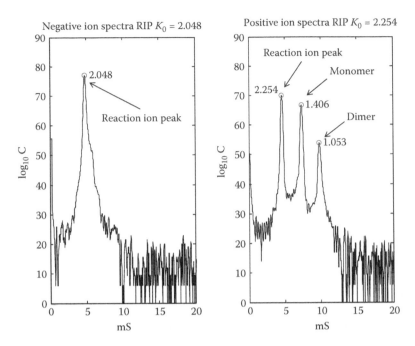

FIGURE 16.26   An IMS plasmagram showing reduced mobilities (K_0) for the reaction ion peak (RIP) in negative mode (left) and the RIP, monomer and dimer peaks in positive mode (right).

signal processor, it can be seen from Figure 16.26 that the IMS plasmagram can be processed just like any other spectrum, so long as you are mindful of the underlying physics.

16.3.5 DETECTING LARGE BIOLOGICAL MOLECULES

Molecular size and composition are the big differences between what is commonly called chemical detection and biological detection. It is not about what elements are present, the molecules that make up proteins, cells, and tissues are all pretty much made from the same things, carbon, oxygen, hydrogen, phosphorous, and nitrogen. DNA is really just combinations of four *nucleotides* called adenine (A), thymine (T), guanine (G), and cytosine (C). Each nucleotide can bind with itself while A binds also with T and G binds with C. So there are four basic "bricks" to life. Combining any three of these four bricks leads to a possible $64 = 4^3$ combinations, but if you remove redundancies it is less than half of that. A group of three DNA nucleotides is called a *condon*. The 20 amino acids that make up life are described as a sequence of anywhere from one to six condons. Combinations of these 20 amino acids make up proteins, which can be very long molecules, thousands of amino acids long, with very complex shapes and functions. The shape, function, and sequence are all tied together in a way which allows these huge molecular structures to bind with one-another, change shape and function, and disassociate in response to subsequent bindings. This forms a fascinating basis for biological signaling within a cell as well as signaling between cells. DNA can replicate itself and be used to replicate more amino acids which in turn form new proteins, cells, and tissues. All of this complexity comes from only five elements [18].

Protein molecules are so big that they tend to ball-up on themselves exposing only part of the amino acid sequence to external stimuli. In a cell there are "rope-like" proteins called trans-membrane proteins (TMP) which extend out of the cell wall and act as a sensor for the cell. For example, one of these TMPs may be designed to signal when sugars are available by binding to the sugar on the end outside the cell, changing shape along the rope-like structure exposing a different part of the amino acid sequence, which in turn binds to RNA inside the cell, then DNA in the cell nucleus, and

RNA back out to the cell wall proteins, which signals them to become porous and let the sugars in for use. Afterward, other biosignaling commences to reverse the process to keep the sugars inside for use. This is a very oversimplified (for signal processing engineers) description of the cell metabolic process. It is how cells respirate to take in energy sources and expel wastes through a controllable differentially permeable cell wall. Another example of biosignaling is *bacteriorhodopsin*, a fascinating protein in the eyes which captures light energy and moves protons across cell walls to trigger a response to light. You are looking through bacteriorhodopsin as you read this book! Similar signaling occurs at the ends of nerve cells (dendrites) where mechanical stresses (from touch or sound) trigger an electrochemical signal to the brain. Chemical reactions (taste or smell) can also trigger a neural signal as part of olfaction. Biosignaling is perhaps the most profound process in all of modern biology and that is why there is such strong research interest at this time. However, once you can see the complexity of how amino acids are expressed within proteins and that this is not static, but part of a vast complex biosignaling and control system, it becomes clear that we will not be able to uniquely identify a pathogen by detecting the presence of a given molecular bond or resonance. Just to identify a single protein, we need to know not only which amino acids are present, but in what order and how many of each, and the molecules are so large that they are coiled up on themselves, exposing only a portion of the protein sequence. When an exposed part of the protein binds with something, the tangled ball reorients and a different section of the protein sequence is exposed.

We are interested in identifying a *pathogen*, that is, something potentially harmful to us, or something we care about, that is biological. By biological we mean it can replicate itself and coexist within us. Generally speaking, pathogens do us harm and our immune systems routinely mark pathogens with *antigens*, and destroy them before they replicate to such an extent that they do damage. We build up immunity against repeated infections by keeping copies of those marker sequences that recognize a past pathogen so that very rapid replication of antigens for destruction can be dispensed to kill off the pathogen if it reappears. The most deadly pathogens are the ones which go unrecognized until it is too late and the immune system cannot catch up and fight the disease. It is largely a numbers game where if the pathogen can out reproduce the immune system the host will soon lose the fight. A pathogen will kill you if it interferes with a key body function, such as breathing or retention of fluids, or if it produces biotoxins, which are chemical wastes that are not biological, but are complex toxins that kill cells, tissues, or impair the nervous system. By the time you recognize the symptoms of a cold or flu, you are already infected. We would love to have sensors that can recognize these pathogens in real time, but our own bodies cannot even do that.

Biological agents can be in the form of a live (vegetative) cell or virus, or as a compact spore capable of surviving long periods without water or nutrients and then be revived later when water is available. *Life is tenacious.* When a cell becomes a spore, whatever proteins that are on the surface of the cell will also be on the surface of the spore, which will tend to cluster with other spores because the removal of water will likely leave or enhance the electric charge or dipole moment of the spore. The usual way one identifies a pathogen is to isolate it and grow it in a controlled environment where one can test the shape, function, and even DNA sequence of the pathogen for verification. This takes a lot of time and expertise. But there are some things that can be sensed in near real time.

The first biological thing to test is the particle size [19]. If particles of any type are in the 2–5 μm range, they pose an inhalation hazard because they are small enough to get past the nose and throat but large enough to get caught in the alveoli sacs of the lung. This usually requires some fine silica powders to keep the particles from clumping together. Particles in this narrow size range with dispersant powders can be immediately deemed both hazardous and man-made. Natural pathogenic spores are carried in the air in clumps of several spores so that the particle size is more diversified. To measure the size of the aerosol particles, one usually uses a particle impactor which moves all the particles around a curve in an air flow where the heavier/larger particles will be separated by centrifugal force. Another option is to use forward Mie optical scattering in a collimated light beam where particles of a particular size will scatter light to a specific detector.

Whether harmful or not, no one wants to breath something in that can get caught deep in their lungs under any circumstance. Simple paper masks are all that is needed to keep hazardous particles out of the respiratory system while hand washing will greatly limit ingestion of the pathogens. The only other way is for the pathogen to enter the body through a cut or sore, but this can be easily protected with a bandage. These are all the things your mother explained to you as a child, and 99% of protection is just that simple.

The second thing associated with biological pathogens is fluorescence. The proteins on the surfaces of spores and cells are long chains of amino acids with quasi-periodic patterns within. Energizing the protein with high-energy photons, say in the UV range, will often excite fluorescence down into the visible range. This is not true for all proteins, cells, and tissues, but most cells have such a diversity of surface proteins that something will fluoresce under UV light. This indicates that the material is likely biological, but not guaranteed so. There has been much interest in examining the fluorescence spectra of various pathogens, and there are characteristic patterns in the fluorescence for a particular pathogen. However, there are 10s of millions of bacteria types out in the environment, most of which are either harmless or are the ones we are immune to. Therefore, expecting the fluorescence spectral pattern to be a unique pathogen identifier is very doubtful.

The most promising technique for automatic identification of pathogens is to create *fluorophores* which bind to unique markers on the pathogen, just like the immune system does. Fluorophores are compounds which contain a fluorescing component that is quenched by the surrounding binding acids. These binding sites can be either nucleic acids or amino acids in a specific spatial order that bind to a particular part of the target pathogen. When the binding occurs, the fluorophore opens up exposing the fluorescing part. By using an array, or *assay*, of these fluorophores which bind "lock and key" to characteristic parts of the pathogen, the pattern of fluorescing can be used to help identify the pathogen. This follows the same philosophy as our immune system, but is a crude facsimile. But instead of T-cells marking and white blood cells destroying the pathogen, the fluorophore binds and changes shape to turn on a particular fluorescence that can easily be detected. A number of these fluorophore markers can be tested at once in a *fluoroimmunoassay*, where the binding agents are antigens (antibodies) to the specific pathogen. These are amazing techniques that are currently in very rapid ascension for rapid detection of harmful bacteria and viruses. There are also techniques such as electrospray IMS and Raman coupled with spore impactors which are very promising for specific pathogen detection. For the signal processor, there is really very little that can be used in real time except spore size and fluorescence, but there is a tremendous amount of pattern finding to do when the proteins and DNA of the pathogen are sequenced. This is a new area called "bioinformatics" which involves generating and analyzing these sequences to unlock the function of the underlying proteins in pathogens.

Biological sensing is quite an exciting field, but unfortunately it is a long way from where we need to be for real-time automatic detection with low false alarm rates. The sensor technology is just not ready yet. We can expect real-time biosensors for detection of respiration hazards, and in particular fluorescing respiration hazards. But the complexity of techniques such as fluoroimmunoassay involving wet chemistry will likely not be fully automated for unattended operation in the near future. The data from such analysis, is the fluorescence of the various binders, which could be an image, and array of fluorescence spectra, or a simple matrix of true–false decisions. In Chapter 17, we will outline pattern recognition techniques which could be used to fuse this data into a decision on what pathogen is present.

16.4 NUCLEAR RADIATION SENSORS

Detection of nuclear materials is perhaps the most worrisome of threats to society because the consequences of radioactive pollution, or the much worse nuclear explosion, are so dire. Isotopes occur naturally in all elements. An isotope is a variant of an atomic element having more or less neutrons than the element listed in the period table. The atomic number (the number of protons) stays the same in the listing, but the atomic mass does not. For example, carbon has an atomic number of 6 for its six protons and electrons. It is the six protons that make the element carbon. So an isotope of carbon with seven neutrons would be denoted as ^{13}C, and eight neutrons would be ^{14}C. Over 99% of naturally occurring carbon is ^{12}C, 1% is ^{13}C, and trace amounts (1 part per trillion) are ^{14}C, which decays into ^{14}N with a half life of over 5700 years. Most isotopes decay through a chain of isotopes to a stable (final) state. Since all life absorbs naturally occurring carbon until death, the decay the radiation from ^{14}C can be used to estimate the time of death. The type of radiation emitted is called β-radiation and is mainly electrons with a typical energy of 0.1565 MeV. One of the useful facts about radioactive isotopes is that they radiate unique energies which can be measured with a nuclear spectrometer. Not all isotopes are radioactive [20].

What are the units of nuclear radiation and what levels are harmful to life? The basic unit is the Becquerel (Bq) which is the number of atomic disintegrations per second (dps), meaning the number of isotope atoms that decay into their final isotope state per second, emitting radiation as they transition. The original unit of radiation is the Curie (Ci), which turns out to be about 3.7×10^{10} Bq. These units describe the *nuclear activity*. The *absorbed dose* of radiation is a more direct energy per mass metric. 1 Gray (Gy) is defined as 1 Joule of radiation energy per Kilogram of absorbing material. An older unit of this is the rad, where 1 Gy equals 100 rads. Now, here is where it gets a little more complicated. Some types of radiation are more harmful to life than others. The *dose equivalent* (DE) adds a metric for the type of radiation with a quality factor Q. The DE is the absorbed dose (Gy) times Q where $Q = 1$ for x-rays, γ-rays, and β radiation (electrons), and $Q = 10$ for neutrons and α-radiation (neutron and a proton or a He atom with no electrons). Neutrons and α particles make other elements radioactive, which is why they are so harmful to the body if absorbed. However, α particles are the easiest to shield against as even a thin sheet of paper will block them. Neutrons are the most difficult to shield, because in part they are neutral particles. The DE unit historically has been the rem (Roentgen equivalent man) and the international standard is the Sievert (Sv), where 1 Sv = 100 rems. What is a Roentgen (R)? It's an old unit of ionizing radiation weighted according to the harm it does to man named after the German physicist Wilhelm Röntgen, the first to produce x-rays which won him the Nobel Prize in Physics in 1905. 1 R is about 2.6×10^{-4} C/kg (charge per mass) and about 500 R will kill most people in five hours. The average dose we each get is about 23 μR per hour. The Sv is the preferred international unit. A simple metric to remember is "*1 Gy will make you gray (and hairless).*" A dose of 5 Gy at one time will lead to death in about two weeks. Medical radiation doses are typically measured in mGy, such as an x-ray (1.4 mGy), CT scan (8.0 mGy), or a selective CT scan (25–30 mGy). Radiation treatments for cancer total 20–80 Gy, but in doses of 1–2 Gy. The radiation ionizes the proteins and DNA, literally killing cells which divide quickly, such as those in the gut and those that make hair, and hopefully, cancer cells. One of the ways radiation harms life is to damage the DNA sufficiently to prevent division. Since cancer cells are already mutated, the theory is that these cells may be more susceptible to lethal radiation damage than healthy cells. Medically, it depends on the type of cancer and the state of health of the patient to determine if radiation therapy is warranted.

Measuring radiation is very interesting from a signal processing point of view. Nuclear radiation is defined in terms of quantum particles such as proton-neutron pairs (alpha radiation), electrons (beta radiation), high-energy photons (gamma radiation), and neutrons. One of the first detectors was the Geiger counter which is essentially a gas tube where the resistance drops when a particle passes through the gas. It is not clear how subatomic particles pass through things until you consider the geometry and dimensions involve. Atoms with their nucleus and electron orbital are not drawn to scale. If the electron cloud occupied the Grand Canyon (miles deep and wide) the nucleus would be a raspberry suspended in the middle of the canyon. Go inside the nucleus and you can repeat the process where the nucleus is the Grand Canyon and the protons and neutrons are little raspberries dispersed evenly throughout the canyon. There is lots of space for subatomic particles to pass through without hitting or even interacting with any other particle. This is most true for neutrons due to their lack of charge. But alpha particles (a proton and neutron pair, of helium without its electrons), is big, charged, and usually does not propagate too far before being captured. That is why α particles are relatively easy to shield. β particles (electrons) create ions of the molecules they hit, which is a good thing for IMS sensors and a bad thing for DNA and proteins which may become damaged from the ionization. γ-rays, like x-rays are photons of electromagnetic energy and they too will ionize molecules they strike and are absorbed into. Even though intuition might lead one to treat a γ-ray as a spherically spreading wave, for nuclear reactions it is treated as a particle.

The earliest radiation detector was the *Geiger Counter* (GC), which uses a gas-filled tube with a high voltage across it such that when a particle passes through the gas, it causes the release of electrons and ions from the gas molecules. The electron–ion pairs are accelerated in opposite directions by the electric field, collide with more molecules, and release more electron–ion pairs in a process known as a Townsend avalanche. This multiplies the amount of charge sensible as a current between the anode (–) and cathode (+) in the tube causing a brief conduction through the gas as the event occurs. Each event can be counted, integrated to represent a flux, and/or presented as an audible "click" on a loudspeaker giving the user an indication of the level of energy radiated. Counting the particles over a long time creates a radiation dosimeter, which is useful because a long exposure to low levels of radiation can have the same damage as a short exposure to high levels of radiation. The GC does not try and quantify the level of energy for each particle event, just to capture and count the events. The gas mixture and electronics can be optimized to maximize sensitivity within a particular energy range and provide a crude estimate of the level of radiation [21].

To breakdown the particle events according to their respective energy levels, a *proportional counter* is made by lowering the plate voltage and optimizing the gas mixture so that the collected charge is proportional to the energy of the particle. This type of detector has multiple "channels" or bands of energy where the number of particle events in each band is tallied independently making it useful for measuring x-rays and γ-rays separately. Another version of this is the *scintillation counter*, where a crystal or gas fluoresces to produce detectable photons that are proportional to the particle energy. These photons are generally quite weak, so the signal needs to be amplified using a phosphor to convert the photons to electrons, and then an electron multiplier tube to boost up the number of electrons for charge detection. A charge amplifier completes the system to produce a reasonable voltage proportional to the particle energy. These types of counter have a limited number of energy bands, or channels as they are called, allowing the relative energies to be estimated. However, the charge integration/amplification time is vulnerable to multiple particles "piling up" to be mistaken as a single, much higher energy particle, so this type of detector can be in error in high radiation fields.

A Gamma Ray Spectrometer (GRS) is used to measure the particle energy over a wide range and with very high resolution. The detector is usually a cooled hyperpure Germanium crystal doped as a semiconductor, and reversed-biased so that a large charge-carrier depletion zone is created. When a radiation particle collides with the atoms of the detector lattice, electron–hole pairs are generated and are swept by the electric field to the diode terminals where they appear after amplification as a voltage spike with amplitude proportional to the particle energy. The signal processing on the back end of the GRS is very straightforward and interesting. The detector voltage is sampled at a very high rate and the samples are sorted and counted in bins that correspond to energy levels of the particles. A GRS can have thousands of channels thus allowing it to identify particular isotopes based on the energy of the radiation given off. Sampling the diode at a high rate lessens the chance of particles piling up into mistaken high-energy particles, but in high radiation fields this possibility should always be a concern. The energy bands for the channels can range from a few keV to over 20 MeV (1 eV is 1.602×10^{-19} J) and they may not be (but usually are) of identical energy bandwidth. While each isotope has a unique dominant energy radiated by its particles, interpretation of a GRS spectra is not that simple. Protons and neutrons emitted may activate other materials leading to unexpected isotope lines. Also, a phenomenon called Compton scattering can occur where the γ-ray loses some of its energy from an inelastic collision where an electron is ejected (β radiation). Interpretation of the GRS requires careful analysis and the maximum SNR possible. Usually GRS measurements take several minutes to integrate enough energy in each channel to be useful. Also, comparison with background radiation spectra is essential to really determine if radioactive material is present or not, so the survey of the area is also critical.

To the signal processor, the GRS output is just another spectrum, but each channel or bin is a histogram of voltage events. Statistical analysis is essential to determine the background noise level and the signal from a radioactive source. Gamma rays are typically above 10^{19} Hz (over 100 exaHz), have energies over 100 keV and wavelengths less than 10 pm (smaller than the atoms they emerge from). This is another reason why they should be thought of as a particle rather than a spherically spreading electromagnetic wave. Particle detection is a function of the distance from the source, the cross-sectional area of the detector, the number of isotope disintegrations per second (dps or Bq), and the quantum efficiency of the detector. Let us say we have a 50% efficient detector of area 0.01 m^2, and a radioactive source 10 m away with 10^6 Bq of 1 MeV particles. The background is about 4×10^{-7} centisieverts per hour (cSv/h). The particles can go in any direction, so the fractional flux through our detector is $10^6 \times 0.01/(400\pi)$, or about 8 particles per second. Our sensor would report roughly 4 particles at 1 MeV each per second, on average. Assuming a human holding the sensor has 1 m^2 area and weighs 100 kg, the dose is 8 MeV per second per kg, or 12.8×10^{-19} J/skg, so even after 1 hour (3600 seconds) the dose would be 4.6×10^{-15} J/kg or 4.6 femto-Gy which is well below the background radiation in the environment. Using a quality factor of $Q = 1$ for 1 MeV particles, 4.6×10^{-15} Gy is 4.6×10^{-15} Sv for 1 h, well below the background of 4×10^{-9} Sv/h. In other words, our sensor will only detect background noise in this situation. These "back-of-the-envelope" calculations are absolutely necessary for the signal processor to make sense of the GRS data and to understand what kinds of sources are detectable and at what distance. There are also shielding effects which also prevent particles from getting from the source to the detector requiring very detailed and complex modeling t accurately predict detection ranges. However, using basic assumptions of background levels and source energies, one can quickly determine the SNR and what level of detection performance is possible.

16.5   MATLAB® EXAMPLES

While this chapter covered an exceptionally broad range of sensor and transducers Table 16.2 show only three m-scripts were used to detail all the transducer figures. This is because the technique of modeling transduction using equivalent electrical circuits is enormously important and rarely covered in signal processing texts. The sensor signal and noise models in Section 16.1 are also important, but straightforward extensions of the SNR models described in ROC curves and detection (Chapter 12). The descriptions of chemical, biological, and nuclear sensors are quite terse compared to what one can find elsewhere. Whole textbooks have been written on each type of detector. Our goal is to explain as concisely as possible the underlying physics, chemistry, and biology for each class of sensor, and to report on what the corresponding sensor data contains. Of these sensors, the acoustic/vibration sensor has the most complexity from a signal processing perspective.

Loudspeakers are all around us and are an amazing invention that has survived since the very first electroacoustic experiments. It has been reported that Alexander Graham Bell's work was

TABLE 16.2
MATLAB m-Scripts Used for Chapter 16 Figures

m-script	Figure(s)
loudspeaker.m	16.7–16.9, 16.11
pzt.m	16.14–16.17
mic.m	16.19, 16.20

focused on a machine that could help the deaf hear by plotting the vibrations from his carbon-needle microphone. It must have been an exciting moment when they realized they could replace the plotting pens with a loudspeaker coil, extend the wires, and have a communication system. Wired Morse code had already been established with transmitting switches and receiving coils. The new transducers just opened up the bandwidth on either end to the available bandwidth on the wires. Many others in the United States and Europe also arrived at the same concept around the same time, but Bell gets the credit, mainly because his patent was first, and his patent survived extensive legal battles that defined patent law in the United States and even abroad. Studying the Bell patents is perhaps the best way to learn the essence of patent and invention law.

The "loudspeaker.m" script shows how a tool like MATLAB can really shine for signal processing. Back in the 1970s, we used stacks of Fortran cards to generate a frequency response using complex algebra, and another program to plot those numbers on a piece of graphing paper to develop the design. It was painful and took a lot of time to get it right. Now, an m-script can read nearly as the equation in the text reads and a simple function call makes the plot. It only takes a few minutes to put the script together. On line 4 a vector of frequencies is defined, scaled to radians on line 5, and carried through the equations for the input impedance and pressure response from the text to make the plots. We only show the closed-box design, but there is also a vented box design with ports for extending the low-frequency response and matching the response to a 4th order Butterworth or Chebyshev polynomial [6–8]. However, this alignment is significantly more complicated and involves a range of bass port sizes and lengths as well as box sizes. One way to save work was to make multiaxis plots with the computer allowing one to look up intermediate parameters and make an accurate guess at the best cabinet size and port dimensions. This was not included here, even though it is much fun to do, because the chapter is already quite long. The equations for the dynamic microphone and geophone are quite straightforward and follow the closed-box loudspeaker.

The script "pzt.m," models an accelerometer, force actuator, receiving hydrophone, and transmitting hydrophone using the Mason-equivalent models for an electromechanical system. This technique is extremely useful for modeling section of any mechanical structure so long as the section is small compared to wavelength in cross section. It is okay for the section to be long. It is really a method of modeling waveguide sections where plane waves move in the waveguide. The tricky part of the accelerometer is the compression bolt and its influence on the response, which is not obvious. Also, for hydrophones, the face of the transducer is usually covered in layers of rubber, both to protect the transducer from corrosion and to better couple the mechanical wave energy from the higher density metal/ceramic to the water. These impedance matching layers transition the impedance more gradually and also help dampen the structural resonance.

The script "mic.m" models a condenser microphone, the most common microphone in use today. The trick here is the role of the vent and shunt resistor in this type of acoustic sensor. The vent prevents static saturation from barometric pressure changes but also limits the low-frequency response. The shunt resistor also plays a role in the low-frequency response, making the sensor a very high impedance sensor. As discussed in Section 15.1, a simple amplifier can drive a low impedance cable into an instrumentation amplifier some distance away to keep noise interference to a minimum. When the vent and diaphragm get a bit larger, it affects the directivity response of the microphone, creating a cardioid pattern (heart shaped) which is actually a dipole plus a monopole response. If there is no cavity behind the microphone it will have a pure dipole directivity response, which is desirable for some applications.

16.6　SUMMARY

Chapter 16 was one of the more challenging chapters to write in this book because of the enormous amount of detail which had to be omitted to be concise. The goal was to familiarize the student signal processing with sensors of all major types, including the built in signal processing usually done to maximize the SNR of the raw sensor output. Simple transducer signals were presented in terms of a dc-like signal in noise and then the concept of a channel was presented for modeling any changes to the sensor's SNR from the point of sensing to the point of processing the sensor signal. The channel could be a cable, a radio link, or recorded medium such as magnetic tape. The effect of modulation, both amplitude and angle, are presented to show how the sensor SNR is modified by the modulation. For angle modulation, both narrowband FM in the form of phase modulation and wideband FM are presented where the channel SNR can be improved by increasing the bandwidth of the transmitted signal. This discussion explains a generally important relationship between signal power and bandwidth for controlling the SNR. There are a near limitless list of simple sensors that can fit in this description of the sensor SNR, the transmission channel, and the signal processing to enhance SNR at the data acquisition point, such as simple filtering and averaging.

The transducer discussion in Section 16.2 summarizes the conversion of mechanical forces and velocities to voltage and current and models the sections of mechanical structures that are spring-like as capacitors, mass-like as inductors, and damping as electrical resistors. This allows the mechanical components of a transducer as well as any structures contributing to the transducers response to be modeled accurately using analog circuit theory. The tricky part of this technique is to understand how the transducer constituent equations relative force and velocity to voltage and current couple into the structure. Sometimes electrical impedance is inversely proportional to mechanical impedance (or directly proportional to mechanical mobility), as is the case with the loudspeaker. However, the application of these electrical analogy circuits allows a rapid and precise assessment of how the transducer works and how various mechanical components affect the transducer's frequency response.

Section 16.3 discusses chemical and biological sensors, how they work, and what their data look like. Chemical sensing ranges from simple sensor outputs like those presented in Section 16.1, to full spectra representing the molecular bonds, and pseudo-spectra in the form of an IMS plasma-gram. There are also gas and liquid chromatographs, where peaks at specific times represent various compounds, and mass spectrographs, where peaks represent molecules of various masses. There are even combinations of two or more spectrographic techniques. The goal of this discussion is to familiarize the signal processor to the major classes of sensors, how they work, and what the data looks like. In particular we focus on those sensors which can operate unattended and in near real time. For biological sensors, one can detect aerosol particle size and fluorescence, and with some sophisticated automation, perhaps perform an assay analysis of binding antigens to help identify the agent, but this is both expensive and time consuming. Breaking down the pathogen into key proteins and their respective amino-acid sequences, we can see where modern biology and the emerging field of bioinformatics is heading, and hopefully new and useful real-time sensors will follow.

Section 16.4 discusses nuclear radiation and its effects on living organisms as well as the common types of counters and spectrometers in use. Foremost, the signal processor has to have a grasp of what the natural background radiation is in terms of the international standard units and accepted guidelines for background radiation. there is much fear and misinformation about radiation levels and the harm it causes. For example, people may worry about the x-rays they get from a dentist or the radon levels in their basement and not be concerned about the CT-scan, or long high-altitude plane ride they just took. People tend to think medical radiation treatment for cancer is safe, yet these doses are by necessity near lethal levels. Nuclear sensors can be in the form of a simple counter, a dosimeter which integrates the counts, a proportional counter which can measure low- and high-energy radiation separately in a limited number of bands, or a spectrometer that can estimate energy levels with such high resolution that the isotope sources can be identified. Different types of radiation harm life differently where protons and neutrons are weighted 10 times more dangerous than γ-rays.

The higher the energy level of the γ-ray, the higher the frequency and the higher the damage. Understanding this, we have to be concerned that the radiation flux does not overwhelm the sensor doing the counting of particle events. If the sensor gets saturated, particles will "pile up" and be counted at an erroneous higher energy level. This is alleviated by sampling at higher rates and reducing the sensor flux through moderator shields or by using a smaller sensing cross section. To the signal processor, one should not only be familiar with the easy to measure background levels, but also the saturation point to avoid erroneous detection of the most harmful particles. For really high energy particles, such as fast neutrons (energies over 1 MeV), there is a really good chance that the detector will not even capture the neutron. Most neutrons detected are thermal neutrons with energy much less than 0.025 eV. Fast neutrons become thermalized fairly quickly in the environment. It takes a special detector to measure neutrons, usually a scintillation proportional counter, but it is difficult to separate the signals from γ-rays and the bigger particles of neutrons and protons. The most important thing for the signal processor is to be familiar with the conversions, expected background levels, and the particular radiation sources of interest in order to processing the signals in the appropriate context.

PROBLEMS

1. A sensor has a 5 V output with a standard deviation of 0.5 V based on raw voltage data sampled at 100 Hz and low-pass filtered for antialiasing at 40 Hz with a sharp cutoff. If we assume the noise is spectrally white, what will the SNR be if a 5 Hz LPF is used?
2. An FM message signal with power $S_m = 1$ and noise power $\eta B_s = 1$ is modulated by a carrier $A_c = 1$. What must the frequency deviation D be to get a channel SNR of 6.8?
3. If one could replace the magnet on a loudspeaker with another which produces a magnetic field in the voice-coil gap 10 times greater, what happens to the driver resonance and damping?
4. A MEMS accelerometer has sensitivity of 0.1 V per g and an electronic noise of 500 nV RMS. What is the SNR in dB for a 0.001 g 5 Hz sinusoid?
5. A pair of microphone transducers in a 5 cm³ air-filled cavity are exposed to an acoustic signal. Mic A responds with 10 times more voltage than Mic B. A current of 20 ma RMS is applied to Mic B and Mic A responds with 1 v RMS at $\omega = 1000$ rad/s. What is the sensitivity of Mic A in V/Pa at $\omega = 1000$ rad/s?
6. A woofer with a resistance of 5 ohms and an inductance of 2 mH is used in a loudspeaker system with a 1 in. tweeter with a LPF of 1 kHz for the woofer and 1 kHz for the tweeter. The inductance of the woofer impedance is interfering with the crossover. Design a shunt network to cancel the inductance from the woofer input impedance.
7. The tweeter in problem 6 has a resistance of 5 ohms and a resonance at 900 Hz where this resonance is interfering with the crossover filter. Design a shunt network to cancel the resonance.
8. A particular absorption resonance of a molecule was found at a wavelength of 1.5 μm using FTIR. If a 785 nm laser is used for Raman spectroscopy, what is the wavelength of the Stokes line for this resonance?
9. Collimated laser light is found to scatter in a forward direction with an angle θ off an aerosol particle of diameter d approximately as sin θ $=1.2 \lambda/d$. What is the range of scattering angles for particles that are a respiratory hazard using a laser with wavelength 680 nm?
10. Convert a radon background of 13 pC/L to Sv for a 100 kg man with volume 0.1 m³ for a 1 hour period. Assume that the radon emits an α particle at 5.6 MeV.

REFERENCES

1. J. Fraden, *AIP Handbook of Modern Sensors: Physics, Designs, and Applications,* New York, NY: AIP, 1993.
2. K. S. Shanmugan, *Digital and Analog Communication Systems*, New York, NY: Wiley, 1979.
3. D. F. Stout and M. Kaufmann, *Handbook of Operational Amplifier Design*, New York, NY: McGraw-Hill, 1976.
4. L. L. Beranek, *Acoustics*, Chapter 3, New York, NY: The American Institute of Physics, 1986.
5. F. V. Hunt, *Electroacoustical Transducers*, New York, NY: The American Institute of Physics, 1982.
6. R. H. Small, Direct-radiator loudspeaker system analysis, *J Audio Eng Soc*, June 1972, 20, pp. 383–395.

7. A. N. Thiele, Loudspeakers in vented boxes, Parts I and II, *J Audio Eng Soc*, May 1971, 19, pp. 382–392; June 1971, pp. 471–483.
8. R. H. Small, Vented-box loudspeaker systems, *J Audio Eng Soc*, June 1973, 21, pp. 363–372; July/August 1973, pp. 438–444; September 1973, pp. 549–554; October 1973, pp. 635–639.
9. L. Mark, *The Complete Beatles Recording Sessions*, First Hardback Edition, EMI, 1988.
10. G. M. Sessler and J. E. West, Condenser microphones with electret foil, *JASA*, Nov 1963, 35(11), pp. 1878–1878.
11. L. E. Kinsler and A. R. Frey, *Fundamentals of Acoustics*, 2nd ed., New York, NY: Wiley, 1962, pp. 327–329.
12. K. E. Kaissling, Physiology of pheremone reception in insects, *ANIR*, 2004, 6(2), pp. 73–91.
13. R. J. Sibley, R. A. Alberty, and M. G. Bawendi, *Physical Chemistry*, 4th ed., New York, NY: Wiley, 2004.
14. Q. Cui and I. Bahar, eds., *Normal Mode Analysis*, Boca Raton, FL: Chapman & Hall, 2006.
15. J. M. Thijssen, *Computational Physics*, 2nd ed., Cambridge: Cambridge Press, 2007.
16. D. Dwight and D. Allara, Surface Enhanced Raman Spectroscopy (SERS) substrates exhibiting uniform, high enhancement stability, US Patent 7,450,227, Nov 11, 2008.
17. G. A. Eiceman and Z. Karpas, *Ion Mobility Spectroscopy*, 2nd ed., Boca Raton, FL: Taylor & Francis, 2005.
18. C. K. Mathews, K. E. van Holde, and K. G. Ahern, *Biochemistry*, 3rd ed., New York, NY: Addison Wesley, 2000.
19. C. S. Cox and C. M. Wathes, eds., *Bioaerosols Handbook*, Boca Raton, FL: Lewis, 1995.
20. M. F. L'Annunziata, *Radioactivity: Introduction and History*, Amsterdam, Netherlands: Elsevier, 2007.
21. D. N. Poenaru and W. Greiner, eds., *Experimental Techniques in Nuclear Physics*, Berlin: de Gruyter, 1997.

17 Intelligent Sensor Systems

There will always be a debate about what is "smart," "intelligent," or even "sentient" (which technically means having the five senses) in the context of artificial intelligence in computing systems. The term "sentient" in particular is used to describe very high-order metal processes such as self-awareness, awareness of one's soul, self-objectivity, and so on, which are all well beyond the scope of anything discussed in this book. However, it is plausible to compare smart sensors to the preprogrammed and sensor-reactive behavior of insects. It is a safe argument that insects lack the mental processing to be compared to human intelligence but, insects have an amazing array of sensor capability, dwarfed only by their energy efficiency. A fly's life probably does not flash before its eyes when it detects a hand moving to swat it, it simply moves out of the way and continues its current "program." It probably does not even get depressed or worried when another fly gets spattered. It just follows its life function reacting to a changing environment. This "insect intelligence" example brings us to define an intelligent sensor system as having the following basic characteristics:

1. Intelligent sensor systems are adaptive to the environment, optimizing their sensor detection performance, power consumption, and communication activity.
2. Intelligent sensor systems record raw data and extract information, which is defined as a measure of how well the data fit into information patterns, either preprogrammed or self-learned.
3. Intelligent sensor systems have some degree of self-awareness through built-in calibration, internal process control checking and rebooting, and measures of "normal" or "abnormal" operation of its own processes.
4. Intelligent sensor systems are reprogrammable through their communications port and allow external access to raw data, program variables, and all levels of processed data.
5. An intelligent sensor system can not only recognize patterns, but also can predict the future time evolution of patterns and provide meaningful confidence metrics of such predictions.
6. Intelligent sensor networks employ information authentication (IA) and organize their information to facilitate automated data mining by hierarchical information processes.

The above characteristics are a starting point for defining the smart sensor node on the wide-area network or intra-network (local network is not accessible by the global Internet) for integrating large numbers of sensors into a control system for production, maintenance, surveillance, or planning systems. Smart sensors provide *information* rather than simply raw data. Consider Webster's dictionary definition of information:

Information: A quantitative measure of the content of information, specifically, a numerical quantity that measures the uncertainty in the outcome of an experiment to be performed.

Clearly, when a sensor provides only raw waveforms, it is difficult to assess whether the sensor is operating properly unless the waveforms are as expected. This is a pretty weak position to be in if an important control decision is to be based on unusual sensor data. But, if the sensor provides a measure of how well the raw data fits a particular pattern, and if that pattern appears to be changing over time, one can not only extract confidence in the sensor information, but also the ability to predict how the sensor pattern will change in the future. Smart sensors provide information with enough detail to allow accurate *diagnosis of the current state* of the sensor's medium and signals, but also *prognosis of the expected future state.* In some cases, the transition of the pattern state over time will, in itself, become a pattern to be detected.

There are some excellent texts on pattern recognition [1,2] which clearly describe statistical, syntactic, and template type of pattern recognition algorithms. These are all part of an intelligent sensor system's ability to turn raw data into information and confidences. We present here a brief discussion of how pattern recognition, signal feature selections, pattern tracking (prognosis), and automated network architectures are functional blocks of a smart sensor system. One of the most appealing aspects of Schalkoff's book [1] is its balanced treatment of syntactical, statistical, and neural (adaptive template) pattern recognition algorithms. A syntactical pattern recognition scheme is based on human knowledge of the relevant syntax or rules of the information making up the pattern of interest. A statistical pattern recognition scheme is based on knowledge of the statistics (means, covariances, etc.) of the patterns of interest. The neural network (or adaptive template) pattern recognition scheme requires no knowledge of pattern syntax or statistics.

Suppose one could construct a neural network to add any two integers and be right 95% of the time. That would be quite an achievement although it would require extensive training and memory by the network. The training of a neural network is analogous to memorizing addition tables in primary school. Indeed, the neural network began as a tool to model the brain and the reinforcement of electrochemical connections between neurons as they are frequently used. Perhaps one of the more interesting aspects of our brain is how we forget data, yet somehow know that we used to know it. Further, we can weight multiple pieces of information with confidence, such as the likelihood that our memory may not be accurate, in assembling the logic to reach a decision. Clearly, these would be extraordinary tasks for a software algorithm to achieve, and it is already being done in one form or another in many areas of computer science and signal processing.

A scientific approach to the addition problem is to memorize the algorithm, or syntax rules for adding any two numbers, and then apply the rules to be correct all the time. Using 2's compliment arithmetic as described in Section 1.1, electronic logic can add any two integer numbers within the range of the number of bits and never make a mistake. That's the power of syntax and the reason we have the scientific method to build upon and ultimately unify the laws of physics and mathematics. *The only problem with syntactical pattern recognition is that one has to know the syntax.* If an error is found in the syntax, ultimately one organizes experiments to quantify new, more detailed rules. But, to simply apply raw data to rules is risky, unless we change the data to information by including measures of confidence as well.

Statistical pattern recognition is based on each data feature being represented by statistical moments such as means, variances, and so on, and an underlying probability density function (PDF) for the data. For Gaussian data, the mean and variance are adequate representations of the information if the underlying data is stationary. If it is moving, such as a rising temperature reading, a Kalman filter representation of the temperature state is the appropriate form to represent the information. In this case both the temperature reading and its velocity and acceleration have mean states and corresponding variances. The pattern now contains a current estimate and confidence, and a capability to predict the temperature in the future along with its confidence. Combining this representation of information with scientifically proven syntax, we have the basis for "fuzzy logic," except for the fuzziness that is not defined arbitrarily, but rather by the observed physics and statistical models.

Neural networks are clearly powerful algorithms for letting the computer sort the data by brute force training and are a reasonable choice for many applications where human learning is not of interest. In other words, we cannot learn much about how or why the neural network separates the trained patterns or if the network will respond appropriately to patterns outside its training set. However, the structure of the neural network is biologically inspired and the interconnections, weightings, and sigmoidal nonlinear functions support a very powerful capability to separate data using a training algorithm which optimizes the interconnection weights. Note that the interconnections can represent logical AND, OR, NOR, XOR, NAND, and so on, and the weight amplitudes can represent data confidences. The sigmoidal function is not unlike the inverse gamma function used in Section 12.2

to model the probability of a random variable being above a given detection threshold. The biologically inspired neural network can carry not only brute-force machine learned pattern separation, but can also carry embedded human intelligence in the form of constrained interconnections, and weights and sigmoidal functions based on adaptive measured statistics processed by the sensor system.

The anatomy of intelligent sensor systems can be seen in the sensor having the ability to produce information, not just raw data, and having the ability to detect and predict patterns in the data. Some of these data patterns include self-calibration and environmental information. These parameters are directly a part of the syntax for assessing the confidence in the data and subsequent-derived information. As such, the smart sensor operates as a computing node in a network where the sensor extracts as much information as possible from its data and presents this information to the rest of the network. At higher levels in the network, information from multiple smart sensors can be combined in a hierarchical layer of "neural network with statistically fuzzy syntax" which combines information from multiple intelligent sensor nodes to extract yet more information and patterns. The structure of the hierarchical layers is defined by the need for information at various levels. All of what has just been stated is possible with twentieth century technology and human knowledge of straightforward physics, signal processing, and statistical modeling, and can be seen in the Internet search engines of the twenty-first century. Given that we will learn about our world and our own thoughts from these intelligent sensor networks, it would appear that we are about to embark on an incredibly fascinating era of mankind, especially in automated linguistic algorithms.

Perhaps the biggest challenge for intelligent sensor networks is how to organize the vast amount of data generated so that it can be found and used in proper context. We call this "the data mountain" and with inexpensive storage vast quantities of documents are being stored somewhere on the Internet every second. The various search engine sites on the Internet run unattended programs called software agents of "ebots" that troll the computers with websites for documents, scan them for important key words, and build a huge dictionary to enable rapid retrieval of links to the documents in response to a query by a user of the search engine at some later time. While these search engines and ebots have become very sophisticated, the documents do not offer much detail to highlight the information in the text except for a few key words and phrases. Document formats such as hyper-text markup language (HTML) allow a generic format for computers to share documents with lots of formatting options for pictures, video, equations, and links to other documents on the network. These documents can also embed program scripts such as JAVA, Adobe Flash, or Active Server Pages (ASP) that aid in processing transactional information while supporting network security. These programs usually access data using a format called eXtensible Markup Language, or XML. XML documents are actually open database documents where a schema document defines all the element names, what data can go where, enumerations, formats, and so on. XML is ideal for organizing the information from intelligent sensor networks. The sensor data in XML can support a live web page with meters and gauges for the sensor raw data or processed information, and it can feed network-based programs for hierarchical information processing. There are lots of ways to implement this architecture, but there are important emerging features that net-centric sensor networks and processing will all require and will be presented here.

Section 17.1 presents the three main techniques for associating information with known patterns. There are clearly many approaches besides statistical, neural, and syntactic pattern recognition and we hope that no reader gets offended by the lack of depth or conglomeration of these different techniques. Our view is to adopt the strong aspects of these well-accepted techniques and create a hybrid approach to the pattern recognition problem. Section 17.2 discusses features from a signal characteristic viewpoint. It will be left to the reader to assemble meaningful feature sets to address a given pattern recognition problem. Section 17.3 discusses the issue of pattern transition over time, and prognostic pattern recognition, that is, prediction what the pattern maybe in the future. Prediction is the ultimate outcome of information processing. If the underlying physics are understood and algorithms are correctly implemented, if the data are collected with high SNR and the feature information is extracted with high confidence, then our intelligent sensor system is ought to be able to provide us with reasonable future situation predictions. This is the "holy grail" sought after in the development of intelligent sensor systems and networks. Section 17.4 presents the important structures of the sensor network information processing architecture. This relatively new technology is in a constant tug-of-war between new convenient networks services and the threat of interference from hackers, crackers, and viruses. There are also logistical issues in keeping software updated and validated across the network. If the sensor processing is designed robustly, the hardware resources of the network, the sensors, and the storage formats, become somewhat static, supporting long useful life, low cost, and predictable interfaces. The intelligent algorithm part of the enterprise becomes a "network ebot process," supporting thousands of sensors and organizing information automatically at many levels. We will call this an intelligent agent, but it is really a component of what is referred to as "cloud computing" where the user's computer is just a display system and the actual computing occurs on servers across the network. Section 17.4 will provide a brief discussion of the Internet and its hardware, how sensors can exploit this connectivity, and the mostly likely successful architectures for automatically processing networked sensor data.

17.1 AUTOMATIC TARGET RECOGNITION ALGORITHMS

Automatic target recognition (ATR) is a term most often applied to military weapon systems designed to "fly the ordnance automatically to the target." However, ATR should also be used to describe the process of detecting, locating, and identifying some information object automatically based on objective metrics executed by a computer algorithm. For example, a tornado ATR would look at rotation in Doppler radar, regional pressures, winds, temperatures, and satellite images. The ATR algorithm is actually many algorithms working together to support the decision process. But, before one can use a computer algorithm to recognize a target pattern, one must first define the pattern in terms of a collection of *features*. Features represent compressed information from the signal(s) of interest. In general, features should have some physical meaning or observable attribute, but this is not an explicit requirement. A collection of features which define a pattern can be defined statistically by measuring the mean and covariance, or higher-order moments, of these features for a particular pattern observed many times. The idea is to characterize a "pattern template" in terms of the feature means and to experimentally define the randomness of the features in terms of their covariance. The feature data used to define the feature statistics is called the *training data set*. The training data can also be used to adapt a filter network to separate the pattern classes. The most popular type of ATR algorithm for adaptive pattern template separation is the *adaptive neural network* (ANN). The ANN is often a very effective algorithm at separating pattern classes, but the ANN does not provide a physical meaning or statistical measure to its internal coefficients. This leaves ANN design very much an art. The last type of ATR we will consider is the *syntactic fuzzy logic* classifier. The difficulty with the syntactic classifier is that one must know the syntax (feature physics and decision rules) in advance. Fuzzy logic is used to tie the syntax together in the decision process such that lower confidence features can be "blended out" of the class decision in favor of higher confidence data. Our syntactic ATR exploits *data fusion* by enabling the fuzzy combination of many types of information including metrics of the information confidence.

17.1.1 Statistical Pattern Recognition

Statistical pattern recognition is perhaps the most straightforward ATR concept. Patterns are recognized by comparing the test pattern with previously designated pattern classes defined by the statistics of the features, such as the mean and covariance [1]. For example, one might characterize a "hot day" pattern class by temperatures averaging around 90°F with a standard deviation of say 10°. A "cold day" might be seen as a mean temperature of 40°F and a standard deviation of 15°. These classes could be defined using human survey data, where the individual circles either "hot day" or "cold day" at lunch time, records the actual temperature, and surveys are collected from, say 100 people over a

12-month period. The 36,500 survey forms are collected and tabulated to give the mean and standard deviation for the temperature in the two classes. Given the temperature on any day thereafter, a statistical pattern recognition algorithm can make an objective guess of what kind of day people would say it is. The core to this algorithm is the characterization of the classes "hot day" and "cold day" as a simple Gaussian distribution with mean m_k and standard deviation σ_k for the kth class w_k.

$$p(x \mid w_k) = \frac{1}{\sqrt{2\pi}\sigma_k} e^{-(x-m_k)^2/2\sigma_k^2} = \frac{1}{\sqrt{2\pi}\sigma_k} e^{-1/2(x-m_k/\sigma_k)^2} \qquad (17.1)$$

Equation 17.1 describes the probability density for our observed feature x (say the temperature for our example) given that class w_k is present. The exponent term in parenthesis on the right in Equation 17.1 is called the *Mahalanobis distance* from x to m_k. Clearly, when this distance is small the feature data point matches the class pattern mean well. This density function is modeled after the observed temperatures associated by people to either "hot day" or "cold day" classes for our example. Why have we chosen Gaussian as the density function? Well for one reason, if we have a large number of random variables at work, the central limit theorem proves that the composite representative density function is Gaussian. However, we are assuming that our random data can be adequately described by a simple mean and variance rather than higher-order statistics. It is generally a very good practice to test the Gaussian assumption and use whatever density function is appropriate for the observed feature data. The answer we are seeking from the statistical classifier is given an observed feature x, what is the probability that we have class w_k? This is a subtle, but important difference from Equation 17.1, which can be solved using Bayes' theorem.

$$p(w_k \mid x) = \frac{p(x \mid w_k)p(w_k)}{p(x)} \qquad (17.2)$$

Integrating the density functions over the available classes, we can write Bayes' theorem in terms of probabilities (note the use of capital P for probability rather than probability density).

$$P(w_k \mid x) = \frac{p(x \mid w_k)P(w_k)}{p(x)} \qquad (17.3)$$

The probability density for the feature x can be seen as the sum of all the class-conditional probability densities times their corresponding class probabilities.

$$p(x) = \sum_k p(x \mid w_k)P(w_k) \qquad (17.4)$$

Equations 17.3 and 17.4 allow the user to *a priori* apply the likelihood of each individual class and incorporate this information in the *a posteriori* probability estimate of a particular class being present when the observed feature x is seen.

Figure 17.1 shows three Gaussian PDFs with respective means of −2.0, 0.0, and +3.0, and respective standard deviations of 1.25, 1.00, and 1.50. This figure also shows a density function for all three classes of data to be used to determine if the feature is an outlier of the observed training data. This context is important because Equation 17.3 will give a very high likelihood to the nearest class, even if the feature data is well outside the range of the associated feature data used during training. To illustrate this and other points, three feature values of $A = -2.00$, $B = 1.00$, and $C = 15.0$ are considered for classification into one of the three given classes. Table 17.1 summarizes the classification results. The data point at $A = -2.00$ is clearly closest to class 1 both in terms of probability and Mahalanobis distance. Furthermore, the feature data point −2.00 (the * on the x-axis of Figure 17.1) fits quite well within the training data.

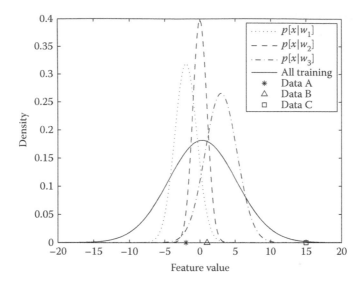

FIGURE 17.1 Three PDFs describe three individual classes in which data points A, B, and C are to be classified while the solid curve shows the probability density for all the classes in the training data.

It is tempting to simply use a small Mahalanobis distance as the metric for selecting the most likely class given the particular feature. However, the real question being asked of the ATR is which class is present given *a priori* feature likelihoods and the feature data. Case 2 where the data point in Figure 17.1 is $B = 1.00$, one can see in Table 17.1 that the highest probability class is class 2, while the smallest Mahalanobis distance is class 3. Each of the three classes is given a 33.33% likelihood for $P(w_k)$.

The third case shows the utility of maintaining a "training data" class to provide a measure of whether the feature data fits into the accepted data for class training. In other words, is this feature data something entirely new or does it fit within the range of the data we so cleverly used to train our classifier (determine the means and standard deviations for each of the classes). For the third feature data point of $C = 15.0$, Table 17.1 shows that class 3 is the best choice of the three classes, but the data are not part of the training data set, as indicated by the low 0.0145 probability for the "training

TABLE 17.1
Classification Results

Data	K	$P(w_k \mid x)$	$(x - m_k)^2/\sigma^2$	$(x - m_k)/\sigma$	Result
$A = -2$	1	0.8066	0.0000	0.0000	Class 1
	2	0.1365	4.0000	2.0000	
	3	0.0569	4.9383	2.2222	
	Training data	0.8234	0.3886	0.6234	
$B = 1$	1	0.1071	3.6864	1.9200	Class 2
	2	0.5130	1.0000	1.0000	
	3	0.3799	0.7901	0.8888	
	Training data	1.0000	0.0000	0.0000	
$C = 15$	1	0.0000	118.37	10.879	Outlier
	2	0.0000	225.00	15.000	
	3	1.0000	28.444	5.3333	
	Training data	0.0145	8.4628	2.9091	

data" class. This low probability coupled with the large Mahalanobis distances to all classes clearly indicates that the feature data is an outlier corresponding to an "unknown" classification.

Multidimensional PDFs are used to describe a number of feature elements together as they apply to a pattern class. Use of multiple signal features to associate a pattern to a particular class is very effective at providing robust pattern classification. To demonstrate multifeature statistical pattern recognition, we begin by defining an M-element feature vector.

$$\bar{x} = [x_1, x_2, \ldots, x_M] \tag{17.5}$$

The means of the features for the class w_k are found as before

$$\bar{m}_k = [m_{1,k}, m_{2,k}, \ldots, m_{M,k}] = \frac{1}{N} \sum_{n=1}^{N} \bar{x} \tag{17.6}$$

but the variance is now expressed as a covariance matrix for the class wk

$$\Sigma_k = \frac{1}{N} \sum_{n=1}^{N} [\bar{x} - \bar{m}_k]^H [\bar{x} - \bar{m}_k] = \begin{bmatrix} \sigma_{1,1,k}^2 & \sigma_{1,2,k}^2 & \cdots & \sigma_{1,M,k}^2 \\ \sigma_{2,1,k}^2 & \sigma_{2,2,k}^2 & \cdots & \sigma_{2,M,k}^2 \\ \vdots & \vdots & & \vdots \\ \sigma_{M,1,k}^2 & \sigma_{M,2,k}^2 & \cdots & \sigma_{M,M,k}^2 \end{bmatrix} \tag{17.7}$$

where $\sigma_{i,j,k}^2$ is the covariance between features i and j for class w_k.

$$\sigma_{i,j,k} = \frac{1}{N} \sum_{n=1}^{N} (x_i - m_{i,k})(x_j - m_{j,k}) \tag{17.8}$$

The multidimensional PDF function for the feature vector when class w_k is present is

$$p(\bar{x} \mid w_k) = \frac{1}{(2\pi)^{(M/2)} |\Sigma_k|^{(1/2)}} e^{-(1/2)[\bar{x} - \bar{m}_k] \Sigma_k^{-1} [\bar{x} - \bar{m}_k]^H} \tag{17.9}$$

where it can be seen that the determinant of the covariance must be nonzero for the inverse of the covariance to exist and for the density to be finite. This means that all the features must be linearly independent, and if not, the linearly dependent feature elements must be dropped from the feature vector. In other words, a given piece of feature information may only be included once.

Ensuring that the covariance matrix is invertible (and has nonzero determinant), it is a major concern for multidimensional statistical pattern recognition. There are several ways one can test for invertibility. The most straightforward way is to diagonalize Σ_k (it is already symmetric) as done in an eigenvalue problem. The "principle eigenvalues" can be separated from the residual eigenvalues (which can be too difficult in matrix inversion) as part of a singular value decomposition (SVD) to reduce the matrix rank if necessary. However, an easier way to test for linear dependence is to simply normalize the rows and columns of Σ_k by their corresponding main diagonal square root value.

$$S_k = \begin{bmatrix} 1 & S_{1,2,k} & S_{1,3,k} & \cdots & S_{1,M,k} \\ S_{2,1,k} & 1 & S_{2,3,k} & \cdots & S_{2,M,k} \\ S_{3,1,k} & S_{3,2,k} & 1 & \cdots & S_{3,M,k} \\ \vdots & \vdots & \vdots & & \vdots \\ S_{M,1,k} & S_{M,2,k} & S_{M,3,k} & \cdots & 1 \end{bmatrix}, \quad S_{i,j,k} = \frac{\sigma_{i,j,k}^2}{\sqrt{\sigma_{i,i,k}^2 \sigma_{j,j,k}^2}} \tag{17.10}$$

For completely statistically independent features both S_k and Σ_k are diagonals and full ranks. The normalized matrix S_k makes it easy to spot which feature elements are statistically dependent because the off-diagonal element corresponding to feature i and feature j will tend toward unity when the features are linearly dependent. The matrix in Equation 17.10 really has no other use than to identify linearly dependent features which should be dropped from the classification problem. However, a more robust approach is to solve for the eigenvalues of Σ_k and apply an SVD to identify and drop the linearly dependent features.

Consider a simple 2-class 2-feature statistical identification problem. A training set of hundreds of feature samples is used to determine the feature means and covariances for each class. The 2D problem is nice because it is easy to graphically display the major concepts. However, a more typical statistical pattern problem will involve anywhere from one to dozens of features. The 1–σ bounds on a 2-feature PDF can be seen as an ellipse centered over the feature means and rotated to some angle which accounts for the cross correlation between the features. For our example, class 1 has mean $m_1 = [7.5\ 8]$ and covariance matrix

$$\Sigma_1 = \begin{bmatrix} \cos\theta_1 & -\sin\theta_1 \\ \sin\theta_1 & \cos\theta_1 \end{bmatrix} \begin{bmatrix} 3.00 & 0.00 \\ 0.00 & 1.00 \end{bmatrix} \begin{bmatrix} \cos\theta_1 & \sin\theta_1 \\ -\sin\theta_1 & \cos\theta_1 \end{bmatrix} = \begin{bmatrix} 1.2412 & -1.3681 \\ -1.3681 & 8.7588 \end{bmatrix} \tag{17.11}$$

where $\theta_1 = -80°$ and a positive angle is counterclockwise from the x_1 axis. Class 2 has mean $m_2 = [7\ 7]$ and

$$\Sigma_2 = \begin{bmatrix} \cos\theta_2 & -\sin\theta_2 \\ \sin\theta_2 & \cos\theta_2 \end{bmatrix} \begin{bmatrix} 2.00 & 0.00 \\ 0.00 & 1.00 \end{bmatrix} \begin{bmatrix} \cos\theta_2 & \sin\theta_2 \\ -\sin\theta_2 & \cos\theta_2 \end{bmatrix} = \begin{bmatrix} 3.6491 & -0.9642 \\ -0.9642 & 1.3509 \end{bmatrix} \tag{17.12}$$

where $\theta_2 = -20°$. The columns of the rotation matrices on the left in Equations 17.11 through 17.12 can be easily seen as eigenvectors. The columns of the covariance matrix on the right are not linearly dependent, but are simply correlated through the coordinate rotation. Clearly, given the symmetric covariance matrix on the right, one can determine the principal variances and rotation orientation. In the three-feature problem, the 1–σ bound on the Gaussian density function is an ellipsoid with three principal variances and two rotation angles. The reader is left to his or her own mental capacity to visualize the density graphics for four or more features, but one can clearly see the connection between eigenvalue analysis and feature independence. For the case of linearly dependent features, one would compute a near-zero eigenvalue on the main diagonal of the principal variance matrix.

Figure 17.2 depicts 50 samples from each of our two classes and the 1–σ ellipses for the Gaussian density functions. The case shown has the two classes overlapping somewhat, which is not desirable but often the case for real statistical pattern recognition problems. The two classes are defined statistically based on the measured means and covariances associated with each class through analysis of a training data set. One can then carry out an eigenvalue analysis to determine if any of the features are linearly dependent and need to be eliminated. The ideal situation would be for the density functions for each class to completely separate in the feature space. In other words, we desire to have large distances between classes and small variances around each class. Given a situation where we do not have a wide class separation, the issue of class discrimination becomes important. Following Equations 17.3 through 17.4, we can estimate the probability of class w_k being present given the feature set $\bar{x}$.

However, we also want to reject outliers which are well outside the range of the entire training samples for all classes. To do so, we derive a new function based on the PDFs of the classes and total training data.

$$p^{\max}(w_k\,|\,\bar{x}) = \begin{cases} \max\{p(\bar{x}\,|\,w_1), p(\bar{x}\,|\,w_2)\} & \forall\ p(\bar{x}) > 0.01 \\ 0 & \forall\ p(\bar{x}) \leq 0.01 \end{cases} \tag{17.13}$$

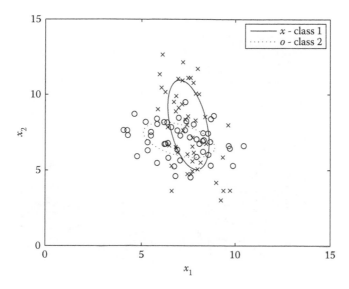

FIGURE 17.2 Samples of a two-feature two-class pattern data set showing the 1–σ ellipses for the Gaussian PDFs for each class and some of the samples plotted in the feature space.

The function in Equation 17.13 is not a probability density, just a function to help us define the decision boundaries for our classes. Because we deal with only two dimensions, this approach is highly illustrative but not likely practical for larger numbers of features. At the boundaries between the two class density functions and at the ellipsoidal perimeter where the training density falls below 0.01 (about a 3–σ Mahalanobis distance) there will be abrupt changes in slope for $p^{\max}(\overline{x})$. Therefore, we can apply some straightforward image processing edge detection to highlight the decision boundaries. Now let us create an "edge" function as follows:

$$p^{\text{edge}}(w_k \mid \overline{x}) = \left| \Delta^2 p^{\max}(w_k \mid \overline{x}) \right|^{(1/8)} \tag{17.14}$$

where the eighth root is used to "flatten" the function toward unity–again to enhance our ability to see the decision lines, which are analytically very difficult to solve. Figure 17.3 shows the results of Equations 17.13 and 17.14. The dark ellipsoidal rings are simply the areas where the class density functions have a near zero second derivative spatially. The bright lines clearly show the boundaries between classes, which in this case overlap, and the boundary for defining outliers. Figure 17.4 combines these edges with some logarithmically spaced contours for the function in Equation 17.13. Depending on where the feature combination $[x_1 \ x_2]$ falls on this map, the statistical classifier will assign the sample to class 1, class 2, or outlier. In addition to making this decision, the algorithm can also provide probabilistic metrics for the confidence in the decision which is very valuable to the data fusion occurring in an intelligent sensor system.

17.1.2 Adaptive Neural Networks

ANNs are a very popular approach to separating classes of data automatically [2]. The concept is based on a biological model for neuron cells and the way electrochemical connections are made between these fascinating cells during learning. In the brain as connections are made between neurons at junctions called *synapses* to create neural pathways as part of learning, chemicals are deposited which either inhibit or enhance the connection. Each neuron can have from 1000 to 10,000 synapses. The human brain is composed of approximately 20 billion neurons. The number of

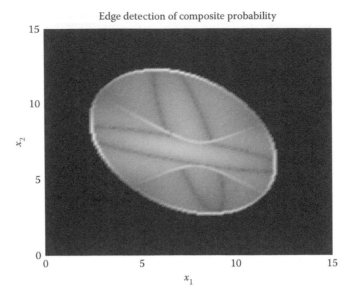

FIGURE 17.3 Thin bright lines of this eighth root Laplacian of the composite density shows the approximate boundaries between the two classes as well as the overall boundary for the data training set.

possible interconnections and feedback loops is obviously extraordinary. Even more fascinating, is the fact that many of these connections are preprogrammed genetically and can be "forgotten" and relearned. While philosophically and biologically interesting, we will explore these issues no further here and concentrate specifically on the "artificial neuron" as part of the most basic ANN based on the *generalized delta rule* (GDR) for adaptive learning. The reader should be advised that there is far more to ANNs for pattern recognition than presented here and many issues of network design and training are beyond the scope of this book [3].

The basic artificial neuron is seen in Figure 17.5 where each of N inputs are individually weighted by a factor η_{jk} and summed with a bias term to produce the intermediate signal net_j. During the

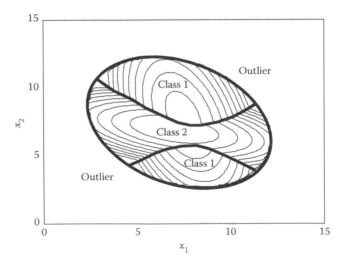

FIGURE 17.4 Outliers as well as the class overlap is easily seen from a simple edge detection of the composite density, excluding any samples outside of 3–σ from the total training set mean.

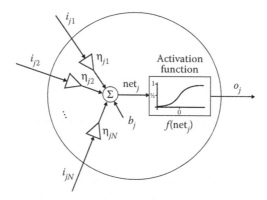

FIGURE 17.5 The "*j*th neuron" show a general set of *N* inputs, each with weight η_i, a bias term b_j, and the activation function $f(net_j)$ to monlinearly limit the output o_j.

learning phase of the neural net, adaptive algorithms will be used to optimize the weights and bias at each node for every pattern of interest.

$$net_j = \sum_{k=1}^{N} \eta_{jk} \, i_{jk} + b_j \tag{17.15}$$

The bias term in net_j is like a weight for an input which is always unity. This allows the neuron to have an output even if there are no inputs. The most typical designs for an artificial neuron "squash" is the signal net_j with a nonlinear *activation function*. This limits the final output of the neuron to a defined range (usually 0 to 1) and allows for a transition rather than a "hard clip" of the output between a zero or one state. By avoiding the "hard clip" or step response, and by limiting the range of the neural output, more information can be preserved and balanced across the outputs from other neurons such that no one pathway dominates. The most common activation function is the *sigmoid*.

$$o_j = f(net_j) = \frac{1}{1 + e^{-\varepsilon net_j}} \tag{17.16}$$

The output of the jth neuron o_j given in Equation 17.16 and seen in Figure 17.5 is a nonlinear response of the inputs to the neuron i_{jk}, the weights η_{jk}, and the bias b_j. The parameter ε in Equation 17.16 is a gain factor for the nonlinearity of the activation function. As ε becomes large even a small positive net_j will drive the output to unity or a small negative net_j will drive the output to zero. The choice of ε, the activation function, as well as the number and size of the hidden layers is up to the designer to choose. This coupled with the difficulty of a complete signal analysis in a neural network fuels the skeptic's criticism of the approach. However, when applied appropriately, the ANN is a valuable and practical tool.

The optimization of the weights and bias terms for each neuron are calculated using error back propagation and the GDR. The GDR is very similar to the least-mean square (LMS) adaptive filter algorithm seen in Section 10.2. In fact, the man acknowledged with developing the LMS algorithm, Prof. Bernard Widrow of Stanford University, is also a pioneer in the development of ANNs. The back propagation algorithm is a little more subtle to understand. Consider the very simple network for classifying two features, x_1 and x_2, into two pattern classes, w_1 and w_2, seen in Figure 17.6. Each of the numbered circles represents a node seen in Figure 17.5. The two network outputs o_4 and o_5, are ideally [1, 0] when the features are from class 1, and [0, 1] when the features are from class 2. To make this happen, the weights and bias for each node must be adjusted to reduce, if not minimize

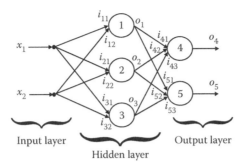

FIGURE 17.6 A simple two-feature two-output class neural network showing the 3-node hidden layer and the inputs and outputs of each node as part of a backpropagation algorithm using the GDR.

the error at the network outputs. It is therefore straightforward to apply an LMS-like update to the weights and bias of each output node, but what about the hidden layer(s)? Back propagation of the error is used by taking the output layer node error, multiplying it by the weight between the output layer and the particular hidden layer node, and then summing all of the back propagated errors for that particular hidden layer node. The process continues using the output error for all the nodes in the hidden layer closest to the output layer. If additional hidden layers exist, the back propagation continues using the back propagated error rather than the actual output error. The weights and bias for each node are adjusted using the GDR until the output and back propagated errors becomes small, thus converging the neural network to a solution that separates the pattern classes of interest.

For the pth pattern class, we follow the nomenclature of the literature, we designate the node inputs i_{jk}^p and output o_j^p and the "training output" for class p as t^p. There are several philosophies about how the train the network, and controversies with regard to overtraining and extendibility of the network result to data outside the training set. For example, one can train the net to identify only pattern p, and when testing for pattern p use those specific weights and biases. Therefore, each class is tested using its optimum weights. Another more efficient approach is to train one set of network weights and biases to discriminate all the patterns of interest. This can be seen as more robust because the classes are directly compared. For an output node, the training error squared is seen to be

$$e_j^p = \left| t_j^p - o_j^p \right|^2 \tag{17.17}$$

For a linear FIR filter, this error is a linear function of the filter coefficients. Therefore, the squared error can be seen as positive definite surface which has one minimum reachable by a series of weight adjustments, in the direction opposite to the gradient of the squared error with respect to the weights. However, herein lies perhaps the biggest inconsistency of the neural network derivation: the error surface is rarely known and generally has multiple minima due to the nonlinear response of the sigmoids. Ignoring this point of lack of rigor, the neural network enthusiast is generally given to point out the success of the algorithm at separating the classes. *We have no way of knowing whether the network is the optimum network, whether a different number of nodes and hidden layers will perform better, or whether the training of the network is complete.* One just has to accept this lack of mathematical rigor and optimization to move on and make use of this curious and useful algorithm. The gradient of the error is

$$\frac{\partial e^p}{\partial \eta_{jk}} = \frac{\partial e^p}{\partial \mathrm{net}_j^p} \frac{\partial \mathrm{net}_j^p}{\partial \eta_{jk}} \tag{17.18}$$

We can apply the chain rule again to break the error gradient down even further as

$$\frac{\partial e^P}{\partial \eta_{jk}} = \frac{\partial e^P}{\partial o_j^P} \frac{\partial o_j^P}{\partial \mathrm{net}_j^P} \frac{\partial \mathrm{net}_j^P}{\partial \eta_{jk}} \qquad (17.19)$$

The gradient in Equation 17.19 is broken down into its components. The gradient of the squared error with respect to output (Equation 17.17) is

$$\frac{\partial e^P}{\partial o_j} = -2 \left| t_j^P - o_j^P \right| \qquad (17.20)$$

The gradient of the output with respect to the weighted sum of inputs (Equation 17.16) is

$$\frac{\partial o_j^P}{\partial \mathrm{net}_j^P} = f'\left(\mathrm{net}_j^P\right)$$

$$= \frac{1}{1+e^{-\varepsilon \mathrm{net}_j^P}} \frac{\varepsilon e^{-\varepsilon \mathrm{net}_j^P}}{1+e^{-\varepsilon \mathrm{net}_j^P}}$$

$$= o_j^P \left(1 - o_j^P\right)\varepsilon \qquad (17.21)$$

and the gradient of the summed inputs and bias (Equation 17.15) is simply

$$\frac{\partial \mathrm{net}_j^P}{\partial \eta_{jk}} = i_{jk}^P \qquad (17.22)$$

The sensitivity of the pattern error on the net activation is defined as

$$\delta_j^P = -\frac{\partial e^P}{\partial \mathrm{net}_j^P} = 2\left(t_j^P - o_j^P\right)\left[o_j^P\left(1-o_j^P\right)\varepsilon\right] \qquad (17.23)$$

and will be used for back propagation to the hidden layer(s). The weight adjustments using the GDR are simply

$$\eta_{jk_{t^+}} = \eta_{jk_{t^-}} + 2\mu\delta_j^P i_{jk}^P \qquad (17.24)$$

where the parameter μ is analogous to the LMS step size, but here it is referred to as the "learning rate." If the feature inputs to the neural network are scaled to be bounded with a ±1 range, it is a fairly safe bet that choosing learning rate less than, say 0.10, will yield a weight adjustment free of oscillation and allow the network to converge reasonably fast. The learning rate does effect the "memory" of the network during training such that a slow learner remembers nearly everything and a fast learner may forget the oldest training data. There is an interesting human analogy as well where many forms of mental retardation are characterized by slow, highly repetitive learning, and excellent long-term memory, while high mental capacity is often accompanied by very fast learning and surprising mid- to long-term memory "forgetfulness." Decreasing the learning rate (increasing the memory span) is advisable if the feature inputs are noisy. The node bias term is updated as if its corresponding input is always unity.

$$b_{jt^+} = b_{jt^-} + 2\mu\delta_j^P \qquad (17.25)$$

Each neuron receives a set of inputs, weights each one appropriately, sums the result in net_j, and then passes the linear result through the activation function (the sigmoid in our example) to produce the neuron node output. At the output layer of the net, this neural network output is used for class decision. For a node in a hidden layer, the neuron output is passed to the inputs of many other nodes in the next layer toward the output layer, if not the output layer. The pattern sensitivity for a hidden layer node is calculated by summing up all the pattern sensitivities times their corresponding input weights for all the nodes in the next layer toward the output layer so that its output is passed as input.

$$\delta_j^p = \sum_{n=N_1}^{N_2} \delta_n^p \eta_{nj} \tag{17.26}$$

In other words, if the output of node 15 is passed to nodes $N_1 = 37$ through $N_2 = 42$ in the next layer toward the output layer, the pattern sensitivity for node 15's weights and bias updates is found from the impact its sensitivity has on all the nodes it affects in the next layer. This is intuitively satisfying because the weight and bias adjustments near the feature input side of the network are specific to the feature inputs but affect most of the network outputs, while the weights and biases near the output layer are specific to the trained network output and encompass most of the feature inputs.

To summarize, given a trained neural network with k pattern classes and M features, one presents the M features for an unknown pattern observation to the input layer, computes the node outputs layer by layer until the network outputs are complete, and then chooses the largest output node as the class for the unknown pattern. Usually, the input features and trained outputs are bounded by unity to simplify the output comparison and net training. To train the neural network, the following procedure is acceptable:

1. The number of input layer nodes equals the number of features. There is at least one hidden layer with at least one more node than the input layer. The number of output nodes is equal to the number of patterns to be identified.
2. The weights and biases for each node are randomized to small values. Neglecting to randomize will lead to symmetry of the network weights and no real convergence. Small values are chosen to avoid "swamping" the sigmoidal functions or having one path dominate the network error. The inputs and outputs are bounded by unity to simply set the learning rate and class selection from the maximum output node.
3. During training, the input features for the given class are presented to the input layer and the network outputs are computed layer by layer (Equations 17.15 through 17.16).
4. The "training error" is calculated for the output layer from the difference between the desired training output and the actual network outputs. The pattern sensitivity (Equation 17.23) is calculated for each output layer node.
5. The weights and biases for the inputs to the output layer nodes are adjusted (Equations 17.24 through 17.25).
6. The pattern sensitivity for the hidden layer nodes are calculated (Equation 17.26) and the corresponding weights and biases are adjusted (Equations 17.24 through 17.25).
7. Steps 3–6 maybe repeated several times for the same input features until the error gets small, and/or steps 3–6 are repeated for a large number of known sample patterns constituting the pattern training set.

Figure 17.7 gives an example with the network output for class 1 shown as a surface plot where the input features form the x and y coordinates of the surface. There are two Gaussian data sets, class 1 with mean $\langle 5, 10.0 \rangle$ and class 2 with mean $\langle 10.0, 5.0 \rangle$. The neural network depicted in Figure 17.6 is trained with samples from class 1 normalized to the interval $\langle 0.0, 1.0 \rangle$ and the output for class 1 (node 4) set to unity and class 2 (node 5) set to zero. The training process is repeated with feature data from class 2 and the network output for class 1 zero and class 2 unity. The weights and biases

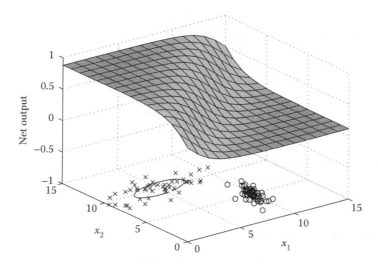

FIGURE 17.7   Neural network output "surface" for class 1 (*x*) showing the scatter and Gaussian ellipse superimposed for both classes on the bottom where class 2 (*o*) is deselected in the class output surface.

for the network are then fixed and the output for class 1 is evaluated for every combination of features to produce the surface plot in Figure 17.7. For illustration, the Gaussian training data sets are superimposed over the surface. Clearly, the network does a good job separating the two classes. A more complex network with more hidden layers and more nodes in the hidden layers could separate the classes with even greater deftness. This is more of an issue when the classes are closer together with larger standard deviations such that many of the training samples overlap. Surprisingly, the neural network still does a good job separating overlapping classes as can be seen in Figure 17.8 where the class 1 mean is $\langle 8.0, 9.0 \rangle$ and class 2 mean is $\langle 7.0, 6.0 \rangle$. The decision is clearly less confident due to the overlap, but it still appears reasonable by most metrics.

A much more sophisticated neural network (more layers, nodes, and more extensive training) might very well completely separate the training set classes with a decision boundary which winds

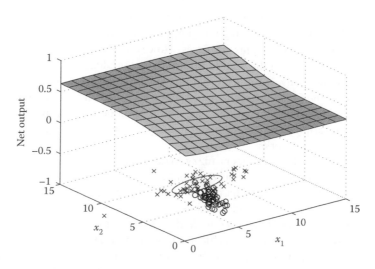

FIGURE 17.8   Neural network output "surface" for class 1 (*x*) when the two classes overlap significantly in the training data but still gives a useable classification capability.

its way between the samples in the overlap area. Therefore, if the training data is not random but rather simply complex, the sophisticated neural network with many layers and nodes should do an amazing job at memorizing which feature combinations go with which class. The converged network provides a fast nonlinear filter where the feature are inputs and the class decisions are outputs. This is of a particular value when high-quality training data is available, the class boundaries are far more complicated than Gaussian boundaries, and one is fairly confident that the actual feature data during usage is well represented by the available training data. Many scientists and engineers who have used neural networks can be described as enthusiasts because they are witnesses to the power of this adaptive nonlinear process to effectively separate complex classes. However, the extendibility of the network performance to data covering a range outside the training data is, like with statistical classification, questionable. To address the problem of extendibility, one can use fuzzy logic and inferencing networks to build in human knowledge into parts of the network.

17.1.3 Syntactic Pattern Recognition

Syntactic pattern recognition is used when one can establish a *syntax*, or natural language to describe the way various pieces of information fit together in a pattern [2,4,5]. Detection of the various information pieces can be done using statistical or neural network-based algorithms, but now we add a layer of logic where pattern data and confidences are fused together. Rather than insisting that all class patterns be completely separate, we allow conflicting pattern estimates to coexist for the benefit of subsequent data fusion to make a more balanced, informed decision. If all the information is binary (true or false), straightforward hard logic (and, or, if, then, else, and other operations in software) can be used to construct a network for processing information. Given the human knowledge in the logic network, good performance can be expected outside the training data. If not, the flawed logic can be identified and corrected. You may remember this from grade school as *the scientific method* and it has worked so well for humans over the last few centuries that one should certainly consider employing it in computer artificial intelligence. *The main difficulty is that one must know the syntax.* Given the syntax rules, one can also employ the ability to discount information with low confidence in favor of higher confidence information and balance combinations of required and alternate information through weights and blending functions. Ultimately, the syntactic logic must have physical meaning to humans. This last point is quite important. For syntactic pattern recognition to reach its full potential in ATR, we humans need to be able to learn from flawed logic and correct the syntax to account for what has been learned. A flow diagram for the syntactic classifier can look very much like a neural network, but the weights, biases, activation functions, layers, and interconnections all have physical meaning and purpose. In a well understood syntactic classifier, no training is required to adjust weights or biases. However, it may be prudent to automatically adjust weights in accordance with the feature information confidence so that unreliable (e.g., low SNR) data can be automatically and gracefully removed from the ATR decision. Syntactical classification algorithms can be found in commercial software for speech and handwriting recognition, decision aids for business logistics, autopilots, and even for product marketing. But most important, syntactical rules can be scripted and layered hierarchically to construct automated situational awareness processes from sensor networks.

The first algorithm needed to implement fuzzy logic is a *fuzzy set membership function* which allows one to control the transition between "true" or "false" for set membership. This transition can be anywhere from a step function for "hard logic" to the sigmoid given in Equation 17.16 where epsilon is small [5]. We call this a blend function $\beta(x)$. The blend function allows one to control fuzzy set membership. Why would one need to do this? Consider that we are combining, or fusing, multiple pieces of information together to achieve a more robust decision on the current pattern, or situation of patterns. Various pieces of information with individual confidences are combined where we want no one piece of information to dominate. And, we want a flexible outcome depending on the quality and availability of information. By blending or smoothing the decision thresholds, we can design the amount of "fuzziness" or flexibility desired in the data fusion and situational awareness algorithms. The following blend function allows a great deal of design flexibility.

$$\beta\,(a,b,c,d,x) \begin{cases} = b, x \le a \\ = \dfrac{1}{2}[d + b + (d-b)\sin\left(\left\{\pi\dfrac{(x-a)}{(c-a)} - \dfrac{1}{2}\right\}\right), & a < x < c \\ = d, x \ge c \end{cases} \tag{17.27}$$

A plot of the blend function in Equation 17.27 is seen in Figure 17.9 for values $a = 1.5$, $b = 0.9$, $c = 4.5$, and $d = 0.1$. This simple function allows one to easily set the maximum "confidence" for the blend function, the minimum "confidence", and the transition points for a sinusoid between the maximum and minimum. Furthermore, the blend function in Equation 17.27 offers a great deal of flexibility than the sigmoid of Equation 17.16 in controlling the transition from one state to the next.

Figure 17.10 shows an example of class pattern membership function derived from two blend functions to describe the average overweight person in terms of body fat. This model is clearly a matter of opinion that the people between 25% and 50% body fat are the average overweight person. If you are below 25% or over 50% you do not belong to this class. So the metrics for defining the "fuzzy set membership" are not set automatically by some optimized algorithm, but are set by human knowledge which makes the syntax to behave in a desired way. This is a key concept for the signal processor to accept when it comes to syntactic classifiers. *We design the syntax so the algorithm will behave according to our rules.* We make the rules. We can use objective information like probabilities or the output of ANNs as parameters in the syntax if we wish, but we do not have to. The fuzzy set membership function is our making, and the underlying fuzzy logic is also our making.

The next operator needed is a *fuzzy AND function* which will allow fusion of multiple pieces of information, each weighted by a corresponding "supportive" coefficient. We will use the same weight symbol as for the neural net η_{jk} (kth weight at the jth AND node) to keep terminology simple. The fuzzy AND function is used when one has the condition where any single information measure

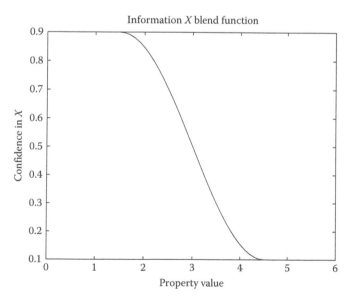

FIGURE 17.9 A blend function or "sigmoid" is used to map property value "x" to a confidence "y" according to a fuzzy set membership function.

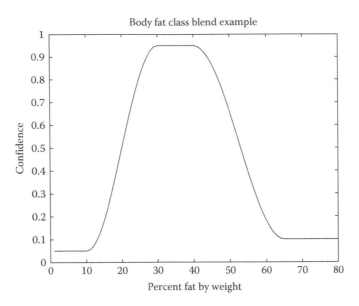

FIGURE 17.10 A "double blend" function is used to map an example of a property of % body fat to a class of "typical overweight people" designated by the user.

false (zero) will cause the resulting decision to be false (zero). If the supporting weight η_{jk} is close to unity, the corresponding information can be seen as required while if the supporting weight is small, the information is merely not important to the decision. To insure that lesser important information does not cause the AND to always produce a near false output, the fuzzy AND function is defined as

$$\mathrm{AND}(\bar{X}_j, \bar{\eta}_j) = \left[\Pi_{i=1}^{N} \left(1 - \eta_{ji} + \eta_{ji} x_{ji} \right) \right]^u$$

$$u = 0.1 + 0.9\, e^{-0.3(\kappa - 1)}$$

$$\kappa = \sum_{i=1}^{N} \eta_{ji}$$

$$\bar{X}_j = [\, x_{j1} \quad x_{j2} \quad \cdots \quad x_{jN} \,]$$

$$\bar{\eta}_j = [\, \eta_{j1} \quad \eta_{j2} \quad \cdots \quad \eta_{jN} \,] \tag{17.28}$$

where u and κ control an activation function on the AND in case the weights η_{ji} do not sum to unity. If the weights sum to something less than unity ($\kappa < 1$), then $u > 1$, which will tend to flatten out the AND making it harder for all the xji to lead to a positive unity output from the AND. When the weights sum to something greater than unity, it will have the opposite effect in that the AND will become steeper and flip to a positive unity outcome more easily. Note how if the supporting weight is small, the corresponding term in the product does not become zero causing the result of the fuzzy AND to be false. Also note that if the sum of the supporting weights is large, the exponent u will tend toward 0.1. This has the effect of making the result of the fuzzy AND compressed toward unity if true, and zero if false. These "behaviors" are very flexible and at the designers discretion to achieve the desired overall fuzzy set membership behavior.

Finally, we need a fuzzy OR function for cases where we are fusing information together where if any one of the pieces of information is true, the result is true. The intuitive mathematical operator to represent this is a Euclidian norm of each of the inputs to the fuzzy OR node times its corresponding supporting weight.

$$E_j = \sqrt{\sum_{i=1}^{N}(\eta_{ji}\, x_{ji})^2} \tag{17.29}$$

The output of the OR can be greater than unity, and re-blended using the single or double blends seen in Figures 17.9 and 17.10 if needed.

The inputs and supporting weights are typically bounded between zero and unity. But, for any range of inputs and weights, the maximum value of any input-weight product is guaranteed to be less than or equal to the Euclidean norm. This is a reasonable place for the transition point of the blend function between true and false. If only one input is used, a step-like blend function is reasonable because no OR'ing is actually happening. As more inputs are used, we want the blend to span a wider range since more inputs increases the likelihood of a true result. Note that all inputs and supporting weights are unity, the Euclidean norm is simply $\sqrt{N}$. Therefore, our fuzzy OR function is described using a Euclidean norm in a blend function as given in the following equation:

$$\mathrm{OR}\,(\bar{X}_j\,,\bar{\eta}_j) = \beta\,(a,b,\sqrt{N},1,E)$$
$$u_j = \max\{(\eta_{ji} x_{ji})\};\ i = 1,2,\ldots,N\}$$
$$a = \begin{cases} 2u_j - \sqrt{N}; & N \le 4 \\ 0; & \text{else} \end{cases} \tag{17.30}$$
$$b = \min\{(\eta_{ji} x_{ji})\};\ i = 1,2,\ldots,N\}$$

The blend transition is symmetric about u_j only for $N \le 4$. If more than four inputs are used, we start the transition at zero and bring it to unity at $\sqrt{N}$, but it is less than 0.5 at u_j. This is largely the user's choice. The blend implementation of the fuzzy OR keeps the output bounded between zero and unity.

When all the fuzzy logic inputs and outputs are bounded gracefully to a range between zero and unity, negating, or the NOT operator is a simple matter of subtracting the confidence (input or output) from unity. Now we have the capability to build all the usual logic operators AND, OR, NOT, NAND, NOR, XOR, and so on, but in "fuzzy" format. Applying the blend function when the transition is rather sharp, it can be seen as adding back the "crisp" to the logic. We build the logic to represent the known syntax and information. The fact that blend points, weights, and min/max values are set by the user should not imply a guess (at least not in all cases), but rather a placeholder to embed the physics, the scientific method has taught us by derivation. We can correct the logic based on observation and new physics built into the problem. In some cases it may be desirable to use some machine training and parameter optimization as seen for the neural network as an input to the logic. However, the real power of syntactic pattern recognition is in possessing the correct knowledge that leads to the right syntax. Algorithm behavior and syntax are linked, so the correction to the behavior carries a corresponding correction to the logic, or highlights an ambiguity in the logic.

Finally, we can even employ a way to resolve a case where two or more inputs, representing the same general information, are conflicting thus giving rise to a state of confusion. Identifying a confusion pattern internal to the syntactic fuzzy network is extremely interesting and useful. For example, if A is true, AND B is false, AND C OR D is false, AND E is true, AND B AND C are not confused, THEN pattern Z is present. Measuring and using a confusion metric as part of the pattern recognition problem brings in a level of "self-objectivity" to the artificial intelligence. This can be an extremely important step in rejecting false alarms. Figure 17.11 shows a confusion blend function based on $\beta\,(-1, 2\ 1, 0, dx_{ji})$, where dx_{ji} represents the difference in confidence (difference in input or

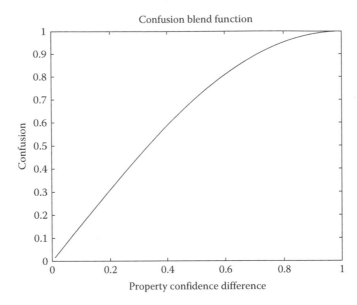

FIGURE 17.11 A confusion metric is derived by mapping the difference in confidences between two separate confidence measures as a way of detecting when the logic is inconsistent "or confused by the data."

output values) for two pieces of data representing the same information. When this difference is significant, the information is conflicting, giving the syntactic algorithm confusion to deal with. The nonlinear blend function is useful in completely controlling the emphasis on confusion based on the information conflict.

To illustrate the power of syntactic pattern recognition, we present several spectrograms of bird calls seen in Figures 17.12 through 17.14. Several useful features can be extracted and sequenced from the spectrograms as seen in Figure 17.15. However, in order to correctly identify the bird calls,

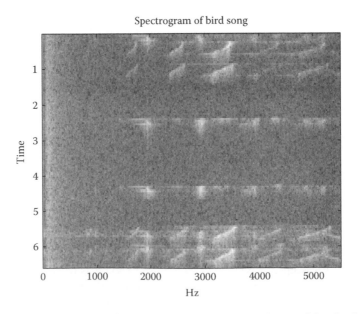

FIGURE 17.12 Spectrogram of a recording of a Blue Jay singing showing a quick pair of downward chirps in the first second followed by a wider harmonic bursts at 2.4 and 4.2 s.

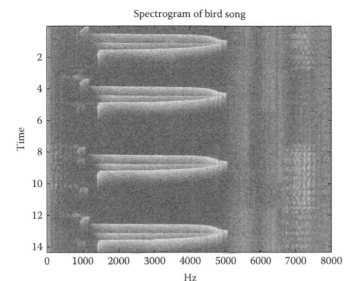

FIGURE 17.13 Dramatic spectrogram of the song of a Screaming Phiha showing a quick 1 kHz burst, then a chirp up to 5 kHz, then another chirp from 1.5 to 5 kHz, followed by a down chirp to 1.5 kHz and a short hold.

we must establish something called a *finite state grammar*. The strings of features must follow a specific order to match the bird call. This *natural language* is established from human application of the scientific method. Making the computer follow the same syntax is a matter of technique. Figure 17.16 shows one possible network of syntactic fuzzy logic to identify the Screaming Phiha bird song from the Blue Jay and Musician Wren. The weights are seen as the numbers next to the signal flow arrows and the boxed letter "B" depicts a blend function. The OR operator is assumed to also have its blend function built in. When the sequence is observed, the confidence output for the

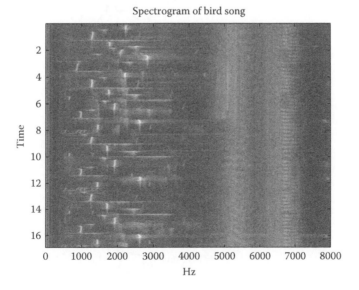

FIGURE 17.14 The remarkable signature of the Musician Wren showing a series of melodic steady "notes" from 1 to 3 kHz which almost repeat as a melodic pattern.

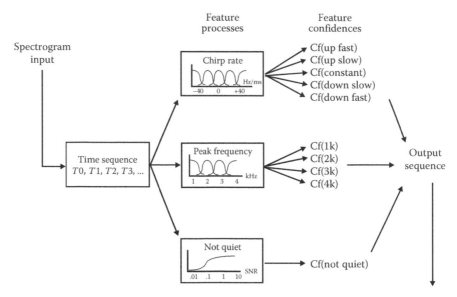

FIGURE 17.15   A block diagram showing the sequencing of features of the bird call spectrogram.

Screaming Phiha will be high. Dropping the "not quiet" confidences for simplicity, the Screaming Phiha confidence at time T, $Cf(T, SP)$, fuzzy logic can be written as

$$Cf(T, SP) = 0.8 * \{Cf(T, constant)\} \wedge 0.9 * [0.5 * Cf(T, 1k) \vee 0.5 * Cf(t, 2k)] \wedge$$
$$Cf(T-1, \quad down\, fast) \wedge \{0.7 * Cf(T-2, \quad up\, fast)\} \wedge Cf(T-3, \quad up\, fast)\} \quad (17.31)$$

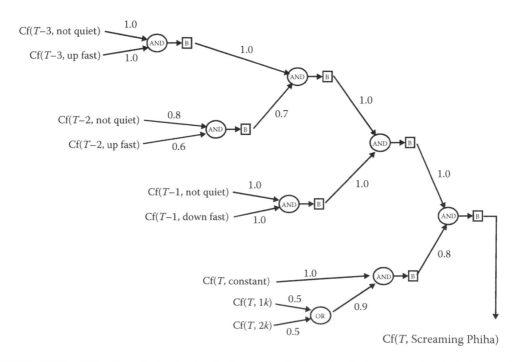

FIGURE 17.16   CINet fuzzy logic showing the features and sequencing that correspond to the inference that the call is a Screaming Phiha.

Where the ∧ symbol depicts a fuzzy AND and blend function and the ∨ symbol depicts a fuzzy OR function with its inherent blend operator. Similarly, the Blue Jay signature seen in Figure 17.12 confidence, Cf(T, BJ), can be written as

$$Cf(T, BJ) = 0.75 * Cf(T, 3k) \wedge \{0.9 * Cf(T, \text{down slow}) \vee 0.75 * Cf(T, \text{constant})\} \quad (17.32)$$

Note that one can invert the "fuzzy true-false" by simply subtracting the confidence from unity. The Musician Wren confidence (signature seen in Figure 17.14) is derived from a series of constant frequencies which are different from time to time. This confidence Cf(T, MW) is a little more complicated to implement, but is still straightforward.

$$Cf(T, MW) = Cf(T, \text{constant}) \wedge$$
$$\{Cf(T, 1k) \wedge [1 - Cf(T - 1, 1k)] \wedge [1 - Cf(T - 2, 1k)] \vee$$
$$Cf(T, 2k) \wedge [1 - Cf(T - 1, 2k)] \wedge [1 - Cf(T - 2, 2k)] \vee$$
$$Cf(T, 3k) \wedge [1 - Cf(T - 1, 3k)] \wedge [1 - Cf(T - 2, 3k)]\} \quad (17.33)$$

The output of the syntactic classifier is all of the pattern class confidences. One can choose the one with the highest confidence, say above 0.5, or choose none if the confidences for any one class are too low. The power of the syntactic technique is that the algorithm is completely a general way to implement many forms of human logic on a computing machine.

Inference engines that generate logic from rules are the basis for many older powerful languages such as LISP and PROLOG. The main challenge for the programmer is that one has to master the syntax of the problem. In other words, for a solid solution, one needs to explicitly follow the physics and use credible signal processing techniques to determine the data confidence going into the syntactic classifier. The information "supportive weights" and choices for blend, AND, and OR functions are more a function of the required logic than the dynamic data quality. One other important technique in classification is the *hidden Markov model*. A Markov chain is a sequence of states of something where the state transition is defined using a probability. When a classifier is constructed based on a sequence of pattern class decisions, often achieving a finer or more specific resolution with each sequential decision, it is said that the classifier is based on a hidden Markov model. This technique is widely used in speech and handwriting recognition, and interpretation. The difficulty is in defining the transitional probabilities. The syntactic classifier technique described here can also be based on a hidden Markov model. Furthermore, its inputs could be generated from statistical classifier metrics or the outputs from a neural network.

Combining fuzzy logic and the inference engine idea leads to a type of pattern recognition called a Continuous Inference Network, or CINet. The CINet is a supervisory-classifier in that it can combine the class outputs of other classifiers along with human knowledge of fuzzy set membership functions and blends, but it does this starting with the conclusion (the inference), and working backward to the inputs. This is the opposite of the order in which we normally work through a problem. The typical signal processing chain is to take the signal, filter it to improve the SNR, apply objective mathematical metrics to obtain useful features of the signal, and see which pattern the features fit the best. The CINet processes in a complete reverse order. Starting with the inference, the fuzzy logic measures if there are consistent set of inputs for the inference to be true, and reports the confidence on a 0.0–1.0 scale. This turns out to be very easy to program a computer to do, and intuitive for a human to examine the logic as a "flow chart" that maps the inputs to the inference in a straightforward manner. It is little clumsy to write the CINet logic in Equations 17.31 through 17.33, but that does not translate into a complicated computer program, in fact it is generally very simple and executes extremely fast. Speed is important because we would like to evaluate a large number of inferences at all levels of an overall scenario or situation. This way the output of one CINet can be the input of another hierarchical CINet, so that the basic sensor information can be assembled layer by layer into complex scenarios. These logic trees can be evaluated, tested, expanded, and scrutinized to ensure that the best information is used and ambiguities are managed.

Consider a civil defense application regarding a fire evacuation scenario where there are sensors for wind, temperature, smoke, traffic, and access to databases for traffic, topography, and recent weather patterns. If there is a drought condition, high temperatures, and high winds, the CINet alerting everyone of the danger is broadcast to everyone in the affected area, by cell phone, text message, desktop icon, radio–television broadcast, and so on. But if there is also smoke detected, the situation is highly contextual depending on where a given person is located. Some people need not leave and should not go on certain roads to keep traffic down. Others should leave immediately by a specific route, still others should wait and then evacuate, and so on. How does everyone know what to do? Your cell phone or desktop has CINets looking out for your interests, as well as the interests of those around you and advises you of how to be safe without endangering others. The rules can be sorted out ahead of time. The sensor and situation data can be made freely available via the wireless network. As smart phone gets more and more powerful, we should expect applications like this to emerge. Civil defense has been largely ignored since the 1950s because the top threat were nuclear weapons, so huge that you could not escape if you wanted to. The terrorist events of September 11, 2001 and thereafter have changed the needs of Civil Defense, and the sensor and network technology is ready to support these new needs. The government needs to help define the rules, which are essentially policies. For example, if a small nuclear device or dirty bomb explodes, running toward your car and getting stuck in traffic will get you killed. Going underground to avoid the radiation is the best idea. If a large chemical fire threatens the area with toxic fumes you do need to find a way to get upwind of the smoke. If it is late at night and the fumes are heavier than air, you need to get up high if you cannot get out, and so on. There is no way to have one central program tell thousands of people what to do, but there is a way for distributed CINets to properly interpret the situation and give the correct guidance, following public policy, to protect themselves without harming others. This is what makes the CINet so interesting for sensor networks of the twenty-first century.

17.2 SIGNAL AND IMAGE FEATURES

Features are distinguishing artifacts which can be assembled to define a useable pattern for classification of a signal. However, a better description of a feature is *an information concentrator*, which makes the job of detecting a pattern that is much easier. For example, a sinusoid can have its information concentrated into one complex number (magnitude and phase) with a corresponding frequency by an FFT. If in the time domain, the signal is an impulse or burst waveform, there is no point in doing an FFT since it will only spread the signal over a wider space making feature detection more difficult. For an image with regular patterns across the focal plane, the spatial domain (the viewable image itself) is the obvious domain to compute features, since they are already concentrated by focusing of the lens-aperture system. The wavenumber (2D FFT) domain obviously offers advantages of spatial processing, filtering, and so on, but not necessarily feature detection, unless the viewable image consists of regular periodic patterns or textures which are well represented by a few FFT bins.

Our discussion will focus first on 1D signal features, such as those from acoustic, vibration, or electromagnetic sensors, or sensor arrays. Our fundamental signal is a "delta function-like" impulse, which can be integrated into step, ramp, quadratic, and higher-order functions. These signal classes are very important to identify for control systems as well as other applications. Next, we consider periodic signals, starting of course with the sinusoid but extending into impulse trains. The impulse train can be seen as one case of a linear superposition of harmonically related sinusoids. Another signal class is a signal distorted by a nonlinear operation, which also generates harmonics from a single sinusoid, and more complicated difference tones when several non-harmonic sinusoids are present. These nonlinearities can be detected using higher-order spectra as seen in Section 6.1. Finally, we will examine amplitude and frequency modulated sinusoid features.

Modulation of the amplitude (AM), phase (narrowband FM), or frequency (FM) of a high frequency radio wave "carrier" is a fundamental method to transmit information (see Section 16.1) [6]. Other modulation schemes can also be used such as generating digital MLSs (a random sequence which repeats precisely) to modulate amplitude, phase, or cause "frequency hopping." These latter "spread-spectrum" techniques (see Section 14.3) are the basis for secure digital communications and allow many information channels to share the same frequency bandwidth. Rather than pursuing signal intelligence gathering techniques, we will focus more on the natural occurrence of AM and FM modulation in rotating machinery. Fatigue condition monitoring of machinery is becoming economically viable, thanks to low cost microprocessors and networking equipment. As this technology becomes proficient at predicting the remaining useful life (RUL) for machinery as well as the current failure hazard, enormous amount of capital are to be saved by industry as well as the military. While predicting the remaining life is useful, just determining when maintenance is needed before a catastrophic failure usually justifies the added cost of monitoring.

It is well known that a signal function can be uniquely defined if its value and all its derivatives are known for one input data point. It can also be known if one has a large number of samples of the function at different input values, allowing the function to be approximated accurately using least-squared error techniques or Fourier series. From physical modeling of the processes which give rise to the signal of interest, one can develop an understanding of the physical meaning of a parameter change in the signal function. So, given that one needs to detect some physical event, signal source type, or condition, the corresponding signal function parameter can be converted into a signal feature. In Section 17.3 we will show how to track the feature changes and make useful predictions for when failure should occur.

17.2.1 Basic Signal Metrics

Signal statistics and state (derivatives, integrals, etc.) are perhaps the common way to characterize a signal. The most basic time-domain 1D signal features are the mean, variance, and *crest factor*, which is usually defined as the ratio of the peak to RMS (standard deviation or square-root of the variance) signal parameters. However, the mean, standard deviation, and crest factor are generally the result of some degree of integration of the signal over a finite period of time. By integration we mean the sample interval over which the average is computed for the mean and standard deviation, and the interval over which the peak factor is obtained for the crest factor. Statistical measures can also include higher-order moments such as skewness and kurtosis (see Section 7.2.7). Yet another statistical signal feature is its histogram, which measures the number of occurrences of a particular value of the signal. This is used quite often in image signals as well as 1D signals to characterize the PDF of the signal. There are a number of well documented distributions that can approximate a given signal by setting just a few parameters of the PDF, or its frequency-domain equivalent, the characteristic function.

Signal characterization by derivative/integral relationships is a very generic way to extract primitive signal information. Consider the primitive signal functions given in Figure 17.17. One can clearly see going down either column the effect of integration. These integrals can have fairly simple features starting with whether they are "impulse-like," "step-like," "ramp-like," and so on. Some useful impulsive signal features are: which integral produces the step function; the width of the impulse; the height of the step; the slope of the ramp, and so on. Together, these features can provide a set of primitive descriptors for an impulsive-like signal. For example, the popping sound made when speaking the letter "p" is recorded in the top waveform in Figure 17.18. The first integral shows a wide Dirac-like delta function rather than the actual popping sound. This is somewhat typical for acoustic impulsive sounds which tend to be zero mean (unless there's some sort of explosion). The characteristics of the step and ramp functions (second and third integrals) should distinguish the "pop" waveforms from other impulsive sounds. The integral is used to generate other features with the assumption that the high frequency components of the sound are not important for

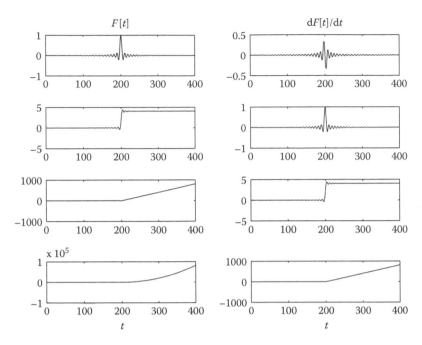

FIGURE 17.17 Primitive signal functions starting with the Dirac delta function (upper left) showing subsequent integration as one proceeds down the plots, and differentiation in the right column of the plots.

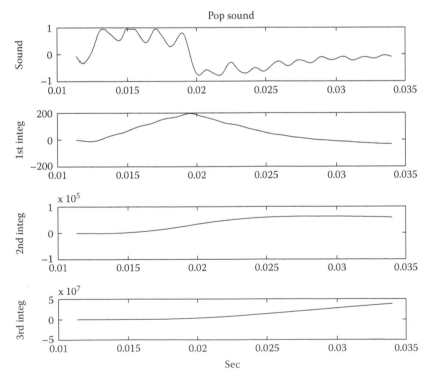

FIGURE 17.18 The "popping" sound of the letter "P" is used to generate some impulsive signal primitive features for analysis of the type of waveform showing the first integral giving a wide delta function.

classification. This is certainly not universally true. However, signal integration is inherently less susceptible to noise than signal differentiation, and is generally preferred.

17.2.2 PULSE-TRAIN SIGNAL MODELS

Periodic signals have unique harmonic patterns which can be identified. Applying derivatives and/or integration to periodic signals is not very interesting in general because the integral or derivative of a sinusoid is yet another sinusoid. Clearly, the periodic signal features are best concentrated in the periodogram of an FFT. This is straightforward enough for single sinusoids, where the amplitude and phase are sufficient descriptors. But for periodic waveforms in general, there is much going on which can be parameterized into features. Consider the following generic digital impulse train waveform.

$$y[n] = \frac{1}{\beta_p} \frac{\sin\left(\frac{[\pi n - \tau]}{N_w}\right)}{\sin\left(\frac{[\pi n - \tau]}{N_r}\right)}, \quad \beta_p = \frac{N_r}{N_w} \tag{17.34}$$

To convert Equation 17.34 to a physical time waveform, simply replace n by $t = nT_s$ where T_s is the sample interval in seconds, and τ is a small delay to avoid dividing by 0 (it makes the waveforms more well-behaved).

The response of this impulse train, which repeats every N_r samples and has peak width N_w samples, is very interesting depending on the bandwidth factor β_p. If β_p is an even integer, one has an alternating impulse train as seen in the top row of plots in Figure 17.19. The bottom row of Figure 17.19 shows β_p chosen as an odd integer, giving rise to the even harmonics in the bottom right

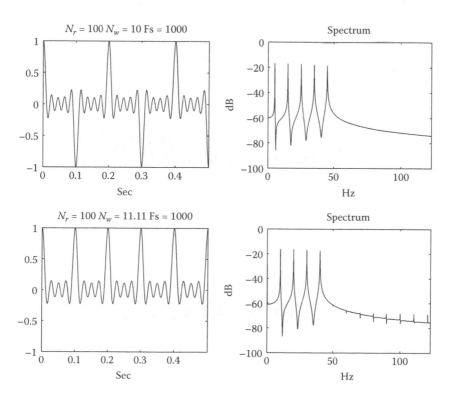

FIGURE 17.19 Even and odd harmonic series are generated by the appropriately repeated impulse trains.

plot. In both cases, one sees either even or odd harmonic multiples up to the β_pth harmonic where every other harmonic is exactly zero. This has to do with the integer relationship between N_r and N_w. When β_p is irrational, the function in Equation 17.34 is not realizable without spurious glitches. However, a nice result is given even if both N_r and N_w are irrational but β_p is an integer. Note that as N_w becomes small and β_p large, the number of harmonics becomes large. As β_p approaches 2, our pulse train becomes a sinusoid and at $\beta_p = 1$ the waveform is a constant dc-like signal. To avoid aliasing, N_w must be greater than or equal to 2. In Figure 17.19, the top pair of plots have $N_r = 100$ and $N_w = 10$, giving an even $\beta_p = 10$, thus causing alternating time domain peaks (Fs is 1000 Hz) and odd harmonics in the frequency domain. The bottom pair of plots have $N_r = 100$ and $N_w = 11.1111$, giving an odd $\beta_p = 9$, thus causing even harmonics in the frequency domain. A surprising number of periodic signals can be characterized in terms of even or odd harmonics.

17.2.3 Spectral Features

The Fourier transform provides the spectral envelope of the signal defined by the magnitude frequency response or a discrete number of frequency–magnitude points defined by a series of narrowband peaks. This envelope can be fit to a polynomial where the zeros provide a basis for signal recognition, or one can simply assemble the peak heights and frequencies as a pattern feature set for identification. Harmonic sets can be separated by a straightforward logic algorithm or cepstral techniques. The dominant Fourier coefficients naturally serve as a feature vector which can uniquely identify the signal.

Figure 17.20 compares the Fourier spectra of the spoken vowels "*e*" (top) and "*o*" (bottom). The narrowband peaks are detected as a means to minimize the number of data points to work with. One can clearly see from the graphs the higher harmonics in the 2–3.5 kHz band which the mouth and nasal cavities radiate for the "*e*" sound. One can also see a difference in low frequency harmonic structure due to the acoustic radiation impedance at the mouth and its effect on the response of the

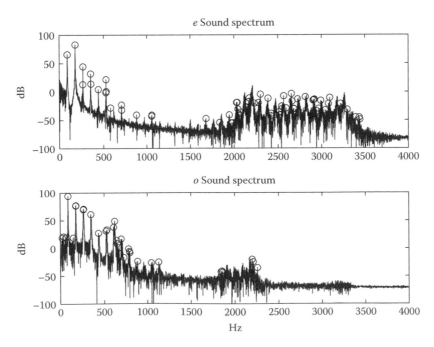

FIGURE 17.20 Spectral peak detection of speech for the sounds "*e*" (top) and "*o*" (bottom) showing the filtering effect of the mouth and lips plus a different structure to the fundamental tones.

throat and vocal chords. The details of this wonderfully complex sound generation model could allow one to build a sensor system which can identify a particular individual (speaker identification). Or, one could put together an "envelope" function for the overall spectral shape to simply identify the speech sound, or *phoneme*. There are over 40 such phonemes which make-up the majority of spoken languages. Commercially available and surprisingly affordable speech recognition software detects the sequence of such phonemes and connects them together using combinations of fuzzy logic, statistical detection, and hidden Markov models. Text is then produced which best matches the speech recognition results. An even more interesting artificial intelligence problem is developing a way for the computer to understand and respond correctly to the spoken sentence.

The log-amplitude of the Fourier coefficients is a great choice for feature elements because a wide dynamic range of signal can be used to parse the feature space. The phase of the Fourier coefficients is also important, but only if this phase can be made time invariant from one FFT data buffer to the next. For example, the processed Fourier transform could represent a cross-spectrum with a particular signal or a transfer function, or coherence. For the phase to be meaningful for an ordinary periodic signal, the FFT buffer size and sample rate should be synchronized with the harmonic of interest to produce a meaningful phase. Otherwise, the FFT phase will not likely be ergodic enough to converge with some averaging. However, one must have a linear, time invariant environment for the Fourier envelope to represent something meaningful physically. This approach to generating signal features is so intuitive and straightforward that we will not pursue its explanation further here. But its usefulness and importance are only overshadowed by its simplicity of application. This is probably the easiest way to generate signal features in an objective way.

17.2.4 Monitoring Signal Distortion

Distortions in signals can be characterized using higher-order spectra. This is because a transducer or system filter nonlinearity can cause frequencies to modulate each other, generating sum and difference frequencies as well as signal harmonics. Signal distortion is a very interesting and important identification problem. Signal distortion can be present for a wide range of signal levels, or only when the signal exceeds some loudness threshold, such as what occurs in amplifier "clipping" where the output voltage is limited to a fixed maximum range. Generally speaking, signal distortion effects always get stronger with increasing signal amplitude. This is because small signal level swings with a nonlinear input–output response can easily be linearized over small range.

A useful way to parameterize nonlinearity is the bi-spectrum. Recall from Section 7.2, Equation 7.46, that an efficient way to compute the bispectrum is to first compute the FFT, then detect the dominant narrowband peaks, and finally directly compute the bispectrum on the relevant combinations of those peaks. Recall that

$$C_3^x(\omega_1, \omega_2) = E\{X(\omega_1) X(\omega_2) X^*(\omega_1 + \omega_2)\} \tag{17.35}$$

The interesting thing about signal nonlinearity is the "phase coupling" between frequencies generated by the nonlinearity and their principal linear components. This is because the nonlinear signal generation is occurring at precisely the same point in time for all frequencies in the time waveform. Therefore, for all applications of Equation 17.35 we have coherence between the $X^*(\omega_1 + \omega_2)$ component and the $X(\omega_1)$ and $X(\omega_2)$ principal waveforms.

As an example, consider the case of two sinusoids.

$$x(t) = A\cos(\omega_1 + \theta_2) + B\cos(\omega_1 + \theta_2) \tag{17.36}$$

The signal in Equation 17.36 is passed through a simple nonlinearity of the form

$$y(t) = x(t) + \varepsilon x^2(t) \tag{17.37}$$

By simply multiplying the signal input $x(t)$ by itself one finds the output of Equation 17.38.

$$y(t) = A\cos(\omega_1 + \theta_1) + B\cos(\omega_2 + \theta_2)$$
$$+\varepsilon\left[\frac{A^2}{2} + \frac{B^2}{2}\frac{A^2}{2}\cos(2\omega_1 t + 2\theta_1)\right.$$
$$+\frac{B^2}{2}\cos(2\omega_2 t + 2\theta_2)$$
$$+AB\cos([\omega_1 - \omega_2]t + \theta_1 - \theta_2)$$
$$\left. +AB\cos([\omega_1 + \omega_2]t + \theta_1 + \theta_2)\right] \tag{17.38}$$

It can be clearly seen in Equation 17.38 that the phases are coupled between the ω_1 and $2\omega_1$ terms, the ω_2 and $2\omega_2$ terms, and the sum and difference frequency terms. The "phase coupling" is very interesting and clearly shows how the bispectrum in Equation 17.35 can detect such coherence between different frequencies in the spectrum. Normally (i.e., for linear time invariant systems), we expect the phase relationship between Fourier harmonics to be independent, and even orthogonal. Thus, we would expect no "coherence" between different frequencies in a linear signal spectrum. But we do expect bispectral coherence between the phase-coupled frequencies for a signal generated using a nonlinearity. Also note that Gaussian noise added to the signal but not part of the nonlinear filtering will average to zero in the bispectrum. The bispectrum is the feature detector of choice for nonlinear signals.

Amplitude and FM of signals is often detected as part of the spectrum of a signal of interest. But the physical features of amplitude modulation (AM) and FM can be broken down and extracted in a number of different ways. AM signals simply vary the amplitude of the "carrier" frequency with the message signal amplitude. It is useful to transmit messages in this way for radio because the use of a high-frequency carrier allows one to efficiently use relatively small antennae (the wavelength at 300 MHz is about 1 m). FM signals can be further categorized into narrowband FM, or PM, and wideband FM. A formal definition of narrowband versus wideband FM is defined from the modulation index, which is the ratio of the maximum frequency deviation to the actual bandwidth of the message signal, or the signal doing the FM. The modulation index can be scaled in any desired way because the frequency deviation of the FM signal from its carrier frequency is proportional to the amplitude of the message signal. But the message signal has its own bandwidth, which maybe larger than or smaller than the bandwidth of the FM, which is proportional to the message signal amplitude. When the FM frequency deviation is smaller than the message signal bandwidth, we call it the modulation narrowband FM. Likewise, when the FM is greater than the message signal bandwidth it is called wideband FM. Wideband FM uses much more bandwidth but offers a significantly greater SNR than narrowband FM. Narrowband FM and AM use about the same bandwidth but the FM signal will have less susceptibility to background noise. For digital signal transmission, the transmitted bandwidth is roughly the maximum possible bit rate for the channel. It is generally on the order or 90% of the available bandwidth because of the need for error correction bits, synchronization, stop bits, and so on.

The general public is most familiar with AM and FM signals from their radio, but AM and FM signals also arise naturally in the vibration (and electromagnetic) signatures of most rotating equipment and electric motors. By applying a little physics and some straightforward communications signal processing, a great deal of useful information can be extracted. Information on machinery condition, fatigue, and failure hazard is worth a great deal of capital for the following (just to name a few) reasons: unexpected equipment failures cause substantial damage to the environment, personal injury, and even deaths; failures of critical machines can cause a sequence of much greater catastrophic failures; time based maintenance (fixing things before its really needed) is on average significantly more costly than condition-based maintenance (CBM). This is because the need for many repairs is actually caused by repairing something else.

A quick analysis of rotating equipment can start with three simple mechanical rules: (1) one cannot make a perfectly round object; (2) even if one could make a perfectly round object, one cannot drill a hole exactly in the center of it; (3) no two pieces of material are identical, respond to stress from heat or external force in the same way, nor will respond to damage the same way. Rolling elements will always have some degree of imbalance, misalignment, and nonconstant contract forces with other rolling elements. These elements will produce vibration, acoustic, and in some cases electromagnetic signals which can be systematically related to the condition of the rotating elements. The contact forces between two rolling elements can be broken down into centripetal and tangential forces. For a pair of gears, the average vibration signal is a harmonic series of narrowband peaks with the dominant frequency corresponding to the gear tooth mesh frequency. The shaft frequency is found by dividing the fundamental gear mesh frequency by the number of teeth in the given gear. The centripetal forces end up on the shafts, propagate through the bearings, and are seen as amplitude modulated vibrations of the gear mesh frequency. The tangential dynamic forces become torsion shear on the shafts and are seen as an FM component of the gear mesh frequency [7]. A typical transmission will have a large number of mesh harmonics and AM and FM sidebands, which can be explicitly attributed to specific rolling elements by a computer algorithm. As the surfaces and condition of each of the rolling elements change, the vibration signatures will also change. Therefore, one can monitor the condition of a piece of machinery through careful monitoring of the vibration and other signatures. Given this capability, one can then track the changes in signature features to predict the number of hours until a failure condition occurs. Feature tracking will be discussed in more detail in the next section.

17.2.5 AMPLITUDE MODULATION

Amplitude modulation signals are common in vibration. Consider a gear mesh frequency of 5 kHz and a shaft frequency of 100 Hz (50 gear teeth). Our modeled AM modulated signal is

$$y(t) = [1 + \alpha \cos(\omega_{sh} t)] \cos(\omega_c t) \tag{17.39}$$

where α represents the amount of modulation for the 100 Hz shaft rate ω_{sh}, component and ω_c represents the 5 kHz carrier frequency component. Once-per-shaft-revolution modulations are seen to be physically due to out-of-round, imbalance, and/or misalignment problems. Figure 17.21 shows the time and frequency domain signals for Equation 17.39 using a sampling frequency of 20 kHz and $\alpha = 1$.

Coefficients for twice-per-rev, three times per-rev, and so on, for the mesh frequency fundamental can be seen as Fourier coefficients for the overall surface shape of the rotating component modulating the sinusoidal fundamental component of the gear teeth Fourier transform. To observe pits, spalls, and cracks in the teeth, analysis is often carried out four times the mesh frequency including the sidebands, if such a high-frequency vibration is detectable. For most turbine-driven transmissions, this type of vibration analysis extends well into the ultrasonic range of frequencies. Equation 14.19 depicts a general, multiple harmonic, and sideband representation of an AM signal

$$y(t) = \sum_{q=1}^{Q} \sum_{m=1}^{M} A_q [1 + \alpha_{q,m} \cos(m\omega_{sh} t)] \cos(q\omega_c t) \tag{17.40}$$

Figure 17.22 shows the time and frequency representation of Equation 17.40 with $Q = 4$ and $M = 5$, and a shaft rate of 350 Hz, which better shows the sidebands on the plot. The A_q coefficients are assumed unity in the response in Figure 17.22 and the $\alpha_{q,m}$ were taken as $1/\sqrt{qm}$ for demonstration purposes. One would expect that a small surface wear would first be detectable up in the sidebands of the higher mesh frequency harmonics, which is the case. Many defects such as spalls, peels, pits, and cracks can be detected early in the high-frequency range of the vibration spectrum. When a crack first occurs, a spontaneous ultrasonic acoustic emission is detectable, which for some

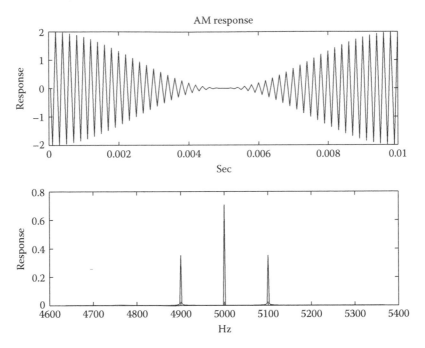

FIGURE 17.21 100% AM modulation of a 5 kHz carrier representing a gear mesh frequency and a 100 Hz amplitude modulation representing the shaft turning rate.

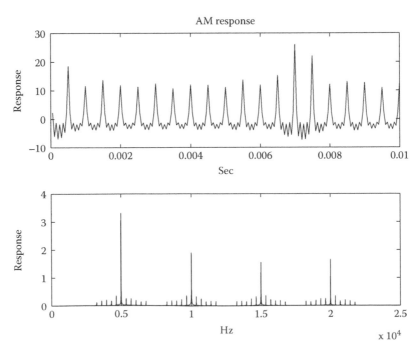

FIGURE 17.22 Time waveform (top) and linear spectrum (bottom) of a multiple harmonic 5 kHz AM signal with 350 Hz sidebands.

materials can also be detected in the electromagnetic spectrum. Complex machines such as helicopter transmissions have been monitored for fatigue at frequencies up to four times the mesh frequency or well into the ultrasonic frequency regime.

17.2.6 Frequency Modulation

Frequency modulation (FM) signals are more complicated but recently are of interest to CBM because the presence of internal cracks in gear teeth and shafts can decrease the mechanical stiffness when the gear is in a particular contact position. As such, as the gear rotates, a dynamic torsion rotation error signal is produced which is seen as an FM component on the gear mesh frequency. Detection of such defects before they reach the surface is of great value because they are precursors which can be used to predict the RUL of the machine. As the crack grows, stiffness decreases more dramatically, the FM increases, and eventually the crack reaches the surface leading to high-frequency features, but perhaps only shortly before component failure. In the case of helicopter gearboxes, a gear failure generally means a complete loss of the aircraft and crew, and the aircraft needs to be flying for the gear to have sufficient applied stress to fail. To complicate matters more, the mere presence of FM does not imply imminent failure. It is when FM and other fatigue features are present and increasing, which provides the basis for mechanical failure prognostics.

FM signals are a little more complicated than AM signals. For an FM signal, the frequency sweeps up and down proportional to the amplitude of the message signal. Consider a very simple case of a carrier frequency ω_c and FM message sinusoid frequency ω_m.

$$y(t) = \cos(\omega_c t + \beta \sin(\omega_m t))$$
(17.41)

The term β in Equation 17.41 is just a constant times the message signal maximum amplitude for what is called *phase modulation*, but is inversely proportional to the message signal frequency for FM. This is because the time derivative of the phase is frequency (or the time integral of the modulation frequency gives the phase). A more general representation of Equation 17.41 is

$$y(t) = \mathrm{Re}\{e^{j\omega_c t} e^{j\beta \sin(\omega_m t)}\}$$
(17.42)

where the right-most exponential is periodic with period $2\pi/\omega_m$ seconds. Therefore, we can write this as a Fourier series.

$$e^{j\beta \sin(\omega_m t)} = \sum_{n=-\infty}^{+\infty} C_n e^{j\omega_m n t}$$
(17.43)

The coefficients of this Fourier series are

$$C_n = \frac{\omega_m}{2A} \int_{-(\pi/\omega_m)}^{+(\pi/\omega_m)} e^{j\beta \sin(\omega_m t)} e^{-jn\omega_m t}\, dt$$

$$= \int_{-\pi}^{+\pi} e^{j(\beta \sin\theta - n\theta)}\, d\theta = J_n(\beta)$$
(17.44)

making a frequency domain representation of our simple sinusoidal FM signal seen as an infinite sum of sideband about our carrier frequency with carrier and sideband amplitudes dependent on the value of Bessel functions of the first kind, $J_n(\beta)$.

$$y(t) = \sum_{n=-\infty}^{+\infty} J_n(\beta) \cos[(\omega_c + n\omega_m)t]$$
(17.45)

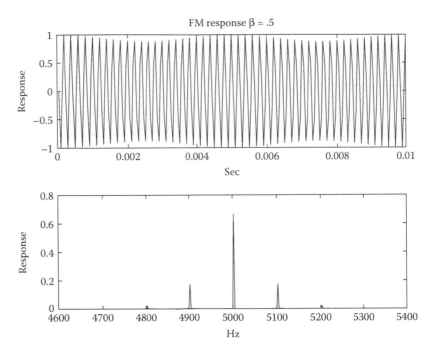

FIGURE 17.23 An FM signal with very small modulation has a barely noticeable frequency shift in the time domain and almost the same bandwidth as an AM signal.

Figure 17.23 shows a fairly narrowband FM signal where the carrier is 5 kHz, the modulation frequency is 100 Hz, and β is 0.5. The sample rate of our digital signal model is 20 kHz, so the amount of modulation relative to the carrier phase is only around π/2: 0.5, or about 30%. This can be seen as a very narrowband FM or PM. Figure 17.24 shows the same carrier and FM, but with a modulation β or 40. In terms of phase shift, the FM can be seen as over 2400%. The maximum rate of this phase change gives the approximate maximum frequency shift of the carrier.

$$\frac{\Delta f}{f_s} = \beta \frac{f_m}{f_s} \approx 20\% \qquad (17.46)$$

The approximation in Equation 17.46 tells us that the carrier frequency is warbling around ±20% of the sample frequency, or 4 kHz, spreading the signal spectrum from around 1–9 kHz. The spectrum seen in the bottom plot of Figure 17.24 shows the spectrum used by the FM as approximately covering this range. Clearly, the greater the modulation, the more wideband FM looks "noise-like" in the frequency domain. The message signal can also be specified as a sum of Fourier components, which complicated the spectrum significantly.

For FM radio, we all know the SNR benefits from personal experiences listening to FM radio broadcasts of music and other information. The reason the received signal is nearly noise-free is because the FM carrier is converted in a nonlinear way into the signal amplitude. At the receiver, the carrier band (typically about 40 kHz wide) is clipped (limited) to remove all amplitude information, and BPF on the edge so the RMS signal is proportional to frequency, which is a relatively simple receiver circuit. Harry Armstrong, one of the great engineers of the twentieth century, developed this technology in the 1930s, when a simple low-cost receiver was needed during the great depression in the US. Unfortunately, he spent the rest of his life fighting to retain the patent rights of his brilliant invention.

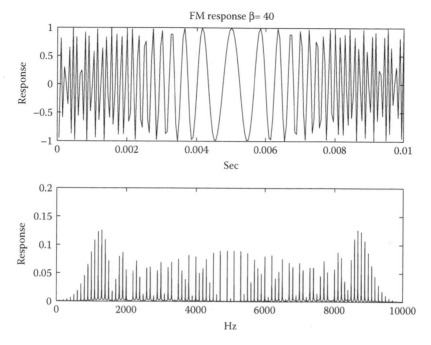

FIGURE 17.24 A wideband FM signal produces a very noticeable chirp in the time domain and a very complex spectrum showing a bandwidth of approximately 2β times 100 Hz = 8 kHz ($\beta = 40$).

17.2.7 DEMODULATION VIA INVERSE HILBERT TRANSFORM

Rotational vibration often contain both AM and FM signals. This poses a significant challenge and opportunity to exploit the features of rotational vibration to measure the mechanical condition or rotating equipment. But how one can tell that the two are apart? The challenge is seen as how to get the AM and FM "message signals" into orthogonal spaces. This is done with the knowledge of the carrier frequency and the Hilbert transform, defined as

$$Y_\mathrm{H}(\omega) = Y(\omega) + Y^*(-\omega) \quad 0 < \omega \le \frac{\omega_\mathrm{s}}{2} \tag{17.47}$$

One can think of the Hilbert transform as a way of representing a real sinusoid as an equivalent complex sinusoid. For example, if one had a Fourier transform of a cosine function with frequency ω, half of the spectral signal will be at $-\omega$ and half of the spectral signal will be at $+\omega$. If the real input signal was a sine function, the half of the spectral signal at $-\omega$ is multiplied by -1 and added into the complex sinusoidal representation. If the input to the Fourier transform is a complex sinusoid, no negative frequency component exists. The Hilbert transform provides a common basis for representing sinusoids for frequencies up to 1/2 the sample rate for real signals, and up to the sample rate for complex time-domain signals. The inverse transform of the Hilbert transform gives us a complex sinusoid, which is what we need to recover the AM and FM signal components.

Consider our combination of AM and FM signal modulation

$$y_{af}(t) = [1 + \alpha \cos(\omega_a t)] \cos(\omega_c t + \beta \sin(\omega_f t)) \tag{17.48}$$

Figure 17.25 shows a time plot (top) and spectrum (bottom of an example combination AM and FM signal) where the AM is 100 Hz with $\alpha = 0.9$ (90% AM), and the FM is 40 Hz with $\beta = 50$ (± 2 kHz modulation) about a common carrier frequency of 5 kHz. By computing the Fourier

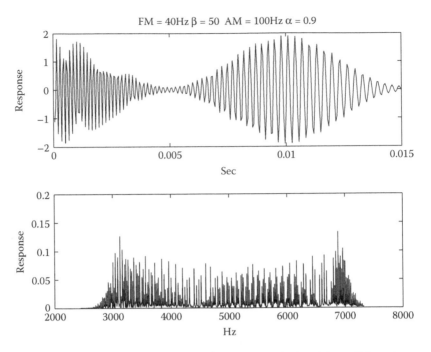

FIGURE 17.25 A combinations of 40 Hz FM with $\beta = 50$ and 100 AM with $\alpha = 0.9$ produces a very complicated waveform and spectrum.

transform of the real signal in Equation 17.48, applying a Hilbert transform, and inverse Fourier transforming, one obtains a complex sinusoid $y_{af}^c(t)$, the real part of which matches Equation 17.48. Therefore, we can demodulate the complex sinusoid by a complex carrier to obtain a "baseband" complex waveform.

$$y_{af}^b(t) = y_{af}^c(t)e^{-j\omega_c t} \qquad (17.49)$$

Or, for digital waveforms sampled at $2\pi f_s = \omega_s$,

$$y_{af}^b[n] = y_{af}^c[n]e^{-j2\pi(f_c/f_s)n} \qquad (17.50)$$

which has the effect of producing a waveform with frequency related to the carrier frequency. The amplitude of this complex waveform is the AM part of the modulation signal, and the phase (unwrapped to remove crossover effects at $\pm\pi$) is the FM part of the modulation. Figure 17.26 depicts the recovered AM and FM waveforms clearly.

There are several important physical insights into the report on this interesting modulation feature extraction example. If the AM is greater than 100%, a phase change will occur when the AM crosses zero. This too can be recovered, but only if one is sure that no FM exists. For very complicated broadband signals as the modulation signal (also known as the message signal in communication theory), one is fundamentally limited by bandwidth and the carrier. If the message signal bandwidth is less than the carrier frequency, there is sufficient signal bandwidth for AM and narrowband FM modulation. The bandwidth of the FM signal is about the bandwidth of the message signal *times* 2 β, which has the potential to cover a very wide bandwidth. In the frequency domain, one can easily select either finite bandwidths or select spectral peaks within a finite band to focus the complex message signal on specific parts of the spectrum. If the carrier demodulating frequency in Equations 17.49 and 17.50 are in error, a slope will be seen in the recovered FM waveform. One

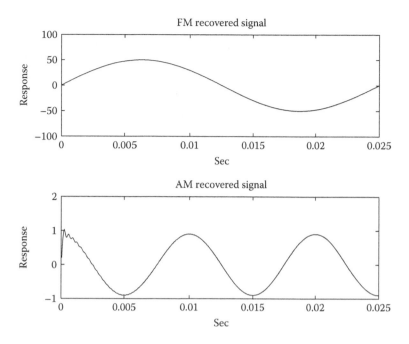

FIGURE 17.26 AM and FM components of the signal in Figure 17.25 recovered using a Hilbert transform.

can adjust the demodulating carrier until this FM slope is zero, to be sure of the correct alignment for a given FFT buffer. Finally, it can be seen that extracting AM and FM signal parameters clearly has the effect of focusing important signal information spread over many time-waveform samples and spectral bins into a few salient parameters for pattern recognition and tracking.

Let us test our modulation separation with a more complicated FM signal. Consider a pulse train where $N_w = 20$ and $N_r = 400$ as defined in Equation 17.34, where the sample rate is 20,000 samples per second. The fundamental of the pulse train is 25 Hz with odd harmonics extending up to 475 Hz. This gives our message signal a total bandwidth of about 1 kHz, which is readily seen in Figure 17.27. The recovered AM and FM signals are seen in Figure 17.28 where one can clearly seen the recovered FM-only pulse train signal. This is a narrowband FM case because the total bandwidth of the FM signal is approximately the same as the bandwidth of the message signal. In Figure 17.29, we increase β to 10 such that now the bandwidth of the FM signal requires most of the available bandwidth of our 20 kHz-sampled waveform with carrier frequency of 5 kHz. Figure 17.30 shows the recovered FM signal which appears correct with the exception of a bias in the modulation frequency. This bias is caused by the angle modulation in our digital signal.

$$y_{af}[n] = \left[1 + \alpha \cos\left(2\pi \frac{f_a}{f_s} n \right) \right] \cos\left(2\pi \frac{f_c}{f_s} n + \beta\, m[n] \right) \tag{17.51}$$

If the FM message signal times β exceeds π in magnitude, the phase of the carrier wraps, causing the bias error in the recovered FM message waveform. Curiously, this did not occur when the message signal was a pure sinusoid as seen in Figure 17.26. To understand why, consider that the frequency shift of the carrier is the derivative of the carrier phase in Equation 17.51. The 40 Hz sinusoid FM message signal in Figure 17.26 with a $\beta = 50$ corresponds to a maximum frequency shift of 2 kHz, which in terms of digital frequency shift is range of 0.7–0.3π. For the case of a non-sinusoidal message signal, once β times the maximum of $m[n]$ in Equation 17.51 exceeds or equals π, a bias error in the recovered FM message waveform is possible due to the higher harmonics of the message waveform. With $\beta = 10$ and a maximum harmonic of 475 Hz in the message signal,

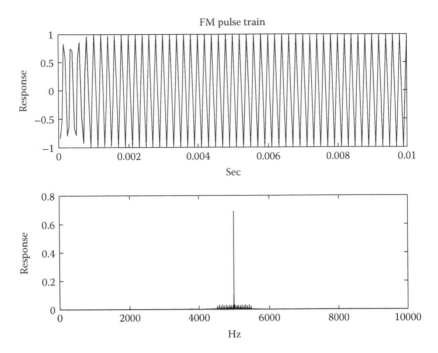

FIGURE 17.27 A narrowband FM signal where the message signal is a pulse train with about 1 kHz total bandwidth and $\beta = 1$ in the FM signal (no AM is used).

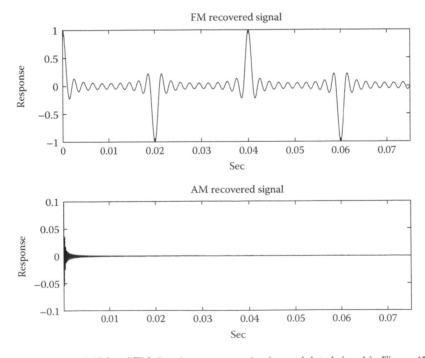

FIGURE 17.28 Recovered AM and FM signal components for the modulated signal in Figure 17.27.

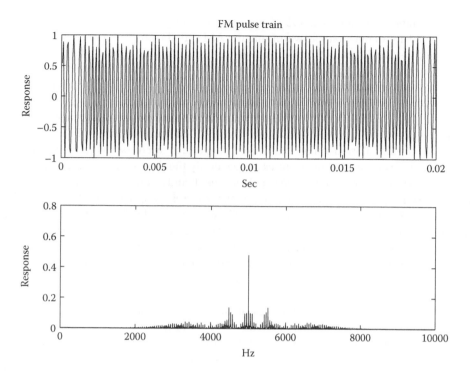

FIGURE 17.29 Wideband FM with the 1 kHz bandwidth pulse train message signal and $\beta = 10$.

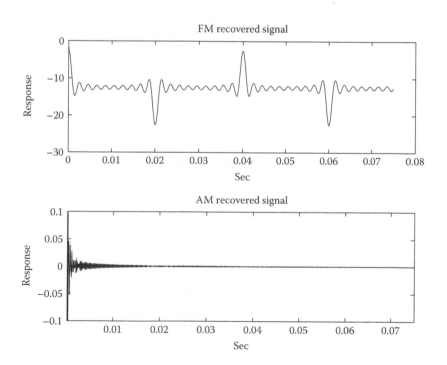

FIGURE 17.30 Recovered AM and FM components of the wideband FM signal in Figure 17.29.

most of the available spectrum (5 $\pm$ 4.75 kHz) is used for our pulse train message signal. If β were increased beyond 15 for our example waveform, the AM signal recovery significantly breaks down, and if increased further beyond 40, neither signals can be accurately recovered.

The 2D features can be extracted from image data as well as other 2D signals. Our discussion will focus mainly on image data, but as the image is broken down into edges and other features, one can easily extend the analysis to any 2D data, or even higher dimensional data. Again, drawing from our philosophy that features should represent concentrated signal information, imagery data can be broken down into two main groups: *focal-plane features* and *texture features*. Focal plane features represent concentrated information on the focused image such as edges, shapes, local patterns, and so on [8,9]. Examples of focal plane features are generally the things we can easily recognize with our eyes and are improved by contrast, edge enhancement, and so on. To help a computer algorithm recognize focused objects, one typically removes the textures by applying first and second derivatives (see Section 5.3) to enhance edges and lines. These lines may then be segmented into sequences of simple functions, like the pen strokes of an artist drawing the object. The functional segments can then be compared objectively to segments for known objects of interest. The functional representation of the object is useful because computing algorithms can be derived to take scale, rotation, and viewing aspect into account for object identification.

Texture features are patterns that are periodic and cover broad areas of the focal plane. Texture features are good candidates for the frequency domain where the broad areas of a periodic pattern on the focal plane can be concentrated into a few 2D FFT bins. For example, inspection of a filled case of 24 beer bottles using an aligned digital picture will normally have a regular pattern of bottle caps. A broken or missing bottle is seen as a "glitch" in the pattern, or aberration of the texture. A simple 2-FFT of the image data and spectral subtraction, and an inverse FFT will quickly reveal the problem before the product is shipped from the factory. Texture patterns can also be applied to a wide range of material inspections including color, reflectance, laser scattering patterns, and so on [10].

One can examine what is known about the physiology of the eye for primates (monkeys, apes, and man) to gain some valuable insight into how biology has optimized our sight for detection and classification of objects. Insect, fish, and other animal eyes generally do not see the way we do. This might be useful for some specific machine vision tasks. Man's eyes and way of seeing (what is known of it) are truly amazing. By examining the psychophysics of the eye/brain operation we find some basic building blocks for the development of 2D image features.

The optical properties of the eye are surprisingly poor [11]. It has a single simple lens with a rather small focusing range controlled by a ring of muscles. The optical aberration can be rather large (as most of us with eyeglasses know all too well) and the aberration over color can be over 2 diopters from red to violet. With little physical correction possible, biology provides the necessary interconnections between receptors to turn this color "defocusing" to an advantage which allows remarkable color recognition over a wide range of ambient light. There are four types of receptors, red, green, and blue *cones* with a significantly overlapping color (light frequency) response, and blue-green *rods* where the outputs are pooled to allow a lower resolution night vision. These retinal receptors are actually situated behind a layer of neural cells which contribute to a loss of acuity but allow more interconnections and additional protection from damage from intense light.

The eye uses tremor and hexagonal variable sampling to enhance acuity. The resolution of the eye is on the order of 30 s of arc (1° = 60 min of arc, 1 min of arc = 60 s of arc), but the eye actually tremors at a rate from 10 to 100 Hz with about the same 30 s of arc. These "matched resolution tremors" have the effect of smoothing the image and removing optical effects of sampling. This is known as "antialiasing" filtering in signal processing and has the effect of spatially low-pass filtering the optical signal. It is interesting to note that during motion sickness, extreme fatigue, mental impairment (say from drugs), or while telling a big fat lie, most humans eyes become noticeably "shifty" (perhaps the brain is too preoccupied). The arrangement of the receptors is hexagonal (honeycomb-like) which can be seen as naturally more optimal to the sampling of a point-spread function which is the fundamental resolution on the surface of the back of the eye to a pin-point of distant light. At the center of the field of view which corresponds to the center of the back of the eye called the *fovea*, the density of the receptors is highest and it decreases as one moves away from the fovea. This variable sampling density, along with the optics of the iris and lens, physically cause the acuity to diminish as one moves away from the fovea. In addition, the optics produce a smoothly diminishing acuity toward the other ranges which acts like a Gaussian window on the iris to suppress sidelobes in the PSF (see Chapter 8). The result of this biology is that objects appear as "blobs" in our peripheral vision and the details are seen once we swivel our eyes around to bring the interesting part of the image over the fovea. This will translate into a useful strategy for image recognition where one scans the entire image with a small window to initially detect the "blobs" of interest, and then subsequently process those "blobs" with much higher resolution.

The structure of the receptors and neurons gives a parallel set of "preprocessed" image features. There are essentially four layers of neurons in the signal path between the rods and cones and the optic nerve. They are the *horizontal cells, the bipolar cells, the amacrine cells, and the ganglion cells*. In humans and other primates, there are additional cells called the *midget bipolar cells* which provide more direct connections to the ganglions and interactive feedback to other groups of receptors. Surprisingly, there are very few blue cones and fewer blue–green rods for low light vision in the fovea area, perhaps to make space for a higher overall density of receptors. However, we "think" we see blue quite readily in our central field of view perhaps because of the poor optical color aberrations of the eye and the feedback interconnections provided by the midget ganglions. The amacrine cells provide further neural interconnections between spatially separated groups of cells which may explain the eye's superb *vernier acuity* (the ability to align parallel lines). Vernier acuity is about six times better (5 s of arc) than visual acuity and requires groups of spatially separated receptors to combine together their outputs. This is also likely to be a significant contributor to the brain's ability to judge distance with stereo vision, another superb human vision capability. From a Darwinistic point of view, without good stereo vision, one does not catch much food nor escape easily from being eaten—thus, these developments in primate eyes are seen as a result of either evolution or divine good fortune. Finally, the ganglion cells can be categorized into two (among many) groups: the X ganglions that have a sustained output which responds in about 0.06 s, decays in about 2.5 s; and the Y ganglions that provide a transient output respond in about 0.01 s and decays in about 0.1 s. Together, the X and Y ganglions provide the ability to detect motion quickly as well as "stabilize" an image for detailed sensing of textures, objects, and color.

So how can one exploit these building blocks to make computer vision possible? First, we recognize that many of the things we "see" are the result of changes, spatially and temporally. Temporal changes are either due to motion or changes in the scene—both of which are of great interest to situational awareness. Spatial changes provide clues for edge detection, lines, and boundaries. In the eye, the time constants of the X and Y ganglions, the spatial interconnections of the amacrine cells, and the receptor responses through the bipolar and horizontal cells provide a strong capability for providing *first and second difference* features. The spatial derivative images seen in Section 4.3 are an excellent example of techniques for detecting edges (such as a Laplacian operator), and the direction or orientation of the edge (a derivative operator). Edge information and orientation can be seen as the primitives needed for object shape recognition. Much of this information is concentrated by the cellular structures and neural interconnections in the eye.

The eye amazingly exploits some of its optical imperfections and turns them to its advantage. From a best engineering practice perspective, one wants as stable and clear an image as possible with constant resolution spatially, temporally, and across the color spectrum. The eye actually exploits tremor and optical aberration to its advantage using the tremor for spatial antialiasing, using the color aberration to feed interconnections and spatially "dither" for, say, the missing blue cones in the fovea area, and uses the blurry images in the off-fovea area to cue motion or image changes to subsequently swivel the fovea into view of the change. The eye distributes its receptors unevenly to meet a wide range of viewing goals. For example, night vision is done at much lower resolution using rods which are pooled together for signal gain in low light. There are practically no rods in the fovea area, which is optimized for intense detail spatially. This is probably where the edges and object shapes are detected through the spatial interconnections and subsequently processed in the visual cortex to associate the viewed shape with a known shape.

Finally, the receptors in the eye have a nonlinear brightness response allowing details in the shadows as well as bright areas of the image. Engineering practice to "linearize" the response of photodetectors on record and playback may not be such a bright idea. This is especially true now that analog circuits are largely a thing of the past. Linearization of the "gamma curve" in television cameras and receivers was very important for the standardization of television using analog circuits of the 1950s and 1960s. Digital systems can allow virtually any known nonlinearity in response to be completely corrected upon playback if desired. There is no technical reason why each pixel in a digital camera cannot have its own nonlinear response and automatic gain control such that the brightness dynamic range of the displayed image is compressed to meet the limits of the display device on playback only. The ability of the eye to see shadows has long been a frustration of photographers using both chemical and digital film. However, such a change in digital image acquisition is so fundamentally different in terms of processing hardware that one might as well suggest the hexagonal pixel arrangement rather than the current square row-and-column pixel arrangement.

Figure 17.31 shows a logical way to layout the processing for computer vision [12]. The reader should note that there are many ways of accomplishing this, but we will use the example in Figure 17.31 as a mean to discuss 2D features. First, we see that the camera output signal is received by a process which corrects any degrading optical aberrations, including color balance, brightness and contrast, but could also include focus, jitter, or blur corrections. The raw image can then be further processed in two main ways: matched field detection and segmentation.

One can carry out a "matched signal" search for any desired objects by correlating a desired scene in the box we call the "fovea window" which scans the entire image for the desired scene. This is definitely a brute force approach that can be made somewhat more efficient through the use of FFT's and inverse FFT's to speed up the correlation process. A high correlation at a particular position in the raw image indicates the presence of the desired subscene there. This is useful for

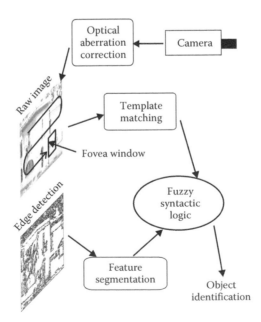

FIGURE 17.31 Block diagram depicting general computer vision using a combinations of template matching and image segmentation.

applications where scale and orientation are not a problem, such as identifying an individual's retinal blood vessel pattern, identifying currency, or scanning a circuit board for a particular component.

When scale and orientation of the object in the image are part of the problem, a process called *image segmentation* is used to essentially convert the image into a maximum contrast black and white image where edges and lines are easy to detect. Use of first and second derivatives allows line segments to be detected and assigned a length and an orientation angle. A cluster of line segments can then be sequenced in a number of ways as if one were drawing the lines using a pencil to reproduce the object. This sequence of segments can be scaled, rotated, and reordered to compare with a known object. A least-squared error technique can be used to select the most appropriate object class. Fuzzy logic can then be used to assemble the primitive objects into something of interest. For example, a cluster of rectangle objects with ellipses protruding from the bottom is likely a wheeled vehicle. The arrangement of the rectangles as well as their relative aspect ratios, number, and arrangement of ellipses (wheels) are then used to identify the vehicle as a truck, automobile, trailer, and so on from a functional description, or *graph*.

Figure 17.32 depicts the process of line segmentation to create a 2D object which can be described functionally to account for scale and orientation. For example, a rectangle can be described as a series of four line segments in a counterclockwise sequence where every other line is parallel. From one line segment to the next, a turn angle is noted where the even and the odd turn angles are equal. The difference between the observed turn angles and a normal angle is due to the orientation and viewing aspect. Virtually any lined object can be segmented into a series of simple functions. These functions can be sequenced in different ways, scaled, skewed, and rotated using simple mathematical operations. Thus, one can systematically describe an object as a graph, or set of segmented functions in much the same way as handwriting can be described as a series of pen strokes. As in handwriting recognition commonly seen in palm-sized computers, given a set of pen strokes, an algorithm can systematically test each known letter segments, but that would be relatively slow. A

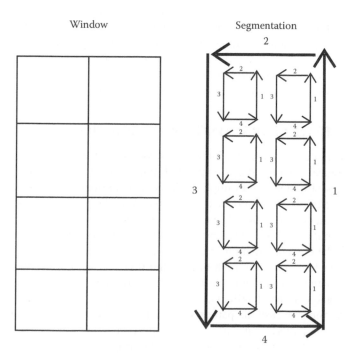

FIGURE 17.32 Sketch of a multipane window and the corresponding line segmentation to describe the window features as a series of line functions capable of preserving scale and aspect.

more efficient process is to apply fuzzy logic to subdivide the alphabet into letters with curves, straight lines, combinations, and so on. In this way, object recognition is similar to handwriting recognition; except for the font set which is much broader.

The image line segmentation can be seen as a hypothesis testing operation. Starting at a point along a difference-enhanced edge, one hypothesizes that the point is on a straight line whose direction is orthogonal to the spatial gradient at the point. The line is extended in either direction and if corresponding "edge" pixels are found, the line continues. At the end of the line, a search begins for a new "edge" to start a new line segment. Segments shorter than some prescribed minimum length are discarded (hypothesis proved false). Several possible hypotheses can simultaneously be active until the final decision is made on the object. Such a final decision may include the context of other objects in the scene and allow obscured objects to still contribute to a correct image classification. This is an example of the power and flexibility of fuzzy logic.

Connecting edge pixels to a curve can be done by either direct least-squared error fit or by employing a Cartesian tracking filter such as an α–β tracker or Kalman filter. Curves belonging to ellipses (a special case of which is a circle) can be described as particle motion with acceleration. By tracking the "particle" motion along the edge pixels, the position, velocity, and acceleration components of the curve can be estimated from the state vector. If the curve is part of an ellipse, this acceleration will change over time and point in a direction nearly orthogonal to the particle velocity direction. The curve can be segmented into sections which are described using specific functions, where the segmentation occurs as the particle "maneuvers" onto a new curve. The analytical representation of lines and curves to describe the salient features of an image allows one to detect objects when the scene is at an arbitrary scale and aspect. This is because our functions for the line and curve segments, and their relative position in the scene, can be easily scaled, skewed, and/or rotated for comparison with a functional description of a known object. The breaking down of an image into edges, lines, and ultimately functional line and curve segments is a computationally-intensive task. It is something that lends itself to parallel processing which is precisely how the brain processes signals from the optic nerve.

We have been careful in addressing signal and image features to stay as general as possible. Every application of this technology eventually ends up with very specific, proprietary, even classified (via government security requirements) features that work in unique ways. The interested reader should consult the books listed in the references for this chapter for more detailed information on image features and algorithms. From our perspective of intelligent sensor system design, it is most important to examine the *strategy* of signal and image features, using physics and biology as a guide.

17.3 DYNAMIC FEATURE TRACKING AND PREDICTION

In this section, we explore a new area for intelligent sensor systems: *prognostics*. Prognostics is the ability to predict a logical outcome at a given point of time in the future based on objective measures of the current and past situation. For example, a physician examines, checks the medical history of a patient, and then makes a careful prognosis of the outcome the patient can expect, and how long will it take to reach that outcome. Intelligent sensor systems can aspire to do the same sort of thing within limits. The limits are determined by our ability to create software algorithms which can accurately measure the current situation, track the changes in situation over time, and predict an outcome with reasonable certainty in both fidelity and time. What is most interesting about prognostics is that most pattern classification is based on a "snapshot" of data. That is, the pattern is assumed to be fixed and is to be identified as such. This "snapshot" pattern recognition is repeated again and again over time, in case the pattern does change for some reason. However, in prognostics, the evolution of the pattern is generally very important. The "trajectory" of the pattern over time is, in itself a pattern to be identified and exploited to improve future predictions.

Take an example of the common cold versus a flu virus. Both can have similar symptoms at times, the path of the sickness is quite significant. A cold is actually a combination of dozens or even hundreds of viruses combined with a bacterial injection. Generally, the body has had time to build up immunity to some of the infection and viruses, but has become overwhelmed. A virus on the other hand, attacks the body within hours often causing severe illness for a day or two before the antibodies gain the upper hand and lead to recovery. So, if one had a "yuk" feature which is zero when one feels great and 10 when one wants to welcome the grim reaper, both cold and virus would have the same feature at various times during the illness. However, the trajectories of the two illnesses are quite different. The cold comes strongly but then takes a while to eliminate while the virus attacks extremely fast and with great severity, but can often stabilize and fade away rather quickly. During the initial stages of a virus or bacterial infection when the severity increases, there is actually a chance that a life-threatening disease is at work. However, one with fever and other symptoms stabilize (stop rising or getting worse), the prognosis becomes much better. During recovery, the rate of improvement can be used to estimate when things will get back to normal.

This example illustrates an important point about prognosis. We know that when things are deteriorating rapidly there is a significant health hazard. But as symptoms stabilize and begin to subside we can be assured that the prognosis is good if we are patient. To make a prognosis, a complete diagnosis of what is wrong is not necessary, only the likely causes of the current situation and their impact on the future progression of the situation are needed. A partial diagnosis with several plausible causes should be used as working hypotheses with corresponding feature trajectories from the current situation to an ultimate outcome feature regime. If any of the possible outcomes presents itself as a significant hazard then corrective action is taken. With machinery health, we cannot expect anything like the "healing" of the body unless we shut down and repair the damage ourselves, although some symptoms maybe reversible with rest (such as over temperature conditions), or resupply. The issue with machinery monitoring is the cost and health hazard of an unplanned failure. Does the airframe crash? Will the ship need a 1000 mile tow? Will the factory make unsafe food or drugs? Will the failure trigger a catastrophe? The technical challenge of machinery monitoring is that the sensor signals have to be put into the context of the loads on the equipment at the time of measurement. This is where CINets can play an important role. Under heavy loads, the machine should be vibrating loudly, but under light loads that same level of vibration indicates a likely failure. One can think of contextual sensing as a hierarchical process that combines objective sensor information with human knowledge and models to make a more sophisticated assessment and prediction. This is where machinery monitoring and prognosis are heading, but the technical challenges are quite formidable.

Currently, most machinery maintenance is done based on a recommended schedule by the manufacturer. For example, automobile engine oil is changed every 3000 miles or 3 months, whichever comes first. But what if you did not drive the car for 3 months? Is the oil not perfectly good? Is it not unnecessary to replace the oil filter? Of course it is! But if the oil is really old there are oil compounds which will chemically break down causing sludge and corrosion, so it is not just the miles driven. Yet *time-based maintenance* is quite common because it is very easy to implement and plan around. However, some machinery is so complex and delicate that every time a maintenance action is done there is a significant chance of damaging some component. The old proverb "if it's not broken, don't try to fix it …" carries a great deal of weight with mechanical systems because doing too much maintenance generally leads to more damage and repairs resulting from the original maintenance action. One way an equipment owner can save on maintenance costs is to simply not do it. Obviously, this looks pretty good for a while until the whole plant infrastructure falls apart. Some plant managers have managed to get promoted because of their "maintenance efficiency" before their lack of maintenance causes a complete shutdown while others are not so lucky. A plant shutdown in big industry such as hydrocarbon production, refining, food production, textiles, and so on, can amount to enormous losses per day and an unplanned, if not catastrophic, equipment failure can lead to many days of shutdown. This is because these industries operate on huge volumes of commodity product with small profit margins. Therefore, a $1 loss of profit may be equivalent to a $100 loss in sales. If the gross production volume falls below a break-even point, they are out of business very quickly.

Some plants are designed to be completely rebuilt after only a few years of operation. Examples are sulfuric acid production, semiconductor manufacture, or military weapons manufacture. An acid plant literally dissolves itself over a relatively short lifetime. Electronic device manufacturing, such as hard drives, memory devices, communications chips, video displays, can become obsolete very quickly. In general, the electronics industry in the twenty-first century has changed from organizations that design and manufacture, to design houses that outsource manufacturing to contract manufacturers. This is a result of globalization and low cost labor in the developing world, but also the ISO9000 standards for manufacturing, so that theoretically, anybody with the appropriate facilities can build a given product. A military production plant is designed to produce a finite number of weapons, bought by mainly one customer (the government), and possibly some more for the government's allies. The factory runs for a while to stockpile the weapons and then is shut down until a war ensues. This creates problems when spare parts are in short supply. The industrial economics is such that these industries have a break-even date for each plant once it enters into production. When a plant continues production past the break-even date, it starts to make a profit on the capital investment, and this profit increases the longer the plant runs. When the product is a commodity such as gasoline, the capital investment is huge with very significant risks, but once a well strikes oil and the gasoline flows to the customer, the profits are significant because of the enormous volumes of production. So how can one increase profits in gasoline sales? The price at the pump and at the crude well is determined by the market. Profits are increased by improving efficiency in production and distribution, both of which have margins affected by technology. Smart sensor networks developed to monitor and predict machinery health are one such technology that will have a profound impact on manufacturing profits in the twenty-first century. This type of maintenance is called CBM and requires a broad spectrum of smart sensors, algorithms, pattern recognition, and prognostics algorithms for predicting machinery health.

The military has a slightly different need for CBM. Most weapon platforms such as ground vehicles, aircraft, and, ships are to be used for a long period of time significantly beyond their designed lifetime. The B-52 bomber is a perfect example which is actually planned to stay in service well into the twenty-first century. Doing unneeded repairs to irreplaceable hardware like a B-52 or CH-47 Chinook helicopter places unnecessary risks on causing fatal damage from the repair, as well as requires unnecessary manpower to execute the repair. In addition, military commanders always want to know as objectively as possible the battle readiness of their forces. Clearly, CBM is an important technology for military operations as well as a means to cut costs and gain more effectiveness. Given the value and long life expectancy of military hardware, CBM using vast networks of smart sensors is in the future. For the Navy, this will have multiplicative effect on reducing the number of personnel on board a ship. If you can eliminate the sailors monitoring the equipment, you also eliminate the sailors needed to clean up, cook, do laundry, and attend the maintenance sailors. This has perhaps the biggest cost impact on the Navy because of the expense of supporting all those personnel on board a ship for extended periods of time. For any branch of the military or emergency responders in the civilian sector automated equipment monitoring support the logistics of keeping everything in working order as well as providing a measure of readiness for the commanders and planners.

Building a CBM system out of networks of smart sensors is straightforward, albeit challenging, but does require a capability in information prognostics. Recall that the difference between data and information is that information is data with a confidence metric and with context to an underlying physical model. Both of these important aspects of information are required for prognostics. The data confidence measures are included in a Kalman filter as part of the measurement error statistics and the contextual physical model is included in the state transition kinematic model. Given this Kalman filter, the "state" of either feature or a "situation" can be tracked over time to produce a smoothed estimate of the current situation and a way to predict the situation in the future. This prediction, based on the current state vector (often contains a "position, velocity, acceleration, etc." elements), into the future is called a linear prediction and it will be accurate if things continue as currently measured. But that is not good enough for some prognostics applications where a great deal is known about the failure mechanisms.

Given a well-understood set of failure trajectories of a state vector, we can apply a pattern matching algorithm to the current observed trajectory and these possible outcome trajectories. In much the same way as multiple hypothesis testing is executed using syntactic fuzzy logic, we can continuously compare the current situational history defined by the observed track to each of the hypothesized trajectories, and use the nonlinear dynamics of the failure progression to determine which path we are on. Given the selected failure mode path, we can do a much better prediction of the RUL and current failure hazard. This requires exquisite knowledge and scientific modeling of the underlying failure mechanisms which is where engineering should and will push the envelope of capability in the future. In the absence of such knowledge of failure mechanisms, we have linear prediction. In both cases we must define a "regime" for imminent failure so that the remaining time and hazard rate can be computed. Defining a regime which signifies unacceptable risk for failure is much easier than defining the nonlinear fault trajectory from the current state to the failure point.

Figure 17.33 depicts graphically the difference between linear prediction prognostics and nonlinear fault trajectory prognostics. Time–temperature plots are given for three hypothetical cases: a healthy power supply; a supply with a defective resistor; and a supply with a failing voltage regulator. The three trajectories are quite different and nonlinear. They are identifiable based on trends and underlying physical modeling of the failure mechanisms. However, at different times along any of the trajectories one or more of the situations can exist as components can and do begin to fail at any time. The prognostics task is to detect the precursors to eventual component failure, and provide accurate prognostics for the current hazard and RUL of the component. "Regimes" are identified on the right edge of Figure 17.33 to identify a set of meaningful situations based on temperature. For example, above 300°C, a fire is assumed imminent. For temperatures above 150°C the components operate out of specification and an immediate repair is warranted. Between 85°C and 150°C, caution is warranted because the situation could quickly deteriorate, but currently is tolerable. Below 85°C, there is no immediate hazard of burning out the power supply. To complicate matters further,

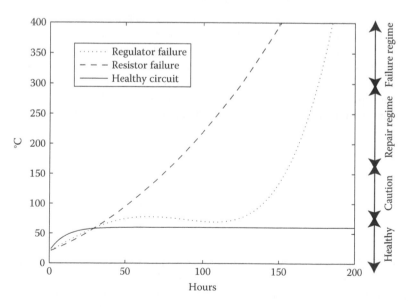

FIGURE 17.33 Temperature readings over time for three types of hypothetical failures of an electronic power supply showing the healthy, caution, repair, and failure regimes.

the power supply temperature will depend greatly on the load which can vary significantly as electronic devices are switched on and off, and catastrophically if an electronic system on the power bus fails, pulling down the power supply with it. With the exception of the latter case, which can often be handled using fuses or circuit breakers, the observed temperature can be expected to have many fluctuations. The job of the prognostic algorithm is to provide an accurate and meaningful prediction in the midst of all the fluctuation noise. This is why a combination of Kalman filtering based on physical modeling and fuzzy syntactic logic is needed to create machine prognostics.

Up until about 30 h, as shown in Figure 17.33, any of the trajectories can exist and the supply is considered "healthy," but a linear prediction alone might prematurely predict a failure or cause an early repair. From 30 to about 130 h the regulator fault trajectory and the healthy supply are not distinguishable, but when the regulator starts to fail, the temperature rises much quicker than the resistor failure. The difference in the nonlinear fault trajectory model is quite useful in improving the prognostics information to the user. We ought not to expect the software at the sensor site t make the full call, but rather to organize and prioritize the information for the human at a hierarchical level. For example, the alert message to the human might look like the following:

Repair of component X requested in 10 h.
Likelihood of 50% failure in 17 h if current situation persists.
Likelihood of 90% failure in 17 h if failure mode is Y.
Likelihood of 35% failure in 17 h if failure mode is Z.
Failure mode probability is Y (80%), Z (50%), and unknown (30%).
Site # reporting all sensors calibrated ±5%.

Note how the sum of the failure mode probabilities is greater than 100%. This is an artifact of fuzzy logic which provides useful statistical measures to be associated with the multiple hypotheses. The human is now put into a desirable position of getting good information to contemplate for the repair decision from the CBM system. The diagnostic information while incomplete is very valuable to the human and may influence the repair decision. Finally, the self-diagnostics at the end of the message ensures credibility of the message data and alert status. This information is straightforward for a smart sensor system to provide and creates a whole new economic basis for industrial production as well as military readiness and efficiency.

Figure 17.34 shows some temperature data for our power supply example. The data are observed every hour and have a standard deviation of 10°C mainly from the load variations. Understanding the variance of the measurements is very important to application of a tracking filter. Recall from Section 11.1 that the Kalman filter has both a *predicted* state vector and an *updated* state vector. The difference between them is that the one-step ahead prediction (T time units into the future) is a little less noisy than the updated state, which has been corrected using the Kalman gain and the prediction error. At any point in time, the predicted state vector can be estimated d steps into the future [13].

$$x'(N+d\,|\,N) = F(N+d)x\,(N\,|\,N)$$

$$= \begin{bmatrix} 1 & T_d & \frac{1}{2}T_d^2 \\ 0 & 1 & T_d \\ 0 & 0 & 1 \end{bmatrix} \begin{bmatrix} x_{N|N} \\ \dot{x}_{N|N} \\ \ddot{x}_{N|N} \end{bmatrix} \qquad (17.52)$$

This can also be done using a new state transition matrix where $NT = T_d$, or simply a one big step ahead prediction. This is a straightforward process using the kinematic state elements to determine the time and position of the state in the future. Therefore, it is really rather unremarkable that we can predict the future value of a state element once we have established a Kalman filter to track the trajectory of the feature of interest. This forward prediction is called a linear prediction. Its limitations are that it takes into account only the current state conditions and assumes that these conditions will hold over the future predicted step.

But, we know that as most things fail, the trajectory is likely to become very nonlinear, making linear predictions accurate only for very short time intervals into the future. To take into account several plausible failure trajectory models, we treat each trajectory as a hypothesis and assess the statistics accordingly. Figure 17.34 shows a linear prediction alongside predictions assuming a resistor or regulator fault. These "tails" are pinned on the end of our Kalman filter predicted state by simply matching a position and slope (rate of change) from the terminal trajectory of our models to the most recent data smoothed by the Kalman filter. This is like postulating, "we know we're at this state, but if a resistor has failed, this is what the future looks like, or if the regulator has failed, then the future looks like this." At 40 h, in our example, one really cannot tell much of what is happening in the power supply. It might be normal or it might be developing a problem. The data points (+) and Kalman filter track (solid line plot) represent our best knowledge of the current condition. The linear prediction at 40 h in Figure 17.34 shows a repair action needed after 100 h, and failure expected after 240 h. The hypothesis of a resistor failure says that a repair will be needed in 80 h and failure is expected after 140 h. The regulator failure hypothesis suggests a repair after 65 h (25 h from now) and that failure could occur after 90 h. Since we have little information to judge what type of failure mode, if any is occurring, the linear predicted prognosis is accepted.

At 140 h, in our example, there is much more information to consider as shown in Figure 17.35. The temperature has stabilized for several days and for the most part, there is little cause for alarm. However, during the last 24 h there has been a slight trend upward. The linear predicted prognosis suggests repair at 200 h (60 h from now) and failure may occur after 400 h. The resistor failure hypothesis suggests repair after 180 h and failure after 300 h. The regulator failure hypothesis suggests repair after 175 h and failure after 210 h. It appears that either the resistor or the regulator is failing, but it is far enough into the future that we do not need to do anything at the moment. Figure 17.36 shows the results of our simulation at 150 h. Failure now appears imminent, but more urgent is the need for a repair starting in a few hours. The linear predicted prognosis suggests repairing after 160 h (just 10 h from now) and failure after 220 h. The resistor failure prognosis suggests repairing even sooner after 155 h and failure at 180 h. The regulator hypothesis suggests

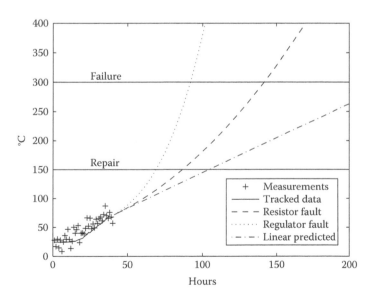

FIGURE 17.34 After about 40 h one cannot diagnose the problem but can make prognosis estimates based on the current state, a hypothetical resistor fault, or a hypothetical regulator fault.

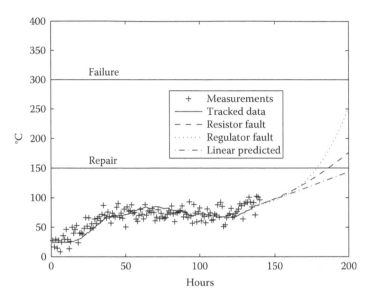

FIGURE 17.35 After about 140 h there is no immediate need for repair and any prognosis is still hypothetical.

repair now and that failure could occur after 170 h, just 20 h into the future. By failure, we mean that the feature will cross the threshold we have defined as the failure regime, where failure can happen at any time. For our example, the regulator failed causing the failure threshold to actually be crossed at 180 h. Even though our prognostic algorithm did not exactly predict the cause nor the precise time of failure, it provided timely and accurate information necessary for the CBM actions to take place. As technology and science improve our ability to model fatigue and perform prognostics, the performance of such algorithms for predictive maintenance will certainly improve. But the basic idea and operations will likely follow our example here.

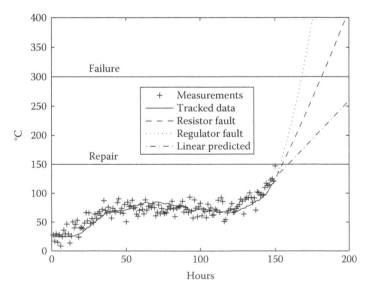

FIGURE 17.36 At 150 h there is a need for repairs based on the state trajectory change where the prognosis for a regulator failure is 12 h and a resistor failure is 30 h.

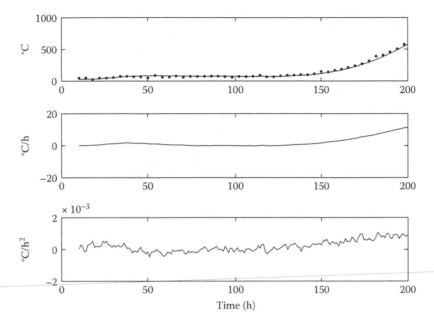

FIGURE 17.37 The Kalman filter states for our power supply example can be used to more accurately predict the statistical rates as well as estimate the type of failure from the temperature rate of growth.

Figure 17.37 shows the results of the Kalman filtering. The measurements and predicted temperature state are seen in the top plot, the middle plot shows the estimated rate of temperature change, and the bottom plot shows the temperature "acceleration." These state components are used to do the forward prediction as shown in Equation 17.52. However, as one predicts a future state, the state error covariance increases depending on the size of the step into the future. The process noise matrix is defined as

$$Q(N) = \begin{bmatrix} \frac{1}{2}T^2 \\ T \\ 1 \end{bmatrix} \sigma_v^2 \begin{bmatrix} \frac{1}{2}T^2 & T & 1 \end{bmatrix} = \begin{bmatrix} \frac{1}{4}T^4 & \frac{1}{2}T^3 & \frac{1}{2}T^2 \\ \frac{1}{2}T^3 & T^2 & T \\ \frac{1}{2}T^2 & T & 1 \end{bmatrix} \sigma_v^2 \tag{17.53}$$

where σ_v is the minimum standard deviation of the "acceleration" element of the state vector. Setting σ_v to a small value make a very smooth track, and also one which is very sluggish to react to dynamical changes in the data. When T represents the time interval of the future prediction step, the state error covariance is predicted according to

$$P'(N+1 \mid N) = F(N)P(N \mid N)F^H(N) + Q(N) \tag{17.54}$$

Thus, the larger the T is (the further one predicts into the future), the larger the state error covariance will be. One can extract the 2×2 submatrix containing the "position" and "velocity" state elements, diagonalize via rotation (see Equations 17.11 through 17.12), and identify the variance for position and the variance for velocity on the main diagonal.

The variance for position is rather obvious, but to get an error estimate for our RUL requires some analysis. Our prediction of the time until our state crosses some regime threshold is the result of a parametric equation in both time and space. Our state elements for position and velocity are random variables whose variances increase as one predicts further into the future. To parametrically translate the statistics to time, we note dimensionally that time is position divided by velocity. Therefore, the PDF for our RUL time prediction is going to likely be the result of the division of two Gaussian random variables (our state vector elements). This is a little tricky but worth showing the details here.

A detailed derivation for the probability density of the ratio of two random variables is presented in Section 7.1. Let the random variable for our predicted RUL time t_F, be z. Let the random variable representing the final "position" prediction be simply x with variance σ_F^2, and the random variable representing the "velocity" be simply y with variance σ_R^2. Our new random variable is therefore $z = x/y$. One derives the pdf by first comparing probabilities, and then, differentiating the probability function for z with respect to z to obtain the desired pdf. The probability that $z \leq Z$, or $P(z \leq Z)$, is equal to the probability that $x \leq yZ$ for $y \geq 0$, plus the probability that $x \geq yZ$ for $y < 0$.

$$P(z \leq Z) = \int_0^\infty \int_{-\infty}^{yZ} p(x,y)\,dxdy + \int_{-\infty}^0 \int_{yZ}^\infty p(x,y)\,dxdy \tag{17.55}$$

Assuming the tracked position and velocity represented by x and y, respectively, are statistically independent, one differentiates Equation 17.55 with respect to z using a simple change of variable $x = yz$ to obtain

$$p(z) = \int_0^\infty yp(x=yz)p(y)\,dy - \int_{-\infty}^0 yp(x=yz)p(y)\,dy$$

$$= \frac{2}{\sigma_F \sigma_R 2\pi} \int_0^\infty y e^{-(y^2/2)\left((z^2/\sigma_F^2)+(1/\sigma_R^2)\right)}\,dy \tag{17.56}$$

$$= \frac{1}{\pi}\left(\frac{\sigma_F}{\sigma_R}\right) \frac{1}{z^2 + \left(\frac{\sigma_F}{\sigma_R}\right)^2}$$

The pdf in Equation 17.56 is zero mean and is known as a Cauchy PDF. It is straightforward to obtain our physical pdf for the time prediction where the mean is t_F.

$$p(t) = \frac{1}{\pi}\left(\frac{\sigma_F}{\sigma_R}\right)\left[(t-t_F)^2 + \left(\frac{\sigma_F}{\sigma_R}\right)^2\right]^{-1} \tag{17.57}$$

The mean of the pdf in Equation 17.57 is t_F and Figure 17.38 compares the shape of the time prediction pdf to a Gaussian pdf where the standard deviation $\sigma = \sigma_F/\sigma_R = 10$. The density functions are clearly significantly different. While the second moment (the variance) of the time prediction pdf in Equation 17.57 is infinite (therefore the use of the central limit theorem does not apply), one can simply integrate the pdf to obtain a confidence interval equivalent to the 68.4% probability range about the mean of a Gaussian pdf defined by $\pm\sigma$. This confidence range works out to be 1.836 times the Gaussian standard deviation. To present the confidence limits for our RUL time prediction, we can simply use 1.836 times the ratio of σ_F/σ_R. Figure 17.39 shows the RUL time based on linear prediction and also shows the 68.4% confidence limits. The straight dashed line represents real-time and shows the actual time the temperature crossed our failure threshold at 300°C was about 180 h. The solid line is the real-time estimated RUL and the dotted line represents the RUL 68.4% confidence limits. Our RUL prediction converges nicely to the correct estimate

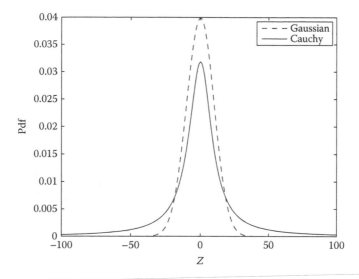

FIGURE 17.38 Comparison of the Gaussian pdf to the Cauchy pdf derived from the ratio of two random variables.

around 140 h, or nearly two days before the failure condition. However, it is only in the range beyond 170 h that the confidence interval is very accurate. This is exactly what we want the prognostics algorithm to provide in terms of statistical data. The plot in Figure 17.39 is known as a *survivor curve* [14].

Figure 17.40 is a 3D rendering of Figure 17.39 showing the RUL pdf at several interesting times for the survivor curve. The RUL pdf is centered over the RUL time prediction. But, for most of the time before 170 h, the statistics on the RUL time are not very good. This is due mainly to the failure threshold being significantly in the future when the trajectory of the temperature plot is fairly flat from around 50–160 h. During this period and situation, we should not expect any reliable prognostic RUL time predictions. The corresponding RUL time pdfs in the period leading up to 160 h are

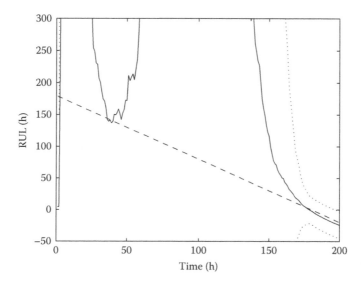

FIGURE 17.39 Using linear prediction, the RUL is predicted statistically and compared to the actual RUL (straight dashed line) showing the 63% confidence intervals (dotted lines).

PDF and survivor curves

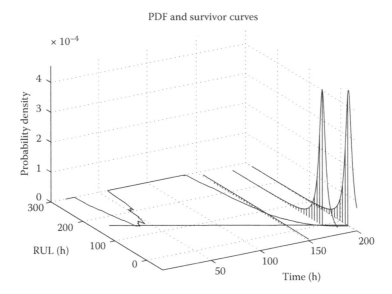

FIGURE 17.40   A 3D version of the survivor curve showing the pdf overlay at interesting times and the integral of the pdf (vertical line shading) used to estimate the probability of surviving as a function of time.

almost completely flat and extremely broad. As the RUL time prediction converges, the corresponding RUL time pdf sharpens. As the temperature feature trajectory passes the threshold, the RUL time becomes negative (suggesting it should have failed by now), and the RUL time pdf is then fixed because we are interested in the time until the failure threshold is crossed and beyond as our failure regime. Again, this is seen as a desirable algorithm behavior.

Consider Figure 17.41 which shows the probability of survival (dotted), which is the integral from infinity down to the current time. When the statistics of the RUL time prediction are accurate (giving a sharp pdf) and the failure time is well off into the future, the survivor probability would be near unity. When the time statistics are not well defined or the RUL time is in the immediate

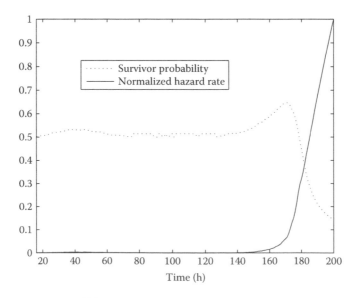

FIGURE 17.41   Estimated probability of survival (dotted line) and the hazard rate (solid line) as a function of time based on the Kalman filter states.

future, the survivor probability is near 50%. Equation 17.58 gives an analytical expression for the survivor rate plotted in Figure 17.41 as the dotted line.

$$R(t) = \int_t^\infty \frac{1}{\pi} \left(\frac{\sigma_F}{\sigma_R} \right) \left[(t - t_F)^2 + \left(\frac{\sigma_F}{\sigma_R} \right)^2 \right]^{-1} dt$$

$$= \frac{1}{2} - \frac{1}{\pi} \tan^{-1} \left(\frac{t - t_F}{\dfrac{\sigma_F}{\sigma_R}} \right) \tag{17.58}$$

In Figure 17.41, the survivor probability is initially around 50% because it is so broad. As the statistics improve, the survivor rate improves. This is desirable because we need assurances that failure is not immediate in the regime approaching the failure threshold. Then the RUL time prediction crosses the current time at 180 h giving a 50% survivor rate. At this point we fix the pdf for the RUL time prediction since we have crossed the failure regime threshold. In other words, we do not want the RUL pdf to start flattening out again from a growing "backward time prediction." However, the survivor rate, which is the integral of the RUL pdf from infinity to the current time still changes as the fixed (i.e., in shape) RUL pdf slides by such that less and less of the pdf is integrated as time progresses. This causes the survivor rate to diminish toward zero as the hazard rate increases. The hazard rate $H(t)$ is defined as the RUL time pdf for the current time divided by the survivor rate and is plotted as the solid line in Figure 17.41.

$$H(t) = 2 \left(\frac{\sigma_F}{\sigma_R} \right) \left[\pi - 2 \tan^{-1} \left(\frac{t - t_F}{\dfrac{\sigma_F}{\sigma_R}} \right) \right]^{-1} \left[(t - t_F)^2 + \left(\frac{\sigma_F}{\sigma_R} \right)^2 \right]^{-1} \tag{17.59}$$

The $(t - t_F)^2$ term in the right-most square-bracketed term in Equation 17.59 is set to zero once the failure threshold is crossed (i.e., the term $t - t_F$ becomes positive). Again, this has the effect of stating that we have met the criteria for passing the failure regime threshold, now how far past that point are we? *The hazard rate measures the likelihood of immediate failure by comparing the current probability density to the probability of surviving in the future.* The hazard rate needs to be normalized to some point on the pdf (in our case, the last point is taken) because it will generally continue to increase as the survivor rate decreases. Clearly, we now have an objective "red flag" which nicely appears when we cross into a prescribed regime we associate with failure.

We have one more metric shown in Figure 17.42 which helps us define component failure we call the *hazard stability metric* [15]. This term is designed to measure how stable the feature dynamics are. When positive, it indicates that the dynamics are converging. This means that if the feature is moving in a positive direction, it is decelerating. If it is moving with negative velocity, it is accelerating. In both stable cases, the velocity is heading toward zero. We can associate zero feature velocity as a "no change" condition, which if the machinery is running indicates that it might just keep on running under the same conditions. Obviously, the most stable situation is when the feature velocity is zero and there is no acceleration. Our stability metric is defined as

$$\Sigma(N|N) = -\langle \dot{x}_{N|N} \ddot{x}_{N|N} \rangle t \tag{17.60}$$

such that a zero or positive number indicates a stable situation. If the feature is moving in a positive direction and is accelerating, or if moving with negative velocity and decelerating, the stability metric in Equation 17.60 will be negative indicating that the feature dynamics are not stable and are

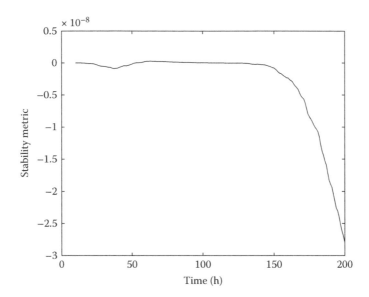

FIGURE 17.42 The hazard stability metric to indicate when the current hazard is getting worse (negative) for converging to a stable rate (positive).

diverging from the current values. Change is not good for machinery once it has been operating smoothly for a time. Therefore, this simple stability metric, along with the hazard and survivor rates, is key objective measures of machinery health.

17.4   INTELLIGENT SENSOR AGENTS

The Internet, and private Intranets, are rapidly dwarfing all other sensor network architectures because of the economies of scale, wide bandwidth available, and flexibility for solving a wide range of sensor network problems including security. Figure 17.43 will help us summarize the basics

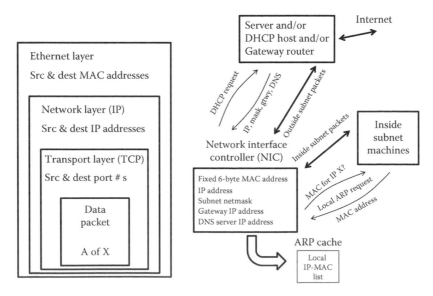

FIGURE 17.43 Basic IP packet layers and network configuration processes showing the wrappers around the data packets and configuration using DHCP.

of how sensors will likely work on the Internet, and the basic process for how a sensor will come online and interface with the rest of the network. The best place to read further on this is from online sources. A quick discussion here will provide a primer for some, review for others, and give us some things to refer to as we get into the details if intelligent sensors and their software agents.

17.4.1 INTERNET BASICS

The first thing to discuss is the various layers that go into Transfer Communications Prototcol/Internet Protocol (TCP/IP) communications [16]. The outer layer, as shown on the left side of Figure 17.43 is called the "Ethernet layer" and contains the source and destination Media Access Control (MAC) addresses on the Network Interface Controller (NIC). The MAC address is a 6-byte unique identification number. When a datagram goes from a computer to a computer, router, server, and so on, the NIC with the source MAC address must have a direct connection to the NIC with the destination MAC address. Devices such as network HUBs simply copy the data packets from the incoming port to all the other ports to facilitate this. A faster device is a switch, which connects the ports based on the MAC addresses listed in the Ethernet layer of the message. This is completely independent of the 4-byte IP address assigned to the computers on each end. How did this outer layer get the name "Ethernet?" It comes from packet radio and follows an analogous protocol as in "messaging through the ether" which dates back to ham radio in the 1950s. The early use of the ARPA net, later DARPA net (Defense Advanced Research Projects Agency) used radio and packet concepts to send digitized seismic waveforms from around the world back to the United States for analysis to see if it was a nuclear event or earthquake. Some very careful and thoughtful engineering discussed how to construct the datagram layers seen in Figure 17.43, and how to implement them, but it is doubtful that the inventors of the Internet (Al Gore had nothing to do with it) could imagine just how well it would all work on the grand scale of operations today.

What happens when you turn on an intelligent sensor connected to this network? There are many ways to set up the network, but we will limit our discussion to the easiest way, and then bring in measures to tighten security and improve performance. The easy way to configure a network device (a node) is to use Dynamic Host Control Protocol (DHCP). The sensor NIC sends a DHCP request to anyone on the subnet (more on subnets later) and a DHCP server (this could be a router or another PC) responds by sending the sensor node NIC its IP address, subnet mask, default gateway IP address, and optionally a domain name server (DNS) IP address, and a lease time for those numbers, typically several weeks or more. The subnet mask may have patterns such as 192.168.1.x where the "x" indicates that all the addresses in the first octet of the IP address are the subnets and any IP address where the last three octets do not match 192.168.1.x are outside the subnet. If an IP is outside the subnet, the sensor NIC uses the MAC address of the default gateway. The gateway router will then find a way to pass the message along (more on that later). If the destination IP is inside the subnet, the sensor NIC goes to a list, called the Address Resolution Protocol (ARP) cache, to find the MAC address that goes with the destination IP address. If the MAC address is not in the list, it sends an "ARP request" to all the nodes in the subnet asking for the MAC address that goes with the known subnet IP address and one node should respond. If the sensor does not have a MAC address to send a datagram to on the first hop, the transmission ends in error. Data transmissions are always between unique MAC addresses.

When the datagram gets to the gateway, the router software opens the Ethernet layer and reads the source and destination IP addresses. A router has *routing tables* that it builds and maintains with MAC addresses of other routers and the IP addresses they can reach. The router then picks one or more other routers to send the datagram to, wrapping the datagram in a new Ethernet layer appropriate for the next hop, and so on. The routers take random turns sending their stored packets over the wire, occasionally colliding with other packets which then need to be resent. The packets that make up a datagram make their way to the destination IP address where the NIC unwraps the Ethernet and Network layers, and sends the data packet on to the destination port number, assuming there is a process listening on that port number. The packets do not arrive in order, some are duplicates, and some may go missing. A missing packet initiates a return request to resend the missing packet(s) until the datagram is complete, and then it is forwarded to the process listening on the destination port. The Ethernet layer can be seen as a letter carrier's mail bag, the Network layer is the letter's envelope, The Transport layer allows for the actual letter to go to a particular person at the address, and the data packet is one page of a possibly multipage letter. This is a very simple and unsecure configuration of the Internet, but explains most of the important concepts.

17.4.2 IP Masquerading/Port Forwarding

In order to manage the burden of a limited (but large) number of fixed IP addresses and to control local network IP addresses using DHCP, a router can act like a broker to handle the sensor node messaging but using the router's IP address instead of the sensor node's IP. This allows the Internet Service Provider (ISP) to manage their own networks independent of the IP addresses in the public which helps them configure and manage network traffic, routers, bridges (routers limited to specific nodes to manage heavy traffic), and most importantly, firewalls (more on firewalls later). It is clear to imagine how the sensor node's packet can go out to the far away server, but how can the response get back past the router that masqueraded the IP address? Well, the router can also change the source port number (the port where the packet originated on the sensor node) to a unique port no one else is using in the subnet. With over 65,000 ports available, it is not difficult to find one not in use during the communication cycle. The router then redirects an incoming packet addressed to it and the masqueraded port back to the sensor node and the correct port. This can happen over and over again as the packets hop between routers on their way back and forth between the sensor node (client) and the far-away server. The *port forwarding* and IP masquerading are not detected (this is called transparent port forwarding) by the client and the server. This makes the Internet very flexible, extensible, and powerful, *but you cannot know where a packet originates from, ever.* This is why there are security threats, viruses, and so on once you are connected to the Internet, or any network using this protocol. But there are ways to manage the risk.

17.4.3 Security versus Convenience

There are dozens of "ports" in a modern PC with network capability to support file transfer (file transfer protocol—FTP), email, remote procedure execution, external log-on sessions, and so on. Every port is a potential security risk in that someone on the network could use a standard port or one of the many unclaimed ports to take control of your PC. The easiest way is to get you to install the malicious software for them. This can happen with any email attachment you open or web page that contains an executable program that requires installation. In the old days of PCs this could be done with a "terminate and stay resident" program that could, say, monitor your keyboard sending every key stroke to a file and sending the file back to the hacker exposing access to your personal information. This is harder to do today because most web browsers do not allow it, or require the user to open up a number of security settings to do so (while advising you that you are making your machine unsecure). Still, every network feature turned on in your computer carries a security risk. The best way to keep those useful network features and keep out hackers is to use a "firewall" gateway to the open Internet, and cut all other connections (security holes) from behind the firewall to the Internet. For example, the firewall could allow incoming traffic to local machines on port 80 to allow web browsing, but shut all the other ports off. That allows the convenience of a network administrator to take over any machine behind the firewall to install software updates, inspect security setting, and help with troubleshooting software, but prevents a hacker from having the same privileges. If other ports need to be open to the public Internet, a reasonable approach is to setup a proxy server outside the firewall with the open ports, and require any users to securely log onto the proxy server to upload programs, download data, and so on. It limits added risk, has similar functionality, but prevents a hacker from getting access to machines and users inside your firewall. There is much more to this than can be explained here, but this simplified description of what is going on highlights most network configurations today.

So how can one get access to a private IP addresses machine behind a firewall? A common example of this is a web-enabled security camera in your home, where the IP address is automatically provided by a DHCP host on your cable modem router, which has an IP address provided by the ISP, and so on. there is no way you can log into your webcam unless you know the IP address, and that can be changed by the ISP, or a power failure in your home and the cable modem router giving the webcam a new address when the power is restored. The webcam needs a port to stream its video, and when an outsider on the network connects to the IP address and port number for the webcam using a standard browser program, the webcam sends an HTML document with a video window and then streams the video to the recipient.

17.4.4 Role of the DNS Server

One way to make this work is to use a third-party service on the public Internet called DNS (Domain Name Service) which maintains a list of IP addresses and ports associated with a web-based name, such as www.mywebcam.dyndns.org (this is called a URL name). The organization "DynDNS" sells web names for a nominal fee like $10 per year, or free for a few months, so long as no one else is using that name. You can use it as a switchyard to redirect traffic from the Internet to site of your choosing. It is a secure server for which you must have an account and password to log into and manage the links. The way to make this service work for your webcam is register the dynamic IP of the webcam every time the camera boots up, and whenever the local IP is changed. When it reaches the DNS server it logs in and sets the last link in the IP chain for the first hop in the route back to the webcam. So long as the routers at each hop maintain this route, a web browser using the web name registered at the DNS server will be redirected properly. The routing tables are maintained automatically and the ISP do not update the assigned IPs through DHCP very often since it is not necessary, but occasionally the IP chain is broken. So to manage that, the webcam reregisters itself from time to time to ensure that the links remain current. Now for the case of firewall, if the webcam uses port 8000 and 8001, the firewall has to open up those ports for packet traffic. Also, if the gateway router is already doing port forwarding for the webcam IP address, we need to stop that from being something the router can change. Usually, you can reserve an IP address for the MAC address of the webcam in the DHCP server at the router and set the port forwarding for the webcam ports to always go to the correct IP, MAC address, and port numbers. This is manually managed port forwarding. This is the reason why the Internet is so robust to a network-wide crash and can largely reboot itself after a localized event. It is also why your favorite website may not be reachable for a few minutes once in a while, or if a hacker launches a "denial of service" attack by having large numbers of PCs with the hacker's "Trojan horse" virus installed, suddenly all wake up and automatically go to the same website jamming all the routers and servers in the process. DNS servers maintain copies of other DNS server's name-IP lists to facilitate the address-finding process and even in the PC configuration there are entries for several DNS servers, but only one gateway router and a DHCP server if DHCP is used. Again, we are only touching the surface here of all that goes on to make the network run smoothly, but the aim is to describe the overall concept so that we can see how intelligent sensors can live on the network.

17.4.5 Intelligent Sensors on the Internet

The sensor as a thin web server is a concept growing more popular as microprocessors become smaller, less expensive, and more powerful. Obviously, the data acquisition system, some computational capability and storage, and the NIC is part of the sensor "box." The sensor raw data can be filtered, calibrated, processed into features, features assembled into dynamic states, and this information can be collected into detections of various levels of complexity and patterns. Such data processing, based on the sensor's limited view of the world, creates a *manifest* of sensor products. Since there is web access, the obvious interface is a fancy web page displaying the sensor data, processed information, a data archive, and with features to support management of the sensor, such as network configuration, software updates, rebooting, downloading data records, and so on. While this makes the sensor information very attractive and exciting for humans to look at, it carries all the security concerns from Internet access and requires manual downloading of data (although a program can interact directly with the web page).

A better approach is to use an existing extensible markup language (XML) schema or create/publish a specific XML schema to organize all the sensor data for use by web pages and automated network programs called ebots (electronic robot agents or an unattended network service). An XML schema is a master document template that defines all the structured elements, their types (numerical, text, etc.), and can even enumerate the allowable ranges for each element and whether the element is required or optional [17]. The sensor manifest describing all the data and features the sensor offers can be part of this XML schema. A person can read the schema to see what is available, translate the data to map to another XML schema if desired, or write an ebot program to automatically harvest the sensor XML data. In addition, the sensor "thin server" can be programmed to connect to another web node and send its data periodically as an XML-formatted message. We use the term "thin" server because the sensor has very limited services compared to a typical PC server. It is really an embedded computer with sensor(s) on one end and communication ports on the other end. The key step here is to organize the sensor data using XML and publish the schema document. The sensor web page can be tied directly to the XML data so that gauges and dials can have their graphics respond "live" to the sensor data. Using the XML data from the sensor, another web page combining many sensors together (a dashboard) can easily integrate the XML from each sensor and update it continuously.

We cannot get into the programming details for each of the language options available for web page display, messaging, and so on, but we can describe the basic constructs. Starting with the web page, there are very sophisticated web page development tools set which allow links to just about anything, but most importantly, graphical parameters such as line widths, angles, colors, and so on, can be linked to an external XML document, rather than the internal XML document the tools use to draw the web page. This literally means that you can draw a fancy gauge and link the line representing the indicator needle to a parameter which is the element of the sensor XML document, and update the needle position continuously making the gauge "live" on the web page. As you might imagine, you can buy graphics packages already setup for this to make beautiful sensor web pages. Furthermore, one can make a dashboard or gauges as another hierarchical web page with all the live gauges. If you click on a particular gauge, you can go to the web page for the corresponding sensor to see more detail, and so on. It is not hard to imagine layers of dashboards and the ability to "drill down" to a particular sensor for greater detail. Figure 17.44 illustrates this concept, the only inconvenience is that sensors may have different XML schema documents (XSDs) and require some translation. XML supports this with eXtensible Style Language (XSL) and a process called XSLT (XSL Transformations) [17]. XSLT is also an XML document and it provides an automatic way to capture XML-formatted data in your schema if the original source data are in a different schema format. This is really the main purpose of XML as a database schema. With each element in the XML document tagged according to the schema, one can then define translators between schemas that themselves are XML documents! While the arrangement of parameter data in XML follows a typical data structure in concept, the size of an XML document is huge compared to its binary counterpart. The advantages of XML are that elements can be empty or missing and the document is still parseable so long as the structure is formed properly (nested opening closing tags are placed correctly). There is also flexibility in updating schemas when new elements are added, and backward compatibility to older XML documents so long as the new schema includes all the old elements. One of the elements in every XML document is a link to the schema document as a file on the network.

FIGURE 17.44 Sensor IP networking showing individual sensor websites with web pages for the gauge, data in XML, and a published XSD easily integrated into a web-based sensor dashboard.

Sensor data can be moved on the network as part of a web page, as a file, or as a message between two ports. Messages between ports can be between multiple programs running on the same PC using the same IP, or separate physical machines with many router hops between them on the Internet. Each of the major programming languages have their specific ways of doing one or more of these operations on the network. For example, in MATLAB the statement "s = urlread('url', 'method', 'params')" reads the content of the "url" address into a string "s" where "method" can be basic web commands such as "get" and "params" is MATLAB cell array of name/parameter value pairs if needed for the action. There is also a "urlwrite()" with the same functionality. If the instrument control toolbox is installed, there is a lot more capability offered. MATLAB also supports COM interfaces and SOAP. So you can do some experimentation with these if you are familiar with them, and MATLAB will likely be adding more and more web capability. Pushing files around the network certainly works, but to make this work really fast one should set up a client–server connection. Think of the server program as a shopkeeper and the client program as a customer. The server sets up the port for listening and when a request comes in from the client, the server responses with the requested information, such as a web page or an XML formatted message (no need to make it a file). So the sensor becomes a server of its data, processed information, pattern detections, and archives on the network. The client to the sensor server could be a web browser displaying the data as graphics, or an ebot capturing the data for other uses and operating as a server for other clients, and so on. All of this is being done today using very straightforward language techniques, but there are still security concerns in our simplified description.

Since we cannot be 100% certain that a given datagram is authentic, how can we at least manage the security of the sensor data and ebot concatenated hierarchical data? There are several popular schemes for developing a trust between two network entities. Figure 17.45 illustrates one approach simple enough to try and explain here yet has the essential concepts for information automation. We have a client who wants data and a sensor who has data and we need to be sure that the data are not

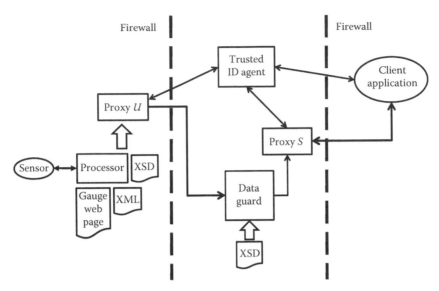

FIGURE 17.45 Sensor networking security using proxy servers, a third-party trusted agent, and the XSD to validate sensor XML information inside a secure firewall.

being spoofed and the client is who they say they are. Generally, this is done using a third party who issues identifications and checks them. Assuming the first time the client and the sensor want to exchange data they are the actual parties, each can receive an authentication key which they can use from then on. If an imposter poses as one or the other with a false key, the "Trusted ID Agent" in Figure 17.45 can detect the problem and alert the one with the good key that the other is not validated. Also, note that the sensor only talks to a proxy server "Proxy U," which means the unconfirmed proxy. This places all the networking burdens on the Proxy U and not the sensor, which we want to focus on sensing and its information processing. The Proxy U has an up-to-date copy of all the sensor's information and real-time data, so it looks like the real sensor to the network. The sensor and Proxy U can encrypt the packets they send each other as another layer of security. The sensor data makes its way to "Proxy S" or the secure proxy in Figure 17.45 through the "Data Guard." The data guard is really important and examines the detailed XSD (XSD in the figure) to validate a match between every required field in the XML document and ensure not a single byte more is present. So at Proxy S we have a copy of the sensor data, but where the information has been assured by the data guard and the trusted ID agent. Proxy S holds good data from an authenticated source. When the client asks Proxy S for the sensor data, Proxy S can check with the Trusted ID Agent to verify the client identity and access authorization and the transaction can be completed. The Data Guard, Trusted ID Agent, and Proxy S have to be protected by a firewall.

17.4.6 XML Documents and Schemas for Sensors

XML is well understood by most people who make web documents and databases, but not to signal processors and engineers. It would require another textbook to provide a more complete discussion [17], but it is very useful to give a quick overview from the perspective of sensor data. Most of the XML books available are really from the database perspective in that the examples are typically names, addresses, or items from a catalog which covers the majority of XML uses today, although we see a growing use as a format for word processing and graphic arts drawings. XML is a human-readable structured database which is restricted to ACSII characters. This restriction means that it can be scanned and deemed safe from executable code and viruses as hidden attachments, since the binary data for these would fold bit fields outside the range of ASCII characters. You can imbed binary data using a conversion called "base64" binary encoding which requires about a 33% increase in the number of bytes. Base-64 works by collecting three 8-bit bytes and dividing the triplet's 24 bits into four 6-bit numbers. These 6-bit numbers cover a range from 0 to 63 which are mapped onto the ACSII sequence "A–Z, a–z, 0–9, +, and /" and so the upper case alphabet has the numbers 0 through 25, lowercase 26–51, the ASCII numbers cover 52–61, and the "+" and "/" cover the range 62 and 63. The ACSII symbol "=" is used for a blank space, one or two of which is needed in case the number of 8-bit data bytes are not a multiple of 3. So even structured binary data can be wrapped in XML and passed through data guards.

The best way to explain XML for sensors is to walk through a simple example. Signal processors are generally knowledgeable of structured languages such as C, C++, VB, Java, and so on, and so the structures aspect of XML will be quite familiar. Table 17.2 provides a short XML example for sensors.

The occasional leading "-" is just something inserted by the editor (Windows Explorer) to help highlight a new element tag. The tag is denoted as "< tag >" and the closing tag as "< /tag >." For an XML document to be valid, all the nested tags must be closed properly. The outer-most tag is called the root, in Table 17.2, it is "< SensorX >" and thus the XML document closes with "< /SensorX >". This is just an example for discussion purposes and the tag names and element text is designed to be as readable as possible. The advantages of XML are that its structure is a database format that human eyes can read and understand and that this structure is basically ASCII characters that are safe to pass across network boundaries. The downside is the document size that can get very large if there are a large number of tags, compared to a binary structure. There is an optional XSD which describes the format and enumerates each element's restrictions and is formatted very nearly the same way, but we will leave that to the XML books to explain, since it is very detailed. What every signal processor needs to know is that one can easily generate a class from a schema document in your software such as C++, VB, Java, and so on. MATLAB is not quite there yet as of 2010, but there are a number of third-party users that offer free toolboxes and/or source script that will take a data structure in MATLAB and write to an XML file and read an XML file into a MATLAB date structure where the structure element names match the element names in the XML document.

The tags in an XML document should be illustrative of what information is in the element (the text between the opening and closing tags), but from a signal processing point of view should be terse enough so that the structure element names are not long dotted lists. In this example, three-letter tag names are used. XML is also supposed to have "meta data," that is, data about the data, such as the length of arrays or the data type or format. This makes reading the XML more robust, so that you can stop before an invalid read. Consider the first element "IOC" which is an abbreviation for "Information Object Collection" in Table 17.2. There is an external "tag name" and so the collection can be associated with a source; there is a time stamp (format shown in the text where capitals are hour–minute–second), and then there is a name for the particular IOC collection, so that multiple collections can be supported of the same or different types. Finally, "noc" is the number of objects in this collection. This metadata allow the collection to be associated with a parent (ACME Sensor), a time, and even have siblings. An object in the collection has a name and a number of coordinates "nco," in/of which each can have a separate name and a number of samples "nsc." Each sample "SMP" has an "x" and a "y" component and a number of corresponding states "st" and state errors "ste." Note how the text of the element is already in MATLAB matrix notation where the semicolon separates each row. This little four-level scheme can cover just about any type of sensor data. The collection concept is very flexible and covers the common situation where very different types of data can all be collected together with the same time stamp, place, sensor, and so on. For example, a LIDAR sampling air pollution could produce multiple absorbance FTIR spectra along a line at different ranges where the line is defined with a GPS coordinate and an elevation and aspect angle. Other objects in the collect could be wind, temperature, and water vapor along the line, and so on, or additional sensors on the ground such as sunlight or ground wind and temperature. These data are "collected" together in the context of the sensor time and place, but may be fused later for other inference-based reasons.

TABLE 17.2
Sensor XML Example

```
<?xml version = "1.0" encoding = "UTF-8" standalone = "no" ?>
-<SensorX>
-<  IOC>
<       tag>ACME Sensor</tag>
<       tim>yyyymmddHHMMSS</tim>
<       nam>Collection ID</nam>
<       noc>1</noc>
-<      OBJ>
<           nam>object ID in collection</nam>
-<          nco>1</nco>
-<          COR>
<               nam>dB</nam>
<               nsc>10</nsc>
-<              SMP>
<                   y>[9 8 7 6 5 4 3 2 1 0]</y>
<                   x>[0 1 2 3 4 5 6 7 8 9]</x>
<                   nst>2</nst>
<                       st>[9 8 7 6 5 4 3 2 1 0; 0 1 2 3 4 5 6
                        7 8 9]</st>
                        ste>[0.1 0.1 0.1 0.1 0.1 0.1 0.1 0.1
                        0.1 0.1;
            0.2 0.2 0.2 0.2 0.2 0.2 0.2 0.2 0.2 0.2]</ste>
                    </SMP>
                </COR>
            </OBJ>
    </IOC>
-<  /BLO>
<       tag>ACME Sensor</tag>
<       tim>yyyymmddHHMMSS</tim>
<       nam>Binary Large Object in xbase64</nam>
<       fmt>binary format description - 16 bit signed in this
case</fmt>
<dat Encodingtype = "base64">OgFzAqkD2wQJBjEHUQhpCXcKewtzDF8NPQ4ND
80PfhAeEawRKRKTEusSLxNhE34TiBN+E2ETLxPrpMSKRKsER4RfhDNDw0PPQ5fDX
MMewt3CmkJUQgxBwkG2wSpA3MCOgEAAMb+jf1X/CX79/nP+K/31/aJ9YX0jfOh8s
Px8/Az8ILv4u5U7tftbe0V7dHsn+yC7Hjsguyf7NHsFe1t7dftVO7i7oLvM/
Dz8MPxofKN84X0ifWX9q/3z/j3+SX7V/yN/cb+AAA=<dat>
    </BLO>
    </SensorX>
```

The format of the XML can be whatever makes sense for the application, but also publishing the schema document means that the XML data can be *validated* in detail.

The last object in the XML example has the tag name "BLO" which stands for "binary large object" also known in the database world as a BLOB. There is no way to read a BLOB by looking at the ASCII representation. The base-64 encoding at least gives a fixed universal ASCII character for each bit pattern where the decoding will produce the exact original bytes, but that is, just bytes. Something else is needed to turn those bytes into integers, floating point numbers, arrays, text, whatever. So the "BLO" in Table 17.2 also has a format statement "fmt" to tell you what to do with the bytes, which in the example are just 100 samples of 16-bit signed integers of a sinusoid. For example, the bytes could be a JPEG encoded image or MP3 audio file. The data are in the element named "dat" and the "dat" tag has what is called an XML attribute named "Encodingtype" which tells the reader to decode the ASCII characters using the base-64 algorithm. There could be other encoding methods, and so this attribute tells the user what method to employ to get the BLOB back from the ASCII characters. XML supports multiple attributes for each element, but we do not use them unless we have to because we do not want the resulting data structure to be cluttered.

XML can even be used to make a very detailed sensor log, or experiment log where multiple sensors with different data structures and report time intervals can be concatenated into a single database. XML is a database format. It is useful for messaging between processes on the network since it can be validated and parsed with links to schema documents. With a big experiment going on with all sorts of sensor data reporting a random times, including network messages, one could append all these XML packages into a single recording. All one needs to do is create a new XML document with the root <log> (or whatever you want to call it). Then take the well-formed <SensorX> document and strip off the root header (<?xml version = "1.0" encoding = "UTF-8" stand-alone = "no" ?>), identify the root (<SensorX>), and closing tags (</SensorX>), and insert this block of text inside the closing tag < /log> in the XML log file. In other words, overwrite the closing < / log> tag and append it back on after the <SensorX > · · · < /SensorX> block. When you load this log into your class, the events may be viewed as a time line, by a sensor, or by the content of the various information sources. No need to make a database, you did it on the fly during the experiment. This is a fundamental lesson for intelligent sensor networks: do not create a data mountain and leave it for someone else to put into a database. It is much better to build the database as you go using your knowledge of how the data will be accessed and used. The XML documents may seem bloated but they can be validated, read by humans, and can create a well-constructed database on the fly. What good is all that data if you cannot get to it quickly?

17.4.7 Architectures for Net-Centric Intelligent Sensors

Without a doubt the twenty-first century will bring faster processors, broader networking bandwidth, and staggering data mountain "ranges." The system has to be constructed from the sensor byte upward with a built-in scheme for finding data and making the connections necessary for extraordinary information processing and situational awareness. We show how well-designed XML documents can fully describe a sensor's capability and the data products available. Using secure proxy servers and agents for verifying client–server identities, we can manage the risk of sensor or client spoofing, and using a data guard we can be assured that the sensor data closely fits the description in the XML schema. The basic principle of the Internet allow large-scale sensor networks to emerge and be manageable both locally and at hierarchical levels following the same basic constructs of validating data and authentication source and destination identities. This also supports legacy sensor data by allowing older schema documents to persist with the older XML data from sensors, and therefore, the architecture should also support new sensors and future data fusion clients by incorporating the required XML schemas and managing the legacy through XSLT. Clearly, the Internet does much more than solve the cabling problem, but with that flexibility comes the sobering reality that there could be billions of sensors on the network transmitting terabytes per second starting in only a few years. Software agent programs can process these data, not to add the documents to the web search engines, but to do real analysis for us, alerting us to situations critical to our security, life, liberty, and the pursuit of happiness. Each person can have his/her own set of ebots looking out for their welfare and interests. This kind of programming is great fun and relatively simple compared to the many algorithms developed and demonstrated in this book. We do not show explicit details because we would have to commit a lot of space to languages such as C++, VB.NET, C#, Java, Flash, and so on, and there are many good texts available for this. The techniques and specific technologies are also still moving fast, much faster than the signal processing algorithms herein. Office software allows web-enabled scripts for automatically gathering data from the web, processing it, and even preparing tables and slides for presentation, sending you a text message at the bar when your report is ready. Too bad writing of this book did not work that way.

17.5 MATLAB® EXAMPLES

The 45 Figures in this chapter used quite a few m-scripts. Table 17.3 summarizes the MATLAB scripts and their Figures for Section 17.1 on classification (ATR) algorithms. There are also several wave files of bird calls used in the CINet analysis. These are widely available on the Internet as MP3-formatted recordings, which MATLAB can read only after converting to the WAV PCM format. For more information on bird songs and software to analyze and compare songs, see www.birds.cornell.edu/brp/software and the Bioacoustics Research Program where they offer free software tools, data, and even MATLAB tools for spectral analysis of a broad range of animals, not just birds.

Table 17.4 summarizes the m-scripts used for various feature types. Most of these scripts are very straightforward and commented. The most interesting are the AM and FM demodulation scripts in "widebandamfm.m" which use the Hilbert transform method for demodulation. It is easy to do in MATLAB and works very well on actual data as well as modeled data.

Table 17.5 summarizes the m-scripts used for feature tracking and statistical prognosis. These are fairly sophisticated in signal processing and graphics. The script "nlfault.m" models the nonlinear fault in the power supply example and the script "kbtrack2.m" uses a Kalman–Bucy tracking filter to generate future predictions and statistical metrics on the feature track, such as the survivor rate and hazard rate.

17.6 SUMMARY

This last chapter was the longest to write of the entire book because of the breadth and depth of diverse information covered. As a result, a significant amount of important approaches to ATR had to be tersely presented, which provides a concise broad view of the area. In Section 17.1 we covered the three main types of decision algorithms used in ATR: statistical, neural network, and fuzzy logic-based classifier algorithms. Statistical recognition is based on the idea that the statistics of a pattern class of interest can be measured through training. Therefore, one can measure how close a particular pattern matches a known pattern class statistically. A pattern to be classified is then assigned to the known pattern class it most closely matches in a statistical sense. A neural network, on the other hand, does not require that one "knows" anything about the pattern classes of interest, nor the "unknown" pattern to be classified. It works by a brute force training of artificial neurons

TABLE 17.3
m-Scripts for ATR

m-Script/Wav	Figure
bayes.m	17.1
mlclass.m	17.2–17.4
neuralnet.m	17.7–17.8
blend.m	17.9–17.11
wavspect.m	17.12–17.14
bluejay.wav	17.12
ScreamingPhiha.wav	17.13
Musicianren.wav	17.14

TABLE 17.4
M-Scripts for Signal and Image Features

m-Script/Wav	Figure
pulseint.m	17.17
readpop.m	17.18
pop3.wav	17.18
pulsetrain.m	17.19
eofft.m	17.20
oooo.wav	17.20
eeee.wav	17.20
amsignal.m	17.21–17.22
fmsignal.m	17.23–17.24
amfmsignal.m	17.25–17.26
widebandfm.m	17.27–17.30

(weights and interconnections) to match a set of known patterns to a particular class output of the network. This amazing algorithm can truly conquer the idea of a simple artificial brain which for some applications may be most appropriate. But its shortfall is that it teaches us little about why it works, and more importantly, what information in the patterns causes the classification. Therefore, there is always a criticism of neural networks about extendibility to new patterns outside the training set. The syntactic fuzzy logic approach captures the best of both statistical and neural algorithms. But the difficulty is that one must know the syntax, design the fuzzy set membership criteria (blend functions, AND, OR, and operations), and test (rather than train) the algorithm on real data to validate the logic. The beauty of syntactic fuzzy logic is that it allows us to use any combination of algorithms to produce inputs to the logic. As a result of testing the logic, flaws are trapped and rewired (recoded in software), and further testing is used to validate the response. Syntactic fuzzy logic embraces the scientific method for model development. It requires the most from us in terms of intelligence, and offers us the ability to refine our own intelligence. We apologize to the creators of those many ingenious algorithms which have been omitted due to limited space, since this is not a pattern recognition book. From the viewpoint of intelligent sensor systems, a basic view of ATR is needed to integrate this technology into the entire intelligent sensor system.

In Section 17.2, we presented some useful techniques for generating signal features in both one and two dimensions. A completely generic summary of feature strategies is given with an emphasis on the concept that useful signal features *concentrate* information. This information should certainly be based on real physics rather than heuristics. But in the case of computer vision, a strategy of following the biology and psychophysics of the primate eye is suggested for developing image features. Vision is extraordinarily complex and is an aspect of human performance in which machines will likely never surpass. But we can train machines to see things, albeit crudely. And the suggested approach is to first determine whether one is interested in a texture or an object. Textures can be matched in the frequency domain while objects may best be identified via image

TABLE 17.5
m-Scripts for Dynamic Feature Tracking and Prediction

m-Script	Figures
nlfault.m	17.33–17.36
kbtrack2.m	17.37, 17.39–17.42
cauchy.m	17.38

segmentation. Segmentation of object shapes leads one to syntactic logic for combining segments into particular objects. This logic can include influence from the presence of other objects or data to suggest a particular scene. Segmentation also offers the ability to account for rotation, skewness, and scaling, which are major algorithm problems for computer vision.

Section 17.3 presents the concept of feature tracking, where the track over time can itself become a feature. This concept considers the evolution of information, rather than just situational awareness in a "snapshot" of data. The concept of prognostics is presented where the algorithm presents where the feature will be in the future, based on current information and reasonable hypotheses. This has significant economic value for machinery systems where maintenance is to be developed based on condition, rather than hours of operation. CBM has the potential for significant economic impact on basic industry as well as increasing reliability of life-critical systems. The prognostics problem is of course helped by as much diagnostic information as possible as well as fault trajectory models for particular cases of damage. This follows closely the medical prognosis we all have experienced at some time. For example, the doctor says "You have a cold. You can expect to feel bad for a few days followed by slow improvement back to normal." Or, the doctor says, "You have liver cancer. I'm sorry but you only have a 50% chance of living more than 6 months." In both cases, the prognosis is based on current information and some projection (model) of how things will go in the future. Even when these predictions are not precisely correct, if they are interpreted correctly, the information is highly valuable. This is definitely the case with machinery CBM as the information regarding several prognoses can be very useful in determining the most economic repair actions.

Section 17.4 discusses the task of placing intelligent sensors on the Internet (or a private Intranet) and how to exploit current network technology to build large-scale networks with manageable configuration and intrinsic data mining capabilities. This is largely supported by using XML-formatted data and published schema documents to allow authentication, automated discovery and integration, and manageable security in information transactions. The specific techniques for network-based programming are only discussed from a high level, due to space limitations here and the fact that much of this technology is fast moving. What one does for network communication will remain stable over time compared to the specific technique for doing communications. Most profound today is the penetration of the Internet into wireless telephone networks and the explosion of computational power in a modern cellular telephone. We cannot begin to discuss the software techniques for phone programming since much of it is still proprietary to the particular networks, but it is clear where this technology wave is going.

The "Intelligent" adjective in the term "Intelligent Sensor Systems" is justified when the sensor has the capability of self-evaluation, self-organizing communications and operations, environmental adaptation, situational awareness and prediction, and conversion of raw data into information. All five of these concepts require some degree of pattern recognition, feature extraction, and prediction. Earlier chapters in the book cover the basic signal processing details of sampling, linear systems, frequency-domain processing, adaptive filtering, and array processing of wavenumber signals. Using these techniques, one can extract information from the raw data which can be associated with various features of a particular information pattern. This chapter's goal is to show the basic techniques of pattern recognition, feature extraction, and situational prognosis, and network communication architectures.

PROBLEMS

1. A pair of features are being considered for use in a statistical classifier. A training set for class A reveals a variance of feature 1 of 2.5 and mean 5, a variance for feature 2 of 5.5 and mean of 9, and a covariance of feature 1 and 2 of −3.2. What are the major and minor axes and rotation angles of the ellipse representing the 1-sigma bound for the training set?

2. A training set for class B reveals a variance of feature 1 of 1.75 and mean 2, a variance for feature 2 of 6.5 and mean of 5, and a covariance of feature 1 and 2 of +0.25. Sketch the ellipse defining class B.

3. Suppose you are given a feature measurement of $\langle 4.35, 5.56 \rangle$ > where 4.35 is feature 1 and 5.36 is feature 2. If the probability of class A existing is 75% and class B is 45%, which class should the feature pair be assigned to?

4. Consider a neural network with two input nodes, four nodes in one hidden layer, and three output nodes. The inputs range from 0 to 1 and the output classes represent the binary sum of the two inputs. In other words, class A is for both inputs being "zero," class B is for one input "zero," the other "one," and class C is for both inputs "one." Train the network using a uniform distribution of inputs, testing each case for the correct class (i.e., if the input is >0.5 its "one," <= 0.5 its "zero").

5. Build the same classifier as in problem 4 but using syntactic fuzzy logic. Select the weights as unity and use identical sigmoidal functions as the neural network. Compare the two classifiers on a set of uniform inputs.

6. A condition monitoring vibration sensor is set to give an alarm if the crest factor is greater than 25. If the main vibration signal is a sinusoid of peak amplitude 100 mV, how big must the transients get to trip the alarm, assuming the sinusoid amplitude stays the same?

7. The harmonic distortion of a loudspeaker is used as a condition monitoring feature in a public address system. To test the distortion, two sinusoid signals drive the loudspeaker at 94 dB (1.0 Pa rms) at 1 m distance at 100 Hz and 117 Hz. If distortion signals are seen at a level of 0.1 Pa rms at 17 Hz and 217 Hz, and 0.05 Pa rms at 200 and 234 Hz, what is the % harmonic distortion?

8. Define a segmentation graph for the numbers 0 through 9 based on "unit-length" line equations. Show how the segmentation functions can be scaled and rotated using simple math operations.

9. The temperature of a process vessel is currently 800°C and has a standard deviation of 5°C. If the temperature is rising at a rate of 10°C/min, where the velocity standard deviation is 1°C, how much time will it take for the vessel to get to 1000°C? What is the standard deviation of this time estimate? (Hint: assume that σ_v (the tracking filter process noise) and T are unity for simplicity.)

10. Compute the hazard and survivor rates as a function of time for the temperature dynamics given in problem 9.

REFERENCES

1. J. T. Tou and R. C. Gonzalez, *Pattern Recognition Principles*, Reading, MA: Addison-Wesley, 1974.
2. R. Schalkoff, *Pattern Recognition: Statistical Structural, and Neural Approaches*, New York, NY: Wiley, 1992.
3. D. E. Rumelhart, J. L. McClelland, and the PDP Research Group, *Parallel Distributed Processing: Explorations in the Microstructure of Cognition*, Vol. I: Foundations, Cambridge: MIT Press, 1986.
4. Z. Chi, H. Yan, and T. Pham, *Fuzzy Algorithms*, River Edge, NJ: World Scientific, 1996.
5. J. A. Stover and R. E. Gibson, *Modeling Confusion for Autonomous Systems*, Volume 1710, Science of Artificial Neural Networks, pp. 547–555, Bellingham, WA: SPIE, 1992.
6. L. A. Klein, *Millimeter-Wave and Infrared Multisensor Design and Signal Processing*, Boston, MA: Artech House, 1997.
7. M. B. Van Dyke, K. Reichard, and D. C. Swanson, Extraction of periodic modulation of rotational speed from roller bearing vibration signals using a modulation template matching algorithm. In *Proceedings NoiseCon '97*, PA: State College, 1997, 15–17.
8. S. Bow, *Pattern Recognition and Image Processing*, New York, NY: Marcel Dekker, 1991.
9. B. Sou_ek and The IRIS Group, *Fast Learning and Invariant Object Recognition: The 6th Generation Breakthrough*, New York, NY: Wiley, 1992.
10. M. Pavel, *Fundamentals of Pattern Recognition*, 2nd ed. New York, NY: Marcel Dekker, 1992.
11. I. Overington, *Computer Vision: A Unified, Biologically-Inspired Approach*, New York, NY: Elsevier, 1992.
12. H. Freeman, ed. *Machine Vision*, New York, NY: Academic Press, 1988.
13. Y. Bar-Shalom and X. R. Li, *Estimation and Tracking: Principles, Techniques, and Software*, London: Artech House, 1993.

14. R. Billinton and R. N. Allan, *Reliability Evaluation of Engineering systems: Concepts and Techniques*, New York, NY: Plenum Press, 1983.
15. J. M. Spencer and D. C. Swanson, *Prediction of crack growth in a tensioned band*, Abstract in 133rd Mtg of ASA, PA: State College, 1997, June 16–20.
16. J. Soloman, *Mobile IP: The Internet Unplugged*, Upper Saddle River, NJ: Prentice-Hall, 1997.
17. D. Livingston, *Essential XML for Web Professionals*, Upper Saddle River, NJ: Prentice-Hall, 2002.

Index

Note: n = Footnote

A

a posteriori error signal, 314
a priori, 322–323
    error signal, 314
    measurement prediction, 325
    state error prediction, 325
A/D conversion, 3
AAC. *See* Apple's Advanced Audio Coding (AAC)
Absolute
    calibration, 552
    threshold level, 379
Absorbed dose, 567
AC-3 (Dolby Digital), 87
Acoustic, 65–69, 70. *See also* Echo; Reverberation
    energy hot spot, 420
    impedance, 204
    large instruments, 76
    lens, 359
    particle velocity measurement, 209
    radiation impedance, 544
    responses of rooms, 398
    reverberation, 90
    room modes, 71
    sensor line array, 226, 227
    underwater arrays, 417
Acoustic and vibration sensors, 530
    α emitter, 562
    APCI, 563
    biological molecules detection, 564–565
    CB sensors, 555
    charge amplifier, 550–552
    chem-FETs, 558
    chemical detection, 556–558
    coherent spatial standing wave, 560
    compression stud, 545
    corona discharge, 562
    exercises, 572
    FTIR, 560
    hydrophone, 542, 543–544
    IMS, 562–564
    MATLAB, 569–570
    MEMS, 549, 550
    microphone, 546
    m-scripts, 569
    nuclear radiation sensors, 566–569
    olfaction sensors, 556–557
    PAS, 560
    piezoceramic materials, 538, 539, 542
    piezoelectric materials, 538–539
    plasmagram, 563
    plasmon resonance, 562
    PVDF, 539

PZT accelerometer, 540–542
    radiation impedance, 544
    reciprocity calibration technique, 552–555
    seismic accelerometers, 545
    SERS, 562
    smoke detector, 562
    spectroscopy, optical absorption chemical, 558–560
    spectroscopym Raman, 560–562
    transducer, 530, 537, 539
Acoustic equation of state, Linearized, 202
Acoustic fields
    farfield plane-wave propagation, 204
    holographic reconstruction, 419
    nearfield, 206
    nonplane wave propagation, 204
Acoustic intensity, 199, 200, 418. *See also* Electromagnetic
        intensity
    instantaneous, 212
    measurement, 209
    from normalized spectral density, 213–215
Acoustic pressure
    statistics model, 387
    wave, 201
Acoustic wave, 397
    equation derivation, 200–203
    speed contributors, 202
Activation function, 585
Active
    intensity, 204, 214
    noise attenuation, 485, 504–519
    noise control, 485, 504, 505
    system identification, 508
Active Server Pages (ASP), 577
Actuator, 3, 552
Adaptive
    beamforming, 441
    digital signal processing, 21
    FIR whitening filter signal flow, 499
    Kalman gain, 325
    Kalman–Bucy tracking filter, 96
    lattice filter, 296, 300
    noise cancellation, 501, 503, 504, 505
    pattern recognition, 125
    Signal Enhancement, 497, 499
    signal processing systems, 188
    signal whitening, 498
    signal-whitening filters, 291
    system identification, 288, 332, 333
    system identification using least-squared error system
        modeling, 266
    system modeling, 21
    tracking systems, 319
    Wiener filter, 501